CITRUS ENTOMOLOGY

CITRUS ENTOMOLOGY

IN THE

MIDDLE EAST

WITH SPECIAL REFERENCES TO

EGYPT · IRAN · IRAK · PALESTINE · SYRIA · TURKEY

BY

F. S. BODENHEIMER

JERUSALEM

Springer-Science+Business Media, B.V, 1951

ISBN 978-94-017-5220-6 ISBN 978-94-017-5218-3 (eBook)
DOI 10.1007/978-94-017-5218-3

Softcover reprint of the hardcover 1st edition 1951

Originally published by Dr. W. JUNK, PUBLISHER, THE HAGUE in 1951.

FOREWORD.

Habent sua fata libelli!

The manuscript of the present book was sent to Amsterdam during the early part of the war, and the drawings arrived there in the first days of the German occupation. The late Dr. W. Junk hid the manuscript during the war in a Dutch fishing village, from where his successor Professor W. Weisbach got it back in good condition after the conclusion of the war. The text was then entirely revised. Some citrus pests of major importance, such as Comstock's mealybug and the silver rust mite, made their appearance in Palestine only after the conclusion of the earlier manuscript. The enforced neglect of the orange groves during the war taught the citrus entomologist some highly interesting lessons, and new insecticides appeared and were introduced into practice, whilst others opened entirely new aspects. Since that time the writer served for three years as entomological expert to the Ministry of Agriculture in Turkey and for one year in the same capacity to the Directorate General of Agriculture in Irak. Fairly representative collections of citrus scale-insects from Iran became available for study. These new experiences have been fully incorporated into the book, whilst many lengthy tables of breedings, as well as the discussion of minor pests have been omitted or cut down.

The classical textbook of H. J. Quayle "Insects of Citrus" gives a world-wide survey of citrus pests dealing briefly with the morphology, the life-cycle, the distribution and the control of the more important citrus pests. The concentration in this book upon a local citrus fauna, of which little was known at the beginning of our work, during 25 years — amongst them ten years of concentrated study, thanks largely to the financial support by the Jewish Farmer's Association of Palestine — permitted to discuss intensively the problems of the ecology and of the population-fluctuations of the more important pests. This intensive discussion of the ecological aspects, as far as our still scanty knowledge permits, including that of the physical ecology of the biotic complex of the citrus pests, is probably the main feature of the book. The establishment of the triple hyperbola-band for many widely spread major pests with the implicated new methods of calculation of life-cycles in a locality of known climate, the application of eco-climograms for a number of major pests, monthly population counts for a number of consecutive years, the discussion of the age of the citrus tree in connection with the infestation and establishment of various pests, the study of the limiting factors — climatic and biotic —, they will all serve as a basis for future corrections and improvements. Also the analysis of the host-relations revealed a number of interesting points.

Another aspect is the detailed survey of our knowledge of the distribution of citrus pests in the various regions of the Middle East. This analysis reveals that each regional citrus fauna is not mainly depending upon the chances of introduction, but to a surprisingly high degree, upon the local climatic conditions. The citrus areas of S. Irak and S. Iran, of the Black and the Caspian Sea regions, of the Mediterranean shores, they all have constitutionally different citrus faunas.

With regard to control, mainly the underlying principles and the present trends have been discussed, and the local experience has been mainly stressed, where it deviates from observations in other countries, such as with regard to the high sensitivity of the mussel scale to HCN-fumigation, the earlier absence of injury due to spraying with oil-emulsions, the working out of sprays efficient against the Florida black-scale, etc.

The bibliography tries to give the most important international literature, as far as ecological observations are concerned, along with the local literature on citrus pests.

Much attention was paid to adequate illustration. Of the **43** plates and the **153** text-figures many contain original material, which is published for the first time, others are taken from not easily accessible earlier publications of the writer, and the rest is borrowed from competent authorities.

Yet it must be pointed out, that most of the Coccid figures were drawn, before Ferris introduced an entirely new standard in this field.

The book is intended as a contribution to the study of animal populations. Physiological laboratory methods, field statistics and ecological analysis have been combined to gain a first insight into some of the factors, at least, which bring about the fluctuations and gradations of the major pests. Physical ecology and host-relations have been specially studied. We have no illusion to have solved the major problems. Yet, without a daring attempt to precise ecological conditions, no final solution can be obtained. Comparison with the gradations of the major pests in other citrus areas, the study of additional factors, and — after the accumulation of a fuller and more complete material — new reconsideration, improvement and correction will be necessary. Many unsolved problems have been indicated. Altogether we hope that our analysis will stimulate further research and a more profound understanding of the causes of the numerical fluctuations of citrus pests in various conditions, which is the final aim of agricultural entomology. The recent intensive development in insect toxicology and in insecticide chemistry is becoming the domain of a new type of entomologists, with special training, entirely different knowledge and entirely different aims. The field grows too large to be still within the grasp of one individual.

In order to help the understanding of the data on physical ecology, as presented here, a few hints may be useful. The methods for calculating the bands of equilateral hyperbolas have been sufficiently exposed in the second chapter of the writer's Problems of Animal Ecology (Oxford, **1938**). The hyperbola, in spite of its theoretical shortcomings, is still the only handy formula for the calculation of the annual number of generations in a given climate and for other ecological applications. The climographs need no further explanation. Their combination with the ecological tolerance of the insect into the eco-climograph has also been adequately discussed in the publication just mentioned.

The scale of activity is the reaction of the animal's activity, as observed, when it is subjected to a slow rise or drop of environmental temperature (starting from room temperature) to the lethal temperature limits of each species. The interpretation of the classes 1 to 8 will be easily understood from the examples given on pages **476** or **552**, for instance. The symbols of the population equation are explained on p. **121** f. More complete data on these and many other methods will be contained in another book on the Principles of Animal Ecology, the second part of which is devoted to methods.

The political events which lead to the foundation of the state of Israel have thoroughly changed many remarks on the economic conditions and outlook given in this book. The very great majority (certainly over **90** %) of the citrus area of Palestine is included in the state of Israel. The only exception are a few groves in the Gaza-Rafia area and near Tulkarem. The actucal conditions in June **1950** are described by Mr. J. D. Ophen, head of the Citrus Department of the Ministry of Agriculture, Hakiryah, Tel Aviv, who kindly put at our disposal the following information:

1. The actual citrus area of Israel is now 140.000 dunam.
2. The export of citrus varieties 1949/50 to various countries was:

Country	Shamouti	Valencia	Grafepruit	Lemons	Total
England.	1.799.147	80.000	580.015	—	2.459.162
Eire.	87.892	32.500	6.500	—	126.892
Belgium	98.638	—	74.014	—	172.652
Holland	202.721	30.000	28.300	—	261.021
Sweden	269.783	94.770	37.707	—	402.260
Norway.	145.000	—	5.000	—	150.000
Denmark	71.444	18.761	6.667	—	96.872
Finland	141.159	15.000	15.000	—	171.159
Poland	109.050	—	6.150	—	115.200
Rumania	21.594	—	390	10.944	32.928
Bulgaria.	—	—	—	3.479	3.479
Yougoslavia	1.500	—	—	4.500	6.000
Czechoslovakia . . .	84.760	—	16.160	—	100.922
Austria	28.300	—	—	—	28.300
Switzerland	5.600	14.000	58.344	—	77.944
France	7.000	4.500	—	—	11.500
U.S.A.	5.161	—	—	158	5.319
Total	3.071.749	292.031	838.749	19.081	4.221.610 boxes

3. Expectation for season 1950/51: Snow and very low winter temperatures caused extensive damage to the orange groves all over the country. Therefore, our total production may remain that of the preceding season or even slightly less. No definite estimates are available, but Valencia and grapefruit were apparently not much affected by the cold.

4. Prospects for new citrus plantations: Amongst other plans, the official general 5 years' plan for agricultural development provides the planting of 100.000 new dunams of citrus. The execution of this plan will, however, be strongly influenced by the results of marketing our citrus fruit in the very next years. Other projects, still in the experimental stage, deal with the introduction of new varieties, in order to expand the citrus season.

5. Mechanization: The very high local costs of production compel our citrus growers to mechanize the work in the groves as far as possible. The Government has put substantial loans, such as five million dollars from an American loan, at the disposal of the citrus growers, to speed up the mechanization of the work. Most growers have already introduced new methods of cultivation in their management. Most conspicuous is the conversion from old ditch-irrigation to various types of sprinkling (artificial rain) by mechanical devices. Thinning of the trees has made it possible in many groves to work with tractors. Weed-killers are applied to assist in introducing so-called "non-cultivation methods". About 80.000 dunams have entered, so far, the scheme of mechanization one way or another."

In the meantime the first provisional regulations of the Plant Inspection Service describing the blames and damages which will prevent export permission for citrus fruit have been published. Insects are concerned in the following paragraphs:

	Highest deviation from cleanness to be passed
10. Clean from silver-mite, thrips, and similar marks	15 %
12. Clean from scale-insects (Shamouti, Valencia, lemons up to 180 fruit per box, grapefruit 112 fruit with no more than 40 scattered scales, greater fruit with maximum 50 scales	10 %
14. A. Fruit not desinfected by 'Decco': free from spots of sooty mould bigger than 3 cm. in diameter, in a quantity to be determined by the fruit inspector	10 %
B. Fruit desinfected by 'Decco': free from spots of sooty mould bigger than 5 cm., in a quantity to be determined etc.	20 %
21. Free from green oviposition spots of *Ceratitis*	50 %
Free from yellow oviposition spots of *Ceratitis*:	
a. in Shamouti up to 28th of February:	
1. when one week in November was warmer than 18° C.	5 %
2. When not so	50 %
b. in Shamouti fruit later than 1st of March	0 %
c. in grapefruit and Valencia	0 %

These somewhat arbitrary prescriptions will doubtless be modified in the light of future experience. As a starting point they are based upon experiences of the past.

Finally the writer wishes to apologize for a number of shortcomings. First, with regard to the morphological descriptions. Apart from the scale insects, they are usually far below the technical standard of modern taxonomy. This is not without intention. It would have been easy to borrow adequate descriptions from various specialists. Yet no one entomologist is, at present, able to determine newly appearing insects or pests of all groups himself. He needs the aid or at least the confirmation of a specialist. The morphological data shall merely give the field-entomologist a first guide to their probable or possible determination.

Second, we regret that the spelling of a number of words is not consequently the same (cf. Acre/Akko, Color/colour, Sulphate/sulfate, or circumgenital = perivulvar = paragenital glands, etc.). This is due to the scattered writing of the text and its repeated rewriting within over 15 years. It would have been unfair to the publisher to insist on strict correction of such inconsequences, which will not disturb the understanding of the text, in the proofs. For this and other minor beauty-faults we ask for the indulgence of the reader.

It is impossible to enumerate here the many persons to whom thanks are due for help of various kinds. Amongst organisations, the Jewish Citrus Growers Federation and especially Mr. Ezra Danin at Hedera, who was an unfailing source of readiness to help, should be mentioned. The scientific co-workers, including my students, their merits are all mentioned on the proper place; yet outstanding amongst them is my debt to Dr. H. Z. Avidov (Klein), to whom I wish to thank for 25 years of happy cooperation and friendship. Long is the list of the technical helpers, who have usually been mentioned in the special papers. Mr. Avigad was a most helpful draftsman. And last but not least my thanks are due to the publisher: to my late friend Dr. W. Junk and to his successor Prof. Weisbach, who undertook in difficult times the production of such a technical book in their usual perfection.

Fully conscious of the many shortcomings of this book, the writer hopes to have produced a sincere attempt to advance applied entomology and to have given a useful contribution for further progress.

Jerusalem, October 1950. F. S. BODENHEIMER.

CONTENTS:

I. GENERAL PART.

II. THE CITRUS PESTS OF PALESTINE. INSECTS AND OTHER ANIMALS, THEIR DESCRIPTION, BIOLOGY, DAMAGE AND CONTROL.

I. GENERAL PART

I. THE HISTORY AND EXTENT OF CITRUS CULTIVATION IN PALESTINE.

A. History: Citrus plants are no natives of Palestine. The original home of the cultivated citrus species is the Oriental kingdom. The bulk of the wild species of the genus occur throughout India, Malaya, S. China, S. Japan and the Malay Archipelagos. The related genera of the same subfamily (Aurantiaceae: 14 genera, about 70 species) live within the same area, a few species only penetrating Africa and E. Australia.

It seems that the Citron (*Citrus medica*) was the oldest cultivated citrus tree spreading westwards. It or its variety the Cederate (citron, etrog: *C. medica ethrog*) was introduced into Palestine between 250 to 200 B.C., where it had reached a certain economic importance by the middle of the second century B.C. Tolkowsky (1938) supposes that the name Peri Etz Hadar in its original meaning refers to cedar cones, and that only after the conquest of Jaffa by Simon Maccabaeus the citron or etrog was used in the ritual of the Feast of Tabernacles. The earliest documentary evidence is a coin of Simon Maccabaeus, struck in 136 B.C. (fig. 1) and a further one a remark made by Josephus in connection with an incident in the early reign of Alexander Jannaeus (103—76 B.C.). Tolkowsky reports that subsequently Caesarea became the principal centre of citron-cultivation. References to its presence are made by many pilgrims of earlier and medieval times until the present time, when its importance is greatly reduced.

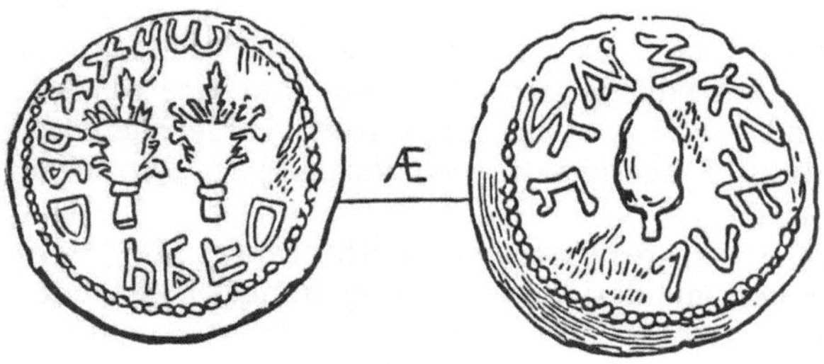

Fig. 1.

The oldest citrus-document from Palestine: A Jewish coin issued by Simon Maccabaeus in 136 B.C. showing a citron fruit (from Tolkowsky).

Apparently, the orange was introduced into Palestine from S. E. Arabia and Irak, perhaps via Syria during the tenth century, if not earlier. During the period of the Crusaders oranges are often mentioned from many places. But whereas Burchardt in 1280, and others mention expressly the sour orange (*Citrus aurantium*) no specific mention is made of the sweet orange (*C. sinensis*) until the 15th century. Yet we agree with Tolkowsky's opinion that it was probably introduced into this country contemporaneously with the sour orange.

This sweet orange introduced in the 10th century or shortly afterwards was undoubtedly the variety called „Beledi" (i.e. local) to-day. It bears small, rounded fruit, produces seeds and but small juice-sacs. Tree and leaves are smaller than in the „Shamuti" variety.

Another variety of sweet-oranges of old standing is the „Fransawi" (De Nice) introduced from France or at least from the Western Mediterranean about 1500. It bears a small, thin-skinned and not very aromatic fruit.

With regard to the famous Jaffa-orange, called Shamuti locally, all reports agree that it is of relatively recent origin. An oral tradition was reported by F. Murad to S. Tolkowsky, that it was brought from China, about 200 years ago, by an American priest whom the Armenian Patriarch had sent on a mission to the Far East. The theory of A. Aaronsohn that it has originated as a hybrid of the Malta orange and some other variety near Jaffa about the middle of the 18th century seems to be likewise erronous. Recent observations of Oppenheim confirm the appearance of common bud variations of beledi on shamuti trees and the rarer appearance of bud variations of shamuti on beledi trees. This makes the origin of the shamuti as a bud-variation of the long introduced beledi very probable and almost certain. The shamuti, proper to type, is locally restricted to an area extending from about Benyamina to about Askalon in the coastal plain. No introduction into other districts or foreign countries has ever succeeded in preserving the special flavour which created the fame of the Jaffa-orange in its proper home.

A series of other orange varieties have recently been introduced into Palestine, none of which have gained any dominant economic importance so far, such as Valencia, blood oranges and Washington navel.

The lemon tree (*Citrus limonum*) has been introduced into Palestine probably contemporaneously with the sour orange or only slightly later. The grapefruit (*Citrus paradisi*) is a recent introduction. Tolkowsky proposes that the Paradise apple or Adam's apple mentioned by medieval travellers from Palestine refers to the pummelo or shaddock, the wild ancestor of the grape fruit. These references, however, may also refer to the citron. Tangerines (*Citrus reticulata*) are also recent introductions.

B. Extent: 85% of the present citrus area is covered with oranges, 12% with grapefruits and 3% with lemons, limes, mandarins, etc. Among oranges the shamuti is the dominant variety. Citriculture is the backbone of the Palestine national economy. It produced more than 50% of the total export of the country before the war. The orange groves around Jaffa which developed about the middle of the 18th century started a modest export in the middle of the last century. They began to develop into economic importance only after this area was considerably enlarged by the impulse given by Jewish colonisation in Judea. The following table gives a survey on the development and economic importance of citriculture for Palestine:

Year	Export of citrus fruits		Citrus area in 1000 dunams*)			Planted in current year
	boxes in thousands	value in 1000 L.P.	Total	Jewish	bearing	
1885	106	27				
1890	200	83		0.9		
1895	260	65				0.9
1900	252	74				
1905	456	112				
1910	854	236				
1913	1609	298	32	10		
1919—20	648	162				
1924—25	1875	541	42	17		
1929—30	2696		110	60		19.5
1934—35	7284	7331	288	153		28.0
1936—37	10776					1.5
1937—38	11416			171	128 (96 full = 8 y. and older)	
1939/40	15310					

*) 10 dunam = 1 ha.

The main citrus-export of Palestine goes to the United Kingdom, about one fourth only being marketed in all other countries. The export of the season **1936/37** was directed as follows:

Export of Palestine Citrus fruits during the season 1936/37.

Country	in thousand boxes	Country	in thousand boxes
United Kingdom	7720	Bulgaria	36
Holland	607	Russia	30
Germany	318	Latvia	30
Sweden	311	Yugoslavia	30
Poland	308	Esthonia	14
France	289	India, Ceylon and Far East	16
Belgium	244	Egypt and Sudan	12
Rumania	220	Various other European countries	12
Czechoslovakia	178	Aden	4
Norway	158	Canada	4
Finland	75	Various other non-European countries	4
Denmark	74		
Switzerland	43		
Austria	39		

Total: 10.776.000 boxes.

It is easy to learn from this table, that the big increase of the citrus area occurred only after 1925. The oldest gardens are very thickly planted and not in a very good condition from the point of view of modern horticulture. Such gardens may still be found near Gaza, or to the north of Acco. The Jewish agricultural pioneer settlers improved conditions since about 1880. But really modern methods and modern management, following in most respect the Californian model, were only introduced and applied in an ever increasing measure since about 1926. Recent Arab plantations have learned very much, indeed, from this progress. But the Arab planter is inclined, as a rule, to believe in closer distances within the grove.

By far the largest part of the present citrus area of Palestine lies in Judea (south of Rehoboth to Petah Tikva), the Sharon (Magdiel) and in Samaria (Nathanya, Hedera, Benyamina). Those plantations laying north or south of this area in the coastal plain are mainly orange groves of the old type and of little production, except some recent plantations laid out in the Haifa Bay. Citrus crops are rather unimportant, as a whole, in the Emek Yesreel (= Plain of Esdralon) and in the Upper and lower Jordan Valley (Lake of Tiberias, Jericho). They are entirely absent in the hills and in the southern steppe and desert, the Negeb.

The economic situation in 1936/37, the worst season experienced for pre-war conditions is described by Poshter: „The average f.o.b. price for oranges amounted to about 6/0. From this must be deducted the costs of packing transportation, storage, loading, charges for advertising and inspection, and the general expenses of the cooperatives, amounting to an average of 3/1. From the f.o.b. price must be deducted the costs of picking estimated at 6 d per box. There thus remains about 2/6 p. as the average income „on the tree". The costs of cultivation are generally estimated at about 10 LP. a dunam. It is more difficult to determine the average yield per dunam. However, the average yield of a grove of medium age and condition may be estimated at approximately 80 boxes of oranges. The costs of cultivation per box thus amount to 2/6 d. It appears, therefore, that for the 1936/37 season the citrus industry in general has realized no net profit which — from an economic point of view — should serve as a cover for interest on investments and as a certain amortisation."

With regard to grapefruit a loss up to 6 d. per box was realized in the same season. Fortunately, the situation was slightly better in 1937/38, but the critical pre-war situation of the Palestine citrus industry is well illustrated by this analysis. It was much aggravated by the second world war.

It is not intended here to give a survey on the management of citrus groves, on which admirable general surveys have been given (Hume, Coit, Oppenheim, etc.) neither on the special conditions of Palestine citriculture which have been described by Yedidyah, Powell a.o. and reviewed partly in the 10th anniversary number of "Hadar", the journal devoted to the citrus industry of Palestine. The few remarks given here just serve as an introduction showing when citrus fruits were introduced into Palestine and what their economic importance is in Palestine. About 4% of the world production is now grown in Palestine, the total productiod being in 1000 cwt.

	Oranges	Grapefruit	Mandarines	Lemons	Cederate, limes, etc.	Total
Spain	9098			510		10,508
France	8		2	2	10	22
Italy	2743		479	2827	198	6,247
Greece	135		42	126	5	308
Malta	3					3
U.S.A.	18324	6275		2684	4	27,287
Brazil	1254					1,254
America (other countries)	2000				200	2,200
Cyprus	92			12		104
Syria and Lebanon						420
Palestine						2,337
Japan	900		4419	220		5,539
Formosa						126
Iran	387					387
Philippines						100
Algeria	387		380	35		802
Tunis	22			6		28
Egypt	1037					1,037
S. Rhodesia	48	2				50
Union South Africa	630	90	94	12		736
Australia	882			130		1,012
Total						66,187

The rapid anual increase of marketable fruit developed into a serious economical crisis. It is worthwhile to consider the position of Palestinian citriculture in the light of these difficulties:

1. Conditions of production are rather favorable as a whole. Damage by frost, unfavorable winds, floods is rather smaller than in most other citrus countries. The citrus area of Palestine is compact and close to its shipping parts. Costs of

production are moderate in comparison with American conditions, but higher than in most other Mediterranean citrus-producing countries (espec. Egypt).

2. The quality of the shamuti orange is superior to all other Mediterranean oranges. This superiority, occasionally contested by competitors, is recognized in the market by higher prices paid for the Jaffa-orange, which is a local and climatic monopoly. Aspects are less favorable for grapefruits where an increasing competition of canned fruit increases the marketing difficulties of the picked fruit.

3. The still progressive tendency of the world-market to increased fruit consumption has by far not yet reached in Europe the N. American level. Even in England the pro-caput consumption of oranges is only about ⅓ of that of the United States.

Even if Russia and Turkey will be able to produce their own consumption it is rather improbable that they will be competitors for export.

Difficulties are to be overcome either by the inclusion of Palestine into the Empire Preference Tax for fruit introduced into the United Kingdom, or by the establishment of trade-agreements with other countries, both measures being barred at present by the rather inconsiderate and highly unjust and injurious paragraph 18 of the Mandate. This view is also held by the Royal Commission (Report, 1937). Mutual competition of different organisations and cooperatives of producers is diminished by the present tendency to merge all Jewish cooperatives into one large, effective organisation. It is hoped to include also the Arab producers in a later stage or at least to arrange a close cooperation with them, as soon as they will start to organise themselves. These cooperatives reduce production costs by common purchase in large quantities, by arranging for pest control on large areas, by organized picking, packing and transport of the fruit, reduction of wastage, enforcing careful shipping etc.

The study of certain problems of irrigation, manuring, pest control, etc. is closely connected with prospects of reduced costs of production or better crops.

The relatively large area occupied in Palestine by citrus-plantations in monoculture is not free of dangers. It depends too much upon annual fluctuations of the world-market. The diversification into mixed farms with citriculture as an important branch is an impending program for the near future. The fruit-fly danger which interferes with many horticultural possibilities and other branches of agriculture, will have to be attended to.

An important step towards reliable grading of the export fruit was the establishment of a fruit inspection service about 15 years ago. This service rejected in 1938/39 374.000 cases or 2.5% of the total export, 320.000 of which were repacked under supervision. The main causes for rejection were:

Cause of rejection of fruit	XI. 38.	XII.	I. 39.	II.	III.	IV
Mechanical injuries	7.4	13.0	24.4	34.2	43.0	27.7%
Waste	0.5	6.5	9.0	22.3	26.0	47.0
Rough skin	26.0	15.0	9.0	6.3	5.4	1.3
Nuqsan (phys. breakdown)	2.0	7.1	12.0	12.2	6.7	5.3
Irregular sizing	7.0	14.0	9.0	6.0	4.3	3.0
Bruised fruit	12.0	11.5	8.1	2.5	0.2	—
Other causes	34.3	16.9	14.3	6.0	6.2	8.0
Sooty mould	2.0	5.0	5.8	7.0	5.5	4.6
Ceratitis capitata	2.4	3.6	3.5	1.4	0.6	2.0
Red scale	1.7	4.1	2.9	0.4	0.3	0.8
Mussel scale	2.3	2.7	1.0	1.4	1.7	0.3
Cryptoblabes gnidiella	2.4	0.6	0.6	0.3	0.1	—

Further data will be given in a later chapter. This table shows sufficiently how far the fruit inspection service had succeeded in preventing damaged fruit from being packed for export. 3 to 9% of the rejected fruit only was due to insect blemishes.

In the same season the official statistics estimated the total citrus area of the country at 300.000 dunams. The war produced profound changes in the local citrus situation. The main export stopped immediately. Many attempts, partly subvented by Governement, to keep most of the existing orchards alive. Most important was the establishment of a Citrus Control Board in October 1940, in order to ensure "control of the areas under citrus cultivation and control of the fruit through the various stages up to and including shipment with the ultimate object to improve the return to the producer by the more orderly handling of the crop in substitution for unregulated individual effort. The ordinance at present does not provide for the control of marketing." Statistical surveys, supervision of the Governement grants by the board, control of marketing, exploration of foreign markets, control of citrus planting under war time restrictions, certain research items (wastage, biological control of *Pseudococcus comstocki*) and suitable advertising were hitherto the main activities of this important body, which is expected to continue its useful work also in the future.

During the war Black and Mussel scales rapidly spread over the whole coastal plain from their hitherto restricted areas. The scale infestation of the neglected orchards was surprisingly low. Neglect of orchards led to reduction of the area. The war time export of citrus in cases and in bulk was suddenly reduced to the equivalent of 2.500.000 cases annually.

The survey of the Citrus Control Board in 1942 stated that the orchard situation was (15 points indicating the best condition):

Condition:	dunam	condition	dunam	condition	dunam
uprooted	25.000	4—6 points	16464	8—10 points	118800
neglected	6.706	6—7.5	25211	10—12	69674
cut back	4.234	7.5—8	18448	12—15	14543

This total of 274.000 existing dunam (after the substraction of the 25.000 uprooted dunam), most in fair condition for revival, was little reduced only in the following year, as Governement grants were then given. In 1943 the area with shamuti oranges was 223, of Valencia 16, and of grapefruit 19 thousand dunam.

It is certain that in agricultural technique as well as in marketing important changes will take place. It will be futile to predict their trends. But pest control will in the near future play a very important part in citrus management. Increase in number of scales, their changing populations in older orchards, which only now will receive proper attention, marketing competition, etc. they will all contribute to increase rather than decrease the control problems.

The prospects of the Palestine citrus industry in the near future are competently outlined by Rokach (1946) in a moderately optimistic analysis as follows: About 240,000 dunam of the pre-war citrus area have been kept alive. The citrus output for 1945/46 is estimated to be 8 million standard boxes (74 cm long, 40 klo net weight). Of this crop 2,500,000 boxes are for export overseas, with the wood for boxes being a limiting factor.

With the improvement of cultivation and recovery from war-neglect, the following prognosis is made:

Season	No. of cases	Approximate weight	Cases for oversea-export
1946/47	10,000,000	400,000 tons	6,275,000
1947/48	12,000,000	480,000	7,925,000
1948/49	14,000,000	560,000	9,325,000
1949/50	15,000,000	620,000	10,225,000
1950/51	17,000,000	680,000	11,125,000

No groves have been planted in the past five years. And, considering the high expenses at present, no extensive planting is expected to take place within the next five years. The analysis assumes that local consumption will rise from 1½ to 2 million boxes within the next five years. A stable export of 375,000 cases is calculated to the neighbouring countries. The citrus by-product factories (juices, jams, essences, etc.) have, in recent years, used about 1 million boxes. With new installations for concentrate juices they can easily raise their annual consumption to 2—3 million boxes. All the mentioned fruit is, of course, not packed in boxes, but sold in bulk. This part of the crop is supposed to rise from almost 4 to almost 6 millions boxes annually. This leaves the quantity indicated by the last column of the table as available for overseas export.

The excellent quality of the Palestine citrus crop with its old established reputation is helpful for regaining the markets. Economic recovery is supposed in the markets to proceed together with our increased export quantities. The citrus export season has extended to nine months by inclusion of citrus fruits other than oranges: "The season now begins in September with the shipment of lemons; grapefruit is already exportable at the beginning of October; Shamuti oranges (Jaffas) are shipped from the end of November until the end of March, when the export of the Valencia type of oranges begins, extending the Palestine season until the end of May."

With regard to competition: Spain, the principal competitor of Palestine in the United Kingdom and on the continent, has not yet recovered from the civil war and has suffered heavily by frosts during the last years. The absence of heavy frosts is just one of Palestine's natural assets. The main citrus export seasons of S. Africa and of Brazil are only slightly overlapping with our shipping season. With regard to American grapefruit and Brazil oranges the fact that they are beyond the sterling area will also reduce their importance.

Governement help for a speedy re-establishment of the citrus industry, reduction of wastage, modernisation and better organisation of picking and shipping, effective grading of the fruit, are needed for the fulfilment of the trend, as outlined by Rokach.

References.

A. Aaronsohn, La production mondiale des Agrumes. Jaffa, 1914.

Annuaire International de Statistique Agricole 1935/36 et 1936/37. Rome, 1937.

J. E. Coit, Citrus Fruits. New York. 1927.

H. H. Hume, Cultivation of Citrus Fruits. New York. 1926.

J. D. Oppenheim, Genetica **9**, 1927, pp. 516—520.

J. D. Oppenheim, Citrus Fruechte. Berlin, 1932.

Dr. Poshter, Bull. Econ. Res. Inst. Jew. Agency for Palestine **2**, 1938, 13—15.

C. H. Powell, The culture of the orange and allied fruits. Pretoria. 1930.

C. H. Powell, Palestine Dept. Agric. Leaflets IV, 9. 1928.

S. Rokach, Palestine and Middle East. **18**. 1946. p. 11—12.

S. Tolkowsky, Palestine Dept. Agric. Leaflets IV, 22. 1930.

S. Tolkowsky, The Gateway of Palestine. London, 1924.

S. Tolkowsky, Hesperides. London. 1938.

H. Viteles, Bull. Palestine Econ. Soc. **3**, No. 1. 1928.

J. Weinberg, Hadar, **1**, 1928, No. 2 and 3.

S. Yedidyah, Citrus Growing (Hebrew). Tel Aviv, 1937.

Hadar 10th Anniversary Number, Tel Aviv 1938.

Report of The Palestine Citrus Control Board (18.10.1940 to 31.7.1942). Tel Aviv. 1943.

2. GENERAL SURVEY OF THE ANIMAL-COMMUNITY OF CITRUS GROVES IN PALESTINE

The citrus groves are a well defined habitat with a poor, but characteristic fauna. The grove itself is a monoculture with a poor animal life. The fauna of the hedges, the weeds, and the intruders from the immediate surrounding biotopes enriches this association somewhat, but most of these intruders do not become real inhabitants of the gardens.

A. VERTEBRATA.

a. Mammals. Large mammals are rare within the orange groves. Small groups of jackals (*Canis aureus*) are met with not infrequently. Fox (*Vulpes vulpes*) and the Mongoose (*Mungos ichneumon*) hide amongst the hedges. The Long-eared Hedgehog (*Erinaceus auritus*) is not uncommon in the groves. A series of rodents are regular inhabitants of citrus orchards, amongst which Tristram's Jird (*Meriones tristrami*), the voles *Microtus guentheri* and *M. philestinus* and the wild House-mouse (*Mus musculus praetextus*) are the more common ones. The injurious Fruit-tree Rat (*Rattus frugivorus*) is dealt with in a later chapter.

b. Birds. The number of species breeding in our orange-groves is relatively poor. The lack of nisting niches for the Rufous Warbler (*Erythropygia galactotes*) as well as for the Sardinian Warbler (*Sylvia melanocephala*) on the smooth trunks of citrus, with high branchings only, explains this partly. It explains also, why the Gracious Warbler (*Prinia palestinae*) builds its nests in young orange gardens only. That lack of nisting opportunities is a dominating factor is easily gathered from the fact that the number of breeding birds increases rapidly when an alley of cypresses, etc. leads through an orange grove. The scarcity of food for seed-feeding species explains the rareness of the Greenfinch (*Chloris chloris*, feeder of Cruciferae seeds) etc. In a properly kept orchard weeds are scarce and rarely grow to maturity. In old citrus groves with the dense cover produced by the crowns of the trees, weeds generally develop poorly. The bird-fauna is very much richer on the hedges of citrus groves, especially if weeds are common on the other side of the hedge.

H. Mendelsohn has recently made quantitative studies on the birds of our orange groves and the following tables are based on his observations, mainly at Mikveh Israel.

It is interesting to follow the bird sequence in a newly planted grove in the coastal plain. During the first two years after plantation the Crested Lark (*Galerida cinnamomina*) still continues to breed, disappearing later on. During the second year the first *Prinia* appear. The Gracious Warblers breed first on the hedges; when the small citrus-trees have grown to bushes they breed in them not infrequently. But as soon as the oranges have grown into trees, they disappear and are again restricted to the hedges. During the fourth or fifth year, *Chloris chloris*, *Cardualis niediecki* and the Olivaceous Warbler (*Hypolais elaeica*) begin to breed. They also begin to nist in the hedges and extend slightly into the orchard. The Palestine Bülbül (*Pycnonotus vallembrosae*) follows and only in the tenth year after plantation the Turtle-Dove *Streptopelia turtur* may be expected.

The main fauna to be met with is given in the following table based on Mendelsohn's observations.

Birds observed in orange groves and on the breeding hedges at Mikveh Israel 1933 to 1935.

P. Orange grove. H. Hedges only. 1—5 = degrees of abundance.

Bird species		month											
		I	II	III	IV	V	VI	VII	VIII	IX	X	XI	XII
Corvus corone sardonius	P H	4	4	4	4	5	5	5	5	4	4	4	4
Chloris chloris chlorotica	P H	3	3	4	4	4	4	4	4	4	3	3	3
Cardualis cardualis niediecki	P H	3	3	3	3	3	3	3	3	3	3	3	3
Serinus canarius serinus	P H	4	4	2								3	4
Fringilla coelebs coelebs	P H	4	4	2								2	4
Passer domesticus biblicus	P H	5	5	5	5	5	5	5	5	5	5	5	5
P. hispanoliensis transcaspicus	H	1	2	2									
Emberiza calandra calandra.	H	2	3	2								2	2
E. hortulana	H			2	2				1	2			
Lullula arborea pallida.	H	1	1									1	1
Anthus trivialis trivialis	H				2					1	2	2	
Motacilla alba alba	H	4	4	2						1	3	4	4
Parus major terraesanctae	H	2	2	2	2	2	2	2	2	2	2	2	2
Lanius nubicus	H			1	1	1	1	1	1	1	1		
Pycnonotus capensis vallembrosae	P H	3	3	3	3	3	3	3	3	3	3	3	3
Muscicapa striata neumanni	H				1	1	1	1	1	2	1		
Phylloscopus sp.	H												
Hypolais pallida elaeica	P H			1	2	2	2	2	2	1			
Sylvia atricapilla atricapilla	P H	1	1	2	2	1			1	2	2	1	1
S. curruca curruca	P H	1	1	2	2	1				2	1	1	1
S. melanocephala momus	P H	1	2	2	2	2	2	2	2	2	1	1	1
		(during summer only in hedges)											
Erythropygia galactotes galactotes	P H				1	3	3	3	3	3	1		
Prinia gracilis palestinae	P H	3	3	3	3	3	3	3	3	3	3	3	3
Turdus ericetorum philomelos	P H	2	2	2								1	2
T. merula syriacus	P H	2	2	1								1	2
Saxicola torquata rubicola	H	2	2	1						1	2	2	2
Phoenicurus phoenicurus phoenicurus	P H			3	1					2	2	2	
P. p. mesoleuca	P H			1						1	1		
P. ochrurus.	H	1	1								1	1	1
Luscinia suecica suecica	P H	2	2	1							1	2	2
Erithacus rubecula rubecula.	P H	4	4	2							1	3	4
Upupa epops epops	P H			1	2				2		2		
Coracias garrulus garrulus	H				2	2			2		2		
Streptopelia turtur turtur	H					1	1	1	1		1		
Coturnix coturnix coturnix	H			2	2				2		2		

The wealth of birds in the hedges as compared with the orange groves is conspicuous. The difference in density is still greater, as will be seen by some observations of breeding density.

1. Typical old orange grove (area 200 × 80 m):

Breeding birds	1934	1936
Chloris chlorotica	1 pair	
Muscicapa neumanni		1 pair
Breeding density per sq. klm. .	63 pairs	63 pairs

2. Typical old orange grove. Area of the same size (200 × 80 m), but including part of a road with some cypresses and acacias. In 1935 the acacias were cut down, which immediately influenced the bird density:

Breeding birds	1934	1936
Chloris chlorotica	2 pairs	2 pairs
Pycnonotus vallembrosae	1	
Hypolaeis elaeica	1	
Prinia palestinae	1	
Erythropygia galactotes.	1	1
Breeding density per sq. klm. .	375 pairs	188 pairs

3. Hedge of orange grove (acacias and casuarinas) 200 × 90 m, including 40 m to each side of the hedge.

Breeding birds	1935
Chloris chlorotica	1 pair
Cardualis niediecki	1
Pycnonotus vallembrosae	1
Hypolais elaeica	2
Prinia palestinae	2
Sylvia momus.	1
Erythropygia galactotes.	1
Breeding density per sq. klm. .	500 pairs

4. Hedge of orange grove (cypresses) 120 × 220 m, being only on the outside of the grove (a great deal of ruderal weeds, partly (in 1937 fully) cultivated with vegetables):

Breeding birds	1935	1936	1937
Chloris chlorotica	4 pairs	5 pairs	3 pairs
Cardualis niediecki	1	2	1
Pycnonotus vallembrosae		2	1
Hypolais elaeica.	1		
Prinia palestinae	4	3	1
Erythropygia galactotes	1	1	1
Parus terraesanctae		1	
Muscicapa neumanni	1	1	1
Breeding density per sq. klm. .	455 pairs	568 pairs	303 pairs

Other birds may appear in other districts of the coastal plain, but the general fauna remains the same. The fauna of the hedges, especially, is much influenced by the vegetation outside the grove.

c. Reptiles. The large Hardoun (*Agama stellio*) is very common on all pumping stations, packing houses and other stone walls within the grove. In the orchard itself the long-toothed Sandy Lizard (*Acanthodactylus schreiberi syriacus*) is very common in the first years after planting, whereas the small striped Scincid *Mabuya vittata* abounds later on. Other lizards which are occasionally met with in the grove are: *Ophisops elegans ehrenbergi, Eumeces schneideri syriacus* and *Chalcides sepsoides.* In the buildings and walls within the orchards geckoes may occur, mainly *Hemidactylus turcicus* and *Ptyodactylus hasselquisti syriacus.* Chamaeleons (*Chamaeleon chamaeleon*) are not infrequently met with in the crowns of citrus trees, where they hunt for flies.

Snakes are not too common in citrus groves and may be met with on the trees, hunting for birds and bird-eggs. *Coluber jugularis, C. najadum, Psammophis schokari* have been found here. In sandy soil lives *Eryx jaculus.* The only poisonous snake, which under very exceptional conditions may be found in citrus groves, is the Palestine Viper (*Vipera palaestinae*). But it is so rare that in orange groves no case of poisoning by this snake has ever come to the notice of the writer.

d. Amphibia. The dominating frog of the citrus orchards is the Tree-frog (*Hyla arborea savignyi*) which succeeds in breeding in small pools during the rainy season. It is quite common on the trees and the small frogs often abound in early spring in the orchards just after the metamorphosis. Where breeding facilities are present in the neighbourhood the Edible Frog (*Rana ridibunda*) may be common. The Green Toad (*Bufo viridis*) occurs everywhere in suitable environment and even *Pelobates syriacus* may be found occasionally.

e. Fish. Fish are absent in the temporary irrigation channels of citrus orchards. But many of the large cisterns or water reservoirs which are connected as a rule with the pumping stations have been artificially populated by a small S. America fish (*Gambusia sp.*). This *Gambusia* has become well acclimatized there and does splendid service in destroying mosquito larvae within those tanks. Goldfish (*Cyprinus carassius auratus*) is sometimes put into those reservoirs, especially in Arab orchards, for the same purpose.

B. INSECTA.

The general insect fauna in citrus groves is extremely poor. The fauna of the citrus trees is limited almost entirely to its pests and to the parasites or predators connected with them. In addition, some insects are attracted to the citrus flowers among which Hymenoptera and a few Diptera dominate, others to the honeydew excretions of citrus scales, where Diptera are more frequently represented. The fauna living on the soil is never a specific one. It is composed of a few very common species and of a varying list of species living either on the ruderal flora or, in young groves, on the residuals of the original flora preceeding the plantation of the orchard, or on the crops cultivated as green-manure. The clean-cultivation of the properly managed orchard, does not allow for many weeds and the few present are not allowed to mature. It is of no interest to give here an extended list of these animals as far as they belong to the normal fauna of ruderal weeds or green crops. All those forms which have any temporary connection with the citrus trees are mentioned in the following chapters.

The most common weeds in our orange-groves are:

Panicum colonum	*Chenopodium murale*
Urtica urens	*Sonchus oleraceus*
Urtica pilulifera	*Malva parviflora*
Capsella bursa pastoris	*Crozophora verbascifolia*
Solanum nigrum	*Citrullus coloquinthus*
Amaranthus album	*Brassica tourneforti*
Polygonum equisetiforme	

In order to give an idea as to the extreme poverty of this fauna, a few groups may be mentioned.

Among the Orthoptera three species only are more or less regularly met with on the trees: Two praying mantids (*Sphodromantis viridis* and *Blepharopsis mendica*) and the grasshopper *Anacridium aegyptium*, which is closely related to tree-inhabiting forms of the savanna. The following other species may be met with on the soil:

regularly	occasionnally
	Ameles heldreichi
Acrotylus insubricus	*Oedipoda miniata*
	Calliptamus palestinae
	Pezottetix meridionalis
Gryllus bimaculatus	*Gryllulus desertus*
Gryllulus burdigalensis	
Pteronemobius tartarus	*Phaneroptera quadripunctata*
	Tylopsis liliifolia
Forficula spp.	*Saga gracilipes*
Labidura riparia	

Coccids are extremely rare. *Pseudococcus lindingeri* has been described from the roots of the weed *Panicum colonum*. Other species of the same subfamily probably occur on roots of other weeds in orange groves, including *Ps. citri*. On some weeds, especially on *Polygonum equisetifolium* the following citrus scales are to be found: *Ceroplastes floridensis*, *Icerya purchasi* and *Coccus hesperidum*.

Aphids abound on the weeds, the more common species being the very polyphagous *Myzodes persicae*, *Aphis frangulae*, *A. medicaginis*, further forms like *Rhopalosiphoninus lactucae* on *Sonchus*, *Dactynotus jaceae* on *Polygonum*, etc.

Sphegids and wasps occasionally build their nests in the packing houses or even among trees.

The following species of ants are recorded:

Dorylus fulvus punicus	*Acantholepis frauenfeldi bipartita*
Crematogaster inermis	*Camponotus compressus sanctus*
C. inermis armatula	*C. compressus thoracicus fellah*
C. jehovae	*C. compressus thoracicus sanctoides*
Messor semirufus ebeninus	*Cataglyphis albicans livida*
M. semirufus intermedius	*C. bicolor nigra*
M. semirufus hebraeus	
M. arenarius	*Paratrechina jaegerskioeldi*
M. rugosus bodenheimeri	*Plagiolepis pallescens ancyrensis*
Monomorium subopacum phoenicium	
Tetramorium punicum lucidula	*Tapinoma israelis*
T. signatum	*T. simrothi phoenicium*

All of the above species are common on light soils.

The insect fauna of the hedges of citrus groves will be discussed separately, as it has a direct bearing to some pest problems of citrus. This is not the case, as far as our present knowledge goes, with the fauna of the ruderal weeds of citrus orchards, none of the species of which has any primary or secondary relation to the genuine citrus-fauna, except in the case of the Citrus Blossom Bug (*Dionconotus cruentatus*).

C. OTHER ANIMALS.

Among the other evertebrate animals of citrus groves, spiders are especially common. A. Shulov made a special study of the Arachnoid fauna in an old and young orange grove at Mikveh Israel during 1937/38. This fauna is poor in species and

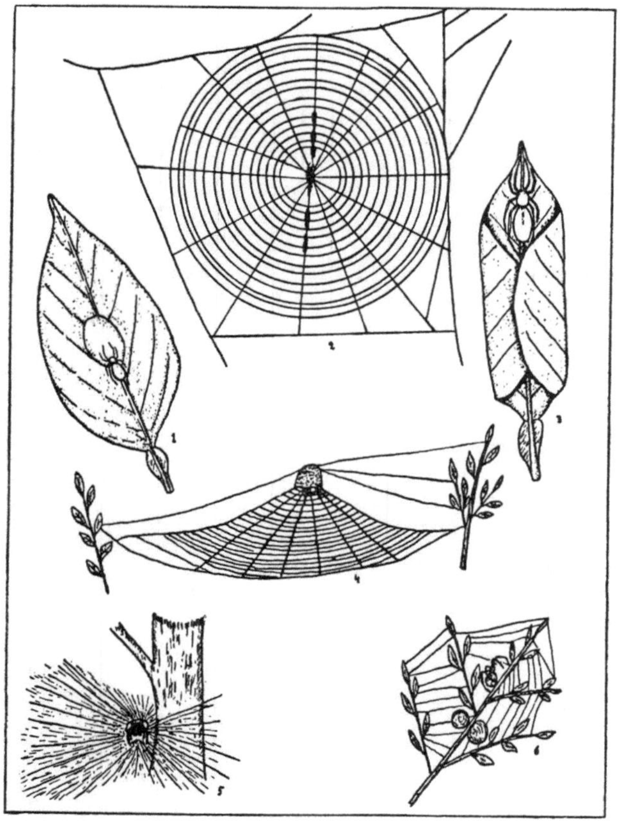

Fig. 2.

Spider webs on citrus trees: 1) Dictyna viridissima with egg-cocoon on leaf; 2) wheel-web of Cyclosa sierrae; 3) Clubiona sp. in a rolled leaf; 4) web of Cyrtophora citricola; 5) Agelena syriaca in its funnel-web; 6) Theridium aulicum with two egg-cocoons in its web (from Shulov).

still more in genera. Besides about 6 species of *Lycosidae*, which live on the soil, all other species live in the crown of the trees. *Agelena bodenheimeri* Shul., only, spins its funnel-shaped webs between the stem and the lower branches. The young orchard has a somewhat richer fauna than the old one (26 against 16 species). Of these species the *Argyopidae* web the typical spider webs, the silvery coloured

	Orchard		Month							
	old	young	V.37	VI	VIII	X	XII	I.38	II	III
Agelenidae:										
Agelena bodenheimeri Shul.	1	2	a.i.	a.i.	a.					
Argiopidae:										
Aranea dalmatica Dol.	1	1	a.i.			a.i.		i.	i.	
A. redii Scop.	1	1		i.		i.		i.		i.
Cyclosa sierrae Sim.	1	2	a.i.p.	a.i.	a.i.	a.i.	a.i.	a.i.	a.i.	a.i.
Tetragnatha minitabunda Camb.. . .	1	2				i.	p.		a.i.	a.i.
Clubionidae:										
Clubiona straminea Cambr..	1	2	p.	i.p.			i.	i.	a.i.	a.i.
Clubiona sp.	1	1							a.	a.
Chiracanthium sp.		1					i.		i.	
Oedignatha sp.		1								a.
Syrisca sp.		1						a.		
Dictynidae:										
Dictyna viridissima Walk.	1	1		a.e.p						
D. puella Sim.		1					a.			
Gnaphosidae:										
Drassodes sp.		1	a.p.	a.						
Zelotes ursinus Cambr.	1	1					i.		i.	i.
Micryphantidae:										
Micryphantis sp.		1							i.	
Salticidae:										
Evophrys clemens Cambr.	1	1								i.
Menemerus indistinctus Cambr. . . .	1	1							a.i.	a.i.
M. flavescens Cambr.		1							i.	
M. bivittatus	1			i.			i.			a.
M. sp. nov. A.		1					i.		a.	a.
M. sp. nov. B.		1								a.
Theridiidae:										
Argyrodes gibbosus Luc.	1				a.e.					
Crustulina sp.		1					i.		i.	
Theridium aulicum Koch.		2							a.	a.
Teutana sp.		1								a.
Thomisidae:										
Heriaeus sp.		1					i.		i.	
Thomisus albus Gmel.	1									
Tmarus piochardi Cambr.		2								a.i.
Uloboridae:										
Uloborus plumipes Luc.	1	1		a.i.						
U. walckenaerius Leth.	1								i.	

1—3 = abundance; e. = eggs; p. = neonate larvae; i. = inadult; a. = adult.

Cyclosa sierrae Sim. and *Tetragnatha minitabunda* Cambr., being especially common and spining their webs always within the crown of the tree. The similar nets of the Uloboridae are always horizontally arranged. The Clubionidae, Gnaphosidae and many Salticidae spin their webs within leaves rolled longitudinally, the Dictynidae on the upper side of leaves or between two leaves, the Theridiidae on the end of the twigs.

None of these species is specifically connected with citrus and they belong to the normal fauna of shrubs and trees of this country. They are not predacious in any appreciable way on citrus insects, but live mainly on Diptera, Psocoptera, Ants, and other insects attracted to the citrus trees by the honeydew, the nectar or the sooty mould. They are decidedly neutral from the point of view of the farmer.

Molluscs are extremely poor in species, the common forms being *Helix joppensis*, *H. pisana* and *H. seetzeni*.

The only group relatively richly represented are soil protozoa. Sandon found 13 species in a sample or orange soil as compared with 9 to 11 species in five other non-irrigated soils. The humidity of the soil during the summer due to regular irrigation is responsible for this increase of species. Earthworms and nematodes have not been studied, both being present, probably in greater numbers than in non-irrigated soils.

References.

H. Mendelssohn, Observations on the population density of birds in Palestine. Thesis. Jerusalem. (1939).

H. Sandon, The Composition and distribution of the Protozoan fauna of the soil. Edinburg and London, 1927.

A. Shulov, Hadar **11**, 1938. p. 206—208.

D. THE INSECT FAUNA OF CITRUS HEDGES.

The importance of the hedges for mammals and birds has been discussed before.

The insect fauna of the hedge-plants, which are indispensable as wind-breakers and which potential sources of infestation are as permanent neighbours deserves special attention.

Eucalyptus hedges are fortunately now abandoned, as they are extremely undesirable from an horticultural point of view. It is not possible to include these tall trees in a fumigation campaign, yet they harbour some of the more important scales, such as *Chrysomphalus aonidum* and *Aonidiella aurantii* and the wind blows their crawlers invariably from the trees to the orchard. The entomologist is indeed ready to give up those Eucalyptus hedges.

A very common hedge-plant is *Acacia farnesiana*. It is a host to quite a number of animals, among which the following abound:

Virachola livia Klg.	*Oxycarenus hyalinipennis* Cta.
Myelois ceratoniae Z.	*Hemiberlesia lataniae* Sign.
Anarsia acaciae Wlgsh.	*Aonidiella aurantii* Mask.
Schistocerus bimaculatus Ol.	*Pseudotargionia glandulosa* Newst.
Bruchus lallemanti Mars.	*Asterolecanium pustulans* Ckll.
Clytus sp.	*Ceroplastes floridensis* Comst.
Himatismus villosus Haag	*Icerya purchasi* Mask.

The infestation with citrus scales is moderate only and none of the other species infest citrus. Contrary to the firm belief of many farmers, *Ceratitis capitata* is unable to develop in the pods of *A. farnesiana* and we have never been able to breed it from many thousands of green pods, collected during many years in such places.

The Prickly-Pear (*Opuntia ficus indica*) is to be found only on some of the oldest citrus orchards as a hedge. Owing to the fact that its fruits are a rare summer host of *Ceratitis*, it is undesirable as a hedge as well as a neighbour of citrus plantations. The Opuntia-scale *Diaspis echinocacti* Bché. is of no importance to the citrus grower.

Solanum coagulans and *Aberia caffra* have recently gained in importance as hedge-plants. Both are infested by *Aspidiotus hederae* and *Hemiberlesia lataniae*. But whereas *Ceratitis* develops only in overmature fruits of *Solanum*, it abounds in fruits of *Aberia* throughout the summer. *Aberia*, at least, is hence an actual danger to the citrus grower and should be exchanged at the earliest possibility for another hedgeplant.

In the Sharon and Samaria-districts Castor-plants (*Ricinus communis*) are common and quick growing wind-breaks. They are infested by a series of insects, the more common of which are:

Grammodes algira L.
Phycita diaphana Stor.
Ephestia calidella Gn.
Acrocercops confluens Meyr.
Crepidodera ventralis Kl.
Retithrips syriacus Myt.
Epitetranychus althaeae Wett.
Anychus orientalis Z.
Nezara viridula L.
Empoasca flavescens F.
Aspidiotus hederae Vall.
Aonidiella aurantii Mask.
Mytilococcus ulmi L.
Icerya purchasi Mask.

The very heavy infestation by Red Scale and Red Mites makes the castor plant a not very desirable intimate neighbour, especially in young orchards. Despite the abundance of *Empoasca* on *Ricinus*, these cicadas have not proved, so far, to be a source of infestation for citrus. The close contact of the castor hedges with the citrus plants and the great number of their rows within the young orchard increase the undesirable features of this hedge plant, which are only partly counterbalanced by its quick growth.

Other hedge-trees, which are decidedly welcome from any point of view, are *Cypressus*- and *Thuya*-trees. None of their common insects is able to infest citrus:

Pachypasa otus Drur.
Brotolomia meticulosa L.
Evetria tesselutana Stgr.
Phloeosinus aubéi Perr.
P. armatus Rtt.
Lineaspis striata Newst.

3. ZOOGEOGRAPHICAL AND ECOLOGICAL SURVEY OF CITRUS PESTS IN GENERAL.

A. GENERAL ZOOGEOGRAPHICAL AND ECOLOGICAL REMARKS ON THE CITRUS PESTS OF PALESTINE.

The list of animals, mainly insects, discovered so far as feeding occasionally or regularly on citrus in Palestine has increased to-day up to 110 species. This number will doubtless still increase considerably in the future, mainly on account of very occasional feeders of no economic importance and on account of new introductions.

Despite the length of this list, from which feeders on nectar, feeders on sooty mould and beneficial insects, living as predators or parasites on the citrus fauna have been omitted, the number of true citrus insects, i.e. of species which are able to develop normally on citrus, is relatively limited. Only 4 species in the following list belong to this group. All other species are occasional or accidental feeders, most of them of no economic importance.

Insects and other animals feeding regularly or occasionally on citrus in Palestine.

1. *Coccidae.*

Aonidiella aurantii Mask.
Chrysomphalus aonidum L.

Mytilococcus beckii Newm.
Mytilococcus gloveri Pack.
Parlatoria pergandii Comst.
Parlatoria oleae Colv.
Parlatoria cinerea Hadd.

Coccus hesperidum L.
Saissetia oleae Bern.
Saissetia hemisphaerica Targ.
Ceroplastes floridensis Comst
Ceroplastes rusci L.
Pseudococcus adonidum L.
Pseudococcus citri Risso.
Pseudococcus comstocki Kuw.
Icerya purchasi Mask.

2. *Borers in trunk and branches.*

Apate monachus F. (Col. Bostr.)

Stromatium fulvum Vill. (*Col. Ceramb.*)
Rattus frugivorus Raf. (Mamm.)

3. *Feeding on leaves.*

Papilio machaon L. (*Lep. Papil*)
Caradrina exigua Hb. (*Lep. Noct.*)
Prodenia litura F.

Euxoa segetum L.

Phytometra ni Hb.

Chloridea obsoleta F.
Chloridea nubigera H.S.
Chloridea peltigera Schiff.
Stenozygum coloratum Klg. (*Rhy. Pent.*)
Toxoptera aurantii Boyer (*Rhy. Aphid.*)
Aphis frangulae Kalt.

Phyllopertha lineata F. (Col. Scarab.)
Phyllopertha arenicola pilosella Rtt.

Labidostoma decipiens Fald. (Col. Chrysom.)
Hermaeophaga ruficollis Luc.
Aphthona pygmaea nigella Kutsch.
Aphthona bonvoulieri All.
Apion miniatum Germ.. (Col. Curcul.)
Psalidium syriacum Mill.
Achradidius creticus Kies.
Strophomorphus panellus Schoenh.
Sitona gressoria F.
Sitona stierlini Rtt.
Schistocerca gregaria Forsk. (Orth. Acrid.)

Aphis laburni Kalt.

Phorodon persicae Sulz.
Aleurotrachelus citri P. & H. (Rhy. Aleur.)
Bemisia tabaci Genn.

Anacridium aegyptium L. (Orth. Acrid.)
Epitetranychus althaeae Wett. (Arach. Acar.)
Anychus orientalis Z.
Bryobia praetiosa Koch.
Helix joppensis Roth (Moll. Helic.)
Helix pisana Muell.
Helix seetzeni Koch

4. *Feeding or sucking on flowers.*
Prays citri Mill. (Lep. Elach.)

Dionconotus cruentatus Brllé. (Rhy. Caps.)

Limothrips cerealium Hal. (Corr. Thys.)
Taeniothrips discolor Ka.
Taeniothrips meridionalis Pr.
Odontothrips robustus Uz.
Thrips major banaticus Ka.
Kakothrips robustus Uz.
Haplothrips tritici Kurdj.
Haplothrips gowdeyi Frankl.
Haplothrips cypriotes Pr.
Hoplandrothrips sp.

Oxythyrea noemi Rche. (Col. Scarab.)
O. abigail Rche.
Epicometis hirta Pda.
Tropinota vittula Rche.
Tropinota squalida Scop.

5. *Insects attacking healthy fruits or loosening their connection with the tree:*
Ceratitis capitata Wied. (Dipt. Tryp.)
Cryptobables gnidiella Mill. (Lep. Pyral.)
Nezara viridula L. (Rhy. Pent.)
Crematogaster jehovae For. (Hym. Form.)
Corvus corone sardonius (Aves)

Heliothrips haemorrhoidalis Bché. (Corr. Thys.)
Thrips tabaci L.
Phyllocoptes oleivorus Ashm. (Acar. Erioph.)

6. *Insetcs feeding on young citrus plants or in branches:*
Gryllotalpa vulgaris L. (Orth. Gryll.)

Gryllus bimaculatus L.

Reticulotermes lucifugus (Corr. Ternit.)

Empoasca bodenheimeri Hpt. (Rhy. Jass)
Empoasca citrea Hpt.
Egropa sp.

Schistoceros bimaculatus Ol. (Col. Bostr.)
Xylopertha picea Ol.
Scobicia chevrieri Vill.
Enneadesmus forficula Fairm.
Sinoxylum sexdentatum Ol.
Zophosis orientalis Deyr. (Col. Tenebr.)
Dendarus syriacus Rche.

Messor semirufus spp. (Hym. Form.)
Tetramorium signatum Men.

7. *Insects feeding on buddings:*
Ephestia vapidella Mn. (Lep. Pyr.)

Tapinoma simrothi phoenicium Em. (Hym. Form.)
Acantholepsis frauenfeldi bipartita Smith
Cataglyphis bicolor nigra André

Adesmia abbreviata Klg. (Col. Tenebr.)
Erodius dejeani Sol.
Anisoplia inculta ferruginipes Pic. (Col. Scarab.)
Tanymecus musculus Fahrs. (Col. Curcul.)

8. *Insects feeding on decaying or dried fruit.*
Drosophila melanogaster Mg. (Dipt. Drosoph.)

Lonchaea sp. (Dipt. Lonch.)

Apterodiscus cernyi Ams. (Lep. Momph.)

Carpophilus immaculatus Luc. (Col. Nitid.)

Carpophilus hemipterus L.
Carpophilus dimidiatus F.

Apion pubescens Klg. (Col. Curcul.).

9. *Root pests:*
Tylenchulus semipenetrans Cobb (Vermes. Nemat.)
Phyllognatus silenus F. (Col. Scarab.)

Only 41 species seem to perform their whole development regularly or occasionally on citrus. But only three of these species seem to be restricted as monophagous feeders to citrus: the Citrus Blossom Moth *Prays citri*, the Citrus Rust Mite *Phyllocoptes oleivorus* and the Citrus Root Nematode *Tylenchulus semipenetrans*. Neither of them is a native of Palestine. They are common in subtropical (and probably also tropical) regions of the world and are undoubtedly introductions. None of the native insects of Palestine have adapted themselves to citrus, so as to be forced to complete their life cycle on this foreign and introduced host.

Among the 41 citrus-feeders we shall first consider the Coccidae. All of the 15 species are of very wide geographical distribution and most of them are extremely polyphagous. Many of them have probably been introduced with citrus at some period or other. The introduction of two species of *Mytilococcus* with Citrus stock from Spain, about 25 years ago, serves as an obvious case of this type and it is of interest to note that one of the species (*M. gloveri*) seems to have died out in recent years, whereas *M. beckii* has rapidly spread over most of the coastal plain within the very last years. *Chrysomphalus aonidum* which immediately after 1918 was restricted to N. Palestine, has by now become firmly established in at least two major centres of the southern district and is rapidly spreading now. Some other species: *Saissetia oleae, Saissetia nigra, Ceroplastes rusci* e.g., have been introduced with other plants and appear occasionally only on citrus, which is never a major host.

Moths like *Caradrina exigua, Prodenia litura, Euxoa segetum*, the three species of *Chloridea, Cryptoblabes gnidiella, Plusia ni* are widely spread in the subtropical and tropical regions of the old and partly of the new world and most of them are exceedingly polyphagous. The same applies to the plant-lice as *Toxoptera aurantii, Aphis frangulae, Aphis laburni, Myzodes persicae;* to red spiders as *Epitetranychus althaeae* and *Bryobia praetiosa;* to the bug *Nezara viridula;* to the Mediterranean Fruit-fly *Ceratitis capitata;* to the thrips *Heliothrips haemorrhoidalis* and *Thrips tabaci* and finally to the Bostrychid beetle *Apate monachus*, in case it should be able to develop in citrus.

The remainder of the citrus-feeders is composed as follows: The Swallow tail *Papilio machaon* was unknown entirely from citrus in Mediterranean countries until about 20 years ago. The first reports from Palestine were followed a few years later by a similar experience in Cyprus. Since then it seems to take to citrus in increasing numbers. Umbelliferae are the known host of this widely spread and well known palaearctic species. *Ruta* was reported as another host for its southern range. The nearest relatives of *P. machaon*, as the tropical *P. demolaus, P. demodocus*, are almost all well known feeders on citrus plants. Citrus is, hence, a potential host of the species and it is rather astonishing that no reports from other Mediterranean regions are available. This lack of reports seems to indicate that is actually does not or not yet infest citrus beyond the area described, as it seems highly improbable that such attacks for many years could evade the entomologist's or even the farmer's eye.

The polyphagous bug *Stenozygum coloratum*, which is restricted to the Middle East, seems to be able to develop successfully on citrus. The trunk boring larvae of *Stromatium fulvum* attack dry wood indiscriminately and no specific preference is given to citrus. The Levante Red Mite *Anychus orientalis*, known so far from Egypt and Palestine only, is limited by climatic requirements, which are discussed in the special part. The Mediterranean moth *Ephestia vapidella* is polyphagous and shows no specific preference for citrus. The same is true for the White Fly *Aleurotrachelus citri*, known hitherto from Palestine and Egypt only, but possibly of wider distribution.

A few general remarks on the factors regulating the numerical fluctuations of citrus insects (mainly Coccids) may be added. The winter-cold is never intense enough to be a limiting factor, but the cool and wet winter is unfavourable. The speed of development is much retarded, the heavy winds and rains create adverse conditions for the settling of the crawlers. The average climate of the summer, especially in the coastal plain where a moderate air-humidity is maintained during that season, is very favourable. But it is anteceded by the heavy spring hamsins (dry desert winds) in April and May which have a rather decimating influence on almost all citrus insects. The absolute lethal temperature of some citrus pests is:

Species	lethal temperature
Aonidiella aurantii (crawlers)	43.0° C
Saissetia oleae (crawlers, Quayle)	46—50
Pseudococcus citri (female)	42.7
Icerya purchasi (large larvae and female) . .	42.2
Toxoptera aurantii.	38.0
Dionconotus cruentatus (adults)	36.7
Ceratitis capitata (adults).	45.8
Heliothrips haemorrhoidalis (adults)	44.2
Epitetranychus althaeae	51.8
Anychus orientalis.	47.4

These low lethal temperatures for most species explain quite easily the enormous mortality following each hamsin period. The favourable days and weeks of spring are far outbalanced by these repeated decimations and, to a minor degree, this is true for September too. Besides the direct killing effect there is a reduction or almost suppression in sexual maturation and reproduction during the hamsin periods, as is described below in detail for the Red Scale.

The fact that the summer is the most favourable season for most citrus pests is obscured by the seasonal population trend. Almost all citrus-coccids, for instance, reach their annual population ebb towards the end of May or early June. It is often difficult to discover even a few scales at that time within the same orchards which four months later are literally encrusted with them. The heavy infestation in autumn and winter should not lead to the wrong belief that they are optimal for those insects. On the contrary, the environmental resistance is always very high at those seasons and the summer climate (July/September) during which two generations develop a highly increasing population is the local optimum. Many illustrations for this general statement are to be found in the following pages. That many pest problems of the citrus grower are not as serious as they might be is due to a large part to those decimating influences of the hamsin winds. Even the conditions in the Upper Jordan-Valley are probably influenced by the scarcity of hamsins. The proverb saying its an ill wind that blows nobody any good, is literally correct in this case.

This mortality caused by the hamsin winds decimates also the beneficial insects and very often to a higher degree than their hosts, the pests. The spring hamsins are injurious to both, the Citrus Fluted Scale *Icerya purchasi* and to its predator, the lady beetle *Novius cardinalis*. But the decimation of *Novius* is heavier than that of *Icerya* and the latter, therefore, increases with a favourable start during summer, excretes abundant honeydew and *Novius*, if left to itself, controls its host population only at the end of autumn, after the damage has been done. In other lady beetles the spring decimation is very heavy and the summer-heat is likewise unfavourable. *Cryptolaemus montrouzieri*, a predator of *Pseudococcus* which proved effective in Australia, in parts of California and on the French Riviera, was unable to become acclimatized because the favourable spring and autumn months are too short to permit the building up of large populations and the rest of the year is decidedly unfavourable. In *Chilocorus bipustulatus* reproduction is almost suppressed during summer and infant mortality is absurdly high. In consequence, it is unable to check the increasing scale population during the most decisive period.

Two native beneficial insects of *Pseudococcus citri* are occasionally rather common and useful, but they are both very sensitive to hamsins and generally to deviations from the vital optimum of temperature and humidity: the small Encyrtid wasp *Leptomastidea abnormis* and the gallmidge *Dicrodiplosis sp.* Another native predator of *P. citri*, the Brown Lacewing *Sympherobius sacer*, is of much higher epidemiological importance, because it is less sensitive than its host.

Quite a number of other beneficial insects are known and mentioned in the following chapters: predacious thrips as *Karnothrips*, predacious bugs as *Triphleps*, predacious caterpillars as those of *Coccidophaga scitula*, many ladybeetles and Chalcidid wasps, etc. But their total influence on the population trend of their hosts seems, as a rule, to be small, except perhaps in mealybugs. There are still many other beneficial insects which deserve closer study: small *Neuroptera* like *Conwentzia*, predacious mites, etc. The influence of entomophagous fungi seems to be restricted to the winter season and does not seem to be of epidemiological importance.

It is theoretically possible and even probable, that an additional host of a parasite or predator does not deteriorate, but on the contrary alleviate the pest situation. This is especially the case when the seasonal population peaks of both hosts are not coinciding. The raised host populations lead to a rapid increase in those of the enemies which are then much more effective. With the seasonal peaks of abundance of both hosts being different, the initial enemy population surviving the population ebb of one host to the following peak is considerably greater.

Bodenheimer and Shiffer (in press) have made a number of analyses of such population trends. In Palestine, following the adaptation of most local enemies of the citrus mealybug on the recently introduced *Pseudococcus comstocki*, this factor seems to have been extremely important in reducing the new mealybug. And since its appearance no important outbreak of *P. citri* has been observed. Now, as both are reduced to very low populations, the way is free for new minor outbreaks of both mealybugs. The importance of such outbreaks depends mainly on the extent of favourable climatic conditions as well as on the population levels which the enemies are able to maintain during the prolonged population ebb of the hosts. In case this level is too low, mass breeding of enemies in order to be ready for any outbreak will remain advisable.

Before parasites are introduced against a pest, not only the question of its viability under the new conditions has to be tested, but also the interspecific competition between the various parasites. The indiscriminate introduction of all available parasites of *Ceratitis capitata* has probably depressed the possible effect, as the two most important parasites are heavily competing among themselves. The same problem has recently arisen in Palestine. The wasp *Clausenia purpurea* was introduced from Japan against Comstock's mealybug. In the meantime, however, most of the local enemies of the citrus mealybug had begun to infest the new mealybug. The local parasites are now the first to attack Comstock's mealybug in every new focus and are subsequently driven out by the competition of the *Clausenia*. No damage has obviously been done, but there remains the interesting question, if the local enemies of *P. citri* would not have done the same work without the new introduction.

In discussing the birds and spiders of the citrus groves, we paid attention to a certain sequence of different groups of animals with the growing age of the orchard.

The same is true for the citrus insects. The following two monthly counts of insects infesting one branch of a young (5 years) and of an old (25 years) orange grove at Hedera, both in 1935, illustrate this point rather well:

Young orchard	month I	II	III	IV	V	VI	VII	VIII	IX	X	XI	XII
Aonidiella aurantii . .	10382	4813	5610	2547	216	26	88	193	483	286	158	158
Coccus hesperidum. . .	—	2	2	2	—	1	2	2	—	—	1	1
Anychus orientalis. . .	—	—	—	1	2	—	—	7	14	14	9	—
Old orchard												
Aonidiella aurantii. . .	11	21	33	15	6	19	52	59	54	84	108	87
Parlatoria pergandii . .	775	533	1303	887	468	230	241	305	185	207	279	351
Coccus hesperidum. . .	—	8	14	12	4	1	2	15	6	1	6	5
Ceroplastes floridensis .	75	28	36	10	10	63	772	956	245	312	432	501
Pseudococcus citri . . .	—	—	—	—	—	—	—	5	—	—	—	—
Icerya purchasi	—	—	—	—	—	—	—	3	2	—	—	—
Heliothrips haemorrhoidalis.	—	—	—	—	—	—	1	—	—	19	4	3
Anychus orientalis. . .	—	6	9	6	5	—	—	8	16	17	—	—

The explosive expansion of the Red Scale population in the young orchard (about 200 leaves) and its absolute as well as relative conctraction on the older trees (about 1000 leaves) is not the only remarkable feature. The annual range (maximum: minimum) was 384 in the young as against 18 only in the old orchard. Both, *Coccus hesperidum* and *Anychus orientalis* are of much higher relative importance on young trees. Most other species are extremely rare in young orchards, except for occasional and local outbreaks of *Toxoptera aurantii*. The older orchard has many additional species, among which abound *Parlatoria pergandii* and *Ceroplastes floridensis; Icerya purchasi, Pseudococcus citri* and *Heliothrips haemorrhoidalis* are always present, but not necessarily in numbers. *Mytilococcus beckii* should be added to those species. Wherever it occurs, it abounds in older orchards only.

Insects to which citrus is not a real host may occasionally become quite injurious. The adult bugs of *Dionconotus cruentatus* which develop on various herbs in the orchard are attracted by the citrus blossoms on which they suck. Extended blossom drop is the consequence of this attack. A timely eradication of the weeds, prevents this damage entirely. Different species of ants are gnawing the bark off citrus twigs either on the root-collar or near the inoculations, but only shortly after planting or after budding. ,,Residual'' insect pests, as the Tenebrionid beetles *Erodius* and *Adesmia,* form another group. When an orange grove is planted, the earlier fauna living there, is deprived of its regular food. The remaining insects are forced to nibble on the young citrus plants which are the only kind of vegetation available. They fail to adapt themselves to the new conditions and consequently disappear within the first few years.

Some insects live on weeds in the orchard as some flea-beetles and are found occasionally nibbling on citrus leaves. Other species behaving likewise are the curculionid-beetles *Sitones* a.o.; they develop on lupine cover crop, which is cut down and ploughed under as green-manure. Most insects of this group appear for a very short period only on citrus, often very locally only and — young plants in seedbeds excepted — do very little damage.

A serious potential danger is the Desert Locust *Schistocerca gregaria* which invades Palestine at regular intervals. Many signs of damage inflicted in the great locust year 1915 were still recognisable in 1925. Luckily, our groves have been spared a repetition of this attack ever since, and it seems almost sure that the country is now prepared to prevent serious damage to any larger extent.

Some of the cases just quoted offer good illustrations of the fact that the accidental flora of the citrus grove may prove to be of some importance. Some of the citrus scales are able to develop on weeds, as *Polygonum* a.o. This is of little importance, but the neighbourhood of summer-fruits even of a few peach-, apricot-, apple-, guava- trees only is a heavy menace in connection with the Mediterranean Fruitfly *Ceratitis capitata*. They enable the breeding of this fly in additional seasons within the same orchard and increase the size of the *Ceratitis*-population which is ready to attack citrus fruits in September. The relative scarcity of summer-fruits in citrus districts is certainly one of the major reasons for the relative low damage of this pest.

A very remarkable trait observed in many citrus-insects is their mono- or oligophagy in Palestine and other Mediterranean regions in contrast to their extreme polyphagy in their original home and other tropical regions. *Mytilococcus beckii, Parlatoria pergandii, Pseudococcus comstocki, Toxoptera aurantii, Heliothrips haemorrhoidalis* are outstanding samples of this kind.

On the contrary; some quite common species of Palestine do not invade citrus in this country by certain ecological reasons. The case of *Aspidiotus hederae*, recently analysed by Steinitz, illustrates this point.

When host conditions are considered, the differential suitability of the various citrus species and varieties should be born in mind. With regard to coccids, red mites, thrips, (*Ceratitis* excepted) the following sequence can almost invariably be observed: citron, sour lemon, grapefruit, sweet lemon and finally sour orange and mandarins. The cause for this differential host resistance has still to be explored. Part of the relatively favourable pest situation of our orchards is undoubtedly due to the hitherto negligible amount of sour lemons among our plantations. This fact should well be born in mind whenever an important increase of our lemon-area is contemplated.

We may conclude that the relatively poor citrus fauna of Palestine is composed almost exclusively of polyphagous species of wide distribution, most of them derived from the original home of citrus, the Oriental Kingdom. The native fauna of Palestine, belonging either to the Mediterranean or to the Saharo-Sindian element, has not been able to adapt itself to the citrus host despite the fact that the citron has been cultivated for more than 2000 years in this country and that wide areas are covered with citrus in monoculture within the last 100 and especially the last 20 years. The only process of „adaptation" is its recent infestation by *Papilio machaon*, which is based on old generic instincts and was prepared by the earlier infestation of a plant of the same family (*Ruta*).

The Coccidae offer a very good illustration for this point. None of the Mediterranean citrus-scales is native, and this phenomenon is of a general importance, as is shown in the following table:

Character of host	Number of species of Coccidae in the Mediterranean regions			
	Total	Mediterranean	Other elements	ubiquists
on 7 native plants	122	54	24	44 species
on 5 introduced plants . . .	51	—	—	51 „

The polyphagous, ubiquistic immigrants have amply invaded the native flora, but no reverse process has been observed so far. Many attempts have been made by the residual fauna, by the fauna of weeds and neighbouring plants to infest citrus, but only a few polyphagous species, mostly of wide distribution, have succeeded so far to invade citrus occasionally. None has established itself permanently on citrus as a permanent major host.

The citrus fauna of the world, and especially within the Mediterranean area, is still in full migration and expansion, aided enormously by human agency and commerce. Now that the importance of strict quarantine measures has been recognised, we should be able to check and prevent further undesirable introductions in the future. The study of the climatic optima and limits as well as host sequence, host-density and of other ecological factors should help in the future to understand the potential dangers of species not yet introduced into this country.

B. CITRUS PESTS OF THE COUNTRIES OF THE MEDITERRANEAN BASIN.

The more important citrus pests of the Mediterranean countries are:

Name	Spain	S.France	Algeria	Italy	Greece	Pale-stine	Egypt
Aspidiotus hederae Vall. . . .	1	2	1	3	1		
Aonidiella aurantii Mask. . .			—	+	3	3	2
Chrysomphalus aonidum L. . .			3	+		3	3
Chrysomphalus dictyospermi Mask.	3	3	3	3	2		
Mytilococcus beckii Newm. . .	3	3	3	2	2	2	1
Mytilococcus gloveri Pack. . .	3	+		+	+		
Parlatoria zizyphi Luc. . . .	1		3	2	2		
Coccus hesperidum L.	2	2	2	2	1	1	+
Saissetia oleae Bern..	2	2	2	2	1	1	+
Ceroplastes sinensis Guerc. . .	—	2	1	2			
Ceroplastes floridensis Comst. .						2	2
Pseudococcus citri Risso . . .	3	2	3	3	2	2	+
Pseudococcus comstocki Kuw. .	—	—	—	—	—	2	—
Icerya purchasi Mask.	3	2	2	2	2	2	+
Ceratitis capitata Wied. . . .	1	1	2	2	2	2	2
Cryptoblabes gnidiella Mill. . .				1		1	
Heliothrips haemorrhoidalis Bch.	+	+	+	1	+	2	1
Toxoptera aurantii Koch . . .	1	1	1	1	1	1	1
Epitetranychus althaeae Wett.	+	+	+	+	+	1	3
Anychus orientalis Z.	0	0	0	0	0	2	2
Phyllocoptes oleivorus Ashm. .	0	0	0	0	0	3	+

No attention was paid in this list to unimportant or occasional pests only and the bonitation (0 to 3) was made for average years. The Mediterranean Fruit Fly (*Ceratitis capitata*) may be extremely injurious in certain seasons or for certain late varieties; in other years, however, its damages are quite negligible. It was therefore qualified only as 2 instead of 3 (highest degree of injury).

The unequal distribution of these species is only partly due to climatic limitations, as is the case with the Levante Red Mite (*Anychus orientalis*). In some cases, the species had not yet spread or has not yet been imported to other countries.

Of special interest are those species which are present all over the area, but invade citrus as host only in some part of it. Besides *Aspidiotus hederae* the polyphagous moth *Cryptoblabes gnidiella* belongs here. The latter is present all over the area, but is recorded from citrus, so far, only from Palestine, S. Anatolia and Italy.

The particular types of distribution of *Aonidiella aurantii*, *Chrysomphalus aonidum*, *C. dictyospermi*, *Mytilococcus beckii*, *Parlatoria zizyphi*, *Ceroplastes sinensis* is discussed in the special part. The economical importance of *Saissetia oleae* and *Pseudococcus citri* seems to decrease from West to East. The damage of *Ceratitis capitata* depends at least as much upon the local host sequence as it does upon climatic conditions in our area.

The following 28 species of Coccidae, the largest and most important groups of citrus pests, are recorded so far from circum-Mediterranean countries:

Icerya aegyptiaca Dougl.
Icerya purchasi Mask.
Pseudococcus citri Risso
Pseudococcus adonidum L.
Pseudococcus comstocki Kuw.
Pseudococcus filamentosus Ckll.
Pseudococcus longispinus Targ.
Phenacoccus hirsutus Green
Coccus hesperidum L.
Coccus tesselatum Sign.
Saissetia oleae Bern.
Pulvinaria floccifera Westw.
Ceroplastes floridensis Comst.
Ceroplastes sinensis Guerc.
Ceroplastes rusci L.
Aspidiotus hederae Vall.
Hemiberlesia lataniae Sign.
Hemiberlesia camelliae Sign.
Pseudaulacaspis pentagona Targ.
Chrysomphalus aonidum L.
Aonidiea aurantii Mask.
Chrysomphalus dictoyspermi Mask.
Mytylococcus gloveri Pack.
Mytilococcus beckii Newm.
Mytilococcus ulmi L.
Parlatoria pergandii Comst.
Parlatoria proteus Comst.
Parlatoria oleae (Colv.)
Parlatoria zizyphus Luc.

A few doubtful species have been omitted from this list. Sixteen of these species are widely distributed polyphagous scales, common on citrus. Eleven other species are known either as widely distributed and polyphagous without having everywhere infested citrus. The position of only *Ceroplastes sinensis* remains doubtful. This polyphagous species has, most probably, been imported from Eastern Asia to Italy on ornamental plants and had become acclimatized in Italy and in the Western Mediterranean; but up till now it has not yet been recorded from Eastern Asia. The assumption of Lindinger that it be a synonym to *C. floridensis* is, of course, incorrect.

The earlier assumption of the writer that a few of these ubiquistic species are originally native to the Mediterranean region, must be abandoned in the light of closer investigation. *Aspidiotus hederae* has very recently extended considerably its area in some Mediterranean countries (as Morocco), and its home is probably in the savannas of Africa or in the drier parts of S. Africa. *Parlatoria zizyphi* is almost certainly of Oriental origin. *Saissetia oleae* was most probably introduced from the African savanna and the origin of *Ceroplastes rusci* is still obscure, but probably similar.

C. ON THE GENERAL DISTRIBUTION OF CITRUS-COCCIDAE.

Somewhat above 100 species of Coccidae have been recorded from citrus in the different regions of the world.

In order to find the endemic species characteristic of citrus of the different zoogeographical zones we must eliminate the ubiquists or widely distributed species, which is more than half of all known forms. Another large group of species to be eliminated comprises those which are restricted to one zone, but are polyphagous and therefore not primarily connected with citrus. The remaining typical citrus-coccidae are as follows (in alphabetical order):

1. *Oriental Kingdom:* 20 species

Aonidiella citrina
Aonidiella comperei
Erium sp.
Geococcus citrinus
Icerya crocea
Icerya okadae
Coccus colemanni
Mytilococcus lasianthi
Monophlebus dalbergiae
Parlatoria sinensis
Phenacoccus iceryoides
Prontapsis vanonensis
Pseudococcus citriculus
Pseudococcus corimbates
Pulvinaria aurantii
Pulvinaria cellulosa
Pulvinaria citricola
Rhizoecus condonis
Tachardia aurantiaca
Takahashia citricola

2. *Australian Kingdom:* 2 species

Pulvinaria ornata
Pulvinaria tecta

3. *Ethiopical Kingdom:* 3 species

Akermes andersoni
Icerya natalensis
Pseudococcus fragilis

4. *Neotropical Kingdom:* 5 species

Akermes punctatus
Coccus deltae
Chrysomphalus albopictus
Chrysomphalus koebelei
Chrysomphalus leonis

5. *Holarctic Kingdom:*

A. Nearctis: 1 species
Chionaspis sassceri

B. Mediterranean region: 0 species

Coccus deltae is probably polyphagous, but no other hosts are yet recorded, as far as the writer is aware. The three *Chrysomphalus* species from Mexico may either by synonyms or polyphagous, but no additional information on the species seems to have been produced since their description. Also *Chionaspis sassceri* and *Akermes punctatus* may easily prove to be either introduced, or not restricted to citrus.

The number of endemic citrus coccidae corresponds well to the natural distribution of the subfamily Aurantiinae, especially if the above remarks are taken into consideration. The Oriental Kingdom again is established as the home of the host by the concentration of endemic citrus coccidae. The few endemic species in Australia and Ethiopia seem to be genuine, whereas in all other zones the presence of any monophagous, specific citrus-coccid is rather doubtful.

The following 22 species of Coccidae are known from more than four major citrus areas:

Icerya purchasi Mask.
Pseudococcus citri Risso
Pseudococcus adonidum L.
Pseudococcus filamentosus Ckll.
Pseudococcus longispinus Targ.
Coccus hesperidum L.
**Coccus viride* Green
**Saissetia hemisphaerica* Targ.
Saissetia oleae Bern.
Ceroplastes floridensis Comst.
**Ceroplastes rubens* Mask.
Aonidiella aurantii Mask.
Chrysomphalus aonidum L.
Chrysomphalus dictyospermi Mask.
**Selenaspidus articulatus* Mask.
Mytilococcus gloveri Pack.
Mytilococcus beckii Newm.
Parlatoria pergandii Comst.
Parlatoria proteus Comst.
Parlatoria zizyphus Luc.
**Prontaspis citri* Comst.
**Pinnaspis aspidistrae* Sign.

These species, of which only the six marked with an asterisk are unknown from citrus within the Mediterranean region, can be considered as the common and therefore most typical fauna of citrus scales. They are all very polyphagous; none is confined to citrus. Of the six species marked with an asterisk, three are present in the Mediterraean region, but on hosts other than citrus, whereas the 3 remaining species (*Ceroplastes rubens*, *Selenaspidus articulatus* and *Prontaspis citri*) are still completely absent.

The information on citrus pests from groups other than Coccidae is still too incomplete to permit a similar analysis. As far as a first survey shows, the results would be in full agreement with those obtained in the Coccidae.

The outstanding character of most cultural immigrants is their considerable euryoecy, their high vitality, which enables them to resist many unfavourable factors, to settle on many hosts (polyphagy), and on these hosts on many parts of the plant (eurymery). They are fit to spread, often with considerable speed, over wide areas. Some are almost ubiquistic, living in cooler climates in xerothermous islands still outdoors, whilst mainly restricted there to greenhouses. *Coccus hesperidum* is a prototype of this ubiquistic group. Other species still remain restricted somehow to one or a few of the broad transversal zones of the earth's climate: to the moderate, the subtropical or the tropical climates. Maskell has pointed out long age for *Icerya purchasi*, that an eurymerous species tends to develop polyphagy. The study of the host-catalogue coccids of the Middle East as well as of most other regions reveals, that almost only cultural immigrants respectively native species which beyond their home have developed into cultural immigrants, show a very pronounced polyphagy. We have however pointed out, that quite a number of these cultural immigrants, in spite of a pronounced polyphagy in their native territory, are restricted in the invasion area to one main, and perhaps a few additional accessory hosts. This restriction may be temporary, yet it may also be the consequence of an unfavourable environment.

Another point which deserves attention is that many cultural immigrants, in spite of their wide dispersal show a definite preference for much more restricted climatic zones than is often realised. The Citrus fauna of the Middle East, which for all practical purposes is entirely composed of cultural immigrants among the coccids, offers a good illustration for this restriction.

Citrus scales of the Middle East.

Species	Greece	Egypt.	Cilicia/ Israel	Colchis	S. Iraq/ S. Iran	Caspian
Icerya purchasi	2	2	2	2	—	2
Icerya aegyptiaca	—	1	—	—	—	—
Coccus hesperidum	1	1	1	1	1	1
C. pseudomagnoliarum. . . .	—	—	—	2	—	—
Pulvinaria aurantii.	—	—	—	2	—	—
P. floccifera	1	1	—	(1)	—	1
Saissetia hemisphaerica;. . .	—	—	—	1	—	—
S. oleae	—	1	1	1	—	—
Ceroplastes rusci	1	—	—	—	—	—
C. floridensis.	—	1	2	—	—	—
C. sinensis.	—	—	—	2	—	1
Pseudococcus citri.	2	2	2	(1)	—	(?)
P. comstocki	—	—	2	—	—	—
P. filamentosus.	—	1	—	—	—	—
P. gahani	—	—	—	2	—	—

Species	Greece	Egypt	Cilicia/ Israel	Colchis	S. Irak/ S. Iran	Caspian
P. maritimus	—	—	—	2	—	—
Phenacoccus hirsutus	—	1	—	—	—	—
Aspidiotus hederae	1	—	—	(1)	—	—
Dynaspidiotus britannicus	—	1	—	—	—	—
Aonidiella aurantii	2	2	3	1	—	1
Aonidiella citrina	—	—	—	3	—	—
A. orientalis	—	—	—	—	3	2
Chrysomphalus aonidum	(1)	—	3	1	—	—
C. dictyospermi	3	—	—	2	—	—
C. pinnulifer	1	—	—	1	—	—
Mycetaspis personatus	—	1	—	—	—	—
Pseudaonidia trilobitiformis	—	—	—	(1)	—	—
Hemiberlesia lataniae	1	—	—	—	—	—
Mytilococcus beckii	2	1	2	2	—	1
M. gloverii	—	—	—	3	—	1
Leucaspis japonica	—	—	—	2	—	—
Parlatoria cinerea	—	—	?	—	—	—
P. morrisoni	—	—	—	—	1	—
P. oleae	—	—	1	1	—	—
P. pergandei	2	1	2	(1)	—	—
P. zizyphi	—	—	1	2	—	—
Total: 36	13	14	14	25	3	9

1 to 3 indicate progressive intensity of infestation.

Of these 36 species the following four are not characteristic, as they are very polyphagous and present everywhere except in S. Irak:

Icerya purchasi	*Pseudococcus citri*
Coccus hesperidum	*Mytilococcus beckii.*

The very low number of Citrus coccids in Irak (*Aonidiella orientalis, Parlatoria morrisoni*) is not due to lack of knowledge, but to the extreme climate and to historical isolation. The following 9 species should be regarded as accidental infestations of polyphagous species: *Hemiberlesia lataniae* in Greece, *Pseudococcus filamentosus, Phenacoccus hirsutus, Icerya aegyptiaca. Dynaspidiotus britannicus, Mycetaspis personatus* in Egypt. *Saissetia oleae* and *Pseudaonidia trilobitiformis* in the Colchis, and *Ceroplases rusci* in the Egean province of Turkey.

Among the remaining 23 species two are characteristic, as mentioned above, for S. Irak and S. Iran. Greece and the Egean coast are characterised by: *Chrysomphalus dictyospermi, Aonidiella aurantii,* and *Parlatoria pergandii.* The Levante coast, from Cilicia to S. Palestine, by: *Ceroplastes floridensis, Pseudococcus comstocki, Aonidiella aurantii, Chrysomphalus aonidum, Parlatoria pergandii.* For Egypt: *Chrysomphalus aonidum* and *Aonidiella aurantii.* For the Caspian coast of Iran: *Aonidiella orientalis, Parlatoria zizyphi, Mytilococcus gloveri, Ceroplastes sinensis, Pulvinaria floccifera.* The very peculiar fauna of the Colchis is mainly composed of recent Oriental immigrants:

Coccus pseudomagnoliarum	*Aonidiella citrina*
Pulvinaria aurantii	*Chrysomphalus dictyospermi*
Ceroplastes sinensis	*C. pinnulifer*
Pseudococcus gahani	*Mytilococcus gloverii*
Pseudococcus maritimus	*Leucaspis japonica*
	Parlatoria zizyphi.

These very great differences of the citrus-fauna of the various citrus regions of the Middle East can not be explained simply on the basis of differential histories of introductions. It is true that in every area are some recent invaders, the definite status of which still awaits elucidation, and some of which have not yet, so far been introduced into the other areas. *Pseudococcus comstocki* and *Parlatoria zizyphi* in the Levante area are such samples. But especially the differences in the following genera: *Aonidiella, Chrysomphalus, Parlatoria, Mytilococcus, Ceroplastes, Coccus,* and *Pulvinaria* are significant and based on the survival value of the coccids in the various climates of the Middle East. Especially the Oriental character of the Colchis with its S. Japanese type of climate is significant, and by no means restricted to the citrus-fauna. This is a striking illustration for the fact that even cultural immigrants of wide distribution show definite zoogeographical areas of preference. We may well say to-day, that the citrus-fauna of any area will be a good indicator to the climatic character of the region.

It is perhaps indicated to remark that information from plant-quarantine services are not reliable with regard to provenience. The U.S.A. quarantine service, the best organized and most reliable service in the world, indicates as intercepted from Palestine between 1926 and 1936 the following pests:

	orange	grapefruit	lemon	cederate
**Aspidiotus hederae*	4	—	—	—
**A. sp.*	1	—	—	—
Aonidiella aurantii	8	—	3	20
Chrysomphalus aonidum	5	1	2	—
**C. dictyospermi*	2	—	—	1
**Selenaspidus articulatus*	1	—	—	—
Mytilococcus beckii	5	1	1	2
Parlatoria oleae	1	—	—	—
P. pergandii	13	1	3	6
**P. zizyphi*	2	—	—	—
Pseudococcus citri	—	—	1	4
**P. kraunhiae*	—	—	—	4
Ceratitis capitata	4	6	—	—
**Hylemyia sp.*	—	—	—	1
**Anobiid*	—	—	—	1
Heliothrips haemorrhoidalis	—	—	—	1
**Paratetranychus sp.*	1	—	—	—

Nine out of these 17 species (marked with an asterisk) have never been found in Palestine on citrus. *Pseudococcus kraunhiae* is the only one which may have migrated to the fruit during the transport. For the other species this is most improbable. It seems therefore that the declaration of provenience was false in the other cases. This is especially probable for cederate fruits, which are used for Jewish ritual purpose in the Feast of Tabernacles and in which Palestine is often wrongly indicated by salesmen as patria.

No conclusions with regard to distribution of pests should therefore be drawn from such lists.

4. SIMPLIFIED SYMPTOMATIC KEY TO THE CITRUS PESTS OF PALESTINE.

The following symptomatic key aims to help the farmer who possesses no special entomological training and is often faced with the symptoms of damage only on the tree or fruit, to find his way to the group of insects which causes the damage in question. Special keys and comparisons of illustrations and of the descriptions will acquaint him better with the species which caused the damage. No explanation for the use of the key seems necessary, besides the statement that the key is generally arranged dichotomic, i.e. selecting an important character for an alternative choice. Both characters of the alternative couple should be compared before a decision is made.

The more important insects of the citrus seedbed and nursery are different from those of the orchards, those of the young orchard from those of the older orchard. In order to avoid unnecessary repetition, only the more specific injuries of the seedbed and nursery are treated separately in addition especially. The general key should be used for scale-insects, aphids, etc.

More detailled keys, based on morphological characters are given for some important orders and families in the special part.

A. *Roots.*

1. Soil particles cling to infected roots, the bark of the rootlet may be easily separated from the inner woody portion, and in heavily infested trees the number of rootlets is very limited. Closer examination reveals the females of a nema, embedded with their heads in the rootlets. The Root-Nema, *Tylenchulus semipenetrans.*
2. Patches of the bark are gnawed off the root-collar, the white wood lying bare on those parts. Ants, *Tetramorium signatus.*

B. *Coccidae* (Scale insects, mealybugs, soft scales, fluted scales).

(As this group may occur on almost any part of the plant, a preliminary key to an easy determination of them is included here). The females are the basis of the key).

1. Small, body covered with a scale. (2)
   Larger animals, without scales. (6)
2. Scale long, mussel-shaped, brown. The Mussel Scale (*Mytilococcus beckii*).
   Scale roundish. (3)
3. Animal yellow to yellow-brown. (4)
   Animal violet. (5)
4. Scale flat, elevated only centrally, translucent, reddish. The Red Scale (*Aonidiella aurantii*).
   Scale elevated, black brown with orange-red central spot. The Black Scale (*Chrysomphalus aonidum*).
5. Scale grey to whitish with yellow-brown excentric spot, ± flat, less than 2 mm. in diameter. The Chaff Scale (*Parlatoria pergandii*).
   Scale gray with greenish-brown subcentric spot, convex, more than 2 mm. in diameter. The Olive Parlatoria (*Parlatoria oleae*).
6. Insect large, segments easily recognisable with a magnifying glass, with wax-secretions (not with wax plates), moving. (7)
   Insect large, no segments recognisable, not moving. (8)

7. Insect dark red with a large white and striated („fluted") ovisac with pink eggs beneath the abdomen. No wax secretions on the back.
The Fluted Scale *(Icerya purchasi)*.
Insect with irregular ovisac, covered with wax threads, with yellow eggs beneath the abdominal end, covered with mealy wax secretions over the whole body, with marginal waxy appendages. (**8**)

8. Marginal wax-appendages short and plump, the posterior too reaching mostly up to 1/6 of the length of the body. The Citrus Mealybug (*Pseudococcus citri*).
Marginal wax-appendages long. The posterior ones may even surpass the length of the body considerably.
The long-tailed Mealybugs (*Pseudococcus comstocki and P. adonidum*).

9. Flat, long-oval, brown. The Brown Soft-Scale (*Coccus hesperidum*).
Highly convex. (10)

10. Insect not covered with wax-plates (11)
Insect wholly armoured with thick wax-plates. (12)

11. Adult female with a distinct H.-shaped crest on its back, dark brown.
The Olive Soft Scale (*Saissetia oleae*).
Adult female with smooth, shiny back; black-brown.
The Shiny Soft Scale (*Saissetia hemisphaerica*).

12. Adult female covered with one apparently homogenous wax-layer (with tubercles, but no distinct plates), 2.8. mm long.
The Florida Wax Scale (*Ceroplastes floridensis*).
Adult female with a central wax plate and 8 distinct marginal plates, 4 mm long.
The Fig Wax Scale (*Ceroplastes rusci*).

C. *Stem, branches and twigs.*

1. Large parts of the bark of branches of single trees only, often including parts of the wood, are gnawed away, especially on lemon trees. Splinters are found abundantly beneath the damage. Rats (*Rattus frugivorus*).
2. Large, shiny, black beetles bore galleries of 4—10 cm length mainly in the branches, of cederate and lemon, preferably, and 1 cm in diameter in May to July.
The Giant Bostrychid Beetle (*Apate monachus*).
3. Small patches of green bark are gnawed away, in spring mainly, on twigs and tender branches, by hornets. The Oriental Hornet (*Vespa orientalis*).
4. The twigs are covered with thousands of small brown mites, which feed on the sooty mould only. The Sooty Mould Mite (*Orbibatula plantivaga*).
5. Large boring galleries in the wood of old decaying or dead trees are caused by Longicorn beetles mainly, espec. *Stromatium fulvum*.
Small boring galleries are caused by Bostrychid beetles (other than *Apate*).

D. *Leaves.*

1. No parts of the leaf are eaten away. (2)
Parts of the leaves are eaten away. (3)

2. Leaves at tips of twigs curled, but not chlorotic. Blackish or green aphids on the lower side of leaves. Aphids, mainly *Toxoptera aurantii*.
Leaves with silvery areas, spotted with reddish black excrements.
Thrips (*Heliothrips haemorrhoidalis*).
Leaves either heavily chlorotic or curled and not chlorotic. Occasional small, black-brown sucking punctures are visible. Leafhoppers (*Empoasca spp.*)
Leaves grayish, mottled or chlorotic. Under the magnifying lense an ample cover of threads and the presence of red mites is recognisable.
Red Spiders (*Anychus orientalis, Epitetranychus althaeae*).
Eggs and larvae of bugs on the leaves, which are occasionally wilting.
Bugs (*Nezara viridula, Stenozygum coloratum*)
Small scale-like, flat-oval insects on the leaves, easy recognisable by their dark black colour and their purely white wax-margin. 1 mm long.
Pupae of the Citrus White Fly (*Aleurotrachelus citri*).
Similar yellowish green ones with white margins belong to the
Tobacco White Fly *(Bemisia tabaci)*.

3. Irregular large patches of the younger leaves, with clean cut margins only, are eaten, from the central parts as well as from the margin.
Caterpillars (*Papilio machaon* or Noctuid moths).
More or less regular large patches with clean cut margins are eaten from the margin of the leaves only. Grasshoppers, espec. *Anacridium aegyptium.*
Irregular feeding patches with ravelled margins. Silvery traces of a dry glue are found on the leaves. Snails, (*Helix* spp.)
Small to very small feeding patches by occasional visitors amongst the
Beetles & Weevils (Coleoptera).
Add: the Silver Rust Mite (*Phyllocoptes oleivorus*).

E. *Flowers.*

1. Visitors of flowers for a short limited period. (2)
Insects inhabiting the flowers. (3)
2. Visitors for gathering the nectar. Bees, espec. the honeybee (*Apis mellifica*).
Visitors for gathering the pollen. Scarabaeid beetles, espec. *Oxythyrea spp.*
Black and red coloured bugs puncture the flowers, which drop in consequence.
The Citrus Blossom Bug (*Dionconotus cruentatus*).
3. Small whitish to pink caterpillars, which web threads, live and feed within the flowers, espec. of cederate and lemon. The Citrus Flower Moth (*Prays citri*).
Small, very slender and very narrow, black or yellow insects, live within the flowers.
Thrips (espec. *Heliothrips haemorrhoidalis*).

F. *Fruits.*

1. Insect larvae in decaying fruit. cf. special key, p. 90.
Insect damage in or on healthy fruit. (2)
2. The contact between fruit-stem and fruit is loosened. (3)
Spoils of the fruit itself. (4)
3. Ants web a small tent around the fruit stem and the end of the latter is often knibbled away. Ants (*Crematogaster jehovae*).
When the June drop is prolonged into late June of July, larvae of the Citrus Mealybug are commonly found beneath the sepal.
(*Pseudococcus citri*) (also *Coccus, Mytilococcus, Chrysomphalus, Thrips* a.o.)
4 Borings of insects in the rind. (5)
Opening or large holes in mature fruits by birds (Corvus) or rodents.
Surface spoils of the fruit. (6)
5. Punctures surrounded by small chlorotic spots on the exposed parts of the fruit. When opened, an enlarged egg-chamber is visible immediately beneath the surface.
The Mediterranean Fruit Fly (*Ceratitis capitata*).
Small punctures, not surrounded by chlorotic spots, but often where two fruits touch and especially where small colonies of mealybugs have settled down. The damage is often followed by local gum-secretion on the fruits surface.
The Honeydew Moth (*Cryptoblabes gnidiella*).
6. Sooty mould covers the fruit, caused by
Soft Scales, *Icerya* and Mealybugs.
Patches of the rind are eaten away and corky groves are formed on its surface.
Cutworms (espec. *Prodenia litura*).
Silvery chlorotic decolorations on the surface caused by
The Greenhouse thrips (*Heliothrips haemorrhoidalis*).
Silvery scaly decoloration on lemons, rusty to black ones on other citrus fruits.
The Citrus Rust Mite (*Phyllocoptes oleivorus*).

G. *Buddings.*

1. Feeding on the lower leaves: by the residual fauna, mainly Tenebrionid beetles.
Tenebrionidae
2. Feeding on the roots and root-collar.
a. „white ants" on the roots. Termites (*Reticulotermes lucifugus*)
b. Black ants on the bark of the root. Ants (*Tetramorium signatus*)

3. Feeding on the bark of the inoculations. Various ants.
Feeding beneath the bark of inoculations.
Caterpillars of moths (*Ephestia vapidella*, *Cryptoblabes gnidiella*)

H. *Seed-beds and nurseries.*

1. Feeding in patches on the leaves by cutworms espec.
Prodenia litura, *Laphygma exigua*, a.o.
The leaves are not eaten, but grow chlorotic, grayish or mottled. Webs and red mites are visible with a magnifying glass.
Red Spiders (*Anychus orientalis* and *Epitetranychus althaeae*)
The leaves at tips of twigs are curled. Citrus Aphid (*Toxoptera aurantii*)
2. The stem, twigs and branches are encrusted with. Red Scale (*Aonidiella aurantii*).
Parts of the twigs and stem are gnawed away.
The Common Cricket (*Liogryllus bimaculatus*)
3. The young seedlings grow chlorotic and die. The roots are cut either by the Molecricket (*Gryllotalpa vulgaris*) or by the large grubs of the Scarabaeid beetle *Phyllognatus silenus*.
The roots are infested with white ants.. *Reticulotermes lucifugus*.

II. THE CITRUS PESTS OF PALESTINE, INSECTS AND OTHER ANIMALS, THEIR DESCRIPTION, BIOLOGY, DAMAGE AND CONTROL.

I. The Israel Orange Country (Phot. Kluger).

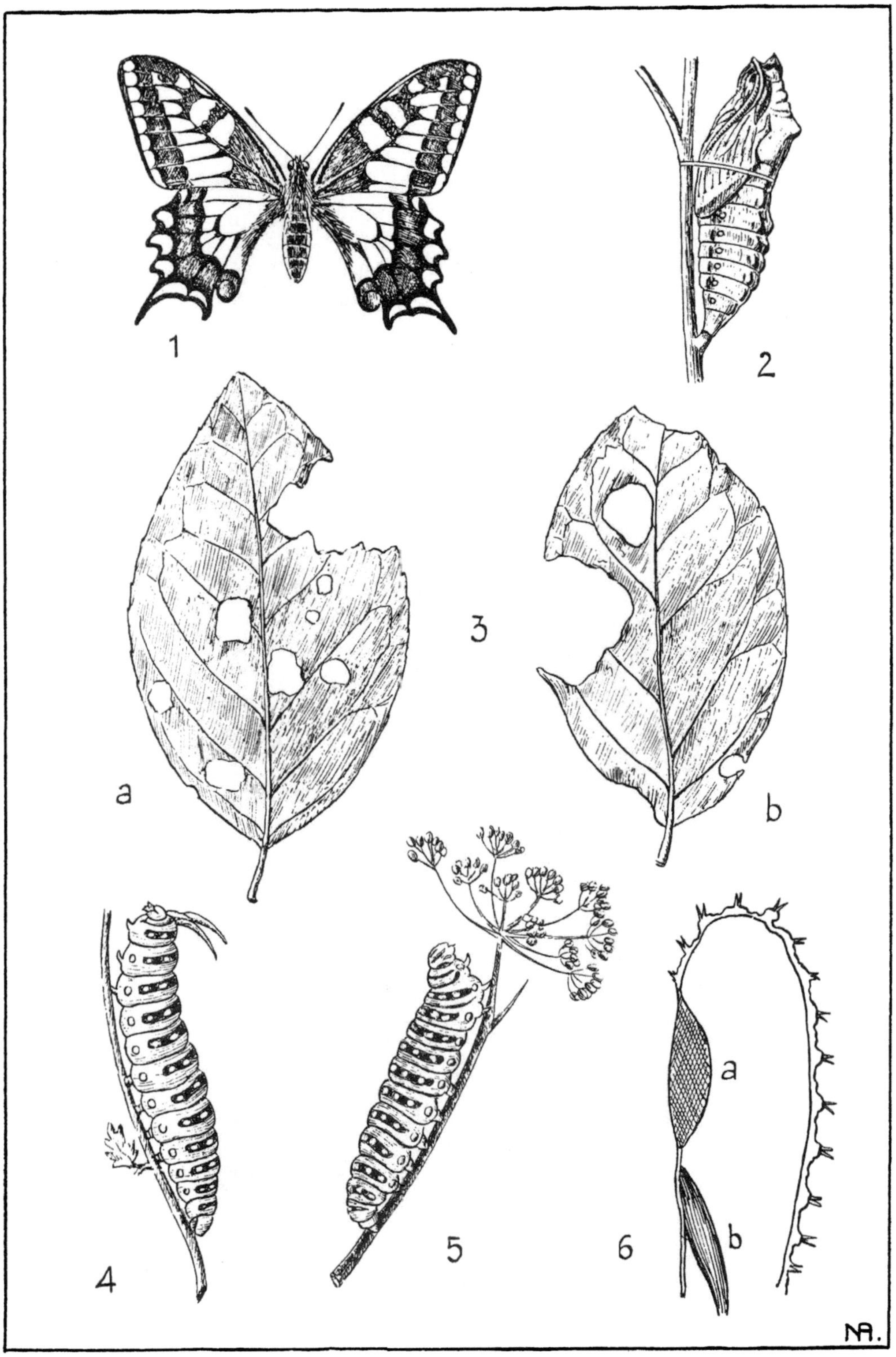

II. The Swallowtail (*Papilio machaon* L.) 1) butterfly, 2) pupa; 3) feeding marks of caterpillar on orange-leaves; 4) caterpillar with neck fork erected; 5) caterpillar (neck fork not erected); 6) cross-section through neck fork (*a.* osmaterium, *b.* retractor muscle).

1. BUTTERFLIES AND MOTHS (*Lepidoptera*).

A. THE SWALLOWTAIL (*Papilio machaon sphyrus* Huebner). (Plate II).

This sulphur-yellow big butterfly with black, and blue markings is readily recognised by its tailed hindwings. It occurs over most of the Palaearctic area and has in Palestine three annual generations. It is on the wing in March, May/June and in August/October. The winter is passed apparently as pupa only. Following complicated nuptial flights they hover quickly over the hosts, laying 1 or 2 eggs at a time. The green caterpillars have black stripes and red spots. When disturbed they protrude two fleshy horns, the neck fork, from the neck. The usual hosts are umbels such as *Daucus*, *Foeniculum*, etc. The weed *Ruta bracteosa*, belonging to another section of the *Rutaceae* than *Citrus*, is another common host in Palestine. Twenty years ago it quite commonly developed here on all species of *Citrus*. The first notice came in XI. 1927 from Magdiel. In 1928 we had reports from Herzlia (V), Gan Hayim (VII) and from Nahlath Yahuda (X.). Since 1929 we have many reports every year from Hedera to Rehoboth. Mr. Morris informed us, that the species has also been observed in Cyprus on citrus since 1933. The closest relatives of our swallowtail, such as *P. demodocus*, *P. demolaus*, a.o. live in tropical regions quite normally on citrus. It is rather peculiar, that no such reports have come from other Mediterranean countries, so far.

Caterpillars of all three generations have been observed on citrus. This proves that the species is now fully acclimatised on this new host. The big feeding patches eaten out of the large yet tender leaves are conspicuous. As only larger trees are infested by a few individuals only on each tree, no real damage is being done. The beautiful butterfly and its equally beautiful larvae are up to date only a decoration of the orchards.

References.

F. S. Bodenheimer, Schaedlingsfauna Palaestinas. Berlin. 1930. p. 12, 163, fig. 43.
P. P. Graves, Transact. Ent. Soc. London. 1925 p. 70—71.

B. THE EGYPTIAN COTTON WORM (*Prodenia litura* F. Noctuidae). (Plate III).

Syn.: P. littoralis Boisduval.

Description: The moth has a wing expanse of 23—38 mm and a body length of 14—18 mm. Forewings dark brown with ochreous lines and bars, running obliquely along the veins, at longer or shorter distances. In the male the ochreous lines are more pronounced than in the female and have an ochreous large band on the lower edge of the forewing. The male, also, has a large bluish or purple area near the apex and a small area of the same color near the base of the forewing. The edge of the forewing is cut by several ochreous bars. Hindwings white, iridescent with a dark brown edge and with brown veins. Thorax and abdomen ochreous to light brown with numerous tufts of hair; head also covered with tufts of light and dark brown scales (Bishara).

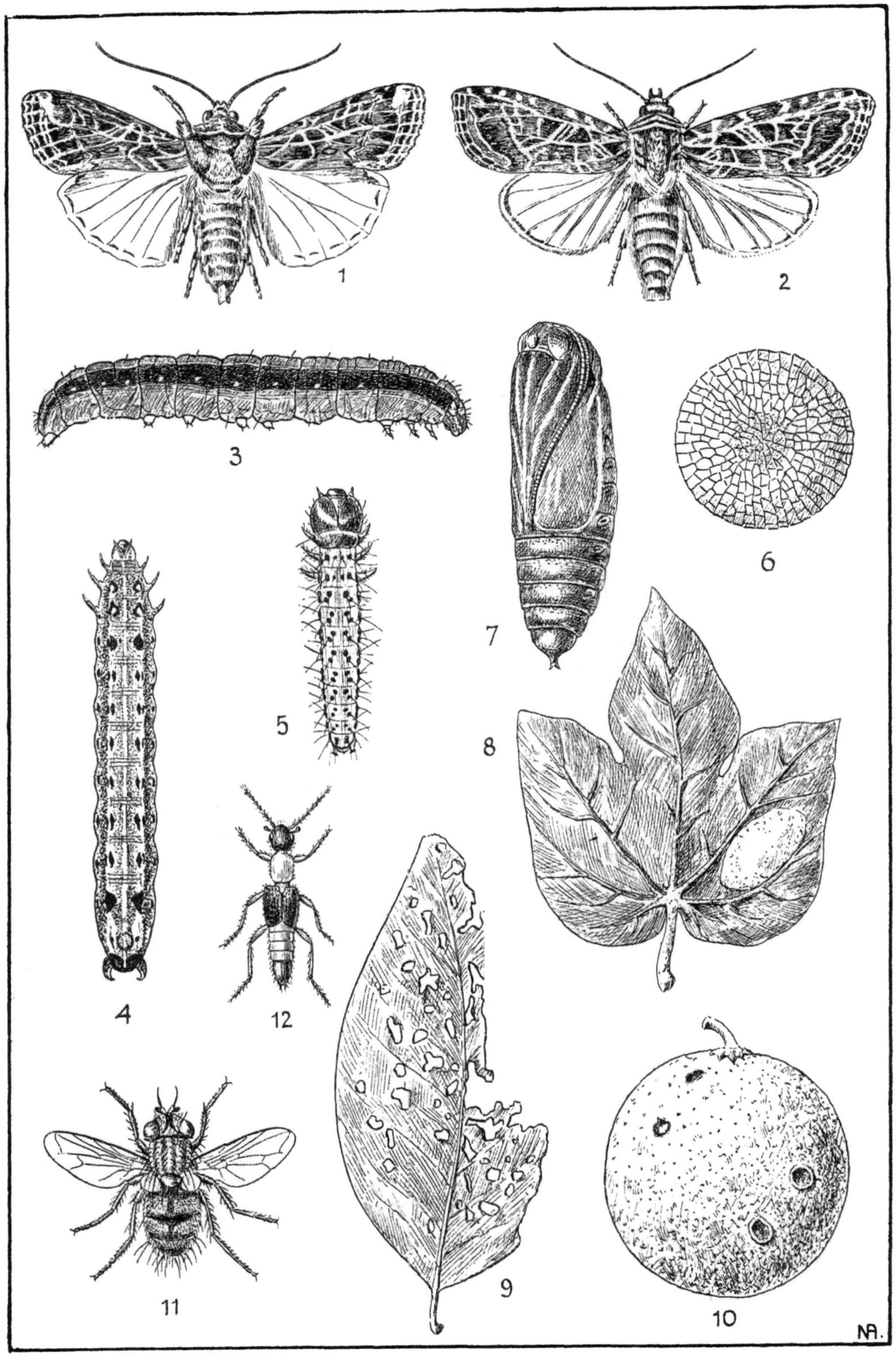

III. The Egyptian Cottonworm (*Prodenia litura* F.). 1) Male; 2) female; 3) and 4) large caterpillar; 5) neonate caterpillar; 6) egg from above; 7) pupa; 8) egg-mass on cotton leaf; 9) feeding marks of young caterpillar on orange leaf; 10) recent and old feeding marks of caterpillars on oranges. 11) *Tachina larvarum* L.; 12) *Paederus fuscipes* Curt. (mainly after Bishara).

The moths are active at twilight and during the night. They copulate soon after hatching and 1—5 days later begin to oviposit, which generally lasts for a similar period. 500—1500, generally over 1000, eggs are laid per female on the leaves or another part of the host plant, in egg-masses, which may contain from about 30 to over 1000 eggs each.

The flat, ribbed eggs are laid in several hundred masses, generally in one, two or three layers. The whole egg-mass is generally covered with ochreous hair from the abdomen of the mother. The originally whitish egg turns black before the hatching.

The young caterpillar has a large, shiny black head. The originally light green body becomes bluish-green after feeding. All its segments are covered with large black tubercles, each tubercle bearing a comparatively long, black hair. The full-grown caterpillar is very variably colored, being generally dark grey, dark brown or almost black, mottled with irregular spots and lines. On the dorsum of meso-and meta-thorax are two black spots with yellow edges; and two large black areas on the first and eighth segments of the abdomen.

Distribution: Tropical and subtropical regions of the old world. In the Mediterranean regions known from the whole Northern Shore of Africa, from Syria, Palestine and Crete.

Life-history: The young caterpillars aggregate in small groups and feed on the lower surface of the host-leaves. After the third moult they disperse and generally pass the day in the soil, climbing on the host plant in the evening, where they feed during the night. The caterpillar goes through six stages.

Pupation takes place in an earth-cell in the soil. The redbrown pupa is 14—18 mm long and the last abdominal segment terminates in two strong hooks.

The following representative data yielded by our breedings will give an idea as to the length of development of every *Prodenia*-generation:

Year	Brood No.	Date of oviposition	Shortest development in days	Date of first ovipostion of F_1
1930	1	10.VIII.	25	4.IX.
	2	4.IX.	32	6.X.
	3	6.X.	38	12.XI.
	4*a*	12.XI.	81	1.II.31
	4*b*	15.XI.	107	2.III.31
1931	1	17.VII.	26	12.VIII.
	2	12.VIII.	23	4.IX.
	3	8.IX.	33	11.X.
	4	12.X.	37	18.XI.
1932	1	25.VIII.	24	18.IX.
	2	18.IX.	28	16.X.
	3	17.X.	38	23.XI.
	4	24.XI.	109	13.III.33
1933	1	23.III.	61	23.V.
	2	22.V.	31	21.VI.
	3	25.VI.	26	21.VII.

Besides the greatly protracted winter generation (3—4 months) and a still protracted spring generation (2 months), development generally takes about one month in summer and autumn. The preoviposition period lasts only 2—5 days, as a rule, and all the eggs are laid within a similar period. About 7 annual generations develop under Palestinian conditions.

The earliest offspring, however, may develop even 8 generations in one annual cycle (cf. 1932/33).

The longevity of the adults changes with the season. Those hatched in the different months lived as follows:

Months	Longevity of moths in days			Average temperature in °C	Average rel. humidity in %
	average	shortest	longest		
I	15.0	10	20	12.4	80.6
II	9.2	4	22	12.9	81.6
III	7.6	2	15	15.1	78.6
IV	6.0	5	7	17.6	74.8
V	6.4	4	9	20.8	79.8
VI	4.0	3	5	23.0	79.0
VII	5.5	4	7	24.3	80.8
VIII	3.5	2	7	25.6	82.0
IX	3.9	2	7	25.0	80.6
X	5.7	3	10	23.6	80.2
XI	6.7	2	11	19.3	78.0
XII	10.1	3	18	14.8	80.7

The scale of activity for the different stages is as follows:

Stage	Month	n	Scale of activity 1	2	3	4	5	6	7	8
Caterpillar (5—10 mm)	VIII	27	4.9	—	15.8	28.4	37.6	50.4	46.1	47.3° C
(20—30 mm)	VIII/IX	14	6.1	—	21.3	28.2	34.3	37.5	46.7	48.4° C
Moths	VIII	16	6.4	—		28.0	37.7	43.4	—	45.4° C

The very young caterpillars are very sensitive to high temperatures; at 47° C they die instantaneously.

Our breedings took place in the laboratory under variable temperatures. The average shortest development (egg to moth) was:

Temperature in ° C	No. of observations	Average	Shortest development in days. Absolute Minimum	At constant temperature (Bishara)
28	1	23		
27	6	24.5	21	24
26	9	26.0	24	
25	11	26.5	24	27
24	3	29.0	25	
23	8	32.0	30	30
22	2	37.5		
21	7	35.3	31	41
20	8	36.4	31	47
19	15	44.1	39	
18	3	50.0	44	62
17	3	82.7	50	
16	1	77		90
15	2	106		
14	3	109.3	102	146

This leads to the hyperbola formula with the following constants:

development zero (c): 10.4° C

thermal constant (Th. C.): 392 days-degrees

This hyperbola agrees fairly well with that of Bishara, gained under constant temperature conditions in Egypt:

c = 11.2° C

Th. C. = 414 days-degrees.

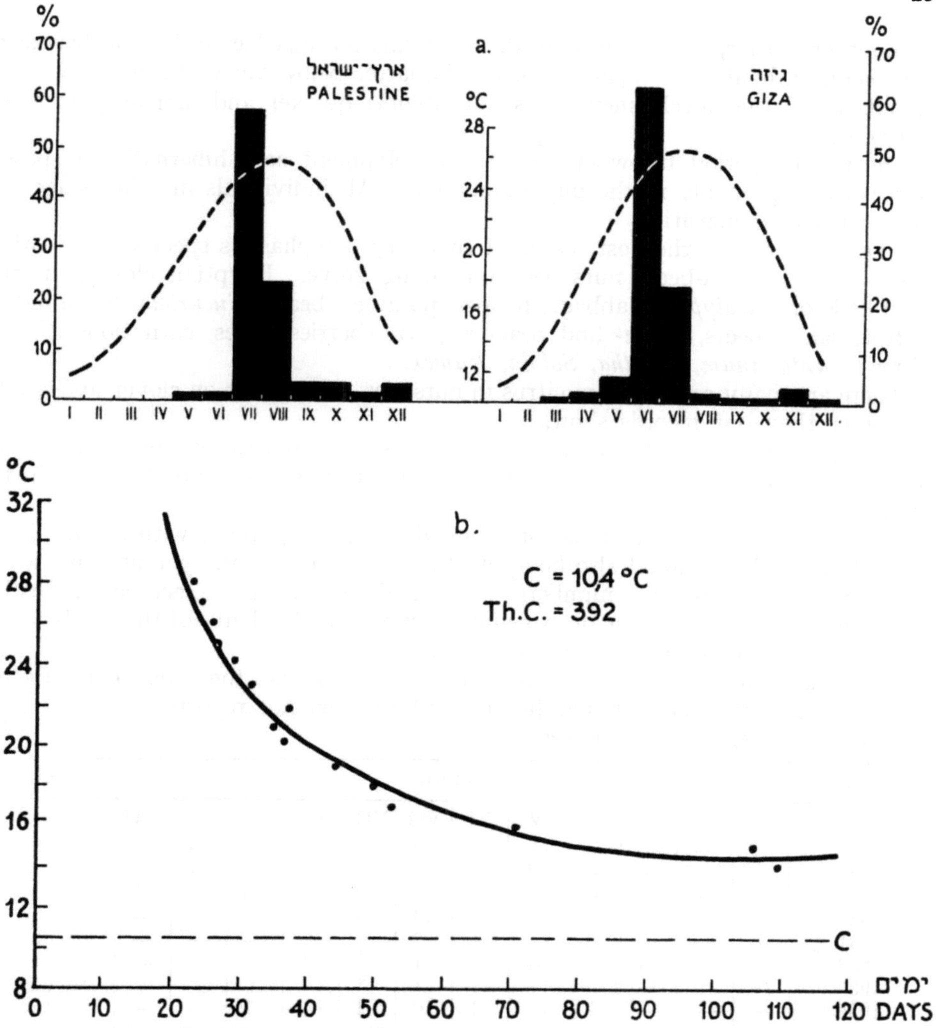

Fig. 3.
Prodenia litura F. A. seasonal population trend in Palestine and Egypt. Black indicates monthly percentage of annual damage, the broken line the average monthly temperature of Tel Aviv and Giza; B. hyperbola of shortest development.

These values run as follows:

Temperature	development in days after	
	Bodenheimer	Bishara
31.2° C	18.9	20.7
26.3	24.5	27.3
21.3	35.6	41.4
16.3	65.3	82.8
14.3	98.0	138.0

The development is hastened at variable temperatures, which phenomenon has been observed by previous authors. The agreement between the hyperbola values and the experimental data of Bishara is remarkable when we consider the paper of another author, who claimed to disprove this formula basing his conclusions on *Prodenia,* without even trying to breed the latter below 29° C. From 29° C on to about 35° C the development does not become quicker and mortality increases considerably.

During the period below or near the development zero hibernation is passed successfully, probably in the pupal stage only. All individuals in other stages die before reaching maturity.

The following are the host plants of this very polyphagous species in Palestine: Citrus spp. in seed-beds, nurseries and young groves, Egyptian clover, lucerne, seed-beds of *Eucalyptus,* cabbage, pepper, potatoes, beans, *Cucurbita,* horse-radish, carrot, salad, beets, apple- and pear-tree, strawberries, roses, carnations, *Canna, Cinnia, Antherinum, Mentha, Salvia, Rumex.*

There are many records from citrus in nurseries. The attack on clover and alfalfa is higher than our records show.

Damage and Control. The damage on citrus is restricted to feeding on the tender leaves. Feeding on the pulp of the young fruit, leaving deep circular holes in the rind later on, is very rare.

In the seed-bed, handpicking, or either dusting or spraying with arsenics, is a radical help, which should also be applied in those parts of the young grove where *Prodenia* appears in large numbers. More modern control is executed by dusting with fluorsilicate of sodium or baryum or cryolithe; by baits of these substances with bran, or of benzene-hexachloride with bran.

Prodenia is, on the whole, more abundant on citrus than the caterpillars of other Lepidoptera, but heavy damage has never been seen nor reported.

Injury was reported as follows:

Year	Month												Total
	I	II	III	IV	V	VI	VII	VIII	IX	X	XI	XII	
1927	2	—	—	—	—	—	3	—	—	—	—	1	6
1928	—	—	—	—	1	—	—	—	1	1	1	—	4
1929	—	—	—	—	—	—	1	—	—	—	—	1	2
1930	—	—	—	—	—	—	14	3	—	—	—	—	17
1931	—	—	—	—	—	—	2	4	—	—	—	—	6
1932	—	—	—	—	—	—	1	—	1	—	—	—	2
1933	—	—	—	—	—	—	1	3	—	—	—	—	4
1934	—	—	—	—	—	1	10	3	—	1	—	—	15
Total	2	—	—	—	1	1	32	13	2	2	1	2	56
Percent	3.6	—	—	—	1.8	1.8	57.1	23.1	3.6	3.6	1.8	3.6	100

The species is practically absent from autumn to summer. 82% of all damage occurred during July and August, with its main toll in July. As fig. 3 shows this agrees well with the seasonal abundance in Egypt. The only difference, the appearance of the peak in June instead of July, is explained by temperature conditions. These seasonal fluctuations are induced by a very high mortality of the young caterpillars. The mortality is as follows:

		I	II	III	IV	V	VI	VII	VIII	IX	X	XI	XII	
Palestine (field)	No. of eggs						900	500	29.150	11.850	1800	900	900	
	No. of moths	—	—	—	—	1.8	1.8	58.3	3.7	3.6	1.8	1.8	3.6	
	Mortality				—		99.8	93.5	99.2	99.97	99.9	99.8	99.6	Average 98.8
Cairo (field)	No. of eggs				500	1000	2500	31.000	11.500	1.000	500	500	1500	
	No. of moths	—	—	1	2	5	62	23	2	1	1	3	1	
	Mortality				99.6	99.5	97.5	99.3	99.98	99.9	99.8	99.4	99.93	Average 99.2
Palestine (laborat.)	Mortality in massbreeding	100	—	—	97.3	99.6	99.5	92.1	95.5	92.9	91.2	99.3	100	Average 96.8

The lower mortality in the laboratory breedings is due to the absence of the extreme high temperatures which prevail in the field at noon. Under optimal conditions and with proper care it is always possible to breed a small number of eggs and caterpillars without losses to maturity. However, such conditions do not prevail in nature, where even the reduction to a 95% mortality in one generation is sufficient to cause a mass-outbreak.

The moths decidedly prefer irrigated crops to non-irrigated ones for oviposition. This is probably the reason for their strong preference for citrus in Palestine, as citrus seed-beds and are well irrigated in the summer. The main oviposition of economic importance takes place in June. The eggs, larvae and pupae depend upon high, but not excessive, environmental humidity. The high winter-humidity even seems to be the main positive factor of bacterial diseases, which are so common in that season.

The young larvae and the eggs, especially, are very sensitive to high temperatures (above 40° C), which kill them after a few hours. This type of mortality is very high during the period of dry and hot desertwinds (hamsins) in May and September. After a heavy hamsin the young caterpillars are frequently seen dead. The high noon-temperatures of August-September are responsible for the breaking down of the population maximum in July.

The fact that the adults prefer irrigated crops accounts, for the species not having been recorded yet as injurious in the mountains, but only in the coastal plain and in Emek Yesreel.

Climatic factors seem to be the main regulators of the seasonal fluctuations of *Prodenia* populations. The influence of parasites seems to be small. A tachinid fly (probably *Eutachina larvarum* L.) and some hymenopterous parasites: *Apanteles ? tenebrosus* Wsm. (Brac.), *Euplectrus ? bicolor* Swed. (Chalc.) and *Angitia* sp. (Ichn.) were bred from the caterpillar only very occasionally. Predators are probably of no major importance. The Staphylinid beetle *Paederus fuscipes* Curt. has been observed. According to observations in Egypt, the common ladybeetle *Coccinella undecimpunctata* L., different bugs, spiders and earworms occasionally pray upon the eggs and young larvae. The epidemiological role of epidemics caused by bacterial diseases etc. remains to be cleared. Those diseases are certainly responsible for a heavy mortality in the winter, though it seems that the climatic conditions of that season would induce a very high mortality in the protracted winter-generation at any rate. This reduction of the insects' vitality is probably the primary factor in the outbreak of these epidemics.

References.

I. Bishara, Bull. Soc. Roy. Ent. d'Egypt. 1934, pp. 288—420.

C. THE LESSER ARMY WORM (*Laphygma exigua Hbn*; Noctuidae.) (Plate IV).

Description: Moth: The forewing is yellowish grey to grey or grey-brown, with two close markings, the one being earshaped and the other orbicular, both of a rusty color. The whitish distal margin is dotted with black dots. The hind wings are pale, whitish grey with a darker fuscous margin.

Body 12—14 mm long, wing expanse 18—30 mm.

The irridescent pearly white or pinkish eggs are flattened, round in diameter and ribbed. Diameter 0.5 mm. They are deposited on the lower side of the host plants in clusters which are covered with hair from the maternal abdomen. The cover hair are grey or white. The clusters contain from two to several hundred eggs each (20—40 usually). The young caterpillars are pale green with a dark need and live gregariously on the parts of the plants above the ground.

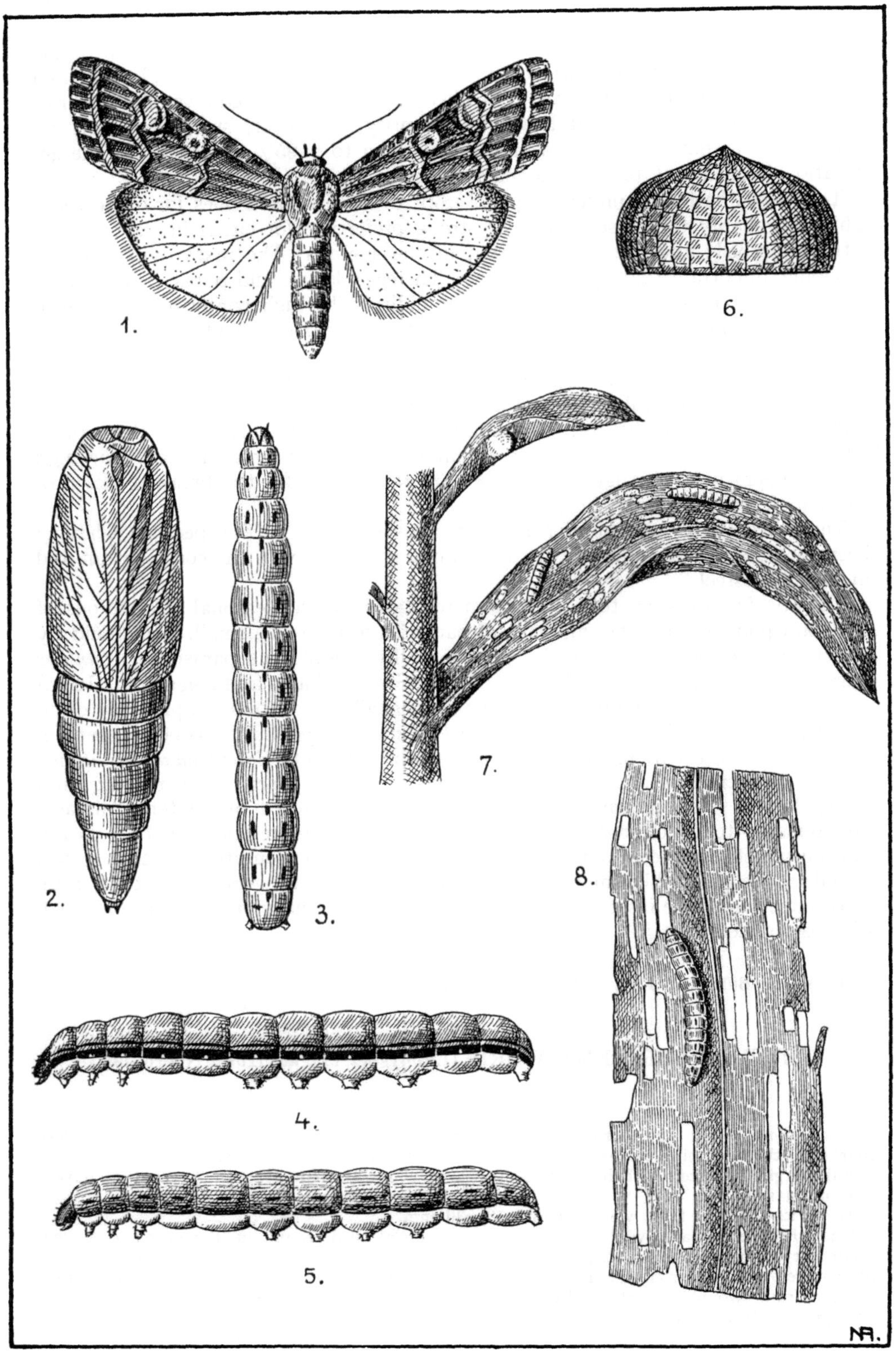

IV. The Lesser Army Worm (*Laphygma exigua* Hbn.). 1) moth; 2) pupa; 3—5) large caterpillars; 6) egg; 7) above: egg-mass; below feeding marks of young caterpillars on maize; 8) feeding marks of full-grown caterpillars on maize (partly after Willcocks).

They feed on the leaves in stripes, always leaving the epidermis of the upper side. ,,On the under side of these leaves a good deal of silken web mixed with black excrement will be found. As they grow older the larvae spread out over the whole plant and migrate to others near-by. They still tend to feed on the lower side. The older stages eat the whole leaf, only the stronger veins remain." (*Willcocks*). The older larvae hide during the day in the soil or in moist shady parts of the host plants. The older larvae are variably colored. However, the lower third of the sides and the ventral surface are always much more pallid than the rest of the body and the transition is sharp. The color of the upper half may be either yellow green with a yellow lateral stripe, or deep with green a black lateral stripe, or brown with short dark brown stripes on every segment or even different. The full-grown caterpillar is 22—30 mm long. There are five larval stages.

The caterpillar finally burrows a few centimeters deep into the soil, where it prepares a smooth-walled cell in which it pupates. The pupa is light red-brown, with the margins of the abdominal segments darkbrown, 15—20 mm long.

Distribution: The species has a Palaeotropical and Holarctic distribution. It has been introduced into the southern states of North America.

Life-cycle: The generations succeed one another, as is the case with ***Prodenia litura***. We have not as many complete breedings as from the just mentioned moth, but from 25.VI.—12.IX.1933, i.e. in 79 days three full generations have been completed. The winter is passed as caterpillars; the last moths on the wing have been seen in mid-December. Our breedings from collected material, our records from injury reports and our catches in light-traps, permit the following reconstruction of its average life-history:

First generation:	March/May	Fifth generation:	August/September
Second generation:	May/June	Sixth generation:	September/October
Third generation:	June/July	Seventh generation:	October/November
Fourth generation:	July/August	Eighth generation:	November/March.

The shortest development (including the short preoviposition period) is about 24 days (3—4 days preoviposition period; 2 days egg-stage; 12—13 days larval stage; 6 days pupal stage). In the autumn it may extend up to 40 days, in the early spring up to 50—60 days. During the winter the development is protracted and takes about 4 months, but no real hibernation or diapause occur. Development simply slows down. This life-history agrees surprisingly well with that of the Egyptian Cotton Worm.

The results of a recent study on *L. exigua* in Anatolia (Steiner) agree with those from Florida (Wilson) fairly well, but permit of more precision. The constants of the hyperbola (based on development-including the pre-oviposition period —of 27.5 days at 26.1° C. and 31.2 days at 23.3° C) are:

Development-threshold (c.) = 11.4° C
Thermal constant (Th. e.) = 378 days-degrees.

The calculation from this hyperbola gives 8.4 annual generations for the coastal plain of Palestine, which figure agrees closely with our breedings at Rehoboth. The calculated number of annual generations is 4.8 for Jerusalem, 9.8 for Tiberias.

Development lasts at	30° C	20.3 days
	25	27.8
	20	44.0
	15	105.0

The adult longevity is extremely short. The life-history in Palestine is illustrated in fig. 5. In Java up to 16 annual generations have been observed, 3 only on the highplateau of Algeria. The emergence, as well as the activity of the moth is mainly nocturnal.

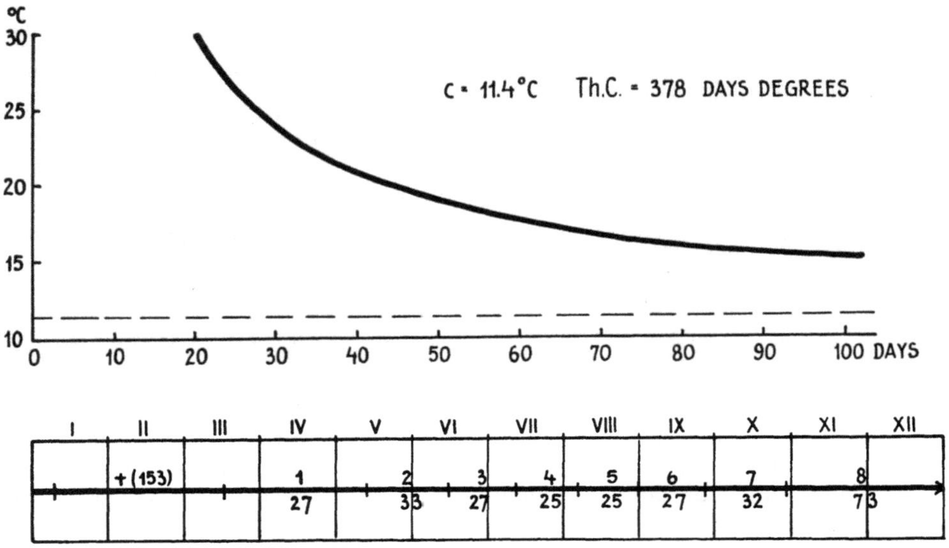

Fig. 4.
Hyperbola of shortest development of *Laphygma exigua* Hb. and maximal number of annual generations at Tel Aviv.

Laphygma exigua occurs throughout Palestine. It is, however, most common in the coastal plain, whence 80% of the cases of injury are recorded, but it is also common in the Jordan Valley and in the Emek Yesreel.

The main host-plants of the very polyphagous species in Palestine are as follows:

	No. of records	
Amarantaceae;	1	*Amarantus caudatus*
Chenopodiaceae	4	*Beta ssp., Chenopodium*
Compositae	2	*Helianthus annuus, Calendula*
Cruciferae	3	*Brassica ssp.*
Euphorbiaceae	1	*Ricinus communis*
Graminaee	6	*Zea mais*
Labiatae	2	*Mentha, Salvia*
Liliaceae	2	*Allium cepa, Asparagus*
Myrtaceae	2	*Eucalyptus ssp.*
Papilionaceae	6	*Trifolium alexandrinum, Medicago sativa, Lupinus termis*
Pedaliaceae	1	*Sesamum indicum*
Rosaceae	2	*Fragaria sp., Pyrus communis*
Rutaceae	9	*Citrus ssp.*
Scrophulariaceae	1	*Verbascum*
Solanaceae	7	*Solanum ssp., Lycopersicum esculentum, Capsicum annuum*
Vitaceae	1	*Vitis vinifera.*

Other herbs and weeds are, no doubt, very commonly attacked, but are not easily recorded since no economic damage is connected with them in Palestine.

From the studies quoted it appears that optimal conditions for rapid mass-increase exist at temperatures above 21° C. Unfavorable temperatures have been experienced above 30° C, the most favorable zone probably being between 24 and 28° C. With regard to humidity it is certain that high relative humidity at low temperatures is extremely unfavorable; bacterial diseases break out quite commonly under those conditions. The importance of humidity seems to decrease with the increase of temperature. Steiner reports summer temperatures of 21.8—23.7° C and 0—22 mm monthly precipitation as optimal in Anatolia, whereas Wilson indicates that in Florida during the months June to September, which are favorable to a *Laphygma* outbreak, temperature ranges from 21—22.5° C and from 126—186 mm rainfall. The fact that eggs which come in contact with water or are temporarily inundated die as a rule (Cunin), is probably connected with the limited outbreak areas in the tropics. Parasites never and nowhere seem to be of any control value. The average number of eggs per female fluctuates as a rule between 300 to 500, but individual maxima up to 1700 eggs per female have been observed.

Damage and control: The species appears in Palestine occasionally in large masses and may become very injurious. In May, especially, when fields of young corn or sesam plants are suddenly attacked, the whole crop may be destroyed in a few days. The feeding on irrigated fields of berseem (Egyptian Clover) and alfalfa is sometimes considerable. The caterpillars may gnaw on the green twigs of the grape vine or in the berries. The fruits of tomatoes and egg plants are damaged. A heavy attack on nurseries may cause considerable damage, if not counteracted. Eucalyptus and citrus suffer especially in nurseries. The leaves of citrus are voraciously eaten from spring to autumn. The moth is known to fly occasionally during the day. However, I was surprised to observe a vivacious flight of the moths over a citrus nursery in Ness Ziona on 15.VII.25 at 1 P.M., probably in connection with oviposition.

The damage is greatest in May. This means that in this month the population level is highest and conditions most favorable. A similar peak is reached again in October. The extent of the damage is not equal every year. The spring of 1926 and the autumn of 1928 were seasons of an especially remarkable injury. The fluctuations of this species depend greatly on climatic factors, no doubt. Unfortunately, nothing is known in Palestine in this respect, except that the caterpillars are especially sensitive to bacterial epidemics during the winter.

Control is not always possible. It is difficult to prevent the damage done to very young fields of mais and sesame even by intensive baiting. Tomatoes are partially protected if grown on wires, because thus they are much less attacked than when growing on the soil.

In citrus nurseries it is easy to prevent serious damage. When *Laphygma* is present in small numbers, it is best to collect them. When in large numbers, hand-collecting is combined with arsenical sprays, or preferably, even, with arsenical dusts. Thus, the damage may be stopped completely within 2—3 days.

Natural enemies of *L. exigua* have not been expecially studied. The number of parasites has never been very high. The following have been bred:

Meteorus laeviventris Wsm. (Hym. Brac.)
M. scutellator var. *unicolor* Wsm. (Hym. Brac.)
Chelonus corvulus Marsh. (Hym. Brac.)
Eutachina larvarum L. (Dipt. Tach.)

Staphylinid-beetles, ladybirds, Anthocorid-bugs etc. occasionally preyed on the eggs and young larvae of *Laphygma exigua*. Diseases of caterpillars are observed during the wet season.

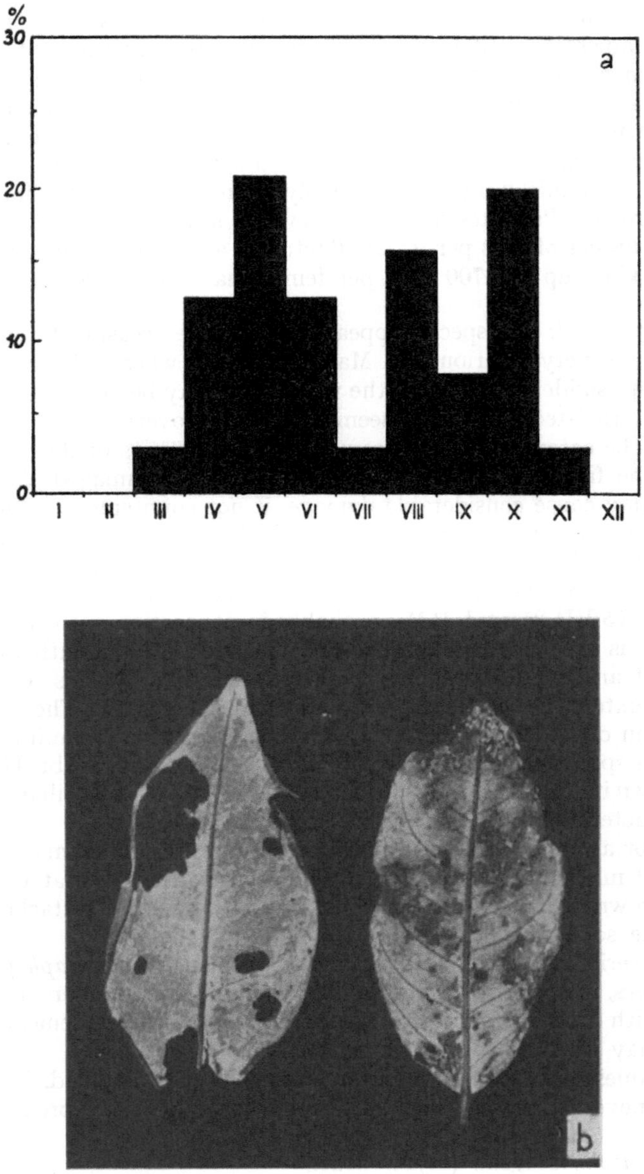

Fig. 5.
Laphygma exigua Hb. a. Monthly distribution of damage in Palestine; b. Feeding marks of caterpillars on citrus leaves (Kfar Saba VI. 1929).

References.

F. S. Bodenheimer, Schaedlingsfauna Palaestinas. Berlin, 1930, pp. 80—81.
R. E. Campbell and V. Duran, Mo. Bull. Dept. Agric. Calif. 18 1929. pp. 267—275.
G. Cunin, Bull. Soc. Hist. Nat. Afrique Nord, 24. 1933. pp. 34—42.
C. J. H. Franssen, Meded. Inst. voor Plantenziekten, Java. Nr. 77. 1930. 24 pp.
P. Steiner, Zeitschr. angew. Ent. 23. 1936. pp. 177—222.
F. C. Willcocks, The Insect and related pests of Egypt II. Cairo, 1926, pp. 139—143, pl. IX.
J. W. Wilson, Florida Agric. Exper. Sta. Gainesville Bull. 271, 1934. 26 pp.

D. OTHER NOCTUIDAE ON CITRUS.

The winter cutworm *Euxoa segetum* Schiff. has apparently three annual generations in Palestine. The spring-generation in April/May is most abundant and causes most of the damage. On citrus only a few cases of feeding have been observed and never in great numbers. In seedbeds occurs occasionally only the cutworm of *Phytometra ni* Huebner.

The American cotton bollworm *Chloridea obsoleta* Fabricius has been observed in Palestine on tomato-fruits, tobacco, berseem and repeatedly on young citrus leaves in nurseries (mainly in VI). No damage to blossoms and young citrus fruits has been observed, which causes them to fall or leaving a large scar on the fully developed fruit. This damage causes at intervals serious damage to the citrus plantations in S. Africa, mainly in spring. Eventually locally the whole crop may be destroyed. The collecting of egg clusters and caterpillars is the only control method. Two other species of the same genus, *C. peltigera* Schiff. and *C. nubigera* Herrich-Schaeffer have been at rare occasions taken in April/May from citrus leaves.

References.

F. S. Bodenheimer, Die Schaedlingsfauna Palaestinas. Berlin. 1930. p. 70, 163, 318.
W. J. Hall and W. K. Ford, Notes on some Citrus insects of South Rhodesia. Oxford. 1933. p. 26—33.

E. THE STUB MOTH OF CITRUS ROOT-STOCK
(*Ephestia vapidella* Mn., Pyralidae).

Description: Moth: Forewing narrow, elongate; fuscous, with indistinct design in white and grey and russet on the basal quarter towards the margin. Ventral side in the female lacking tufts on the anterior margin. Hind-wings nearly purely white. Wing expanse 12—13 mm.

Larva: Whitish to yellowish-white. Head capsule brown. Single hair scattered over the body. When full-grown, it is 8—10 mm long.

Pupa: Light brown in a white cocoon, 7 mm long.

Distribution: Mediterranean.

Life-history and damage: The Citrus stub moth has a series of annual generations. From late summer to spring 2—3 generations were observed. Moths appeared in early September, early November and early April. The number of generations developing from April to September is still unknown.

The species is probably polyphagous. So far, it has been observed on carob and citrus and as a pest only in the plain of Sharon (Herzlia, Kfar Saba, Ramath Hasharon Ramath Gan).

From early August on, the feeding signs of *E. vapidella* may be noticed on the bark of Citrus. Small, pale yellow-brown excrements are being rejected and small drops of gum are exuded on the infested parts. The feeding activity increases noticeably from day to day. The larvae quite commonly derive their food from the stub near the grafting union, often eating half of the basis of the scion and gnawing around it. The scion breaks at such places frequently. The same damage is noticed on twigs, which show callus-formation, occasionally also on the root-suckers, which grow from the root-stock (after unsuccessful grafting). In all observed cases, which

Fig. 6
Damage of *Ephestia vapidella* Mn. On scion: 1. lateral view of damaged scion, 2. view of the removed scion. On citrus stubs: 3. feeding place, 4. excrements protruding from the feeding gallery. 5. cross-section of feeding galleries in the twigs (from Bodenheimer and Klein).

were grafted on the spot, the root-stock was sweet lime. Trees which were planted after having been grafted and treated properly (branches cut at their bases, stubs cut out) were not noticed to be damaged. These stubs, which often project up to 3 cm, dry quickly and the larvae enter at those dried spots. Fresh green wood was not attacked, with the exception of a few attacked root-suckers. When the larvae gnaw around the cutting, they always begin their feeding activity by eating the drying stubs of the root-stocks. They feed during the entire month of August and sometimes up to early September. Later in the year, they are scarcely observed on citrus. Root-stocks 1—2 years old are being attacked. In one large orange grove about 400 trees had to be replaced, in consequence of this damage. When young trees are found dead during the first year following grafting, one should always think of *E. vapidella*.

During January a larva was found boring in a citrus twig, where it managed to create a small groove. It abandoned the groove before pupation. The infested branch died off.

Branches were generally attacked, when the cutting surface was covered with plaster.

Prophylactic control measures are as follows: The projecting stubs should be cut not later than early June, if possible still earlier. The branches should be cut at their bases immediately and the cut surfaces covered with asphalt. A cover of plaster must be regarded not only as ineffective, but as actually attractive to *Ephestia*.

No direct control measure can be recommended. Cutting off the damaged parts of the scion or of the stub, washing with Bordeaux solution and covering with tree-wax, generally weaken the young trees greatly.

References.

F. S. Bodenheimer and H. Z. Klein, Hadar 7. 1934. pp. 9—10.
A good description in: O. W. Richards and W. S. Thomson, Transact. Ent. Soc. London. **80. 1932.** p. 174.

F. THE HONEYDEW MOTH (*Cryptoblabes gnidiella* Mill., Pyralidae). (Plate V).

Description: Moth: Forewings grey-brown, covered with scattered silvery white scales, the veins broadly crimson. A white strongly curved narrow transverse band along one third of the body, and a white serrate wider band near the terminal end. Both bands are occasionally very indistinct. Basis and margins darker than the centre. Hind-wing whitish grey with veins and margin darker. The male antennae are provided with a styliform appendix on the fourth joint; palpi long and thin.

Body 7 mm long, 15—20 mm wing expanse.

Caterpillar: Basic color variable, generally grey brown; head chestnut-brown and pronotum dark brown. Dorsally a median white pinkish and laterally a black-brown longitudinal band. Ventral side pinkish or grey.

The bright red-or yellowish brown *pupa* lies in a white cocoon of silk. 7 mm long.

Distribution: Common in Mediterranean region, Egypt, New Zealand. Probably of a wide distribution.

Life-history and damage: The number of annual generations in Palestine is still unknown, but at least 3—4 generations are to be expected. Not much is known of its biology, either.

The caterpillar is very polyphagous, but apparently never damages healthy fruit, but always such which has been previously spoilt by other animals. Its congener *C. aliena* Swezey from Hawaii is also known to be connected with aphids or mealy-bugs. The honeydew exuded by the latter probably attracts them and wherever the honeydew, leaf, stalk or fruit are attacked and either eaten or pierced. *C. gnidiella* behaves similarly, known to attack cotton balls or grape berries, which have been previously attacked by *Platyedra*, *Polychrosis* or other insects (*Virachola livia* in pomegranates, *Carpomyia incompleta* in Zizyphus etc). Over 100 caterpillars were met in August, 1923 on three spadices of sorghum, where they moved about in slight silken webs, spun by them. After the moths had hatched, the grains were examined. Not one grain was damaged. The caterpillars lived only on the sweet viscous excretions which cover the external cover-leaves of the grains. A few years ago, it was possible to confirm this statement. However, Willcocks reports from Egypt that they may occasionally feed on the immature grains. They are found in the ears of corn, where they seek nourishment in the silk hair and in the tassels. Willcocks observed it feeding in grains, previously damaged by birds.

From August on, it may quite commonly be bred from grapeberries, which have been previously injured by *Polychrosis botrana* (Sargonia, Kiryath Anavim).

It has been found only in those balls of cotton, which were attacked primarily by *Earias* or *Platyedra* (Akko, Sichron Yakob).

The damage to citrus is remarkable. Silvestri probably rightly attributes the pinning of webs in the flowers to a confusion with *Prays citri*. However, manys fruits, which have a hole of about 1 mm diameter, drop in our orange groves, from

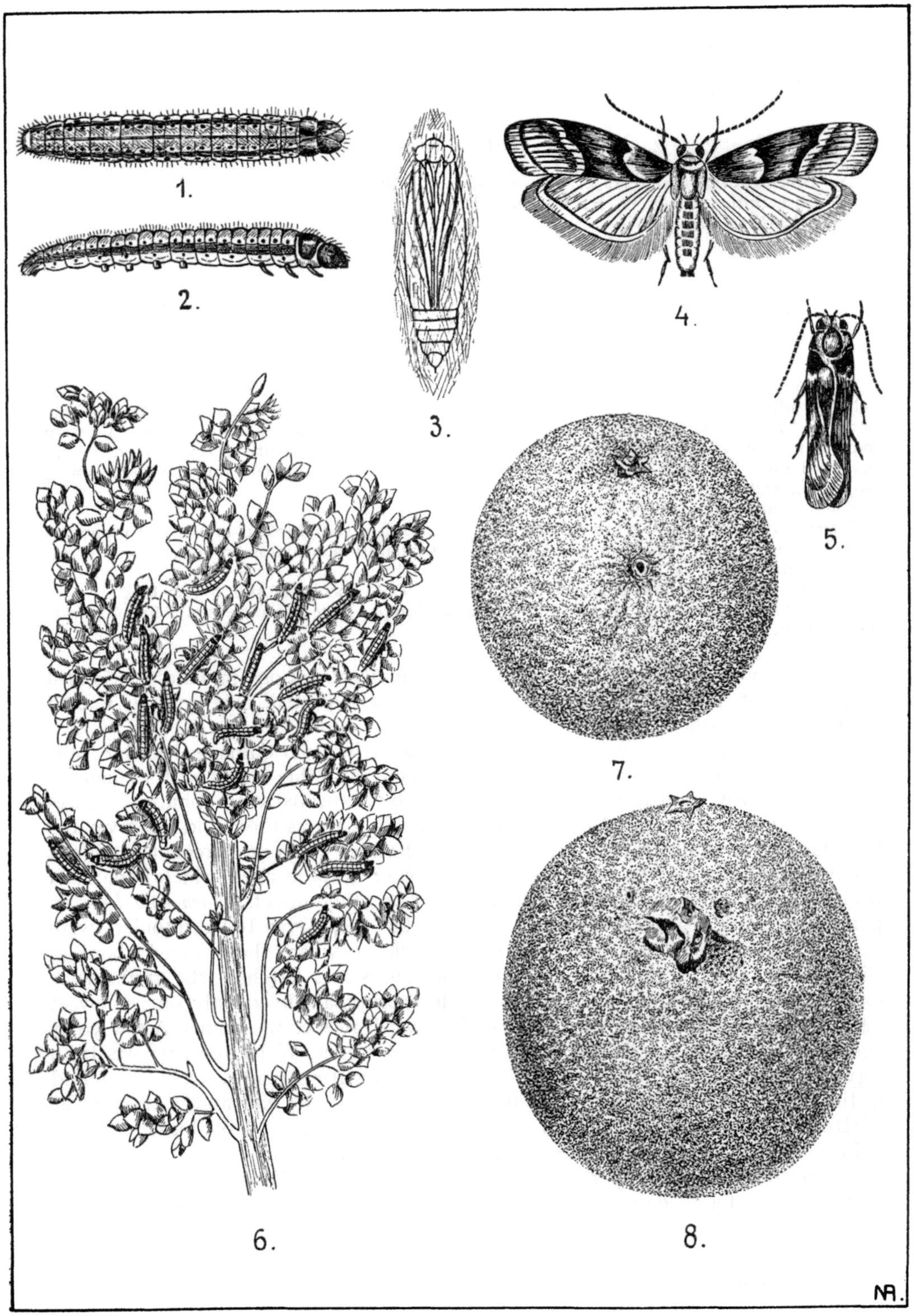

V. The Honeydew Moth (*Cryptoblabes gnidiella* Mill.). 1 and 2) caterpillars from above and from the side; 3) pupa in pupal cocoon; 4) and 5) moth with extended wings and in resting position; 6) caterpillars feeding on sorghum; 7) and 8) feeding marks of the caterpillars with exits and gum-exudation on orange fruits (partly after Bodenheimer).

September on. The layman generally ascribes this damage to *Ceratitis capitata*. Many of the dropped fruits, which have become yellow prematurely, show depressions in the rind of a size of a 5 mil coin. A feeding gallery, mm in diameter, leads from the centre of this groove into the interior of the fruit, where a grey brown pyralid caterpillar, up to 3 cm long, may be found. It can easily be distinguished from the white fruit-fly maggot by its different coloring and by the presence of a black-brown head capsule, which is absent in the latter. Those depressions are met with almost exclusively on those parts of the fruits, where two fruits touch and press each other and where the caterpillar lies on a branch or below, near the petal of the fruit, respectively. In those places it is generally restricted to those where mealybugs had settled previously and remains of honeydew and sooty mould are still recognisable.

The caterpillars are negatively phototactic. Gum is often exuded in the neighborhood of the entrance hole. It is not known yet in what stage the caterpillar is able to pierce the rind of the fruit. It is probable, that the early larvae feed mainly on the honeydew and that only the later stages are able to pierce the fruit at the weakened parts.

The typical seasonal appearance of *C. gnidiella* in the citrus orchard is as follows: In the second half of July the caterpillars enter the fruit either below the stalk of the fruit or at the contact place of two fruits and drops of gum appear in the neighborhood of the entrance hole. In the last days of August a large part of the infected fruit begins to fall off. In October most infested fruits have fallen off. Only very few fruits continue to drop until early November. In some orchards in Hedera the infestation was as follows:

Year	month VII			VIII			IX			X			XI		
	1/10	11/20	21/31	1/10	11/20	21/31	1/10	11/20	21/30	1/10	11/20	21/31	1/10	11/11	21/30
1930			.	.	.	+	+	+	+	+	+	+	(+)	(+)	
1931			.	.	.	+	+	+	+	+	+	(+)	—		
1932		.	.	.	—	+	+	+	+	+	+	(+)	—		

caterpillar infests the fruit + fruit drops

Oranges and grapefruit are equally attacked, the latter sometimes slightly earlier than the former.

In order to get an idea as to what extent *Cryptoblabes* causes the dropping of fruit, counts were made, each on one dunam, at Hedera during 1930.

Date	Fruits dropped		Fruit on the tree, ready to drop	
	Total	owing to *C. gnidiella*	Total	owing to *C. gnidiella*
29.VIII	93	48	47	22
8.IX	45	29	45	45
12.IX	125	91	79	79
19.IX	141	121	73	73
26.IX	192	142	83	83
3.X	145	97	64	64
11.X	148	109	80	80
17.X	107	64	48	48
24.X	123	63	44	44
31.X	88	45	53	53
	1211	809 67%	616	591 96%

This means that ⅔ of the fruits dropping during this season are infested by *Cryptoblabes*. The remaining third are constitutionally weak fruit. Nearly 100% of the damaged fruit, which falls off as a result of the damage, is or was infested by this moth.

In normal years the actual damage is less than 20 fruits per tree, which loss certainly does not justify any intensive control measures.

Another type of damage has been observed once near Kfar Saba. The caterpillar fed heavily in the stubs of Shamuti, grafted on sweet limes. This damage, which is very similar to that caused by *Ephestia vapidella*, appeared at an earlier season, from early June to early July. When the surface of the stubs was covered with asphalt, the larvae were, as a rule, prevented from entering the branch. When the damage is noticed in time, the injured part may be cut out and the surface covered with asphalt.

We may conclude from all the above facts, that *Cryptoblabes* is a serial pest, connected in some way with the presence of honeydew or other sweet substances. Whether this connection originates with oviposition or to supply the nutritional needs of the early larval stages, has still to be decided. In Europe it has been observed on the likewise glabrous flowers of Daphne gnidium. No serious primary damage has ever been reported.

It is very unlikely that collecting the dropped oranges would prove to be a control measure, owing to the fact that by August the oranges are not attacked anymore. It is necessary to follow the life-cycle of *Cryptoblabes* in Palestine throughout the year and only, by studying its full food ecology, may we find practical ways of control.

References.

F. S. Bodenheimer, Die Schaedlingsfauna Palaestinas. Berlin, 1930, pp. 166—168, 251, 262, 277, 301, 302.

F. S. Bodenheimer, and H. Z. Klein, Hadar, 7. 1934. p. 10.

F. Silvestri, Dispense di Entomologia agraria. Portici, 1911, p. 269.

O. H. Swezey, Hawaian Sugar Planters Exper. Sta. Entom. Bull. 6, 1909, pp. 24—26.

F. C. Willcocks, The Insect Pests of Egypt. Vol. II, Cairo. 1925. pp. 67, 145, 161. Pl. X.

G. THE CITRUS FLOWER MOTH (*Prays citri* Mill., Yponomeutidae).

Description: Moth: Grayish brown. Head light brown with black eyes. Antennae tapering slightly toward the tip. Wings long and narrow, broadly fringed, especially the hind wings. Forewings bear numerous irregular markings which are distributed over the wings. Hind-wings light gray, membranous, without spots. Abdomen and thorax light brown. It is 2.8—3.2 mm long. Forewing 4.6 mm long, 1.3 mm broad.

The subelliptical *egg* varies from colorless to light yellow, becoming darker during its incubation. Eggs are laid wingly, usually at night, on flowers or grafting buds. In the Philippine Islands they are also laid on the young fruits. The egg is 0.2 × 0.1 mm.

The full-grown *larva* has 14 segments. Besides the three pairs of thoracic legs, there are 5 pairs of abdominal legs. The subcylindric, semi-transparent body slightly tapers posteriorly and is covered sparsely with fine hairs, which are not visible to the naked eye. The anal segment bears the last pair of legs. The larva is 4.2—5.5 mm long.

The larva pupates in a loosely woven silk cocoon. The anal extremity of the chocolate-brown pupa is fastened to the host by silk threads. It is 4.8—5.5 mm long.

Distribution: The moth is distributed over Southern Europe, Syria, Palestine, India, Ceylon, N.S. Wales and the Philippines.

Life history and damage: In the Philippines the development of the species takes 66 days during the autumn (69 days including preoviposition period) at an average temperature of about 27° C and the species has 5 annual generations. In Italy, Silvestri reports 3 annual generations, the moths being on the wing in April/May, August and October/November. This allows a tentative construction of the hyperbola (69 days at 27° C at Manila; 92 days at 20.5° C at Naples). The constants are as follows:

Development threshold (c) = 1° C
Thermal constant (Th. C.) = 1794 days degrees.

This computed number of annual generations agrees well with the actual observations:

Locality	Thermal sum	No. of generations	
		computed	observed
Manila	9204 days degrees	5.1	5
Jaffa	6672	3.7	—
Naples	5073	2.8	3

Dry heat seems to be unfavorable to the species. Damages being generally observed during the cooler half of the year. *Prays citri* seems to be restricted to *Citrus* spp, but it shows a high variation of feeding types on this host.

The caterpillars are met in January/March quite commonly in the flowers of citron (*Citrus medica*) and sour lemon; but records are obtained from other months

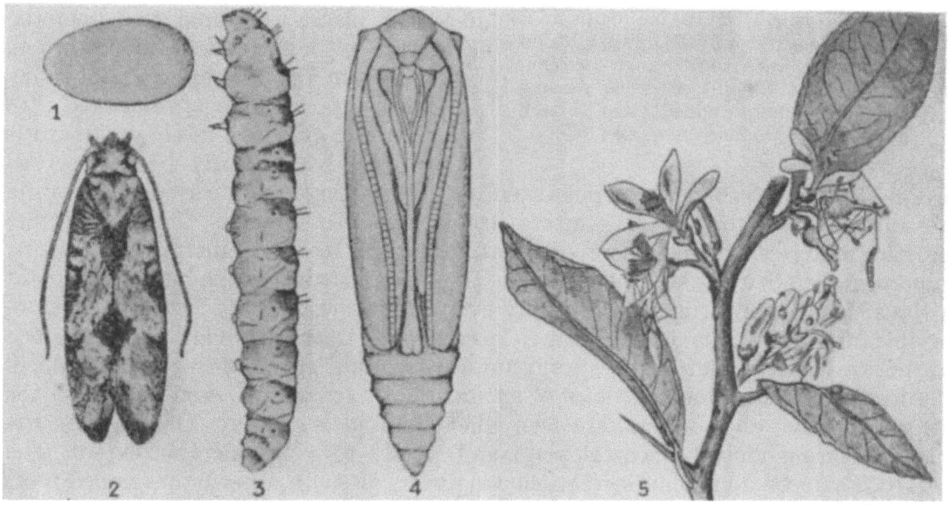

Fig. 7.
The Citrus Flower Moth (*Prays citri* Mill.): 1. egg, 2. moth, 3. caterpillar, 4. pupa, 5. larvae feeding in lemon flowers (1—4 from San Juan, 5 from Penzig).

also from September on. They spin in the flowers and feed on the calyces in the manner, which is illustrated in fig. 7. Similar damage is reported from other Mediterranean countries; in Sicily considerable damage is observed in some years. The caterpillars are most common at the end of January and the beginning of February. The moths are on the wing one month later. Their economic importance is negligible in Palestine, as citron is cultivated to only a very slight extent and only a small fraction of the flowers matures into fruit even under optimal conditions. They undoubtedly occur on orangeblossoms also, but obviously in small numbers only, since they have escaped our actual observation to this day. They are dangerous to lemon blossoms.

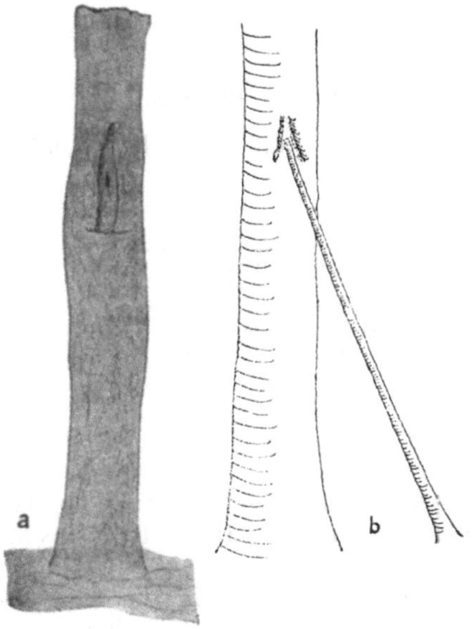

Fig. 8.

A. Caterpillars of *Prays citri* Mill. feeding in grafted citrus buds. No damage is recognisable externally. B. Sweet lemon tree inarched with sour orange seedling. The place of union serves as feeding place of the indetermined citrus inarch moth (from Bodenheimer and Klein).

Another type of damage has been observed in the Philippines (San Juan). The caterpillars live in the rind of the fruit, next to the pulp, of six different citrus species. By feeding, they cause the growth of a gall-like tumor or swelling. The inside tissues of the rind are consumed, before the caterpillar leaves the gall to pupate. Twenty to thirty galls or even more may be found on the surface of one fruit in cases of severe infestation. The swellings are very prominent, often spoiling the shape of the fruit, and the openings in the rind furnish excellent germination places for fungi and bacteria. This type of damage is so far, known to occur only in the Philippines and India.

A third type of damage, connected with injury to the grafting buds in young citrus plantations, has recently been observed in Palestine. The young caterpillars enter the grafting union through the rafia dressing shortly after the fall-grafting (from November onwards). They penetrate into the living cambium, where they eat through a feeding gallery. The grafting union begins to dry out and the bud dies. The caterpillar leaves the bud and pupates in the soil or on the trunk, occasionally even inside the bud. Its feeding is not noticeable from the outside. Only in a later period, upon removing the rafia, one can notice how the bud changes its natural colour and grows brown and wrinkled. The feeding galleries and particles of small bright excrements, eventually also the full grown larva or the pupa, are seen when the bud is cut out. This damage has occurred throughout the coastal plain and locally to a considerable extent. The regrafting can be done only in the following year, often on a newly developed root-sucker. The general development of the tree is thus retarded. Very often, however, this retardation is compensated for after a few years.

So far, control measures against the caterpillars feeding on the blossoms have not been necessary. In a limited outbreak, handpicking of the caterpillars and of the infested blossoms would be advisable, also dusting with cryolithe on lemons.

At the present state of our knowledge, an outbreak can be neither prevented nor controlled. The latter would be too late, in any case.

References.

F. S. Bodenheimer, Schaedlingsfauna Palaestinas. Berlin. 1930. p. 165.
F. S. Bodenheimer and H. Z. Klein, Hadar **7**. 1934. p. 8—9.
J. M. San Juan, The Philippine Agriculturist. **12**, 1924, pp. 339—348.
F. Silvestri, Dispense di Entomologia Agraria. Portici, 1911, p. 278.

H. OTHER MICROLEPIDOPTERA.

a. *The Citrus Inarch Moth.*

An unindentified moth larva feeds on the grafted inarches. The grafting union of recently grafted inarches of sour orange on a diseased sweet lime was attacked by them. Such feeding caused considerable damage in some orange groves. Covering the grafting union with plaster or rafia provided no protection. In some cases the larvae entered even through a cover of asphalt. They eat through galleries and the grafted inarches dry out.

References.

F. S. Bodenheimer and H. Z. Klein, Hadar 7. 1934. p. 10.

b. *Apterodiscus cernyi* Amsel. (Momphidae).

Description: Gradeolar of fore-wing yellow with unclear dark and broad transversal anal band, another, narrower band reaches almost the apex. Wing basis also dark, especially near the costa. The entire pattern unsharp and variable. Fringe yellow with occasional dark scales. Hindwing bright yellow-grey, with fringe of the same colour. Head yellow. Antennae simple the first joint incrassated. Wing-expansion 7—10 mm. (Amsel).

Biology: Zacher observed the caterpillars of this moth, described from the Drogue Bazaar in Cairo. They lived in dense webs on dry orange peels. It is probable that their caterpillar live in similar habitats in Palestine, belonging perhaps to the fauna which attacks dry fruits.

References.

F. Zacher, Entomol. Beihefte Berlin-Dahlem **1**, 1934, p. 107.
H. G. Amsel, Mitteil. Zool. Mus., Berlin **20**. 1935. p. 303, pl. 12, 151.

2. BEETLES (COLEOPTERA).

KEY TO THE FAMILIES OF THE MORE COMMON BEETLES FOUND ON CITRUS TREES IN PALESTINE.

1. Head prolongated into a snout. Tarsi 4-jointed. Antennae kneeled. *Curculionidae*
   Head not prolongated into a snout. **2.**
2. Antennae kneeled, ending in a club, composed of 3—7 lamellae. Tarsi 5-jointed. Common on herbs. *Scarabaeidae.*
   Antennae not ending in a laminated club. **3.**
3. Antennae neither kneeled, nor ending in a club. Tarsi cryptopentamer, i.e. the fifth tarsal-joint is often difficult to be seen. **4.**
   Antennae not kneeled, ending in a 3-jointed club. Tarsi 5-jointed. **6.**
   Antennae filiform, sometimes ending in a club. Fore- and middle-tarsi 5 in number, hind-tarsi 4. **5.**
4. Antennae shorter than half the body. Body short and plump, generally shiny. Living on leaves. *Chrysomelidae.*
   Antennae longer than half the body. Body elongate, pubescent. Develop in wood. *Cerambycidae.*
5. Anterior coxae spherical. Ungui smooth. Generally live on the soil, plump. *Tenebrionidae.*
   Anteriors coxae protruding cones. Ungui dentate. General visitors of flowers and herbs; elongate. *Alleculidae.*
6. Anterior coxae spherical, separated by a prolongation of the prosternum. Tarsi 5-jointed. Elytra not reaching the end of the abdomen. Live on decaying fruit. *Nitidulidae.*
   Anterior coxae protruding cones, which generally touch one another. Tarsi 5-jointed, the first joint being very small or hidden. Elytra reaching the end of the abdomen. Develop in the branches. *Bostrychidae.*

BEETLES IN CITRUS ORCHARDS.

Many beetles live in the citrus groves, yet most of them has no direct interrelation with the crop. And even those few species, which occasionally may cause damage do so to a very restricted limit only. We mention most of these secondary pests mainly, because every year we obtain a number of inquiries about them which is decidedly far beyond their economic importance. The scarcity of beetles living in numbers in dying and dead wood of citrus is the consequence of the relative youth of our extended citrus areas.

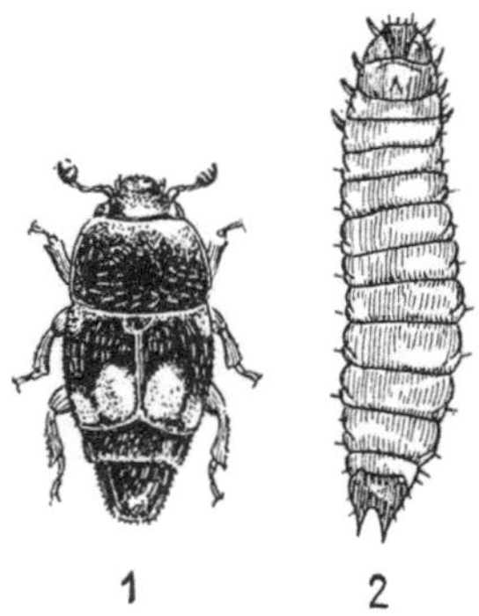

Fig. 9.
Carpophilus hemipterus L. and larva.

Nitidulidae. Three closely related species of *Carpophilus* abound in our orchards (*C. hemipterus* Linné, *C. dimidiatus* Fabricius, *C. immaculatus* Lucas). They are about 3 mm long, brownish, with relatively short elytra. They live and develop in plant decay, such as fruits in decomposition, in all warmer parts of the world. They develop frequently in oranges rotting on the soil, together with *Drosophila.* (Larva see fig. 19). They are of no economic importance, but may become so when in future the citrus by-product industry is more highly developed.

KEY TO SPECIES OF BOSTRYCHIDAE.

1. Scutellum with yellow pubescence. *Schistocerus bimaculatus.*
   Scutellum not pubescent, shiny. **2.**
2. The first 3 tarsal-joints enlarged below, bearing a yellow hair-brush. *Apate monachus.*
   Tarsi simple, the first 3 joints neither enlarged below nor bearing a yellow hairbrush. **3.**
3. Posterior half of pronotum smooth (or with very fine punctation only). The 3 joints of the antennal club decreasing in size. **4.**
   Posterior half of pronotum finely granulated. The 3 joints of the antennal club branched inwards. Elytral slope lunched. *Sinoxylon sexdentatum.*
4. Suture and end of elytra without spines. **5.**
   Suture of elytra in the middle of the slope bispinate, laterally with a small horn. *Enneadesmus forficula.*
5. Frons with long hair arranged in a semi-circle. Dark red-brown to black-brown. 2—4 mm long. *Scobicia chevrieri.*
   Frons simple or only slightly haired. Yellow-brown. 4—8 mm long. *Xylopertha picea.*

a. THE BLACK GIANT BOSTRYCHID (*Apate monachus Fabricius,* 1775).

Description: Shiny black to shiny dark brown. Head pointed downward and its view is barred from above by the large prothorax. In the female, the frons is covered with dense golden hair. The head bears two triangular thorns in its middel. Eyes strongly protruding laterally. Pronotum highly convex with thorns on the anterior angles, especially in the males. Pronotum and elytra highly sculptured. The caudal end in the female with one pointed projection on each side, which is absent in the male. Male 14 mm long, 4 mm broad; female 18 mm long, 6 mm broad.

Larva similar to that of *Schistocerus bimaculatus,* but bigger.

Distribution: Throughout tropical Africa. Africa Minor, Palestine, Syria; West Indies (probably introduced).

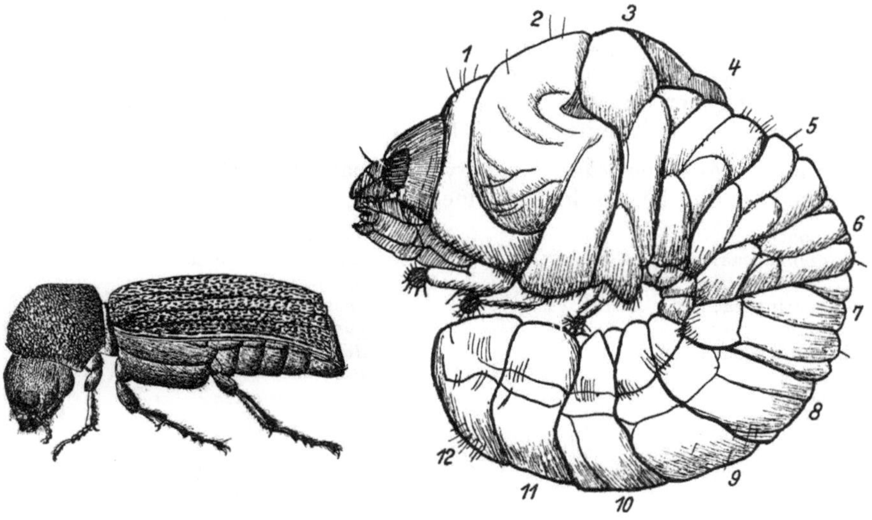

Fig. 10. *Apate monachus* F. and larva of related species.

Life-history: The life-history of the beetle in Palestine is still unknown. It has one annual generation. The adults appear from mid-May to beginning of August. During that season the beetles attack the stems of young trees, or branches of about 3 cm in diameter. The preferred trees are: *Citrus, Grevillea, Sterculia, Casuarina,* all strangers to the flora of the country. The beetle enters the stem or branch at an angle of about 45° and a circular gallery runs in to the interior parallel to the stem and distad. The gallery is 4—17 cm long and in 1 cm diameter, ending in a right angle upwards. Apparently, only the female bores galleries, either in order to reach thereby sexual maturation or to secure a nuptial chamber. Up to four beetles have been met in one gallery. However, the bigger female was always in the most distant part of the gallery and the smaller males entered it later. We have never seen the male making similar galleries. Most probably copula takes place in the interior of the gallery. Such feeding has been reported, as follows:

month	IV	V	VI	VII	VIII	IX
no. of observations	—	11	24	21	4	—

A. monachus is known from the coastal plain only.

Larval development has been described by Aulmann from cotton in Tropical Africa:

,,The larva enters the wood (splint) after having pierced the bark and feeds below the bark around the tree, leaving some parts of the splint untouched. Later, when the larva has grown, it enters the wood in a gallery which surrounds in circular windings the pith of the tree, without touching the latter. Having bored these circular windings in the twigs or branches the larva turns upwards and continues its ways in an irregular gallery, but remaining always below the bark. Now the pith is devoured as well as the wood and the splint. The bark dries on the entrance hole and the twig or branch breaks off by the wind or by its own weight. Cotton is rarely attacked by *A. monachus* probably because its wood is too porous."

In Algeria it has been bred from *Calycotome spinosa*, a leguminous shrub. In Palestine a near relative, *C. villosa*, is fairly common in the Mediterranean territory and should be systematically searched for larvae of *A. monachus*.

Aharoni observed its larva once in citrus in 1914/15. The larvae, which he believed to be those of *Apate*, are reported to have appeared from May to July. The author, however, has seen *Apate*-larvae only once, in the dry stem of *Poinciana regia*, despite the fact that the characteristic and big larva of this Bostrychid can hardly be mistaken. It is quite certain that, as a rule, it does not develop in citrus.

Host-plants: Apate monachus is a very polyphagous beetle. The following hosts are attacked by the beetles in Palestine:

Anonaceae:	*Anona muricata*
Bignoniaceae:	*Jacaranda mimosaefolia*
Caesalpinaceae:	*Bauhinia sp.*
Casuarinaceae:	*Casuarina equisetifolia*
Juglandaceae:	*Juglans regia*
Leguminosae:	*Acacia floribunda*
	Poinciana regia
Meliaceae:	*Melia azederach*
Moraceae:	*Ficus carica*
	Morus alba
Myrtaceae:	*Psidium guayava*
Oleaceae:	*Olea europaea*
Proteaceae:	*Grevillea robusta*
Punicaceae:	*Punica granatum*
Rosaceae:	*Cydonia vulgaris; Prunus persica;*
	Prunus amygdalus; Pyrus ussuriensis;
	Pyrus communis
Rutaceae:	*Citrus spp.*
Sterculiaceae:	*Sterculia sp.*
Tamaricaceae:	*Tamarix spp.*

57 out of 69 reported cases live on introduced plants and the rest mainly on trees with tropical affinities. Very many trees, as coffee, cacao, palms, are attacked

in the Tropics. However, in all cases hitherto observed by us, the attack was the maturation feeding of the immature adults. The late A. Ritchie, who was well acquainted with this beetle in Tropical Africa, imparted to the author in conversation that even there the damage is done by the adults and not by the larvae. Little is yet known on the regular hosts of the larva and further information as to its host-relations is greatly desirable.

The damage done is considerable. Young trees of *Sterculia, Grevillea, Casuarina,* etc. die in a short time after the attack, when 1—3 years old.

Citrus is seriously attacked only locally and occasionally. Citron is always the preferred citrus-species. Serious damage occured ex. gen. at Kfar Saba in 1930 in a citron-grove. The grove in question was 4 years old and was rich in fruit. 60 new galleries per tree were not rare. Large branches of some trees, heavy with fruit, broke off distad of the feeding gallery. Many galleries were observed in trunks, about 10 cm in diameter. The attack on not-inoculated trees of cederate was very small and the neighbouring shamuti-trees were not or only very slightly attacked. On the other hand, young shamuti trees in another grove at Kfar Saba suffered considerably during the same year. Both groves were surrounded by fallow land, on which grew plants of *Calycotome.* The second choice of *Apate* is sour lemon. In July 1932, lemon trees were heavily attacked by this beetle near Hedera. 15 galleries were found in one branch. In this case, as well, the lemon trees grew near uncultivated land, where *Calycotome* was common.

However, serious damage on orange or grapefruit trees is rare and scattered or local cases of *Apate*-feeding never required any special emergency measures in Palestine.

In a 2 years old orchard, planted with grapefruits and shamuti at Hedera, ex. gen., some damage was observed in late June, 1937. In early August about 25 trees were infested by *Apate,* each tree with 4—6 galleries on the average, in the stem, in the primary and in the secondary branches. No external damage was recognisable, but probably later on some of the branches broke during the winter storms.

No satisfactory control measures are known. From May on, the beetles should be looked for, collected and destroyed. Those, beginning to eat galleries, should be destroyed by means of wire. The study of the larval host plants of *Apate* should be the next step. The eradication of these plants (possibly *Calycotome*) in the immediate environments of the citrus grove or of ornamental gardens would then be a radical preventive measure.

Larvae and beetles of *Schistoceros bimaculatus* Olivier develop in the twigs of vine and fruit trees. One we met beetles in an orange twig. Also *Xylopertha picea* Olivier was once found boring in citrus twigs. It normally develops in *Acacia* and *Zizyphus. Enneadesmus forficula* Fairmaire, the grub of which develops in *Acacia tortilis,* once did some small damage to young citrus plants at Jericho. *Scobicia chevrieri* Villa and *Sinoxylon sexdentatum* Olivier both are common polyphagous borers in wild and cultivated trees in Palestine. *Scobicia* was observed repeatedly in twigs and stems of young orange trees, often accompanied by heavy gum secretion from the entrance hole. This feeding may easily be a maturation feeding, such as described for *Apate monachus. Sinoxylon* was only seen by us in dead citrus wood, where it once almost swarmed in an arm-thick dead lemon branch.

ALLECULIDAE.

The black adults of *Cteniopus gibbosus* Baudi, *Omophlus syriacus* Mulsant and *O. lucidus* Kirsch. abound in spring on all kinds of flowers. The larvae inhabit the soil, feeding mainly on the roots of grasses. In Italy damage to citrus flowers has been observed. We have never seen any real damage caused by these common beetles.

TENEBRIONIDAE.

Some of the many black desert- beetles are locally injurious by feeding on fresh inoculations. This occurs, as a rule, in groves where the lower part of the wildling is being stripped of its leaves. The beetles then climb to the inoculation point, where they feed, often destroying the upper part. When the lower leaves remain for some months on the wildling, the beetles feed on them and no damage is done. The feeding of these beetles is observed from April to July. They are residual

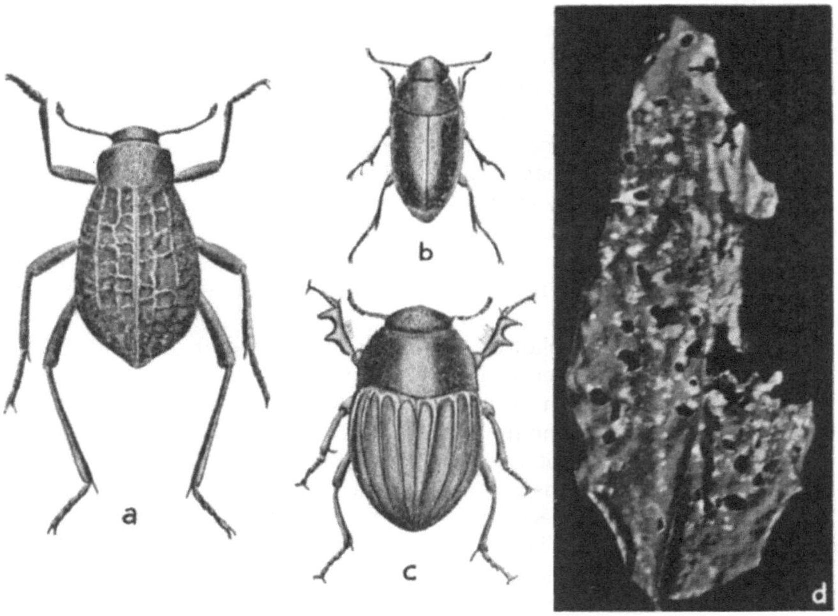

Fig. 11.
Residual pests: A. *Adesmia abbreviata*, B. *Zophosis asiatica*, C. *Erodius duponcheli* (from Bodenheimer). D. Feeding marks of Chrysomelid beetle.

insects, which have lived on the wild flora before the grove was planted. They soon disappear, being unable to maintain themselves on the inadequate food. Such damage was observed by *Erodius dejeani* Solier, *Zophosis orientalis* Deyrolles and *Dendarus syriacus* Reiche. *Adesmia abbreviata* Klug once caused a more extended damage in citrus nurseries and in recently planted orchards in spring 1930 at Pardess Hanna. The attack was easily controlled by tying paper funnels with the broad side down around the stem, about 20 cm above the soil.

Many records on Tenebrionid-beetles reach us every year, mainly in April. The damage is generally restricted to a few trees only. It is never specific on citrus. It merely is an attempt of the pre-citrus fauna to find some food. We have therefore not even collected information on all the species concerned. As preventive control we have advised to leave the lower leaves on the wildling for some time or active prevention as just described for *Adesmia*. When large scale field operation should seem advisable, the following baits may be tried:

60 klo	wheat-bran	or	200 g	wheat-bran
400 g	Paris green		5 g	Paris green
20 g	amyl acetate		10 g	molasses.

Both baits to be wetted with water. It is however highly improbable that in our citrus groves such measures should be asked for. The danger is here usually passed one month after planting or inoculation.

SCARABAEID-BEETLES.

1. Anterior margin of prosternum raised to a lump with long hair. Mandibles visible from above. *Phyllognatus silenus.*
   Anterior margin of prosternum simple. Mandibles not visible from above. **2.**
2. Hind-tarsi with two ungui of unequal length. **3.**
   Hind-tarsi with two ungui of equal length. **4.**
3. Shield of head protracted mouth-like anteriorly. *Anisoplia inculta.*
   Shield of head not protracted, broadly rounded. *Phyllopertha nazarena.*
4. Anterior tibiae with 3 teeth on the external margin. Pronotum with a median carena, of one colour. **5.**
   Anterior tibiae with 2 teeth on the external margin. Pronotum and ventral side with white spots. **6.**
5. White pubescence, lateral margin of elytra not elevated. *Epicometis hirta.*
   White pubescence, lateral margin of elytra elevated. *Tropinota vittula.*
   Yellow pubescence, lateral margin of elytra elevated. *Tropinota squalida.*
6. Pronotum with 3 small white spots only laterally. *Oxythyrea abigail.*
   Pronotum with white lateral bands. *Oxythyrea noemi.*

Most of these *Lamellicornia*-beetles have apparently one annual generation. The adults are active during early spring.

The grubs of the shiny chestnut-brown *Phyllognatus silenus* Fabricius feed on the roots of citrus seedlings sometimes in the seedbeds of citrus, also loosening the soil. Also the beetles remain for some time in the soil. They never were in sufficient numbers to justify control measures. The larvae of *Anisoplia inculta ferruginipes* Pic, *Phyllopertha nazarena* Mulsant, *P. lineata* Fabr. and *P. arenicola pilosella* Reitter develop in the soil on roots of grasses and cereals. Very occasionally beetles are caught nibbling on the leaves or on the bark of citrus seedlings, but not to any extent which could be claimed as damage.

Every spring, some beetles of the *Cetoniini* swarm in large numbers on all flowers, on the pollen of which they feed. On citrus flowers the following species have been recorded:

Epicometis hirta Poda
Tropinota vittula Reiche
Tropinota squalida Scopoli
Oxythyrea abigail Reiche
Oxythyrea noemi Reiche.

Their size fluctuates between 5 to 13 mm length. The beetles abound especially on uncultivated steppe and dunes. In areas under concentrated cultivation they do not appear in such numbers. Farmers often accuse them of having caused up to 40 or 50% of the flower-drop in citrus orchards on light soils.

This mass-appearance of large beetles on the flowers of citrus causes some alarm amongst the farmers every year. It was therefore desirable to determine to what extent damage is caused by the beetles. Miss S. Avisohar studied this question in our laboratory.

During March and April 1935 *Epicometis vittula* was quite common on citrus flowers at Rehoboth. All observations on flowers of citrus and lupines showed that under normal conditions only the pollen is eaten and no other part of the flower is in any way damaged. However, sometimes all the pollen of one plant is devoured. Lupine flowers withered rather quickly after the visit of the beetles. Damage to the pistil or to the filaments has never been observed. These observations were confirmed by laboratory experiments, performed by Miss Avisohar. In one series of experiments parts of the flowers were given to beetles in petri-dishes:

1) whole flowers
2) all parts of the flower, but separated from each other.
   In both cases only the pollen was eaten.
3) pistils, still losely connected to the calyx.
   The pistils remained untouched and closely connected to the calyx.
4) flower-leaves only.

The leaves which remained untouched in the earlier experiments were eaten in absence of pollen. Whereas in the experiments 1—3 no difference between orange, grapefruit, and lemon-flowers was observed, in the fourth group the flower-leaves of lemon were preferred to those of the other citrus species. Repetition with flowers of *Dianthus*, *Anthemis*, *Matthiola* and *Lepidium draba* gave the same results. In nature, the beetles are sometimes seen inserting their heads deeply into the flower for some hours and even between the pistil and the calyx. This probably leads to the common belief that they eat the „heart" of the flower or damage the ovary, the fruit to come. Repeated inspection often showed that no damage whatsoever was done and that the connection of the pistil with the calyx was not weakened to any considerable extent. Sometimes, they feed holes in the flowers of papilionaceous plants or in the flower-leaves of roses and other plants to get to the pollen. In this way they are injurious to ornamental plants, but not to the open flowers of citrus.

Whatever said above for *Epicometis vittula* is true also for *Oxythyrea abigail* and probably for all other beetles of this group. Filaments, pistil and calyx are not eaten, flower-leaves only in hunger or in order to penetrate to the pollen. The movements of their legs may do damage also to ornamental flowers, especially in cases where no free access to the pollen exists, as is the case in papilionaceous plants or where the flower-leaves almost cover the pollen (rose, carnation). It is quite conspicuous that these beetles neither damage the flowers of citrus nor weaken the connection between the pistil and the calyx. The farmer has therefore no need to worry about control measures.

These observations are confirmed for *Tropinota hirta*, which feeds only on secretions on the basis of the flower and on the exine of the pollen, whereas the pollen itself is not digested. The pollen is transmitted on the hairs of the cuticula and no traces of feeding was observed on flowers of cherries, apples and rape (Fuchs).

The beetles appear in the spring. They were recorded as follows:

Month	I	II	III	IV	V	VI
no. of records. . . .	—	7	10	18	3	—

The grubs develop in the soil and sometimes cause damage to the roots of plants in ornamental gardens.

The species of *Amphicoma* (*A. vulpes* F., *A. vittata* F., *A. hyrax* Tqui., *A. papaveris* Sturm etc. *Glaphyrini*) have similar habits and life history. It is probably a coincidence that they have not been met with yet on citrus. They have about the same size as the similar Cetoniini, from which they are easily distinguished by the green or violet colour of the dense pubescence on their pronotum.

CHRYSOMELIDAE.

The ore-green beetles of *Labidostomis decipiens* Faldermann with pale yellow-brown elytra appear in April and May, feeding on leaves of vine and apricots. We have twice received samples from Beer Tovia (V. **1931**) and from Givath Brenner (V. **1934**) nibbling on leaves and twigs of citrus, on both occasions to a very limited extent only. It was presumably a residual insect of the original fauna (Vineyard before citrus grove).

A few flea-beetles (*Hermaeophaga ruficollis* Lucas, *Aphthona pygmaea nigella* Kutsch., *A. bonvoulieri* All.) only occasionally are met with on citrus. The normal host of the former species is *Crozophora tinctoria*, of the two *Aphthona* ruderal milkweeds (*Euphorbia*). Their nibbling on citrus is insignificant. We note however the fact, that in some reports it is stated, that they were observed on citrus seedlings, after the weeds had been destroyed. All our records from the southern coastal plain are from August to October.

CERAMBYCIDAE.

The larva of the big yellow-brown *Stromatium fulvum* develops in all kinds of dead wood. We obtained the beetle twice only from dying citrus trees. With the aging of our citrus groves its number will increase. Also other Cerambycid, Bostrychid and other beetles will then appear, such as *Rhesus serricollis* Motsh. observed by us in Cilicia.

CURCULIONIDAE.

Weevils are found feeding on citrus only accidentally. The small red weevil *Apion miniatum* Germar nibbles on leaves, its black congener *A. pubescens* Kirby on small, dropped oranges in spring. The later species develops on clover. On leguminous undercrops for green manuring develop two species of *Sitones* (*S. gressoria* Fabricius and *S. stierlini* Reitter). When this crop is cut, many of the weevils pass on to the citrus trees. Inspite of their large number, which often causes anxiety, they have never done any real damage. Occasional nibbling has also been observed on citrus leaves by *Psalidium syriacum* Miller, *Achradidius creticus* Kiesewetter, *Strophomorphus porcellus* Schoenheer and *Lixus sp. Tanymecus musculus* Fahrs., once collected from a recently planted orchard, is a residual pest.

References.

J. Aharoni, Hadar. 1. 1928. No. 2. p. 8—10.

G. Aulmann and W. La Baume, Die Fauna der Deutschen Kolonien. V, 2. Schaedlinge des Kaffees. Berlin. 1911. p. 5—9.

F. S. Bodenheimer, Die Schaedlingsfauna Palaestinas. Berlin. 1930. p. 160—163.

F. S. Bodenheimer, Animal life in Palestine. Jerusalem. 1935.

W. H. Fuchs, Ztschr. Pflanzenkrankheiten u. Pfl.schutz. 43. 1933. pp. 563—565.

3. ANTS, WASPS AND HONEYBEES (HYMENOPTERA).

A. ANTS (FORMICIDAE) CONNECTED WITH CITRUS-GROVES.

None of the ants connected with citrus-groves are regular or important citrus pests. Some species frequent the honeydew producing scales and mealy-bugs, others damage young plants or fruits occasionally.

The damage they cause is apparently never specific. Short information with regard to those species observed hitherto in connection with citrus-groves will be given below.

KEY TO THE MORE COMMON ANTS MET ON CITRUS TREES IN PALESTINE.

1. Cloacal orifice round, terminal, surrounded by a fringe of hairs. Abdominal pedical consisting of a single segment; no constriction between the second and third segments. Nymphs rarely naked, most frequently enclosed in a cocoon. (Formicinae) **3.**
   Cloacal orifice in the shape of a slit. Nymphs always naked. **2.**
2. Abdominal pedicel consisting of a single segment. Aromatic smell. *Tapinoma phoenicia.*
   Abdominal pedicel consisting of 2 segments. (Myrmecinae) **7.**
3. Antennae 11-jointed. **4.**
   Antennae 12-jointed. **5.**
4. Epinotum bidentate. Scale of pedicel incised or bidentate. Workers with well developed ocelli. *Acantholepis bipartita.*
   Epinotum unarmed. Scale of pedicel not incised from above. Workers without ocelli. *Plagiolepis ancryensis.*
5. Antennae inserted some distance behind the clypeus. *Camponotus compressus.*
   Antennae inserted near the posterior margin of the clypeus. **6.**
6. Maxillary palpi very long; its fourth joint nearly twice as long as the fifth. *Cataglyphis nigra.*
   The fourth joint of the maxillary palpi only a little longer than the fifth. *Paratrechina jaegerskioeldi.*
7. Postpetiole articulated to the *dorsal* surface of the following (*Crematogaster*) **8.**
   Postpetiole inserted at the *anterior* end of the following segment. **9.**
8. Epinotum with spines. *Crematogaster jehovae.*
   Epinotum unarmed. *Cr. inermis.*
9. Antennae 4-jointed clavate (clubbed). *Messor semirufus.*
   Antennae 3-jointed. **10.**
10. Epinotum smooth, unarmed. *Monomorium phoenicium.*
    Epinotum dentate. **11.**
11. The first 2 joints of clavus nearly twice as long as large. *Tetramorium signatus.*
    The first 2 joints of clavus nearly as long as large. *Tetramorium punicum.*

ANTS IN CITRUS ORCHARDS.

Of the many ants in citrus orchards some are without interest to the trees, such as the common harvester ant *Messor semirufus* André. All others cause occasional and local damage only.

The regular attendance of mealybugs and soft scales in connection with the sweet honeydew of these coccids gives these later a certain protection against their natural

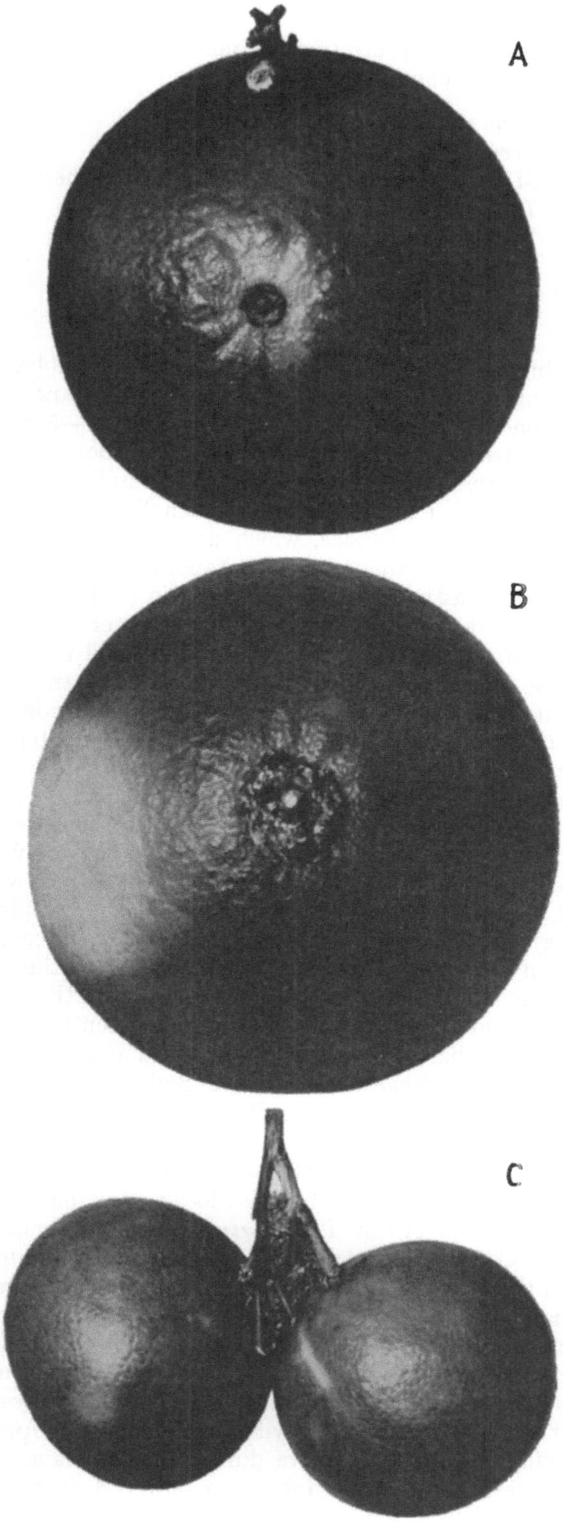

Fig. 12. Damage of *Crematogaster jehovae* and citrus mealybug on grapefruit at Dagania (VIII. 1935): A, B. Damage of stalk of fruit and feeding marks C. view of tent (from Klein).

enemies. Apparently so far this protection has not been of major importance in this country. The main species concerned are:

Crematogaster inermis Mayr
Crematogaster jehovae Forel
Monomorium subopacum phoenicium Emery
Tetramorium punicum Smith
Tapinoma simrothi phoenicium Emery
Camponotus compressus Fabricius
Plagiolepis pallescens ancyrensis Santschi
Paratrechina jaegerskioeldi Mayr

The large black-brown *Camponotus compressus* even assembles the coccids into real stables. In a recently planted orchard at Rehoboth, for instance, in late November the root collars of sone trees were laid bare (5 cm deep, 1—5 cm from the tree). The ants had transported about 25 young *Icerya* to the root collars in these places and attended them eagerly. There also was found a caterpillar of *Cosmolyce boeticus* (*Lycaenidae*), which exudes a sweet secretion much liked by the ants.

The red-brown *Crematogaster jehovae*, which usually nests in the wood of trees or in the stems of thistles and other herbs, has caused some local damage to grapefruits at Dagania, which has been described by Klein (1936). A few hundred trees were heavily infested by the citrus mealybug in the summer of 1935. Fruits dropped heavily in July. The brown ants swarmed over the infested trees, especially in the neighbourhood of the mealybugs. Just bases of the above the sepals white configurations obscured the calyx (Fig. 12). These were tents consisting of whitish threads, fragments of insects and hard, green plant fibres. These tents had a diameter of 15 to 20 mm, a height of about 10 mm. When these tents were removed, ants were seen attending to the honeydew of the *Pseudococcus*. Wherever two fruits touched one another, the tents were larger and higher (diameter: 20 mm, hight: 35 mm). Most of these fruits would drop upon the slightest touch. The calyx had been mostly nibbled away and thus the connection of the fruit with its stem was weakened. Fig. 12, fig. 2 shows this nibbling clearly. The nibblings are included in the tent. All the tents contained mealybugs. The damage caused by the ants was not smaller than that caused by the mealybugs. With the disappearance of *Pseudococcus* from the trees in late autumn, the ants left the trees as well.

Damage is inflicted on recent inoculations or on recently transplanted seedlings, when patches of the bark are eaten by ants, such as *Tapinoma simrothi phoenicium* Emery, *Acantholepis frauenfeldi bipartita* Smith or by *Cataglyphis bicolor nigra* André. This kind of damage is usually restricted to a few days following inocolation respectively transplantation. As a rule 3 to 5, occasionally up to 20 trees are attacked in one orchard. From among 25 cases of such attack about 90% occurred in May, the rest in June. More serious, but also localised damage was only suffered from the small, shiny black-brown *Tetramorium signatus* Menozzi, an endemic species. In May 1926 600 seedlings of sour oranges were transplanted at Ein Harod from the nursery to the grove. The season was unusual for such transplantation to take place. The small ants swarmed immediately about the trees and nibbled large irregular patches of bark off the roots, root collars and lower stems, often on many places. The damage ceased about 10 days after the transplantation, as soon as the seedlings began to take root. About 200 trees had to be removed and replaced. Opening of the root-collar and covering it with rags previously dipped into kerosene was quite a successful emergency measure during the few days of active danger. In a neighbouring vineyard, the same ant was a powerful enemy of the injurious termite *Reticulotermes lucifugus Rossi*. A similar damage of *T. signatum* occurred in late May 1931 at Kfar Yeheskiel. Serious damage occurred only in spots of heavy soil which was badly prepared and hard as stone. 6% of the trees were killed by the ants in such patches. The feeding was restricted to the subterraneous parts of the plants. Even in mid-june a few ants were still observed gnawing on the rind. Some more trees had to be replaced by healthy seedlings. The generally bad condition of the seedlings, the unsuitable season of transplantation, the insufficient preparation of the soil and delayed irrigation have doubtless greatly contributed to the extent of the damage. No damage was observed on light soils or on properly prepared soil.

Ants have been observed twice feeding in citrus flowers (Petah Tikwa, XII 1926; Kfar Atta 24.III.1931). As we have received neither samples of the ants nor of the flowers, we are unable to decide if any damage has been done. We have observed. *Tapinoma israelis* Forel and *Monomorium phoenicium* Emery in citrus flowers, without any damage being recognisable. They were probably in search for nectar.

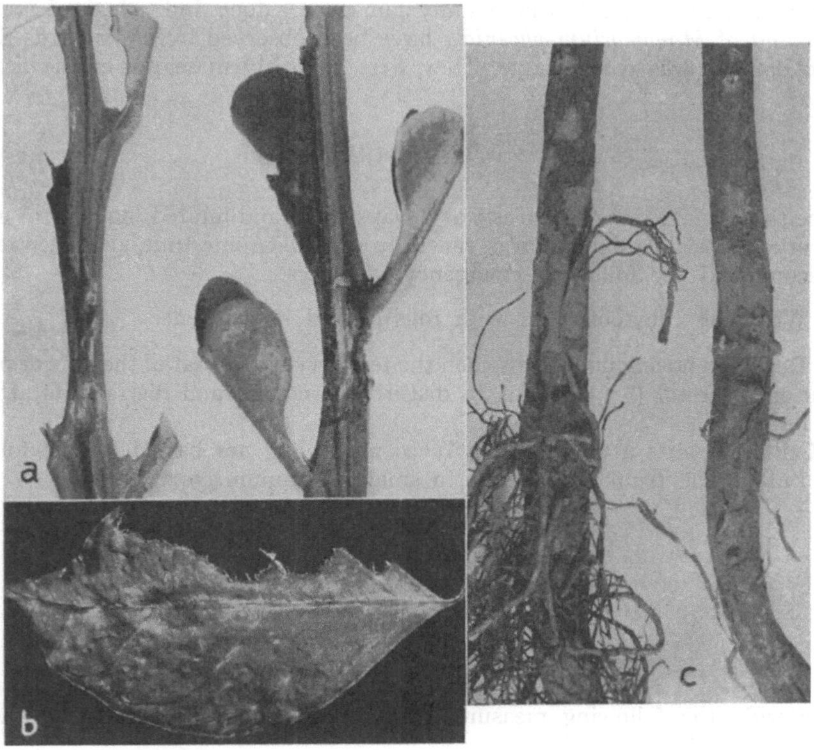

Fig. 13.
Feeding marks of ants: A, B. on fresh inoculations and on leaves of grapefruit (Rehoboth V. 1929), C. of *Tetramorium signatum* Men. on root collar of just planted nursery stock (Ein Harod V. 1929).

DAMAGE.

The injuries caused by ants in citrus orchards are manifold.

The regular attendance to Mealy-bugs, Soft-scales, etc. in connection with the sweet excretions of these coccids gives them a certain protection against their natural enemies. However, it does not seem that this side of the damage done has ever been of major importance in our country. This attendance becomes more disagreeable when connected with the construction of tents and nibbling on the fruit stem, as in the case of *Crematogaster jehovae.*

Serious damage is inflicted on recent inoculations or on recently transplanted seedlings, when patches of the bark are eaten. It is still unknown whether this feeding is provoked by the aromatic essences in the bark or whether it aims to obtain access to moisture. In any case, this type of damage (except the one attack described from *Tetramorium signatus*) always remains restricted to a few days following transplantation or inoculation. As a rule, 3—5, occasionally 10—20 trees are attacked in one orchard. It is to be regretted that in most of these cases no ants have been collected and sent in for determination. From among about 25 cases of attack observed, nearly 90% occurred in May, the rest in June.

Ants have twice been observed feeding on citrus flowers. (Petah Tikva, XII, 1926; Atta 24.III.1931). Neither the injured flowers nor the ants have been sent in. We are unable to judge as to whether any damage was done. Individuals of *Tapinoma israelis* and of *Monomorium phoenicia* have been observed by us in citrus flowers, but no damage was recognizable. They were probably in search of the nectar.

CONTROL.

Since the damage done to citrus was always small and limited, no control method was worked out. The damage was generally reported some time after it was done. We recommend the following emergency measures:

1) Wrapping the root-collar with rags dipped in kerosene.

2) Repeated hoeing up of the soil on the territory of the nest of the ants or pouring a solution of cresol (5%) over it to disturb the colony and distract its attention for a few days.

In California ants attending heavily to mealy-bug are baited by the following liquid put on the trunk of the tree in small aluminium cups (Quayle):

11 pints water
¼ ounce tartaric acid
⅓ ounce benzoate of soda
12 pound granulated sugar
$3^2/_5$ ounces chemically pure sodium arsenite
2 pounds strained honey

In Florida the following measures are recommended for preventing the ants from climbing the trees (Watson):

1) Banding the tree with tow worked in a mixture of corrosive sublimate and lard and then allowed to dry.

2) A circle of air-slaked lime in a cleared space around the trees (no ant nests to be included within the circle!) prevents the ants from crossing the lime, as long as it remains dry and the wind does not blow it away. (For further compounds see Watson, 1926).

In the last years ants have aroused attention in connection with the possible protection of mealybugs before ennemies. This problem is still under research.

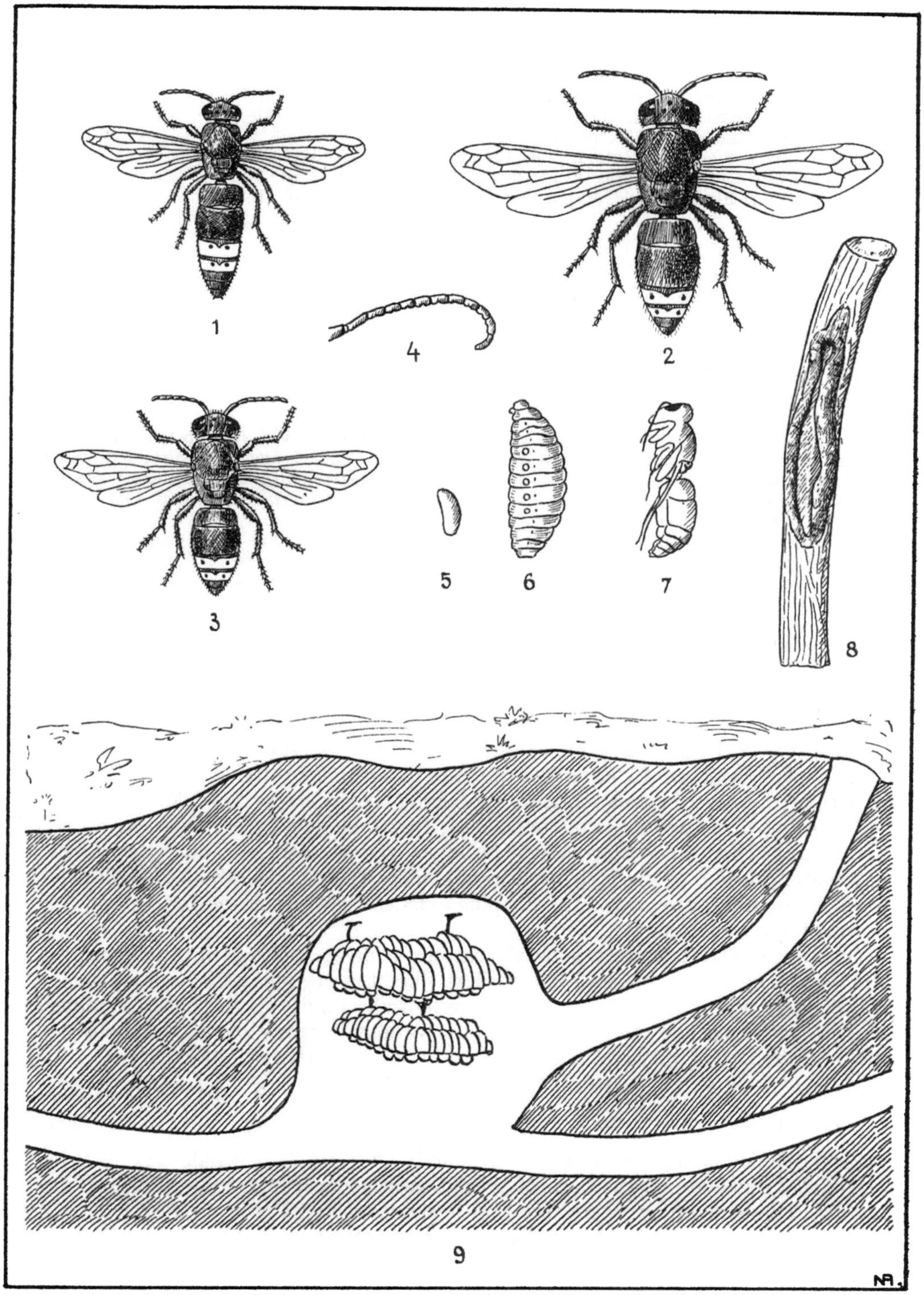

VI. The Levantine Hornet (*Vespa orientalis* F.) 1) worker; 2) queen; 3) male. 4) antennae of male; 5) egg; 6) full-grown maggot; 7) worker pupa; 8) healed damage following bark injury by Vespa on young citrus; 9) sketch of a nest.

References.

F. S. Bodenheimer, Schaedlingsfauna Palaestinas. Berlin, 1930. pp. 83, 94, 176.
F. S. Bodenheimer, Animal life in Palestine. Jerusalem, 1935, pp. 257—266.
F. S. Bodenheimer and H. Z. Klein, Zeitschr. vergl. Physiol. **11**, 1930, pp. 345—385.
H. Z. Klein, Zeitschr. wiss. Insektenbiol. **25**, 1930. pp. 45—52, 141—156.
H. Z. Klein, Hadar, **9**, 1936, pp 42—43.
C. Menozzi, Memor. Soc. Entom. Ital. **12**, 1933, pp. 49—113.
H. I. Quayle, California Agric. Exper. Sta., Berkeley, 1932, Bull. 542.
J. R. Watson, Florida Agric. Exper. Sta. Gainesville. 1926, Bull. 183.

B. THE LEVANTINE HORNET (*Vespa Orientalis* F.)

The Levantine Hornet is a very common insect in Palestine. It is easily recognised by its size and the yellow and red-brown color.

Description: The hairy, red-brown head is enlarged behind the eyes; the area between antennae and clypeus yellow; maxillae yellow with black margins. Thorax and legs hairy, red-brown. Wings reddish-brown. Abdomen red-brown, but slightly darker than head and thorax; a small yellow band on the first and two larger ones on the third and fourth tergit; two small yellow triangles laterally on the second tergit.

The casts may be distinguished as follows: The antennae of the male is 13-jointed, with two small humps on the ventral side of each joint. Its abdomen is composed of 7 segments and is 13—15 mm long. The female (and worker) have 12-jointed antennae with no humps on the lower side of each segment. Their abdomen has 6 segments only and is 11—12 mm long in the worker. The female is larger than the worker.

Body-length of male 24—28 mm; of female 26—34 mm; of worker 19—23 mm.

Distribution: S. & E. Mediterranean, S. W. and C. Asia, Egypt, Sahara, Abyssinia, Madagascar.

The nests are generally built in the soil in Palestine, They consist of 2—6 stores of combs. These nests lack the paper cover, which generally surrounds the combs of wasps. Occasionally they build also in corners of windows or among the mud walls of Arab villages. In Egypt locality the latter seems to be even preferred and if part of the nest remains exposed, the grey paper cover is built (Mellor).

The colonies of the hornet die out during November, i.e. before their earth-nests are destroyed by the winter rains. The males and workers die; only the females hibernate. The end of the breeding season is induced by the death of the old queen which lives only one year, and the stopping of breeding is responsible for the breaking down of the social contact.

Observations in Jerusalem showed that the females appeared in numbers from the beginning of May on, and the workers since the beginning of June. From the end of June onwards, females became rare and all activity of the hornets stopped with the end of November. Statistical observations in many parts of the country yield the following relative index of activity per observation and month:

Month	I	II	III	IV	V	VI	VII	VIII	IX	X	XI	XII	Total
No. of hornets	—	—	3	6	8	20	22	22	**58**	**55**	30	—	223

Single females were observed even in March. On 6.III.24 a female was observed at Nahalal, gnawing on the bark of young carobshoots. From June to August the number increased, the workers appeared in numbers, but the peak of activity and population density was reached in September and October. In November the population was rapidly contracting. From the end of November to the beginning of March all workers have died. The males also died before winter, after having copulated with the young females. The hibernating females may be seen occasionally on flight during nice, sunny days; but certainly no regular activity takes place in winter. This annual curve of activity runs parallel to the breeding curve of the nest.

A few counts will illustrate this seasonal development:

Nest Locality and date	A Jerusalem 4.VII.37	B Ganigar 26.VI.37	C Jerusalem 1.X.37	D Jerusalem 5.X.37
no. of combs	1	2	2	2
no. of cells	70	709	651	931
no. of eggs	29	96	54	54
larvae	18	97	106	106
workers pupae	13	270	56	108
sexual pupae	—	—	70	23
workers	9	91	226	226
females	1	1	30	13
males	—	—	10	10
total population	70	555	552	540
adult population	10	92	266	249
empty cells	10	246	365	640

Nest B is almost contemporaneous to nest A. The warmer climate in Emek Yesreel is the reason for its much further advanced development.

The diurnal activity is concentrated in early spring and in autumn on the warm hours of the day. During the summermonths the main activity is in the early morning and late afternoon hours. The activity as correlated to the environmental shade temperature and as observed on some observation plots thouroughout the year, is:

Shade temperatures	15	17	19	21	23	25	27	29	31	33	35	37°C
no. of observations	3	5	12	30	44	57	39	32	17	10	4	

The range of 21—31° C embraces 80%, that of 23—29° C 55% of all observations.

The scale of activity was:

1.	Beginning of cold torpor:	8.0° C
2.	Only weak movements:	12.2
3.	Crawling with interruptions:	20.0
4.	Beginning of normal activity:	23.3
5.	,, ,, high activity:	27.1
6.	,, ,, irritation:	35.4
7.	,, ,, heat paralysis:	41.9
8.	Instantaneous heat death:	44.0

The hornets are true heliothermic insects, the body temperature of which is always above 30° C as long as it is exposed to sunradiation. In accordance herewith the normal temperature of the breeding hornet nest ranges from 30 to 35° C (Janet, Himmer). When this upper limit is surpassed, a state of irritation begins in the nest and the workers transport water drops into the nest, the evaporation of which lowers the nest temperature. The extended interval from 35.4 to 41.8°C., the beginning of heat paralysis, is very important for the hornet. It enables it to transport the water without danger and to create a more suitable microclimate in the nest. Normal activity (running) starts in nature about 22° C shade temperature, and 27—25° C will be the optimal temperature for the normal flight. Below 16° C and above 37° C no activity was observed.

The hornets are very omnivorous feeders. They are a most troublesome pest of honeybees, with they prey on or their way to the nest. They even enter the nest in order to pillage the honey. They attack ripe fruits, of all kinds, on the tree and feed on them (grapes, quinces, pears, etc.). Dates can only be saved by covering them with a paper bag. They are attracted to the honeydew of aphids and coccids, as well as to animal cadavers and human excrements. They gnaw the bark of young trees and branches, which they probably use mainly for the construction of their nests. A real mass outbreak of hornets occurred in 1934, when considerable damage of this type was reported from Central and Eastern Emek Yesreel, from Sichron Yakob and Rehoboth. The debarked plants were especially: *Citrus, Cydonia, Tamarix, Pyrus, Ceratonia, Poinciana, Prunus* and *Vitis*.

Of citrus, only young trees are damaged, up to 3—4 years after the inoculation. The bark is eaten mainly off the trunk, wich they bare in large patches. Gum is exuded from the damaged parts. Its colour is whitish to bright brown and it flows slowly downwards. But also young leaves are occasionally devoured. The sap dripping from the wounds of the tree is eagerly drunk by the hornets.

On other plants (carob, *Tamarix, Prunus*, etc.) the bark of the branches is preferred to that of the trunk. However, last year's bark is always preferred to that of the present year.

The attack in 1934 started from the end of July. The damage became remarkable from mid-August and lasted until the end of October.

The only reliable control is the destruction of the nests by dusting with calcium cyanide. In the evening, when all the workers have returned to their nests for their night's sleep, the entrance of the nest should be closed and 30—60 g calcium cyanide dust blown into the nest with the aid of special pumps. One treatment is sufficient to destroy the nest. This control should be made not later than early August and its effect is radical only when all the nests of the immediate neighbourhood are carefully marked and treated. The dust is issued gratis to bee-keepers from the Veterinary Officers. Mashed liver with thallium sulfate is a favoured bait.

During the mass-outbreak in 1934 it was repeatedly reported that after the destruction of the nests in September the hornets remained on the trees during the night. One quince grove at Ein Harod was so heavily attacked in this way, that the growers left the fruit on the tree for fear of the hornets. Breeding is stopped in any way by the destruction of the nest. The extensive damage and the difficulty to control the hornets in that year were mainly the result of the delay in the beginning of the anti-hornet campaign.

References.

F. S. Bodenheimer, Schaedlingsfauna Palaestinas. 1930, pp. 400—401.

F. S. Bodenheimer, Zool. Anzeiger **102**. 1933, pp. 135—140.

J. E. Mellor, Bull. Soc. Roy. Entom. d'Egypte 1927, pp. 80—97.

(Anonymous) The Hornet and Methods of Combating it. Dept. of Agric. Pal. Series VII, No. 1 (1933) 6 pp.

C. THE HONEY-BEE (APIS MELLIFICA SYRIACA BUTT. REEP.) AND THE QUESTION OF BEE-KEEPING WITHIN THE ORANGE GROVE.

The problem of the interrelations between the honeybee and citrus plants is rather complex and has not yet found a satisfactory solution until this day. Webber (1930) describes the situation: „Though all commercial citrus varieties have showy white flowers, develop an abundance of nectar, are regularly visited by bees and are excellent honey plants, no fixed economic relation between bees and fruit production is recognized. Citrus fruits are not grown for their honey and yet honey-production is an important secondary industry. The source of wealth which the rich bee pastures of citrus groves represent should not be minimised. Questions at issue are 1) whether bees are of value in a citrus grove from the fruit production standpoint, 2) if of no value to crop, do they do no harm to the crop so that they may be permitted to gather the honey without loss?"

The rich, not easily accessible honey source and the morphological structure of the male organs of citrus flowers seem to indicate a close coaction with insect-pollination, especially by the very frequent visitor, the honey bee. And ample pollination by honey bees is undoubtedly a very common fact.

A few observations in orange groves in the coastal plain may illustrate the frequency of visits of the honey bee to citrus flowers.

The number of visits per 200 flowers (one quarter of the total number of flowers on each tree) on one tree of an outside row and on one from the fifth internal row in an orange grove was noted at Petah Tikva (17.III.36):

Meteorological observations					No. of visits p. hour		Visits per flower	
Hour	Temp. °C	R.H. %	Clouds	Wind	External row	Intern row	Nectar n × Time	Pollen n × Time
7—8	15	70	+	—	—	—	7× 4 sec.	—
8—9	16	70	++	(+)	108	102	7 11	4× 4 sec.
9—10	20	60	++	(+)	252	152	24 10	11 4
10—11	23	50	+	(+)	216	24	19 19	9 3
11—12	23	50	+	(+)	276	126	20 15	8 4
12—1	23	50	++	+	192	84*	29 12	8 4
1—2	21	50	+	+	42	54*	23 10	11 4
2—3	20	60	+	(+)	186	42*	16 13	6 4
3—4	19	60	+	(+)	138	48*	20 7	6 3
4—5	19	80	(+)	(+)	96	—	14 6	7 4
5—6	16	80	—	—	—	—	—	—
			Total		1506	632	179	70
			Mean				11.2	3.5

The duration of each visit fluctuated with the time of day, especially in the case of the nectar-collecting bees. It was remarkably constant in the pollen collectors. On an average each nectar visit lasted 2—3 times longer than the pollen visit. The average number of visits per flower per day was 7.5 on the outside tree, 3.2 on the internal tree.

In order to gain preliminary information on the effect of distance from the apiary on the frequency of visits of bees, a series of observations were made on 29.III.36 in an orange grove at Givath Hashloshah between 10 and 12 a.m. (26—29 °C; 49 per cent R.H.):

Distance from apiary m.	Visits per 100 flowers and day (7 hours)
50 (interior)	454
300 (exterior)	330
300 (interior)	196
700 (exterior)	50

* In shade

In the height of the flowering season the distance from the apiary is an important factor in the frequency of the visits.

These facts do not prove that cross-fertilisation is necessary for the development of the citrus trees, nor that self-pollination is inhibited by the bee. Many observations have shown that self-pollination occurs quite frequently in citrus-flowers. It may be that different varieties show a different behaviour in this respect. Oppenheimer's experiments seem to indicate that a lower percentage of fruits stay on the tree till the time of ripening (i.e. a lower crop-expectation, but not a lower set-percentage) as well as a lower percentage of seeds occur in self-pollinated (i.e. within the same citrus variety) fruits of the shamouti-orange:

Year	Type of pollination	Average no. of seeds	crop	June drop
1933	self pollination	0.4 per fruit	13%	70%
	pollinated with Temple Orange. .	0.1	30	0
1934	self pollination	1.4	4	60
	poll. with Clementine	5.5	10	40
	poll. with King's Orange	9.0	15	46

But no generalisation is yet possible. Further experiments of Oppenheim and Oppenheimer show that cross-pollination increases the number of seeds per fruit in most crossings, whereas in a few others no influence was recognisable. Whereas seediness is not a desirable character in the shamouti orange, no interplantation of different citrus-varieties is advisable.

Self-pollination of the flower (by its own pollen) occurs quite frequently during visits of honey bees. An agreement exists between horticulturists that pollination by wind is insignificant, where-as self-pollination within the same flower occurs quite often. The direction of wind is of no direct importance with regard to pollination, except perhaps as far as it influences the activity of honey-bees.

In many citrus-varieties a surprisingly high percentage of sterility of pollen has been observed by a number of observers, whereas in other varieties it is rather low. In some citrus varieties it has been conclusively demonstrated that normal setting and development of fruit takes place when pollination of any kind, even self-pollination, is experimentally excluded. Yet the question if all citrus varieties behave this way, and if all varieties can dispense not only with normal fecundation, but also with the stimulation caused by pollination and the growth of the pollen tubes (even if no normal or successful fertilisation follows) is still under study.

The problem how far the size of the citrus crop is correlated to the number of visits of honey-bees to the flowers cannot yet be definitely settled.

Webber concludes for California and Florida that in his experience neither has the presence of bees increased nor their absence decreased the normal citrus crop. Orchards with several citrus-varieties are not more fertile than those in which one variety has been planted.

In Malta the opinion prevails ,,that orange groves in which bee-hives are kept are always more productive than those in which there are no bee-hives. and therefore our gardeners are glad to see the bees busily humming on the orange trees in bloom. It is not easy to establish accurately by experiment what is the actual increase of production due to fertilization by bees, but is is generally believed to be 15—20% and may be much more in certain circumstances. The honey obtained from the orange grove, though fine in quality and very light in colour, is lacking in flavour, and on this account it is often allowed to remain in the hive to be used by the bees, in order not to have to feed them artificially for a considerable part of the year." (Borg).

C. Oppenheimer concludes for Palestine: „We do not consider the introduction of apiaries into citrus groves as indispensable for increasing crops, but they are harmless as far as danger of increasing seeds is concerned, always provided that the grove does not contain trees of certain species or varieties (as Valencia, Baladi, Duncan, Yussuf Effendi, Sour Orange), pollination with which tends to increase the number of seeds."

The late S. Yedidyah agrees with Webber's conclusion. In three consecutive years trees were covered at Mikve Israel with muslin cages and bee-hives were placed under some of them. The results were as follows:

Tree	year	with or without bee-hive in cage	number of fruits per tree	weight of fruit in kilos	Average no. of seeds per fruit
I	1934	—	290	69	1.8
I	1935	+	549	100	4.5
I	1936	+	323	72	1.6
II	1934	—	377	91	2.0
II	1935	+	686	130	4.9
II	1936	—	210	52	0.5
III	1934	+	418	86	5.0
III	1935	—	424	97	1.8
III	1936	—	439	93	1.8
IV	1934	+	275	53	5.0
IV	1935	—	428	95	1.1
IV	1936	+	461	96	1.0
	Average	—	361	83	1.5
		+	452	89	3.7

The increase in fruit and weight of crop on the trees with apiaries is not constant and even the increase in seeds per fruit is not without exceptions. Other factors besides fertilization by bees also seem to influence seediness. Offering no definite solution these experiments confirm the general trend of results and show that no damage is inflicted by the bees, but that much more and more definite experience is needed before any conclusion as to the positive benefit the citrus-crop derives from the introduction of apiaries is drawn.

Similarly, limited experience of Fujita in Japan states that bees sometimes open flower buds to take honey, this not being injurious as it hastens the flowering and stimulates the setting of fruit. The surface of the very young fruits is sometimes slightly injured, but the wounds disappear during growth.

On the other hand, no damage is caused to citrus crop by honey-bees. The nectar, once secreted, does not benefit the citrus plant anymore. Its collection and consumption by the visiting honey-bees does not affect the plants' metabolism. The pollen which is collected is anyhow surplus pollen, even if the pollen should prove to be necessary or useful for the setting and maturation of the fruit. Neither mutilations of the flower nor transmission of diseases are known to be connected with visits by honey-bees.

The following conclusions seem therefore to be justified for Palestine:

1) The introduction of apiaries into the citrus groves is of no definite benefit to the citrus crop. The theoretical side of this problem is not yet satisfactorily solved, but it seems certain that insect- and especially honey-bee fertilization is not needed for a normal setting of fruits, despite the apparently strongly adapted morphological structure of the citrus-flower for insect-pollination.

2) No damage is apparently connected with the introduction of apiaries into the citrus orchard, except under very special conditions which are rare in Palestine (mixed plantation of certain citrus species or varieties).

3) Bee-keeping is a natural secondary industry of citrus-growing and may gain in importance, when growing marketing difficulties will enforce a more intensive and more diversified form of management of citrus-gardens. But even if the citrus grower is unwilling to undertake the exploitation of this natural by-product of the citrus grove, it offers an important basis to the professional bee-keeper.

The Palestinian honey-bee belongs to the Syrian Race of the European Honey-bee (*Apis mellifica syriaca* Butt.-Reep.), which is spread over the Levant-coast and

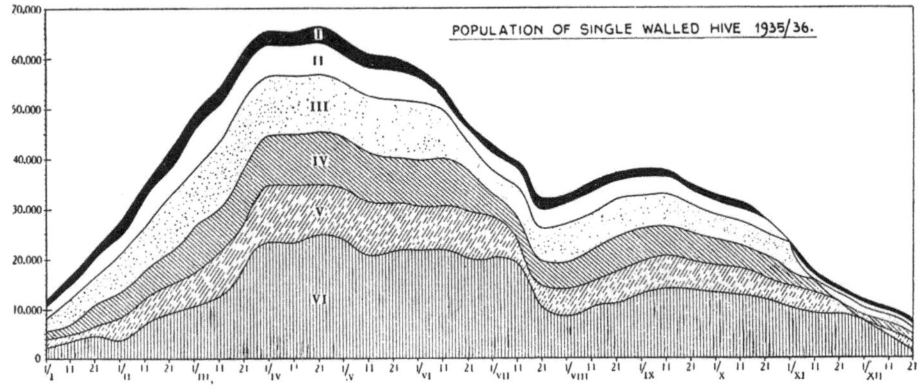

Fig. 14.
Seasonal population trend of a colony of honeybees in Kiryath Anavim (1935/36): I eggs, II larvae, III sealed brood, IV nurse-bees, V house-bees, VI field-bees (from Bodenheimer and Ben Nerya).

closely related to the Cyprian bee. It is a dark variety, swarming speedily and possesses a bad temper, which makes bee-keeping sometimes rather unpleasant. Italian and Carniolian queens were recently introduced. Italian queens are now bred locally, Mrs. G. Wertheim has made a biometrical study of the Palestine honey-bee in our laboratory. The following table compares her results with those obtained by Alpatov for other races:

Characters	Palestine	C. Russia	Ukraina	Italy	Egypt
Tongue (length)	6.41±0.03	6.12	6.55	6.23	5.55
forewing (length)	8.97±0.03	9.47	9.36	9.05	8.19
,, (width)	3.01±0.01	3.17	3.19	3.17	2.75
hooks on hindwing	23.47±0.47	20.70	21.08	21.51	20.50
3rd tergit (length)	2.29±0.02	2.35	2.27	2.17	2.15
yellow stripe of 3rd tergit (length)	1.23±0.04			1.08	1.28
1. tarsal joint (length)	2.16±0.02	2.13	2.10	2.08	1.97
femur (length)	3.40±0.01	3.36	3.29	3.23	3.03
tibia (length)	2.74±0.004	2.74	2.68	2.65	2.42
cubital index	50.81±2.21	67.57	50.79	47.05	50.81

Alpatov has recently shown conclusively that the Egyptian race belongs to the „Rassenkreis" of the African bee and must be named *A. unicolor fasciata* Latr.). The Syrian and Cyprian races were generally regarded as a result of crossings between lighter European espec. Italian bees and the dark Egyptian bee. This conception must now be abandoned because the Syrian bee undoubtedly belongs to the „Rassenkreis" of the European *Apis mellifica* L. The value of the cubital index as a racial and geographical character, which has recently been greatly stressed by Alpatov, seems to have been overestimated.

The number of total flights made from all field-bees within one hive in 1935/36 at Kiryath Anavim (Near Jerusalem) was:

Month	no. of total flights: in thousands	no. of total flights: in percentage of annual activity
I	219	3.3%
II	404	6.2
III	1198	18.3
IV	1214	18.5
V	1047	16.0
VI	583	8.9
VII	417	6.4
VIII	752	11.5
IX	334	5.1
X	225	3.4
XI	41	0.6
XII	100	1.5
Total	6534	99.7

Considering that in the coastal plain the seasonal trend is somewhat earlier we may fairly assume that at least 30—40% of the total annual flight activity is concentrated on the season of citrus flowering. Honey is collected almost only in

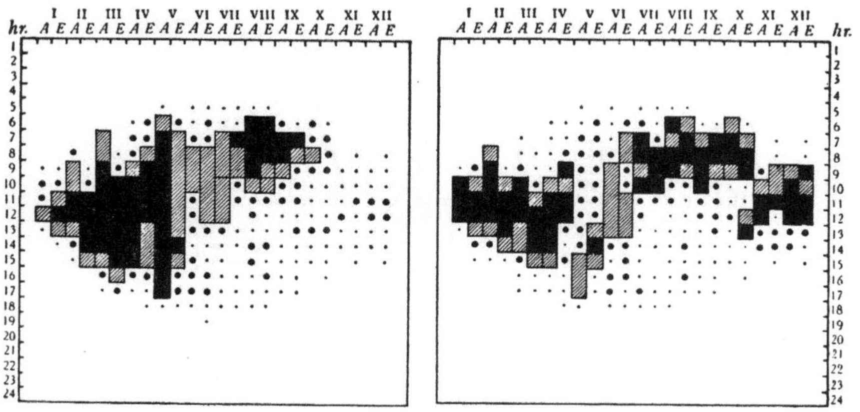

Fig. 15.
Diagram of seasonal and daily activity of a bee-hive at Kiryath Anavim, measured by the number of bees leaving the hive every hour (from Bodenheimer and Nerya). Abscissa: month; ordinate: hours of day. A. Absolute annual activity. B. Relative annual activity (total daily activity for every month = 100). Black: over 3000 flights per hour (or over 15% daily activity); striated: 1500—3000 flights (10—15%); big dot: 600—1500 flights (5—10%); small dot: 1—600 flights (0.5—5%); white: no activity.

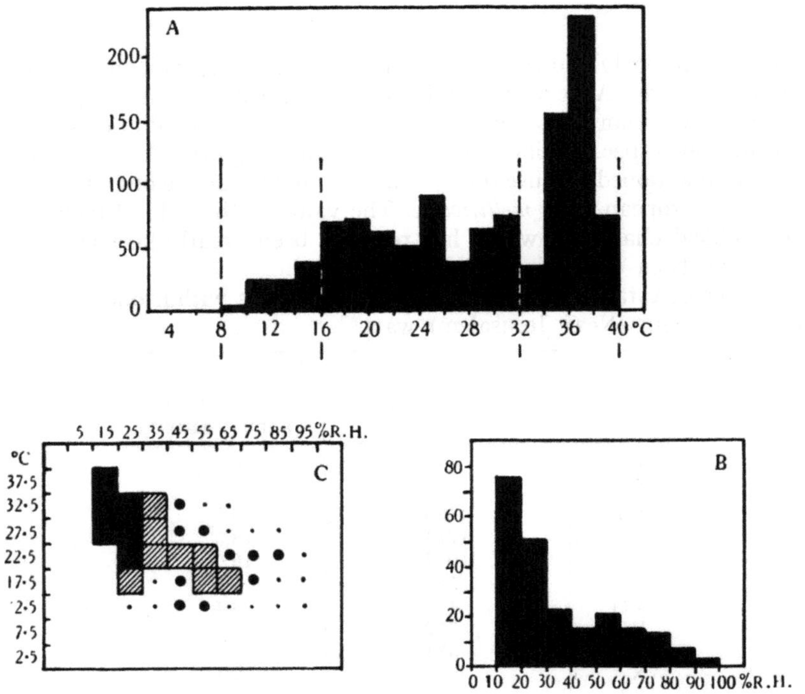

Fig. 16.
Correlation of environmental temperature and humidity with the flight activity of the honey-bee: A. temperature and activity, B. rel. humidity and activity, C. combination of temperature and humidity and activity (from Bodenheimer and Nerva).

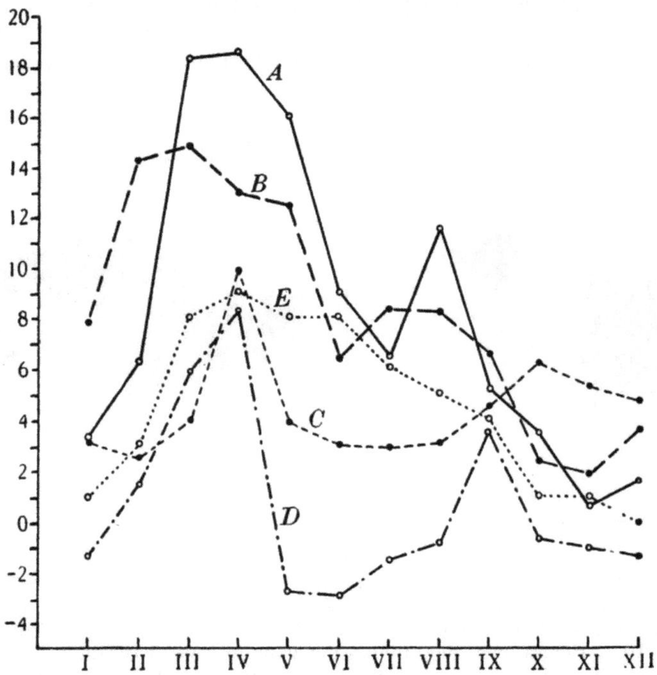

Fig. 17.
Various seasonal trends within the bee-hive: A. percentage of flight activity, B. of total egg-production, C. no. of full honey combs, D. change of weight of hive (in klo.), E. no. of nectar plants (from Bodenheimer and Nerya).

February to April and, to a minor degree, again in September. It follows that orange blossoming and seasonal field activity, (as measured directly by the number of flights and indirectly by the weight increase of hive) agree in Palestine to a high degree.

The influence of climatic factors on bee activity is as follows: Temperatures in Palestine may be divided into 5 groups:

1) Below 8° C no activity was observed at all.
2) From 9 to 16° C activity is below normal (22 flights per minute)
3) From 16 to 32° C range of activity is optimal (66 flights per minute)
4) From 32 to 34° C activity is reduced (34 flights per minute)
5) Between 34 and 39° C rise of activity (watertransport) (153 flights per minute)

High air humidity depresses the activity of the honey-bee, whereas low humidity stimulates it:

10	20	30	40	50	60	70	80	90	100%	Rel. Hum.
76	51	23	16	21	16	14	7	3		flights per minute

The sesonal trend of egg production is likewise closely correlated to the orange-blossom (always considering that the data given here were obtained in the hills and are therefore retarded with regard to the coastal plain). The egg-production was in a normal single-walled hive:

month	Total	Average per 10 days period			Daily average during month
		I	II	III	
I	17,695	4,780	5,850	7,065	571
II	28,940	9,500	9,500	9,940	1,034
III	33,105	11,990	9,740	11,375	1,068
IV	28,330	9,390	9,810	9,130	944
V	27,930	9,335	9,275	9,320	901
VI	13,710	4,605	4,030	5,075	457
VII	18,825	5,570	5,685	7,570	607
VIII	18,390	6,355	6,110	5,925	593
IX	14,110	5,170	4,575	4,365	470
X	7,185	3,990	3,195	0	232
XI	4,060	1,100	1,320	1,630	135
XII	7,875	1,270	1,630	4,975	254
Total	220,155				603

The total population trend is illustrated in fig. 14. In its major progressive part, which lasts four months, from January to April, it follows a logistic curve of the formula:

$$y = \frac{66{,}285}{1 + e^{2.8462 - 0.0464x}}$$

The almost continuous egg-production and breeding throughout the whole year, the vegetation period being rather limited, is rather a heavy drawback for the Palestinian bee-keeper, because if the honey is collected through autumn and winter, heavy artificial nutrition is needed.

With regard to the citrus-crop from the point of view of the bee-keeper, the following figures are due to Mr. A. Ben Nerya.

The main orange crop in the coastal plain of Palestine occurs from the 15.III. to the 25.IV. It is a very selected honey of white colour, very tasty and is highly appreciated on the market. The average crop of orange honey obtained from a normal hive is 15—18 kilos whereas the most intensive migration adds in the average only up to 12 kilos from other flowers. This means that the citrus-flower is *the* fundamental crop of the bee-keeper who would find it extremely difficult to manage without it. The general management of the apiary is intended for a maximal utilisation of the citrus blooming. Stimulation feeding of the queen begins in mid-January and the brood-room of the nest is subsequently enlarged in preparation for a maximum field-bee population during the orange season. Fig. 14, concerns conditions in the hills, where the seasonal population peak is reached later. The genuity of orange honey is easily confirmed by microscopic examination, which reveals the presence of a large number of typical citrus pollen (fig. 18) within it. It is widely oval, with rough exime and 4 deep groves. An end view presents the picture of a perfect yellow cross. Colour: deep yellow; size: 24 μ long.

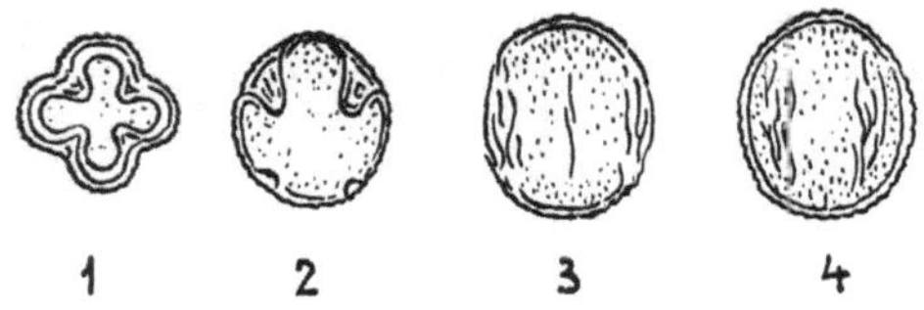

Fig. 18.
Citrus pollen from honey (from Zander).

References.

F. S. Bodenheimer and A. Ben-Nerya, Ann. Appl. Biol. **24**, 1937, p. 385—403.
F. S. Bodenheimer, Quart. Rev. Biol. **12**. 1937. pp. 411—422.
J. Borg, Cultivation and diseases of fruit-trees in the Maltese Islands. Malta, 1922, p. 56.
K. Fujita, Agric. and Hortic. **9**, 1934, p. 1413—1414, 1616—1618 (Japanese).
C. Oppenheimer, Hadar, **8**, 1935. p. 261—267, 291—296. (Further references to botanical literature).
H. I. Webber, Calif. Citrograph **15**, 1930, p. 304, 322—323.
S. Yedidyah, Citrus Growing Tel Aviv 1937 (Hebrew). p. 77—78.
E Zander, Pollengestaltung und Herkunftsbestimmung bei Blütenhonig. Berlin. 1935. pp. 205—206.

4. FLIES (DIPTERA).

A. THE MEDITERRANEAN FRUIT FLY (*Ceratitis capitata* Wied.) (Plates VIII, IX, X).

Synonym: 1824 *Trypeta capitata* Wiedemann
1825 *Petalophora capitata* Macepi
1829 *Trypeta citriperda* Mac Leay
1842 *Trypeta hispanica* de Brême.

1. DESCRIPTION AND DISTRIBUTION.

Description: A handsome fly, well characterised by the peculiar color of body and wings, the latter being also broad and short, and by the large lancet shaped, frontal appendage of the male. 4—5 mm long; ovipositor of the female extended 1.2 mm; wings 4.5 mm.

„Frons yellow, getting much darker towards the base of the antennae, where it is dark brownish and bearing some blackish erect hairs in the middle; in additions the male only possesses 2 curious spatulate appendages at the end of 2 chaetae, which are silver grey in colour; face and cheeks yellow; vertex yellow, somewhat shining, with the ocellar triangle brownish-black; proboscis and palpi yellow, bristly; antennae with the 2 basal segments reddish-brown and the third segment usually yellow; arista blackish.

Thorax black on the disc, very shiny, with very characteristic light grey markings which are due to a fine pulverulence of that colour; the black shining areas are as follows: the upper margin of the disc, the centres of the humeral calli, the two large upper corners, 2 small isolated rounded spots in a straight line with the suture, 2 much larger isolated spots below these and the 2 large inferior corners; the humeral and prealar calli are whitish, except the centre of the former, as stated above, black; the pleurae are whitish and entirely covered with a delicate white tomentum; the mesopleurae possess a tuft of strong erect hairs, which in the male are black and in the female white; there are also some white strong hairs on the pteropleurae and sternopleurae, but these are constant in both sexes; scutellum very shiny black like the thorax with a pale yellow band all round the base which is emarginate on both sides and not touching the extreme base; mesophragma dark yellowish-red; the pubescence on the disc and scutellum is short, adpressed and shining pale yellow; it is scarcer on the black areas and on the scutellum it is not adpressed; all the bristles are black (except the scp yellow).

Abdomen orange-yellow with 2 reddish-brown bands which appear grey owing to a pale whitish pulverulence; the first of these bands occupies the lower half of the second segment and the second band the lower two thirds of the fourth segment; the pubescense is fairly dense, short, erect and blackish and on the lateral margins and apex long and bristly; female ovipositor short, broad, flattened and reddish-yellow.

Legs reddish-yellow with a row of rather long yellow bristles on the hind tibiae and the hairs on the upper side of the front femora are black in the male and yellow in the female.

Wings characterised by their breadth, the deep curve in Cu_2 and by the acute inferior angle of first M_2 cell. The wing markings although almost indescribable are nevertheless characteristic, being hyaline with black, brown and ochraceous markings with somewhat faded tints as shown in the figure." (Efflatoun, 1925).

The *eggs* are smooth, shining white, long elliptical, but (*oftener*) convex on the dorsal side and flat or slightly concave on the ventral side when freshly laid. The micropylar region at the anterior egg-pole is distinctly tubercular.

0.9—1.1 mm long, 0.2—0.25 mm broad.

Maggot creamy-white. Shape elongate, conical anteriorly, sub-cylindric posteriorly.

Antennae 2-jointed with the basal joint longer than broad and the apical joint narrower but rounded at the tip. Maxillary palpi small, bearing 3 minute sensoria. Cephalo-pharyngeal skeleton with oral hooks, each furnished with a strong, fairly well curved tooth and not possessing a distinct preapical tooth although the lower margin of the mandibular sclerite is feebly notched; hypostomal sclerite very small and weak, triangular, with the posterior triangle narrow and elongate. Oral lobes convex, each possessing 9—10 raised, transverse laminae. Anterior spiracles with from 8—10 lobes. Posterior spiracles each with 13 slits which are about 3 times longer than wide; distance between the stigmata equal to less than 1½ times the length of one

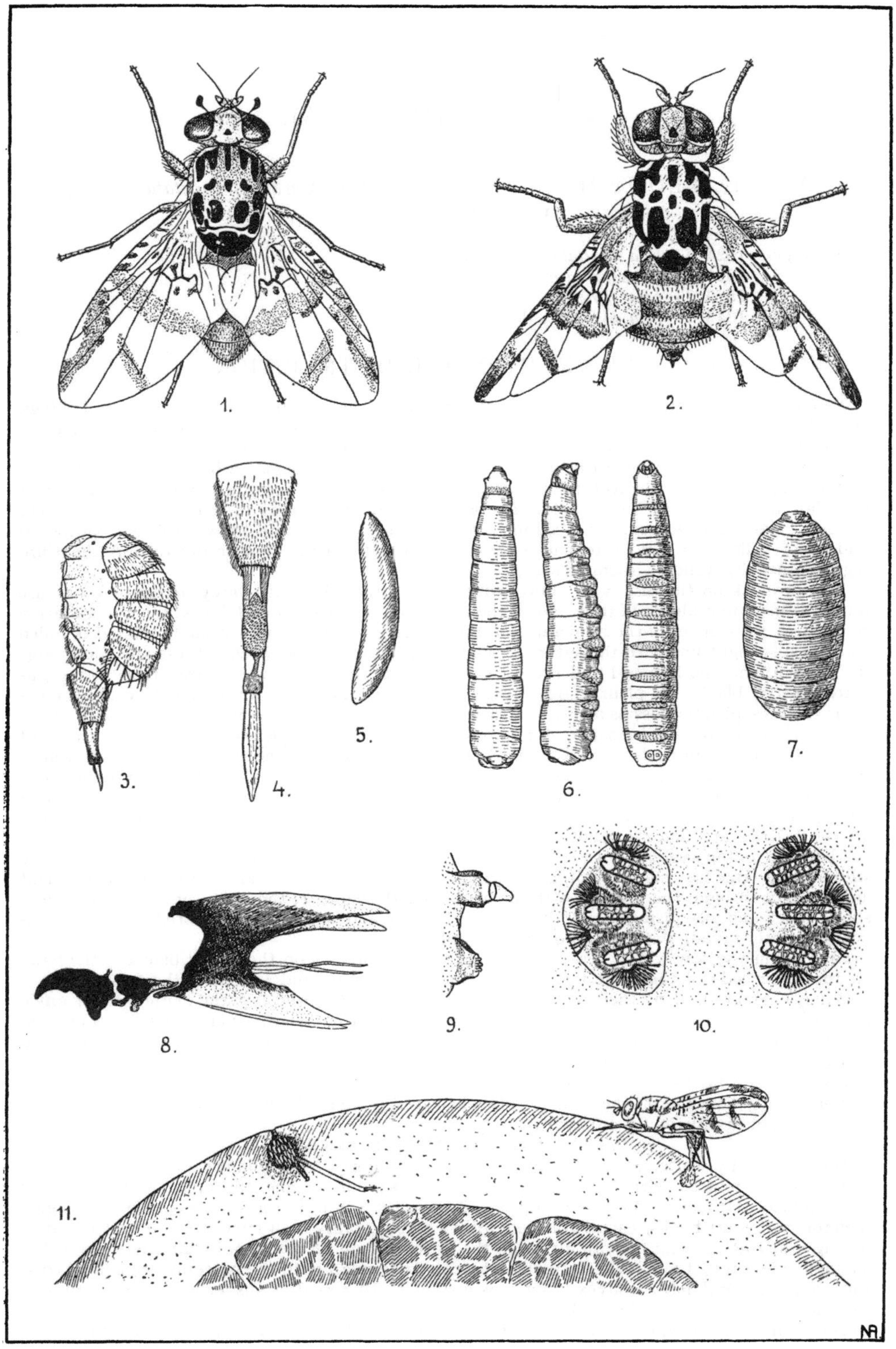

VII. The Mediterranean Fruit Fly (*Ceratitis capitata* Wied.). 1) male; 2) female); 3) and 4) ovipositor of female seen laterally and from above; 5) egg; 6) maggot as seen from above, from the side and from below; 7) pupa; 8) mouth hooks of fullgrown maggot; 9) antennae and maxillary palpi of full grown maggot; 10) posterior spiracles of full grown maggot; 11) ovipositing female and egg-cavity with larval gallery in orange rind. (partly after Constantino, Efflatoun and Woglum).

slit. Interspiracular processes very large, greatly branched, with very long and somewhat inflated processes or branches, rounded at their extreme tips. (Efflatoun, 1927).

6.8—8.2 mm long; 1.5—2.0 mm broad.

Pupa dark testaceous in colour, elongate elliptical and a little more convex on the dorsal side than on the ventral, especially on the posterior half. 4—4.3 mm long; 2.1—2.4 mm broad.

DIFFERENTIAL DIAGNOSTIC OF CERATITIS-MAGGOTS FROM OTHER INSECT-LARVAE INFESTING CITRUS FRUITS.

The more common insect-larvae met in decaying oranges in Palestine can be distinguished by the following characters (partly after Lewis):

1. Larvae with legs and with a distinct head. Larvae of beetles (*Carpophilus*).
Maggots without legs and without a distinct head. **2.**

2. Last segment of posterior end extending backward finger-like. About 5 mm long, when fully grown. Maggots of *Drosophila*
Posterior end blunt. About 8 mm long, when fully grown. **3.**

3. Anterior third long and slender. Posterior spiracle with 1 slit only. Maggots of *Lonchaea.*
Anterior third plump and conical. Posterior spiracle with 3 slits each. Maggots of *Ceratitis.*

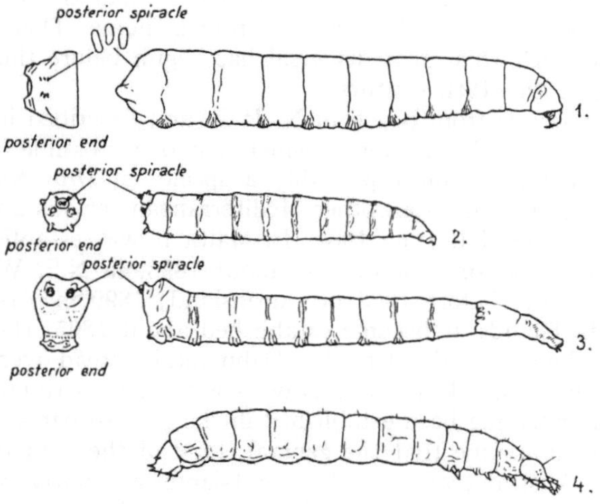

Fig. 19.
Insect larvae from decaying citrus fruits: 1. *Ceratitis capitata*, 2. *Drosophila melanogaster*, 3. *Lonchaea sp.*, 4. *Carpophilus hemipterus* (after Woglum).

It is rather important to be able to determine these main groups of insect larvae, scavenging in decaying citrus fruit. Not every maggot is a fruit fly maggot, but many of them are secondary scavengers continuing the decay started by *Ceratitis* or by moulds.

Distribution: Today, the Mediterranean fruit fly is distributed throughout all the continents. The localities in which it is permanently established are as follows:

Europe: S. Spain, S. France, Italy, Sicily, Malta, Dalmatia, Albania, Greece, Turkey. (Intercepted in England in 1917 on oranges from Sevilla).

Asia: Palestine, Syria, Turkey (littoral), Cyprus, ? Java (1915). Definitely absent, so far, in the Syrian Desert, the whole of Irak, E. und S. E. Anatolia.

Africa: Egypt, Lybia, Tunis, Algeria, Morocco, Atlantic Islands (Madeira, Canaries, Cap Verde, Azores), West Africa (Dahomey, S. Nigeria, Congo). S. Africa (Cape Colony, Natal, S. Rhodesia, Delagoa). East Africa (Brit. East Africa, Portuguese East Africa, Uganda). Eastern Islands (Madagascar, Mauritius).

Australia: W. Australia, N. S. Wales, Victoria, Queensland. (In New Zealand and Tasmania eradicated).

America: Brazil (Rio de Janeiro, Sao Paolo), Argentine (Buenos Aires). (Florida: eradicated). Bermuda Islands.

Pacific: Hawaiian Islands.

The type-specimen was collected „in the Indian Ocean" (1817). It probably hatched on board of a ship from fruits collected at one of the African ports. Tropical Africa is no doubt the original home of *C. capitata*, probably the primaeval forest of W. Africa. Silvestri (1914) considers the northern limit of its natural area as 8° N. latitude, but he does not make it clear whether the western part only or the entire tropical Africa should be regarded as its original habitat. It is certainly distributed by now throughout Africa, except the deserts and the dry savannas.

In N. Africa it was first recorded from the Atlantic Islands (Azores Madeira, Cape Verde) in 1829; 1859 from Algeria and in the second half of the 19th century from S. Africa; Egypt 1904. These late records do not prove a late appearance of the species, but there are reasons to believe that human traffic has considerably promoted the dispersal of the fly from its tropical home. There is no indication available as to whether this great dispersal has begun before the end of the 18th or the beginning of the 19th century.

The fly bred by Redi from peach in N. Italy and described in De Generatione Insectorum was not, as the author considered it once possible (1928. p. 321), the Mediterranean fruitfly, but most probably a species of *Drosophila* (certainly not *Ceratitis*). The oldest records from other Mediterranean regions are: S. Spain 1842, Italy, 1863, Sicily 1878, Palestine 1904. In Malta it was established about 1875.

In W. Australia it became established about 1893, in N.S. Wales about 1895, about 1907 in Victoria, before 1903 in Queensland; 1899 in Tasmania.

In the Bermuda Islands it became established about 1865. Its first appearance on the Hawaiian Islands was about 1907 at Oahu and it spread over the other islands during the following years. It reached Hawaii with ships arriving from Australia, about 10 years after having been established in the eastern parts of that continent.

It is reasonable to assume that the general trend of the data from the Mediterranean countries, Australia and the Pacific Islands are fairly reliable. The first appearance of this fly could hardly have been overlooked for more than a few years, considering its rather high economic importance.

In Palestine it was first reported by Compere in 1904 from oranges and pomegranate in Jaffa and Jerusalem. But its immigration and acclimatisation may have easily occurred a few decades earlier. Today, it is distributed throughout Palestine, except for the deserts.

The dispersal of *Ceratitis* takes place mainly trough human transport in infested fruits (ships, railway, other transports). The adults can disperse over two kilometers and perhaps more, especially in the direction of the wind. They disperse regularly into neighbouring orchards, but outbreaks are build up within the area concerned. No migrations or long-distance flights take place, as a rule. Long distance travels of infested fruit swimming in the sea connot take place, as Marlowe and Hong (1934) have demonstrated. By strict quarantine measures the fly can be safely prevented from settling in new continents. The very efficient quarantine service of the United States illustrates this clearly. Countries, hitherto free from *Ceratitis*, are therefore highly justified in using strict quarantine measures against the introduction of this fly. This was the case in 1937 with Jaffa-oranges in Ceylon, where a whole consignment was thrown into the sea.

The eradication of the Mediterranean fruitfly in Florida. The nursery inspector Goodwin, whilst preparing a grapefruit from Orlando, Fda. for his evening meal on 6.IV.1929 discovered some maggots in the fruit. He submitted them for determination to an entomologist. This event led to the discovery of an intensive infestation of several citrus groves around Orlando with *Ceratitis capitata* Wied. The State Board of Florida, the Federal Plant Quarantine Administration and the Bureau of Entomology were mobilised. On the 9.IV. a conference was held in the office of the Secretary of Agriculture and two days later authorities, such as C. I. Marlatt and A. C. Baker, arrived in Orlando. The opening of an eradication campaign was immediately decided upon and the following activities were immediately organised: inspection of groves throughout Florida, destruction of all host fruits in infested properties, spraying of poison baits in infested as well as in protective zones, destruction of all potential summer hosts in order to obtain a summer period with no available alternative fruits in the citrus belt, removal of all fruits of potential hosts in the infested area prior to the age at which they are liable to *Ceratitis*-infestation, and strict plant quarantine measures. The Florida National Guard was mobilised and engaged for over one year in road patrols in enforcement of the quarantine laws. The campaign was declared officially as successfully completed in late December 1930.

During this period 100 gardens in 20 counties were declared infested, ranging from one infested fruit or tree to almost complete infestation of the garden. Identification of flies and maggots was made by competent specialists throughout the campaign. In April 1929 all potential host fruits one mile around every infested zone were destroyed, since September 1929 only those on infested property. During 1929 almost half a million boxes of citrus fruits, 50.000 bushels of vegetables, 27.000 bushels of other minor hosts were thus destroyed. Kerosene traps, at one time as many as 12.645, were invaluable indicators. In some instances they gave the first evidence of *Ceratitis* in new foci. President Hoover, the Secretary of Agriculture and congress sent special committees to Florida and full publicity was given to the eradication campaign.

The headquarter of the campaign was established at Orlando. The number of trained inspectors (for scouting, for traffic inspection, for supervision of the eradication work) and of labourers engaged in the campaign was:

Month	1929		1930	
	inspectors	labourers	inspectors	labourers
January			695	1656
February			755	1480
March			750	1452
April	137	416	56	175
May	189	3650	70	117
June	202	4520	425	783
July	245	5262	642	939
August	243	5647	670	508
September	261	5310	656	502
October	299	3452	384	124
November	683	1807	428	46
December	569	1647	216	—

A research station under A. C. Baker was locally installed and a small staff was later transferred to Honolulu (Hawaii) for the continuation of the research, after the success of the eradication campaign in Florida.

The infestation in the central foci was very heavy. Photographs show orchards, the soil of which was literally covered with fallen citrus fruits. We have never seen such heavy damage, but the reports are beyond doubt. This extensive damage was certainly one of the main causes for the energy with which the eradication campain was launched. Over 70% of the fallen fruit was usually infested with fruitfly maggots. At one time up to 10 larvae per grapefruit were a common average.

At its greatest extent the area marked for the eradication campaign covered 40 million dunam (15.000 sq. miles). All evidence tends to show that the infestation was very recent at the date of its discovery. In Florida 500 eggs per female were regarded as the average, the development in summer (including the preoviposition period) lasted 34 days only. Most Florida oranges and tangerines have a fairly thin rind. The ovipositor of the female fly is thus able to pierce through the rind and to lay the eggs into the flesh of the fruits, avoiding thus the high nutritional mortality in thick-rinded citrus fruits. The warm and humid summer of Florida offers definitely favourable conditions for adult survival, in contrast to the unfavourable dry summers of the Mediterranean region. Favourable climate, low nutritional mortality, high egg production and short development in summer account easily for the rapid spread and the enormous damage by the pest in Florida.

At least 118 Florida plants were successfully infested by *Ceratitis* in laboratory experiments. Yet in nature only the following few plants were found infested, apart of citrus:

Casimiroa edulis	*Eugenia edulis*	*Eugenia uniflora*
Malpighia glabra	*Armeniaca persica*	*Passiflora incarnata*
Psidium cattleianum	*Psidium guajava*	*Pyrus communis.*

This list is small and *Passiflora* was the only wild plant infested. This small host list in nature also points to the recent introduction of the pest. This early discovery and the immediate start of the eradication campaign are doubtless responsible for the rapid and complete success of the work. A few years or even months latter, the number of hosts and the area of infestation would have made eradication very difficult and perhaps even impossible under normal budgetary limitations.

The number of discovered foci during the campaign was in 1929: 364 in April, 318 in May, 185 in June, 64 in July, 8 in August, 1 in November, and in 1930: 1 in March, 1 in July. No new foci were furthermore discovered and in late 1930 full victory was declared. The appropriations and expenditures during the entire period were:

381.000 Dollar	from Florida State Funds
6.471.000	from Federal Funds
6.852.000	total expenditure.

The energy, simplicity and the speedy success of the Florida fruitfly eradication campaign make it a classical case in the history of applied entomology.

2. LIFE CYCLE.

The hatching flies are sexually immature. Fig. 20 shows the maturation of the ovaries. The length of the preoviposition period depends mainly on temperature. No oviposition has been observed below 15 to 16° C. In our breeding the preoviposition period lasted at

16—17° C	121 (71—163) days
20—24	29 (17—47)
26—28	18 (2—36)
29—30	27 (3—39)

Averages or shade temperatures are of little value, as the fly is heliothermic, except during noon hours. If exposed to sun rays for some hours the day, the preoviposition period may be much shortened. The shortest preoviposition period of 2 to 3 days was rare, on figs and on *Casimiroa*. Night temperatures and special conditions of larval nutrition must have much contributed in these cases. The lengthening

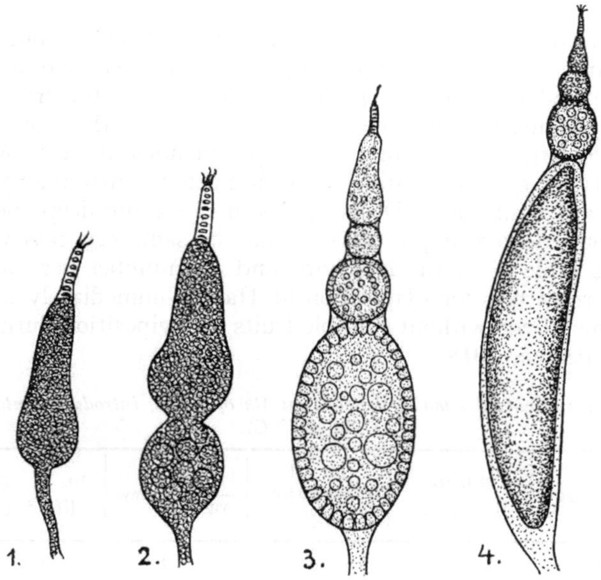

Fig. 20.
Maturation of ovaries of *Ceratitis capitata* Wied.: 1, 3, 8 and 10 days after the emergence of the female at about 21° C (from Back and Pemberton).

of the period above 29° C is significant. The maturation of the males is slightly shorter. The sex ratio in breeding is approximately equal, but of a great range seasonally and locally in populations caught in clensel traps. The low percentage of females in the months of high populations, i.e. 35% in June/July indicates, as do other experiments, that gravid females are less attracted to the bait than are the males. At Dagania the sex-ratio of 2543 bred individuals as compared with the trapped flies (about 100.000) was 49.8% of all flies (Klein 1943). The percentage of females was:

Month	I	II	III	IV	V	VI	VII	VIII	IX	X	XI	XII
no. individuals bred	110	242	346	415	155	775	154	120	65	20	7	73
breedings	51	51	50	49	54	51	47	53	51	(60	71)	41%
traps	51	50	78	56	35	32	30	24	50	67	64	57%

Before copulation the male seeks to attract the female by curving and raising its last abdominal segment, then bending its apex, swelling it into a pin-head, while the venter is drawn back half of its length and the abdomen puffed out laterad (Martelli). As sexual stimulation we must regard the emanation of a stale mucous smell, emanated at that stage by the males. The females are not attracted to the smell, but apparently stimulated by it (Back and Pemberton). This smell is not emanated below 16° C. The normal copula lasts in summer from a few minutes to over one hour. Constantino (1930) describs its duration as 2 to 3 hours. His observations must have been made in the shade at fairly low temperatures. The female often repulses the approaching male. Copulation is repeated by both sexes throughout life. In winter copulation is restricted to the warm noon hours, when temperature is higher than 17° C. The first copulation at Dagania occurred in summer 3 to 13, in autumn 6 to 26, in winter 20 and more, in spring 5 and more days after hatching.

When ready to oviposit, the females look over the fruit by touching it with the proboscis in search of a more mature (soft) or injured area, or of a punctured hole. They curve the abdomen and insert the ovipositor into the fruit under rythmic abdominal movements. One to nine eggs are now laid within two to ten minutes. The fruit is then left. The puncture is not conspicuous at first, but subsequently shows various types of discoloration in its immediate surroundings. The primary egg-cavity is about 1 mm deep. Deeper holes (up to 4 mm deep ones are observed) are the result of repeated ovipositions within the same egg cavity.

The following table gives the longevity and egg number per day of 1) females which were offered fruits for oviposition at Hawaii immediately after copula and 2) of females which lived without suitable fruits for oviposition during some months (Back and Pemberton, 1918):

1. *Oviposition of 9 individual females, hatched on 4th of April, introduced into apples on 14.IV (22—27° C).*

Female no.	no. of eggs	no. of life-days	no. of oviposition days	$\frac{\text{life}}{\text{oviposition}}$ days	no. of eggs / life-days	no. of eggs / oviposition days
1	191	72	30	2.4	2.7	6.4
2	34	46	10	4.6	0.7	3.4
3	86	54	14	3.8	1.6	6.1
4	314	95	40	2.4	3.3	7.9
5	622	152	106	1.4	4.1	5.9
6	312	81	49	1.6	3.9	6.3
7	426	119	71	1.7	3.6	6.0
8	405	147	77	1.9	2.8	5.3
9	3	69	1	69.0	0.0	(3.0)
Average	266	93	34		2.4	5.9
2. *Oviposition of 6 individual females, hatched on 28th of February, introduced into apples on 1.VII.*						
1	162	156	20	7.8	1.0	8.1
2	97	164	21	7.3	0.6	4.6
3	0	135	0	—	—	—
4	76	156	12	13.0	0.5	6.3
5	12	146	3	14.7	0.1	4.0
6	43	136	6	22.7	0.3	7.2
Average	83	147	10		(0.5) resp. 3.1	6.0

Oviposition remained high from the second to the 12th decade, with the peak in the 4th to 8th decade:

Decade	1	2	3	4	5	6	7	8	9	10	11	12	13	14
eggs per ♀-day	1.4	2.9	3.2	3.8	4.8	5.0	3.8	4.8	2.9	1.5	3.0	3.0	2.3	0.9

The maximum number of eggs per female-day observed was 21, whereas 2.4 (1—4) eggs per fermale-day at 20° C and 5.4 (3—9) at 28° C were the average observed by Back and Pemberton. 1 to 3 ovipositions per female-day can be expected under favorable conditions. 30—300 eggs (B. and P. observed as many as 926 eggs in one egg-chamber) found in one egg-chamber are the result of many ovipositions by generally more then one female.

While Back and Pemberton estimate the maximum egg number per ♀ at 800, this figure is certainly very exceptional. 266 eggs deposited per female (3—622) under optimal conditions were the average, considering the high, and often early mortality of adult populations. Under unfavorable conditions this number is about halved or even smaller. We, therefore, assume

under optimal conditions	500	eggs per female
under favorable conditions	300	,, ,, ,,
under unfavorable conditions	100	,, ,, ,,

Of great interest are the data of the second series, which show that the 6 females having had no opportunity to oviposit for a period of 4 months, deposited 83 eggs on the average during the next month, with an oviposition rate of 6.0 eggs per oviposition-day or 3.1 eggs per day of the oviposition period. These values are identical with those of the normal females. This fact shows that the total egg number is reduced only by the shorter duration of the oviposition period. With regard to the epidemiology of the fruit fly these delayed ovipositions after a prolonged absence of suitable host fruits are of paramount importance for the maintenance of the fly within the same region or even orchard.

Virgin females start normal oviposition at the regular time. Back and Pemberton proved, however, in extended experiments that none of these unfertilized eggs develop. In our own experiments we have neither observed oviposition or even puncture of virgin females. No mature eggs were found in their body. Their longevity was not changed in comparison with paired females, in October. About half of about 500 females, which died a natural death after a prolonged period of oviposition in the laboratory, contained no mature eggs, the other half 5 to 17 eggs with a maximum of 72 ones. Females supplied with food but without fruit suitable for oviposition lived also in our experiments noticably longer in summer than fertilised females with oviposition hosts. This difference was not apparent in winter.

Adult longevity is a very important factor in the ecology of *Ceratitis*. The long life-span of the adult flies is of great importance for the maintenance of the species during periods of absence or scarcity of suitable hosts.

Adult flies lacking food or water are short-living. In the experiments of Back and Pemberton, the flies did not survive the fifth day:

day	1	2	3	4	5
no. of living flies	482	151	13	2	0

Flies hatched in mid-October lived at 8—15° C and were fed with honey-water from 28 to 75 days (44 days on the average) (Verguin). In 5 series of 129 flies, at about 25° C, the survival in our experiment was as follows:

day	1	2	3	4	5	6	7	10	12	13	15	16	18	21
no. of surviving flies.	123	117	102	82	34	8	7	6	5	4	3	2	1	0

The heaviest mortality fell between the third and sixth day. The longevity of flies which are normally fed on honeydew, sugar-water, plant-saps, etc., is much longer. But while the great majority of the flies die during the first 2 months of their adult life, a few individuals always survive for a rather extended period. And, interested as we may be in these few individuals from an epidemiological point of view, we should never forget that they represent only a small minority of their generation and that the average longevity of their generation is much shorter.

At lower temperatures, especially if they are connected with heavy rains, mortality grows considerably and below 5° C the adult fly cannot live beyond a period of more than 1—2 weeks.

The average longevity observed has been 28—40 days in summer, 28—65 days in winter in W. Australia (Newman) and about 25 days in Victoria (French).

Extreme longevity observed was as follows:

Locality	longest longevity	author	remarks
S. Africa	111 days	Lounsbury	on apple
W. Australia	112	Newman	
Hawaii	315	Back and Pemb.	at 15.5° C. in refrigerator
Hawaii	230	Back and Pemb.	at about 24° C.
Palestine	207	Klein	at Dagania in November

Some details of the mortality given by these authors will be of interest:

Flies surviving (%)	n	days 1	10	20	30	40	50	60	70	80	90	100	110	120	130
1. W. Australia	200	100		90	80	60	50	30	17	8	4	2	0.5		
2. Hawaii	500	100					8					—			
	250	100							90	71	24	16	7	1	—

or among 800 individuals, survived at Hawaii

days	1	125	135	147	161	171	182	194
survivors (%)	100	19	10	7	4	2	1	1 (last individual died at the age of 230 days).

It is therefore obvious that an interval of 3—4 months in the absence of suitable host fruits must reduce the endemic fruit fly population to a minimum. Heavy infestations following such an interval indicate that an intermediate host must be present in numbers in the near neighbourhood.

The longevity of the adults in laboratory breedings at Dagania was:

Sex		I	II	III	IV	V	VI	VII	VIII	XI	X	XI	XII	
female	n	30	13	5	17	78	39	57	101	86	173	26	44	
	M	47	30	6	25	21	23	22	6	14	54	99	78	days
	max.	96	83	11	45	90	65	63	40	49	189	206	156	,,
male	n	31	9	2	24	56	31	51	113	82	151	33	34	
	M	40	26	34	23	19	21	23	5	20	51	81	65	days
	max.	115	94	64	85	66	63	46	45	150	193	207	142	,,

A tentative hyperbola has been calculated from these data. This has, however, no pretensions to great precision. The data are few and important factors, such as relative humidity, sexual and oviposition conditions, etc. are not included. The constants of the hyperbola are: c = 16.3° C, Th.C = 301 days-degrees for average, Th. C. = 963 days-degrees for maximal longevity (For the later the data are especially questionable, as the scattering of the complete mortality in the few individuals is large). This hyperbola yields:

Temperature	Average longevity	Maximal longevity
35° C	16 days	52 days
30	22	70
25	35	111
20	81	260
18	177	566

The threshold well agrees with the observed lower limit of normal activity in *Ceratitis*-flies.

The difference between the average and the maximal longevity is of highest epidemiological importance, as it gives an indication to the number of flies which are able to survive at seasons when no suitable hosts for oviposition are available. Whilst the average winter longevity was 2 to 3 months, the last survivors lived up to 6 to 7 months. In summer the corresponding data are one respectively two months. The May hamsins and the excessive heat in August are both extremely unfavourable for survival, in nature possibly even more than in these breedings. During longer hamsin winds the flies die in the preoviposition period, and pupating maggots almost all die in the early pupal stage. No oviposition has been observed during hamsin periods.

Klein points out the importance of these observations for the potential survival of *Ceratitis* in the Upper Jordan Valley: Some adults which emerge in October survive until mid-April. The flies from their last ovipositions in March hatch in early May. The latest flies of that generation hatch in early August and live to mid-September. Their latest offspring hatches in early October and these flies hibernate. This means that in the complete absence of citrus fruits in winter the species can maintain itself locally on figs, grapes and dates. The recent introduction of Surinam cherry, white Sapota, and guava create still more favorable conditions. With the recent diversification of citrus fruits, *Ceratitis* is even able to maintain itself on citrus fruits alone. Infested grapefruits are found from October to June, as a few fruits are always left by the pickers in the orchard. When the flies hatch in early July from these last maggots, they can survive until September. Maggots in grapefruits have been found in October. Yet clementine, sour orange and tangerine ripen even earlier than grapefruit. Two more generations would appear in November and in April.

In this respect it is of high importance that, whilst lack of fertilisation does not markedly lengthen the life of the female, this is definitely the case when fertilised females are kept without suitable fruit for oviposition. The experience of Honolulu as well as of Dagania show this clearly. Survival chances in absence of seasonal fruit rotation are thus even higher than under normal conditions.

Whilst four to five annual generations develop in nature in the Upper Jordan Valley Klein (1932) obtained 6 to 7 in breedings, taken always from the first patch of eggs from which adults of the following generation were obtained: From parents hatched on 20.V. developed the first adults of each of the following generations and oviposited: 26.V., 27.VII., 18.VIII., 26.IX., 1.XI., 28.II., 3.V.

But the fate of the various ovipositions of one and the same female may be very different, especially towards the winter. The offspring of female no.135, for instance:

Oviposition no.	1	2	3	4	5	6	7	8
Date.	27.X.	4.XI.	13.XI.	16.XI.	18.XI.	21.XI.	28.XI.	12.XII.
development days. . . .	23	28	—	32	58	56	55	57
temperature °C	22.0	20.0	—	18.7	17.3	17.2	17.0	15.6

In autumn the first ovipositions of the same female develop in less than half the time of the last ones. The temperature relation of pre-adult development at Dagania was (in days):

Temperature	average	minimum	maximum
16° C	57	46	75
18	48	32	58
20	35	28	39
22	34	23	44
24	24	18	31
26	18	13	24
28	17	13	24
29.5	17	13	25

Pupal mortality has been little studied, so far. It has been obviously underestimated as an ecological factor. Excessive soil humidity, especially if combined with low temperatures or frost, has a very unfavourable influence on pupal survival in the soil. Most pupae die under such conditions. In very humid soil the maggots or pupae often die from an unknown disease. Maggots which pupated during hamsin conditions, often shrunk and died.

Pupae which were bred under normal laboratory conditions and the maggots of which had previously developed in various fruits showed the following mortality at Dagania (Klein and Paker 1943):

Host	Mortality	Host	Mortality	Host	Mortality
grapefruit	29%	*Eriobotrya*	—%	plums	11%
Valencia orange	33	apple (Hashebi)	14	grapes	32
casimiroa	12	guava	8	tomato	15
banana	5	*Carissa*	58	green pepper	18
		fig	18		

These data indicate that pupal mortality is much higher than usually assumed and that it deserves a detailed analysis.

The fly is positively phototactic. Even in the hottest summer days it is seen in the orchard as it seeks the shade of the trees as protection against strong sun-radiation, but always on the brightest parts of the tree. It practically never enters the shade interior of plantations or forests. This habit leads to a heavy infestation of the outside rows in dense plantations, such as older orange groves, whereas in loose and sunny plantations, as those of peaches and apricots, no such preference is observed.

One typical count in an older orange-grove at Petah-Tikva yielded the following results:

	no. of trees	no. of fruits infested	no. of infested fruits per tree
Outside row	108	479	4.4
Second row	88	73	0.8
Transversal row	26	28	1.1
Diagonal row (2 outer trees excluded)	24	8	0.3

The intensity of attack on the four outside did not give conclusive preference for any side of the compass.

The flies are not only positively phototactic, but also heliothermic. On an early morning or on sunny days, from autumn to spring, they utilise the sun rays to raise their body temperature. At noon in summer, they may seek protection from direct sun radiation in the halfshade of the external parts of the trees.

Strong chemo-taxis towards smells is important in connection with control measures.

The maggot tends, as a rule, to move immediately after having hatched into the interior of the fruit, in the pulp of which it develops. The physiology of the maggot has never been studied. Which senses lead it into the interior of the fruit? To what degree may the maggot live anoxybiontically within the fruit? Neither has its nutrition or digestion been studied, except for the symbionts which are present in the intestines of maggot and fly.

In the normal course of events the full grown maggots leave the fruit. They are strongly negatively phototactic and burrow into the soil. When exposed to light at this stage, they begin to crawl restlessly and 2—3 minutes later begin to jump up to 15 cm high and to a distance of 10 cm. Before jumping, they bend their body in a high curve, bringing the mouth parts into contact with the much contracted posterior end of the body. From this position they jerk in a flat, occasionally in a high bow. After about 10 minutes they stop jumping, being tired out. They crawl and jump alternatively. Every distinct increase in light intensity stimulates them to a new series of jumps.

They enter the soil about 5—15 cm deep, where they pupate. The emerging flies may pierce even through a layer of 45 cm of loose soil. In the experiments of Newman (1924) the following percentage of flies succeeded in reaching the surface:

soil texture	Depth		
	15 cm	30 cm	45 cm
dry, loose sand	84	—	4
soil	74		5
clay	96		
wet, loose sand		24	0
soil		28	0
None, emerged in wet, hard clay, and in moist, rammed sand			

Dry, soil hard, such as desert clay, inhibits the existence of the species as well as very moist and heavy soil. These factors are responsible for a strong reduction of the fly population, even if all other conditions should be favorable.

3. HOST RELATIONS OF C. CAPITATA.

A. General remarks: Host conditions influence the epidemiology of the Mediterranean fruitfly in many ways.

1) *The sequence of hosts* is of paramount importance, because it determines the number of adult females ready to oviposit in a maturing suitable crop of fruits. Any interruption of the food chain for some months unfavourable climatic conditions, as low temperature and heavy rains or high temperature and extreme drought, shorten the longevity of the adult flies and the number of females ready to attack the new crop is small to very small. The duration of the interruption, its season, its complete or incomplete interruption are the main factors which determine the epidemiological importance of the interruption of the host sequence.

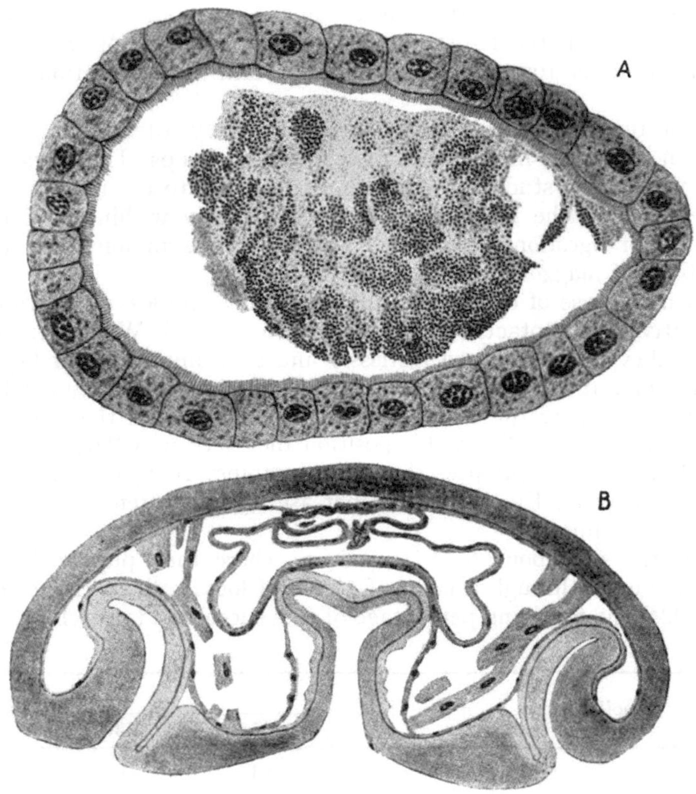

Fig. 21.
Bacterial symbionts of *Ceratitis capitata* Wied.: A. Bacterial lump in mid-intestine of adult, B. within the ovipositor a bacterial lump enters from the rectum into the vagina (about 500 ×; from Buchner).

2) *The accessibility of the host* is determined mainly by the abundance of the host, its distributions in space and its distance from the micro-locality studied. Most accessible are generally the most abundant fruits, but fairly rare hosts can be heavily attacked when growing on places which are especially attractive to the flies. This

is the reason why distribution in space is of such importance. Suitable fruits abounding on the soil of a shady forest remain practically free from fruitfly attack, even if climatic conditions are optimal, as the fly does not enter the interior of shady forests. Accessibility is, therefore, by no means a purely physical conception, but a biological one, in which animal behaviour is quite as important as the physical environment. The distance of the ripening crop from the place of the development of the earlier fly generation determines its accessibility to a high degree. Accessibility is much higher in or against the direction of the prevailing winds than in

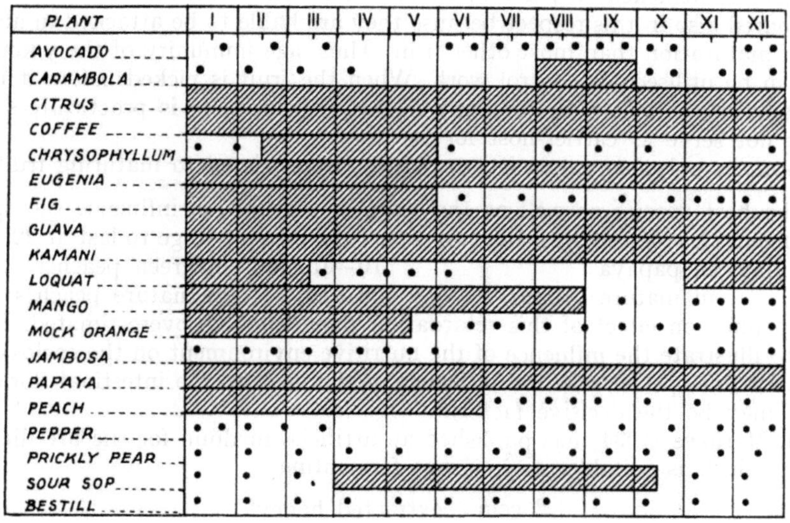

Fig. 22.
Seasonal host sequence of *Ceratitis capitata* Wied. in a block of house gardens at Hawaii. (From Pemberton).

the other directions of the compass. The quantity of a crop determines also its degree of mass-attraction for the fly, provided other climatic, behaviouristic, etc. circumstances are favorable.

3) *The suitability of the host* is not of less importance. Egg and larval mortality is fairly high in some fruits, practically zero in others. The same fruit very often possesses different degrees of suitability in its different stages of maturation. The suitability of the hosts liable to be attacked in Palestine will be dealt with in detail later on. One example only may serve here to illustrate the importance of this factor.

When 100 adult flies are present (50 ♀ : 50 ♂, 500 eggs per ♀ under optimal, 300 under favorable and 100 under tolerable conditions), and in one locality or season 100 flies attack nearly mature peaches, in another thick-skinned oranges, the increase of flies in the next generation under other optimal conditions will be as follows:

host	nutritive mortality	optimal environment	favorable environment	tolerable environment
peach	0%	25.000	12.500	5.000
orange	99.5	1.250	625	250

In order to obtain identical numbers of flies in the filial generation, the original population must be 20 to 200 times as large in the orange grove as in the peach grove. Or, 20 to 200 flies in an orange grove are the economic equivalent of one fly in a peach grove.

4) *The suitability* of any host is by no means absolute. Besides the large range of resistance which we often observe among the different varieties of one host, the age of the fruit is of no less importance. It is a general rule that unripe, green and hard fruits are unfit for *Ceratitis* infestation and that in most hosts the fruit is liable to be attacked only a very short period before reaching maturity. Peaches are favoured also in this respect because they are liable to be attacked in an earlier stage of maturation than most other fruit. Thus age immunity of the young fruits can often be utilised for control work. When the fruit is picked before it is liable to fruitfly infestation in the orchard or in the field, it 1) is practically safe and 2) does not serve as carrier-host for the fly.

C. capitata, polyphagous as it certainly is, is restricted to maturing fruits only.

5) The host species as well as the stage of maturation influence the speed of development. Back and Pemberton report the maggot-stage to last at 22—25° C

5—20	days in papaya	10—15	days in green peaches
19—20	in mature lemons	6—10	in mature peaches.

The population effect of this retardation should not be overestimated. It again serves to illustrate the influence of the nutritive environment on the ecology of the species. In this special case the influence of the host on the intestinal flora of the maggot may be the decisive factor.

R. H. Marlowe (1934) has published an artificial medium for the breeding of *C. capitata*, which is suitable, as it is not fermenting:

2 ccm. „extracted honey"
4 ccm. crushed papaya
3 ccm. water
1 gr. brown sugar
0.16 gr. agar agar.

This mixture is to be autoclaved for 20 minutes at 15 lbs. pressure.

B. *The Mediterranean fruitfly in citrus:* Citrus fruits are rarely a suitable host of *Ceratitis*, yet the damage caused to it is nevertheless sometimes considerable. Back and Pemberton (1915) were the first to point out the very high mortality occurring in the rind of citrus fruits. They observed in Hawaii:

Variety	no. of fruits examined	% of empty oviposition punctures	Mortality in the rind in %			Surviving larvae in pulp
			eggs	larvae	total	
Grapefruit	8222	25	90.5	9.3	99.8	0.0%
Lemon	991	67	76.1	21.0	97.1	0.0
Lime.	1054	46	43.1	55.6	98.7	1.7
Kusaie Lime . . .	838	0	47.6	51.8	99.4	0.0
Sweet orange . . .	3635	17	19.0	79.0	98.0	0.5
Sour orange . . .	2357	71	71.7	17.3	89.0	9.8
Shaddock No. 1. .	1155	36	3.8	95.3	99.1	0.3
Shaddock No. 2. .	303	25	13.5	18.1	31.6	10.8
Chinese orange . .	1039	1	3.7	46.5	51.2	41.5

This same type of mortality occurs in Palestine too. The above table allows for the following analysis (excluding the non-commercial: Chinese orange and Shaddock no. 2):

In 256 fruits there were 2716 punctures (= 10.6 punctures per fruit; 43.6 which were empty). Of the 18672 fruit fly individuals contained in these fruits (which had been kept long enough to pass through the entire egg-stage) the mortality has been as follows:

	dead		alive		dead in % of the total
eggs	11.507	100.0%	—	0.0%	61.6
maggots in egg-cavity	3.166	96.2	124	3.8	17.5
maggots in rag of rind	3.387	95.3	167	4.7	18.1
maggots in pulp	1	0.3	320	99.7	0.0
	18.061		611		97.2

This is not the final mortality because the great majority of the maggots still alive in the egg cavity or in the rag of the rind are bound to die before reaching the pulp. Once having reached the pulp they are relatively safe.

Savastano (1914) explained the practical immunity of lemon fruits, by claiming that the high percentage of acid in the lemon pulp is the protective agent. Back and Pemberton (1915) transplanted young maggots from other hosts into the pulp of green and ripe lemons, where the maggots developed with no higher mortality than in other citrus fruits. These experiments have been repeated by the author in 1932 with the same success. That acidity in the pulp is not a primary factor in lemon resistance, seems thus to be established.

Mr. D. Oppenheim of Rehoboth was kind enough to determine for us the acid content of orange rind in Palestine. The measurements yielded the following results (average of 3 titrations each):

date	in rind of	
	healthy orange	orange with *Ceratitis* puncture
27.IX.1925	2.398%	
11.X.	2.134	2.975%
25.X.	1.824	2.685
30.X.	1.793	
18.XI.	1.465	
24.I.1926	1.037	

It is not certain, considering the small amount of measurements made, whether oranges with higher acid content in the rind are preferred for oviposition or whether the puncture possibly retards the lowering of acid content. But this result makes it still less probable that the acid content of the rind is a primary factor in the mortality.

First importance must be ascribed to the etheric oils of the rind. The first conclusive experiments were performed by Back and Pemberton. From two batches, each containing 1600 eggs, one was covered with the fine spray of squeezed orange rinds, the other left untouched. Only 163 maggots hatched from the first batch as against 1313 from the control. And from among the 163 maggots all died either

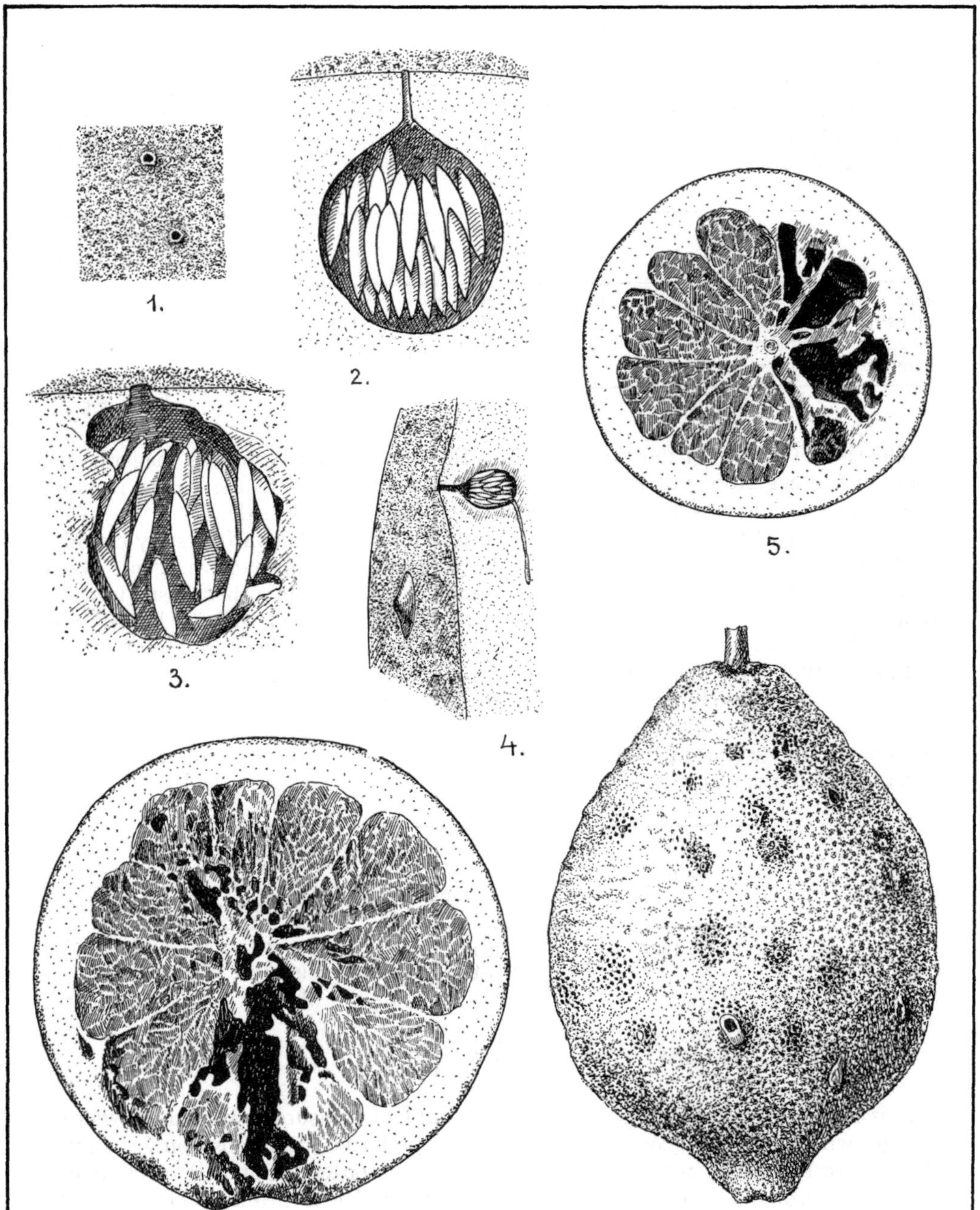

VIII. Damage caused by *Ceratitis capitata* Wied. in citrus fruits. **1**) orange with exit holes through which full-grown maggots have left the fruit; **2**) and **3**) egg-cavity in peach immediately and **36** hours after oviposition; **4**) egg-cavity in orange rind, showing the slightly elevated crater around the puncture; **5**) grapefruit after attack of fruitfly maggots; **6**) grapefruit after attack of fruitfly maggots combined with bacterial decay; **7**) discoloration on lemon fruit around the punctures. (partly after Back and Pemberton).

before having entirely left the egg-shell or before having crawled much more than 0.6 cm.

That high density of oil glands in the rind interferes even with oviposition has been demonstrated by Rivnay (1935) on citron. In each of 9 fruits of younger fruits with a high density of oil glands and of older fruits with a decreased number of oil glands per area, the following observations were made:

Series	no. of oil glands per sq. cm.	egg cavities		no. of eggs
		without eggs	with eggs	
young fruit	108 (60—160)	4	4	9
old fruit	50 (35—75)	10	69	680

In the grove no oviposition took place in a fruit having more than 65 oil glands per sq. cm. and only 10% of the attacked fruit had more than 50 oil glands per sq. cm. (average 30—40). In higher concentrations of oil glands oviposition occurs under enforced conditions only. When having the choice, the older fruit is always preferred, as the above mentioned experiment illustrates, where one young and one old fruit were placed in each cage. The number of eggless puncture holes in the orchard was only 8% out of 318 holes as against the 41% among the 565 holes in the laboratory. This indicates that a repellent, probably the odour of the etheric oil, inhibits oviposition and that in nature mature fruits are preferred. Before October no puncture was observed in citron in the grove.

Another additional mortality factor is gum secretion. Quite frequently gum oozes from the cells into the egg-cavity, floods and fills it and the eggs drown. Eggs completely embedded in gum are not unfrequently found.

The mortality of eggs and larvae within the rind proceeds along the following steps: Almost all those eggs which are exposed to contact with etheric oil, liberated by the female ovipositor in the preparation of the oviposition cavity, die as eggs, especially if the contact has been made, as usually is the case, in the earliest stages of embryonic development. The young maggots trying to pierce the rind in order to enter the pulp meet with 3 types of resistance: The most important one is again the lesion of oil glands which liberate etheric oil deadly to the younger maggots too. This danger sets in when the maggot tries to pierce the wall of the egg-cavity. The second factor which may be combined with the first one is the often high elasticity and mechanical resistance offered to the larva by a thick rind. The stronger this resistance, the longer the maggot remains in the dangerous zone of the rind, the stronger must be its movements in order to pierce the rind and the higher the chance of the lesion of an oil gland. The third factor, the intensity of which depends mainly on the thickness of the rind, is gum exudation. The boring gallery of the burrowing maggot is often filled with gum and at its end, before reaching the pulp. The asphyxiated dead maggot lies embedded in the gum. Whereas maggots develop quite normally within the pulp, it is different with eggs which may be deposited within the pulp in very thin-skinned citrus fruits. The eggs are often still subjected to a high mortality brought about either by excessive moisture or by lack of air. Back and Pemberton observed a 84% mortality of this type in 85 fruits of Chinese oranges.

The more or less heavy infestation within the pulp of mature citrus fruits may, except in the case of the very thin-skinned varieties, be explained partly as a consequence of persistent attack. We know that the batch of eggs deposited by the female at one time fluctuates between 3—9 eggs. Up to 300 eggs in one puncture-hole are the result of repeated ovipositions of the same female into the same puncture, which has been observed by us as well, and of other females preferring to deposit their eggs in already prepared oviposition holes. This use of already existing punctures continues sometimes over a considerable period, whereas in other cases pathological changes within the hole interfere with it (gum-exudation, proliferation, etc.). The maggots hatching first die as pioneers in piercing the first steps of the tunnel and the maggots hatching later continue to pierce through. Finally, group after

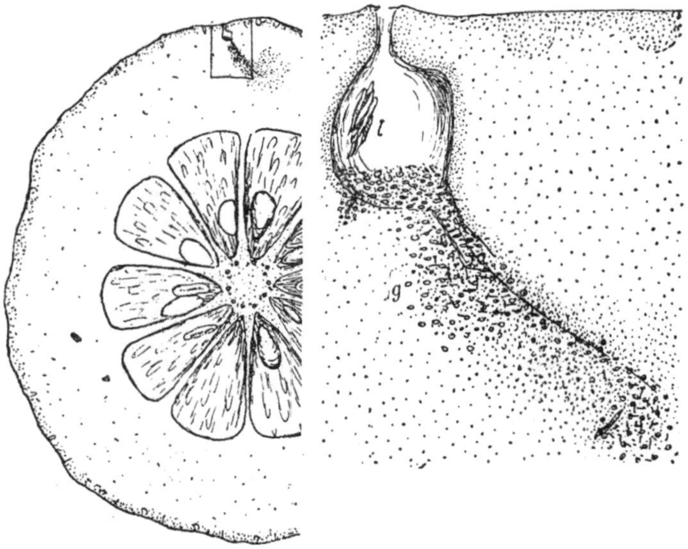

Fig. 23.
Cross-section of a citron-fruit with an oviposition hole of *Ceratitis capitata*. The oviposition cavity is shown at right enlarged: l dead larvae, others which succeeded to burrow into the rind are drowned in gum (g; from Rivnay).

group having died, some late individuals succeed in reaching the pulp and in developing normally. The percentage of the pioneers dying is always very high and depends on the thickness, texture and density of the oil glands of the rind. In most species the mortality fluctuates between 95—100%. Most citrus fruits cannot therefore be claimed as suitable hosts, especially as they are being picked in a well-cared orchard before the optimum season for infestation has arrived. Besides the entomological indications, there exist commercial seasons in Palestine advocating the enforcement of the end of the picking season at about mid-March. But at any season and in any stage of maturation the development in the rind of citrus fruits is accompanied by a very high mortality of eggs as well as of maggots. Ripe and overripe fruit grows much less resistent, as the physical structure of the albedo offers no serious resistance and gum excudation season.

Back and Pemberton (1915) describe a crater-like excrescence surrounding the puncture hole very often in Hawaii. The author has never seen these galls in Palestine, where the mouth of the puncture is, as a rule, only slightly alone the general surface-level of the fruit or even slightly sunken.

The susceptibility of the different citrus fruits shows a wide range. It is determined by the three main factors mentioned above. The classification of the susceptibility following species is impossible since the different varieties of fruits often show a wider range of varieties with regard to rind-structure than that occurring within the same species.

Sour lemon and citron belong to the most immune species. Lemon was regarded as absolutely resistant to fruit fly attack, until Quayle (1914) reported from Sicily of having collected 15 lemon fruits heavily infested with maggots. However, all these fruits showed signs of having been injured previously either mechanically or by fungi. Under normal conditions, green or early ripe fruits are sometimes heavily punctured, but the density of oil cells and a hardening of the neighbouring wall-cell of the egg-cavity by growth, joined with thickness and rough texture of the rind, practically interfere with any infestation. Back and Pemberton never observed hatching flies either from enforced oviposition or from fruits inspected in the field. Martelli (1914) did not succeed in getting any successful development in lemons under enforced laboratory conditions. Punctures have been observed in Palestine repeatedly, but no development. Lemons are therefore to be regarded as practically immune in Palestine.

The same is true, as has been mentioned before, for citron. Among 4768 eggs, only 111 (= 2.3%) hatched; among the 10.604 maggots found, only 17 were alive (Rivnay, 1935). The fate of 100.000 eggs laid within the rind of citron is therefore as follows: 2300 maggots hatch, from which 99.8% die shortly after hatching and 4.6 maggots only survive. But none of the observed maggots had pierced through more than mid-way of the rind. It is almost certain that all these maggots would have been found dead were the inspection made a few days later. The citron in Palestine is even more immune than the lemon. However, as oviposition occurs frequently, the loss may occasionally be great, as the ritual prescription for the symbolic fruits of the tabernacle feast requires fruits without any blemish. Even an empty *Ceratitis* puncture disqualifies the fruit from being used. Fortunately, the festival enforces very early picking (August/September) before punctures become common. The structure and condition of the rind of our local variety offers at that season a complete guarantee against successful development of the fly within the fruit. How far this conclusion is correct for other varieties of citron remains to be seen.

Whereas grapefruits show the highest percentage of immunity in the list of Back and Pemberton, this is certainly not the case in Palestine. The reason is that these authors observed mainly a fairly thick-skinned variety with a tight texture of the rag. The shaddocks, which — partly at least — showed a fairly heavy infestation in the pulp in Hawaii are also thick-skinned, but in part of their varieties the rind texture is not tight at all and the burrowing is fairly easy to the maggots. The general type of grapefruits cultivated in Palestine is far more thin-skinned and they are liable to infestation as soon as they mature. Early picking of this fruit is essential to prevent damages, at least in heavy fruitfly years. Especially in the Upper Jordan Valley the total egg and larval mortality in grapefruit is rather low, often as low as 50 to 60% only. The infestation in the grove was (Klein and Paker 1943):

Month	No. of punctures per fruit	Punctures on picked fruit				Dto. fallen fruit % with maggots
		with no eggs	with dead eggs	with living eggs	with maggots	
X	6	57	36	8	0	0
XI	6	25	15	60	5	0
XII	7	27	46	27	11	62
I	6	55	37	8	5	67
II	3	76	17	7	11	29
III	2	77	15	8	15	68
IV	1	22	13	65	32	91

The infestation in two grapefruit groves at Dagania was at two picking times in **1941**: 17.3% in late March, 57.3% in late April. Infestation is especially high in autumn and early winter. From 100 eggs developed in the average of all seasons in grapefruit: **43** maggots, **31** pupae and **26** adults. The thin layer of oilglands and the looseness of the white rind are the main reasons for this low nutritive mortality.

Oranges behave very differently. The common shamuti orange is fairly well-protected by its thick skin. Only when overripe or when injured mechanically or by fungi, the fruit flies may develop to a certain extent. Most of the September ovipositions do not lead to a successful development. But when the fruit matures, the weather being still sunny and favourable in December and January, some infestations are successful. About 100 punctured and dropped fruits which were collected in Petah Tikva at the beginning of January 1928 led to the development of 15 flies in two fruits only during the second half of February. A certain amount of fruit cropping will be caused from September to December by fruit fly punctures and other types of damage will be discussed later separately. But Rivnay, Nadel and Littauer (1938) established the fact, that maturing fruit *on healthy, vigorous trees* punctured by *Ceratitis* in autumn does not drop, as a rule, but remains on the tree for a long period. However, no heavy attack is to be expected in the spring, as a rule, when picking is ended in mid-March.

There exist, or existed, a few orange-groves in which a very thinskinned variety called ,,Franzawi'' was grown. This variety is liable to heavy infestation because a great part of the punctures go right through the skin and oviposition occurs here between rind and pulp or very near the inner border of the rind. Heavy infestations have repeatedly been observed here as early as late autumn and again in early spring. Such varieties are unsuitable for growing in this country and no thin-skinned orange-variety should be planted extensively before its liability to fruitfly development has been studied throughout a series of years.

A special problem presents the Valencia-orange, a late variety which matures from March onwards and which could extend the orange season for two additional months. Unfortunately, it is attacked heavily by *Ceratitis*. The late Mr. Yedidyah (1936) has recently studied this problem at Mikweh Israel during three consecutive seasons. Seven trees were chosen for experiments with baits, the results of which will be discussed later on, and every type of damage was registered. Dropping caused by fruitfly attack occurred as follows:

month	IV		V			VI
decade	11—20	21—30	1—10	11—20	21—31	1—10
spring 1934	—	—	—	—	50	98
,, 1935	—	55	101	61	269	—
,, 1936	110	205	345	298	181	16

In bad fruitfly years the oranges began to drop as early as mid April, but in **1934** only in late May. This depends on the climatic variation of the season, which influences the maturation of the fruits as well as the development and numbers of the fruitfly. The total damage done to 7 trees in the 3 seasons was as follows:

	no. of fruits dropped	Picked fruit		Total fruit	% of damage
		infested	sound		
1934	153	172	931	1258	25.8
1935	320	1186	191	1697	88.8
1936	1175	54	34	1263	97.6

The relatively thin skin combined with the fact that, for late export, the fruit has to be kept as long as possible on the tree in a season when climatic factors are favorable and the fruitfly population in the orange belt fairly high, bring about disastrous results. It is obvious that in fruitfly years Valencia late is exposed to extreme infestation. Even if the damage may not always be as severe as it has been in the two bad fruitfly springs of 1935 and 1936, it is nevertheless still high in other years. In 1941 the percentage of infestation of the picked fruit at Hedera was: 0.2% in late May, but 78% in late June. Washington Navel oranges are infested when approaching maturity.

With regard to sour oranges, commonly grown in Florida, Back and Pemberton report that they are as severely attacked as the sweet orange, but on account of the looseness of the rind and rag their pulp becomes infested more easily.

Clementines are among the first citrus fruits liable to be infested under Palestine conditions. The tangerine Yussuf Effendi is only slightly infected.

Limes are not grown in Palestine yet and the Chinese orange has a very thin skin, so that all eggs are practically laid below the rind; hence the large percentage of survivors. Other oranges approaching the tangerine type as the thin skinned Japanese Satsuma are liable to heavy fruitfly attack, which may prove to be inhibitive against their introduction, if no efficient fruitfly control is found.

Fig. 24 and the following table illustrate the conditions of citrus rinds for Palestine.

	Oil cells per sq. cm.	Yellow rind + white rag	texture of rag + = tough — = feeble
1. Orange: Shamuti (small) . . .	94	1.6+4.5 = 6.1mm	+
2. Shamuti (large) . . .	149	2.0+ 8.9=10.9	—
3. Valencia	135	1.8+ 3.1= 4.9	+ +
4. Navel	140	1.9+ 5.0= 6.9	+ +
5. De Nice	175	2.1+ 4.0= 6.1	+
6. Lemon: green	115	1.6+ 4.1= 5.7	+ +
7. just ripe	79	1.5+ 3.0= 4.5	+
8. overripe.	58	1.4+ 3.0= 4.4	—
9. Grapefruit:	67	2.0+ 5.2= 7.2	— —
10. Pomelo:	74	1.4+12.0=13.4	+
11. Tangerine: Clementine	103	1.7+ 1.2= 2.9	— —
12. Satsuma	58	1.8+ 3.8= 5.6	— —
13. Citron:.	46	2.8+18.0=20.8	+ +
14. Kumquat	67	1.2+ 3.7= 4.9	+
15. Tangelo	134	2.9+ 3.6= 6.5	—
16. Poncira trifoliata	243	1.5+ 2.0= 3.5	+ +

The nutritional mortality of *Ceratitis* in citrus fruits is easily demonstrated by simple experiments, performed in the laboratory course of animal ecology at our University:

1) Maggots of *Ceratitis* (or if absent: full grown maggots of *Drosophila*) are put in 3 groups of 10 into small petri-dishes, the bottom of which is covered with moist filtering paper. The first group is sprayed with water, the second with the sap of squashed immature lemons, the third with etheric oils squeased from fresh citrus rind. At the end of 1—2 hours all maggots are well in the two first groups, while all of the third are dead or moribund. This proves that citric acid is harmless, but the etheric oils of the rind are deathly. The distribution of the oil glands is then inspected.

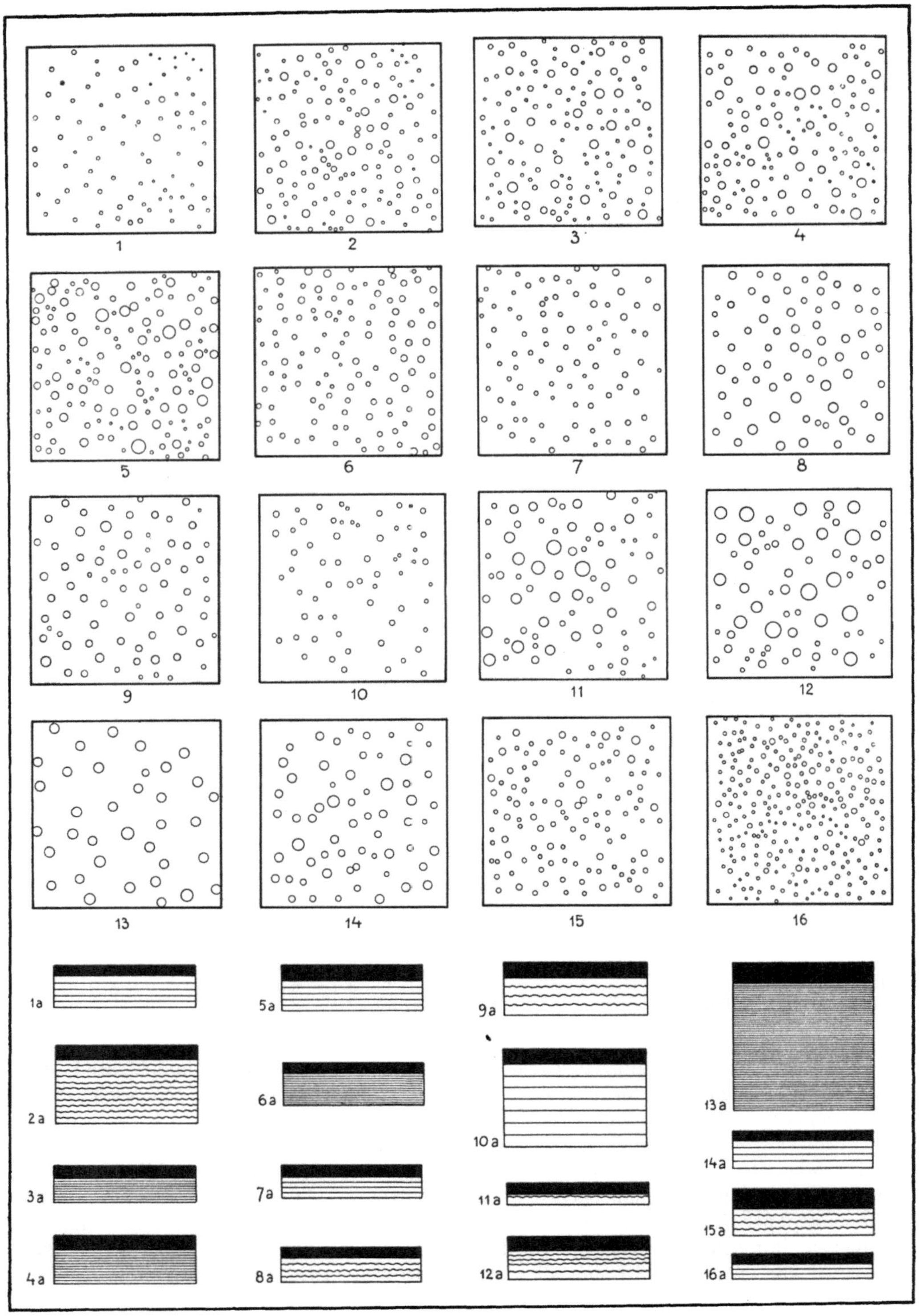

Fig. 24.

Distribution of oil glands and structure (thickness and elasticity) of rind in various citrus fruits in Palestine. No. 1 to 16 indicate the abundance and distribution of oil cells in 1 sq. cm. of peel: 1—5: Oranges. 1. shamuti (normal export size) ,2 . large shamuti fruit, 3. Valencia, 4. Navel, 5. De Nice. 6—8: Lemons. 6. green fruit, 7. maturing fruit, 8. overripe fruit. 9. grapefruit, 10. pomelo, 11. clementine, 12. Satsuma tangerine, 13. citron (etrog), 14 kumquat, 15. tangelo, 16. *Poncira trifoliata*. No. la to 16a show the corresponding thickness and the ratio of rind to rag layer in the peel, the horizontal lines indicating elasticity of peel.

2) The *Ceratitis* punctures of a number of fruits which 3 days previously were kept at 30° C are opened with the aid of a scalpel and the egg chamber and larval galleries inspected. All eggs present are dead (living eggs would have hatched in 3 days at 30° C) and the length and number of galleries, the number of living and dead maggots and the reason for their death (elastical resistance of rag or gum secretion) are noted. The result of a class of students was, e.g., as follows: Among 130 punctures 71 were empty. In the 59 remaining punctures 117 dead maggots are formed in the oviposition chamber resp. in the rag and 15 dead eggs. Not one maggot succeeded to pierce the rag and to enter the fruit-pulp.

The relative importance of etheric oils, thougness of rag and gum secretion, which vary in intensity and sequence in different fruits and at different seasons, can easily be studied in this way.

C. *The mediterranean fruit fly in other hosts.*

Peaches (*Prunus persica*) are the most preferred host in all countries where *Ceratitis* appears. The statement that in Hawaii peach-growing has been abandoned since no peach matures in the lowlands due to fruitfly infestation is correct for many regions. Peaches are not only the most susceptible fruit, but also help to raise the *Ceratitis* population considerable even when a few peach trees only are present. This is of paramount importance when no other suitable hosts are at the disposal of the fly during August/September. Peaches should be rigourously eliminated from the citrus belt. In all Mediterranean regions, as well as in S. Africa, the fruitfly is the major pest of this crop. Nectarines are about as susceptible as peaches.

The earliest varieties in the coastal plain, as well as in the hills, of Palestine are often only moderately or very little attacked. But the later varieties, including the most delicious ones, are always attacked to some extent. The strong infestation has greatly reduced peaches in the coastal plain. In the hills there are years with a very low climatic factor and a very poor host-sequences, during which even the late varieties are only moderately attacked. But in those years *Anarsia lineatella* often takes the place of the fruitfly. In other years, as in autumn 1936, the damage was so heavy that not one uninfested fruit of the more selected varieties could be collected. In a fruit garden in Jerusalem, in the summer of 1930, from 4 fruits collected in early August, from 9.VIII—20.IX hatched 70 flies, i.e. 17.5 maggots per fruit. An infestation of 15—30 larvae per fruit is quite common in late peaches, but much higher figures have been observed.

Green peaches are rarely attacked before maturation. The only way to protect the more valuable varieties is to wrap each individual fruit, in mid to late July, in a paper bag.

In the experimental garden at Kiryath Anavim the varieties of peaches showed, in the heavy fruitfly year 1936, a high correlation of intensity of infestation with date of maturity (not with date of flowering):

Variety	Date of flowering	Beginning of maturity	Intensity of attack
Mayflower	25.III	end of VI.	+
Philipskling	9.III	end of VI.	++
Cote d'Azur	8.III	20.VII	+++
Alberta	8.III	28.VII	++++
American	15.III	24.VII	++++
Lovel	9.III	20.VIII	++++

Nectarines are relatively free, despite their late maturation:

Variety	Date of flowering	Beginning of maturity	Intensity of attack
Cardinal	25.III	1.VIII	+
Hanwick	15.III	10.VIII	+
Champion	15.III	10.VIII	+

The correlation of intensity of infestation to date of maturity of the peaches in the mountains is based upon the late appearance of the second *Ceratitis*-generation in the hills. In the coastal plain, at Mikweh Israel for instance, all varieties were 100% infested at an early stage, long before the peaches were mature.

The clensel traps at Kiryath Anavim yielded (average per trap):

Year	June		July			August	
	10—20	20—30	1—10	11—20	21—31	1—10	16—31
1937	2	6	19	5	2		
1938	16	—	12	31	6	7	17
1939	10	5	7	2	3	2	

At Mikweh Israel Klabi suffers 80 to 100% damage, all other varieties 100% in a heavy fruitfly year. They were infested as immature fruits in May/June, whilst maturing in late May/June.

Apricots (*Prunus armeniaca*) are likewise a very favoured host. Especially the later varieties are heavily infested. There is scarcely an inhabitant in this country who has not yet met with the maggots in this popular fruit.

The early apricots in the hills, belonging mostly to the palefleshed varieties Muskatavi and Losi, are fairly free from *Ceratitis*. But the later varieties and those of the coast plain suffer heavily. Both, the late and very large fruits of the variety Amerikani, which are grown with irrigation, and the small fruits of Klabi, which are grown near Ramleh, have bitter kernels and are generally sold still green, are strongly infested. Mr. Zisik estimates the normal infestation of the former variety at 80%, of the latter at 30—60%. Rarely more than 3—5 maggots are bred from one fruit.

Late apricot varieties were heavily attacked at Kiryath Anavim in 1936, not so the early varieties:

Variety	Date of flowering	Date of maturation	Intensity of attack
Muskatavi	14.III	2.VI.	+
Losi	8.III.	5.VI.	+
Titton	12.III.	6.VI.	+
Royal	3.III.	14.VI.	+
Blenheim.	3.III.	14.VI.	+
Moorpark.	12.III.	16.VI.	++
Amerikani	12.III.	23.VI.	++++
Grosseillier	27.III.	23.VI.	++++
Gross válla	26.III.	23.VI.	++++
Prec. Alexandre.	26.III.	23.VI.	++++

Plums (*Prunus spp.*) are relatively resistant to fruit fly attack. Mature fruits of all kinds of plums were readily infested in the laboratory at Hawaii. Excessive moisture of well ripened fruits, especially of the large and juicy Japanese plums, causes, following Back and Pemberton, a high mortality among young larves. Plums are reported to be attacked in heavy fruit fly years only in S. Africa.

In Palestine, we observed for many years, only rare cases of *Ceratitis*-infestation of plums. The reason was the small number and the economic unimportance of plums at that period. In the meantime plums are gaining an increasing importance and Japanese plums are often heavily infested. At Mikweh Israel the infestion of various varieties was:

Variety	Infestation	Fruit mature	Fruit at infestation	Degree of infestation
Combination	VI	late VI	almost mature	10—15%
Ogden	VI	late VI	almost mature	5—10
Wickson	VII	late VII	immature	10—20
Myrobalan	VII	mid-VII	almost mature	80—100
Mariposa	VII	late VII	immature	20—25
Kelsey	VII	late VII	almost mature	40—60

We cannot maintain, therefore, our earlier opinion that plums are a relatively safe fruit in the citrus belt.

Pears (*Pyrus communis*) are again a very suitable host. In Hawaii pears are generally and badly infested. In Italy some early varieties with small and juicy fruits suffer most, in S. Africa others. In Jerusalem, 6 flies hatched in late August 1930 from one pear. In an orange grove in Petah Tikva 31 flies hatched from 4 pears in August/September. Pear growing has much expanded in the last decade in almost all cultivated districts of the country. A few data on infestation by fruitfly were obtained by Klein at Mikweh Israel:

Variety	Infestation	Fruit mature	Fruit at infestation	Degree of infestation
Beurre Giffard	VI	mid-VI	almost mature	10—15%
Sukkari	VI	mid-VI	almost mature	15—20
Ein Harodi.	VII	late VII	immature	—25
Superfine.	VII	late VII	immature	—40
Bartlett	VIII	late VIII	immature	—25
Artas	VIII	VIII	immature	1943: 90

Apples (*Pyrus malus*) are much less susceptible to fruit fly attack than pears. They are sometimes heavily attacked near orange groves, etc. From 14 apples picked from a few trees in an orange grove in Petah Tikvah in mid-June 34 flies hatched during July. But in a large apple orchard at a distance of several kms. from orange groves *Ceratitis* was very rare in 1925. An old farmer at Rehoboth who originally had two neighbouring orchards, one of apples and one of oranges, always complained about a heavy fruitfly infestation in both groves. After having followed our advise to eliminate the apples the fruitfly damage among the oranges has decreased considerably.

Yet with the expansion of apple cultivation the importance of the fruitfly has increased and is today by no means anymore depending on the neighbourhood of orange groves. At Mikweh Israel the infestation is low, inspite of the neighbouring orange groves. Only the late varieties, such as Calville de San Sauveur, are infested to some extent. Queen Pippin and Rome Beauty, also late varieties, are much less infested. At Ein Harod apples are considerably attacked by *Ceratitis*.

Quinces (*Cydonia vulgaris*) are fairly susceptible. As a late fruit they are important for the maintenance of the host sequence. They are readily infested.

Figs (*Ficus carica*) are susceptible to *Ceratitis* attack only while ripening. The milky acrid juice of the immature figs prevents the fly from ovipositing, but soft and ripe fruits are heavily attacked. However, the different varieties show a different intensity of attack. In Hawaii 36% of the fruits were infested with 9.5 larvae per fruit (Mason, 1931). From Italy, N. Africa and S. Africa heavy attacks have been heard of.

We have bred it repeatedly from figs in the hills during August/September, but never to any large extent. The common maggots of figs in Palestine belong to another fly, *Lonchaea aristella* Beck. The importance of the fruitfly as fig pest has not yet been thoroughly studied in the coastal plain; it is probably slight. This small percentage is important rather for the maintenance of the host-cycle. The infestation of edible fig as well as that of sycomore fig (*Ficus sycomorus*) should be studied therefore for that district. In the Upper Jordan-Valley the former is a host to *Ceratitis* from April to July and Grünberg regards it as a potential pest of importance.

Locally figs are heavily infested. Klein reports a 100% infestation from Mikweh Israel, together with *Lonchaea aristella*. Apart of the morphological difference between both maggots, they are easily separated. The maggots of *Ceratitis* live in the interior of the fig, those of *Lonchaea* only beneath the rind.

Loquats (*Eriobotrya japonica*) are a favoured host in Hawaii and Quayle (1931) observed it at Valencia. We have observed the fly only once in this fruit in Palestine and are still ignorant as to the role it plays as fruitfly carrier. The economic damage effected to this crop is certainly neglegible.

Pomegranates (*Punica granatum*) have been reported to be infested from Hawaii, Palestine and Algeria. These few observed infestations were apparently subsequent to another primary injury of the fruit. The very hard and resistant skin of the healthy fruit is inaccessible to the ovipositor of *Ceratitis*. It was never punctured even under enforced conditions in the laboratory.

Mangos (*Mangifera indica*) are on a whole a very favored host of the fruit fly, a very different degree of resistance having been observed in different varieties. Whereas the common seedlings are often infested to 100% in Hawaii, other varieties showed a wide range of infestation (Mason, 1931) from 6 to 10% in seven varieties. It was in Manila 14.0, Hawaiian 35.3, French 58.8.

Elimination of the strongly succeptible varieties is recommended by Mason for Hawaii and a material reduction of the fruitfly population is expected by the removal of this main supporting summer host in the lowlands. In Palestine fruits are sometimes clean, sometimes heavily infested. At Rehobot 5 to 8% of the fruits, mainly overripe ones, were infested in July/September. At Mikweh Israel slight infestations in September only were noted.

C. Oppenheimer (1939) has carried out en extensive research of fruitfly infestation in various varieties of young mango trees for eight years. The average infestation at fortnightly inspections was in percentages:

Year	1938	1939	1940	1941	1942	1943	1944	1945
damage	5	6	8	8	7	1	4	1

Sprayings with sodiumfluosilicate were executed in 1938 (8 sprays), 1939 (3), 1941 (10), 1942 (8). Since 1943 immature fruit was picked for marketing. This procedure is apparently, perhaps except in years of heavy fruitfly infestation a sufficient prevention of damage. The thick-skinned varieties were only very occasionally infested in the mixed groves. This does however not mean, that they are safe when planted alone. In thin-skinned varieties, fruit left to full maturity and softening on the tree tends to be highly infested. Seedling fruits were infested at 3 to 17%. The early ripening fruits till about mid-September are heavily infested. In 1943 and 1945 when the fruit was picked early, this infestation is less. Mid-season fruit was least infested, with an increase towards the end of the season. This increase depends on the fact that fruits from the top of the trees are picked in a riper state than the rest. These careful experiments of C. Oppenheimer show that early picking, perhaps also selection of varieties, may offer safe measures for mango growing, even in the citrus belt.

Guavas (*Psidium guayava*) are another favoured host. At Hawaii 62% of the fruits are attacked with eleven maggots per infested fruit. At higher elevations of these islands wild guava is the only host of *Ceratitis*. Guavas are often heavily infested in Palestine. 70 to 100% of the fruits were infested at Mikweh Israel in September/October, when approaching maturity. They are early picked, when a few punctures, without eggs however, occur. This fruit is well *marketable*.

Avocado (*Persea americana*). All avocado varieties are possible carriers of the fly, although the infestation is so slight that it is not considered in Hawaii as a pest. Among 1270 fruits of 12 varieties, 5 were infested in the orchard. Under enforced conditions a 4 % infestation only was obtained. Picking while the fruit is hard eliminates most chances of infestation. Of 1300 eggs deposited 54% failed to hatch in consequence of the oiliness of the pulp (Mason 1931, Willard 1927). At Mikweh Israel the variety Duke was infested, when immature in July (mature late July to late August). Caliente is slightly infested in August (mature in August/September). Varieties maturing in winter are not infested. At Rehoboth the fruits of Duke, when not protected by paper bags, were completely infested.

Bananas (*Musa cavendishii, M. sapientium*) are never attacked while green. The exudation of tannin-laden sap prevents the female from ovipositing. 30.000 fruits ready for shipment did not show one puncture; 1000 prematurely ripened fruits in the field did not yield one maggot, despite an abundant fly population in the field in Hawaii. On certain cooking bananas a few flies can be bred very exceptionally (Mason, 1931). No case of infestation has been reported from Palestine in nature, whilst ripe bananas are easily infested in the laboratory.

Papaya (*Carica papaya*) is so rarely infested that it can scarcely be regarded as a host, as long as the fruit is picked whilst green. Oviposition in the green fruit is prevented by the exudation of the milky acrid juice, which fills any oviposition hole immediately. The young maggot is unable to develop in green fruits, whereas among 171 overripe fruits at Hawaii, 23% were infested. Early picking will therefore be a sufficient safeguard, when this crop will leave the experimental stage in Palestine. It was regularly observed in overripe fruits, during the summer at Rehoboth.

Persimmon and Kaki (*Diospyros spp.*) are heavily attacked in Hawaii, Italy and Egypt. In Palestine only a few trees exist, to which little attention has been paid hitherto.

Sour Sap (*Anona muricata*) is liable to be heavily attacked as soon as it matures. Up to 50 flies may hatch from a single of these small fruits. Its cultivation is still very restricted in Palestine.

Dates (*Phoenix dactylifera*) have only once bred *Ceratitis* at Hawaii. Two adults hatched from a bunch of diseased dates which were abnormally moist. A few were bred at Dagania from dates, which certainly is not a regular host there.

Grapes (*Vitis spp.*) are a possible, but quite rare host. Back and Pemberton observed it once infested (*V. labrusca*) in Hawaii and Lounsbury reports it as a secondary host from S. Africa. Its discovery in Almeria grapes in Spain led to the prohibition of the import of Spanish grapes into the U.S.A., but a thorough inspection in Spain and Italy failed to show it as a regular host of *Ceratitis*. About 86% of eggs laid into grapes under forced conditions in Hawaii failed to hatch, „doubtless owing to a lack of oxygen." (Willard, 1927). Also this lost has been discovered recently to be infested in the Upper Jordan-Valley. It is not important enough to be a pest, but it permits fruitfly development from June to August.

At Dagania 10% and more of the Es Salti variety was infested quite regularly. Dattier de Beyrouth was much less attacked. We have no observations from the main vineyard areas of Palestine. We have pointed out on another occasion that the host situation at Dagania is quite exceptional.

A few other fruits of recent introduction are: Casimiroa was heavily infested. Feijoa and *Psidium cattleyanum* a little only, *Pytanga* heavily at Mikweh Israel and Dagania, very little at Rehoboth, where a less favourable host sequence exists.

Another group of hosts of the Mediterranean fruit fly are the hedge-plants and wild shrubs growing in the neighbourhood of the groves. The following species may be mentioned: *Prickly pear* (*Opuntia ficus indica*) is of common occurrence in the citrus belt. It is important because of its long fruiting season. The failure to find infestation in its fruits often leads to the belief that it is not liable to be attacked at all. From 22 fruits of a shrub, 30 m from an orange grove in Petah Tikwa, we bred 18 flies in September/December 1930. This is certainly not a strong infestation. In a number of enforced breedings fruit of prickly pear was not infested. In Italy heavy attack is reported (Constantino). It should not be allowed as a hedge-plant in the citrus belt.

Aberia caffra had become quite a popular hedge-plant for orange groves in Samaria. Unfortunately its fruits have proved to be very susceptible to *Ceratitis* infestation and it should therefore be strictly eliminated from the citrus belt. Infestation starts as soon as may and lasts until winter.

Solanum coagulans is another hedge-plant of orange orchards in Samaria. The skin of its fruits is too hard to be pierced by the ovipositor of the fly. But the overripe fruits are liable to infestation; hence it cannot be regarded as a desirable hedge-plant.

The pods of the common hedge-acacia (*Acacia farnesiana*) have proved in many experiments not to be liable to infestation even under enforced conditions.

The fruits of the shrubs of *Lycium europaeum* which are quite common among the shade of other hedge-plants, are rarely liable to be infested. The two cases in which we observed a very slight infestation are probably exceptional and it is doubtful whether any attention should be paid to this host.

With regard to vegetables, there is one report on tomatoes (*Lycopersicum esculantum*) having been attacked 15 years ago in Palestine. We have no other report, despite the fact that under enforced conditions tomatoes are readily infested. Also Mason reports from Hawaii, that no field infestation has become known to him. Eggplants (*Solanum melongena*) and green pepper (*Capsicum annuum*) which are both liable to fruitfly attack at Hawaii to a minor degree have never been observed here as hosts. In years of heavy infestation attention should be paid to these vegetables, not because *Ceratitis* is bound to be of any economic importance for these crops, but because of their importance as fruit fly carriers. Pepper was readily accepted in laboratory breedings at Dagania, eggplants never. In none, field infestation was observed.

The flies bred from melons, cucumbers, squash and onions belong to other species. *Ceratitis* has not yet been recorded from them in Palestine.

The host sequence of *Ceratitis* in the mixed plantations of Mikveh Israel during 1936 was as follows:

In May apricots showed a 100 per cent infestation and *Eugenia uniflora* was heavily infested. The early maturing plum Wickson was only slightly infested (3%) in July, whereas of the later variety Kelsey 30% were injured. Myrobalan, also, was heavily infested since late June. Apple and pear were also attacked in July, guava and *anona* in September. During the same month the fruits of an ornamental tree *Tevethia neriifolii* (Apocyn.) were infested in the Botanical garden.

Adults hatched from shamuti in April, from Valencia-oranges in June, from Clementine in December. Washington Navel was infested in November. In localities with such a favourable host sequence as at Mikveh Israel each year favourable to *Ceratitis* necessarily causes heavy damages to various fruits.

The real host sequence in a given locality is not necessarily that of the available fruits. H. Z. Klein has compiled the real host sequence for two localities in different areas of Palestine, yet both are exceptionally rich in potential hosts of many kinds. These data are based on actual oviposition or infestation with maggots observed in the orchard:

Month	Dagania	Mikweh Israel
I	Grapefruit	—
II	Grapefruit	—
III	Grapefruit	Citrus spp.
IV	Grapefruit	Citrus spp., espec. Valencia.
V	Grapefruit, Valencia, Pytanga.	Valencia, Pytanga, apricots.
VI	Grapefruit, Valencia, Pytanga, Plum Beauty, grape (Es Salti).	Valencia, apricots, peaches, pears, apples, plums.
VII	Grapes (Es Salti, Dattier de Beyrouth) Casimiroa.	Plums, pears, avocado, Casimiroa.
VIII	Grapes (Es Salti), Casimiroa, fig.	Peaches, apples, pears, fig, walnut, avocado.
IX	Guava, dates.	Guava, anona, mango.
X	Guava, grapefruit, Washington Navel, Shamouti, clementine.	Quince, Citrus spp., espec. clementine.
XI	Grapefruit, Washington Navel, shamouti, clementine.	Feijoa, Citrus spp.
XII	Grapefruit, blood orange.	—

Klein and Paker (1943) discuss this actual host sequence in connection with their extended observations at Dagania. Quite a number of fruits which were eagerly accepted in laboratory breedings, and which occurred within the area under observation, were not found infested in nature. The fly was bred for 10 months the year in green pepper and in tomatoes. Casimiroa was accepted for two months longer in the breedings than observed in nature. Bananas were accepted in breedings for three months, yet they were never found on this host in the orchard. *Spondias cytherea* and apples were accepted for two months in the breedings, but not infested in nature.

In the following fruits oviposition or oviposition holes were never observed in nature or in breedings at Dagania: tangerines (Yussuf Effendi), green lemons, papaya, anona, *Doryalis hebecarpa*, *Passiflora edulis*, the grape varieties Chasselat and Muscat Hambourg, pomegranate, *Ficus sycomorus*, egg-plant, prickly pear. This behaviour is quite irregular, as pomegranate, egg-plant and prickly pear, for instance, are known hosts from other regions.

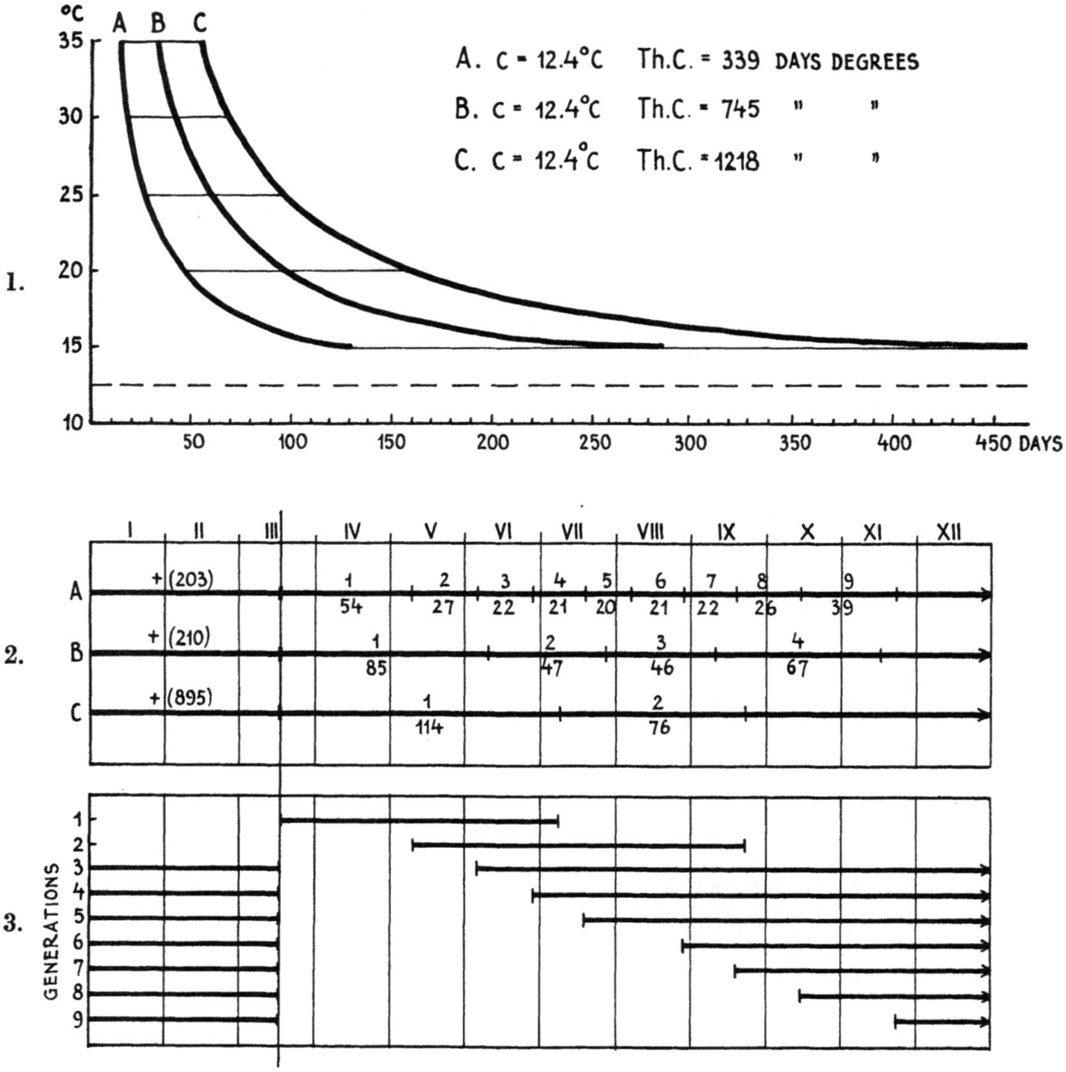

Fig. 25.
Physical ecology of development of *Ceratitis capitata* Wied. at Dagania: 1. The hyperbola band, 2. the local life-history, as calculated from the three hyperbolae, 3. the overlapping of the generations.

Comparative observations of oviposition in nature and selection from various fruits offered together in the laboratory showed that from September onwards guava is the preferred fruit. Among plums Kelsey is preferred to all other varieties. The other easily accepted hosts have just been enumerated. Most interesting is the often repeated observation, that the host in which the mother developed herself had under orchard and laboratory conditions no influence on the host preference of here own oviposition.

4. PHYSICAL ECOLOGY, DISTRIBUTION AND DENSITY OF FRUIT FLY POPULATIONS.

A. *Annual number of generations, local life-history and world chart:*

The author published a hyperbola for *Ceratitis capitata* (1924 and 1926), with separate hyperbolas for each stage of development. The constants for development from egg to fly were:

development threshold (c) = 13.5° C
thermal constant (th c) = 250° C.

Data published recently by Mc Bride (1934) as well as our own experiments permit the calculation of the hyperbola for total as well as for partial development, including the preoviposition period, with a higher degree of accuracy. The values for the corrected hyperbolas are as follows:

	egg	larva	pupa	total	preoviposition period	total
development threshold	10.5	9.8	9.7	9.9	16.5	12.4° C
thermal constant	24.3	117.8	182.3	321	45.0	339 days-degree
duration of stage at 35° C	1.0	4.7	7.2	12.9	2.4	15.3 days
30	1.2	5.8	9.0	16.0	3.3	19.3
25	1.7	7.7	11.9	21.4	5.3	26.7
20	2.6	11.5	17.7	31.8	12.8	44.6
15	4.4	22.7	34.3	51.4	—	—
12	9.7	53.6	79.1	142.4	—	—

This development zero at about 10° C for egg, larva and pupa is empirically confirmed by the extended observations and experiments of Back and Pemberton (1918). The lower empirical limit for oviposition is between 15—16° C. The oviposition zero corresponds less with the average environmental temperature, as well as with the length of the preoviposition period, because of the heliothermic behaviour of the fly.

The quickest development from oviposition to oviposition, however, does not permit the recognition of the true epidemiological number of generations. Oviposition is fairly scattered and lasts for some months. In order to obtain a full picture we again calculate the hyperbolas B and C from the data at our disposal (cf. adult longevity).

The constants for the three hyperbolas are as follows:

	A	B	C
development threshold	12.4	12.4	12.4° C
thermal constant.	339	745	1218 days-degrees
35° C	15.3	33.0	53.9 days
30	19.3	42.5	69.2
duration of development at 25	26.7	59.1	96.7
20	44.6	98.0	160.3
15	130.4	286.5	468.5

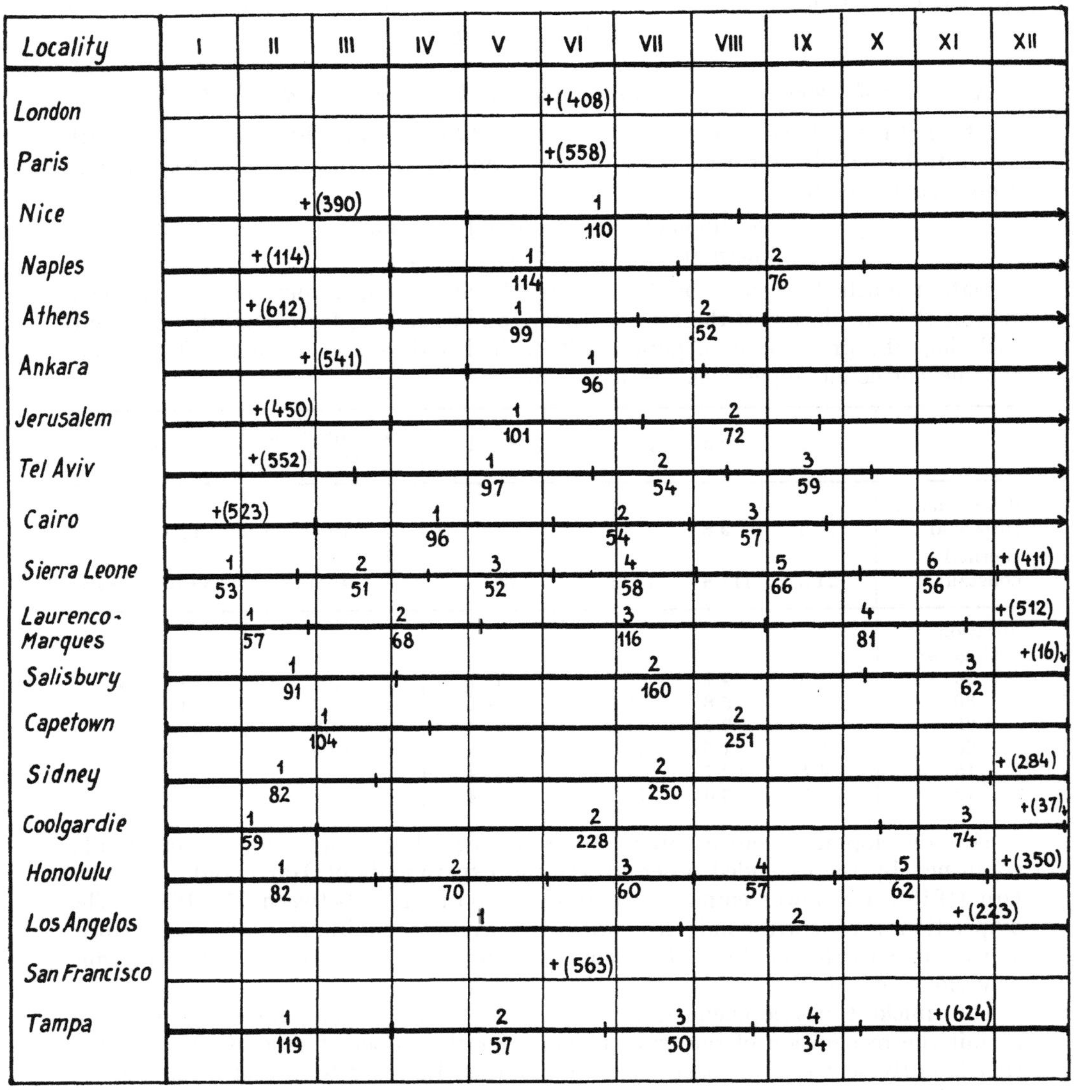

Fig. 26.
World chart of life-cycle of *Ceratitis capitata* Wied.

The values of hyperbola B may be regarded as representing the epidemiological average of annual generations. A world-chart has been calculated for the life-history of *C. capitata* in different parts of its area of distribution (fig. 26), which is based on these epidemiological averages.

The life-history of *C. capitata* can be computed in this way for any place within or without its actual area of distribution. Some of these life-cycles (Tel-Aviv, Jerusalem) are illustrated by fig 27. In evaluating these charts, one has to consider that some adults may live much longer than indicated by hyperbola C. In spite of the small number of those long living flies, their epidemiological importance for the maintenance of the species is great. The prospects for the presence of such survivors are especially high when only the host sequence is unfavorable, but climatic conditions favorable.

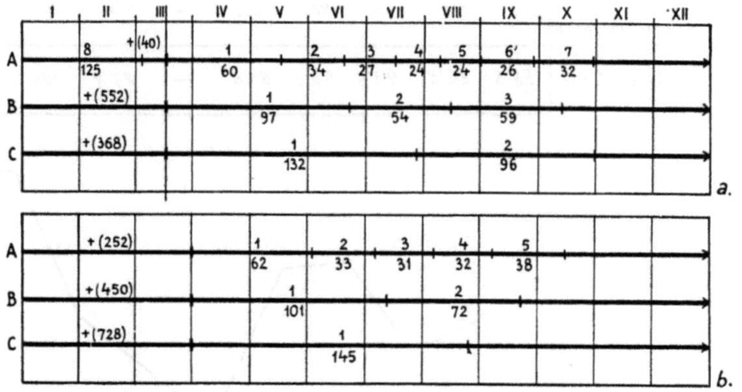

Fig. 27.
Local life-cycles as calculated from the three hyperbolae at Tel Aviv (above) and at Jerusalem (below).

The development potential of one pair of flies at different climates, assuming an average egg-number of 300, resp. 100, eggs per ♀ and an equal sex-ratio, during one year, is as follows epidemiological average (hyperbola B):

Number of annual generations (B)	Localities	Annual potential increase	
1	Paris, Nice (San Francisco)	100	300
2	Naples, Capetown (Los Angeles)	5000	45000
3	Athens, Jerusalem, Salisbury	250000	6750000
4	Tel Aviv	12500000	1012500000
5	Jericho, Tiberias, Lourenco Marques, Honolulu, Tampa	625000000	151875000000
6	Sierra Leone	31250000000	22781250000000

Due to climate as well as host conditions, only a small fraction of this potential is realized. The cooperation of these factors has been exposed before. This development potential can never realize, as the environment would grow too small for it. The maximum population level which could be maintained is that of a high population pressure of adult fly, which would enforce the immediate discovery and utilisation of any fruit fit for oviposition. This would result in an overoviposition of every fruit. Conditions would be similar to those of *Lucilia*-flies, which discover any carrion of their district due to a high population pressure and try to benefit from every milligram of its meat for reproduction despite heavy inter- and intra-specific competition. Hitherto, no region is known for *Ceratitis* where such conditions prevail for any prolonged period.

Environmental resistance for the two assumed cases would be 99.7 and 98.0% respectively, should the population be stable. The actual resistance may be computed for the two statistical series known to us (m = f, $R_0 = 99.2$, e = 250). The symbols in this as well in all following population equations indicate m: percent-

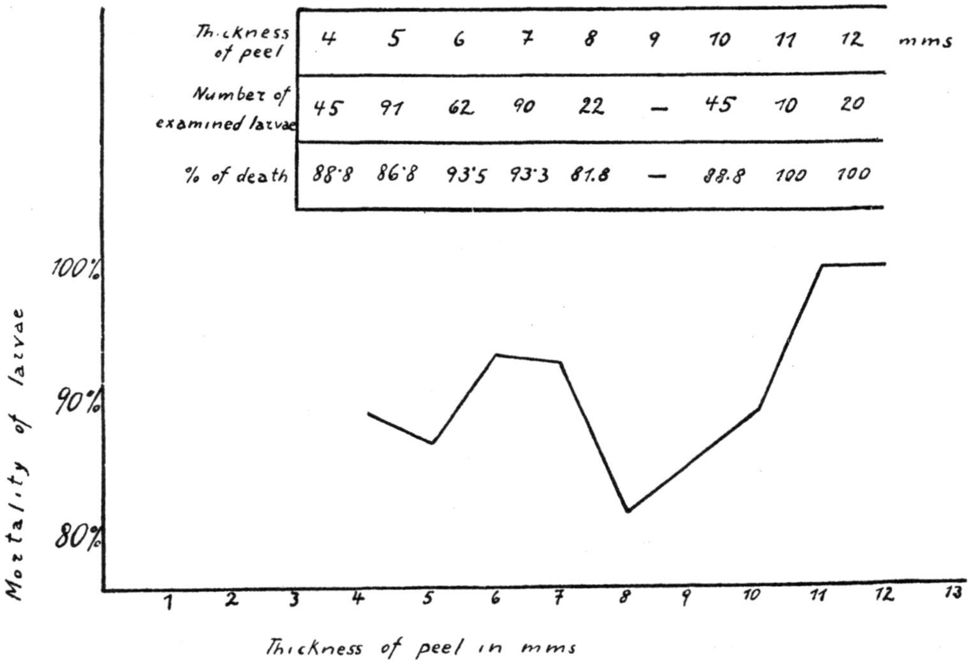

Fig. 28.
Seasonal trend of *Ceratitis*-populations in the Upper Jordan Valley (from Grünberg).

age of males, f: of females in the population; e: eggs per female. R: Environmental resistance, expressed as the change of population size from one the to following generation $\left(\frac{P_1}{P_2}\right)$. R_0: Environmental resistance in a stable population, with no change in the size of the populations in the following generations.

At Honolulu:

date	1.I.	19.III.	27.V.	27.VII.	22.IX.	28.XI.
population	176	74	275	727	152	132
R	98.93	99.66	97.03	97.88	99.82	99.27
a $\frac{P_1}{P_2}$	1.3	0.4	3.7	2.7	0.2	0.9

At Rehoboth:

date	1.IV.	21.VI.	15.VIII.	12.X.
population	7	—	—	217
R	99.95	?	?	?
a $\frac{P_1}{P_2}$	31.0	?	?	?

It is improbable that at Rehoboth the fly population could have maintained itself within the orchard. It is also entirely impossible that its population level could have increased on the spot 31-fold during an interval of 8 months or three normally developing generations. Hence, we must assume that they have developed in the neighbourhood in summer-fruits, at least during 2 generations.

Data illustrating the actual increase were obtained on various hosts at Dagania (Klein and Paker 1943):

Month	II/III	IV/V	VI/VII	VII/VIII	VIII/IX	IX	X/XII
Aver. temperature	17	23	27	30	30	27	21
increase factor	4	12	17	10	10	27	14
n	5	4	7	3	2	13	9

The increase factor indicates the number of flies bred from one pair of *Ceratitis* during their entire reproductive period (divided by two). The maximal increase was 90 (i.e. 180 adult flies from one female). The average increase factor on various fruits was:

Host	n	increase factor	Host	n	increase factor
guava	11	23	grapefruit . .	7	8
pepper	4	17	banana. . . .	2	8
Kelsey plum . .	2	12	casimiroa . . .	2	6

Physical, host and parasite conditioned mortalities partake in this empirical resistance. It is of interest to note that at a 100-fold increase of the population R_2 would be still 20.0, at a 100-fold decrease 99.992.

Physiological constants for *Ceratitis* are:

Scale of activity.

Scale of activity.	1	2	3	4	5	6	7	8
Temperature	8.0		17.0	(25.5-28.0)	29.1	38.6		45.8°C

At 13.4° C the cooled flies reawakened to sluggish activity.

The preference temperature in 38 readings was 30.27° C, the preference zone embracing 28 to 33° C (68% of all observations).

B. *Physical ecology of the Mediterannean fruitfly and its value for the bonitation of a given locality.*

The study of the life-history combined with the increase potential and the local host-sequence, gives us only the frame of the seasonal fluctuations and of the general abundance of *C. capitata* in a given region. No exact physiological studies have yet been made as to the mortality of the various stages of the fruit fly at different combinations of temperature and humidity.

The main empirical facts are as follows:

Below 10° C, no development takes place and mortality is complete when these conditions prevail over a sufficiently long period (about 4 months). When about

2000 eggs were exposed to varying temperatures from 3.5—19.5 °C in apples, less than 0.3% were found still alive after about 85 days and those would also have died in all probability (Back and Pemberton, 1918). 11 days at 0° C kill all eggs (Mason and Mc Bride, 1934).

During cold and wet weather the pupa, and not the adult, carries the species over the unfavourable season. Pupal mortality is high in cold and wet winters too, whereas a mild and relatively dry winter favours an outbreak in the following spring by reducing pupal mortality in the winter.

Mating and oviposition are restricted to temperatures above 15.5° C but probably to much higher body temperatures, considering their heliothermism. Above 25° C, no sun-exposure is needed for normal adult activity.

Adult longevity is influenced decidedly by air humidity. Extreme drought, even for a relatively short period, is detrimental. Prolonged, uninterrupted rains are also unfavourable to adult activity.

Gjullin (1931) classified the climatic requirements of *C. capitata* based mainly on temperature and rainfall, as follows: The optimal zone (of normal abundance) is limited by temperatures of 15.5° C (60° F), in combination with precipitations exceeding 125 mm during December/February, or of 13° C (55° F) with lower rainfall during these months.

The favorable zone (of occasional abundance) is limited by a monthly temperature of not less than 3° C (37° F) combined with precipitations not exceeding 125 mm during December/February, or 6° C, with higher rainfall.

The zone of mere existence (or of possible abundance) is limited by —1 to +2° C (30—35° F) mean monthly temperature, in proportion to the precipitations in December/February. Areas with —1° C have less than 125 mm precipitation during these months.

In this way the regions of the northern hemisphere may be classified in a quite useful way. This classification proves to be surprisingly useful in getting a first survey on the suitability of a major region for fruitfly infestation. However, this first pioneer step in evaluating the physical ecology of *Ceratitis* has still many drawbacks. The most important of them is the impossibility to evaluate suitable estimates for regions with summer-months of no rainfall. Also, as to how far irrigation can replace the unsuitable conditions of dry summers. And an even more important drawback is the impossibility to compare, in those regions, the climograms for different years in order to analyze the reasons for serious outbreaks in some special years.

We therefore attempted to substitute relative humidity for rainfall. The following zonation seems to serve as a suitable basis to predict fruitfly abundance, as far as it is conditioned by physical ecology.

Relative humidity has been introduced into the study of fruitfly ecology for the first time. The American students in Hawaii neglected this factor as, in the environment they studied, humidity never seems to fluctuate far from the optimal conditions and light rainfalls, even in the dryer months, counteract the seriousness of deviations of an even medium extent. However, the decisive and destructive influence of prolonged hamsin periods on the adult flies in May cannot be disregarded. It is a fundamental assumption of ecology that factors which are conspicuously influential at higher or extreme intensities are important, but in a lesser degree, at less extreme intensities, whenever deviating from the optimum values. The importance attributed to humidity, significant in the life cycle of the fly as co-determining the adult longevity, is therefore justified, especially when temperature conditions allow quick development and the drought or humidity periods are sufficiently prolonged to effect one or two full generations, at least.

The different zones of abundance are limited (always excluding the area of the preceding areas) by:

Zone	temperature-range in ° C	humidity range in % of relative humidity
A. Optimal.	16—32	75—85
B. Favourable	10—35	60—90
C. Non-favourable. . . .	2—38	40—100
D. Impossible.	<2—40<	<40

In this way we are able to obtain a bonitation of any region as to what degree it is reliable to outbreaks or serious damages effected by the Mediterranean fruitfly. The presence of 1 to 3 consecutive months in the impossible zone inhibits any permanent occurrence of damage in the locality concerned. In the non-favourable and the favorable zone the population density will, as a rule, be very low. Mass-appearance and danger begin where favorable or optimal conditions exist for some consecutive months, at least. In counting every point of the climogram which falls within the favorable zone as 1, every one of the optimal zone as 2 and summing the positive values of all months together we obtain the intensity index of *C. capitata* within the given climate, resp. locality. Later tables reproduce a series of such bonitations which show a fair approach to actual conditions.

A series of corrections can be made to improve this bonitation.

Rainfall of more than 125 mm combined with low temperatures (below 15° C) is detrimental, especially to the pupae in the soil and to the adult flies. The number of such months should be indicated. Prolonged lack of rains (irrigation lacking) is detrimental to the fly, directly by its influence on humidity and indirectly by its influence on vegetation. The biological unit for measuring the importance of any factor to which the fly is exposed is the generation. The speed of development depends mainly on temperature. A special indicator consigns the speed of development (from egg to egg) in the actual month, as follows:

		1 generation is		
>10 °C	= 0	∞	days	
10—15	= 1	over 50	,,	(no oviposition)
15—25, > 32	= 2	50—29	,,	
25—32	= 3	29—18	,,	

Of greater importance, however, than both these factors are the host-conditions. Abundant supply of favorable and suitable fruits for oviposition and development throughout the physically favorable season is optimal. The conditions are less favorable when the hosts are present during a shorter period only or when they are unsuitable, i.e. if eggs or larvae show a high nutritional mortality during development. Conditions grow uncomfortable when the sequence of ± suitable fruits is interrupted for a ± prolonged season and when hosts grow scarce in number. Wherever these conditions become even more extreme, the host factor is very unfavorable. The bonitation with regard to the host factor should never be made for a whole country or region, but must be determined for each locality separately. In the oasis of Jericho, ex. gen., host conditions change within a few metres from fairly to very favorable in the fruit-groves of the oasis to extreme unfavourable in the desert. The complete local bonitation for a given locality will be explained by some examples:

1) An isolated fruit garden in Jerusalem, in which apricots, peaches, prunes, apples and pears are grown, during 1932:

	I	II	III	IV	V	VI	VII	VII	IX	X	XI	XII	
Climate:	8.0	9.9	13.0	16.7	19.8	23.8	24.4	25.0	23.4	23.2	15.8	9.3 °C	temperature
	80	60	61	50	36	38	41	41	53	45	60	75 %	rel. humidity
	47	127	7	10	—	—	—	—	11	34	19	7 mm	precipitation
Bonitation	0	0	1	0	0	0	0	0	0	0	1	0	damage factor: 2
		—1											rain-factor: —1
	0	0	1	2	2	2	2	3	2	2	2	0	generation factor
	0	0	0	0	0	3/3	3/3	3/3	0	0	0	0	host factor

2) For the year 1926, the following bonitation would be correct in the same garden:

Climate:	8.9	9.4	11.2	17.0	22.0	23.8	23.2	22.4	21.7	20.9	16.3	11.7 °C	
	79	78	70	48	43	48	65	85	67	75	59	68 %	
	96	113	59	12	—	—	—	—	—	—	8	96 mm	
Bonitation	0	0	1	0	0	0	1	2	1	2	0	0	damage factor: 7
		—1											rain-factor: —1
	0	0	1	2	2	2	2	2	2	2	2	1	generation factor
	0	0	0	0	0	3/3	3/3	3/3	0	0	0	0	host factor

In 1932, physical ecology is not favorable to an outbreak in any month of the year, when fruits were available. The host sequence is very unfavorable, because of the long interruption from September to May. The rain-factor is only slightly unfavorable during one of the winter-months. The extreme draught from May to August is very depressive to the adult fly population. The number of generations during the favorable host period is only 1—2. Very slight damage at the utmost may therefore be expected only during August in late peaches.

The situation was different in 1926. 4 consecutive months were climatically favorable, two of them optimal. The rain-factor was only slightly unfavorable during one winter-month. The number of generations during the suitable host period was only 1—2. Host-sequence remained very unfavorable because of its long interruption, but optimal conditions prevailed during three consecutive months, two of which coincided with favorable to optimal climatic conditions. During July a sufficiently high fly population could have been built up to cause serious damage to the most suitable host, peaches, in August. Beyond the garden, where practically no suitable hosts were present, the presence of the species was impossible in both years.

2. A fruit-garden in Jericho, which contains some trees of apricots, oranges, pomegranates, guavas, papayas, etc. in 1935:

	I	II	III	IV	V	VI	VII	VIII	IX	X	XI	XII	
Climate:	14.3	15.0	18.8	22	29.2	30.6	30.4	31.2	29.4	26.6	19.1	16.2 °C	
	67	75	59	44	34	46	44	48	54	56	68	68 %	
	27	43	1	—	—	—	—	—	—	17	15	5 mm	
Bonitation	1	1	0	0	0	0	0	0	0	0	1	1	damage factor: 4
													rain factor: 0
	1	2	2	2	3	3	3	3	3	3	2	2	generation factor
	1/1	1/1	1/1	2/1	2/1	2/1	2/1	2/1	2/1	1/1	1/1	1/1	host factor

The damage factor is 4 and the favorable months are four consecutive winter months. The rain-factor is very favorable. Extreme drought rules from April to August, but is slightly improved by irrigation. The host sequence is fairly favorable, but during February and November only oranges are available, and they are an unsuitable host. Only slight damage to winter fruit should thus be expected, despite the very favorable generation factor. In the neighbouring desert where practically no hosts are present the presence of the species is impossible.

Considering the intensity of effect on the fruit fly population, the montly factors for climatic and host conditions should be multiplied by the generation factor. The sum of the monthly products would serve as final bonitation factor (under Palestine conditions, where the rain-factor is negligible). The size of the single monthly bonitation factor indicates the local probabilities of damage in the year concerned. The final host factor should be combined as the sum of suitability and accessibility (abundance of host and distance from the locality concerned). Other factors absent, only the climatic factor is inserted.

The final bonitation for the years just analyzed would be as follows:

	I	II	III	IV	V	VI	VII	VIII	IX	X	XI	XII	Year
Jerusalem 1932	0	0	1	0	0	0	0	0	0	0	1	0	2
Jerusalem 1926	0	0	1	0	0	0	12	24	1	2	0	0	40
Jericho 1935	2	4	0	0	0	0	0	0	0	0	4	4	14
Tiberias 1934	4	4	0	1	0	0	0	12	12	8	0	0	41
Tel Aviv A1932	4	4	8	1	1	1	1	1	12	8	8	0	49
Tel Aviv A1934	4	4	8	2	2	2	2	2	12	16	8	4	66
Tel Aviv B1932	4	4	8	8	8	8	12	12	12	8	8	0	92
Tel Aviv B1934	4	4	8	16	16	16	24	24	12	16	8	4	152

The last two lines (B) assume an uninterrupted host sequence (2/2) from April to August with the other conditions remaining equal to those assumed in the two preceeding lines. Such estimates, which include all the important local factors, can be calculated only for localities the climate and host conditions of which are known. The final annual sum in itself does not replace the study of the detailed analysis. In assuming 10 as the lower damage limit, one must consider that in Tel Aviv, e.g., in 1932 the value 12 in September follows five consecutive months of 1, which have reduced the fly population considerably. Whereas in 1934 two consecutive values of 12 and 16 (September, October) follow five months of 2, which means favorable conditions for a longer time, following a much smaller decimation of the fly population during the interval.

3. In an orange grove at Tiberias, the following conditions prevailed during 1934:

	I	II	III	IV	V	VI	VII	VIII	IX	X	XI	XII	
Climate:	12.4	10.9	17.7	18.7	23.1	28.1	27.9	27.3	26.9	23.1	20.2	13.9	°C
	69	77	57	65	57	59	59	61	64	65	57	61	%
Bonitation	1	1	0	1	0	0	0	1	1	1	0	0	damage factor: 6
													rain-factor:
	1	1	2	2	2	3	3	3	3	2	2	1	generation factor
	1/3	1/3	1/3	0	0	0	0	1/3	1/3	1/3	1/3	1/3	host factor

Climate was favorable in January, February and April, not very unfavorable in May to July, growing again favorable from August to October. Rain-factor was favorable the drought never extremely low. The generation factor was favourable. Host sequence was not very unfavorable, as only during 4 months no host was present, but the host during the other months was not very favorable. No other suitable hosts being present in the neighbourhood (apricots, prickly pear, guavas, figs, etc.), a serious damage could not be effected. However, in the presence of other suitable fruits within or near the garden the damage may become quite considerable, especially in autumn and early winter, even in the orange grove.

4. An orange grove in the neighbourhood of Tel Aviv in 1932 and 1934 respectively permits the following conclusions:

1932	I	II	III	IV	V	VI	VII	VIII	IX	X	XI	XII	
Climate:	11.9	13.2	15.5	17.7	20.1	24.2	26.2	26.9	25.1	23.8	19.0	13.5	°C
	70	67	64	65	73	71	74	73	69	72	60	55	%
	131	96	8	3	—	—	—	—	1	38	58	43	mm
Bonitation	1	1	1	1	1	1	1	1	1	1	1	0	damage factor:
	—1												rain factor: —1
	1	1	2	2	2	2	3	3	3	2	2	1	generation factor
	1/3	1/3	1/3	0	0	0	0	0	1/3	1/3	1/3	1/3	host factor
1934													
Climate:	12.8	11.7	16.3	18.5	21.6	24.9	26.7	27.1	25.7	23.2	19.2	14.4	°C
	76	79	71	80	80	77	79	79	74	75	73	80	%
	165	152	10	11	1	—	—	—	—	1	13	445	mm
Bonitation	1	1	1	2	2	2	2	2	1	2	1	1	damage factor: 18
	—1	—1										—1	rain factor: —3
	1	1	2	2	2	2	3	3	3	2	2	1	generation factor
	1/3	1/3	1/3	0	0	0	0	0	1/3	1/3	1/3	1/3	host factor

In Tel Aviv climatic conditions are favourable to optimal all year round. In 1932 they were favourable, except for December. The rain factor is only slightly unfavorable in January. No extreme drought decimates the adult fly population seriously. The generation factor is very favorable from March to November. Host sequence is fairly favourable, being interrupted for five months. The main host is not very suitable. No high fly population can be present in Sep-

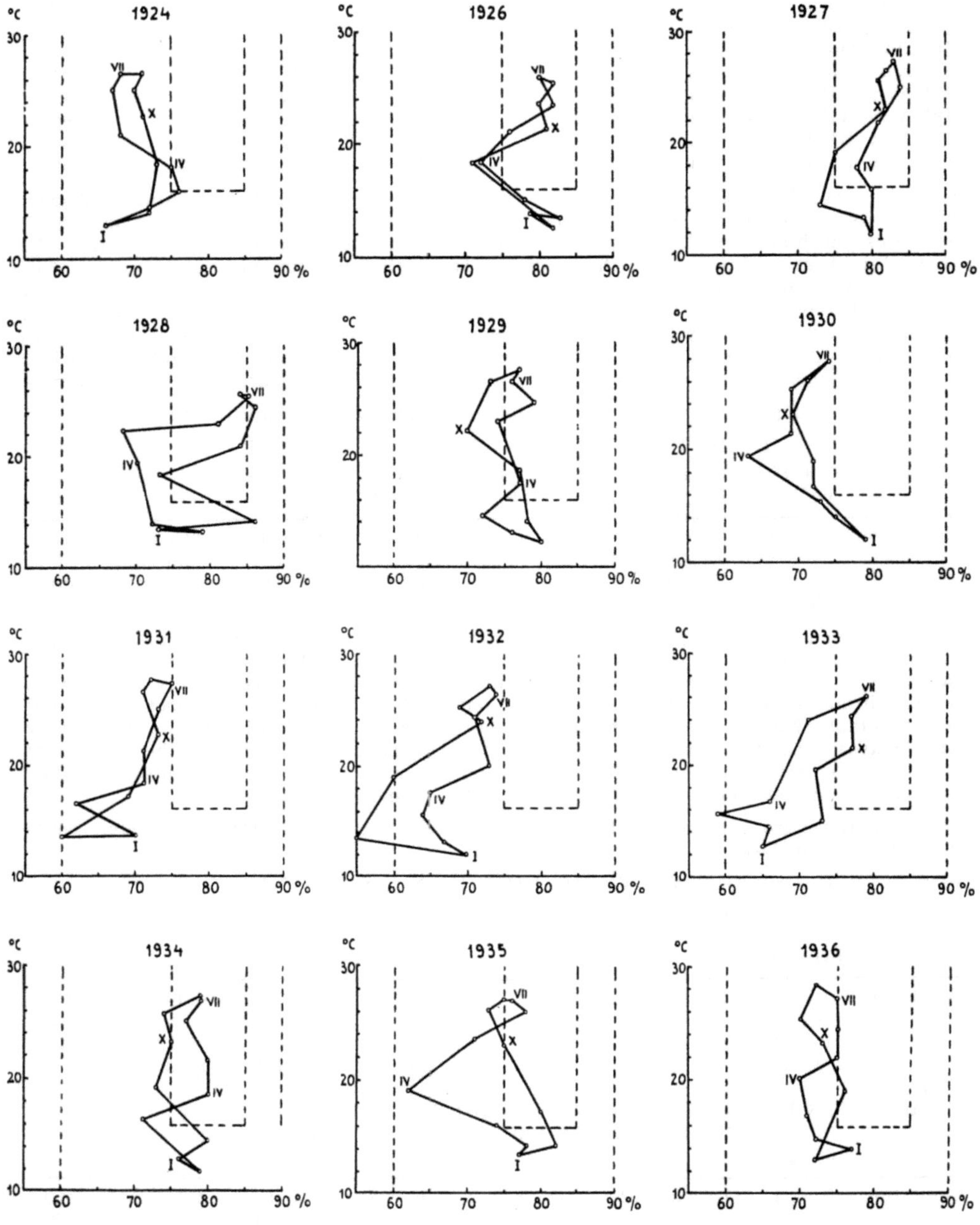

Fig. 29.
Monthly ecoclimograms for Ceratitis at Tel Aviv from 1924 to 1936.
(Broken lines: limits of optimal and favourable conditions).

tember, after the host interval, but damage of late fruit may be expected from late March on.

In 1934, conditions are rather different. Whereas host- and generation-factor remain constant, the rain factor grows more unfavorable, being slightly unfavorable in January and February, fairly unfavorable in December. The high precipitations in December affect only the following year. Climatic conditions, however, are far more favorable: they are optimal from April to August, very close to optimal in September and optimal again in October. No interruption occurs in December. The optimal climatic conditions in spring and summer permit the survival of a relatively high number of flies, which should be able and has actually been able to attack citrus fruits in September to a considerable extent. These conditions are improved, as far as the fly is concerned, by the fact that some other fruits are almost always available within the garden or in its neighbourhood, throughout spring and summer. Apricots, prickly pear, figs, Lycium, etc. are among the more common ones. The quantity and distance of these intermediate fruits determine the intensity of the attack on oranges in autumn. An interrupted sequence of suitable hosts in the neighbourhood would have resulted in serious damage in 1932, but in extremely serious damage in 1934. The intensity of attack, the host factor being 2/2 or 3/3 in April to August, may well be imagined.

The following series of bonitations for Palestine localities is appended here:

	I	II	III	IV	V	VI	VII	VIII	IX	X	XI	XII	Total	Opt.	Fav.	N.Fav.	Imposs.
								Tel Aviv									
1923	1	1	2	2	1	1	2	1	1	1	1	1	15	3	9	—	—
1924	1	1	2	2	1	1	1	1	1	1	1	1	14	2	10	—	—
1925										1	1	2					
1926	1	1	2	1	2	2	2	2	2	2	1	1	19	7	5	—	—
1927	1	1	2	2	2	2	2	2	2	2	2	1	21	9	3	—	—
1928	1	1	1	1	1	2	2	2	2	2	1	1	17	5	7	—	—
1929	1	1	1	2	1	2	2	2	1	1	2	1	17	5	7	—	—
1930	1	1	1	1	1	1	1	1	1	1	1	1	12	—	12	—	—
1931	1	1	1	1	1	1	2	1	1	1	1	1	13	1	11	—	—
1932	1	1	1	1	1	1	1	1	1	1	1	1	11	—	11	1	—
1933	1	1	0	1	1	1	2	2	2	2	1	1	15	4	7	1	—
1934	1	1	1	2	2	2	2	2	1	2	1	1	18	6	6	—	—
1935	1	1	1	1	1	2	2	2	1	2	2	1	17	5	7	—	—
1936	1	1	1	1	2	2	2	1	1	1	2	1	16	4	8	—	—
								Jerusalem									
1922	0	0	0	0	0	0	0	0	0	0	1	1	2	—	2	10	—
1923	0	0	0	0	0	0	0	0	1	0	0	1	2	—	2	9	1
1924	0	0	0	0	0	0	0	0	0	2	2	1	7	2	3	7	—
1925	0	0	1	1	0	0	0	1	0	0	1	1	5	—	5	7	—
1926	0	0	1	0	0	0	1	2	1	2	0	1	8	2	4	6	—
1927	0	0	1	0	0	0	0	0	1	0	1	1	4	—	4	7	1
1928	0	0	0	0	0	0	0	0	1	1	1	1	4	—	4	8	—
1929	0	0	1	0	0	0	0	0	1	0	0	0	2	—	2	10	—
1930	0	0	1	0	0	0	0	0	0	1	1	1	4	—	4	8	—
1931	0	0	1	0	0	0	0	0	0	0	0	1	2	—	2	10	—
1932	0	0	0	0	0	0	0	0	0	0	1	0	1	—	1	9	2
1933	0	0	1	0	0	0	0	0	1	0	0	1	3	—	3	8	1
1934	0	0	1	0	0	0	0	0	1	1	0	1	4	—	4	6	2
1935	0	0	0	0	0	0	0	1	1	1	1	1	5	—	5	7	
								Jericho									
1925	1	0	0	0	0	0	0	0	0	0	1	1	3	—	3	8	1
1926	1	1	1	0	0	0	0	0	0	0	0	1	4	—	4	7	1
1927	1	1	1	0	0	0	0	0	1	0	1	2	7	1	5	6	—
1928	1	1	0	0	0	0	0	0	0	0	1	1	4	—	4	6	2
1929	1	1	1	0	0	0	0	0	0	0	0	1	4	—	4	8	—
1930	1	1	0	0	0	0	0	0	0	0	0	1	3	—	3	9	—
1931	1	1	0	0	0	0	0	0	0	0	0	1	3	—	3	9	—
1932	1	1	0	0	0	0	0	0	0	0	0	0	2	—	2	7	3
1933	1	1	0	0	0	0	0	0	0	0	0	1	3	—	3	7	2
1934	1	1	0	0	0	0	0	0	0	0	0	1	3	—	3	9	—
1935	1	2	0	0	0	0	0	0	0	0	1	1	5	1	3	7	1

C. *Fruit fly bonitation in different climates.*

We have pointed out before that a complete bonitation of the fruitfly population, including the important host-factor, is possible for local conditions only. The following regional bonitation is based on climate only. The following analysis is based on this assumption.

a) *The fruitfly in the humid tropics:* The Hawaiian Islands are the classical locality of *Ceratitis* research. The conditions in the lowlands were repeatedly described. McBride (1935) now published the first results of a study on *Ceratitis* at higher elevations. His stations were at an altitude of 200, 250, 1000 and 1800 feet. Citrus and mango were present at the lower stations, whereas at the higher ones mango fruits were present only for 2—3 months. The fly-populations decreased during the 3—4 months following the absence of this only fruit hand reappearance at these higher levels is due to flies moving in from the lower levels. The mean annual temperature decreases 0.56° C. for every 200 feet of elevation, parallel to a corresponding decrease in *Ceratitis*-abundance at higher elevations. The average number of generations, calculated on this base, as is follows:

Altitude	Mean annual temperature	Average no. of generations	
		A	B
Honolulu	23.5° C	11.6	5.4
250 feet (80 m)	23.0	11.2	5.1
1000 (330 m)	20.7	8.8	4.0
1800 (600 m)	18.5	6.5	2.9
4500 (1500 m)	13.3	1.0	0.4

Unfortunately, exact data on temperature and humidity were not available for these stations. We, therefore, illustrate the influence of altitude by calculations on Java, which is still free of fruitfly infestation.

The bonitation will be as follows:

Locality	altitude	no. of generations		Bonitation				Bonitation factor	months below 12.4°C
		A	B	0	F	N	I		
Batavia	8 m	14.3	6.5	9	3	—	—	21	—
Bandung	730	10.3	4.7	10	2	—	—	22	—
Tosari	1735	3.7	1.7	10	2	—	—	22	—
Pangerango	3023	—	—	—	12	—		0	12

Fig 30 illustrates this bonitation by means of climograms. Conditions are extremely favorable throughout the lowlands of the Malay-Asian region, as the bonitation for Singapore and Manilla show:

Locality	no. of generations			Bonitation				Bonitation factor	months below 12.4 °C
	A	B	C	0	F	N	I		
Singapore	14.8	6.7	4.1	12	—	—	—	24	—
Manila	15.1	6.8	4.2	8	4	—	--	20	—

Bonitation ranges generally from 19—24.

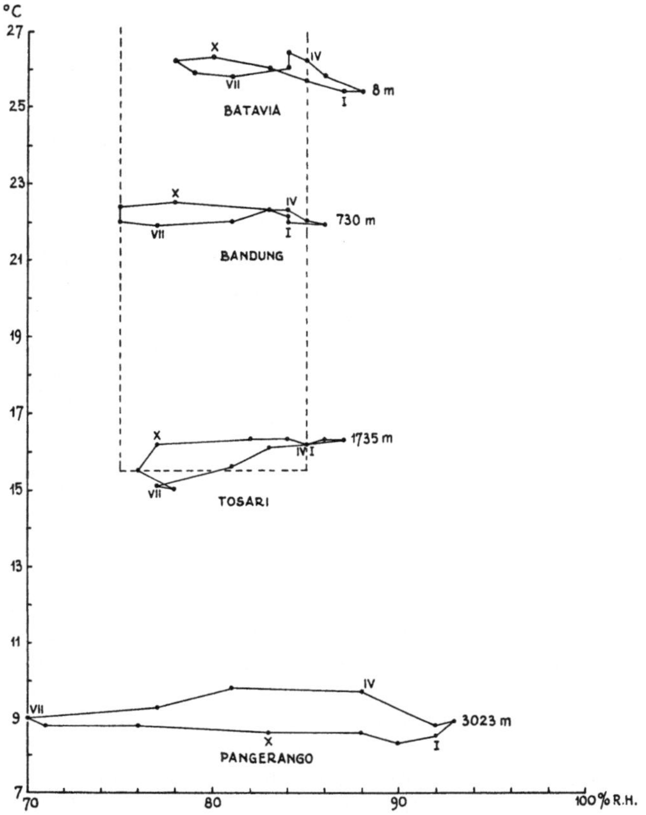

Fig. 30.
Ecoclimograms for *Ceratitis* at various altitudes at Java.

b) *The fruitfly in desert, steppes and cool climates:* The climates of 3 different regions, representing the desert (Saharo-Sindian), the steppe (Irano-Turanian) and a moderate northern climate (Euro-Siberian) will now be analyzed:

No optimal combination is ever obtained in the Irano-Turanian region, the number of favorable months being low, and at least 2 are impossible. No permanent damage and heavy outbreaks are to be expected in this region:

Locality	no. of generations			Bonitation				Bonita-tion factor	months below 12.4 °C
	A	B	C	O	F	N	I		
Damascus	6.7	3.1	1.9	—	1	9	2	1	4
Mosul	8.9	4.1	2.5	—	2	6	4	0	4
Ankara	3.7	1.1	1.0	—	—	9	3	0	6
Budapest	2.5	1.1	0.7	—	7	2	3	0	7

In the true desert, fruitfly cannot permanently exist because of lack of host sequence and of dry air. In cases where host sequence, as well as humidity conditions, are favorable (cf. Cairo and Jericho), existence and even high infestation are possible, but:

Locality	no. of generations			Bonitation				Bonitation factor	months below 12.4 °C
	A	B	C	O	F	N	I		
Bagdad	11.9	5.4	3.3	—	4	5	3	0	2
Khartoum	16.8	7.6	4.7	—	—	3	9	0	—

Ceratitis may still successfully invade the southern parts of the Euro-Siberian region. The comparison between Paris and Madrid (northern points of permanent habitat) and Berlin and Pinsk, is of interest:

Locality	No. of generations			Bonitation				Bonitation factor	months below 12.4°C
	A	B	C	O	F	N	I		
Madrid . . .	4.2	1.9	1.1	—	3	9	—	3	6
Paris	1.5	0.7	0.4	—	6	6	—	6	7
London . . .	1.2	0.6	0.3	—	5	7	—	5	8
Berlin. . . .	1.9	0.9	0.5	—	5	4	3	0	7
Pinsk	1.7	0.8	0.5	—	5	2	5	0	7

The high number of months in the impossible zone prevents any permanent settlement in the last mentioned regions, the low annual thermal sum in S. England.

c) *The fruitfly in the Mediterranean regions:* There is no impossible zone throughout the Mediterranean region, most of the months being, as a rule, in the favorable, others in the not favorable and a few in the optimal zone. Bonitation fluctuates from 7 to 15. The warmer parts of the region are more favourable. The winter rains rarely suffice to suppress the wintering pupae and the summer drought is rarely severe and not long enough to exterminate the adult population.

The following list contains also a series of African bonitations, which confirm the general conclusions hitherto reached.

Climatic bonitation of a number of localities for *Ceratitis capitata*.

Locality	no. of possible generations			no. of months				climatic bonitation	no. of months below 12.4° C
	A	B	C	O	F	N	I		
Paris.	1.5	0.7	0.4	—	6	6	—	6	7
Lyon.	1.8	0.8	0.5	—	7	5	—	7	7
Nice	3.3	1.5	0.9	—	7	5	—	7	6
Genoa	4.6	2.2	1.3	—	9	3	—	9	5
Naples	4.6	2.1	1.3	—	9	3	—	9	4
Palermo . . .	5.5	2.5	1.5	—	12	—	—	12	3
Valencia . . .	5.3	2.4	1.5	—	11	1	—	11	3
Malaga. . . .	6.6	3.0	1.5	—	12	—	—	12	1
Lisboa	4.0	1.5	1.1	—	10	2	—	10	3
Las Palmas. .	8.0	3.6	2.2	3	9	—	—	15	—
Athens	6.1	2.8	1.7	—	6	6	—	6	4
Famagusta . .	8.1	3.7	2.2	1	11	—	—	13	2
Tel Aviv . . .	8.2	3.7	2.2	2	10	—	—	14	—
Jerusalem. . .	5.6	2.5	1.6	—	3	9	—	3	3
Jericho	12.0	5.5	3.3	—	4	8	—	4	—
Tiberias . . .	10.7	4.9	3.0	—	5	7	—	5	—
Alexandria . .	7.7	3.5	2.3	3	9	—	—	15	—
Cairo.	7.7	3.5	2.1	2	8	2	—	12	2
Khartoum . .	16.8	7.6	4.7	—	—	3	9	0	2
Mongalla . . .	14.5	6.7	3.9	6	3	3	—	15	—
Sierra Leone .	14.3	6.5	4.0	—	—	—	—	—	—
Lourenco Marques . .	10.2	4.7	2.8	7	5	—	—	19	—
Salisbury. . .	6.6	3.0	1.8	4	4	4	—	12	—
Capetown. . .	4.2	1.9	1.2	—	—	—	—	—	—

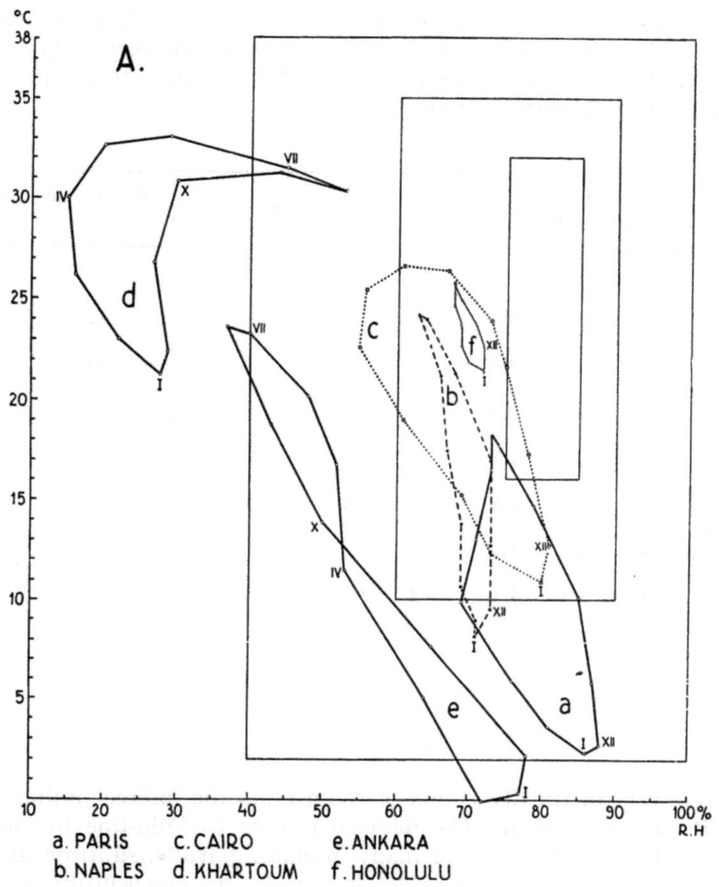

Fig. 31.
Ecoclimograms in correlation to suitability of various regions for *Ceratitis:* a) Paris, b) Naples, c) Cairo, d) Khartoum, e) Ankara, f) Honolulu (from Bodenheimer).

d) *The potential danger of the Mediterranean Fruit fly in U.S.A.* is of interest in connection with the damage experienced recently in Florida and the strong quarantine measures maintained in California against this pest. In dividing the regions analysed into 2 groups, a) hot climate permitting the development of 10 generations and more (hyperbola A) and b) with a cooler summer, the importance of relative humidity is dominating: At Tampa optimal conditions prevail, which are even more favourable than the climate of Hawaii, whereas at Yuma with an average relative humidity of 38% not less than 9 months are in the impossible, the 3 others in the unfavourable zone.

The dry interior of California (Fresno) is likewise too dry. A passably high air humidity is maintained only during the cold winter months. No outbreaks are to be expected with 1 favourable and 4 impossible months. In Central California at

San Francisco all months are in the favourable zone, but temperature is too low to permit many generations. The damage would probably be higher than at Paris, where 6 months are in the non-favourable zone. At Los Angeles damage would be quite considerable, whereas at San Diego it would be very heavy indeed. The damage in California would decrease from the Pacific shore to the interior parallel to the decrease of relative air-humidity.

Locality	No. of possible generations			Bonitation				Bonitation factor	months below 12.4° C
	A	B	C	O	F	N	I		
Honolulu (Hawaii). .	11.9	5.4	3.3	—	12	—	—	12	—
Tampa (Florida) .	10.4	4.7	2.9	2	10	—	—	14	—
Yuma (Arizona) .	10.4	4.7	2.9	—	—	3	9	—	—
Fresno (Calif.) . .	6.2	2.8	1.7	—	1	7	4	1	3
San Francisco (Calif.) .	1.6	0.8	0.5	—	12	—	—	12	4
Los Angeles (Calif.) . .	4.9	2.2	1.4	—	9	3	—	9	—
San Diego (Calif.) . .	4.0	1.8	1.1	4	8	—	—	16	—

5. DAMAGE.

The Mediterranean fruitfly is one of those pests the indirect influence of which surpasses its actual damage. The damage it does to Palestine by preventing or interfering with the cultivation of many promising fruits, subtropical and others, in our citrus belt is serious indeed. The danger of large monocultures could be most easily overcome by planting horticultural by-crops in this belt. But unfortunately most of the latter mature from spring to autumn, when a fairly high fruitfly population is present. In this way citrus growing is a danger to such crops as guavas, mangos, peaches, etc. and vice versa. *Ceratitis capitata* must be regarded as a factor of imminent importance for horticulture in all semitropical and tropical countries and its presence or absence may mean the entire difference between a prospering horticulture and one fighting hard for its existence. Its presence always either limits the choice of crops to a far higher degree than climatic limitations alone would do or it decreases considerably the annual crop revenue of one or, usually of many fruits.

The damage begins with oviposition, when the rind of the fruit is pierced. This puncture and the larval gallery open the way for moulds and bacteria, which in turn induce decay. The fruit may fall off the tree before the maggots are ready tc pupate. Citrus fruit drop after oviposition of *Ceratitis* only, when they were weakened by other reasons, such as lack of irrigation in autumn, light and poor soil. These and all other causes which provoke premature yellowing of the fruits attract also oviposition of the fruitfly. In spring fruits may drop when overripe. Other types of damage will be discussed later.

Not in all the cases is the infestation recognisable from the outside. Apples, peaches or oranges may be heavily destroyed within the fruit, before the damage becomes recognisable to the casual observer. The oviposition puncture is always the earliest sympton, and usually develops some discoloration. In oranges and other citrus fruits the puncture is surrounded by a yellow patch on the green fruit, whereas in loquat the puncture and its environment remain green, even when the fruit grows yellow. Sunken areas without discoloration or with brown decay develop in mature peaches or pears. In prickly pear the puncture is discolored to dark brown which is not connected with decay, but probably with the proliferation of cork cells.

Ceratitis conditions in Hawaii are representative of those met with in most countries of the humid, resp. semi-humid tropics. They are described by McBride (1934) as follows:

„The Mediterranean fruitfly infests more than 75 species of fruits, which are distributed throughout the islands and which furnish a constant supply of suitable food. The abundance of these host fruits, however, varies with the season, and the fly population is largely determined by the quantity of available host fruits. During periods when host fruits are plentiful, the fly population builds up quickly, while a decrease in the number of fruitflies, which however, comes 3 to 5 weeks after the host fruits disappear. At elevations above 1500 feet (= 489 m), fruits of the wild guava are available for a period of only 2 to 3 months each year, and then disappear. In these areas the fruit flies are present in decreasing numbers during the first 3 or 4 months following the absence of this host fruit. Mature guava fruits first appear at the lower elevations and gradually thereafter to the higher elevations. The fruit fly population also moves forward and upward with the maturing fruits. Reappearance of the flies at the higher elevations is due to flies moving in from the lower levels."

There is no doubt that in the lowlands of Hawaii the Mediterranean fruitfly has effected a most serious and permanent check upon the horticultural development. The same is true of the Bermuda Islands.

The situation is slightly different in countries of a Mediterranean type of climate.

Constantino (1930) reports that in S. Italy, where July to October are favorable, the damage fortunately is not repeated every year with the same intensity. But here, as well as in S. France, heavy damages and even full destruction of the peach and apricot crops occur from time to time.

Conditions at Valencia are described by Quayle (1931): The fly disappears for the winter about the 1st of January. It reappears about June 1st. In April and May, hundreds of acres of orange trees still carry their full quota of matura fruit without one of those being attacked, as the fly has not yet appeared from its winter quarters. It may become injurious to other crops in the summer and autumn and there is no doubt that it has restricted the cultivation of peaches considerably.

In Egypt serious damage is restricted to fruit and house orchards which have a specially favorable host sequence. Damage is very limited wherever the host sequence is unfavorable. But here also *Ceratitis* has induced horticultural limitations.

Lounsbury and Mally, of South Africa, consider the fruit fly as one of the greatest drawbacks to the development of the fruit industry in Cape Colony, and have stated that during certain favorable seasons almost entire areas of apricots, figs, pears, plums, apples and quinces were infested. Following Quayle (1931) „the highest quality of peach is grown in spite of the fly. The greatest care is taken in the frequent picking up of fallen fruit and, in oranges, in addition the prematurely coloured fruits were seen to be picked off. This prevents fruit fly infestation or destroys it (beside the control of the false Codling moth). No fruit is allowed to accumulate either in the orchard or about the packing houses and all infested fruits are properly disposed of."

Also in W. Australia its damages are sporadic, being more permanent in the mild coastal area, where in 1914 the ravages of the fly were so widespread that it became a question as to wich was to exist — the grower or the fly (Newman, 1924).

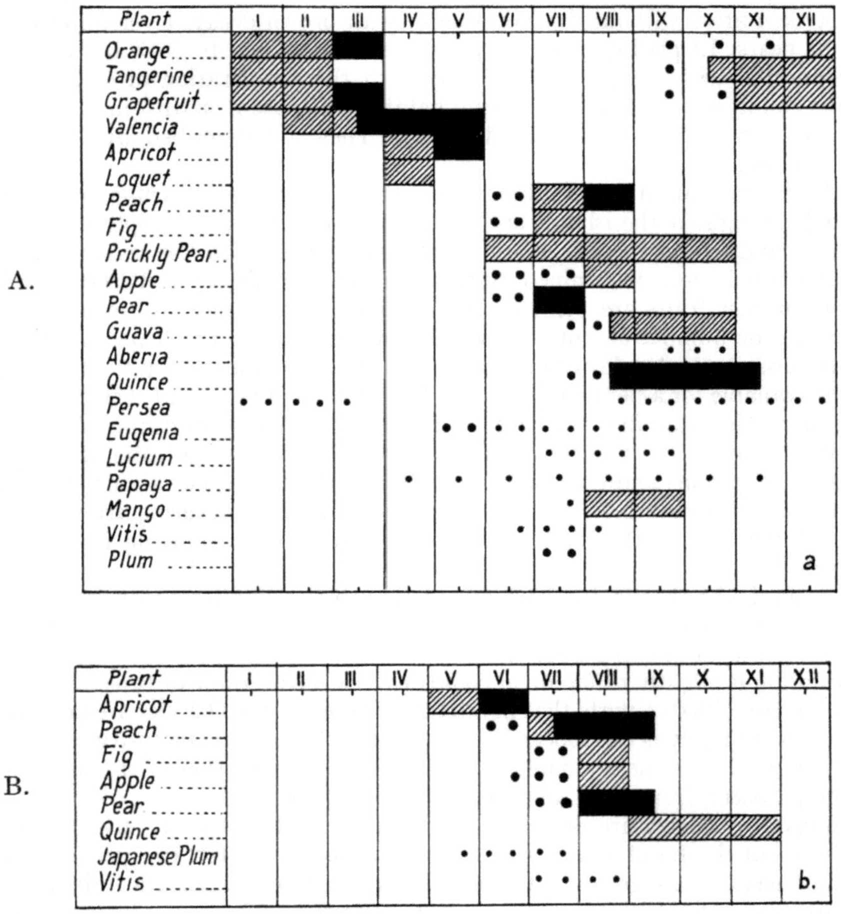

Fig. 32.
Seasonal host sequence for *Ceratitis capitata* Wied. A. in the coastal plain of Palestine, B. in the Judaean hills.

In regions of a Mediterranean climate damages may be high and permanent, wherever the temperature is high throughout the year (wintergeneration up to 3 months at most) and host sequence favorable. Its damage is more limited and only sporadically high in regions with a long winter interruption when the fly population is considerably reduced. But the presence of the fruitfly restricted horticultural development even in these regions, especially as far as peaches are concerned.

Paris may serve to illustrate life conditions of *Ceratitis* along its northern border. „Aux environs immédiats de Paris la présence de Ceratitis fut observée à de nombreuses reprises et periodiquement, souvent à plusieurs années d'intervelle, dans un même jardain, sur les fruits d'un même arbre, ce qui démontre sans contradit l'existence d'un foyer permanent. Vu pour la première fois en 1898, . . . hiverment dans la region de Rambouillet, *C, capitata* existe bien aujourd'hui à l'état endémique dans la région parisienne." (Balachowsky et Mesnil 1935). Its damage seems to be negligible there in spite of its existence and no serious outbreaks, not even sporadic ones, have yet been reported.

Palestine: Oranges and grapefruit begin to be infested in September. The yellow punctures, easily recognisable, are called 'samra' in Arabic. Only very few of the eggs laid at that period develop into mature maggots. Most of them die as eggs in the rind, the rest as very young maggots soon after hatching. Only in clementines and thin-skinned oranges (as in the variety known as Franzawi) the number of successfully developing larvae is higher even in autumn. The same is true of especially thin-skinned grapefruits.

In normal years new infestation ceases about the beginning or middel of December, when the number of ovipositing females decreases suddenly. Development is very slow in the winter, the full cycle from egg to adult lasting about 4 months in the coastal plain. The samra found in this season originates from oviposition in the preceding one.

Hatching and maturing of the flies of this generations are generally at their peak in the middel of March. Fortunately, it has become customary, owing to imperial customs policy in the last years, to terminate the orange-season about the middle of March. In the normal orange grove this generation is picked before having matured and does not find any suitable citrus fruit for infestation. In mild winters the interval is often much shorter and the beginning of the spring infestation earlier.

Before discussing the damage done to citrus crop in detail, some ciphers may be of use. In a typical, monocultural orange grove at Rehoboth the damage caused by *Ceratitis* was 27.7% of the total ,brarah' (cull) and 3.5% of the total crop in 1935/36. The damage caused by *C. capitata* was

in month	no. of boxes absolute	(per 150 fruits) percentage
X/XI	53	100.0% picked off the tree
XI/XII	46	3.9 because of samrah
XII/I	27	2.3
I/II	3	0.5
II/III	—	—

The Mediterranean fruitfly was responsible for the following share of the citrus-cull in four seasons in a grove at Rehoboth (Klein):

Ceratitis share of total cull	XI	XII	I	II	III	% of total crop	% of total cull
1934/1935	—	—	—	—	—	7.0	—
1935/1936	89	37	12	1	—	3.5	27.7%
1936/1937	17	1	0.2	—	—	0.6	3.7%
1937/1938	—	—	—	—	—	0.02	0.2%

The fruit dropped prematurely owing to fruit fly punctures or for other reasons is not included in this count.

The catch (4 traps) baited with clensel thoughout the year 1936 in the same orchard was, as follows:

date	no. of flies	date	no. of flies	date	no. of flies
1.I.	2	17.V.	1	26.	137
9.	1	5.VI.	3	2.XI.	128
16.	—	18.	—	22.	3
26.	—	11.VII.	—	12.XII.	49
5.II.	1	25.	—	2.I.1937	33
16.	1	9.VIII.	—	18.I.	6
3.III.	—	10.IX.	3	2.II.	—
12.	—	20.	50	16.II.	—
25.	10	26.	123		
2.IV.	4	14.X.	—		
16.	1		(no clensel)		
26.	2	22.	217		

Or reduced to months:

Month	I	II	III	IV	V	VI	VII	VIII	IX	X	XI	XII
No. of flies	1	2	10	7	1	3	—	—	176	354	131	82

Other statistics show the following infestation of the total crop (of the total culls in brackets):

		on Shamuti	on grapefruit
Hedera	1936/37	5(35)	2(9)
	1937/38	0.05(3)	0.05(4)
Mikveh Israel	1936/37	0.6(6)	0.02(0.3)
	1937/38	0.04(0.5)	0.02(0.3)

Klein (1940) resumes the damage since 1934 as follows:

	Coastal plain	Inner plains
Autumn 1934	+++	+++
Early 1935	Valencia ++	Fruit early picked
Autumn 1935	++	++
Early 1936	Half of the culls caused by *Ceratitis*, 50% of the fruit infested.	

Early spring 1936 Late Valencia and Fruit late on the trees: +++
Autumn 1936 to 1938 only damage to Valencia, excepting early winter 1937 when locally heavy damage was observed.

The monthly catches on three plots near Hedera calculated for 110 bottles each were in 1937/38:

Month	III	IV	V	VI	VII	VIII	IX	X	XI	XII	I	II
Plot A.	—	—	194	797	497	365	290	159	212	490	31	24
Plot B.	12	28	193	688	298	259	182	122	182	295	—	6
Aberia-hedge	—	—	3	962	969	747	777	618	1032	57	36	—
Total	12	28	390	2477	1764	1371	1249	899	1426	842	67	30

The above data show:

1) The early samra is transportable and marketable, but reduced in value. Its export possibilities are restricted in consequence of quarantine laws of certain countries (cf. Ceylon). It must be picked prematurely and this again lowers its value. 53 out of 3536 boxes (= 1.5%) were reduced in value in this way.

2) During September and early October of some years, a large part of the punctured fruit drops, even if no oviposition has taken place. The injury caused to the fruit by oviposition leads to its fall at this period, during which the fruit seems to be especially sensitive.

3) From early December on, the samra is not anymore included in the export fruit. From this period on, it belongs to the 'brarah' (cull). This 'brarah' was 2.1% of the total crop during the season **1935/36**. The fruit inspection service is very rigorous in preventing samra fruit from being exported. This attitude is entirely to the benefit of the grower, for one decaying fruit inside the box may cause, during the long transport, the total destruction of all the fruits in the box.

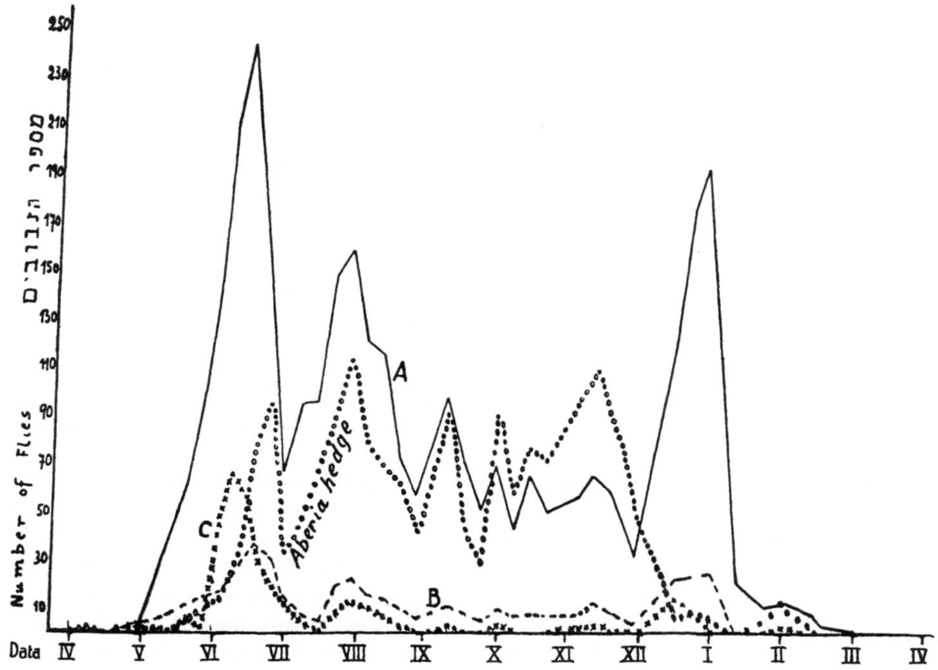

Fig. 33.
Fluctuations of Ceratitis capitata in orange groves at Hedera (A—C) and in Aberia-hedge (D. from Klein).

4) The fruit continues to drop throughout winter, but especially in the spring, especially when infested with maggots. During heavy winter-storms, the punctured fruit drops easier than normal fruit.

In the autumn infestation local differences occur in the intensity of attack among different orchards or even within the same orchard. These differences are explained by Rivnay (1936) as due to early maturing fruit which is more attractive to the ovipositing females. Some of the reasons for early maturation are

1) *late* spraying (better: too late spraying!) with oil emulsions (against red scale),
2) trees planted in too light, poor and dry soil.
3) insufficient irrigation, or the exploitation of the moisture in the ground for cypress and eucalyptus trees,
4) failing trees.

"All these factors should be removed in order to avoid an early attack" of *C. capitata*, says Rivnay. But whereas the early maturation only concentrates the damage, but does not increase it, no real control can be gained in this way, except for the benefit that the orchard gains from any thorough sanitation.

With regard to the intensity of the spring attack, our statement needs an amplification. When mature citrus fruit remains on the tree longer than mid-March, its infectation will always be heavy, occasionally enormous. This is not only obvious from our own experiences in earlier years, but is illustrated most clearly by the experiences of Yedidyah (1936) with Valencia late at Mikveh Israel. The number of flies entrapped on 7 trees was in 3 successive seasons:

month	III		IV		V		VI	Sum
days	1—15	16—31	1—15	16—30	1—15	16—31	1—15	
1934	—	—	—	14	117	508	347	986
1935	—	32	166	269	727	1571	318	3081
1936	—	500	1005	1085	2156	2190	191	7127

The corresponding infestation during these seasons was:

	dropped fruit	picked fruit		Total fruit	% infested
		infested	sound		
1934	153	172	1074	1399	23.2
1935	320	1186	191	1697	88.8
1936	1175	54	34	1263	97.6

This shows the total destruction of the crop for 1935 and 1936. The conclusion to be drawn is that no Valencia late should be planted and the export season should not be prolonged any longer than mid-March, until our knowledge with regard to fruit fly control has advanced much further.

Rivnay, in his above mentioned paper, has his doubts as to whether the common opinion that the proximity of summer-fruits is of great importance for the intensity of the autumn attack of the fruit fly within the orange-grove is justified. He correlates the infestation with the state of maturity of the fruit only. Our statistics on the annual fruit fly catch, which was made in the same orchard to which Rivnay refers, show that the autumn catch is about 35 times bigger than the spring catch (after a cold and wet winter). However, whereas we must assume a high mortality throughout a 4—5 months interruption of the host sequence, i.e. a strong reduction of the spring population level, we find a 35-fold increase. This indicates that the fruitfly must have developed, reproduced and increased repeatedly throughout the summer in other hosts than citrus and have reentered the orange-grove when ready for a new infestation. This experience, again, demonstrates very clearly the paramount importance, which summer fruits in the neighbourhood of orange groves posses for the intensity of autumn infestation.

The influence of alternating hosts available in the orchard has been well brought out by some observations of Klein (1945):

1. In two plots of one orchards at Hedera clensel traps were put in three consecutive years. Plot A was adjoining a hedge of *Aberia caffra*, plot B well within the orchard. The number of flies per clensel trap was:

Plot	Month	V	VI	VII	VIII	IX	X	XI	XII	I
A	1937	2	**6**	4	3	2	2	1	4	—
	1938	4	7	2	6	**24**	20	14	2	—
	1939	4	4	8	9	12	2	1	1	—
B	1937	2	**7**	2	1	—	—	—	—	—
	1938	6	**16**	1	1	3	4	3	—	—
	1939	3	9	2	1	1	2	—	—	—

The average number of flies per trap was almost twice as much in plot A as compared to plot B. The seasonal trend differed much also. In B the peak was in June and flies were always few in autumn. In A the June peak was small as compared to the autumn peak. The fruits of *Aberia* ripen in summer. In 1937 they were carefully collected and taken away. For this reason the seasonal trend is identical in both plot during that year. No fruits were collected in 1938, which lead to an exceptional concentration of *Ceratitis*-flies in the neighbouring rows of orange trees. The collection of the fruit was begun late in 1939, which led to a steep drop of flies in October.

2. In an orange grove (A) at Mikweh Israel close to the deciduous fruit tree collection (B) similar trappings were arranged:

Plot	Month	V	VI	VII	VIII	IX	X	XI	XII	I
A.	1937	**30**	6	1	—	4	7	8	1	—
	1938	3	3	4	8	14	**66**	4	2	—
B	1937	19	42	47	18	**73**	52	2	—	—
	1938	13	**78**	35	39	14	4	1	—	—

The fruit rotation was: apricot in June, Mirabolan and *Casimiroa* in June/July, Kelsey plum and peaches in July/August, guavas in September/October. There is no doubt that the neighbourhood of the deciduous fruits had considerably increased the number of *Ceratitis* in the orange grove. The antibatic trend of the fly populations in both seasons suggests migrations in both ways.

The following table contains some information on the phenology of *C. capitata* in Palestine. It registers for some localities and years the first appearance of the fruitfly in clensel traps:

Locality	1937	1938	1939	1940	1941	1942
Dagania (orange grove)		28.III.	31.III.	14.III.	—	
Beth Alpha (orange grove)		10.IV	3.IV.	11.III.	—	
Hedera-Nasat (orange grove)	28.IV.	7.IV.	7.IV.	25.III.	12.II.	
Hedera-Meretz (orange grove)			24.IV.	12.IV.	27.III.	
Mikweh Israel (orange grove)	30.IV.	7.V.	13.IV.			
Mikweh Israel (fruit garden)	5.V.	13.V.	12.V.		late IV.	late IV.

In the warmest area (the two first localities) a few flies are caught throughout the winter months. The data indicate the sudden increase in early spring. In 1941 no interruption was observed at all.

The above data are arranged in two couples of fairly closely neighboured gardens. The fairly large and significant differences between the two groves of each locality are surprising. These data strengthen the conception of small scale migrations as a common feature in the fruitfly.

And in 1943 Klein and Paker conclude for Dagania: "If for instance a citrus grove is in the vicinity of a deciduous fruit orchard, the danger to both is considerably increased. . . . In a small area at Dagania a school grove, an acclimatisation garden and a vineyard were established on adjoining lands. In this small block where fruit is available at all seasons of the year, the damage by the Mediterranean fruitfly to the vines has mounted steadily from year to year. This infestation of the grapes raises the population level of *Ceratitis* in the neighbouring citrus groves."

We have a few other counts on the actual damage caused by the Mediterranean fruitfly:

	n	month	% of *Ceratitis* from total picking	author
Petah Tikva	11	I/III.1926	1.4(0—3.4)	Bodenheimer (1930)
Petah Tikva	7	I/III.1927	0.3(0—1.2)	,,
Southern coastal plain . .	3	X.1930/III.1931	0.6(0—0.8)	Reichert and Hellinger (1932)

However, in all these counts, neither were all the fallen fruits included into the count, nor those samra-fruits which were exported. The fruit inspection service was very lenient in those early years and infested fruit was exported in quantity, especially in the first half of the season. In this way, most of the damage is included in the estimate, but no conclusions as to be degree of infestation should be drawn.

We have recorded the damage done by decay following fruitfly development within the fruit. Reichert and Hellinger (1932) placed fruit in which actual development had taken or was taking place in cold storage and a 100% *Penicillium* wastage ensued. Since the decay of one fruit causes the decay of many neighbouring fruits during the long journey, it is obviously impossible to permit any fruit showing any signs of infestation to be exported from the end of November on.

Keck (1934) has recently shown that in the orchard *Ceratitis* is an active carrier of the common green mould of citrus fruits (*Penicillium digitatum* Sacc.), which it inoculates into the fruits while ovipositing. "Since the green mould organism produces quite an area of soft rot within 3 days after inoculation, it follows that if the female should cause infection at the time of oviposition the development of the organism would soften the rag of the rind and in this way aid the larvae in gaining entrance to the pulp. Several tests have been made in an attempt to determine whether this is the case and, although a higher percentage of infestation occurred in some tests, results have been inconsistent and a definite statement cannot be made at this time."

Observations on fruit-dropping of citrus in connection with infestation by *C. capitata* and green mould (*Penicillium digitatum* Sacc.) were made in spring and autumn 1936 at Rehobot by one of our students, Miss R. Ben-Ami, who has repeated some of the experiments of Keck, with the following results:

In the autumn, the green immature fruits showed an equally high resistance to the infestation by both *Ceratitis* and *Penicillium*. In the spring, when fruits were mature, especially after mid-March, and when air humidity was high, both were dangerous to the fruit, when acting separately or together. These conclusions were confirmed later independently by Rivnay, Nadel and Littauer (1938).

Fruits dropped:

1) when infested by both *Ceratitis* and *Penicillium*,
2) when infested by *Penicillium* only,
3) when infested by oviposition punctures of *Ceratitis*,
4) when infested by maggots of *Ceratitis*.

Fruits remained on the tree:

1) when infested by oviposition punctures of *Ceratitis* only, and
2) occasionally, when infested by maggots, even after the latter dropped to the ground for oviposition.

All these observations were confirmed by studying dropped fruits as well as fruits on the tree, infested by either *Ceratitis* or *Penicillium* separately, or by both together, by cultivating the mould on a special medium and by microscopic examination. In autumn fruits seem to be immune as a rule against green mould-infestation. At that period they are also extremely resistant against fruitfly. The only two cases where green mould was observed were connected with earlier infestation by *Ceratitis*: in one case maggots were within the fruit; in the second case no maggots were present, but in the centre of mould infestation of the already yellow fruit was an oviposition hole of the fly.

The role of the fruit fly in dispersing green mould on its ovipositor is yet dubious, but experiments are continued in this direction.

Fruitfly infestation in the Upper Jordan Valley has recently been studied by Gruenberg. It was exceptionally heavy in 1935/36; as in the rest of the country, when *Ceratitis* often caused 25—50% ot the total cull and about 10% of the total crop was damaged.

Sex-ratio and percentage of gravid females was in the trapbottles in 1936, as follows:

month	I	II	III	IV	V	VI	VII	VIII	IX	X	XI	XII
% of ♀	67	75	31	49	55	56	34	single flies		78	68	66
% of gravid ♀	73	54	35	16	68	73	32	only		70	60	52

Klein obtained at Hedera in 1937/1939 (A) and at Dagania in 1938/1940 (B):

Plot	Month	I	II	III	IV	V	VI	VII	VIII	IX	X	XI	XII
A.	% of females	45	57	50	41	49	51	47	41	40	43	42	50
	% of gravid females	77	100	67	67	55	61	42	42	47	63	55	60
B.	% of females	51	50	78	56	35	32	30	24	50	67	64	57
	% of gravid females	57	62	73	69	36	48	81	—	78	73	69	61
	Total catch	167	30	267	1270	69928	27063	884	17	70	2120	3188	344

In the seasonal trend of population density of *Ceratitis capitata* in the Upper Jordan Valley April to June is the peak of the infestation, which drops nearly to zero during July/September. Gruenberg's conclusion that crops other than citrus are more or less safe from fruitfly attack during July/mid-September and are therefore not objectionable, is not shared by the author. Even if unfavourable climatic conditions (espec. low air humidity) are extremely unfavourable to the fruitfly, it is certain that any additional host increases the population attacking citrus in September. In addition the air humidity is sometimes quite high in summer and in such years considerable damage has to be expected. Until our knowledge on *Ceratitis* populations and their fluctuations is much further advanced, the rigid interruption of seasonal host sequence during a period as prolonged as possible seems to be the best cultural policy in areas subjected to fruit fly attack during any season of the year.

There is no need to discuss in detail the damage done to other fruit crops in Palestine, this having been done in the chapter on host-relations. The restrictions, enforced by *Ceratitis* are still far more serious to out horticultural development, than the actual damage done.

There are heavy fluctuations in the abundance of *C. capitata* and correspondingly, in the damage caused by it. We tried to analyze the physical and host factors which are connected with these fluctuations. However, all year round statistical observations, made by the same method, are needed in many places, in many seasons and under different host conditions before any final conclusions can be drawn.

6. CONTROL.

A. NATURAL ENEMIES AND BIOLOGICAL CONTROL

Only few parasites have been bred in Palestine so far. The first one obtained repeatedly, from the pupae of *C. capitata* is, a not yet fully determined species fo *Trichopria* (*Planopria*).

This Proctotrupid is related to *T. capensis*, which has been bred by Silvestri from pupae of *C. capitata* in Cape Colony. In his case of heavy infestation, almost only, of only males emerged from the fly puparia, whereas in our case the percentage of females in slightly infested puparia is quite normal. It is of great interest to study whether and how far population density of the larval stage influences the sex in this species.

Description: Body red-brown, antennae larger distally in the female, cylindrical in the male. 1.5—1.7 mm long. Oviposition occurs in the early pupa oı the fruit fly. As in most Proctotrupids the time of active development is very short. The percentage of infestation is never very high in Palestine.

Another hymenopterous parasite of *Ceratitis*, the small Chalcidid *Pachyneuron vindemniae* Rondani, bred from fruit fly, is preserved in the collection of the Department of Agriculture, Jerusalem. This is a normal and common parasite of *Drosophila*-flies. Its development on other flies (*Lonchaea:* Silvestri, *Ceratitis:* Mimeur in Morocco; Dept. of Agric. in Palestine) is only occasional and exceptional. Both species, just mentioned, give very little promise to develop into parasites of economic importance.

Biological control in other countries: The first attempts at biological control were made by Silvestri, who in 1909 imported a small Chalcidid wasp, *Syntomosphyrum indicum* Silv. into Italy from India. These wasps enter the infested fruit through the oviposition hole or another injured spot of the rind and deposit their eggs in groups in the posterior part of the host maggot. Then they leave the host until about 200 eggs have been laid. Two to three days later the endophagous larvae hatch and after 9—10 days they pupate within the *Ceratitis*-pupa. They pierce the puparium of the host on their way out. In summer, the development is as short as 2—3 weeks only. Its introduction into Italy did not succeed. Insufficient number of hosts throughout the greater part of the year, low winter temperatures and lack of attempt at breeding it permanently in a central laboratory are among the main reasons for this failure.

In July 1925 Silvestri left Italy for Tropical and S. Africa, as well as Australia, for about one year, in search of fruit fly parasites, which were to be introduced into the Hawaiian Islands. Living adults of *Opius humilis*, *Dirhinus giffardii* and *Galesus silvestrii* from South Africa, and *Diachasma tryoni* from Australia were brought to Honolulu. This material served as the starting point of the biological control of *C, capitata* in Hawaii. The perseverance shown in the later stages of this control is not less remarkable than the results of Silverstri's classical exploration trip. Since 1926, four species of parasites, at least, may be regarded as well established. The absolute fluctuations of parasitisms and the relative ones between the various parasites at Hawaii were (14000 to 88000 counts per year):

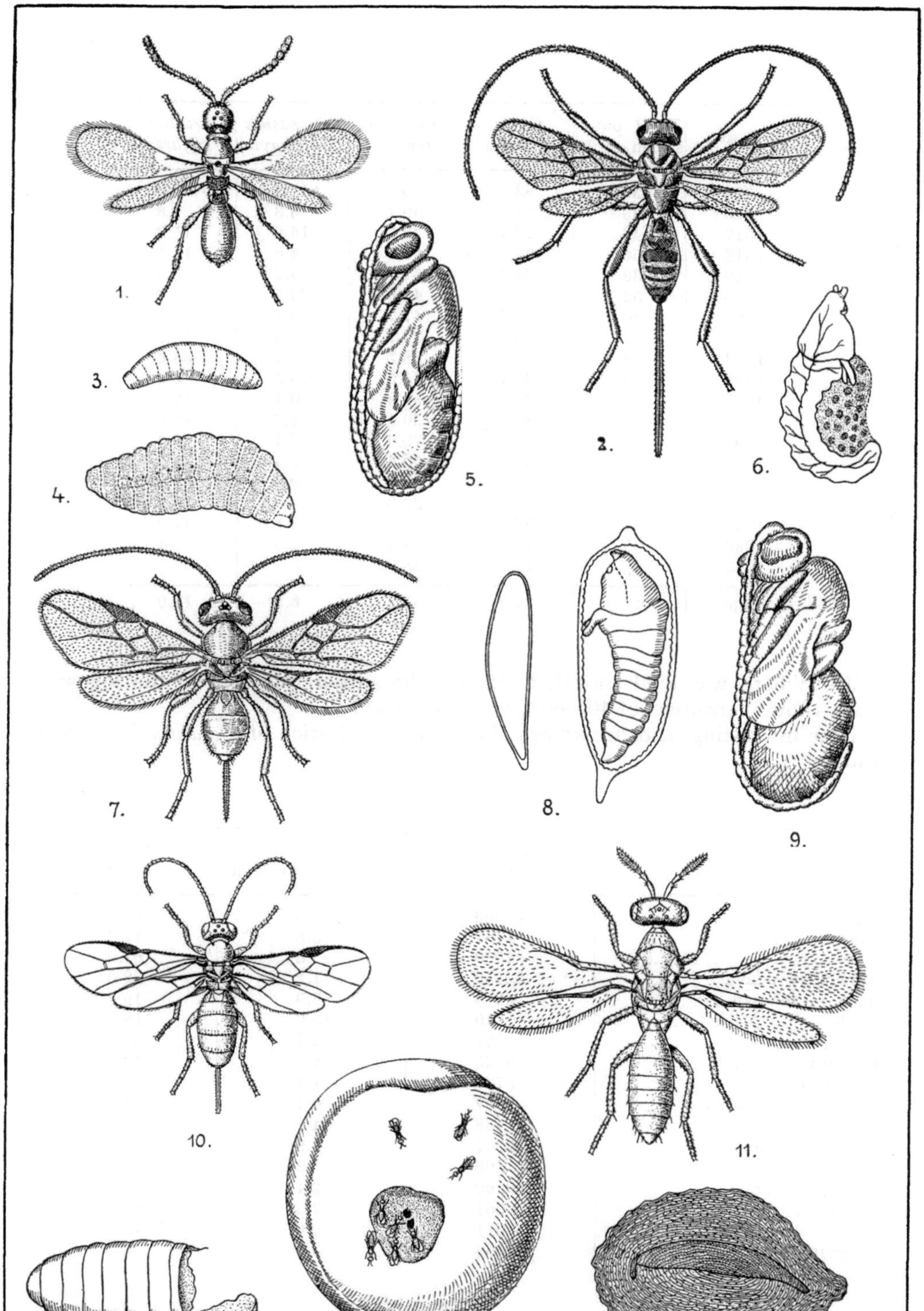

IX. Important parasites of *Ceratitis capitata* Wied. 1) *Trichopria* sp.; 2)-6) *Diachasma tryoni* Cam.: 2) female, 3) third-instar larva, 4) mature larva, 5) pupa, 6) neonate larva, showing right main tracheal trunk with branches and characteristic position and some of egg serosal cells clinging to ventral surface; 7—9) *Opius humilis* Silv. 7) female, 8) egg one and three days old, old, 9) pupa. 10) *Opius concolor* Sz., female; 11) *Syntomosphyrum indicum* Silv., female; 12) peach infested by *Ceratitis* with holes through which females of *S. indicum* infest the maggots of the pest; 13) empty puparium of *Ceratitis*, left by *Diachasma tryoni*; 14) egg of *Diachasma tryoni*, encysted and killed in the pupa of an unsuitable host (Melon fly). The length of the encysted capsule is 0.55 mm (mainly after Pemberton, Back, Willard and Sivestri).

Year	Total parasitism in %	*Opius humilis*	*Diachasma tryoni*	*Diachasma fullawayi*	*Tetrastichus giffardii*
1915	38	83.1	0.8	15.5	0.6
1916	33	51.8	40.1	6.3	1.8
1917	48	26.4	42.7	15.6	15.3
1918	56	22.2	62.0	4.6	11.1
1919	38	24.6	51.3	4.2	19.9
1920	52	18.1	43.6	23.6	14.7
1921	56	13.5	47.7	9.7	29.1
1922	47	9.7	71.5	5.3	13.5
1923	43	9.5	28.2	3.7	58.6
1924	45	32.1	45.9	15.1	6.9
1925	46	31.8	35.3	0.4	12.5
1926	38	26.8	50.1	4.2	18.9
1927	46	17.9	70.0	2.4	9.6
1928	31	36.9	50.6	2.5	9.9
1929	40	31.9	59.4	0.8	7.9
1930	25	54.2	26.9	—	18.9
1931	35	56.6	29.1	3.1	11.2
1932	33	40.4	47.3	3.9	8.4
1933	58	45.0	48.0	—	7.0
Average	42	31.0	47.3	6.8	14.9

Mr. Willard was kind enough to inform the author that also in the following years a total parasitism of 40—60% was maintained.

Some interesting facts are known about the parasitation of *Ceratitis* in various fruits.

Host	No. of fruits collected	Average no. of *Ceratitis* per fruit	Total % of parasitism
Indian almond	34066	5.5	56.0
Mango	1787	3.5	12.3
Coffee	4080	.6	67.0
Strawberry guava	22133	1.2	62.3
Black myrobalan	3373	6.9	19.3
Peach	10	15.6	18.7
Satin-leaf.	801	4.7	53.3
Rose-apple	4225	11.5	48.3
French cherry.	5725	1.4	57.2
West Indian medler.	2287	0.4	4.0
Kamani	438	4.9	0.3
Yellow oleander.	2367	1.9	60.1
Carambola	106	.2	—
Chinese orange	40260	2.4	22.6
Guava	4051	7.2	46.5
Orange.	611	6.0	15.4
Tangerine.	869	0.9	—

The parasitism is thus much higher in smaller fruits than in the bigger ones, which are not as easily accessible to most parasites. In most of the small fruits the economic damage has been greatly reduced.

Those biologists who like to refer every regulation in abundance to parasites may

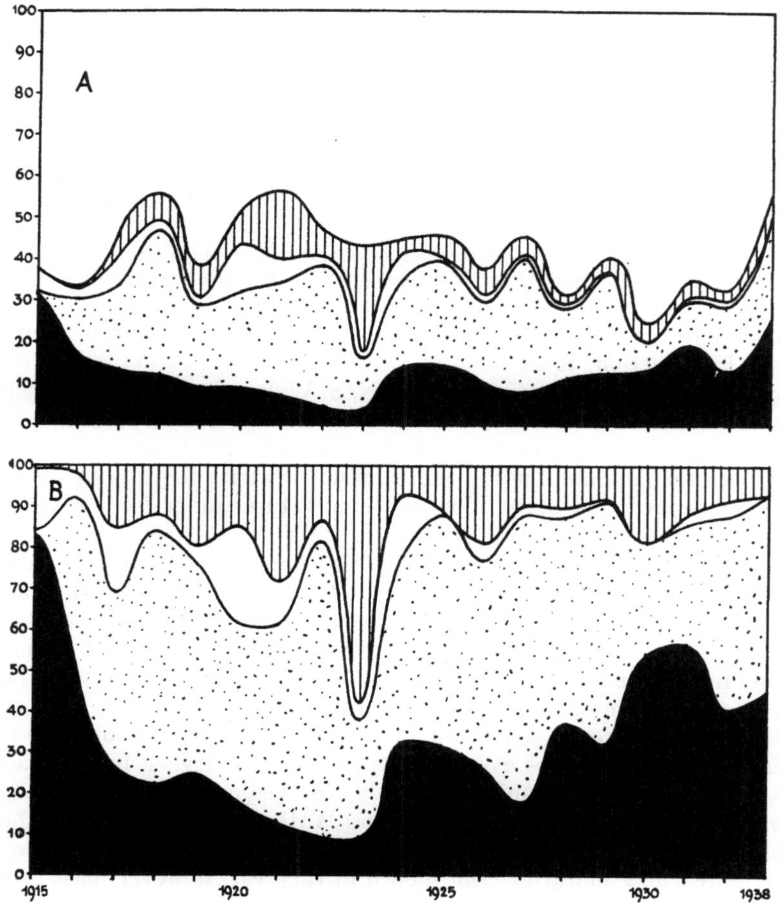

Fig. 34.
Fluctuations of the populations of the four parasites of Ceratitis capitata in Hawaii from 1915 to 1934. A. Absolute percentage of parasitation. B. Ratio between the parasites (all parasites = 100). 1. Opius humilis, (black) 2. Diachasma tryoni, 3. D. fullawayi, 4. Tetrastichus giffardii. (striated).

be interested in the comparison between the seasonal population trend of *Ceratitis* and its parasites in Honolulu:

month	I	II	III	IV	V	VI	VII	VIII	IX	X	XI	XII
Ceratitis (% of absolute population)	2	2	2	3	8	17	32	16	4	4	5	5
Parasites (% of absolute population)	1	2	2	2	7	19	31	16	2	6	7	5

The complete parallelism of both series arouses suspicion if the parasites are the primary regulating factor. Parasites have their real importance in nature, but not the exaggerated one often attributed to them.

Description: 1. *Opius humilis Silv.* The *female* is ochraceous, wings hyaline with brownish veins, legs brown. Head hardly more than ⅓ wider than long, hairy. Antennae 35-jointed, a little longer than the body. Ovipositor almost straight, a bit shorter than the abdomen. 2.6 mm long, ovipositor 1.2 mm.

Development stages illustrated on plate IX. Egg 0.45 mm long when laid; 0.8 mm long 2—3 days later. Larva 12-segmented, 4 stages. Pupa 2.3 mm long; may be easily distinguished from that of *Diachasma* by its short ovipositor sheath.

2. *Diachasma tryoni* (*Cam.*) The *female* testaceous-ferruginous with black brown abdomen. Wings slightly infuscated, with the stigma and veins brown. Head a bit broader than the thorax, about ¼ wider than long. Antennae 45-jointed, longer than the body. Ovipositor about as long as the body. 3.5—4.5 mm long, ovipositor 4.5 mm.

Development stages illustrated on plate IX. Egg 0.45 mm long when laid, 0.65 mm when mature. Larva 14-segmented, 4 stages, up to 3 mm long. Pupa about 4 mm long, striking by its long ovipositor sheath.

3. *Diachasma fullawayi Silv.*The *female* has yellow ochraceous wings, hyaline with fulveous veins and stigma. Antennae 30-jointed. Ovipositor somewhat shorter than the abdomen. 3.5 mm long, ovipositor 1 mm long.

The ecology of the two most important of these parasites, from a point of view of parasite introduction, is as follows:

Diachasma tryoni (Cam.) originates from N. S. Wales, the climate of which resembles greatly that of the Mediterranean region. On the coast, at Sidney e.g., the monthly average temperatures fluctuate between 11—22° C and rainfall is about 1200 mm, with the maximum in the winter, from February to July. In the lowlands, towards the interior of the state, monthly temperatures range from 7—26° C, total annual precipitation from 250—600 mm.

The female deposits 5—25 eggs daily, one at a time, up to 50—150 eggs altogether. Most of the eggs are deposited during the first week. The last larval instar of the host is preferred, but oviposition has also been observed in second instar larvae. At about 25° C its adult longevity is on an average about 15 days, about 25 days being the maximum. In the laboratory, they live as long as 56—70 days, without ovipositing. Unmated females produce males only.

D. tryoni has the longest ovipositor of all Braconid parasites of fruit flies and may therefore reach maggots even in deeper layers of bigger fruits, inaccessible to *Opius*. A diapause which is quite frequently observed in the full grown maggots of *Diachasma* is of importance. Pemberton and Willard (1918) observed the length of the last larval stage to be from 6 days to over a year. "A certain proportion of the larvae reaching maturity each month of the year, however, pass into a dormant stage and may remain in this condition for from a few weeks to several months, or occasionally a year." This percentage ranged from 0.3% in the summer to 52% in the winter. At 24.5° C in the summer about 1% of the larvae entered diapause, whereas another portion when placed in a refrigerator at 17° C showed a 68% diapause. Cool temperatures are certainly an important factor in inducing diapause. But drought, food shortage and other unfavorable environmental conditions have the same effect. Diapause larvae may survive even prolonged periods of unfavorable conditions. It is therefore most probable that *D. tryoni* would develop successfully in our coastal plain.

Opius humilis Silv. originates from Constantia (Cape Colony) where the climate also resembles greatly that of the Mediterranean region. At Capetown temperature ranges from 12—21° C and rains amount to 838 mm, with the maximum in winter (May to August). 100—250 eggs are laid by the female one at a time, 5—30 eggs daily. The adult longevity is 3—4 weeks, the main oviposition generally terminating after two weeks. However, longevity up to 100 days and more has been observed in females. All parthenogenetic eggs develop into males. The ovipositor is rather short and restricts the efficiency of the species considerably. *O. humilis* is known to be somewhat more resistant to cold than its host, so that it should be able to maintain itself in any cold region where *Ceratitis* is able to persist. No diapause is known from *Opius*.

The following constants, based on the data published by Pemberton and Willard (1918), have been calculated:

	Opius A	*Opius* B	*Diachasma* A	*Diachasma* B
development threshold (c)	11.9	6.9	11.7	9.0°C
thermal constant (th. c.)	191	405	238	366 days-degrees

The development (including the first half of the oviposition period for the two parasites) lasts as follows (hyperbola B):

	Ceratitis	*Opius*	*Diachasma*
35° C	33.0	14.4	14.1 days
30	42.5	17.5	17.4
25	59.1	22.3	22.9
20	98.0	30.9	33.3
15	286.5	50.0	61.0

The comparative number of annual generations at different localities (hyperbola B) would therefore be at:

	Ceratitis	*Opius*	*Diachasma*
Honolulu	5.4	**14.7**	**14.4**
Capetown	1.9	**8.6**	7.7
Sidney	2.4	9.1	**8.1**
Paris	0.7	3.9	3.1
Palermo.	2.5	9.2	8.2
Alexandria	3.5	11.3	10.5
Valencia.	2.4	8.9	7.7
Los Palmas	3.6	11.5	10.8

Considering all these indications the prognosis for a possible introduction must be regarded as rather favorable. Probably both of them are able to exist in any climate in which *Ceratitis* lives. It is, however, probable that both species have high humidity requirements. Their behaviour and mortality at relative humidities below 70 and 60% respectively should be studied. Further, taking into consideration the climate of their original home-places, the chances of a good prognosis are especially high in the coastal districts of the warmer parts of the Mediterranean region. Fairly high air humidity, mild winters and not too hot summers are combined in Palestine.

Opius humilis and *Diachasma tryoni* act antagonistically in Hawaii, as the maggots of *D. tryoni* devour all other maggots, including those of their own species, wherever they meet them. Pemberton and Willard (1918) maintain that *Opius* is more fertile, has a shorter development, a longer adult longevity and is not inhibited in winter development by diapause. *Opius* alone is superior to *Diachasma* and is restrained by the cannibalistic habits of the latter's larvae. ,,Knowing the capacity of *O. humilis* for parasitizing from 80 to 90 per cent of the larvae of the fruitfly in favorable localities, such as the large Kona coffee-belt, the writers here maintain that detrimental resultst to a certain extent have arisen from the liberation in Hawaii of parasites other than *O. humilis* that attack the larva in the fruitfly. The total parasitism has simply been reduced in value to that of a parasite of secondary value."

This statement has a relative value only. Even if *O. humilis* is slightly superior to *Diachasma* in some respects, it has also its drawbacks. In Palestine, where small fruits are only very secondary hosts of *Ceratitis*, the longer ovipositor of *D. tryoni* is of great importance. *Opius* may prove unsatisfactory under our host conditions. The easiness with which *D. tryoni* can be induced to enter diapause, is also a decided advantage helping the species to survive unfavorable seasons. Therefore, if it as a question of selecting only one species for the Mediterranean regions, preference should be given to *Diachasma tryoni* which has a greater chance of successful introduction.

The authors just mentioned are certainly correct in their statement, that if two parasites compete on the same host and one has a parasitism capacity of 60%, the second of 40%, there is no guarantee that their combined efficiency will surpass 60%. But even if no higher percentage of parasitism is gained, a double guarantee for that 60% is given. If both parasites show a different range of climatic tolerance, one of the parasites will in all probability remain active. A thorough knowledge of the physical ecology of these parasites will enable us to choose with greater wisdom and safety the beneficial insects to be introduced into a foreign country.

The earlier experiences of Hawaii up to date were recently published by Willard and Mason. Whereas *Opius humilis* was originally the most successful of the introduced parasites, *D. tryoni* lately controlled *Opius humilis*. Also in the monthly fluctuation, where *Opius humilis* reaches its peak in April, it is from May on superseeded by *D. tryoni*. Whenever both larvae meet within one host-maggot, *D. tryoni* eliminates *O. humilis*. The high degree of multiple parasitation of fruit fly maggots in coffee-berries in the Kona district e.g. has eliminated *O. humilis* as a control factor, its place being taken by the other species.

Texture of skin and thickness of pulp of various host fruits seems to be a governing factor in the degree of *Opius*-parasitation, by rendering the host maggot less accessible to the parasites in certain fruits.

	% of total parasitation										Av. No. of Ceratitis per fruit	
	1924	1925	1926	1927	1928	1929	1930	1931	1932	1933	1916/24	1925/33
Coffea arabica	64	—	64	58	22	—	—	40	48	82	0.4	0.1
Terminalia catalpa	52	54	46	53	44	44	29	42	43	69	7.9	3.7
Psidium cattleyanum	44	28	18	40	32	32	53	0	—	64	1.5	0.9
Citrus sp. (Chinese Orange)	13	10	8	9	5	10	7	4	7	11	2.4	1.2
Mangifera indica	17	17	20	13	20	19	27	25	14	6	5.9	6.5
Psidium guajava	12	15	23	15	12	19	9	8	4	5	8.2	4.9
Mimusops elengi	—	1	6	0	1	—	3	0	2	1	3.7	3.1
Thevetia neriifolia	45	16	13	7	9	16	25	24	19	8	4.0	2.5
Citrus aurantium	23	12	6	10	—	14	0	24	31	10	3.0	3.0
											37.0	25.9

Coffee: large seed, skin easily punctured, and pulp very thin so that any maggot can be reached by the *Opius* larvae. Parasitation always high, and locally occasionally up to 95—100% parasitation. (Similar *Terminalia catalpa*). *Psidium cattleyanum* and *Chinese Orange:* small fruits of the same size, but *Psidium cattleyanum* has very thin skin and soft pulp, Chinese Orange skin thicker and more difficult to puncture. Consequently parasitism much higher in former species.

Mimusops: fruit also small, but hard cover difficult to puncture by a few parasites. Parasitism low, despite its high infestation by fruitfly.

Mango and Guava of equal size, but Guava pulpy throughout and Mango contains a large seed covered with a thick pulp. Many *Ceratitis* maggots feed in the Guava centre and are out of reach for the parasites, while the large seed of Mango forces the maggots to feed nearer the surface, where they are subjected to parasitism. Higher percentage of parasitism in Mango, although the though texture, etc. of the skin in some Mango varieties probably decrease the rate somewhat.

In Orange and *Thevetia* fruits the per cent of *Tetrastichus giffardi* is more important than that of *Opius*, contrary to the general experience. *Opiinae* larvae are undoubtedly eliminated to a large part from those fruits by the larvae of *T. giffardi*.

The decrease of average infestation per fruit in the last 10 years, although due largely to parasitic control, was to some extent, at least, caused by other environmental factors affecting the fly.

Balachowski et Mesnil (1935) recently report that *Opius concolor* Silv., a well known parasite of the Olive fly (*Dacus oleae*), has been bred in the region of Sous in Morocco from *C. capitata* too, but in very small numbers only. It is quite possible that also other olive-fly parasites will attack occasionally the Mediterranean fruitfly. However, one should record the experience of Pemberton and Willard (1918) who found that the eggs of *Ceratitis* parasites, when deposited in the maggots of the Melonfly generally become encysted and die, only a small percentage succeeding in developing normally (pl. IX). *Opius concolor* is regularly present in the coast plain and in the Mediterranean hills of Palestine, and is a not very efficient parasite of Olive flies. Counts from three different seasons at Petah Tikva yielded the following results:

	Dacus	*Opius concolor*						Total	Total parasites
		VI	VII	VIII	IX	X	XI		
in ciphers	348	—	3	18	6	9	—	36	168
in percentages	69	—	—	—	—	—	—	7	31%

Other *Ceratitis* parasites in Morocco, recently mentioned by de Lepiney et Mimeur (1932) include *Pachyneuron vindemmiae* Rond. This is a common parasite of *Drosophila* maggots and its hatching from *Ceratitis* pupae is rather exceptional. The predacious maggots of the Anthomyid fly, *Phaonia trimaculata* Bché, were also occasionally met in infested peaches. Whenever parasite introduction is planned in Mediterranean countries, *Opius concolor*, *Trichopria* sp. and the other known endemic parasites of Trypetid flies should be studied too.

Mr. Willard, Federal Entomologist at Honolulu, was kind enough to send us a large shipment of parasitized pupae of *C. capitata* in the winter 1925/26. These were transported in the cool-room of a mailsteamer from Honolulu to Alexandria and from there by ordinary mail to Tel Aviv. The shipment was about 44 days on the way. 17 specimens of *Opius humilis* were hatched during the transport. They arrived dead and no further parasites hatched.

We did not receive any further shipments of parasites. But very interesting results have been obtained in Spain in the meantime. The following parasites reached Valencia from Honolulu (via San Francisco—New York) living in June, 1933:

	♂	♀
Diachasma tryoni	46	102
Opius humilis	4	5

Of this stock, the following developed in the laboratory, until September 1933;

	♂	♀
Diachasma tryoni	63	16
Opius humilis	167	6

The extremely low percentage of females is of great interest and, so far, unexplained. No further reproduction took place and the attempt ended unsuccessfully. Clemente (1934) proposes to choose Los Palmas as the place of a second experiment which by its higher and more moderate annual temperature seems to be more fit as a first acclimatization station for these parasites.

The continuation of the Spanish experiment and its further results should be carefully followed. The difficulties involved in shipment have been overcome. Present day's aviation reduces them, especially in winter, to a bagatelle.

After careful consideration we came to the conclusion that the introduction and establishment of at least two of the most successful fruit fly parasites of Hawaii into the countries of the Southern Mediterranean region has a favorable prognosis. This establishment does not yet include any forecast as to their control-value against *C. capitata* here. The high success of these parasites is established in the lowlands of Hawaii, where a large number of highly suitable hosts succeed one another throughout the year. Conditions are very different at the medium and higher elevations. Many suitable hosts are inaccessible to the flies, because they grow in the interior of the dark and shady woods. And in the sunny woods of wild guava, suitable hosts are present only during a season of 3—4 months. The importance of parasitism is here greatly reduced. This may be the fate of fruitfly parasites in Palestine. In the rich and beautiful ornamental parks of Alexandria and Cairo, or in horticultural centres, such as Mikweh Israel some permanent and recognisable parasitism may occur. This means, wherever fruitfly damage is normally low because of the unfavorable long interrupted host sequence no remarkable reduction will occur. Wherever it is high because of favorable host sequence and climatic conditions, a certain percentage, probably even a high one, of the flies will be parasitized, but both *Ceratitis* as well as its parasites may maintain themselves on a high population level.

Whereas the practical success of parasite-introduction should not be overestimated, it is highly advisable not to neglect any opportunity of studying and promoting such a possibility.

Fruit-fly parasites have been introduced about 1937 from Hawaii into Egypt. If these parasites should be established in Egypt, every attempt should be made to introduce them into Palestine too.

B. TECHNICAL CONTROL.

1. *Clean cultivation:* All conclusions to be drawn from the host relations are included in this chapter.

"The records show that Hawaii is able to grow an abundant crop of fruits for home use, and in some cases for export, with little loss caused by the fly. The growing of more resistant varieties of some of the favored hosts, elimination of nonessential host fruits, and a program of control through orchard and dooryard sanitation, together with the use of poisoned baits, would probably result in largely removing the insects from the lowlands of Hawaii."

This Hawaiian program of Mason (1931) includes all measures to be applied in Palestine too. Whereas it seems to be practically impossible to interfere with the growth of summer fruit trees in the citrus belt, without the help of a law which would be decidedly unpopular, the following measures should be rigorously enforced in this district: No subtropical summer fruit liable to be seriously attacked by *C. capitata* should be grown, neither as plantation nor as single trees, before sufficient experimental data (Agric. Exper. Sta. Rehoboth or Government Exper. Gardens) with regard to more or less fruitfly resistant varieties or to the safe handling of these fruits have been gained. (See later).

All susceptible hosts among hedge plants, wild or ornamental shrubs (*Aberia caffra, Opuntia ficus indica, Lycium europaeum,* etc.) should be rigorously suppressed. In many cases it will be within the jurisdiction of the local Village Councils to enforce such measures.

All those summer fruits which are even slightly exposed to fruit fly infestation should be early picked and packed. Regular visits to the orchard must make sure that no late fruit matures on the tree.

Peaches should not be planted at all within the citrus belt. Intensive propaganda should be maintained against the plantation of single summer fruit-trees liable to fruitfly attack within the orchard or in the private garden. The benefit and pleasure gained from these trees does not counterbalance the damage they do by serving as an additional host in the season in which the maintenance is most difficult for the *Ceratitis* population.

Strict adherence to these rules, which are possible and reasonable under the present conditions of Palestine horticulture, is an important step in support of any other control measure.

2. Infested fruit, especially after it has fallen off, should be regularly eliminated from the orchard. However, this seems easier than it really is. Difficult as it is to educate the farmer to do such regular collecting of the infested fruit, the question as to how to dispose of these fruits is still more difficult. The old recommendation of deep burial is generally abandoned. Whereas the cover of soil should be at least half a meter thick and then covered by a layer of lime, many adults will always hatch because of careless execution of the work or through cracks in the soil. Burning of the collected fruit must be thorough. Superficial burning of the upper layers of the heap only is of no use.

Desirable as the permanent elimination of the infested fruit may be, the difficulties involved in its effective disposal are too many to make a permanent and reliable execution possible. This disposal must be executed daily in order to fulfil its aim, as otherwise many or all maggots leave the fruit for pupation in the soil and remain on the spot, even after the fruits are removed. Furthermore, these procedures have to be performed in every orchard, also in the house gardens. Daily burning of the collected fruit may be practical in small orchards.

For all these reasons, it is practically impossible to attain a sufficient fruit fly control in this way. The removal of fallen, infested fruit should be recommended as valuable supplementary remedy, but cannot be relied on as a primary or only control method.

3. *Chemo-taxic control by baits:* The only way of intensive control at our disposal to-day is baiting. The flies are lured into a poisonous liquid sprayed on the tree, or into a trap, in which they are caught. This control is based on the chemo-taxic reactions of the fly. In view of the importance of this question, it is regrettable that only empirical experiments have thitherto been made, but no systematic study of the chemotaxis of the fly. Also differential reactions with regard to age and sex should be studied. At present, we may only state that sugary liquids, liquids with substances in fermentation, kerosene and certain industrial brands, as Clensel, are attractive. Of these, kerosene ex. gen. is useless for control purposes since practically only males are attracted by it.

The main qualities to be demanded from an efficient fruitfly bait are (Ripley and Hepburn, 1929):

1) High odour intensity, surpassing the odour attraction of the fruit, luring the flies in their search for food or for oviposition hosts.

2) The evaporation of the attractive substance must be slow, in order that the bait should remain attractive as long as possible (3—4 weeks preferably).

3) The efficiency of the attractive substance should not be changed substantially by the evaporation or addition of water.

4) All substances used in the composition of the bait should be inexpensive and easily obtainable.

Spraying was the first application of chemotaxic control. Thanks to the efforts of Mally, especially, the sprays have been developed, in South Africa first, since 1904. The following are some of the most recommended sprays:

1. Mally's improved spray:	Sugar	6 klo
	Lead arsenate (paste)	500 g
	Water	100 l.
2. Newman's spray:	Molasse	10 klo
	Lead Arsenate (paste)	1 klo
	Sap of oranges (or other fruits)	70—90 fruits
	Water	100 l.
3. Recommendation of		
Palestine Governement:	Sugar	16 klo
(superseeded now)	Lead arsenate (powder)	150 g
	Water	100 l.

Later recommendations of the Palestine Government (1939) are four formulas, all with 10 liter water and 1 kilo sugar, and in addition, either 40 g of lead arsenate, 30 g of sodium fluosilicate, 60 g of copper carbonate or 80 g of copper carbonate. K. Schweig (1944) recommends for deciduous fruit trees two cover sprays with lead arsenate (0.6%) six and four weeks before the fruit reaches maturity.

Into our own experimental series Dr. H. Z. Klein introduced tartar emetic. The formulas compared were:

a) Tartar emetic 45 g, sugar 225 g, water 10 l.
b) Copper carbonate 33 g, sugar 200 g, water 9 l (at Hedera).
Copper carbonate 40 g, sugar 400 g, water 9 l (at Dagania).
c) Sodium fluosilicate 11 g, sugar 420 g, water 7½ l.
d) untreated control parcel.

Every tree was sprayed with 200 ccm. once per week. The beginning of the control was directed by the initial increase of the clensel bait traps (late February in Dagania, early March in Hedera. The results in 1941 were (% infested fruit):

Formula	Hedera			Dagania	
	Grove A	Grove B		Grove A	Grove B
Host	Shamuti	Shamuti	Valencia	Grapefruit	
Picking date		30.VI.	30.VI.	30.IV.	30.IV.
Bait a. . .	40	23	13	20	20
b. . .	60	48	30	53	74
c. . .	92	71	60	42	37
Control d .	100	100	100	100	100

Continued experiments with tartar emetic proved its superiority over other baits. Weekly sprayings (6 in 1941, 7 in 1942, 4 in 1943) according to the appearance of the flies in the clensel traps and to the picking date in a grove of Kelsey plums at Mikweh Israel (in 1941 the entire grove was sprayed, hence no control plot):

Year	1941	1942		1943	
	sprayed	control	sprayed	control	sprayed
no. of fruits.	1680	6516	7258	6174	5130
% *Ceratitis* infestation	5.5	26.2	14.3	32.2	14.6

The bait with tartar emetic was for some years our preference recommendation.

Small scale experiments with DDT (of two different proveniences) in 1945 on a garden of Kelsey plums at Mikweh Israel gave rather promising results (Klein). The following four series were arranged:

No.	Treatment	quantity per tree	% infested fruit	Total crop (klo)
1.	DDT: a Cover-spray 5%.	9 l	32	260
2.	DDT: b dto.	9 l	15	330
3.	DDT: a Bait-spray 5% (with sugar). . .	¼ l	47	200
4.	DDT: a Dusting 2%	150 g	60	250
5.	Control plot	—	86	180

The control was performed twice within the last 14 days before picking. In the two series with full cover sprays the damage was reduced to 16—21% of that of the control plots. Dusting and baiting was less successful in these preliminary experiments.

Spraying should always be applied by sprinkling lightly the sunny side of the trees and should be renewed at intervals of 7 to 10 days.

The disadvantage of spraying lies in the frequent development of heavy sooty mould on the sugary residues on the leaves. The sugar content of the spray should therefore be as low as possible. Also, the content of lead arsenate should be as low as possible in order to reduce the danger of burning the leaves. There is no doubt that spraying can be still enormously improved and attempts should be made in centrally situated laboratories ro obtain a still better spray-formula. Fruit-fly infestation of citrus was rather low in the last years. We give therefore the results of these years' control experience against fruitfly infestation of Kelsey-plums and of guavas:

Fruit		Kelsey plum		Guava	
Year	Remedy	Treatment	% clean fruit	Treatment	% clean fruit
1947	DDT 5%	D : 4 × 7	95	D : 7 × 7	79
	DDT 5%	D: 3 × 10	80		
	Untreated control	—	79	—	7
1948	DDT (10%) 2% + spreader	S : 5 × 7	78	S : 7 × 7	74
	DDT 10%	S : 5 × 7	92	D : 7 × 7	95
	Untreated control	—	87	—	7

S = spraying; D = dusting; 4 × 7 = 4 treatments at 7 days interval.

Our present recommendation is therefore: Dustings with DDT 10% at intervals of seven days.

The Severins at Honolulu (1915) have shown conclusively how favorable results can be obtained by spraying. The average daily catch of 10 kerosene traps was as follows:

Before spraying		292 Ceratitis
After the first spraying	15.IX—13.XII	16
,, ,, second ,, ...		3
,, ,, third ,, ...		1
,, ,, forth ,, ...		4
,, ,, fifth ,, ...		6
10 days after last spray		6
15 ,, ,, ,, ,, ...		12
20 ,, ,, ,, ,, ...		21
25 ,, ,, ,, ,, ...		18
30 ,, ,, ,, ,, ...		23
35 ,, ,, ,, ,, ...		24
40 ,, ,, ,, ,, ...		34

Comparative experiments on the efficiency of spraying and baiting yielded the following results with regard to the percentages of infested fruits in the orchard (A when applied during the preoviposition period; B when applied later; Newman):

Spraying		Trapping		Control
A	B	A	B	
12%	30%	5%	18%	89%

These experiments prove sufficiently the possibility of obtaining a large reduction both af the fruit fly population and of the damage. The ample use of unspecific poisons in the spray is a decided disadvantage. Beneficial insects as hymenopterous parasites, Tachinid flies and honeybees are also attracted and poisoned. The same is true of most trapping baits, with the exeption of Clensel which seems to be specifically attractive to *Ceratitis*.

In earlier times *trapping* has been done in tins. Now they are generally replaced by glass-jars. The most convenient form has been described by Ballard (1933): "A glass bottle about 12 cms high and 26 cms maximum circumference, with the base blown into the form of a truncated cone, the apex of the cone being open to permit the entry of flies into the bottle. The mouth of the bottle if closed with a cork or stopper and round the neck is a wire and hook for hanging it in a tree. The bait, which is a liquid, is poured into the bottle to about half the height of the conical based and is renewed weekly at the time when the trapped flies are removed."

Preliminary experiments (1926) with formulas

450 g wheat bran	or	250 g molasse
30 g molasse		extract of 4 mature oranges
60 g borax		2 l water
2 l water		

yielded entirely negative results. Recent results obtained with other substances are compiled in the following series.

Bait		Ceratitis per week and bottle	% of ♀	author
1.	Kerosene.	204.4	0.3	Severin 1915
2.	Clensel.	15.6	70	Newman 1931
	Pollad and borax . . .	3.9	74	
3.	Vine vinegar 20% . .	2.1	72	Ballard 1933
	30% . .	2.2	68	
	Country vinegar 20% . .	0.4		
	30% . .	0.7		
	40% . .	0.7		
	Molasses 10% . .	12.0	57	
4.	Bran and borax	54		Cyprus Dept. of Agr.
	Clensel 1 : 20.	100		
	1 : 30.	67		
	1 : 40.	77		

Among the substancess tested, fermenting substances, kerosene and clensel seem to be specially attractive to the fly. From the many experiments, we compile those of Constantino (1930) and of the Severins (1914): The latter measured the attractive powers of different substances as compared with kerosene (= 100), unfortunately for the males only.

Aromatic and ethereal oils, like these of oranges, lemons, eucalyptus, have no, or a very low, attractive power. Syrups are not attractive, either. But those of alkermes, kimmel, maraschino and *Ribes* (a 10% dilution in water) had a slight attractive power.

Cocos oil and fish or whale oil have no attraction. Mineral oils show a high range of variation from O-103: naphtha, benzene and kerosene ranging high, heavy oils and highly refined ones low.

Fermenting liquids, as diluted molasses, vinegar, bran in fermentation, etc., have the greatest attraction.

The aim of successful trapping is to catch the females during the preoviposition period which lasts from 5 days in summer to over 2 weeks in winter and spring under our conditions. Granted, the various baits have a very different degree of attraction for both sexes, it is still of further importance to investigate as to what percentage of the females is attracted before, during, or only at the end of oviposition. Ballard (1933) reports from his experiments in Egypt that most of the females had not laid their eggs or were unfertilized.

Good vinegar, diluted 1 : 3 with water, fermenting bran, 50—75 g per liter of water, are mixed together. This mixture, after having stood for 1—2 days, is now widely being used in Mediterranean countries. Fermenting bran seems to be superior to any other of these substances, except clensel. Clensel should be used in a 3—5% concentration.

The number of bottles indicated as necessary in order to receive effective control is generally 1 bottle for every orange tree (2 in the exterior rows) and at least 2 bottles for every peach tree. The bottles must be refilled every 5—6 days, in order to maintain their attraction. In summer they should be hung in half-shade, but always in the exterior parts of the tree, in winter in the sunniest part.

Experiments with all these baiting bottles have been performed in Palestine. None of the studied substances are able to prevent heavy infestation of the fruit, when population pressure is high. In September 1934 the author visited the Government garden at Paphos (Cyprus). Despite the fact that at least 2 bottles with clensel were hanging on each tree and that hundreds of *Ceratitis* flies were caugth in every bottle, flies were seen to oviposit quite abundantly on every fruit, which were literally covered with punctures. This indicates that in the presence of a high population pressure of *Ceratitis* none of the bait-traps in use are able to prevent serious damage. An additional and very striking proof therefore, we find in the results obtained by Mr. Yedidyah (1936), which we have described before. Better results should be expected in the case of medium and light infestation. The results of a one year campaign are never conclusive, because of the marked fluctuations in the abundance of the fly from year to year. Furthermore, only experiments on a sufficiently large scale (50—100 dunams minimum) will allow economical conclusions. Such experiments have not yet been performed in Palestine, neither has the season for baiting been determined. Traps should be available from two weeks before the normal beginning of infestation until 4—6 weeks after the removal of the crop; in orange groves on the coast from the beginning of September to early May, in the fruit orchards of the hills from April to September. But on the whole is seems advisable to maintain some traps throughout the year as indicators.

Gruenberg, who has experimented for some years in the Upper Jordan Valley, comes to the following practical recommendations for this district:

Fruitfly control should be confined to the seasonal peaks of infestation (late September/early December and March). Trapping bottles within the orchard should be used as indicators of fruitfly abundance and beginning of control measures. Vine Vinegar 25% proved to be the most suitable baiting fluid, Clensel being less sensitive as indicator. Besides destruction of infested fruit and early picking, spraying and trapping is recommended, the first being the cheaper and a thoroughly satisfactory method.

Expenses for treatment (pre-war) of one dunam (40 trees) of citrus grove by spraying and trapping methods.

Method	hours of work	quantity of material	approximative no. of treatments per season	total expenses per season	remarks
Trapping bottles	2	½ l Clensel	14	Clensel 700 mils 3½ days labour	½ gallon Clensel 200 mils
Spraying	1	65 g Copper Carbonate	12	Copper Carbo- 200 mils Sug. or Mol. **200 mils** 1½ days lab. 400 mils	780 gr. of copper carbon. 100 mils per lb. molasse

The formula for the spray-fluid is:

Copper carbonate	65 grams	per dunam
Molasses	400 ,,	
Water	7.600 ,,	

No sooty mould development has been observed in the Upper Jordan Valley in connection with spraying, so far. It is possible that copper-carbonate deserves to be preferred to the somewhat cheaper sodium fluosilicate, as it is a fungicide and for this reason possibly counteracts sooty mould development.

4. *Various control methods:* Control by low temperature has often been discussed,

especially in the American literature. Back and Pemberton state that no stage of *Ceratitis* is able to survive refrigeration:

4.5—7.2° C	for 7 weeks
0.5—4.5	for 3 weeks
0.0—0.5	for 2 weeks.

However, it should be born in mind that cold storage serves as control only as long as superficial punctures are made. As soon as the larvae have entered the pulp of oranges, or penetrated their rind, these oranges are liable to be attacked by various moulds during cold storage.

Adult flies lived in cold-storage rooms in Palestine at 5—7° C from 4 to 11 days. At decreasing temperatures from 9 to 2° C the last flies died after 19 days. In another experiment, the last flies were dead after 7 days under temperatures fluctuating between 2 to 11° C (Klein and Paker 1943).

No stage survives submergence in water for longer than 7 days; however, no practical use can be made thereof.

Covering the fruits of special commercial value may help. Dates and valuable peaches may be wrapped in paper bags during the early stage of maturation. This procedure gives full protection to the fruits.

5. *Strict quarantine measures* can be employed to the benefit of the countries hitherto uninfested. The American quarantine service is a model of succesful and efficient prevention of such introductions.

In conclusion we may summarize:

A wise choice of fruits, not permitting an uninterrupted hostsequence within the same district, is of paramount prophylactic importance. Neither should the planting of single trees with fruit ripening at another season than the main crops be permitted, nor that of any hedgeplants, which are suitable fruit fly hosts. We admit however, that the continued study of subtropical fruits by Ch. Oppenheimer has done much to disperse exaggerated conceptions. Where the fruit is picked early, where relatively little infested varieties and species are concerned, where dropped fruit is collected, most of the subtropical fruits are well compatible with citrus growing. The greatest objection remains against the host par excellence, the peach. Recent developments of control, especially through DDT, which may eventually be replaced in a near future by other synthetic or refined by synergistic insecticides, open new prospects of a subtropical horticulture relieved from the nightmare of fruitfly years.

The daily destruction of infested fruit is an important supplementary remedy for any systematic control.

Spraying (or trapping) with one of the indicated substances should be performed carefully throughout the dangerous season, and some time before and after. Further improvements are to be expected with regard to both methods; until then the present prescriptions should be carefully applied.

With regard to late peaches or other crops of special commercial value it will pay to envelop every fruit during an early stage in a paper bag.

Considering the manifold physical and chemical properties rendering many fruits and varieties naturally resistant to fruitfly attack, it is the duty of the horticulturist to select those protected varieties carefully and to promote their growth in preference to others, liable to be heavily infested. There may even be work for future plant-breeders along this line.

Serious steps should be taken, eventually by inter-state agreements, to introduce the natural parasites of *Ceratitis* from over the world, especially *Diachasma tryoni* and *Opius humilis* from Hawaii, into the Mediterranean region. Even if the parasites cannot be expected to ensure a permanent control, they at least may lower considerably the danger and damage caused by *Ceratitis.*

References.

E. A. Back and Pemberton, Bull. U.S.A. Dept. Agric. No. 536 (profess. paper). 1918.
E. A. Back and Pemberton, Bull. U.S.A. Dept. Agric. No. 640. 1918.
E. A. Back and Pemberton, Journ. Agric. Res. 3. 1915. pp. 311—330, pp. 363—374, 5. 1916. pp. 793—803, pp. 657—666; 251—260.
A. Balachowsky et L. Mesnil, Les insectes nuisibles aux plantes cultivées. Paris, 1935. pp. 242—253.
E. Ballard, Bull. Soc. Roy. d'Entom. d'Egypte. 1933. pp. 8—13.
F. S. Bodenheimer, Bull. Soc. Roy. Entom. d'Egypte, 1924. pp. 149—157.
F. S. Bodenheimer, Schaedlingsfauna Palaestinas. Berlin, 1930. pp. 120—130.
F. S. Bodenheimer, Observations on citrus insects in many parts of the world. Tel Aviv, 1932.
F. G. Clemente, Bol. Patol. Veget. y Ent. Agr. **4**, 1930. 20 pp.; **7**, 1934. 11 pp.
G. Compere, Monthly Bull. Dept. Agric. California, **1**, 1912. pp. 842—845.
G. Costantino, Boll. Labor. Zool. generale ed agraria, Portici, **23**, 1930. pp. 237—322.
H. Efflatoun, Mem. Soc. Roy. Entom. d'Egypte, **2**, fasc. 2. 1925.
H. Efflatoun, Bull. Soc. Roy. Entom. d'Egypte, 1927, pp. 17—50.
C. M. Gjullin, Ecology **12**, 1931, pp. 248—258.
A. Grünberg, Bull. Ent. Res. 1938
W. J. Hall, Bull. 45. Min. Agric. Egypt. 1924. p. 13.
H. Z. Klein, Hadar **10**, 1937. pp. 193—195. **11**, 1938. pp. 203—204.
H. Z. Klein, Masike hedadar be Eretz Israel. Tel Aviv. 1940.
H. Z. Klein and M. Paker, J. A. P. Agric. Sta. Res. Sta. Bull. 32. 1942.
H. Z. Klein, A. Shnitzer and M. Paker, Hassadeh, **22**, no. 3. 1942.
H. Z. Klein and N. Yavolsky, Hassadeh. 24, no. 7. 1944.
J. de Lepiney et J. M. Mimeur, Mem. Soc. Sci. Nat. Maroc. **31**, 1932.
C. W. Mally, Agr. Journ. Cape Good Hope, **28**, 1904, pp. 627—662.
C. W. Mally, Union S. Africa, Dept. Agric. Leaflet No. 183. 1915.
R. H. Marlowe, Journ. Econ. Entom. **27**, 1934. p. 1100, **28**, 1934, pp. 914—919.
G. Martelli, Boll. Zool. e Entom. Agr. Portici, **9**, 1914, pp. 161—164.
A. C. Mason, Proceed. Hawaiian Entom. Soc. 1931, pp. 163—178.
A. C. Mason, Journ. Econ. Entom. **27**, 1934, pp. 891—897.
A. C. Mason and O. C. McBride, Journ. Econ. Entom. **27**, 1934, pp. 897—902.
O. C. McBride, Proceed. Hawaiian Ent. Soc. 1935, pp. 99—108.
L. J. Newman, Dept. of Agric. W. Austrilia, Bull. 122. Berth. 1924.
L. J. Newman, Journ. of Agric. Western Australia, 1931, pp. 316—318.
C. Oppenheimer, Gidul etze pri subtropiim be eretz Israel. Tel Aviv. 1939.
C. E. Pemberton and H. F. Willard, Journ. Agric. Res. **12**, 1926. pp. 285—296.
H. J. Quayle, U.S. Dept. Agric. Bull. 134, 1914.
H. J. Quayle, U.S. Dept. Agric. Circular, E. 282. 1931.
J. Reichert and E. Hellinger, Hadar, **5**, 1932, pp. 287—292.
E. Rivnay, Journ. Ent. Soc. S. Africa. **4**, 1941. pp. 166—176.
E. Rivnay, Hadar, n, 1935, pp. 49—52; **9**, 1936. pp. 134—137.
E. Rivnay, H. Nadel and F. Littauer, Hadar, **11**, 1938. pp. 227—231.
H. H. P. and H. C. Severin, Journ. New York Entom. Soc. 1914.
H. H. P. and H. C. Severin, Journ. Econ. Ent. **8**, 1915, pp. 329—338.
F. Silvestri, Hawaii Board of Agric. and Forestry Div. Entom. Bull. 3. 1914.
L. A. Stronge, Rept. Chief Plant Quarantine and Control Administration (for 1929/30). Washington 1930, 1931.
J. Verguin, Revue Zool. Agric. et Appl. **27**, 1928. p. 141, 149, 175.
H. F. Willard, Proceed. Hawaiian Entom. Soc. 1927, pp. 505—515.
H. F. Willard, Journ. Agric. Res. **25**, 1923, pp. 1—8.
H. F. Willard and B. C. Mason, U.S. Dept. Agric. Circular 439. 1937. 17 pp.
F. C. Willcocks, The more important economic insects and mites in Egypt. Cairo, 1922.
R. S. Woglum, Handbook of Citrus Insect Control for 1929. Los Angelos. 1929.
S. Yedidyah, Hadar, **9**, 1936. pp. 167—171.
P. K. Yonge, Report State Plant Board (1925/30). Sta. Plant Board Florida. Bull. 15. 1931.

C. OTHER FLIES IN CITRUS FRUITS.

The ubiquistic pomace fly (*Drosophila melanogaster* Meig.) and some of its congeners are the most common larvae found in decaying citrus fruits. The female is attracted to oviposition by the smell of fermentation. No oviposition occurs into intact citrus fruits. The eggs and larvae develop quickly, in about 5 days at 25° C, the pupal stage lasting three more days. The longevity of the flies was in one strain, for instance, 46 days on the average, but up to 83 days at 27° C (Pearl). The whitish slender maggot has a very elongate conical shape and can be easily determined by the characteristic caudal protuberances as well as by the shape of the posterior spiracles (fig. 19).

As the pomace fly infests only previously damaged fruits in which decay has already started, it cannot be regarded as a pest at all. It is not specific to rotting citrus fruit, but develops on the yeast flowering in all vegetable decay.

Other dipterous larvae in decaying citrus fruits belong, apart from other species of *Drosophila*, to a species of *Lonchaea* and to some Muscid flies, such as *Musca vicina* and *Chortophila cilicrura*. The occurrence of the latter in decaying fruit is quite accidental.

5. BUGS, LEAFHOPPERS AND APHIDS (*Rhynchota I*).

A. THE ORANGE BLOSSOM BUG (*Dionconotus cruentatus* Brullé, Capsidae).

Description: Oblong. Head broad, eyes large and protruding. Pronotum trapezoidal. Body, legs and the 4-jointed antennae black. The elytra have a broad dark red longitudinal stripe towards the margin. The central half of the elytra is black.

Body 11 mm long, 13 mm including the elytra; 3 mm broad. It is easy to distinguish this species from all other relatives by its distinct colours and design of the elytra. In the similar *Lopus infuscatus* and others neither are the colours as pure, nor the pattern as distinct.

Eggs oblong oval, yellow. 0.8—1.0 × 0.4—0.7 mm large. They are deposited in wet soil, near the hosts.

The *larvae* have 5 instars. The sheds of wings and elytrae become visible in the third larva.

First instar: Colour of dorsal side blackish brown with redbrown marginal stripes. Dorsal side of abdomen reddish and glassy. Antennae brown and claws grayish yellow. 1.8 mm long. Second instar: Colouration similar to first instar. 1.2 mm long. Third instar: Body dark brown to black, in the center of the dorsal side yellowish and glassy. Antennae grayish brown; sheds of elytra wine-red; claws yellowish and glassy. 2.6 mm long. Fourth instar: Body black. The last antennal joints yellow-brown. Margin of the dorsal side of abdomen and elytra red. 3.3 mm long. Fifth instar: Colour as in fourth stage. 5.0 mm long.

Distribution: Southern Europe, Cilicia, Syria and Palestine. In Palestine more common in the coastal plain than in the hills.

Life-history: The first larvae hatch in late December, but not contemporaneously. In early years the first adults appear even at the beginning of February, but generally in its second half. They then become scarce and disappear entirely in the coast-plain until the middle of March. In the hills their development is delayed for about one month. The larval development lasts 5—6 weeks. During the night, they hide among the herbs. They become active after sunrise, mainly on the shadyside of the leaves. They are fairly active in all stages. When approached, the bug flies away or drops to the ground, feigning death. The adult females lived 19 days in captivity. During copulation, which takes place at noon, the bugs are with their faces averted. The bulk of the eggs is laid in the stalks of the hosts in the second half of February. They enter diapause through spring, summer and autumn. The heavy winter-rains break the diapause and they begin to develop in December, contemporaneously with the development of their host.

Dionconotus cruentatus is physiologically adapted to a life in winter. Its scale of activity is as follows:

Scale:	1	2	3	4	5	6	7	8
°C:	—2.8	0.0	1.8	16.0	26.0	32.3	35.5	36.7° C

A few spiders are minor enemies of this bug.

Host plants: *Dionconotus cruentatus* is a very polyphagous species. Actual sucking is easily discovered, because the bug leaves behind typical chlorotical discolorations. The following are several of the more common hosts observed:

Borraginaceae	*Anchusa*
Compositae	*Pallensis spinosa, Echinops sp.*
Euphorbiaceae	*Euphorbia helioscopia*

Gramineae	Wheat, barley, etc. The leaves are often entirely covered with the sucking marks of the bugs.
Leguminosae	*Anagyris foetida*
Liliaceae	*Asphodelus microcarpus; Urginea maritima*
Malvaceae	*Malva nicaeensis*
Polygonaceae	*Emex spinosa*
Rosaceae	*Rubus discolor*
Rutaceae	*Citrus spp.*
Scrophulariaceae	*Verbascum sp.*
Umbelliferae	*Eringium creticum.*

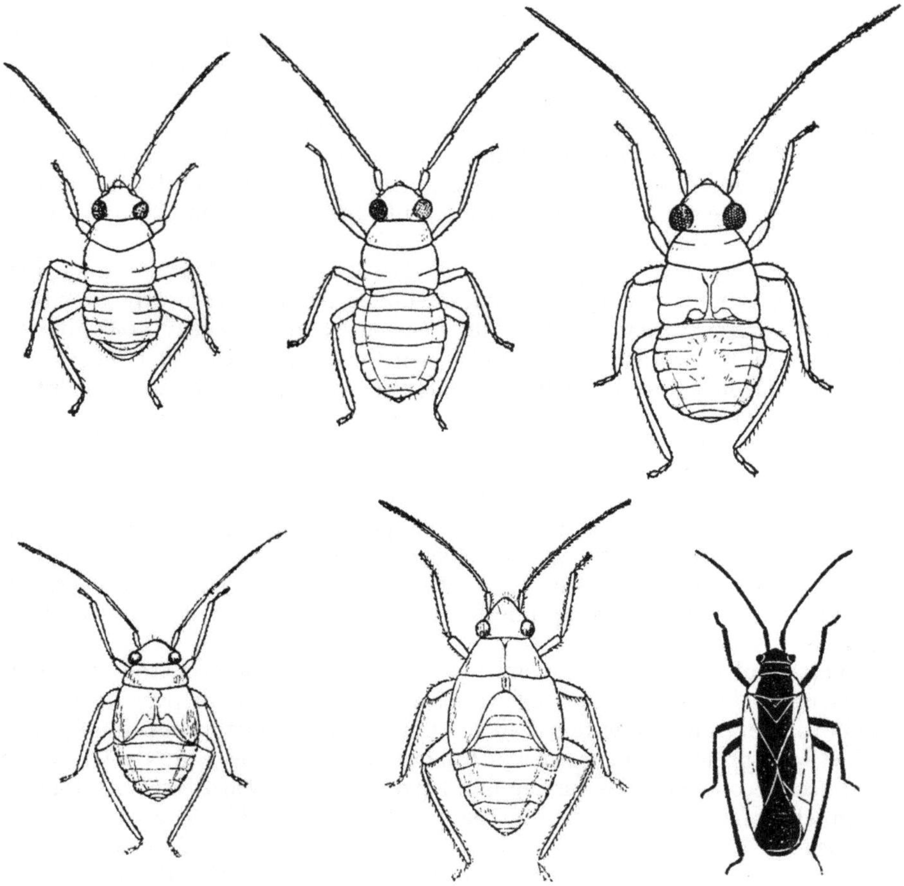

Fig. 35.
The Orange Blossom Bug (Dionconotus cruentatus Brllé.): larval stages 1—3 five times, larvae 4—5 two and half times more enlarged than the adult bug (from Bodenheimer).

The herbs and grasses here mentioned are real host-plants. It is not so with citrus. We have never observed any larvae on this plant, but the adults are attracted by the smell of the orange-blossoms. They suck on them and the blossoms consequently drop.

Damage: The damage done to cereals is relatively negligible, but locally serious damage is done to oranges. A typical case of such injury may be reported from a grove on heavy soil (hamra) at Petah Tikva:

13.III.1924. Soon after sunrise *Dionconotus* begins to fly about. It is then also engaged in other activities, but is fully active only during the warm hours of the day. Many of the bugs leave the herb-stratum and fly towards the orange blossoms, whereas the large majority has already died. Under some trees may be found over 2000 small unopened orange blossoms, with only 200—3000 flowers remaining on the trees. These are extreme cases, though the number of trees which have lost ½—⅔ of their blossoms is quite considerable. The feeding puncture of the bugs can be easily discovered on the dropped blossoms, because transparent sweet drops are exudated from it, to which *Bibionidae,* bees and others insects are attracted. When the early flower bud is pierced by *Dionconotus,* the buds dry up and drop after 1—2 days, with their stalks. The damage appears in patches within the grove. The crop in the more infested parts yielded only ⅓ of the expected crop (owner's estimate). A second series of flower-buds develops in place of the dropped blossoms and these buds form late fruits — about 50 per tree — of a lower quality, with leathery epidermis. Bugs which were put in a muslin bag on flowering orange twigs produced the typical injuries to blossoms and buds, about 2 days after the beginning of the experiment.

This damage is known, so far, only from the Jaffa-district (Jaffa Sarona, Petah Tikva). Weeds abound in all heavily attacked orchards. They are mostly low situated regions, half swampy from December to February and not easily accessible. The bugs develop on the weeds and the adults visit the orange blossoms later, to which they are strongly attached. In 1939 similar damage was observed by us in the Antalya district of Cilicia.

Control: Clean cultivation of the orchard is the only reliable control. The whole grove must be carefully weeded by mid-January, at the latest. No *Dioconotus* bugs develop then in it, and only some few individuals may perhaps immigrate from the immediate neighbourhood. Ever since this practice has been carefully observed in the concerned groves at Petah Tikva (since 1925), no further damage had ever occurred. I have seen the accounts of one owner, who reports a tripling of the revenue of one orchard, which was heavily infested for some years, after this system of clean cultivation was introduced.

When clean cultivation has been omitted by negligence or other reasons, the control of the larvae is important. Dusting with commercial brands of pyrethrum, derris (rotenone) and even with fine flower of sulphur, gives satisfactory results. All stages are killed, except for some insects just moulting. A simple hand-duster will do. We therefore recommend the dusting of all spots of the orchard threatened and the herbs on which *Dionconotus* larvae actually develop in numbers, carefully, in the early morning, from the end of January to mid-February on. The dusted plants are not attacked by other larvae and adult bugs, and when visiting them by chance, leave them immediately.

Direct control of the adult bugs is not advisable. They are too active and fly away when approached. Therefore, every winter, careful attention has to be paid to the development of *Dionconotus* in those localities, where the damage is known to occur.

References.

F. S. Bodenheimer Hadar **1**, 1938, Nr. 11—12. pp. 12—13; Schaedlingsfauna Palaestinas. 1930, pp. 169—172.

F. S. Bodenheimer, Ziraat Dergisi (Ankara). March 1941, p. 2-3.

B. THE GREEN STINK BUG (*Nezara viridula* L.; Pentatomidae).

Description: Adult. Body oblong oval; generally bright green, but variation reaches a yellowish stramineous colour. Dorsally more punctuated than ventrally; slightly convex dorsally, more convex ventrally. Head flat, and emarginate. Eyes rounded, dark green to brunescent, ocelli prominent. Antennae 5-jointed. The hexangular pronotum is emarginate and deflected anteriorly. The manubrium of the large scutellum reaches well beyond the middle of the insect. The light brownish rostrum is 4-segmented and reaches to the hind coxae. The legs are pale green, hairy on tibia and tarsus. The sexes are easily distinguished by the shape of their abdominal ends. These are notched in the male, but entire in the female. The male is smaller and narrower than the female. 20 mm long, 10 mm broad.

The *eggs* are deposited on the lower surface of a leaf in rags, consisting of regular rows. The eggs, 30—100 in number, are glued together and to the leaf by a colourless viscid substance. From above they resemble the form of a honeycomb.

The shape is cylindrical, rounded at the lower end and flattened above. The dorsal region covers the egg as a cap, which is thrust off by the emerging larva. The latter possesses a T-shaped egg-burster for this purpose. After some days, the original creamy white colour changes into a pinkish yellow.

The five *nymphal stages* are very dissimilar — colour and drawing — from the adults. Besides this their antennae are only 4-jointed and their tarsi have only 2 joints (3 in the adults). Two colour varieties of nymphs frequently occur together in the fourth and fifth instars.

The first instar nymph is broadly oval; thorax quadrangular. The rostrum reaches the hind coxae. Abdomen with two visible dorsal odoriferous glands. The colour is yellowish brown, with black markings on thorax and abdomen; and a white patch along the sides of the base of the abdomen. 1.4 mm long.

The second instar piceous, two white oblong areas extending over basis of abdomen; 4 light brown areas indicate the position of the dorsal glands on abdomen. 2.3 mm long.

The third instar has black head and thorax with yellowish areas along the margin of the head and down the centre of the thorax. Abdomen dark brown with a regular pattern of yellowish or white spots. 4.5 mm long.

In the fourth instar the light form has head and thorax pale green, with black markings, abdomen slightly darker than head and thorax, with a pattern of yellowish or white markings. The centre of the abdomen has 3 distinct dark areas, legs brownish. In the dark form the head is black with 2 yellow markings at its side. Thorax black with orange-coloured marks along the side. Abdomen dark red or brown, central markings black, covered with numerous yellowish spots. Antennae black, legs brownish black. 6.2 mm long.

Fifth instar: Light form. Head, thorax, wing-buds pale green with black margin and other black markings. Abdomen yellowish green, darker towards the centre with red areas surrounded with black. Numerous other yellowich spots, lateral margin of abdomen bordered with black. Antennae green, with terminal segments black, legs greenish brown. Dark form: with head and thorax black; margins of head and thorax with orange markings. Abdomen dark brown to black, with numerous white spots and pinkish margin. Antennae and legs black. 9.5 mm long.

Life-history: The incubation period of the eggs follows a hyperbola with the following constants: development threshold (c) = 12.2° C
thermal constant (th. c.) = 74 days-degrees.

The mortality of the eggs is fairly low. Unparasitized egg-clusters hatch generally to about 100%.

We have no experimental data on the total duration of larval development in Palestine. It is reported to last 97—98 (53—141) days on the average in the two non-hibernating generations in Western Australia (Newman), 66 days in the summer-generation in Cape Colony (Heerden) and 35 days in June in Florida (Drake). Three annual generations are reported from both regions. We might expect at least 4 annual generations in the much warmer climate of Palestine. A tentative hyperbola for the larval development is gained from the data: 53 days at 24.2° C, 71.5 days at 21° C (average from 77 days at 22° C and 66 at 20.5° C), with the constants c = 11.6° C th. c. = 668 days degrees.

This gives for total development (egg + larva) a constant of c = 11.7° C and th.c. = 772 days degrees.

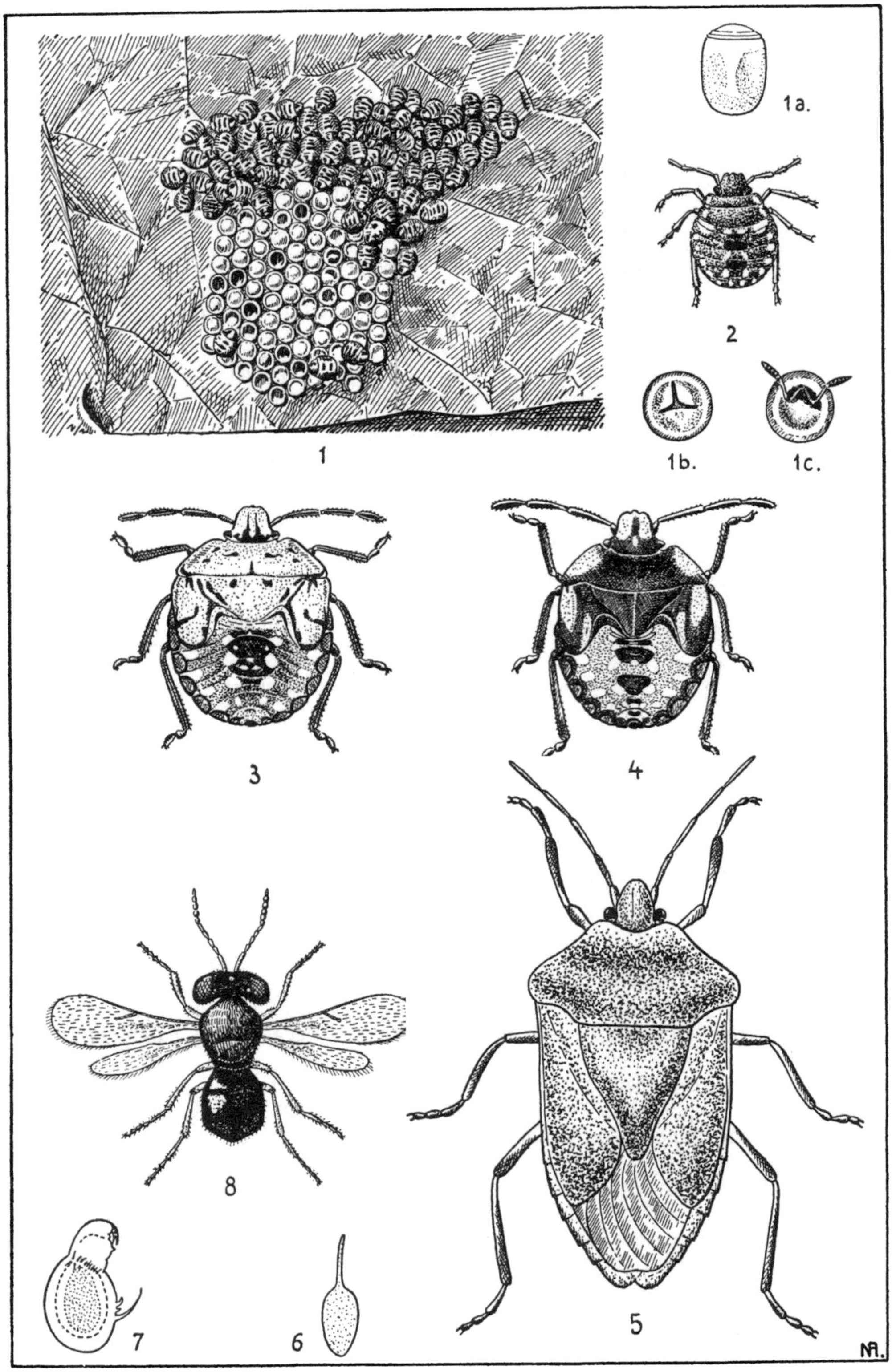

X. The Green Stink Bug (*Nezara viridula* L.) 1) Neonate larvae aggregated near the empty egg-shells; 2) third larval instar; 3) fifth larval instar (bright form); 4) fifth larval instar (dark form); 5) adult bug; 6—8) *Microphanurus megacephalus* Ashm., the common egg-parasite of the Green Stink Bug. 6) egg; 7) first stage of larva, 8) wasp (6—8 from Priesner and Kamal).

The sum of effective temperatures is for

Perth (W. Australia):	2118	days degrees	= 3 annual generations
Simonstown (Cape Col.):	2133	,, ,,	= 3 annual
Tel Aviv (Palestine):	2886	,, ,,	= 4
Florida (Orange Zone):	3255	,, ,,	= 4

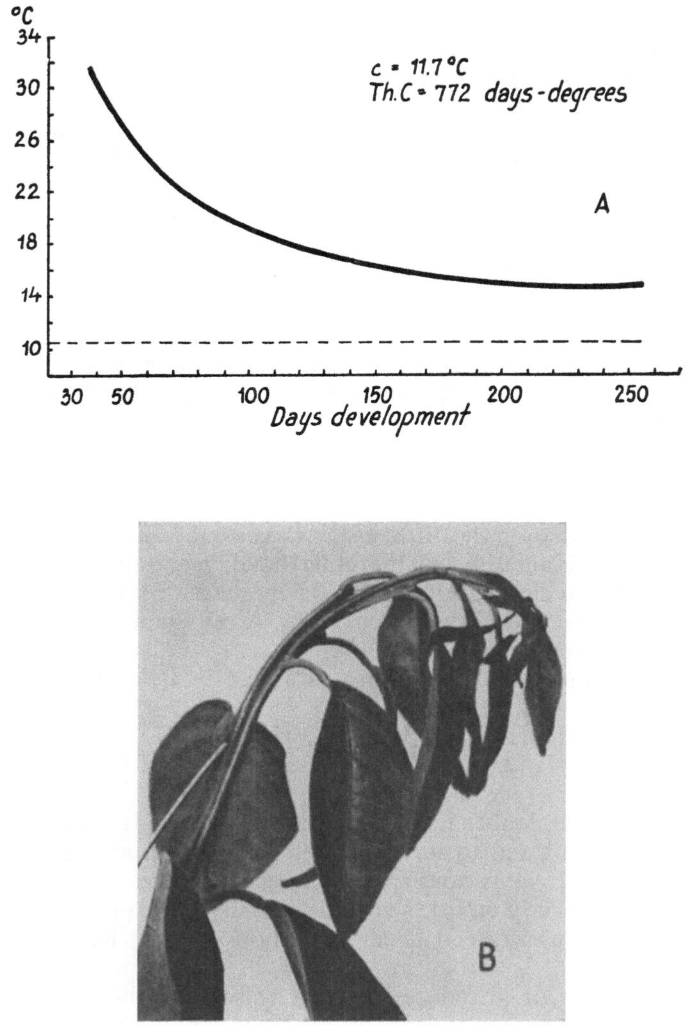

Fig. 36.
Nezara viridula L. A. Hyperbola, B. Wilting orange shoot following sucking of the bug.

This is about the average, not the shortest development. This calculation is verified for Egypt, where 3 generations (excepting the hibernating generation) develop (Kamal).

These generations overlap to a certain degree. The first larval stages develop quicker than the later ones. Heerden reports the difference between the shortest and slowest development among individuals of the same egg-cluster to be up to 15 days.

The newly emerged nymphs usually remain clustered on the empty egg-mass until after the first moult. Then they cease to be gregarious and scatter over the host plant, where they are found mainly on the growing tips. They suck in the early morning or late afternoon of the hot season and rest in the shade at noon. The wingless larvae are rather quick moving insects, which may traverse distances of dozens of meters quite easily.

The adults suck on the fruits or succulent tips of many plants. They do not use their wings readily and probably never over long distances. They were seen flying from herbs to neighbouring trees and on similar short flights only.

The number of males and females is about equal. As in most bugs, the two partners face opposite directions during copulation and the larger female often draws the male after herself. In nature, the act is generally repeated several times and lasts some hours, up to over one day. The exact longevity of the adults is not known, but in the summer it should not be under 1 month, during the winter not under 3—5 months.

The preoviposition period lasts 25—35 days (Heerden). One female oviposits several egg-clusters from 1 to about 5, each of 30—100 eggs (in Palestine nearly 40—50 on the average). There may be a few days to a few weeks interval between the different ovipositions.

The adults pass the winter hidden in crevices, etc. below leaves. They do not undergo a true hibernation, but are semi-quiescent. They may be seen active on every nice, sunny winter-day. Newman even points out, that they require food during the winter and that the fifth instar larvae may during the winter slowly develop into adults. Heerden, on the other hand, maintains that under natural conditions these larvae die out. During winter the mortality is very heavy and this is the most serious check of *Nezara* populations. A mild and humid winter is apparently decidely unfavorable.

Our records of its seasonal appearance as a pest in Palestine are as follows:

Months	I	II	III	IV	V	VI	VII	VIII	IX	X	XI	XII	Total
No. of records			1	3	7	19	11	3	2				47

Nezara viridula is, no doubt, a very polyphagous feeder. It is reported to be even occasionally carnivorous and to suck on insect eggs. From among its many hosts, some leguminous plants, as various species of beans, vetches, etc. seem to form the main hosts. It sucks eagerly on fruits or seeds of many other plants, which it injures, without being able to undergo on them normally or without heavy mortality complete development.

Damage: By sucking, it introduces its toxic saliva into the plant tissues. Young plants, when attacked, wither and finally die. It invariably prefers the young and most tender twigs and the vegetation points. We have observed such damage on orange-trees, castor-plant and tobacco.

Attacked fruits often drop, as has been observed with tomatoes, egg-plants, cucumbers. In other cases the fruit shows discolorations or deformations. The punctured tissue stops growing and hard dry changes are found on the punctured part of the fruit (apricots, melons). The berries of grape-vine shrink and dry prematurely. The puncture on Citrus remains visible as a black dot, surrounded by a chlorotic field. Our oranges seem to be protected from dropping by the thick diameter of their peal. Thin-skinned oranges and tangerines, when damaged, invariably drop, as is reported from Florida, where the bug is a troublesome citrus pest.

It seems desirable to study its food-ecology. Which are its favoured host-plants? On which host-plants may it develop, undergoing, however, a heavy mortality? On which plants may it suck occasionally without being able to perform its whole development (especially in the younger stages) on them? Can it perform the total cycle on any plant, only on twigs and leaves, or is — at least in some stages (maturation of eggs, for example) — the sucking on the fruit or seed indispensable?

The reports in Palestine concern the following plants:

Citrus	4	Tobacco	5	*Cucurbitaceae*	3
Tomatoes	9	*Vitis*	2	*Ricinus*	1
Eggplants	2	Apricots	1	*Beta*	2

The injury occurs mainly from May to July.

It sucks on the young sprouts of citrus, causing them to wilt (Fig. 36b). The discoloration of the fruit had just been described. When it sucks on the fruit of young trees, the fruit drops heavily. This phenomenon has been observed in Hedera, in May 1931. The bugs were collected in the early morning by hand-picking.

Nezara viridula is known as a carrier of fungal and virus diseases, besides the direct damage it does. Its activities should, therefore, be carefully watched, wherever it appears in numbers.

It seems that the dry summer is unfavorable to their mass increase. Outbreaks might, therefore, be only occasional and local. In the event of an outbreak, hand-picking of egg-clusters and bugs may be applied, where modern insecticides are not available.

In Florida, where heavy damage occurs during the hot and humid summer, they are shaken into large nets from the tree. If certain leguminous crops (as green manure or cover-crop) remain too long in the orchard, an undesirable mass-increase of *Nezara* may develop in the orchard. They should be cut or ploughed at the earliest convenience.

A small black Proctotrupid wasp (*Microphanurus megacephalus* Ashmead) is a common egg-parasite of *Nezara*. In W. Australia its entire development lasts 9 to 10 days only in summer. In summer it is common in Palestine. Parasited egg-clusters are usually 100% infested. Yet we never found more than at most one third of the egg-clusters parasitised in one place. Where and when the host is rare, the parasite is very rare, but increases quickly when the host becomes abundant. A high crowding of egg-slusters of *Nezara* is needed for effective biological control by *Microphanurus*. The parasite is known from the West Indies, U.S.A., Palestine, Egypt, the Sudan, and has recently been introduced into W. Australia for biological control and is posted there to the farmers on request.

References.

F. S. Bodenheimer, Schaedlingsfauna Palaestinas. 1930, pp. 172—173.
C. J. Drake, Quart. Bull. Sta. Plant Board Florida **4**, 1920, pp. 41—94.
M. Kamal, Bull. Soc. Roy. Ent. d'Egypte. 1937, pp. 175—207.
L. J. Newman and B. A. O'Connor, Journ. Agric. West. Australia **11**, 1934, pp. 101—112.
H. Priesner, Bull. Soc. Roy. Ent. Egypte 1931, pp. 137—139.
P. W. van Heerden, Annals Univ. Stellenbosch, **11** A 1933, No. 17. 24 pp.

C. THE VARIEGATED CAPER BUG.

The Variegated Caper Bug (*Stenozygum coloratum* Klug, *Pentatomidae*) is a short oval bug with shining dark blue, orange and white colouration. Its normal host is the caper bush (*Capparis spinosa*), but it has also been collected by us from tobacco, fig-trees, grape vine, cauliflower and raddish. Its egg clusters and larvae have been found repeatedly in June/July on citrus leaves (Hedera, Petah Tikva), but never caused any damage.

D. LEAFHOPPERS.

a. Empoasca bodenheimeri Haupt. Description: 3.0—4.0 mm long; green, without any design. Recognized by its genital-characters.

Male: The genital-plates surpass the anal segment. They are flat, nearly parallel sided, somewhat bent outwards at the end. The lateral view shows these plates sloping from the dorsum. Their anterior front is covered with fine and long hairs. Long, parallel hairs cover the distal margin. The lateral lobes of the anal segments are triangular.

Female: No genital-valve. The last sternit shows an angular incision and the basis of the ovipositor is visible. The two vaginal ridges touch their longer distant half with their interior margins. They are covered with conspicuously long and fine hairs, among which short and strong bristles are noticed on the distal margin.

The arrangement of the sensory bristles and the topography of the face in the larval stages s illustrated in fig. 37. The antennae of the larva are exceptionally long.

b. Empoasca citrea Haupt. Description: 3.0—3.5 mm long. Green, with indistinct whitish designs on vertex and pronotum: Two lunular stains on the vertex and 2 circular stains in front of them. Two irregular spots behind the middle of the anterior margin of the pronotum and one on each side. The distance between the eyes is broader than the posterior eye-margin. The antennae are nearly as long as vertex, pronotum and scutum put together.

Male: The genital plates surpass the anal segment slightly. They join boat-shaped and are bent upwards towards the anal tube. A series of bristles is observed at their basis along the median line and turns towards the interior margin. The lateral lobes of the anal segment are triangular and possess a row of bristles on their inferior margin.

Female: The genital valve is long, very round at the end. The vaginal-ridges do not join on the inner side and the ovipositor is free and visible from its basis to the top. They have short bristles on the distal half, 2 of which are on the interior margin and 5 form a row along the mediane line.

c. Tettigometrid species. Description: Colouration mainly shiny black-brown. The sides of the pronotum, the centre of the thoracic dorsum between the wing sheets, an irregular longitudinal stripe on the mid of the abdomen and the lateral edges of the fourth and fifth tergit are whitish. Vertex and front are faint, the clypeus shiny. The latter is provided, before its upper margin, with a short conical process, which can be seen from above, as the margin of the vertex protrudes only very slightly. Almost the entire body is covered with short bristles, which are especially dense on the head. The second antennae-link is twice as long as the first one and surpasses the eyes. Behind the eyes are thick ridyes. The pronotum is only slightly broader than the head with the eyes. The back of the abdomen is elevated in a flat-obtuse angle. The tergits protrude in angles on both sides. The anal-tube runs obliquely upwards. The sternites are white.

Damage: Nothing is known so far on the life-cycle of those leafhoppers. No permanent damage has ever been reported from *Homoptera* on citrus. Some reports on occasional infestations may be quoted in detail.

Petah Tikva, 2 years old orange grove of 50 dunams with many weeds at the time of observation on 11.VII.1929. Adult green leafhoppers (*Empoasca citrea* Hpt.) were common on citrus leaves, only few larvae being present. The infested leaves grew yellow, their margins rolled downwards, wilted and dropped. On their lower side were small black spots (faeces!) and small brown or black sucking punctures. Identical faeces and sucking punctures were observed abundantly on the common weed *Portulaca oleracea*. The succulent leaves of *Portulaca* neither rolled nor grew chlorotic, but grew stunted. The damage was especially heavy in 10 dunams of weak

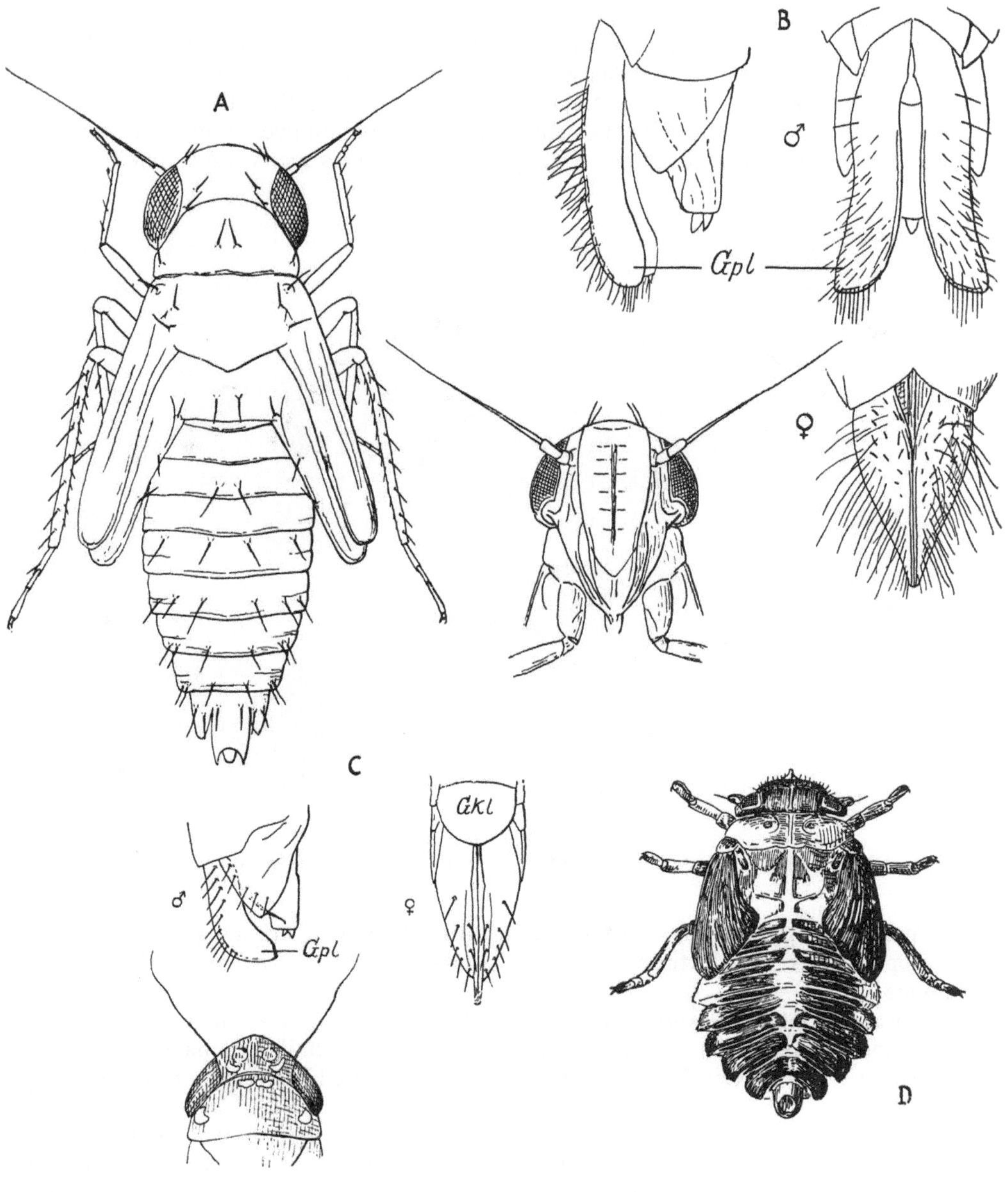

Fig. 37.
Homopterous citrus pests: A, B. *Empoasca bodenheimeri* Hpt. A. larva, B. larval head from below, male and female genitals; C. head, male and female genitals of *Empoasca citrea* Hpt.; D. Tettigometrid larva (Gpl. genital plates; from Haupt).

and badly developed citrus-trees, where *all* young shoots are chlorotic and their yellow colour is obvious from afar. In the other parts of the grove only isolated chlorotic shoots occur. The manager informed us that these trees were weak ones which had been early grafted (as were all the trees on the 10 dunams mentioned). Many leafhoppers occurred on trees grafted later and with fully developed leaf-crowns. On shaking the branches many leafhoppers jumped about, but not one chlorotic leaf was seen on them. Among the older leaves many were rolled, but no

Fig. 38.
Rolled orange leaves following attack by *Empoasca* sp. (Hedera VII. 1934).

recent sucking punctures were visible. In all neighbouring orange groves with well developed trees the leafhoppers were likewise observed, without inducing any damage.

A fortnight later, we were consulted by Dr. Eichenberg, the manager of a large young citrus plantation at Kfar Saba, with regard to similar damage caused by *Empoasca bodenheimeri* Hpt. The leafhoppers abounded on citrus, but relatively little damage was observed.

The blue-black larvae of a *Tettigometrid* were first observed in small numbers in Juli/August 1929 at Hedera. It was observed again in the same locality sitting in some numbers, mainly on the fruit, of one tree late in April 1932. The small red brown young larvae were quite abundant on a few flowers in Mikveh Israel in April 1938.

No serious damage has ever been observed on healthy trees and a general policy of keeping the orchard fit is the best prophylactic. We have repeatedly observed some curling of the leaves in young orchards, which in certain years is more abundant than in others (fig. 37). Clean cultivarion of the orchard is an important preventive measure. None of these leafhoppers seems to be a genuine citrus-insect. The occasional infestations originate from the weeds within the orchard and it is not improbable that they belong to the "residual" fauna, which tries to maintain itself for some time in the recently planted citrus groves, without being able to establish itself permanently there.

In case some control should be found advisable on weak trees, spraying or dusting with derris or pyrethrum preparations should be attempted.

References.

F. S. Bodenheimer, Die Schaedlingsfauna Palaestinas. Berlin, 1930, pp. 175—176.
H. Haupt, ibidem, pp. 419—422.

E. WHITE FLIES

THE CITRUS WHITE FLY (*Aleurotrachelus citri* Priesner and Hosni; Aleurodidae).

Description: Female: Head and thorax more or less dark brown, thorax partly orange, eyes black to purple, abdomen pale yellow with grey markings on the first three segments only, laterally and narrowly; a large grey patch round the anal orifice, and the last sternite grey. Legs pale yellow (coxae shaded with grey), antennae pale but very sligthly infumated towards tip. Fore wings remarkably coloured: clear hyaline, with three large patches (fig. 39); one along cubital area, a second, obliquely x-shaped, about the middle of the fore wing, and a third near tip, forming a regular upright cross; principal vein darkest where it passes the x-spot; hind wing with but faint darkenings, thought the principal vein more distinctly shaded than in the fore wing. Thorax width about 350 μ.

Mid tibiae, length 252, tarsal joint 1: 70, joint 2: 75, hind tibiac 350 μ, tarsal joint 1: 98, joint 2: 77 μ. The pointed apex of the abdomen narrower than in *Siphoninus granati*.

Male: Somewhat smaller, coloration of body, legs and wings as in the female, abdominal shadings more pronounced, the last two segments with claspers grey, also some infumation on the preceding segments, dorsally.

Pupal case: When mounted, the pupal case is oval, 858—892 μ in length, 627—693 μ in width, greatest breadth the middle; eye spots small; margin of pupal case crenulated, 11 to 15 teeth of it cover a space of 0.1 mm; teeth rounded, indistinctly serrate; abdomen with a well marked trachea-like area along the middle, which is provided with small dorsal tubercles on segments 2 to 6, the segments 1 and 7 are free from tubercles; number of tubercles varying from 2 to 8; there are 5 or 6 in most cases, arranged in one or two cross rows; the specimens from pomegranate leaves have less tubercles, on an average; trachea-shaped area, on each segment, projecting into lateral lobes which are always rather blunt though they are somewhat different in shape, according to be state of pressure on the preparation. Vasiform orifice, length, 56—60 μ, width 49—53 μ; operculum oval, blunter at base than apically, length 50—58, width 35—42 μ; it almost completely fills the orifice. Bristles on pupal case inconspicuous; pairs of anteopercular pores 38—46 μ distant.

The living pupal case, when attached on the leaf is black; shining, somewhat convex, the trachea-like area forming a slight keel; a regular close fringe of white "wax" (of 70—85 μ width) surrounds the puparium; there is never any wax secretion on the upper surface of the puparium." (Priesner and Hosni).

Life-history and damage: The pupal cases of this species have been discovered on the lower side of the leaves of one orange tree at Mikveh Israel, in November 1935. This orange tree is in the immediate neighbourhood of the Botanical Garden of that Agricultural School. The species is endemic in Palestine and has been collected on Zizyphus near Jericho by Prof. Priesner and the author. It is endemic in Egypt also, from where it was described. Its main host plant there is çitrus (lemons, mandarines, oranges, bitter oranges, Adalia lemons), but other hosts are Pomegranate and *Zizyphus*, *Lawsonia inermis*, *Psidium guayava* and *Dodonea viscosa*. It is well distributed over Egypt, but heavy attack was observed only in neglected orchards in Upper Egypt, where it caused leaf drop. In all other cases districts of Egypt it is of no economic importance.

In Palestine this Aleurodid, was observed in small numbers on orange leaves in Petah Tikva, in 1923. In the case of Mikveh Israel, also, only 1—2 individuals per leaf were observed. The black pupal case with its white fringe, about 1 mm in length, is sufficiently obvious not to escape the eye of the trained entomologist, were it is common over the country. It is therefore probable that in the recent case the White fly passed over on citrus from other trees of the Botanical Garden, with which it was introduced from Jericho or Egypt. It has not spread, since, in the citrus orchard at Mikveh.

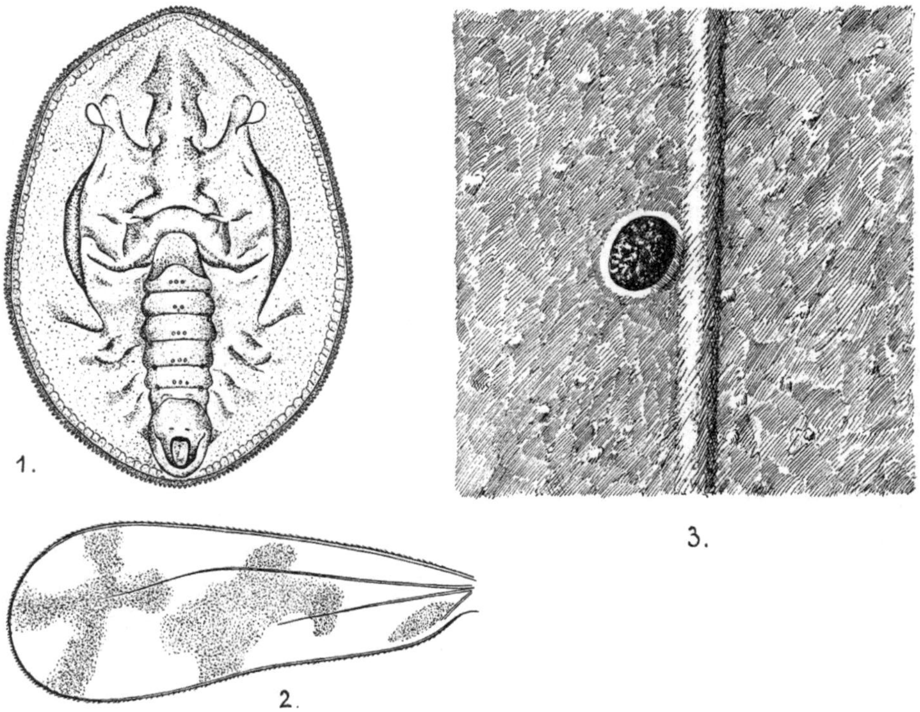

Fig. 39.
Aleurotrachelus citri P.a.H.: 1. ventral view of pupa, 2. wing of adult, 3. pupa on leaf (after Priesner and Hosn).

References.

H. Priesner and M. Hosni, Tech. Bull. 145. Min. of Agric. Egypt, Cairo, 1934.

THE TOBACCO WHITE FLY (*Bemisia tabaci* Gennad.; Aleurodidae).

Synonymy: *Aleurodes tabaci* Gennadius.
Bemisia gossypii Misra and Lamba.
Bemisia hibisci Takahashi.

This small yellow whitefly (1.0 × 0.3 mm.) is widely scattered over the Palaeotropics, from the Pacific Islands to Nigeria. In Mediterranean countries it is recorded from Egypt, Palestine and Italy.

Description of pupa (or fourth larval stage): Body elliptic, slightly convex, deep yellow, 0.6 × 0.5 mm. Antennae very short, but more conspicuous than in former larval stages. Margins dentate. Submarginal area not distinctly separated from the dorsal disc. Along the later 7 pairs of long bristles, plus a longer pair of caudal setae (0.1 mm.). Thoracal spiracles folded and not conspicuous. Vasiform orifice triangular and elongate (0.007 × 0.005 mm). Lingula with two short, thin setae, subequal in lenght to the operculum.

Bemisia tabaci is a rather polyphagous insect, being recorded in Palestine alone from over 50 hosts, wild and cultivated, belonging to 17 plant-families. Damage occurs mainly on tomatoes, eggplants, cucumbers and melons, but also on:

vegetables: potatoes, beans and cabbage;
forage crops: cowpeas, clover;
green manure plants: *Cajanus indicus, Sesbania, Crotalaria*;
ornamental plants: roses.

The Tobacco White Fly is, apart of the damage caused by wilting and by sooty mould, known as the carrier of the following two virus diseases (in the nomenclature of K. M. Smith): Tobacco virus 10 and cotton virus 1. One of these is observed in Palestine on tomatoe seedlings. Occasional local damage has been reported from the Upper Jordan Valley in older citrus plantations. The heavy infestation of the leaves, mainly on their lower surfaces, is followed by intensive sooty mould development subsequent to strong honey-dew excretion. This damage can be prevented, when the infestation is early observed, by dusting with DDT.(5%).

The life cycle and the ecology of this pest has been well studied by my coworker H. Z. Klein (1946, 1949). The physical constants of development (from egg to first oviposition) are:

$$C. = 12.0°C.$$
$$Th. C. = 225 \text{ days-degrees.}$$

Up to 11 annual generations (Hyperbola A) develop in the coastal plain, up to 15 ones in the Upper Jordan Valley. Average adult longevity ranges from two weeks in summer to eight weeks in winter. The average number of eggs per female is about 50, the sex-ratio almost equal.

B. tabaci grows injurious in Palestine mainly in the warmer valleys of the interior in humid summers. Whenever the average air humidity drops below 60% for two or three spring or early summer months, no damage to crops occurs. But when the hamsin winds are missing or weak, and air humidity remains high during that season, a mass-outbreak of the pest has to be expected in late summer and autumn. Only then it is liable to become a local and minor citrus pest.

References.

H. Z. Klein, Knimath Ash-Hatabak be-Eretz-Israel. Tel Aviv. 1946 34 pp.
H. Z. Klein, The Tobacco White Fly in Israel. Agric. Res. Sta. Bull. Rehoboth (in press).

F. APHIDS.

THE BLACK CITRUS APHIS (*Toxoptera aurantii Boyer*; *Aphididae*).

Synonymy:
1841 *Aphis aurantii* Boyer.
1843 *A. camelliae* Kaltenbach.
1857 *Toxoptera aurantiae* Koch.
1880 *Aphis coffeae* Nietner.
1891 *Ceylonia theaecola* Buckton.
1906 *Toxoptera theobromae* Schoutenden.
1914 *T. aurantii limonii* Del Guercio.
1921 *Aphis coffeae thomensis* De Seabra.

Description: Apterous, viviparous female. Body broad oval, slightly convex with small lateral tuberculi on the prothorax and on the first and seventh abdominal segment. Back bare with irregular polygonal sculptures. Lower side with transversal rows of short hair. Antennae only little shorter than the body. Rhinaria with short their corona. Proboscis reaching up to the third pair of coxae. Siphunculi moderately long and thin, about cylindrical. Cauda club-shaped, 3/4 of length of the siphunculi. Legs well developed, with short spines.

Body dark brownish to greenish black, on the abdomen reddish brown between the siphunculi. Cornicles, cauda and the anal plate almost black. Eyes large and black. Antennae: 2 basal segments black, the others pale with 4 black rings on the apex of the third to sixth segment. Legs pale; tarsus, tibial and femoral apices black-brown; hind femora black.

Body length	2.07 mm	Body width	1.17 mm
Antennae length	1.71 mm	Siphunculi	0.31 mm

Cauda 0.23 mm.

Alate viviparous female: Body with small lateral tuberculi on the prothorax, first and seventh abdominal segment. Integument with polygonal structures, rhinaria and proboscis, as in the apterous form. Antennae slightly longer than the body. Siphunculi long and thin cylindrical; the club-shaped cauda half as long as the siphunculi. Wing with normal *Toxoptera*-venation, subjected to some variation, much longer than the body; furca of media I very large; with 2—3 adhering clubs.

Legs with short spines.

Body and thorax black, abdomen dark to black brown, around the siphunculi red-brown. Eyes large, black. Antennae pale, with 4 dark rings on the third to sixth segment. Legs pale; tarsus, tibial and femoral apices, hind-femora and siphunculi black. Wings hyaline, often tinged with yellowish brown; stigma large and black, veins yellowish brown.

Body length	1.70 mm	Body with	0.94 mm
Antennae	1.76 mm	Cauda	0.16 mm
Siphunculi	0.31 mm	Wing-expanse	6.10 mm

The neonate larvae are ferruginous and grow darker as they mature, until they become dark brown in the last stage.

The *male* which was apparently unknown and does not seem to appear at all in tropical climates, appeared in early summer in the breedings of Rivnay. Dr. F. Laing — British Museum confirmed its identity. Later experience with *Brachycolus brassicae* has since shown, that dispersed light in the laboratory induces the appearance of sexuales, which do not occur in nature (Elze, Bull. Soc. Fuad Ier d'Entom. d'Egypte 1944). We wish to point out that no sexuales were ever found by us in the material collected in nature of *Toxoptera.*

The very complicated synonymy of the species was caused by the wide range of host-plants as well as by the high variability of the species which concerns not only the colour (from black and dark brown to blackish green), but also morphological structures, as the relative length of the antennal joints of cornicles and proboscis, in the number of the caudal hairs, even in wing-venation and in the antennal sensoria. The banded antennae and the marked specific sculpturing of the cuticula serve to ascertain the identity of the species. A transition of the variable characters, just mentioned, has been observed in experimental transfer of *T. aurantii* from one host to another. The repetition of such experiments under controlled conditions and accompanied by biometrical measurements is desirable.

Distribution: The Black Citrus-Aphis is almost tropico- and subtropicopolit. It is recorded so far from *):

Europe: Spain, Italy, Sicily, Malta, Turkey (in greenhouses).
Africa: Egypt, Morocco, Erytrea, Uganda, Kenya, Tanganyika Territory, Nyazaland, French and Belgian Congo, Cameroon, S. Nigeria.
Asia: Palestine, Syria, Anatolia (incl. Black Sea region), Caucasus, Transcaucasia, India (Pusa, Bombay, Madras), Ceylon, Malaya, Indochina, China, Formosa, Philippines, Malayan Archipelago.
Australia and Polynesia: Queensland, New Zeeland, Samoa, Hawaii Isl.
America: California, Florida, C. America, the West Indies, Argentine, Brazil, Chile, Venezuela.

---

*) Thanks are due to Dr. F. Laing—British Museum for help rendered in compiling the distribution and synonymy of the species.

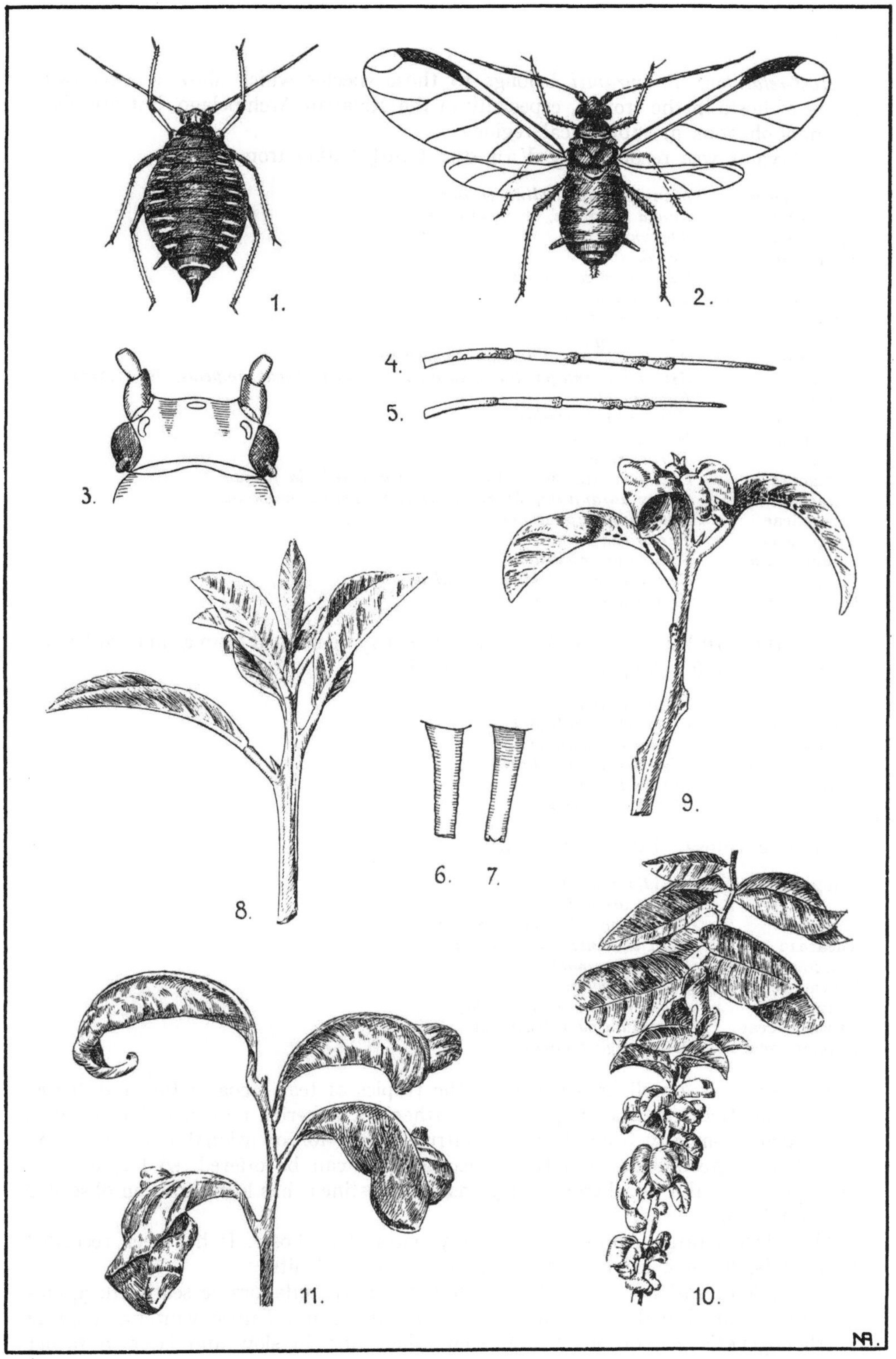

XI. The Black Orange Aphis (*Toxoptera aurantii* Kalt.)1) apterate viviparous female; 2) alate viviparous female; 3) head of alate viviparous female; 4) and 5) antemnae of alate resp. apterate female; 6) and 7) siphunculi of alate resp. apterate female. 8) and 9) show ng the effect of early infestation of *Toxoptera* on the growth of the young shoots (8: normal, 9: infested); 10) later development of young twig after early *Toxoptera*-infestation; 11) damage on end of shoots some months later, showing the curling of the old leaves.

Host-relations: T. aurantii belongs to those species which show a very wide range of hosts in the tropics, especially in the Malayan Archipelago, but are oligo- to monophagous in subtropical regions.

In Java it was recorded by Van der Goot (1941) from:

Anacardiaceae	*Mangifera sp., Odina wodier*
Anonaceae	*Anona muricata, A. squamosa*
Apocynaceae	*Apocynea sp.*
Araliaceae	*Panax sp.*
Ericaceae	*Vaccinium varingifolium*
Hamamelidaceae	*Distylum stellare*
Leguminosae	*Butea frondosa*
Loranthaceae	*Lorantus sp.*
Meliaceae	*Aglaia odorata, Lansium domesticum*
Moraceae	*Artocarpus integrifolia, Cudrania javanica, Ficus ampelas, F. hispida*
Myrtaceae	*Eugenia sp.*
Polygonaceae	*Antigonon Leptopus*
Proteaceae	*Helicia attenuata*
Rosaceae	*Pyrus malus*
Rubiaceae	*Cinchona ledgeriana, Coffea arabica, Gardenia florida*
Rutaceae	*Citrus aurantium, Murraya exotica, Evodia aromatica*
Sapinadeae	*Nephelium lappaceum*
Sterculiaceae	*Theobroma cacao*
Symplocaceae	*Symplocos sessiliflora*
Theaceae	*Camellia thea, Schima noronhae*
Urticaceae	*Trema amboinensis*

That this host list is far from being complete may be learned from a small collection at Peradeniya (Ceylon, Van der Goot, 1921):

Ulmaceae	*Celtis cinnamomea*
Leguminosae	*Cynometra cauliflora*
Myrtaceae	*Eugenia mooniana*
Flaccourtiaceae	*Flaccourtia ramontchi*
Guttiferae	*Mesua ferrea*
Plumbaginaceae	*Plumbago capensis.*

or from Formosa (Takahashi, 1923);

Rubiaceae	*Arsidia sieboldi*
	Coffea arabica
Rutaceae	*Citrus spp., Murraya exotica*
Theaceae	*Thea chinensis, T. caudata*
Flaccourtiaceae	*Scolopia crenata*
Salicaceae	*Salix warburgii*
Moraceae	*Ficus retusa, F. foveolata*
Caprifoliaceae	*Viburnum odoratissimum*
Euphorbiaceae	*Macaranga tanarius*

T. aurantii is a well known pest in the tropics of tea, cocoa, coffee and citrus. But in the Mediterranean regions, in Southern S. America and in California it is not known from any other host but citrus except for occasional infestations on camellia, in greenhouses mainly. No explanation can be offered, so far, for this monophagous habit in subtropical regions. In Palestine it has hitherto been observed only on Citrus.

The Black Orange Aphis attacks many parts of its hosts. It has been recorded from buds, flowers, leaves, tender twigs and young fruits.

Life-cycle in Palestine. Not the winter is the most unfavorable season to aphid-life in Palestine, but the dry and hot summer and autumn. During winter and spring parthenogenetic generations follow each other, first in slow and later in rather rapid succession. When the average temperature surpasses 22—23° C in April and May, i.e. when the maximum surpasses 40° C, for many hours during the dry hamsin winds, the spring peak of *Toxoptera*-populations rapidly decreases. During the summer only a careful search in orange groves shows that small scattered colonies of viviparous females are present and maintain the continuity of life wherever young leaves are present. The latter are rare, however, in summer. The normal cycle is therefore an uninterrupted chain of parthenogenenic generations, But, as mentioned before, alate males have been discovered in the beginning of the unfavorable season. No morphologically different forms of females were observed and all attempts to obtain copulation of the normal females with these males were in vain and no eggs were observed. The sexual generation does not seem — all facts considered — to play an actual part in the maintenance of the cycle during the unfavorable season; if it occurs in nature at all.

The development of the apterate female is rather quick.

The shortest development of 6 days is at 25° C, but even at 15° C maturity is reached after 19 days. These figures combined with those of adult longevity permit the construction of the following hyperbolae:

Hyperbola		A	B	C
Development threshold (c)		7.5	6.5	5.9° C
Thermal constant (th. c.)		120	258	535 day-degrees
Duration of development	30° C	5.3	11.0	22.2 days
	25	6.8	13.9	28.0
	20	9.6	19.1	37.9
	15	16.0	30.4	48.8
	10	48.0	73.7	130.5
Number of generations at	Tel Aviv	36	19	10
	Nice		14	
	Palermo		15	
	Cairo		20	
	Sierra Leone		28	

$B = A + \frac{C - A}{3}$, as half of the offspring is born at the end of the first third of the adult life-span.

No complete development is observed above 30° C.

During the parthenogenetic cycle apterate females develop under favorable, in increasing number of alates under unfavorable environmental conditions. Their fate is apparently decided during the early larval period.

Many students have dealt with the problem of development of the alate ovivi-parous females and many factors, as length of day (light), temperature, crowding, humidity, precipitation, condition of host-plant have been shown to influence this transformation under controlled conditions. Evans has just shown in unpublished experiments that in the cabbage aphid it is dependent on the content of nitrogen and sugar in the cabbage-leaves. Rivnay (1937) obtained an increased production of winged forms in *T. aurantii* under crowded conditions or on wilted host-plants, but no influence of temperatures ranging from 14 to 32° C was observed. He maintains that air-humidity and water content of the host are the causal factors inducing the formation of winged forms through a disturbance of the water balance of the insects. Evans (1938) has protested against this "sweeping statement."

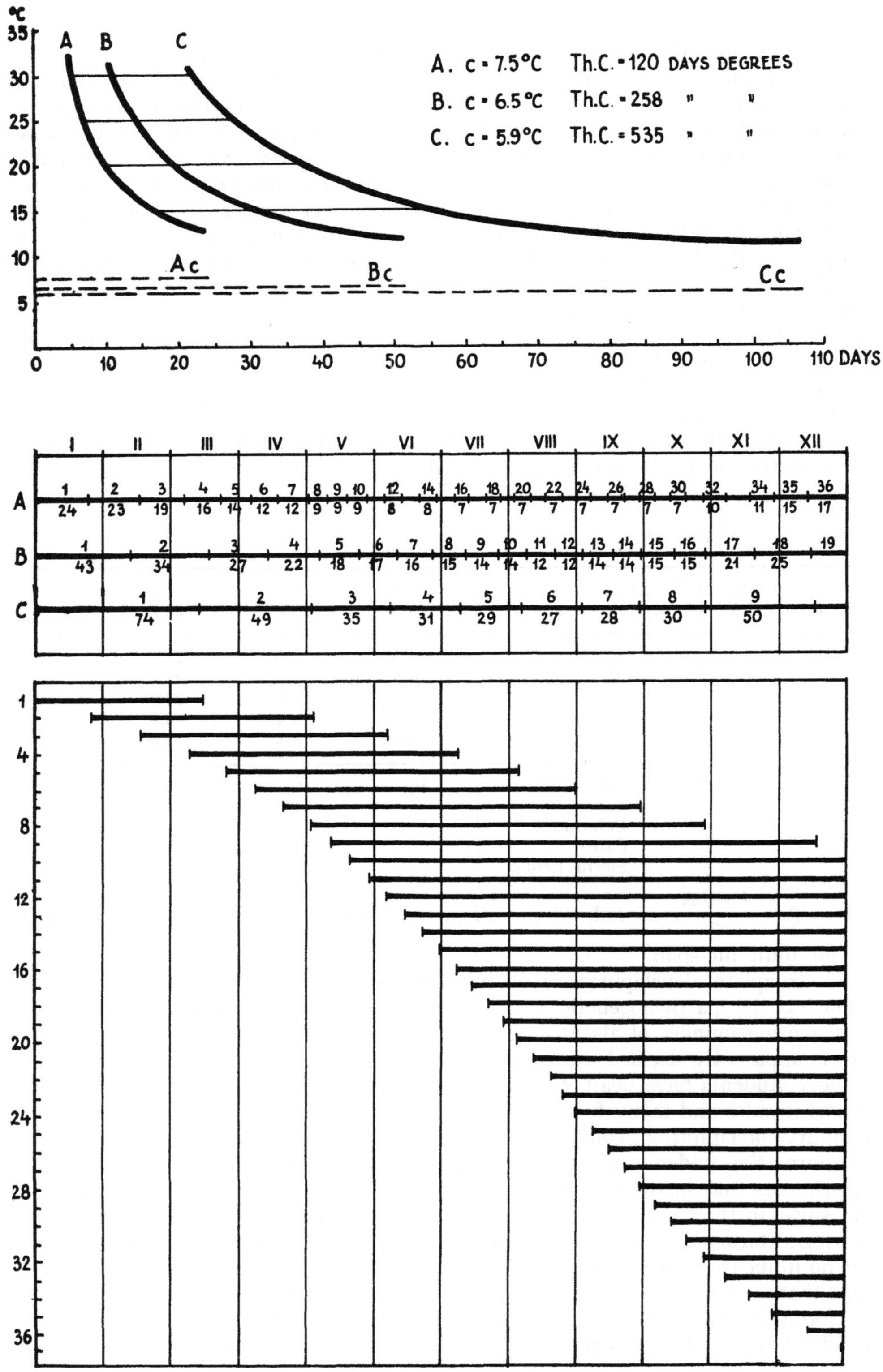

Fig. 40.
Life-history of the Black Citrus Aphis (Toxoptera aurantii Fonsc.) at Tel Aviv. 1. The hyperbola-band. 2. Local life-history, as calculated from the three hyperbolae. 3. The sequence and overlapping of the generations.

Host condition	Humidity	Development of young larvae into apterates	alates	n
Normal	10% R.H.	68%	32%	137
,,	96	91	9	158
No food	humid	80	20	101
,,	70	56	44	100

It is hardly probable that *one* factor alone is the only physiological factor affecting wing-development. But it is quite possible that in any specific environment one ecological factor is temporarily dominant. The general physiological conclusion justified from the accumulated information seems to be that almost any ecological factor (temperature, light, humidity, and chemical composition of host) when growing sufficiently unfavorable, induces wing formation in a quite unspecific way. Host condition and low air humidity, espec. the dry hamsins, seem to be prevalent factors in Palestine, inducing the appearance of alate females in the spring.

The reproductivity of *T. aurantii* is low in cool weather and depressed at temperatures above 25° C. The nativity in one series of breedings was:

Temperature	n	Number of larvae born per ♀
29° C	3	8 (5—10)
28	12	29 (15—42)
26	12	25 (9—38)
25	6	65 (56—72)
24	3	67 (56—85)
23	4	63 (45—85)
22	15	62 (29—85)
18	7	42 (42—58)
16.5	3	66 (62—69)
15	9	51 (31—75)
14	2	42 (39—45)

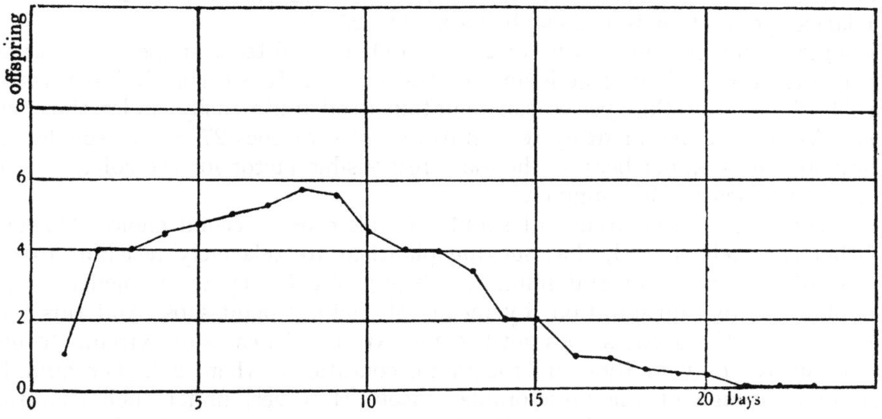

Fig. 41.
Daily larval production of apterate viviparous females of *Toxoptera aurantii* at 22° C. (after Rivnay).

Between 20 and 25° C the number of larvae produced surpasses 60 per female. The adult longevity ranges at those temperatures between 3 and 4 weeks. More larvae are born during the first half of this period (up to 7 larvae per day) than in the second half (fig. 40). The decrease of total reproductivity at higher temperatures has recently been well explained by Smith, who demonstrated that at higher temperatures the intensity of food-intake cannot keep pace with the speed of maturation and that at high temperateurs the females grow mature in a conditon of undernutrition.

It is quite certain that besides temperature, other factors (humidity, host-condition, early history of generation) influence the reproductivity. The influence of chemical composition of the host plant upon aphid-fertility has recently been most neatly demonstrated by Evans.

Seasonal fluctuation and environmental resistance: The importance of the single factors of abiotic and biotic environmental resistance in connection with the seasonal fluctuation of the aphid is rather complex.

The observer has the impression, at the very start, that the heat and dryness of the summer are the primary factors which are responsible for the summer-reduction of the aphid populations. And this impression grows considerably, upon following the effect of the spring hamsins, the hot and dry desert winds of that season. Abundant Aphid populations are reduced to scattered relics after a prolonged period of these winds. And the small, scattered summer colonies, which we find in many species, exist only on shady places. But even the largest reductions of numbers caused by hamsins last only for a very short time. In spring 1935, ex. gen. a ten day hamsin reduced the woolly apple-aphis practically to zero, in the neighbourhood of Jerusalem. But two to three weeks later the trees were covered again by this pest as if no destruction had occurred so short a time ago. This species persists quite abundantly on the lower side of the tender apple-twigs, throughout the summer. Aphids are very sensitive to high temperatures. At 27° C and higher no individual larva of *Toxoptera aurantii* developed successfully in the thermostat. The instantaneous heat death occurs between 33° C and 40° C in different species. 31—32° C is the preference temperature for several species. All these temperature constants are relatively low. The range of normal activity was between 7 and 32° C.

The shortest development of *Toxoptera aurantii* was observed at 25° C and the total larval production is highest between 22—25° C.

Although a sudden temperature of 27° C is detrimental to most species, a variable temperature of a 27° C average is not necessarily fatal. It is certainly less favorable than 23—25° C, but the species can maintain itself quite easily under those conditions. As our average monthly temperature rarely reaches 27° C. we should have to eliminate the summer heat as the main responsible factor for the collapse of the aphid populations in the summer.

The dryness of the environment should be taken as a second choice. However, the belief is widely spread, that sucking plantlice are relatively resistant to fluctuations of the environmental humidity, being able to replace immediately any losses of water by transpiration, by means of sucking plant-saps. And this holds good, as long as the dryness does not become extreme. In cases of extreme dryness, the lice survive only in those microclimatic conditions, where a higher humidity is retained throughout the unfavorable season. However, under such conditions they may thrive. Consequently, the low humidity cannot be considered as the most important responsible factor. But we should bear in mind, that the combined effect of light, temperature and dryness is necessarily much greater than that of every single factor alone.

Of much greater general importance is another factor, to which students of aphid biology have paid only slight attention. This factor is nutrition. The entomologist who breeds an aphis, always tries to supply the coming generations and new breedings with suitable food, which consists invariably of young shoots of the host. If he should attempt to breed an aphis on old, hard leaves of the host, he would be faced with the same results, which were ours, when infecting old citrus leaves with *T. aurantii*. In our experiment, all the larvae left the leaves and no development took place.

The same is true of other species and this phenomenon is in conformity with the well known sensitiveness of gall-forming aphids to a very short period of the leaves, which they infest. They die without having produced gall formations or offspring, when the larva is only several hours or days older than required. We can make the statement, that the abundance of *T. aurantii*, *Aphis calotropidis*, *A. frangulae*, *A. laburni* and many other species depends proportionately upon the presence of suitable food. The aphids are abundant whenever young sprouts abound and are extremely scarce when the latter are practically absent. This food relation is of greatest importance in Palestine and is here certainly of much greater importance in regulating the number of aphids than the effect of climatic factors combined. The mere fact that *Aphis frangulae* has its peak of abundance in summer is sufficient proof thereof. Suitable host conditions — the Cucurbitaceae are cultivated in summer — are the cause and the unfavorable climatic conditions do not interfere with its pullulation.

The aphids are one of the staple foods of nature and have countless enemies. The most important among them are *Aphidiinae* (*Lysiphlebus* sp.), *Syrphidae* (*Lasiophticus pyrastri*), *Coccinellidae* (*Coccinella 7-punctata*), *Cecidomyidae* (*Guerciobremia toxopterae*) and *Neuroptera* (*Chrysopa vulgaris*). Their decimation is enormous. But these enemies multiply only at a dense population level of the Aphids and practically disappear with the reduction of the Aphids. Despite their intensive decimation activity, they are of much lesser importance than the food factor. With the disappearance of Aphids in June, we find sometimes whole aphid colonies, which show the inflated appearance caused by the parasitism of *Lysiphlebus*. The voracity of the larvae of some ladybirds, syrphids and chrysopids is great, but their entire influence is restricted to the period of highest pullulation of the host, when the potential reproduction of the latter is enormous. They are rare or absent during the population ebb.

The importance of the single factors of the total environmental resistance for Palestine should therefore be ranged as follows:

Status of the host-plant
Enemies
Climatic factors.

This does not reduce the very important part played by climatic conditions in the seasonal population trend of *T. aurantii*. The temperatures of February to April are of decisive influence on the start of the spring pullulation. Cool weather delays it, warm weather helps to hasten it.

The following analysis of the outbreak in spring **1936** will explain this (Rivnay, **1936**). In early February and March, unusually heavy infestations of the young shoots was observed everywhere and caused much trouble amongst the farmers, but the outbreak broke down during April. In a normal year the population rise in February would have been very slow, the population rising in March to the level reached in **1936** in early February and would have slowly gone down until May. Fig. **42** shows that January and February were warmer than normally, and the warmer weather lasted until late April, when normal temperatures set in again. In II.1937 in 3 nights the minimum dropped below 0° C, in 4 nights to

3—4° C and in the rest of the nights to 6—8° C. The maximal daily temperature fluctuated from 14° C to 17° C, only on a few hamsin days early in the month it rose to 20° C. In II.36 the minimal temperature dropped in 6 nights to 4°C, generally to 6—8° C, and in a few nights to 10° C only. The maxima ranged from 18 to 24° C. The warm weather stimulated also an earlier shooting of leaves and en earlier flowering. The anticipated heat in February and March together with the anticipated shooting of the host caused a protracted period of optimal increase, which was ended by heavy hamsins in April 1936, when the maxima jumped up to 42° C.

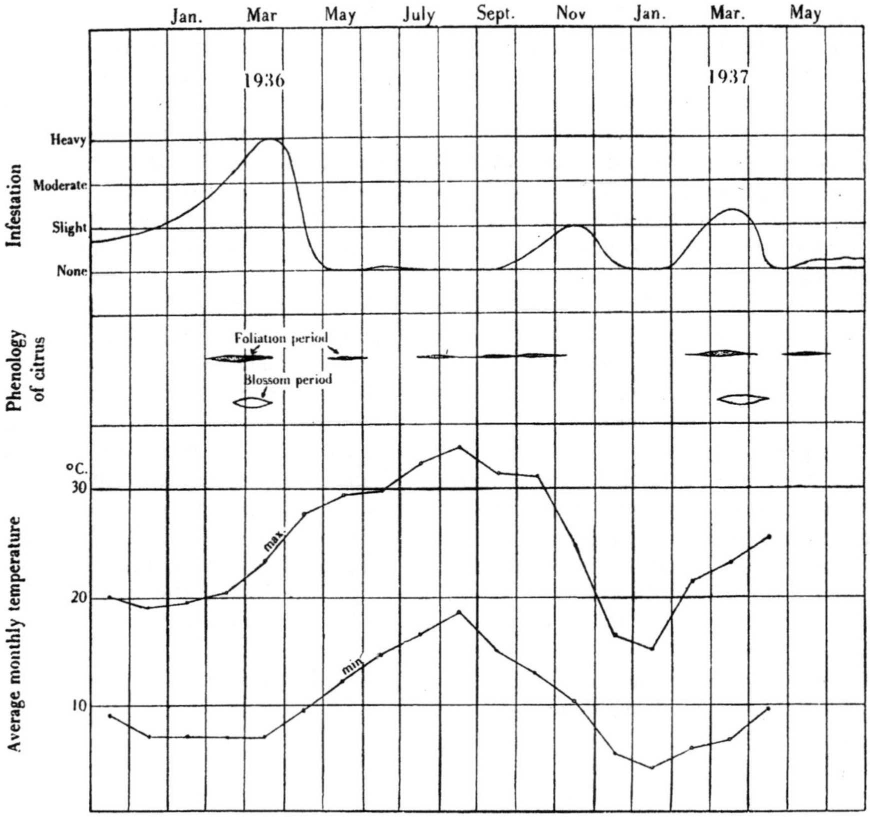

Fig. 42.
Seasonal population trend of *Toxoptera aurantii* in a normal (1937) and in an outbreak year (1936; from Rivnay).

Damage and control: Considering the polymerous feeding habits of the Black Orange Aphis the damage is rather limited. Young leaves are commonly attacked but damage is mainly restricted to young orchards. It is rare that blossoms and flowers are attacked intensively enough to suffer and damage to the young tender fruit, which falls in consequence, has never been observed yet by the writer. From Italy it is reported by Silvestri as a common type of injury. All parts, heavily infested, are arrested in their growth, as is shown in plate XI. In normal years the damage is moderate and becomes insignificant in older groves. Only those tender leaves are attacked, where the females settle on the lower side. The foliage curls and twists, offering in the interior of the rolled, cup-shaped leaves a perfect protection to the aphids. In cases of slight and late infestation, only partial curling of the leaves occurs. The fully grown, hard leaf is left as it is unsuitable as a nutritional substratum. On young trees it may, in years of outbreaks, destroy the whole new top on every branch.

Despite the heavy toll, which numerous natural enemies levy on the aphid populations, they are insufficient to prevent outbreaks. It is difficult to apply technical control measures owing to the crowded condition of the leaves and to the protection offered to the aphids by them. Spraying with or with any other aphicid in suitable concentration them. Spraying with nicotine-sulfate (40%; 1 : 1000) is advisable only when the late winter is extraordinarily warm. Spraying can then be performed in a dry interval in February, before the curling of the leaves has started and even then only in young orchards and nurseries. For the latter, it is an advisable precaution, whereas even in young orchards, the farmer generally prefers to take the risk, until it is too late to do anything else.

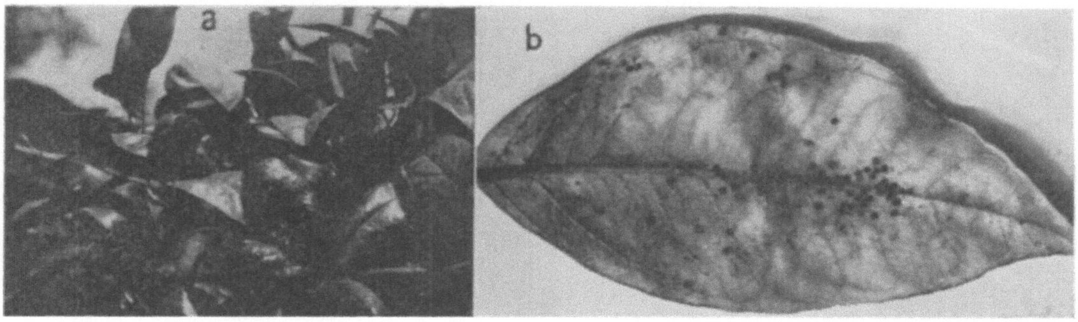

Fig. 43.
Orange leaves infested by *Aphis laburni* Kalt. A. rolled leaves; B. small colonies in early July of parasitised colonies of the aphid (from Bodenheimer).

G. OTHER APHIDS ON CITRUS.

A few other plant-lice have been occasionally observed on citrus in Palestine, none causing real damage and calling for control measures. The very polyphagous light green *Myzodes persicae* Sulzer is found occasionally on citrus leaves, which slightly curl. It is a pest on tobacco and some fruit-trees. The equally polyphagous greenish melon aphis *Aphis frangulae* Kaltenbach sometimes develops small colonies on citrus leaves. It causes serious damage in summer on melons and cucumbers. The black *Aphis laburni* Kaltenbach, which usually infests leguminous herbs and crops, was found twice on citrus leaves, which they curled.

References.

F. S. Bodenheimer, Schaedlingsfauna Palaestinas. Berlin, 1930.
F. S. Bodenheimer, Acti VI. Internat. Congress. Entom. 1935. Madrid, 1940. p. 49—58.
E. O. Essig, Pomona College Journ. Entom. pp. 601—603, fig. 196.
E. O. Essig, Mthl. Bull. St. Comm. Hortic. **1**, 1912, p. 131—132, fig. 45.
R. Mayne, Insectes attaquant le Cacoyer au Congo Belge. London, 1917. p. 49.
E. Rivnay, Hassadeh, 1936, pp. 290—294.
E. Rivnay, Bull. Ent. Res. **28**, 1937, pp. 173—179.
E. Rivnay, Ann. Appl. Biol. **25**, 1938, pp. 143—154.
A. F. Swain, Univ. of California Public. Entom. **3**, 1919, p. 129.
R. Takahashi, Dept. of Agric. Formosa Rept. 4. 1923, pp. 93—96.
F. V. Theobald, Bull. Ent. Res. **4**, 1914, pp. 329—332; **7**, 1917, p. 337—342.
P. Van der Goot, Contrib. Faune des Indes Neerlandaises **1.2**, 1914, pp. 76—81, 294.
P. van der Goot, Spolia Zeylanica **11**, 1921, p. 72.

6. SCALE-INSECTS (RHYNCHOTA II).

A. KEY TO DETERMINATION OF THE COCCOIDEA (ADULT FEMALES), INFESTING CITRUS IN THE MIDDLE EAST.

Key to families.

1. Body always covered with a real scale furnished with one or two exuviae. Small, very rarely exceeding 3—4 mm. Legs absent; antennae tuberculate: rostrum consisting of a single segment. Abdomen divided into 4 movable cephalic segments (= preabdomen) and a caudal series of fused inflexible segments forming pygidium, bearing plates and lobes on the hind-margin. Armoured Scales, *Diaspididae*.
   Body not covered with real scale bearing one or two exuviae 2.

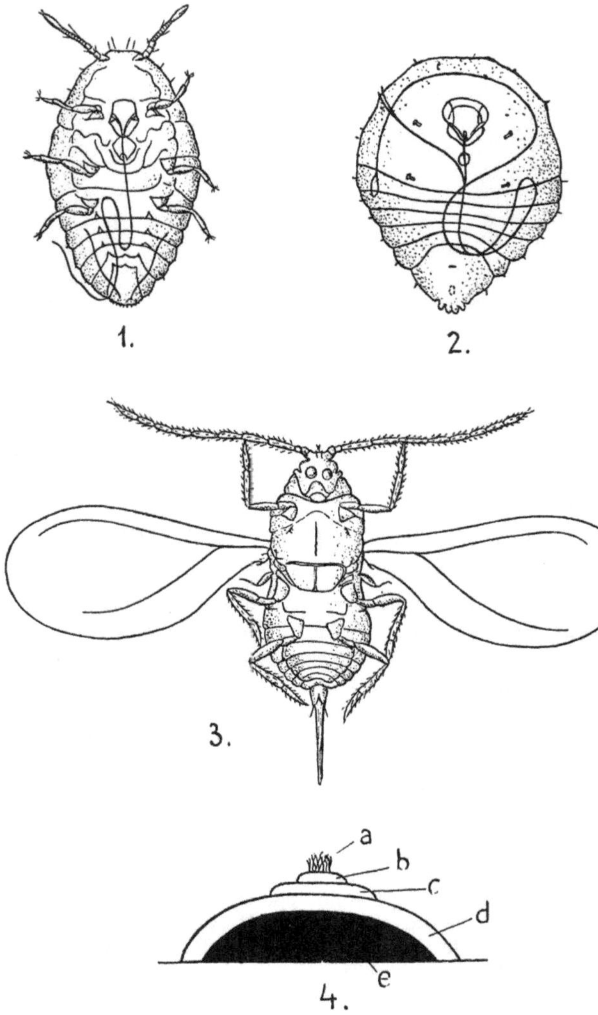

Fig. 44.
Stages of an armoured scale (*Aspidiotus hederae* Vall; *Diaspididae*): 1. crawler; 2. adult female, 3. adult male, 4. cross-section through scale of adult female: a. waxcap, b. first larval exuvia, c. second larval exuvia (or scale), d. third scale, e. female's body (from Bodenheimer).

2 Body not segmented, generally highly convex (only *Coccus hesperidum* flat to slightly convex). Without locomotion in the adult stage. Legs much reduced. Abdomen with anal cleft at caudal end with pair of moveable opercula at cephalic end of cleft. Rostrum consisting of single segment. Antennae 7—8 jointed. Soft Scales, *Coccidae*.
Body always segmented. Adults with fully developed legs and with locomotory power until the beginning of oviposition; never with anal cleft. 3.

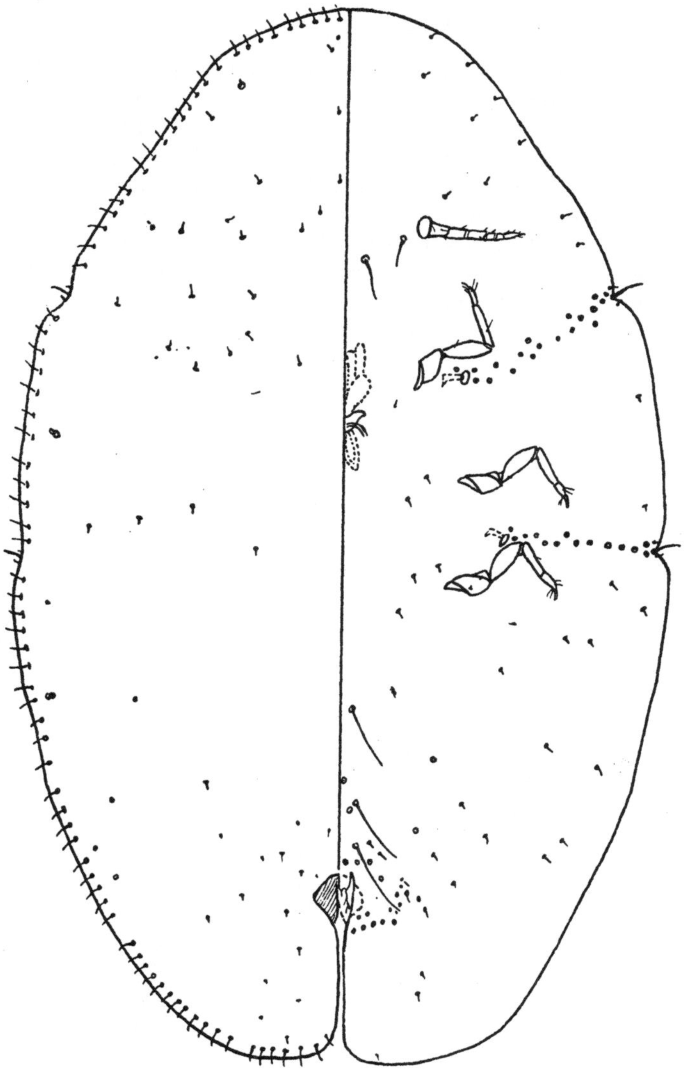

Fig. 45.
The Brown Soft Scale (*Coccus hesperidum* L; *Coccidae*). Left: dorsal, right: ventral. Pay attention to the anal cleft and to the (striped) operculum (after Steinweden).

3. Insects large with large, white, distinct ovisac at caudal end. Antennae 10—11 jointed. Rostrum consisting of 2—3 segments. Abdominal stigmata present. Anal orifice without a setiferous ring. Fluted Scales, *Margarodidae; Monophlebinae*.

Insects covered with whitish secretions („mealy bugs") and with marginal wax-processes; egg-laying into an unformed ovisac of wax-threads. Antennae 7—9 jointed. Rostrum consisting of 2 segments. Anal orifice with anal ring and setae. (Fig. 47). Mealybugs, *Pseudococcidae*.

Fig. 46.
Adult of *Icerya purchasi* Mask. on ovisac (from Bodenheimer).

Key to the species.

Diaspididae (adult females).

1. Second pygidial lobes usually bilobate (or scale elongate). Anterior spiracles with disc pores. Antennae with 2 or more setae. Macroducts two-barred (*Diaspidini*) 12.

Second pygidial lobes always simple. Anterior spiracles very rarely with disc pores. Antennae with one seta only, as a rule. Macroducts one-barred. Scale always roundish. (*Aspidiotini*) 2.

2. Pygidium with paraphyses from the base of present or obsolete lobes (more than a sclerotisation of the folds of a pore or a pore furrow). The paraphyses are sometimes small, but always conspicuous. 3.

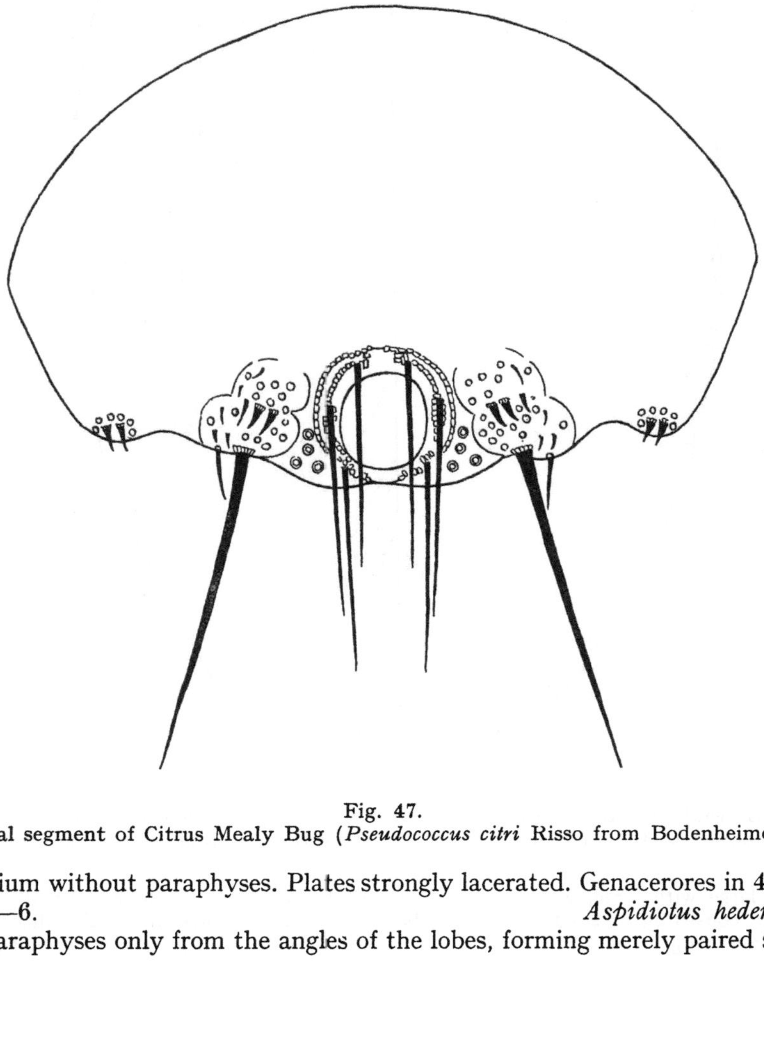

Fig. 47.
Anal segment of Citrus Mealy Bug (*Pseudococcus citri* Risso from Bodenheimer).

Pygidium without paraphyses. Plates strongly lacerated. Genacerores in 4 groups: 8—9: 5—6. *Aspidiotus hederae* Vall.

3. Paraphyses only from the angles of the lobes, forming merely paired supports

for the lateral margins of the intersegmental poriferous furrows. Anal opening large, plates well developed. The median lobes only conspicuous. 4.

At least one paraphyse arises from about the centre of at least one interlobal space (usually between second and third lobes). Frequently with paraphyses beyond the third lobe. Anal opening small to moderate. 5.

4. Genacerores in four groups: 2—10 : 3—8. *Hemiberlesia lataniae* Sign.

Genacerores wanting. *Hemiberlesia camelliae* Sign.

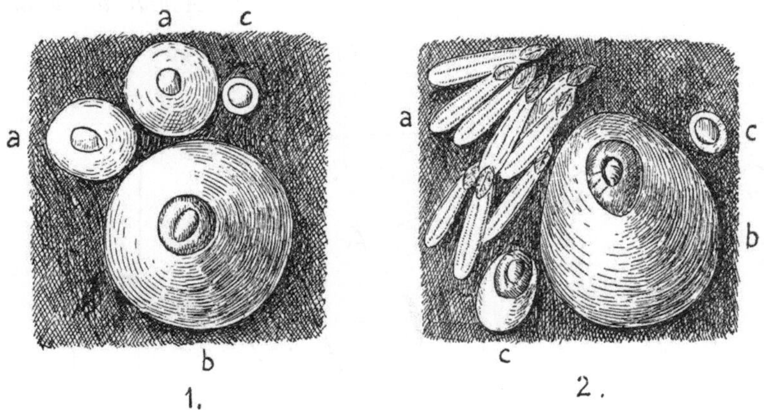

Fig. 48.
Scales of 1. the Aspidiotine *Aspidiotus hederae*, Vall. 2. the Diaspidine *Sasakiaspis pentagona* Targ. a. male puparium, b. female scale, c. first larval scale (from Bodenheimer).

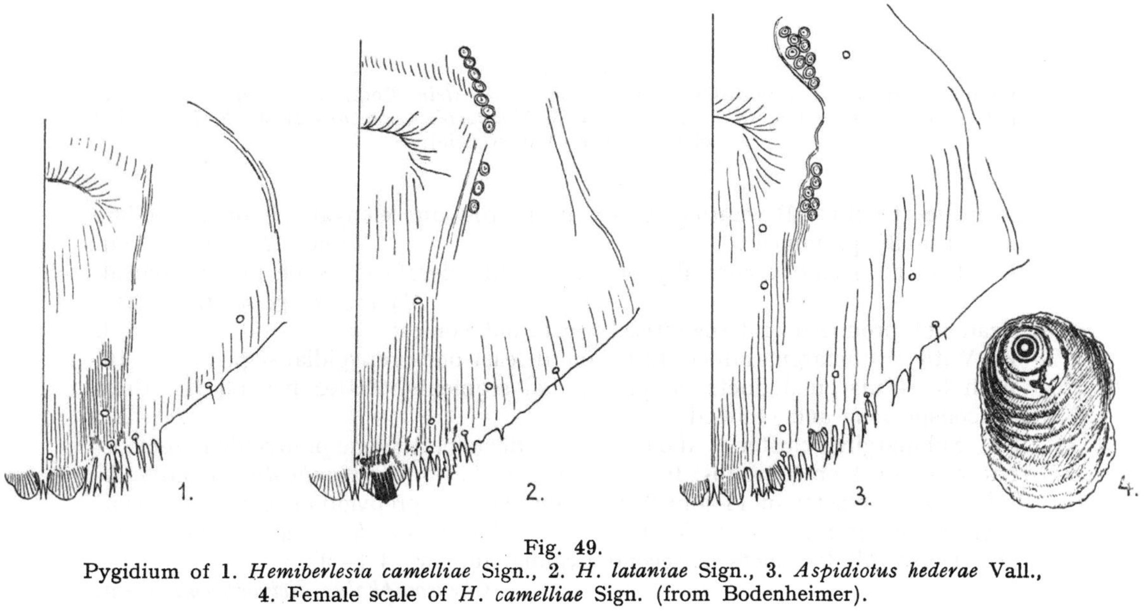

Fig. 49.
Pygidium of 1. *Hemiberlesia camelliae* Sign., 2. *H. lataniae* Sign., 3. *Aspidiotus hederae* Vall., 4. Female scale of *H. camelliae* Sign. (from Bodenheimer).

5. Prosoma much swollen and strongly sclerotised, often reniform and embracing pygidium. 6.

Prosoma, if sclerotic or swollen, not reniform. 8.

6. Genacerores in 4 groups: 1—7 : 1—5. With submarginal clusters of macroducts on three prepygidial segments. *Aonidiella orientalis* Newst.

Genacerores and prepygidial submarginal duct clusters wanting. 7.

7. With sclerotic structure B on ventral part of pygidium. Sclerotic structure A circular. Fourth lobe-like process usually inconspicuous. *Aonidiella aurantii* Mask.

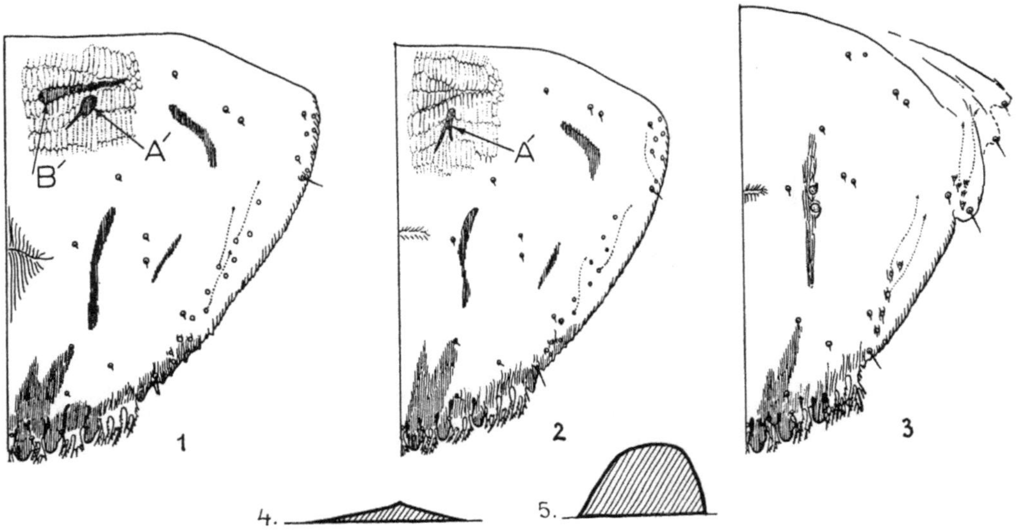

Fig. 50.
Pygidial structures of 1. *Aonidiella aurantii* Mask., 2. *A. citrina* Coq., 3. A. *comperei* McKenz., Female scale (cross-section) of 4. *A. aurantii*, 5. *Mycetaspis personatus* Mask. A, B pygidial scleroses (after Mac Kenzie).

Sclerotic structure B wanting A, usually narrow and elongate. Fourth lobe-like process usually prominent. *Aonidiella citrina* Coq.

8. Head at maturity conically produced and sclerotised. Scale highly conical. *Mycetaspis personatus* Mask.

Head not produced and sclerotised. Scale not conical. 9.

9. With a submarginal duct cluster on at least one prepygidial segment. 10.

With 3—4 marginal ducts on prepygidial segments. Plates beyond the third lobes conspicuous and clubbed. 11.

10. Submarginal cluster of about 15 macroducts on the one prepygidial segment. Genacerores in 4 or 5 groups: 0—1 : 4—10 : 4—8. *Chrysomphalus aonidum* L.

Submarginal cluster of 11 and 9 ducts on the two prepygidial segments. Genacerores in four groups: 3—4 : 3—4. (*Chrysomphalus bifasciculatus* Ferr.)

11. Dorso-tubular ducts of second pygidial furrow: 8 (—9). *Chrysomphalus dictyospermi* Morg.

Dorso-tubular ducts in second pygidial furrow: 15 (10—18).

Chrysomphalus pinnulifer Mask.

12. Adult females entirely enclosed within second exuvia, scale elongate and white, with first exuvia only (cryptogyrous). 13.

Adult female not cryptogynous, scale with two exuviae. 14.

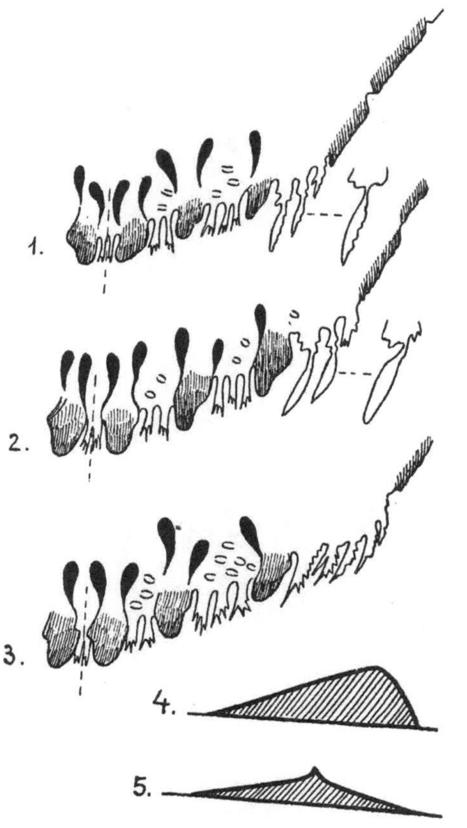

Fig. 51.
Pygidia of 1. *Chrysomphalus dictyospermi* Morg., 2. *C. pinnulifer* Mask., 3). *C. aonidum* L., Female scale (cross-section) of 4. *C. aonidum*, 5. *C. dictyospermi* (after Balachowsky).

13. Dorsal pygidium with about 16 sclerotic patches arranged in two rows. Lobes slender, tricuspidate, plates slender. Genacerores in 5 groups: **34**:13:21.

Leucaspis japonica Cock.

Dorsal pygidium with no sclerotic patches. First lobes slender, entire, plates very flat, broadly rounded, not fimbriate as in preceeding species. Genacerores in 5 groups: 14 : 12 : 12.

Leucaspis hydrangeae Tak.

14. Median lobes large and at basis yoked by an internal sclerosis.

Sasakiaspis pentagona Targ.

Median lobes not yoked at base. 15.

15. Marginal pygidial macroducts with orifice transversal and parallel to margin and with heavily sclerotised rim. Pygidial margin beyond the lobes with a continous series of broad fimbriate plates which extend to the margins of the prepygidal segments. Scale roundish. 16

Marginal pygidial macroducts with the orifices diagonally or longitudinally. No fimbriate plates on prepygidial segments. Scale elongate. 20.

16. No sclerotised gland tubercles on the prosoma and with an ear-like membranous lobe each side laterad of anterior spiracles.

Parlatoria zizyphus Lucas.

With at least a few sclerotised gland tubercles on prosoma, not with ear-like lateral lobes. 17.

17. Fourth lobe sclerotised, toothlike, notched. 18.

Fourth lobe absent or not sclerotised. 19.

18. With one pair of orifices of dorsal macroducts in to area between the anterior genacerores (ventral), another pair each side just covering them from above.

Parlatoria morrisoni Bdhmr.

No orifices of dorsal macroducts in the said area.

Parlatoria pergandii Comst.

19. Pygidium usually with four plates between third and fourth lobes. All pygidal lobes large, prominent and deeply notched on the outer margin.

Parlatoria oleae Colv.

Pygidium usually with three plates between third and fourth lobes. 20.

20. Plates between third lobe and site of fourth lobe broad at base and tapering towards apex. Dorsal macroducts present in area close to anus.

Parlatoria cinerea Hadden.

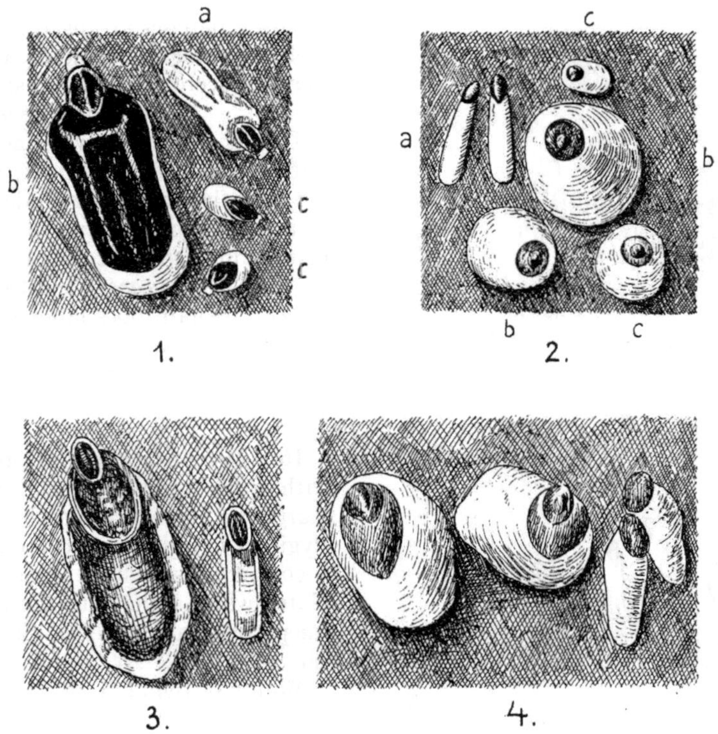

Fig. 52.
Scales of 1. *Parlatoria zizyphi* Luc., 2. *P. oleae* Colv., 3. *P. proteus* Curt., 4. *P. pergandii* Comst. (a—c as in fig. 48; from Bodenheimer).

Plates between third lobe and site of fourth lobe broad at base and expanding toward apex. No dorsal macroducts in paraanal area. *Parlatoria proteus* Curt.

21. The three prepygidial abdominal margins with chitinous spur. Median plates somewhat shorter than median lobes. Scale very narrow.

Mytilococcus gloveri Fack.

Prepygidial abdominal margins without distinct spurs. Median plates about twice as long as median lobes. Scale elongate mussel-shaped.

Mytilococcus beckii Newm.

COCCIDAE.

1. Adult females highly convex, the sides and back covered by a thick, comp act wax-layer. 2.

Adult females flat or convex, but not covered by a compact wax-layer. 3.

2. Females covered by a central and eight marginal wax-plates. 4 mm long. Antennae 6-jointed. *Ceroplastes rusci* L.

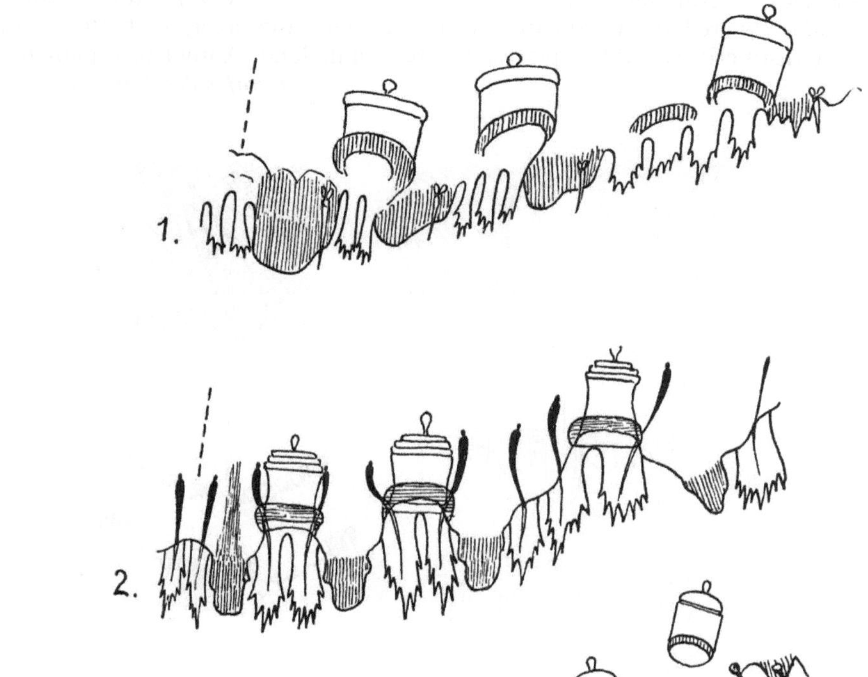

Fig. 53.
Pygidia of 1. *Parlatoria oleae* Colv., 2. *P. zizyphi* Luc., 3. *P. pergandii* Comst., 4. *P. proteus* Curt. (from Bodenheimer).

Females covered by a central and six marginal wax-plates. 4 mm long and very highly convex. Antennae 7-jointed. *Ceroplastes sinensis* D.G.

Females covered by an apparently homogenous wax-layer, with tubercles, but without distinctly separated plates. Up to 3 mm long. Antennae 6-jointed. *Ceroplastes floridensis* Comst.

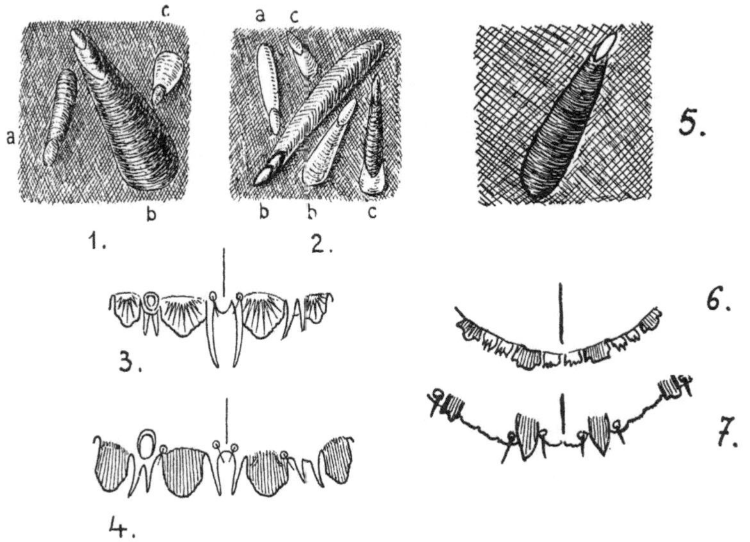

Fig. 54.
Scales of 1. *Mytilococcus beckii* Newm., 2. *M. gloveri* Pack. 3. middle of pygidial margin of *M. beckii*, 4. of *M. gloveri*. *Leucaspis hydrangeae* Tak. 5. female scale, 6. pygidial margin of second larva, 7. of adult female (from Bodenheimer).

3. Mature females producing a big, white ovisac. 7.

Mature females not producing a conspicuous white ovisac. 4.

4. Females flat or moderately convex only. 5.

Females highly convex. 6.

5. Females asymmetric oval, light to dark brown. *Coccus hesperidum* L.

Females symmetrically elliptic, greyish. *Coccus pseudomagnoliarum* Kuw.

6. Females with a distinct H-shaped crest on the back. *Saissetia oleae* Bern.

Females with smooth back, without such crest. *Saissetia hemisphaerica* Targ.

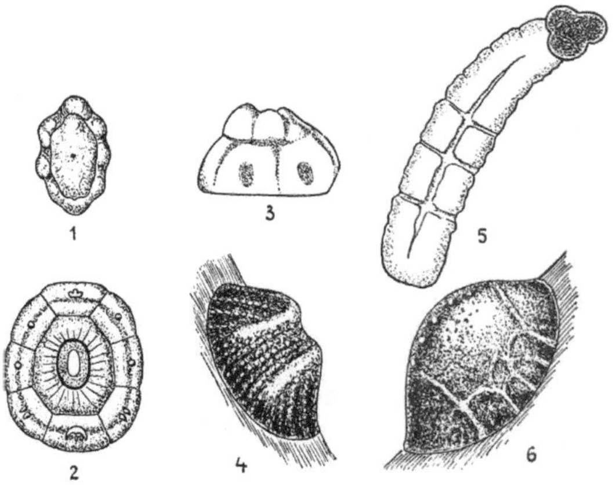

Fig. 55.
Adult females of Citrus Soft Scales: 1. *Ceroplastes floridensis* Comst., 2. *C. rusci* L., 3. *C. sinensis* Guerc., 4. *Saissetia oleae* Bern., 5. *Pulvinaria floccifera* West. with ovisac, 6. *Saissetia hemisphaerica* Targ. (from Bodenheimer).

7. Ovisac short, two to three times as broad as the female.
Pulvinaria aurantii Ckll.
Ovisac long and narrow, not distinctly broader than the body of the female. 8.
8. Female reddish-brown with transparent greenish-yellow margins, flat.
Pulvinaria floccifera West.
Female pale yellow to greenish with irregular black markings, moderately convex.
Pulvinaria psidii Mask.

MARGARODIDAE.

1. Body dark red. *Icerya purchasi* Mask.
Body orange-yellow. *Icerya aegyptiaca* Dougl.

PSEUDOCOCCIDAE.

1. Antennae of female 9-jointed. Body reddish, covered with mealy secretions, no marginal wax-filaments. *Phenacoccus hirsutus* Gr.
Antennae of female 8 (or 7-jointed). Marginal wax filaments conspicuous. 2.
2. Anal cerarii on heavily chitinised area. 3.
Anal cerarii not on heavily chitinised area. 4.
3. Caudal wax filaments as long or longer than body. Disc pores on last cerarius clustered and touching one an other. Circumanal setae surpassing caudal setae. *Pseudococcus adonidum* L.
Caudal wax filaments robust, at most as long as half the body length. Disc pores of last cerarius scattered, not touching one an other. Circumanal setae not surpassing the caudal setae.
Pseudococcus gahani Gr.
4. Caudal wax filaments very short, not reaching one fourth of body length. 5.
Caudal wax filaments robust, surpassing half the body length. With a few scattered translucent pores on the hind limbs (much less than in *P. maritimus*). Last cerarius on slightly chitinised area only, with disc pores not in cluster as in *P. adonidum*, but less scattered than in *P. gahani*.
Pseudococcus comstocki Kuw.-group.

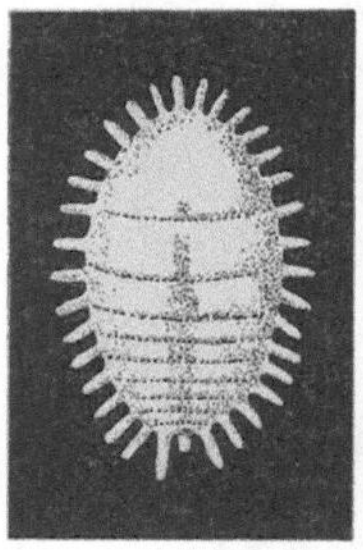

1

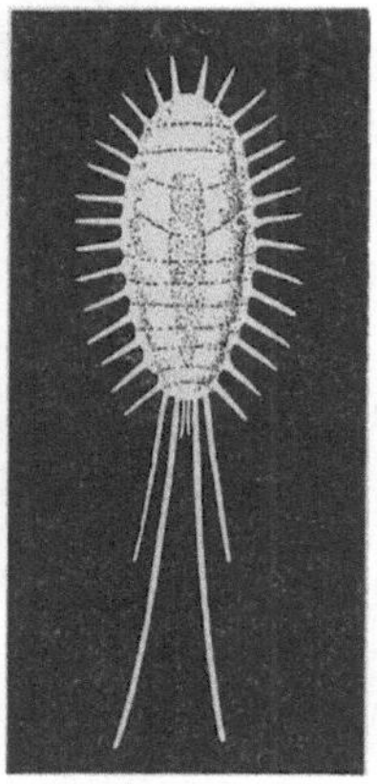

2

Fig. 56.
Females of 1. *Pseudococcus citri* Riss. and 2. *P. adonidum* with wax-filaments (from Bodenheimer).

5. Hindlegs covered with many translucent pores. Hindtibiae usually markedly dilated. Disc pores on last cerarius clustered. Caudal wax filaments slender, three times as long as lateral ones. *Pseudococcus maritimus* Ehrh.
Hindlegs without translucent pores, hindtibiae not dilated. Disc pores on last cerarius scattered. Caudal wax filaments robust, at most twice as long as latera ones.
Pseudococcus citri Risso.

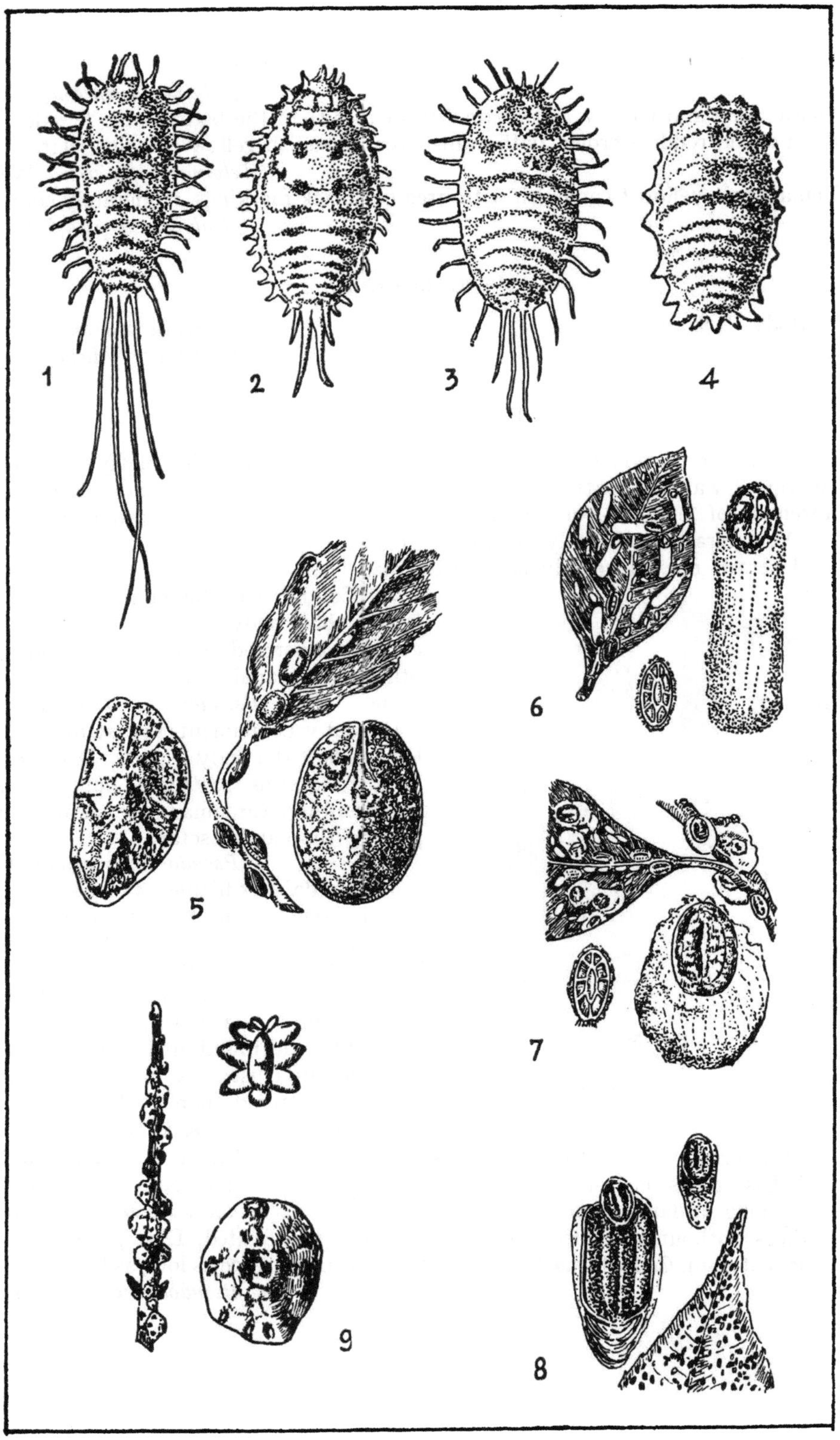

XII. Citrus scale insects from the Black Sea region of the Caucasus: 1) *Pseudococcus adonidum* L., 2) *P. gahani* Green, 3) *P. maritimus* Ehrh., 4) *P. citri* Risso, 5) *Coccus hesperidum* L. (left) and *C. pseudomagnoliarum* Kuw. (right), 6) *Pulvinaria floccifera* Westw., 7) *P. aurantii* Ckll., 8) *Parlatoria zizyphi* Luc., 9) *Ceroplastes sinensis* Guerc. (after Gogiberidze).

B. GENERAL REMARKS ON THE BIOLOGY OF THE ARMOURED SCALES (*Diaspidae*).

A few general remarks on the life-history of the armoured scales will be welcome to the non-specialist of this group.

The life-cycle is markedly different in both sexes. Some of the species are oviparous, others viviparous. Perivulvar glands (genacerores) are found as a rule in egg-laying species. In many of the oviparous species, the development of the embryo has already proceeded far, by the time the eggs are laid. Only the very latest period of embryonic development is passed after oviposition; such species are called ovivivi-parous.

The hatching larva has six well developed legs and two normal antennae. These crawlers enter the bristles of their mouth-parts into the plant-tissue and settle down. Soon after having successfully settled, they secrete a whitish cap and the first scale is formed beneath this first protective cover. Both the free-living crawler and the larva beneath the first stage scale belong to the same stage, the first larva. Only after some days or weeks, the first moult occurs, after which the legs disappear, and the antennae are much reduced, but the pygidial characters, i.e. the chitinous plates and lobes on the anal margin of the last abdominal segment develop. These pygidial plates and lobes are a help to the larva and adult female in their locomotion around their fixed beak. They may also be of help in the distribution of the eggs below the scale. After some time, the second female larva moults again and changes into the adult female. The female has no legs and very rudimentary antennae only. It is often pear-shaped or long-oval just after moulting, but when the growing eggs fill its abdomen, the latter often grows very large. It may obtain a reniform shape, the pygidium protruding from the centre of the "renal" concavity (*Aonidiella*). The female turns, at this stage, as often occurs in parasites, into a mere ovisac. Almost all armoured scales go through a relatively long preoviposition period (often as long or longer than the total larval development). Eggs or larvae are then deposited, never all at a time, but successively, a few every day for a period of some weeks to some months. The oviposition activity is highest — as a rule — at the beginning and — roughly speaking — half of the eggs are laid at the end of the first third of the oviposition period. After the end of oviposition, the females die rather quickly. A total longevity of 4—8 months is quite usual among the females of armoured scales; egg- and larval development are only a fraction, generally less than a third, of the total longevity (except if those stages pass the winter).

The early development of the male is very similar, in most cases morphologically still indistinguishable at first from that of the female during the egg and the first or even the second larval stage. The second larva often shows definite sexual dimor-phism. Before turning adult, it passes two other moults: into a first nymph or prepupa with short antennal- and wing-sheads and into a second nymph or pupa with longer antennal- and wing-sheads and a distinct stylus. In both stages no food is taken. Each of these stages lasts, as a rule, a few days only, and then the adult male hatches. The latter is characterized by unfunctional mouthparts, extremely reduced venation of the forewings, halteres instead of hindwings, peculiar eyes which are not homologous to facett-eyes and a long abdominal appendix, the stylus, sheating the penis. The delicate males, which seem to be invariably vesperal or nocturnal in their activity, live from 2 to 5 days only and die.

Copulation takes place within this short adult life-span, soon after the second moult of the female. The male dies long before the first eggs are deposited. It is still extremely difficult to say as to how the male is able to insert his stylus within the female vagina. The female has in the species with rounded scale no fixed position beneath the scale to be noticed from the outside. It seems most probable that the female leaves behind a fissure, when the building of the third scale is finished. The male is perhaps attracted by the smell emanating from this fissure (large antennae!) and inserts there the stylus.

Many species are able to develop parthenogenetically, others definitely not. In the first group we find either normal sex-ratio or females only. No case of pure male-production has yet come to our knowledge. In some species, as in *Aspidiotus hederae,* two biological races occur, one with obligatory copulation and the other with parthenogenetical development, as a rule.

The formation of the scale needs some explanation. As soon as the crawler begins to suck on its host plant, it simultaneously starts to excrete coarse wax threads from a few body glands located mainly near the anterior margin. These threads become entangled and form a sort of whitish cap over the body of the crawler. The secretion of the permanent scale begins beneath this first protective cover. The threads are very much finer than the first ones and cover the insect more compactly. The formation of the scale proceeds from the posterior to the anterior part of the body. After the larval body is completely covered, two horn-like projections are still being formed at the anterior end, serving the antennae, in most of the elongate scaled *Diaspidinae.* This scale now becomes more compact and thicker from day to day.

The moulting is different in different scales. In the case of *Aonidiella, Chrysomphalus, Aspidiotus,* etc., the rent in the old skin occurs along the margin of the body, so that the dorsal half becomes incorporated in the dorsal scale and the ventral half of the ventral scale. The rent in the skin of *Mytilococcus* occurs on the ventral side in a more or less irregular line a short distance from the margin. At the anterior end this is between the rostrum and the antennae; on the sides, between the legs and the body margin; and at the posterior end, just anterior to the pygidium. The insect first emerges, presumably at the anterior end and later pulls the posterior end forward and downward, leaving the old skin directly above it. The exuvium, therefore, is entirely in the upper or dorsal part of the scale covering. (Quayle). Antennae, pygidium and legs are easily recognisable in the dorsal scale, after the latter has been treated in potash.

When the first moult is finished, being supported by convulsive movements of the insect's body, the scale is extended. In the round-shaped scales the larvae move turning themselves around their own beaks and always excreting on the posterior end, adding and extending the scale. As no legs are present anymore, the movements probably occur with the support of the strongly chitinized pygidial margin. The exuvia is solidly incorporated into the scale. In the elongate scales, the female moves and extends backwards only. After the second moult, the second exuvia is likewise inserted in the scale and the scale of the adult female is extended in the same way as in the second stage. It is often overlooked that a ventral scale which sometimes is very thin and sometimes rather thick is formed in a similar way.

In the male development only the first skin is incorporated into the scale covering. The narrow, elongate covering, the male puparium, is completely formed during the second larval stage and three moults occur beneath this scale, the exuviae being accumulated at the posterior end of the puparium.

The metabolism of the armoured scale is still fairly unknown. The intestinal tract is interrupted, no connection existing apparently between mid- and hind-intestine. The secretory glands fill a rather large part of the body during growth and are probably not only secretory but also excretory. All the food sucked from the plant diffuses into the haemolymph. No defaecation occurs at all. The wax, respectively scale secretion, must therefore perform both functions (defaecation and excretion). This process has never been analyzed and is the more remarkable as all the near relatives (other coccids, aphids) are rather extremely heavy defaecators of honeydew. All armoured scales live on cell contents of their hosts, which are rich in proteins, whilst the heavy honeydew defaecators live on the saps of the plant vessels, which are extremely poor in proteins.

C. THE (IVY OR) OLEANDER SCALE (*Aspidiotus hederae Vall.*)

Synonymy: 1829 *Chermes hederae* Vallot
1833 *Aspidiotus nerii* Bouché
1869 *A. ceratoniae* Signoret
1869 *A. hederae* Signoret
1869 *A. limonii* Signoret
1880 *A. oleae* Colvée and many more.

Description: Scale of adult female subcircular, flat to moderately convex, of young female circular and whitish, later yellowish to pale brown, the exuviae central and yellow. Ventral scale delicate, adherent to the plant. 1.5—2.5 mm.

Scale of male puparium flat, variable in form being usually ovate, white. Exuviae central, yellow, 1.0—1.5 mm long.

Adult female broad pear-shaped (young) to subcircular (old), broadest about the middle. Rostral loop long. Antennal tubercle furnished with a long curved hair. Colour light yellow, with the pygidial margin darker. No parastigmatic glands. The free abdominal segments distinct in young individuals, but obscured when the body becomes distended with eggs.

Three pairs of well developed variable lobes on the pygidial margin: the median one is the longest, the second somewhat smaller, the third small, thin and pointed. The first lobe is notched on both the inner and outer sides, the other two indistinctly notched on the outer side. Plates prominent and branched. Pygidial formula runs as follows: P_1, L_1, $2P_2$, L_2, $3P_3$, L_3, 7—$8P_4$.

The anal opening is circular, at a distance of three times its diameter from the median lobe. Circumgenital glands generally arranged in four groups:

0 — 2			0
4—18	4—18	generally	8—13;
2—13	2—13		5—10

Marginal gland orifices rather large. Dorsal pores arranged in 3—4 indistinct rows.

Male: bright yellow, eyes black, legs and antennae yellow-brown. Body short and relatively stout for the family. 0.8 mm long (+ 0.3 mm stylus).

Egg: long oval, pale yellow, 0.15—0.2 mm long; 0.1 mm broad.

Neonate larva: long oval, yellowish, 0.3 mm long.

Distribution: The oleander scale is probably the most ubiquistic of all armoured scales. To-day, it is of general occurrence throughout the world in tropical and subtropical climates. In more moderate climates it is still common in greenhouses. Following Balachowsky (1931) its original home is S. Africa, where many of its nearest relatives lives. This is certainly as incorrect as the earlier assumption of the present writer, that the Levant be its home. The revision of Ferris (1941) has reduced the genus to at most 12 species of Palaeotropical origin. The actual home of *A. hederae* is entirely obscured by its wide distribution by human agency. Wherever it occurs, it is also very abundant.

Host-relations: A. hederae is one of the most polyphagous of all armoured scales. Hundreds of hosts are recorded from over a hundred different plant families. It is of interest to note that the same plant is not attacked everywhere with equal intensity. Whereas olive fruits are heavily attacked and badly deformed in Palestine, Africa Minor and the interior of California, no such damage is known in S. Europe. In Palestine, the preferred hosts are *Nerium oleander, Acacia longifolia, Ceratonia siliqua* and *Melia azederach.* It is often injurious to ornamental plants and occasionally to olives.

Citrus as a host of A. hederae: Heavy damage to lemons is reported from Sicily, Italy, Cyprus.

Other records from citrus, which are probably more or less occasional only, are as follows:

Greece Koroneos, 1934), S. France (Balachowsky, 1932), Africa Minor (Balachowsky, 1932), Morocco (Lepiney and Mimeur, 1931), Chios, Spain, Syria (Lindinger, 1912), S. Africa (Brain), Florida (Merril and Chaffin, 1923), California (Essig, 1915).

In Palestine and Egypt it has not yet been recorded from citrus. Lemons and citron are heavily attacked in all regions, whereas oranges and grapefruits show a very slight infestation.

The problem of the non-infestation of citrus fruits by the oleander scale seemed to deserve special study, which has been made by H. Steinitz in our laboratory:

1. Crawlers normally settled on leaves of sour orange, formed, with few exceptions, the first scale and died shortly afterwards. A second infestation did not succeded.

2. Crawlers settled normally on leaves of sweet lemon, formed, almost all, the first scale. Most of them died in the first larval stage; the few which survived the first moult perished soon afterwards. Only one of two infections on twigs of sweet lemon succeeded, but only a few crawlers settled, dying shortly after the formation of the first scale.

3. The crawlers died soon on twigs of shamuti. Only half of the larvae succeeded in forming the first scale and these perished soon afterwards. On leaves, about ⅔ of the crawlers succeeded in forming the scale, but they also died soon. Conditions are better on shamuti fruits, where development is partly completed. Some of the males hatch, others die as adults beneath the puparium. The mortality of second larvae is high in the female and increases in the adult females, which die before reaching maturity. Only two out of nine females contained young eggs, three eggs being the maximum. Scale and females are small, the first reaching an area of 105 (relative size).

4. Conditions for development are better on grapefruit twigs than on the twigs discussed above. Some larvae reached the second stage, but development is extremely slow. At an age of 54—60 days, when the larvae developing on the leaves were more or less full-grown, a larva on the twig just passed the first moult. Leaves of grapefruit are a better substrat than even the fruits of shamuti. Mortality is relatively low in the first larval stage, increases in the second larval stage and most of the females die before reaching maturity. Most of the males hatch. The maximal number of intraabdominal eggs was four. These were of normal size and started embryonic development. The scale of the female was flat, their area covered by them reaching 187. Most individuals die in the first and second larval stages even on grapefruit-fruits. The maximum of intraabdominal eggs was 16. The scales were somewhat more convex than those on the leaves and grew to cover an area of 275. At least one living larva of a daughter generation was observed. Males hatch almost all.

5. On lemon-fruits the larval mortality is considerably lower than in the cases hitherto reported and the egg-production is much higher (up to 60 intraabdominal eggs.). The scales are convex and the area rose to 395. Normal development in the daughter-generation was repeatedly observed.

6. On fruits of citron larval mortality was even lower and the egg-production still higher (up to 100 intraabdominal eggs). The scales are quite convex and the area covered by them reached 605.

These experiments may be resumed as follows:

„Infections of twigs lead only into the first stage; on grapefruit only very few individuals survive until after a delayed first moult. On leaves it was possible once to breed the beginning of a second generation (grapefruit). On lemon a few larvae reach the second stage, whereas on sweet lemon and sour orange all larvae die in the first larval stage.

On fruits the adult stage is always reached by a few individuals. On shamuti no maturity is gained by the females. On grapefruit, lemon and citron the beginning development of a daughter generation was observed, which on citron — at least — seemed to be approaching normal reproduction." (Steinitz).

The sequence may be illustrated as follows:

Scale	1	2	3	4	5	6	7	8
twigs	shamuti	lemon	grape-fruit					
leaves		sour lemon sweet orange	lemon		grape-fruit			
fruits				shamuti		grape-fruit	lemon	citron

Steinitz discusses the application of these experiments to conditions in nature as follows: Only grapefruit leaves and citrus fruits permitted the development of at least some individuals of a daughter generation. But leaves of lemon and citron were not tried. Scales living only on the fruit would be exterminated every year with the crop.

However, the conditions of these laboratory experiments were much more favorable to the development of *Aspidiotus hederae* than those in nature. In the laboratory the wind did not blow away the majority of the crawlers. Fluctuations of temperature and humidity were less extreme than in nature and levied a lower toll of larval mortality.

The reproduction on grapefruit leaves and fruits was so small, that not even the mere maintenence of the scale-population was assured. Maintenance and probably even increase of population would be possible on fruits of lemon and citron. The fact that to infestation occurred hitherto under natural conditions in Palestine must be ascribed to the following factors:

1) The (probable but not yet experimentally ascertained) unfavorable conditions on leaves of lemon and citron.

2) The relative scarcity of lemon and citron fruits and the absence of dense plantations of these fruits in Palestine.

3) The scarcity of favoured hosts of *A. hederae* in the neighbourhood of citrus plantations in Palestine.

Life-history: The life history of the oleander scale in Palestine has not yet been worked out. 3—4 annual generations are probably the normal in our coastal plain.

The normal reproduction in Palestine is certainly by copulation. No pure female populations, as those observed by Gabritschevsky and Schrader in Russia and U.S.A., have been seen. The suggestion of Balachowsky (1932) that agglomerations of male puparia indicate the existence of arrhenotokeous parthenogenesis seems premature. However, the percentage of males seems to become more abundant towards the winter.

Among the natural enemies of *A. hederae, Chilocorus bipustulatus* and *Aspidiotiphagus lounsburyi* are occasionally met with, but they do not check the increase of their hosts.

References.

A. Balachowsky, Encyclopedie Entom. Vol. 15, 1932. Paris.
A. Balachowsky, et L. Mesnil, Les insectes nuisibles aux plantes cultivees. Paris, 1935, pp. 475—480.
G. F. Ferris, Microentomology. **6.** 1941. p. 33-70.
H. Steinitz, Hadar. **11.** 1938. pp. 160—163.
F. Schrader, Psyche **36**, 1929, pp. 232—236.

D. THE RED SCALE (*Aonidiella aurantii* Mask.)

Synonymy: 1878 *Aspidiotus aurantii* Maskell
1881 *Aspidiotus citri* Comstock
1881 *Aspidiotus coccineus* Gennadius
1884 *Aonidia aurantii* Targioni
1886 *Aonidiella aurantii* Berlese
1899 *Chrysomphalus aurantii* Cockerell.

Description: Scale of adult female: Nearly circular, often slightly broader than long; with thin flat margins and the central area slightly convex. Pale yellowish grey or pale ochreous. The scale being semi-transparent, the form and colour of the female show through it. Its true character is often visible only at the margin of the scale; this margin is broader in old females. The reddish colour is due to the female beneath the scale. The exuviae are central, orange red or yellow, covered with a thin layer of secretion and the larval exuviae, at their centre, with a small prominant nipple with a concentric ring of wihtish secretion. The second exuviae are pyriform, pale ochreous to orange brown. The greyish white ventral scale is thin at the centre, but thick towards the margin. It is firmly attached to the dorsal scale. The adult female is thus enclosed when the scale is removed.

Diameter of scale 1.6—2.2 mm; of first exuviae 0.3—0.4 mm; of second exuviae 0.8—0.9 mm.

Scale of male pupa: Oblong, but narrower towards the back. Exuviae subcentral. Dull reddish brown, paler towards the margin. Towards the anterior margin exuviae similar in colour and structure to those of the female.

0.9—1.3 mm long, 0.6 mm broad.

Young adult female broadly pear-shaped. During the period of gestation the broad parts of the body extend laterally considerably. The matured female is reniform, the pygidium protruding from the concave side and the whole body is filled with mature eggs. Colour yellow pale, dull orange to brownish with age. The body becomes fairly chitinized during maturity and therefore retains its characteristic shape. Mouthparts relatively small with short rostral loop. Antennal tubercles small with one long, slightly curved spine. Stigmata normal, peristigmatic glands absent.

The pygidium of the adult female is parabolic. Its margin is furnished with three well developed-lobes and well-defined plates. The first two lobes are usually distinctly notched on both margins, the notch on the outer margin of the second lobe being more pronounced. The size of the lobes decreases considerably from the first towards the third. The third lobe is notched only on the outer margin. The first three plates are not very broad at the base, with long projections.The following plates have broader bases and are toothed at their outer edges. The pygidial formula runs as follows:

$$P_1, L_1, 2P_2, L_2, 3P_3. L_3, 3P_4.$$

The dorsal surface of the pygidium bears the anal opening, which is approximately circular and is about 5 times larger in diameter than the median lobe. The orifices of the dorsal tubular glands are arranged in three, partially double, rows on the lateral parts of the pygidium:

group	I	2—3	orifices	between first and second lobe
,,	II	8—10	,,	above the third lobe
,,	III	6—8	,,	above the last plates.

The orifices are oval. These tubular glands are very long and extend throughout the anal segment, some even entering the preanal segment. Certain chitinous thickenings on the pygidial fringe, four in number, are called paraphyses. They are slender, not club-shaped and not very long. Two are connected with each of the second lobes, only one with the first (internal margin) and one with the third lobe (external margin.)

On the ventral surface lies the vaginal aperture in a central position. Circumgenital glands are absent. The length and width of the adult female change with gestation from about 1.0—1.3 × 0.9—1.0 to about 0.9—1.1 × 1.3—1.5 mm.

The specific recognition characters, as recently described by Mc Kenzie (1938), are: "Prosomatic region at maturity strongly sclerotized and strongly reniform, its lobes crowding against

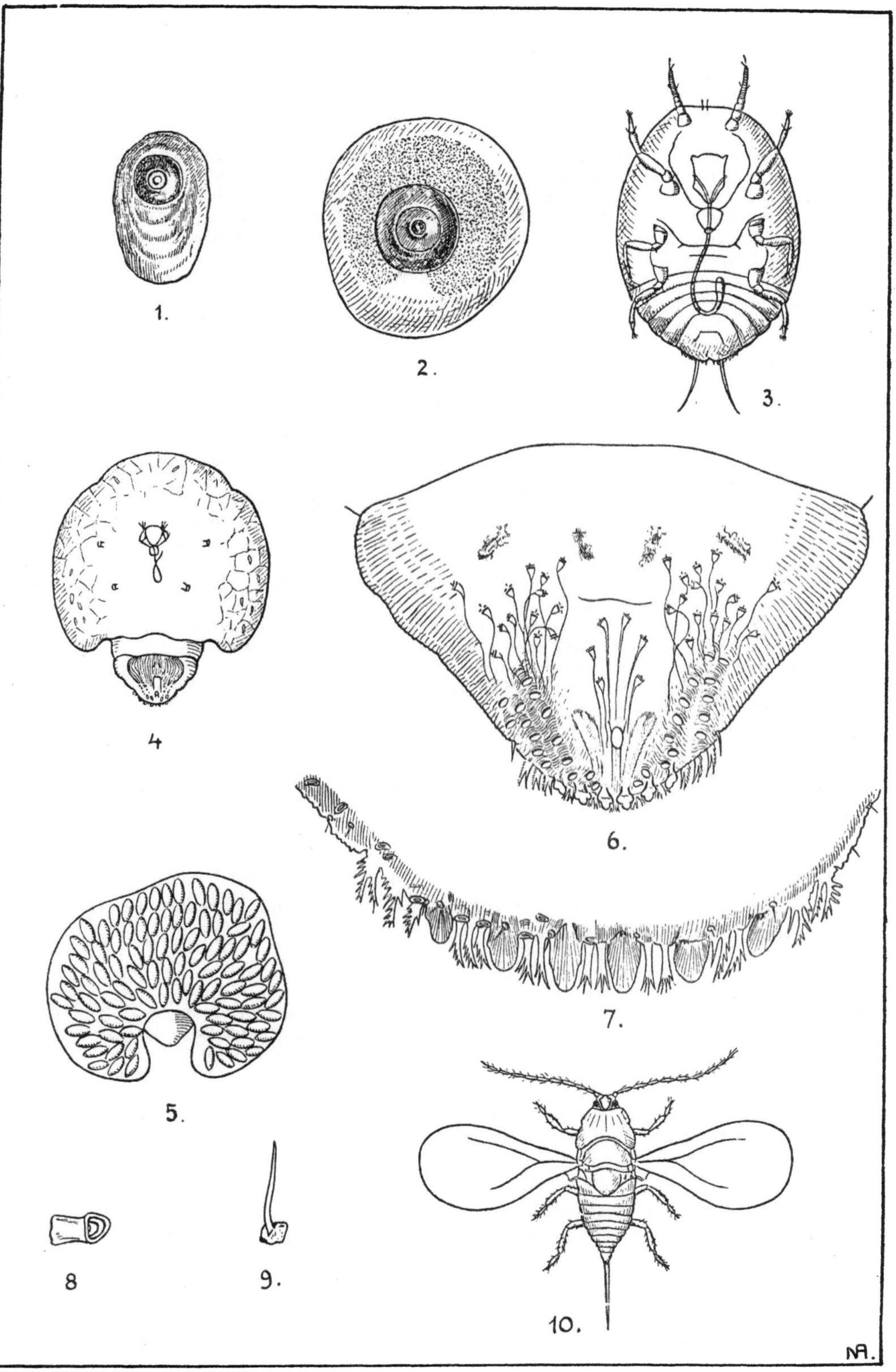

XIII. The California Red Scale (*Aonidella aurantii* Mask.) 1) male puparium; 2) scale of adult female; 3) neonate larva; 4) young female; 5) old female; 6) pygidium of female; 7) posterior margin of female pygidium; 8) anterior spiracle; 9) antennae of adult female; 10) male.

he retracted pygidium and prepygidial segments, perivulvar pores lacking, prevulvar scleroses and apophyses present, paravulvar fold usually sclerotized, three well-developed pairs of pygidial lobes present, the median pair slightly the largest, fourth lobe slightly indented, plates not present beyond the fourth lobe, dorsal ducts arranged in three rather definite rows, those of the first row slightly shorter than the others."

The *first stage larva* is short oval and 0.2—0.25 mm long, yellow. The antennae are 4-segmented, the last segment being annulated and longer. The pygidium has two lobes only, with six plates. 3—4 knobbed hair extend from the tibia, generally reaching beyond the one tarsal segment.

Carmin and Shenkin maintain that the male crawlers are slightly longer, more agile and more prolongated in shape than the female ones. No breedings have been made. First stage larvae were simply biometrically divided into two groups. No other student has been able to distinguish the sexes in this stage. Only individual breedings can confirm this assumption.

The *second female larva* has neither legs nor antennae. The pygidial characters now grow quite distinct. It has about twice the size of the first larva.

The *second male larva* is more elongate (o.7 mm long, 0.4 mm wide). The pygidial characters are as in the female, but two pairs of conspicuous purple eyes are present. The *male prepupa* is orange yellow with dark brown eyes. O.7 mm long, 0.35 mm broad. Dorsal eyes are just posterior to antennal sheats. The ventral eyes are larger and closer together and a little more posterior than dorsal. The sheats of the antennae and wings are visible, and a faint indication of those of the legs. However, all of them are more rudimentary an lie closer to the body than is the case in the next stage. There is no style present, but in its place we find a blunt protuberance which bears two distinct spines. The truncate posterior end with the button at the tip is the most evident character distinguishing this stage from the true pupa (Quayle).

The *male pupa* is of the same general colour as the prepupa. 0.7 m long (+ 0.1 mm style), 0.32 mm broad. The sheats of the antennae, wings and legs are clearly visible and lie close to body along the ventral margin.

The adult male is not much different from other Diaspidind males. It is 0.6—0.8 mm long and has a 1.5 mm wing-expanse. Orange yellow, only the 10-jointed antennae (0.5 mm long), the legs and the thoracical band being very light brown. The lateral pair of eyes situated just laterally of the antennae are dark brown; the larger ventral pair of eyes are closer together and more posterad. The halters are club-shaped.

Distribution and origin: It is impossible to give at the present state of our knowledge a correct and reliable survey on the distribution of the red scale. The reason for this lack of information lies in the fact that until the very last years three species have been confused even by the most competent specialists. Only thanks to the work of Nell (1933) *A. aurantii* Mask. can be morphologically distinguished from the Yellow Scale (*A. citrina* Coq.) and thanks to the studies of Lupo (1936) the first can be differentiated from *A. taxus* Leon. Mc Kenzie (1937) adds *A. comperei* Mc K. from citrus in India and *A. orientalis* Newst (= *Aspidiotus o.N.* = *A. cocotiphaga* Marl.), a very polyphagous species known from India, Ceylon a.o. countries.

Mc Kenzie discovered some sclerotized structures (A. and B. in fig. 50) which offer reliable features for the determination of this group. This key for the Red Scale group runs as follows:

1.	Perivulvar pores present	2
	Perivulvar pores absent	3
2.	Perivulvar pores in two groups	*Aonidiella comperii* Mc Kenzie.
	Perivulvar pores in five groups	*Aonidiella oreintalis* (Newst.)
3.	Sclerotized structures A on ventral part of pygidium present (see fig. 50)	4
	Sclerotized structures A absent	*Aonidiella taxus* Leon.
4.	Sclerotized structure B on ventral part of pygidium present; structure A globular, fourth lobelike process usually inconspicuous	*Aonidiella aurantii* (Mask.)
	Sclerotized structure B absent, an irregular fold present, structure A usually narrow and elongate: fourt lobelike process usually prominent	*Aonidiella citrina* (Coq.)

A. taxus Leon. is of world-wide distribution and all records from hosts as *Taxus* and *Podocarpus* should be referred to this species for the time being. It probably has a still wider host range not yet studied. The true Yellow Scale has been discovered so far only in Japan, California, Texas and Transcaucasia. However, it is quite possible that other records, especially from the Oriental region, will later be referred, to this species which is so far known from citrus only.

All records from Palestine have been attributed to the red scale. The morphological study of the scales of this group should be taken up wherever *Aonidiella* occurs as a pest, considering carefully the now established differences.

Distribution in Mediterranean countries: In Spain the species must be local and very rare. It is mentioned by Lindinger from citrus. However, neither Balachowsky nor the author have encountered it in Spain, nor is it recorded in the reports of the regional Agricultural Experimental Stations of the Spanish Government.

In S. France it is likewise extremely rare. It has been observed on two occasions only on *Evonymus* and *Podocarpus* (Balachowsky). The latter should almost certainly be referred to *A. taxus*.

In Italy it has been reported by Leonardi from citrus at Naples, Portici and Apricena (Apulia), and by Lupo at Fucrigrotta. Lindinger mentions it also from *Diospyros*, palms and *Taxus*, but at least the last determniation belongs to *A. taxus* Leon.

In Greece it is common everywhere, infesting especially citrus and evonymus. However, only 14 hosts are known so far, as compared with 37 hosts known for *Chrysomphalus dictyopermi*, which was introduced only about 10 years ago, whereas *A. aurantii* was described as early as 1881 as doing serious damage to oranges on Chios Island.

In Turkey *A. aurantii* is a very important citrus pest and quite common on other hosts. It is restricted to the Mediterranean lowlands. It is common also in Syria (author's' observation). In Cyprus it is known from *Citrus, Acacia, Ceratonia, Pistacia, Rosa* (Morris, Hall).

In Egypt red scale is common everywhere and there is hardly a garden in the country in which it cannot befound. It does very little damage there (Hall). From Lybia it is not yet recorded, neither from Tunis nor from Morocco. In Algeria it was discovered only on oranges from Misserghin and is in any case extremely local and rare (Balachowsky).

From the Canary Islands it is unrecorded, whereas it is known from Madeira on *Cydonia* (Lindinger).

Africa: It was probably introduced into South Africa at the beginning of the 19th century with citrus trees brought by the captains of the East India Company as presents for S. African inhabitants who entertained them. "This would account for the hold the species had on the old oak trees (often far removed from citrus), on apple and pear, and on the old rose hedgerows, some of which were killed by the pest." Its oldest authentic record is in 1857 near Cape Town. When C. P. Lounsbury came in 1895 as Entomologist to the Cape Colony, Red Scale was one of the first insect pests to which his attention was drawn. It is now widely spread throughout the Union (Brain). Jones gives a similar picture for S. Rhodesia, where it is, as in the whole of S. Africa, the most important citrus pest.

In Tropical Africa it is widely spread: It is common in Tanganyika Territory (Lindinger). In Kenya Colony it was first recorded in 1914. It was introduced in this colony with cirtus consigments and has spread greatly recently (Dry). It is still unknown in other parts, as for example, Paoli (1933) did not mention it in his recent book on pests of Italian Somaliland. Among the Eastern African Islands it is known from Madagascar, Zanzibar and Mauritius (Lindinger, Charmoy).

America: In S. America it is present everywhere in the tropical and subtropical zone. Its presence and importance had often been overlooked because it is relatively rare on citrus, occurring mainly on roses and occasionally on other ornamental plants. It is reliably known from: Argentine, Chile, Paraguay, Uruguay, Brazil, Mexico (Compere, Lupo, Bodkin, Costa, Lima). From the West Indies it is recorded for some islands (Jamaica, Gowdey), but apparently lacking in others, as in Cuba, the scale insects of which have been well studied (Ballou, 1926).

It seems to have been first introduced into N. America about 1879, when its introduction from Australia into California was ascertained. It is by now common in California, Arizona, Texas, Florida (and probably all other southern states); in Ohio, Indiana, New York and other states it is a common greenhouse pest (Comstock, Fernald, Merrill, Sanders, Quayle, Dietz, etc.).

Australia: In Australia it is widely distributed and apparently present, wherever citrus is planted. It was a serious citrus pest as far back as 1870 (Comstock, Froggatt). In Queensland it is a pest mainly in the drier and hotter parts of the state (Summerville). It has also been recorded from New Zealand, Hawaii Islands, Samoa, Fiji Island, Marianna Island and Palau Is. (Fernald, Bodenheimer, Takahashi).

Orient: In the Oriental regions its economic importance is small, but its distribution seems to be general. In India it occurs everywhere, but mainly on cultivated or ornamental plants (Fletcher, Lindinger). In Ceylon it is extremely rare on citrus, but occurs on agave and aloe (Green, Bodenheimer). Further records include Burma (Fletcher) and Siam (Lindinger). In the Dutch East India it is widely spread, but of no economic importance, (Voute). In S. China it is known, but likewise not injurious (Hoffmann), where it seems to be as well as in Japan of less importance than the yellow scale. (Kuwana, Lupo). The area it covers further embraces the Phillippines (Lindinger), Riukiu Is. and Formosa (Takahashi, Clausen). In the latter it is much less common than *C. aonidum.*

The conclusions to be drawn from this distribution with regard to the home of red scale are as follows: N. and S. America as well as Africa certainly are centres of recent introduction. The same is true of the Mediterranean regions and also of Australia. The Oriental region is the home of the entire *A. aurantii-group*, certainly of *A. citrina, A. orientalis* and *A. comperei.*

LIFE-HISTORY.

Longevity: Quayle records the following typical life-history in summer and autumn from California: The females grow adult 40—50 days after having hatched, grow fertilized 10—20 days later. They begin to give birth at an age of about 90 days and live for another 1—2 months until death. In total, the females live 4—5 months, as compared with unfertitlized females which survive 6 months and longer. The males hatch at an age of about 60 days and die within the five following days.

Dry reports from Kenya Colony (certainly in the shade) that at 19—21° C males hatch after 65 days, the first larvae of the following generation after 110 days.

Nell observed 60 days development until first birth and 2 months reproduction period (at 27.8° C) on citrus leaves in California. The results of Bliss and co-workers will be dealt with later as also those of Clark and Friend.

Representative life-histories of red scale in S. Rhodesia, as observed by Jones, are as follows (no. of days):

month of settling	females on fruit			females on leaves			males average development	
	A Average development in days	B Average duration of production period	C Average longevity	A Average development in days	B Average duration of production period	C Average longevity	leaf	fruit
I. 1934	58	20	78	67	23	90	39	34
II.	63	18	81	70	71	141	42	34
IV.	118	19	137	133	29	167	81	68
V.	104	18	122	129	21	150	71	63
VIII.	73	16	89	78	22	100	49	40
IX.	60	23	83	67	32	99	39	34
X.	55	31	86	60	34	94	33	30

From among a great number of individual life histories obtained in our breedings in Palestine a short summary may be given:

Month	II	III	IV	V	IV	VII	VIII	IX	X
Females n			1		2	15	12	22	4
larval development	—	—	30	—	26	22	22	28	34
preoviposition period . . .			118		73	50	54	131	147
Males n	2	4	9	2	5	22	4	2	
larvae	62	62	30	26	22	21	18	30	—
pupae	79	77	44	38	32	31	28	39	

The following table gives the complete life cycle for 50 females at Hedera, from birth to death:

No.	Date (crawler to death)	longevity			Total
		larva	preoviposition period	after first oviposition	
1.	6.III.1931 —12.VIII.1931	28	77	43	148 days
2.	15.VI. —10.IX.	20	35	32	87
3.	15.VI. —15.XII.	23	50	110	183
4.	5.VII. —27.IX.	21	30	33	84
5.	10.VII. —11.X.	14	32	47	93
6.	14.VII.1931— 3.I.1932	19	37	115	171
7.	21.VII. — 2.XII.1931	27	30	77	134
8.	24.VII. —12.IX.	21	33	57	111
9.	5.VIII. —18.I.1932	21	36	109	166
10.	30.VIII. — 3.V.	28	95	123	246
11.	30.VIII. — 5.V.	22	226		248
12.	1.IX. — 8.III.	20	60	108	188
13.	8.IX. — 5.V.	24	81	134	239
14.	8.IX. — 5.V.	24	70	145	239
15.	5.VI.1932 —26.VIII.	22	56	4	82
16.	12.VI. —21.VIII.	24	30	16	70
17.	10.VII. —30.VIII.	23	18	10	51
18.	15.VII. —18.IX.	23	37	5	65
19.	22.VII. —18.X.	21	32	35	88
20.	11.IX. — 2.V.1933	27	38	168	243
21.	11.IX. —25.VI.	32	170	85	287
22.	13.IX. —25.VI.	28	49	208	285
23.	13.IX. —25.VI.	33	133	119	285
24.	13.IX. —21.IV.	30	136	34	200
25.	13.IX. —22.IV.	30	136	27	193
26.	13.IX. — 2.V.	27	204		231
27.	14.IX. — 2.V.	19	79	132	230
28.	25.IX. —14.V.	29	116	86	231
29.	25.IX. —23.IV.	19	146	45	210
30.	25.IX. —28.V.	33	112	70	215
31.	27.IX. —17.III.	29	123	19	171
32.	28.IX. — 2.V.	35	25	146	216
33.	1.X. — 2.V.	27	125	61	213
34.	1.X. — 2.V.	27	121	65	213
35.	3.X. —17.V.	30	116	80	216
36.	16.X. — 2.V.	40	95	63	198
37.	21.VII.1933—18.X.	26	23	63	112
38.	24.IX. — 5.II.1934	38	47	49	134
39.	14.VII.1934—16.X.	27	6	62	95
40.	14.VII. —16.IX.	27	8	30	65
41.	21.VIII. —14.I.1935	60	12	43	115
42.	16.IX.1935 —21.I.1936	65	8	35	108
43.	19.XI. —20.XII.1935	44	12	34	90
44.	23.IX. —23.XII.	40	15	36	91
45.	23.IX. —24.III.1936	54	20	108	182
46.	25.IX. —24.III.	41	20	119	180
47.	17.X. —24.III.	72	42	33	147
48.	29.X. —19.II.	45	42	26	113
49.	4.XI. —16.IV.	91	21	51	163

The development of the generations at Dagania in 1933 was as follows (Klein):

No. of generation	Month of birth	Larval development	Preoviposition period	Total development	°C during development	% R.H. during development	Longevity after first birth	Total longevity
1	I	53	40	93	16	64	57	150 (135—170)
2	IV	24	34	58	23.5	59	88	146 (140—155)
3	VI	19	30	49	28	60	56	105 (110—120)
4	VIII	21	40	61	25	61	32	93 (85 —105)
5	XI	38	50	88	15	67	20	108 (100—115)

A few life-histories on orange leaves in the coastal plain may be recapitulated here, in order to show the seasonal trend of the life-histories of Red Scale:

Month	Larval development	Female		Male	
		pre-oviposition period	reproduction period	Total longevity	larval and nymphal development
III	(28	77)	43	158	75
IV	(30	88)			42
VI	20	35	32	87	33
VII	21	33	57	111	30
VIII	21	34	123	178	27
IX	29	116	86	231	36
X	30	116	80	226	62
XI	91	21	51	163	75
XII	(60	57)			

Development is shortest in July/August, whereas the reproduction period is considerably longer in August than in July.

In many breedings the complete development of some individuals took an abnormally long period. This phenomenon, recorded also by Jones, occurred as a rule in the preoviposition period. At first we thought of the possibility of an imaginal diapause in these cases, though it seems by now that it should be referred to delayed fertilization.

Temperature and development: The first series of experimental breedings published was that by Bliss, Broadbent and Watson in California. The scales were reared in an open, screened and roofed insectary, in which conditions of temperature and humidity approximated those in near-by citrus groves. The light-intensity was, however, lower in the insectary, at least in comparison with the most exposed parts of the tree. A hygro-thermograph in the centre of the insectary supplied temperature records from which means were secured with a planimeter. The breedings were made on fruits as well as on leaves and stems.

Bodenheimer observed the development in the grove on orange leaves only as the twigs were covered with a bag of thick muslin. The scales lived, therefore, in full shade and were never exposed to direct sun radiation. The average local temperature of the field station in Hedera was gained by interpolation from the sum of two-thirds of the monthly average temperature of Tel Aviv, one third of that of Haifa.

Correlation between temperature and length of development.

Development of females in days. Development of males in days.

Temperature	No. of observations	Larval period	Preoviposition period	Total development	Range total of development	Computed values of hyperbola	Deviations of the empirical value for hyperbola	Temperature	Number of observations	Total development	Deviations from the hyperbola.
19.5°C	4	40	83	123	100—135	125	—1.6%	19.5°C	4	49.3	—
20.5	4	30.5	73	103.5	96—114	105	—1.4	20.5	3	48.5	+11.3
21.5	2	30	54	84	72—89	92	—8.7	21.5	13	41.3	—
22.5	7	32	48	80	68—86	81	—1.2	22.5	8	39.6	+2.5
23.5	1	22	44	66	—	72	—8.3	23.5	6	35.0	—2.8
24.5	5	25.6	38	63.6	48—80	65	—2.1	24.5	7	31.3	—8.8
25.5	6	22.4	39	61.4	48—72	60	+2.2	25.5	15	31.7	—
26.5	29	19.7	33	52.7	39—67	54	—.24	26.5	59	29.0	—3.3
27.5	58	22.1	31	53.1	31—72	51	+4.1	27.5	90	29.6	+3.4
28.5	11	21.5	30	51.5	41—64	48	+7.3	28.5	20	27.5	—
	127						—1.2%		225		+0.6%

Jones reared the scales in an insectary in S. Rhodesia. The mean temperatures were computed by taking the average of the daily mean temperatures during each stage. His values show also a good adaptation to the observed values:

Temperature	*Female*				*Male*			
	fruit		leaves		fruit		leaves	
	hyperbola	observed	hyperbola	observed	hyperbola	observed	hyperbola	observed
17.8°C	118	118	138	138	68	68	81	81
18.3	108	104	125	129	62	63	74	71
21.7	69	73	76	78	39	40	45	49
23.0	61	61	67	70	34	36	39	42
23.1	60	60	67	67	34	34	39	39
23.6	57	60	64	67	32	34	37	39
24.3	54	55	60	60	31	30	35	33

The following hyperbolae have been hitherto obtained:

Locality		Threshold of development	Thermal constant	Author
California		12.4°C	570 days-degrees	Bliss
S. Rhodesia	(fruit)	12.3°C	648 ,,	Jones
	(leaf)	12.8	690 ,,	,,
Palestine		14.0	533 ,,	Bodenheimer

Munger and Cresson (1948) showed that fluctuating temperatures (18 to 30° C) did not produce accelerated development as compared with that at constant temperature of the same average.

The body temperature of the Red Scale in its natural environment: The body temperature of a poikilotherm animal is of decisive importance for the speed of all its physiological processes, including development, maturation, respiration, etc. It is therefore necessary to know the body temperature of an insect in its natural environment. The old theory, that a poikilotherm animal has the temperature of its environment is not true. As recent studies of the writer (3) a.o. have shown, there are several distinct types of thermoregulation. One of them, heliothermy is very generally found in diurnal insects. A poikilotherm animal is called heliothermic if its body temperature is rising under the influence of the sun-rays higly above the shade temperature of its environment. This phenomenon was first discovered in insects by the author (4) on the Desert Locust (*Schistocerca gregaria)* and has since been found to be of general distribution. The sun-exposed situation of the red scale in our main citrus districts makes it obvious that during some hours of the day they are fully exposed to the sun-rays and it was necessary to study its body temperature under these conditions. All measurements have been made with the thermocouple method. The fine point of the thermocouple was inserted in the body of the red scale through the opposite side of the citrus leaf and the body temperatures were read from early morning to half an hour after noon.

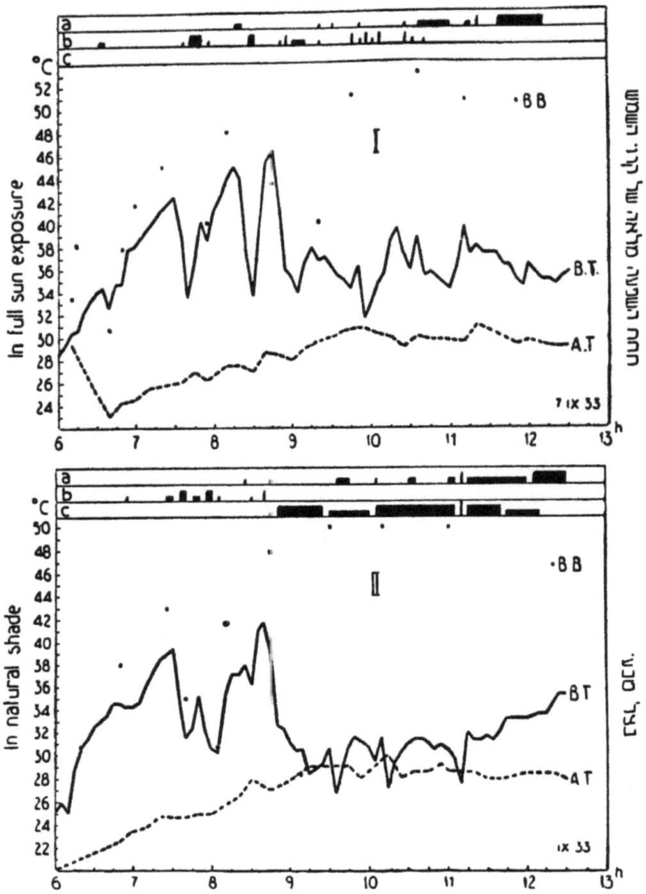

Fig. 57.
Body temperature of Red Scale (*Aonidiella aurantii* Mask.) AT Air temperature in shade; BT Body temperature of Red Scale in the sun; BB Black Bulb thermometer; a wind, b shade of clouds, c natural shades of other leaves; hight of black line in a—c designates the intensity of factors a—c. (from Bodenheimer).

Fig. 57 shows quite clearly, that the average body temperature of the Red Scale in its main environment is higher than in the shady centre of the tree. The average values of readings every five minutes between 6 A.M. and 12.30 P.M. were as follows:

I Locality	II Leaf in sun	III Dto. Clouds	IV Leaf in natural shade	V Leaf in sun (clouds)
Jerusalem (shade)	44.4°C(50) 17.5°C 26.9(29/8)	33.2(51) 6.1°C 27.1(29/8)	33.8(55) 10.2°C 23.6(28/8)	31.7(30) 11.7°C 20.0(3/12)
Rehoboth (shade)	39.4(66) 10.8°C 28.6(7/10)		32.8(67) 5.3°C 27.5(8/9)	
Daganiah (shade)	40.6(63) 12.0°C 28.6(15/9)	37.4(62) 7.9°C 29.5(13/9)	31.2(61) 3.4°C 27.8(17/9)	
Average difference fr. shade-temp.	13.4°C	7.0°C	6.3°C	

Explanation of the table: { 44.4°C (50) 17.5°C / 26.9(29/8) } means: The average body temperature in 50 readings was 44.4°C. The average temperature in the shady centre of the tree was 26.9°C i.e. a difference of 17.5°C as compared with the first value. The observation was performed on the 29.8.1933.

The average difference under natural conditions was 6.3°C in the preferred natural environment (col. IV), but normal development still occurs at average differences from 0.0 to 13.4°C.

The high sun radiation at Jerusalem, which is much higher than at Rehoboth or Dagania, is unfavourable to the red scale. The relative dryness of the air is likewise unfavourable to its development in Jerusalem.

Munger (1948) confirmed these results and points out, that in the sun the temperature of the scale is lower than that of the surface tissues of the host plant.

The hyperbolae (A) for the single stages of the female have been carefully calculated by both Bliss and Jones. The author's observations fit well into the general trend described by both. The results of the Californian writers show a somewhat shorter, those of Palestine a slightly slower development. The figures obtained by Jones form the basis of the calculations given below, mainly because they extend over a whole year and because the values for fruit and leaves are clearly differentiated. The following values are calculated from his data (under equal conditions of exposure to sun-radiation).

A. Development on fruit.

Stage		Larva I	Larva II	Preoviposition period	Total development
development threshold (c)		10.2	12.7	12.4	12.3°C
Thermal constant (Th. C)		182	110	408	690 days-degrees
Length of stage at	30°C	9.6	6.7	21.5	36.6
	25	12.8	9.4	30.0	51.0 days
	20	19.4	15.9	49.7	84.2
	15	39.6	50.4	153.1	240.0

B. Development on leaves.

Stage		Larva I	Larva II	Preoviposition period	Total development
development threshold (c)		11.3	13.8	13.0	12.8°C
thermal constant (Th. C)		182	110	408	690 days-degrees
Length of stage at	30°C	9.7	6.8	24.0	40.1
	25	13.3	9.8	34.0	56.6 days
	20	20.9	17.7	58.3	95.8
	15	49.2	91.7	204.0	313.6

These two groups of observations show that two factors are apt to induce considerable deviations from the simple dependency upon the average environmental temperature: 1) Rise of body-temperature under the influence of sun-radiation (heliothermy) and parallel rise in the speed of development during the hours of exposure. 2) The host. In all cases development was quicker on fruit than on leaves, stem or twigs. The influence of the substratum is greatest at lower temperatures, when the difference in speed approaches 35 cent per, whereas at high temperatures it falls to less than 10 per cent.

The figures obtained for deviations of development by different authors show them to lie between the following duration of days at various temperatures:

°C	Quickest develop-ment	Develop-ment at partial exposure	Develop-ment in shade	S. Rho-desia (Jones)	Texas (Clark & Fried)	Palestine (Boden-heimer)	(Klein)	California (Bliss)
30	26.5	40.1	47.0					32
27	32.1	48.7	56.9		52	53	50	39
25	37.4	56.6	66.3			63	61	45
23	44.7	67.6	79.1	70		74	60	54
20	63.3	95.8	112.3	103		110		75
18	87.7	132.7	155.6	132	130			102
15	207.3	313.6	367.7		93		88	220

At the lower hyperbola values the quickest development may be even quicker than seen from this calculation, provided the individuals are fully exposed to the sun.

It is thus almost impossible to calculate the usual three hyperboles. Following our experience it seems most advisable to make the hyperbola offered by Jones as the base of epidemiological evaluations.

Among the development stages of the female the second stage is as a rule shortest, the first larva following and then the preoviposition period. The first larval stage fluctuates least with temperature, the second slightly more, whereas the preoviposition period is rather sensitive. All students agree in this respect. These facts are well illustrated by some breedings at Rehoboth:

date	Larva I	Larva II	preoviposition	Total
VII/IX.1934	11.3	5.8	27.3	44.4 days
VII/VIII.1935	12.0	9.0	16.9	37.9 ,,
VIII/IX	12.4	9.5	14.8	36.7 ,,
IX/XI.	15.5	13.0	39.3	67.8 ,,

In the male, the first larval stage is longest, following second larva and the two nymphal stages. In July 1935, it lasted 12.0 and 7.8 and 8.2 resp. = 25 days.

The males hatch, as a rule, 1—2 weeks after the female has entered the preoviposition stage. This agrees well with Quale's observation that the females are fertilized 10—20 days after the last moult. The development of the male is also highly dependent upon temperature and host.

The conditions of female development lead to just described the fact that the preoviposition period fluctuates more than any other developmental stage. This is easily seen from the foregoing table.

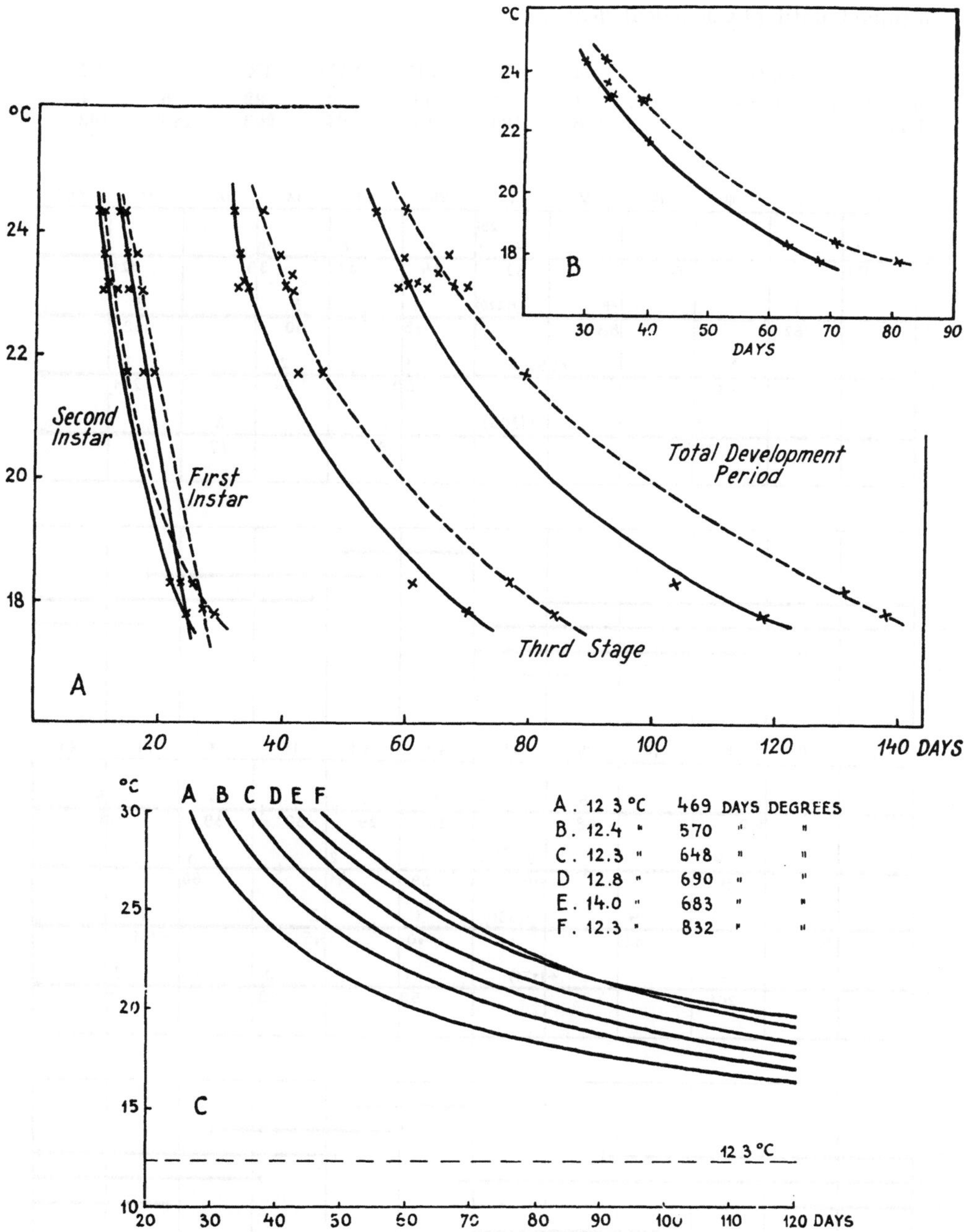

Fig. 58.
Temperature dependency of development of *Aonidiella aurantii* Mask. on fruit (full line) and leaf (broken line) A. female s, B males (after Jones). C. Hyperbolea of temperature correlation of development in *Aonidiella aurantii* as calculated from different authors: A. shortest development. B. on leaves in California (Bliss). C. on fruit in S. Rhodesia (Jones). D. on leaves in S. Rhodesia (Jones). E. on leaves in Palestine (Bodenheimer). F. Slowest development.

Female longevity fluctuates as demonstrated by the breedings of 1935, which were continued until physiological death:

month	III	VI	VII	VIII	IX	X	XI
no of observations	1	4	11	4	22	6	1
longevity	158	106	97	194	202	185	163

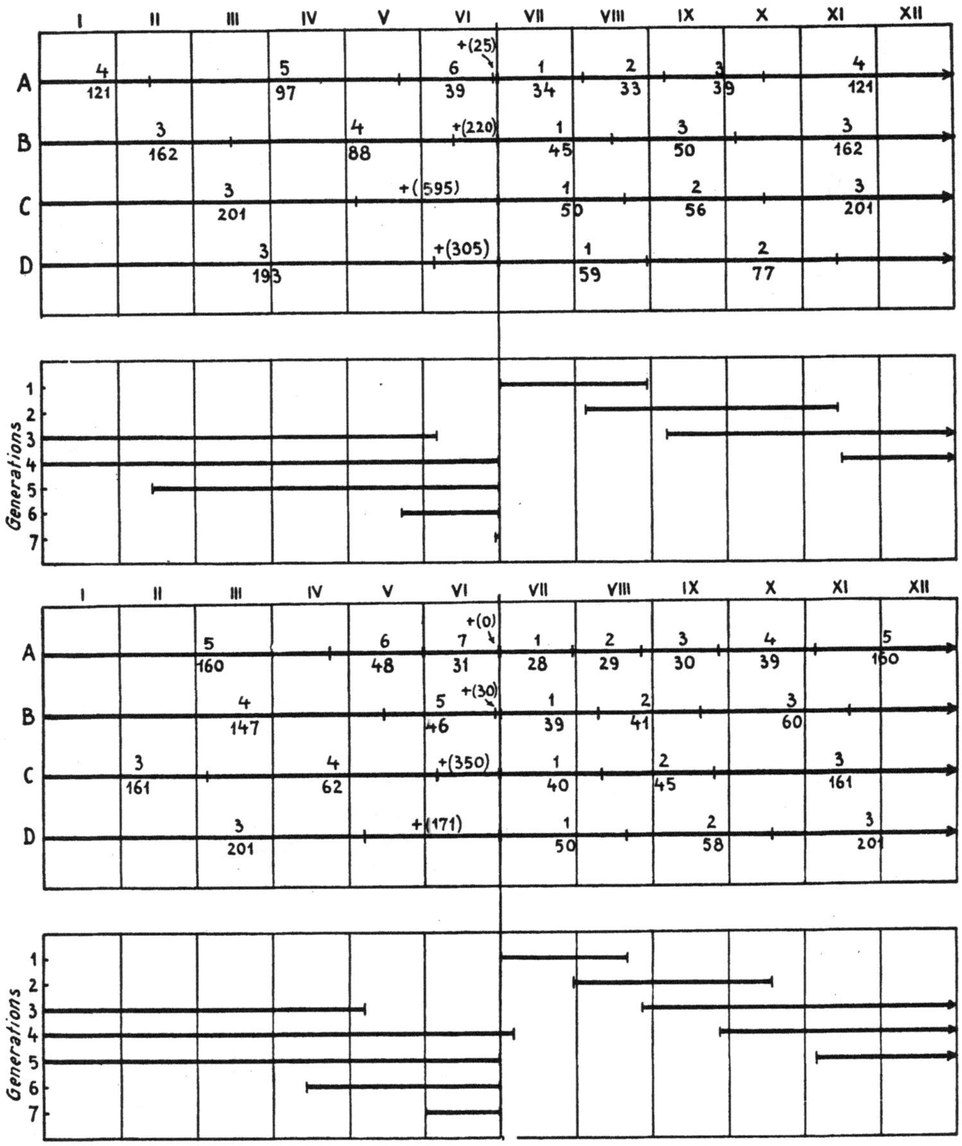

Fig. 59.

Life history, sequence and overlapping of generations of *Aonidiella aurantii Mask.* Above: at Tel Aviv. Below: at Dagania. A—D: A. Shortest development, B. such in the sun, C. such in shade, D. slowest development.

Of this period reproduction lasts about 50 per cent in the hibernating generation, fluctuating in the others between 20 and 40 per cent. In S. Rhodesia this percentage is somewhat lower (Jones):

month	I	II	IV	V	VIII	XI	X
no. of observations	25	22	14	15	18	28	36
% of production on fruit . . .	25	22	14	15	18	28	36%
period of total longevity on leaves	26	50	12	14	22	34	35 days

Number of annual generations: Following the statements of Quayle, three generations are the normal in California, a partial fourth generation often occurring. April, July, October and February/March are the months of main larval appearance at Whittiers. This statement is confirmed by Bliss, who ascribed the partial or full fourth generation to the exposure to sun-radiation. In the hot interior of the Southern California as well as in S. Florida the number of annual generations rises to five.

In the Oriental tropics the annual number of generations may rise from six to eight.

In S. Africa 2—3, in S. Rhodesia 3—5 generations are reported. The prolonged summer period, especially when poor in rains, is very favourable to red scale development. In the shade, the larvae of the first generation after winter appear from middle to end of September. The first generation which develops quickest, gives birth to the larvae of the second generation from mid-November to mid-December. The main larval development of the third generation lasts from mid-January to mid-February, and that of the fourth through the end of March and April. The fifth generation is represented by those scales slowly developing through the remainder of the winter and together with a partial sixth generation gives rise to the first generation of the annual cycle in September. The total number of generations in the shade is five and in the more exposed parts of the tree, an additional one and probably two generations may be observed, making a total of six and a partial seventh, or of seven generations. In the field these generations are not as clearly defined and considerable overlaping occurs (Jones). In analyzing this report of Jones, we find that only four annual generations develop:

generation	1	2	3	4	
date	20.IX.—	1.XII.—	1.II.—	15.IV.—	20.IX.
days	71	62	73	158	

In the coastal plain of Palestine four major generations are the rule. The main periods of larval development may be observed in February/March, May/early June, late July/August, September/October. Considerable overlapping occurs between these generations. Under exposure to sun, a fifth generation may certainly develop.

Klein reports from Dagania (southern shore of Lake of Galilee) 5 annual generations reproducing in January, March/April, June, August and October/November.

All these reports are in close correlation to the computed number of annual generations. Summerville reports 5—6 generations in the drier and hotter parts of Queensland, where the calculated values fluctuate between 5.1 (Brisbane) to 5.9 (Ravenswood) generations. The quickest (A) and slowest (C) development, as well as the epidemiological average for some localities are given below:

Locality	Hyperbola			Months above threshold (in shade)
	A	B	C	
Nice	2.4	1.7	1.4	6
Naples	3.4	2.3	1.9	8
Athens	4.4	3.2	2.5	8
Tel Aviv	6.1	4.3	3.4	12
Jerusalem	4.1	3.1	2.3	9
Dagania	7.0	5.3	4.0	12
Gizah	5.6	3.8	3.2	11
Laurenco-Marques	7.4	5.4	4.5	12
Mazoe	4.6	3.4	2.6	12
Salisbury	4.8	3.5	2.7	12
Capetown	3.1	2.3	1.8	11
Batavia	10.4	7.6	5.9	12
Honolulu	8.7	6.3	4.9	12
Los Angeles	3.6	2.7	2.0	12
Riverside	3.9	2.8	2.1	9
Rio Grande	8.0	5.6	4.5	12
Tampa	7.6	5.5	4.2	12

Biology of reproduction.

Sex-ratio: Quayle obtained 64% of males during winter and spring in California, about 50% during late summer and autumn. Nel obtained in the same country among 521 adult red scales 202 males (= 39%) in the second half of the year; in early spring 46 males among 65 adults (= 71%). In Texas the proportion of males was in 1930 (Clark and Friend):

month	IV	V	VI	VII	VIII	IX	Total
total no. settled	128	58	40	102	349	228	905
% of males	67	60	65	60	49	62	57

A statistical approach could not yield satisfactory results. Whereas the longevity of adult females varies from 28 to over 250 days, that of the adult male is restricted to a few days. The male puparium remains on the host for a much shorter period than the scale of the adult female. This discrepancy leads necessarily to a disproportion in the numerical sex-ratio in favour of the females. Among 400 scales (males: larva II and nymph; female: larva II and adults) per month the percentage of males was as follows:

year	I	II	III	IV	V	VI	VII	VIII	IX	X	XI	XII	Average
1932	—	23	**44**	19	17	**47**	5	**23**	18	25	**42**	40	28%
1933	35	**46**	38	33	23	**48**	24	**36**	13	17	**41**	15	31
1934	9	**23**	8	5	3	**14**	7	**29**	27	**35**	23	35	17
1935	9	26	**27**	4	2	0	**41**	0	30	**33**	24	**36**	23
1936	18	**30**	**30**	15	11	**27**	19	**22**	22	28	**33**	32	24

Similar results were obtained by counts made in the same way by Klein in the Jordan Valley in 1933/34:

Generation	1	2	3	4	5	Average
% of males	25	24	30	32	21	26

The four annual peaks coincide with the main larval development of the four annual generations. During this period only the percentage of males present on the host becomes equal to that of the females or even surpasses it considering the differences in the longevity of both sexes and their scales.

Direct counts in breedings seemed to be the only way to obtain correct values for sex-ratio.

Laboratory breedings did not further the problem greatly. Two large sets of breedings were arranged with the following result:

	male	female	total	% of males
IX.1934	200	402	602	33
VIII/XI.1935	2368	1363	3731	63

The result was contradictory. Observations of scales both from birth and time of settlement of the host throughout development were made contemporaneously. The result of 3 years breedings are as follows:

month	1934		1935		1936		Total	
	male	female	male	female	male	female	male	female
I			1	2	3	3	4	3
II			1	11	1	7	2	18
III			4	12	—	1	4	13
IV	4	—	2	4	—	—	6	4
V	41	28	5	13	—	1	46	42
VI	68	25	—	2	—	1	68	28
VII	23	29	28	15	2	—	53	44
VIII	141	162	37	26	—	—	178	188
IX	103	67	27	26	9	6	139	99
X	50	54	37	49	8	10	104	113
XI	5	16	23	33	2	1	30	50
XII	3	8	4	4			7	12
Sum	447	389	179	197	25	30	641	616
% of males	53		48		46		51	%

In breedings extended over a longer period the sex ratio is equal (51% males absolute average, 49% male relative average). If any seasonal shift may be distinguished, it is as follows:

period	male	female	% males
I—V	62	80	44
VI—VII	114	70	62
VIII—XII	465	766	50

These results agree well with those obtained from breedings by Jones in S. Rhodesia:

period	male	female	% males
I—IV	76	146	34
V—VIII	244	147	62
IX—XII	168	156	52
Total	488	449	52

The male percentage was highest in winter (V—VIII), which fact is in agreement with the results obtained by Quayle and Nel in California. It is possible that the few figures obtained in our breedings for these months are misleading.

Whereas the sex-ratio has now been definitely established as approaching equality, if observed for a period not shorter than one year, the problem of factors controling the seasonal shift in sex-ratio, remains to be solved.

We find some indications in literature hinting at a different distribution of males and females on the upper and lower side of the leaves. This is certainly not the case with regard to red scale. The percentage of females was always more or less identical on both sides of the leaves.

Year	side	I	II	III	IV	V	VI	VII	VIII	IX	X	XI	XII	Total upper	Total lower
1932	upper					80	37	97	72	86	76	71	43	70	
	lower					77	62	93	76	81	73	72	57		74
1933	upper	64	65	69	36	79	81	70	60	94	81	53	96	71	
	lower	57	38	60	55	70	78	74	62	90	82	75	81		69
1934	upper	82	73	96	98	99	80	91	74	59	69	62	71	80	
	lower	91	93	98	89	99	75	94	72	81	72	88	86		85
1935	upper	89	62	71	96	98	95	—	—	61	71	43	76	76	
	lower	89	80	79	96	—	—	—	—	79	81	35	77		77
1932—35	upper	78	67	72	73	86	69	86	69	75	74	57	72	73	
	lower	79	70	79	80	80	72	87	70	83	77	68	75		77

Reproduction and productivity: Copula is essential for reproduction. Thoroughly isolated females, do not oviposit. This has been confirmed by almost all observers. However, the conditions for such an isolation are not always realized. The writer has seen a trained entomologist isolating immature females, the isolation not being sufficient as the cover consisted of wide-meshed mull. No wonder that under such conditions normal reproduction has been frequently observed. The individuals experimented with should be isolated in the second larval stage the latest and the isolation must be thorough, preventing the intrusion of the tiny males. Jones describes that two females of his experimental series gave birth to larvae, but inspection showed that males had entered the cages.

Quayle, Jones and the writer have published the results of extended isolating experiments in all of which the unfertilized females died without having given birth to larvae. Our own experiments have now been increased up to 97 females at almost all seasons with the same conclusions.

Carmin and Shenkin have promoted a rather unbased theory. Assuming that female and male first larvae can be differentiated by their size form, an assumption which has not been proven, these authors state that sometimes either only males or only females are born by one mother, sometimes larvae of both sexes. They thus draw the conclusion that parthenegenesis occurs in the June/July generation. ,,There should also be kept in mind a possibility of seasonal parthenogenesis occurring in some individuals or even in the same individuals at fixed periods.''

In all the breedings in which all stages from larval to adult stage have been observed, males and females have always been present (Quayle, Jones, Bodenheimer). Not one case of parthenogenesis has ever been observed in properly conducted experiments.

Fertilization is not only needed for birth but also for the maturation of the eggs. As Comstock has first pointed out, the unfertilized female may be easily distinguished from the female which has copulated. The former has a greyish appearance due to a very wide greyish or whitish margin. Her scale is rather flat at this period and looks greyish. After copulation the eggs begin to mature, the female's body swells, the scale is elevated and being semi-transparent, the body of the female is seen through it giving it now the reddish tinge. The greyish or whitish margin disappears. Whereas the unfertilized female is free and not connected to the scale, the fertilized one is attached to the dorsal as well as to the ventral scale and may therefore not esaily be separated from it.

The longevity of unfertilized females is extended as compared with the fertilized ones, a phenomenon of common occurrence amongst insects. Jones reports of an experiment in which (in September) two series of females were bred: one in isolation, the other under normal conditions. The latter commenced to reproduce about two months after the beginning of the experiment. Seven weeks later 8 isolated scales were exposed in order to allow access to males and after a period of 20—26 days three of them commenced to produce young. The number that emerged compared favorably with that from normal scales. A similar experience of the author may be mentioned. In August 1934, the cages of isolated females were opened, giving thus access to the males 75 days after isolation, when the corresponding normal females had started for over 30 days to reproduce. All of these gave birth to offspring, the number of which was unfortunately not established.

The second free-living stage of the red scale is the adult male, which flies actively and copulates with the female. In all experiments they proved to be positively phototactic. They are sensitive to very slight differences of light intensity even in diffuse light and speedily react to the brighter light. They are likewise positively geotactic in the light as well as in the dark.

The copula is not yet fully understood. The delicate males are certainly noctural. Experiments on the tactic responses showed that the males are 1) positively phototactic (at least in not too strong a light), 2) negatively geotactic, 3) positively chemotactic towards the females, in a varying degree. The experiments on sexual chemotaxis were performed by crushing some females on paper and putting this paper near the male or at least in the same jar with it. All males used in those experiments were less than one day old. The results were at first very contradictory. Whereas in some cases immediate reaction followed and the attraction proved to be stronger than light- or gravity stimuli, the latter stimuli often proved to be stronger and no attention was paid to the crushed females. Among 10 experiments 3 were strongly positive, whereas 7 cases belonged to the latter type. These experiments were repeated later on with the purpose to establish whether the state of maturation of the female has any influence on this reaction. Assumingly unfertilized females proved to be strongly attractive in ⅔ of the experiments, decidedly fertilized ones never. This tends to show that the male is directed towards the female by her smell. The unfertilized females may secrete a special smell which attracts the male and the secretion of which ceases after the copula. That the search after the female is dominated by smell is in conformance with the structure of the sense organs of the male. The eyes are large, but as far as recent researches show not able to distinguish between shade and light and movements. The antennae which are covered with numerous chemical sense-organs are large and well developed. P. Marchal (Ann. des Epiphyties 2, 1936, p. 481) has recently discovered that *Trichogramma* is attracted to host eggs as long as those are in active development or have died in such state and that attraction ceases when the eggs have entered diapause. He refers this change in attraction to some radiation and suggests that „en dehors d'un sens purement olfactif, l'existence d'un sens radiesthesique localisé dans les antennes et capable de deceler des radiations, peut être de l'ordre de celles de Gurwitsch". In armoured scales it would be difficult to assume such a radiation, as the scale probably intercepts them efficiently.

The question arises how the male is able to detect the female pygidium and vagina beneath the scale into which it has to introduce its long stylus-like penis. After many exclusions which had to be made from preliminary experiments as well as from deduction, the following hypothesis seems to be conclusive with the facts: As soon as the male has entered the stylus beneath the scale the active role passes to the female. The female rises her body on the pygidial lobi and turns round until the stylus touches the vagina. By mutual cooperation, probably, the stylus is now introduced into the vagina and copulation begins. The male dies in the night of copulation, as a rule, and there seems to occur one copula only. An excess of males would therefore be definitely useful to *A. aurantii*. In the males also prevention of copula prolonged longevity.

Birth: A. aurantii is generally described as being viviparous. This fact is confirmed by all Californian and S. African writers. True oviposition has never been observed.

Whereas the majority of all individuals are born as young larves with free legs and antennae, there is a varying number of individuals still enveloped in the amnion, a very thin membrane, which they cast off very soon after having been expelled from the body of their mother. The occurrence of this „amniopary" is by no means uncommon. The duration of this stage is very short, often a few minutes only, and the amnion is so thin and transparent as to be hardly observed. But despite lack of observation, local as well as seasonal fluctuation occur, cases of real vivipary probably always constituting the overwhelming majority.

Some confusion has been caused by the statement of Carmin and Shenkin that eggs are found quite often beneath the scales of *Aonidiella*.

„The eggs burst under the scale of their mother leaving there thin white shells which are of somewhat transparent nature, the larva extrudes from the egg starting with its pygidial side, the antennae are folded at first and then straighten. They may be found under the scale at a time a maximum of four eggs, usually no more than 2 or 3." The author has discussed this statement with Carmin, who agreed with him and permitted an authentic correction, in so far as his observations are fully covered by the just mentioned „amniopary". No real eggs with chorion have ever come to his observation.

We may resume: Vivipary is the rule in red scale. Amniopary has been noticed by every student of the species, but has generally been regarded as extremely rare. This rarity of observation is due partially to the short duration of this passing stage, partially to the difficulty the unexpecting student has in observing it. However, in certain localities and probably seasons this phenomenon is by no means rare, assumingly in summer especially and early autumn.

Egg-maturation and egg-size: The eggs begin to mature in the body of the female immediately after fertilization. We regard eggs as mature when the length axis surpasses 200 μ.

Some dissections of females show the process of maturation (in size classes):

age of female after 3rd moult	date	no. of intra-abdominal eggs	size of eggs	no. of intra-abdominal eggs	size of eggs
20 days	17.X.1935	5	126 × 70 μ	10	224 × 168 μ
20	27.XI.	8			280 × 252
23	26.XI.	12			236 × 224
33	10.XI.	5	210 × 112	12	280 × 140
39	26.XI.	5			224 × 112
41	2.XII.	27			210 × 140

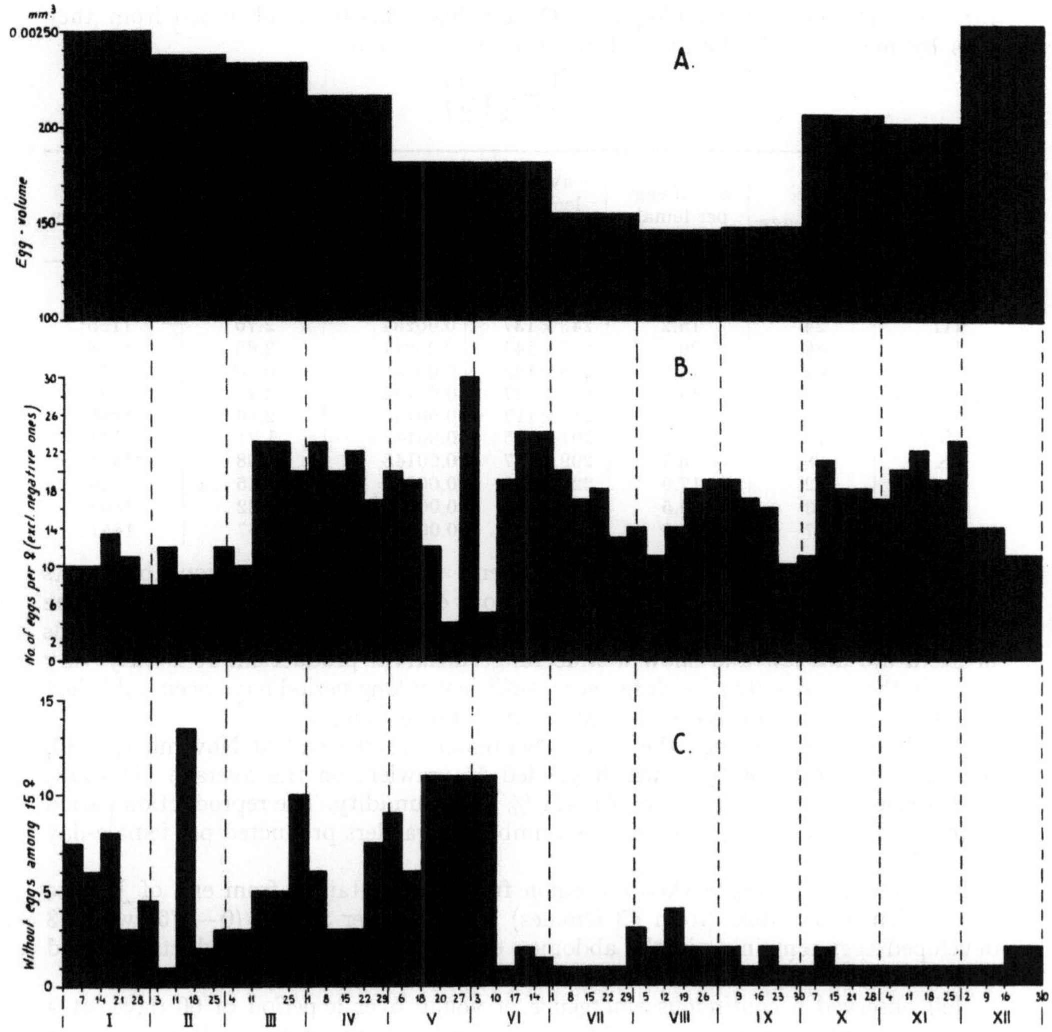

Fig. 60.
Seasonal variation in eggs and reproductivity of *Aonidiella aurantii* Mask. in 1929 in Tel Aviv. A. Monthly average volume of mature eggs; B. Average weekly number of eggs per female (the sterile ♀♀ excluded); C. Average weekly number of sterile females among 15 females.

The number of eggs within the body of the female after the last moult was as follows:

month	0—5	6—10	11—15	16—20	21—25	26—30	31—35	36—40	41—45	46—50	51—55 days old
VIII.1935	—	10	31	39(1L)	16	15	11(1L)				
IX.	3	21	14	12(1L)	13(2L)	17	12(11L)	15(1L)	20		
XII.	—	—	—	1	4	11	28		36	37	23

The first larvae appeared in the summer 19 days after the last moult, in the winter no larval birth was observed in these dissections even after 53 days.

One of the most interesting outcomes of our long continued dissections was the observations that winter eggs are much larger than summer eggs. The average ratio is 5 (December) to 3 (August). Their volume has been calculated from their axis by means of the formula of the rotation ellipsoid:

$$v = \frac{4}{3}\pi\frac{l}{2}\left(\frac{\omega}{2}\right)^2$$

month	% ♀ with eggs	no. of eggs per female	average length × width of eggs	average volume	% posit ♀ × no. of eggs per ♀ × average volume	eggs per 100 females
I	40	10.8	238 × 143 μ	0.00251 mm	1.26	648
II	33	9.3	239 × 138	0.00238	1.48	623
III	24	15.2	243 × 137	0.00234	2.70	1155
IV	39	20.1	238 × 141	0.00217	2.66	1226
V	62	14.2	218 × 125	0.00182	0.98	540
VI	33	19.8	217 × 127	0.00182	2.42	1327
VII	1	18.5	211 × 117	0.00155	2.86	1843
VIII	18	14.2	207 × 115	0.00147	1.71	1164
IX	8	15.3	209 × 117	0.00148	2.08	1408
X	2	17.0	220 × 133	0.00207	3.45	1666
XI	3	19.5	222 × 132	0.00202	3.82	1892
XII	7	15.6	226 × 147	0.00253	3.67	1451

Total young-production: The red scale seems to belong to that group of insects in which every exact study meets with various obstacles, not easily overcome. The productivity of *A. aurantii* belongs to one of the most thorny of these. Indications in literature are few and show a wide range in larval production. It is only very recently that more extensive date over a sufficiently long period have been published and some of these are data of relative importance only.

Quayle counted between the end of September to the end of November, 1910, the offspring of 14 females, which yielded 55 crawlers on the average (34—143) on citrus leaves at 20—23° C and 67—71 % Rel. Humidity. The reproduction period lasted 24 days in the average and the number of crawlers producted per female-day was 2.3.

Bliss and co-workers worked on lemon fruits and obtained from end of July to end of September 1930 (from 43 females) 73 larvae per female (0—176) with 43 developed eggs remaining in the abdomen of the females. The reproductive period lasted 18 days only, but 4 crawlers were born per female day.

Nell observed a continuous emergence of young over a period of 60 days, at a fairly constant temperature of 27.8° C. No exact counts were made on a larger number of females, but in one case 150 young were produced by one female on a lemon fruit. He remarks that there seem to be more young produced on the fruit than on the leaves.

All these observations refer to Californian conditions.

Full records on fertility have been published by Clark and Friend from the Rio Grande (Texas):

A Month	B Temperature °C	C % Rel. Hum.	D Average no. of larvae born per female-day	E Maturation period	F Length of life	G Length of young-producing period	H % crawlers settled	I % males	K Total larval production D×G
XI.1929	16.6	84	1.22	54	192	138			168
XII.	15.3	82	.88	91	217	126			111
I.1930	12.7	80	.52	102	258	156			81
II.	19.3	77	.43	—	—	—			—
III.	18.5	66	.70	123	234	111			78
IV.	24.1	76	1.36	88	172	84	31	67	114
V.	26.8	85	2.18	54	129	75	58	60	114
VI.	26.5	85	2.07	50	94	43	44	65	89
VII. /	28.1	78	1.95	43	93	49	27	60	96
VIII.	29.0	72	1.67	40	99	59	62	49	99
IX.	28.5	70	1.69	43	110	68	—	—	115
X.	24.0	80	1.32				67	62	
XI.	18.5	82	.60						
XII.	14.5	78	.55						

Similar observations have been made by Jones in S. Rhodesia:

month	Average no. of larvae born per female-day on fruit	Total no. of larvae produced per female: fruit	leaves	Reproduction period of female in days: fruit	leaves
I.1934	6.76				
II.	5.54	143	111	33	41
III.	3.29				
IV.	3.25	57	48	20	26
V.	1.43				
VI.	1.71	66	61	48	63
VII.	1.16				
VIII.	2.03	37	41	19	29
IX.	1.78				
X.	3.28	48	45	17	25
XI.	2.56				
XII.	4.40	88	72	27	31
		73	63	27	36

The continuation of our earlier experiments on the total fertility of red scale yielded results very different from those obtained in the first five series.

The total fertility of Red Scale in Palestine is shown in the following table:

The seasonal fluctuation of fertility proved stronger than the difference between fruit and leaves at the same season. In addition, there seems to be a higher fertility on young leaves than on old ones (as observed in experiments no. 9 and 10).

It seems impossible to give an accurate account of the total fertility of red scale. The nearest approach of rough evaluation would be for the optimal range of relative humidity (70—80%):

temperature	15—20	20—25	25—30° C
no. of larvae	50	100	150 larvae per female on leaves and

A No.	B Date	C n	D no. of larvae born per ♀	E max. no. of larvae born	F no. of eggs in female at the end of observation	G total egg production	H total larval production $\left(= D + \frac{F}{3}\right)$	I Average temperature in °C	K Rel. Hum. in %	L Remarks
1	VIII/IX.1931	16	40	46+54	35	75	52	27.5	72	Orange leaves in laboratory
2	XI.1931/I.1932	23	38	97+59	33	71	49	15.1	63	Lemon fruits in laboratory
3	IX/X.1933	28	42	62+62	13	55	46	23.4	72	Lemon leaves in laboratory
4	IX/X.1933	28	58	93+ ?	?	?	>60	22.5	77	Orange fruits in laboratory
5	XI/XII.1933	28	100	153+64	31	131	110	17.3	73	Orange fruits in laboratory
6	VII/VIII.1934	15	89	184+31	18	107	95	27.0	79	Cederate leaves outside, exposed to sun and wind
7	VIII/IX.1934	7	64	105+32	15	79	69	27.1	79	Sweet lemon leaves outside, exposed to sun and wind
8	VIII/X.1934	24	81	113+32	22	103	88	25.9	75	Orange fruit in grove
9	VII/IX.1934	26	149	186+20	19	168	155	26.7	79	Cederate leaves in laboratory
10	VI/VIII.1934	21	188	302+20	18	206	194	26.4	79	Sweet lemon leaves in laboratory

20—25% more on fruit. The maximal fertility of one female hitherto known is 322 in a series where the total average egg-production of 21 females reached 206. In the absence of rains, even at low temperatures, the total production may be higher in those cases where the period of production is protracted. The utilisation of sun-radiation and the consequent rise of body temperature above that of the shade temperature are the causes of such deviations. The rise in production of offspring during July and August is an important factor contributing to the heavy increase of *Aonidiella* at that period.

The positive correlation existing between the total production of crawlers and the duration of the production period has been clearly put to light by Jones (fig. 61). Until about 50 days the total number of offspring rises parallel to the duration of this stage. When this lasts longer, no correlation exists anymore. This occurs in winter, when the individuals with 90 days duration of production may again regain nearly the summer peak.

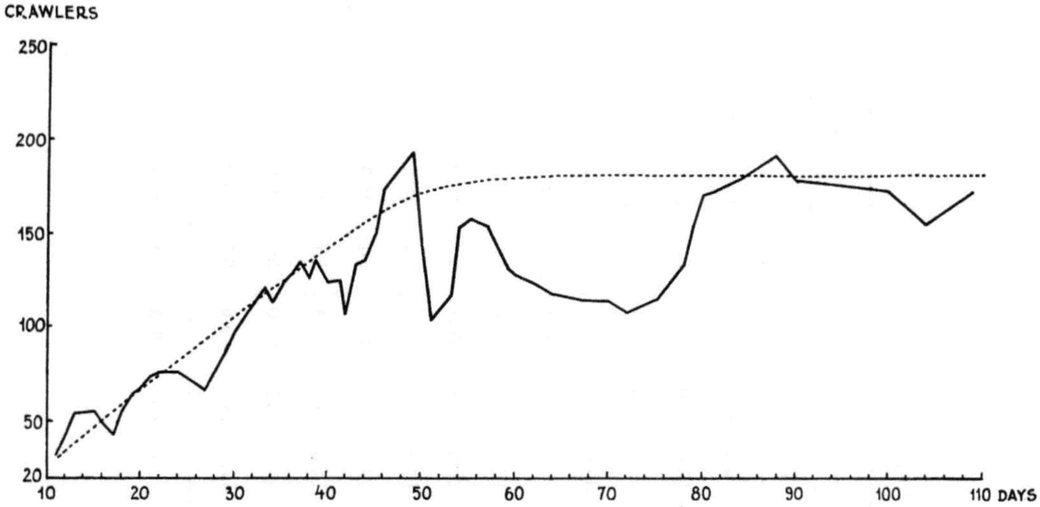

Fig. 61.

Correlation between total production of crawlers and the duration of the reproductive period in Red Scale females. (Recalculated after Jones and smoothed curve introduced).

Relative fertility: Comprehensive series of monthly dissections of mature females have been made in order to obtain the number of matured eggs within the body of the female. Such data certainly do not yield the total eggproduction of the dissected females, but must nevertheless be regarded as relative values permitting comparisons. The results are compiled in the table on page 219.

In order to understand this table the speed of maturation has to be considered too. The peak in spring (25.8 eggs per female in April) coincides with slow maturition in this season, whereas the fewer eggs produced in late summer and autumn develop more speedily. Climatic influences are strong. Many of the eggs are killed during the May hamsins and the depressive effect of these winds is still recognizable by a depression in early June.

Mature females are present on the leaves the whole year round. They are absent or rare on the fruit (except the latest fruits of the old crop) between April and July. They are absent from the green twigs during the hot season (July to September).

Average number of eggs in the abdomen of Red Scale females on sour orange.

Month	on leaves														Total on fruits		Total on green twigs	
	1928		1929		1930		1931		1932		1933		Total					
	n	M	n	M	n	M	n	M	n	M	n	M	n	M	n	M	n	M
I	—	—	36	10.7	52	8.9	60	14.3	60	19.8	75	14.3	283	14.1	71	39.7	60	11.5
II	—	—	40	9.7	45	12.0	60	20.8	60	29.8	60	16.2	265	18.4	60	50.1	135	17.2
III	—	—	34	14.1	53	18.8	44	25.7	60	23.9			191	21.1	60	48.6	60	21.7
IV	—	—	46	20.3	28	30.6	41	25.0	60	28.2			175	25.8	18	33.8	60	25.8
V	—	—	18	16.7	—	—	20	21.1	69	24.7			107	22.7	—	—	60	28.9
VI	—	—	32	18.9	72	17.2	45	25.0	4	9.0			153	19.7	48	39.9	22	20.2
VII	—	—	74	18.6	45	23.8	58	26.4	27	19.0			204	21.5	79	36.7	—	—
VIII	—	—	49	15.1	60	17.5	75	21.8	75	18.6			259	18.6	100	40.6	—	—
IX	—	—	55	15.2	72	16.3	60	19.1	60	22.9			247	18.3	128	34.7	—	—
X	53	13.0	59	17.0	38	14.3	60	16.1	75	22.5			285	18.1	108	40.7	46	19.5
XI	47	14.9	58	19.4	—	—	75	20.0	60	16.8			240	17.9	63	40.1	44	12.2
XII	54	11.1	56	15.6	29	17.8	60	19.0	60	18.0			259	16.2	45	44.4	60	13.9
			557	16.2	494	17.1	658	20.9	670	21.4			2668	18.8	780	39.7	547	18.8

The highest egg number within the females is observed on the green twigs from April to June.

The differential egg-production on fruits on the one hand, on leaves and twigs on the other hand, is well illustrated in these observations. The ratio 39.7 : 18.8 eggs per female points to the higher fertility on the fruit.

This difference was even greater in the Upper Jordan Valley, where Klein obtained the following values from dissections made the whole year round:

Host	leaf	fruit	
grapefruit	17.0	46.0	eggs per female
lemon	14.2	38.4	eggs per female

A drop in the number of mature eggs is seen in August, November, January and June, when larvae appeared in masses. The rise in April can be explained by the fact that the females, after a long hibernation, produce their first larvae only at the end of April.

Rate of young-production: The author has not obtained values for rate of reproduction. Fortunately, the papers of Jones and Clark and Friend permit a discussion of this problem. That the daily production rate shows considerable seasonal fluctuations is obvious from the tables of these authors which have been reprinted above. Fig. 62 illustrates this seasonal trend in S. Rhodesia. It shows that the rate is highest at the highest temperatures and highest air humidities. It further shows that the rate is always higher on fruits than on leaves.

By plotting the numerous daily records of crawlers produced against temperature and humidity, Jones received a thermo-hygrogram of the daily production rate. The basic figures are:

Average numbers of crawlers produced per day at varying relative humidities and temperatures (per female day):

	15.6—18.3	18.3—21.1	21.1—23.9	23.9—26.7° C	sum
80% R.H.	1.0	1.9	4.3	4.8	12.0
75	1.6	2.0	4.7	5.9	14.2
70	1.7	2.6	4.2	5.4	13.9
65	1.7	2.4	3.8	4.5	12.4
60	2.0	2.4	3.1	3.7	11.2
55	1.7	2.4	2.4	3.8	10.3
50	1.5	2.0	2.1	2.7	8.3
40	1.2	1.3	2.0	2.3	6.8
average	1.6	2.1	3.4	4.1	

The optimal rate of reproduction occurred therefore at 24—27° C and 70—75% Rel. Hum. Lower as well as higher humidities reduce the rate of reproduction. The rate of reproduction rises with temperature. It is questionable, whether 27° C is yet the upper limit of this rise, which probably occurs at 28—30° C only. The averages of the single values, meaningless in themselves, indicate these shifts quite distinctly. At lower temperatures the limit of optimal humidity conditions falls to 60 or even 55% rel. hum. The combined influence of temperature and humidity has been illustrated by Jones in fig. 62.

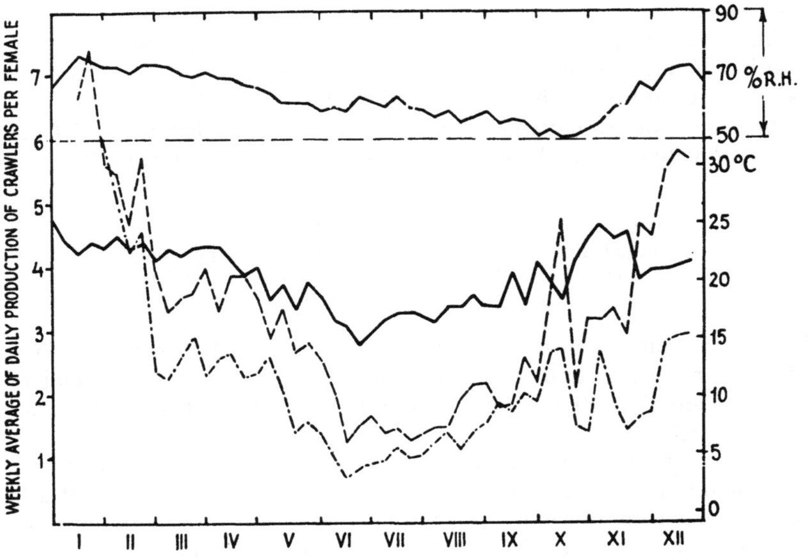

Fig. 62.
Weakly average of daily crawler production per red scale female in relation to temperature and humidity (full lines) on fruit (broken line) and on leaves (dotted line) in 1934 at S. Rhodesia (from Jones).

The effect of climatic factors on the seasonal variation of productivity at S. Rhodesia may be resumed as follows (Jones):

Period	Productivity			Climatic conditions	
	rate	duration	Total larvae born	temperature	Rel. Hum.
1. XII—III (late summer with rains)	+++	++	+++	fairly high	fairly high
2. IV- end V (approach of winter)	++	+	+	decreasing	decreasing
3. V- end VIII (winter)	+	+++	++	low	fair
4. VIII—IX (approach of summer)	+(+)	+	(+)	increasing	low
5. X—XI (summer prior to rains)	++	(+)	+	very high	very low

Remarks: 1. Conditions optimal for productivity, especially in I and II. 2. Rate of productivity slowed down by both climatic factors. 3. Rate of productivity slowest through low temperature, which with a fair humidity prolongs the production period, giving a fairly high total productivity. 4. The slow rate of productivity with low temperature and the limited production period through low humidity results in lowest total productivity. 5. Considerable increase in temperature increases the rate of productivity but the very low humidity prevents a maximum rate and produces the shortest production period, resulting in low total productivity.

For Palestine (coastal plain) a similar table would run as follows:

month	Productivity			Climatic conditions		
	rate	duration	Total larvae born	temperature	Rel. Hum.	rains
I	+	+	+	12.8° C	72%	132 mm
II	+	+	+	13.5	74	99
III	+(+)	+++	++	15.5	71	25
IV	++	++	++	17.9	71	19
V	+	+	+	21.5	72	2
VI	++	+	++	24.5	74	—
VII	+++	+	+++	26.4	76	—
VIII	++	++	++	26.7	76	—
IX	+++	++	++(+)	25.5	74	1
X	++	+++	+++	22.6	73	16
XI	++	++	+++	18.7	70	65
XII	+(+)	++	+++	14.7	73	144

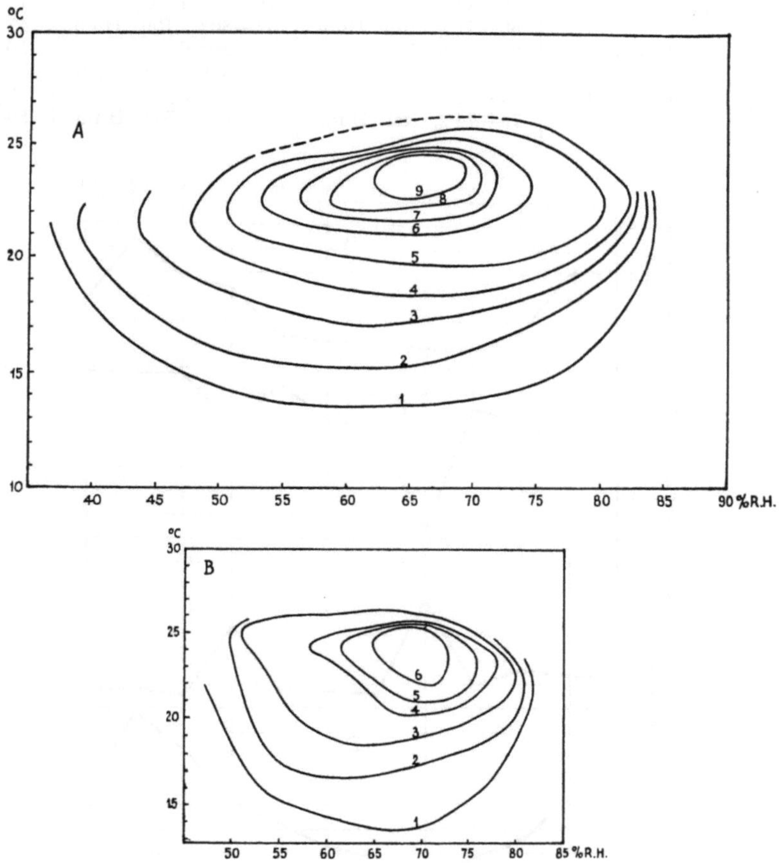

Fig. 63.
Productivity of *Aonidiella aurantii* Mask (per female day) in dependency upon temperature and humidity. A. on fruit, B on leaves (from Jones).

Remarks: In I and II low rate of productivity due to low temperature, short longevity due to low temperature and rains and therefore low total productivity. In III and IV considerable increase in rate of productivity, production period and total productivity as consequence of increase of temperature. In V rate of productivity higher than before, preceding the hamsin winds. These winds reduce the rate of productivity, production period and total productivity to a minimum, the height of which is determined by the intensity of these dry not winds. In VI recovery and strong increase. In VII to IX highest rate of productivity, fair production period and highest total productivity. In IX often depression, when hamsin winds are strong. In X—XII fairly high rate of productivity, longest production period and therefore a fairly high total productivity, contracted to a slight degree by the first rains.

The optimal climographic zones of fertility would be as follows from the present experiences:

Country	Author	Optimum	Favourable	Unfavourable
S. Rhodesia	Jones	20—24.5° C; 58—71% Rel. Hum.	18.3—25.6° C; 52—80% Rel. Hum.	13.9—26.7° C; 47—82 Rel. Hum.
Texas	Clark and Friend	25—27.5° C; 70—75% Rel. Hum.	20—29° C; 60—85% Rel. Hum.	10—30° C; 50—95% Rel. Hum.
Palestine	Bodenheimer	17—25° C; 70—74% Rel. Hum.	15—27° C; 68—80% Rel. Hum.	10%30° C; 50—95% Rel. Hum.

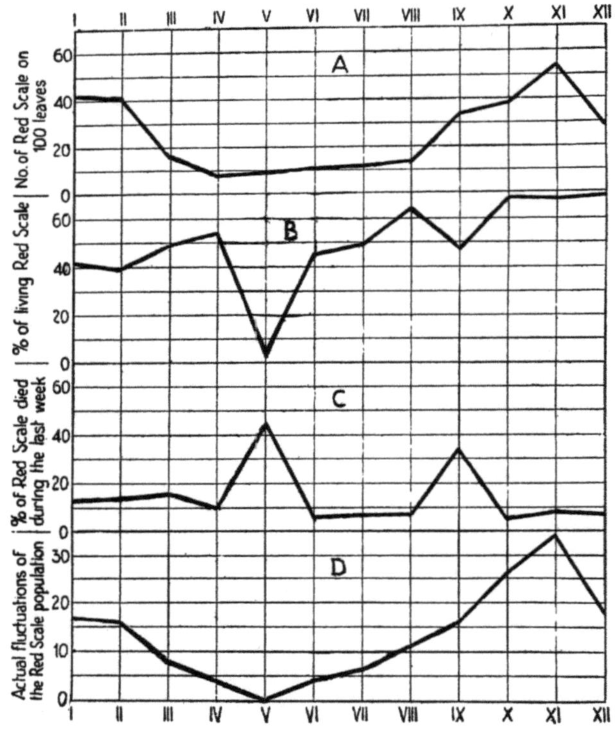

Fig. 64.
Seasonal trend of Red Scale population and mortality in Palestine.

Fig. 65 illustrates these conditions and shows clearly that with higher temperatures the humidity requirements rise as well.

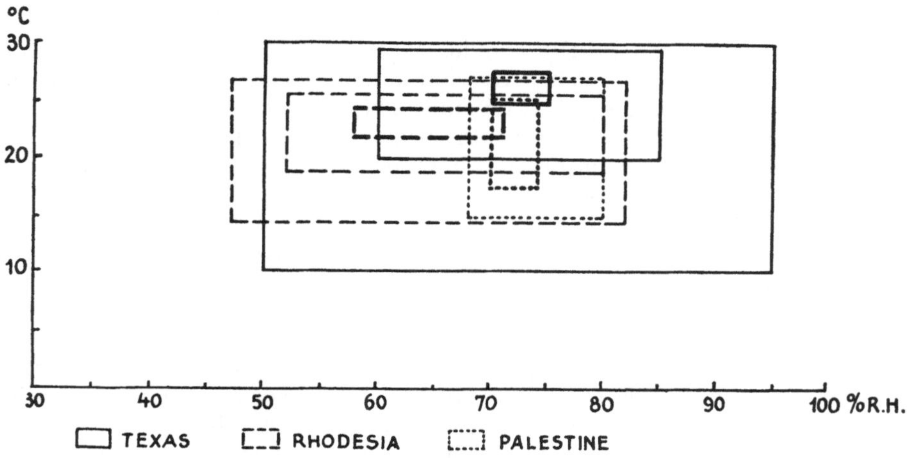

Fig. 65.
Optimal climatic zones of fertility of *Aonidiella aurantii Mask.* in different regions.

THE CRAWLERS.

The young crawler remains, as a rule, for ½ to 2 days below the scale of its mother. The scale of activity of these crawlers ranges (39 series of experiments):

1. Beginning of cold torpor 10.5° C (Average upper limit)
2. Only weak movements of antennae 12.8° C (Average upper limit)
3. Crawling with interruptions 13.8° C (Average upper limit)
5. High activity 39.2° C (Average lower limit)
8. Heat death (instantaneous) 43.0° C (Average lower limit)

Normal activity of the crawlers occurs between 14 and 39° C, 25 to 32° C being the optimum.

The locomotion of the crawlers has not yet been adequately studied. Quayle's experiments (78 cms at 18.9° C on the average, maximum at 32.7° C 278 cms per 2 hours) permit the following calculations:

1/78 (18.9—x) = 1/278 (32.7—x)   c = 13.0° C   (Bodenheimer)
1/104(18.9—x) = 1/279 (32.8 — x)   c = 10.7° C   (Jones)

However, the limited number of experiments and the lack of certainty as to whether exposure to sun-radiation was excluded during the experiment do not make these calculations very reliable. The fact suffices that both values range about the development threshold and are slightly below the lower range of normal activity obtained in our experiments.

The normal crawling activity per 2 hours would therefore be at

35°C	30° C	25° C	20° C	15° C
297	230	162	94	27 cm

The distance covered by the crawlers in 5 minutes varied, according to our observations, from 15 to 38 mm at 25° C. As a matter of fact, only very few crawlers ever surpass by locomotion about ½ m distance from their mothers; as a rule, they remain much nearer. In the laboratory, they show a tendency to settle in the immediate neighbourhood of their birth-place. Their active dispersal in nature is just extensive enough to infest the young sprouts of their hosts. They may even infest the leaves of a neighbouring tree, when the crowns of both trees touch each other. Quayle has proved quite definitely and his experiments have been repeated with similar results by the writer, that active infestation of neighbouring trees by migration passing over the soil is practically impossible. The crawlers may theoretically pass over some metres distance on a smooth surface, but they are decidedly handicapped by a barrier even of a few inches only of sand or soil. The main source of infestation from tree to tree and from locality to locality is the wind, which blows the small crawlers away and eventually brings them to a suitable host. This importance of the wind is certainly higher than realised by Quayle. The speed of reinfestation of thoroughly treated orchards, if the neighbour has not treated his orchard, is sufficient proof therefore.

The *sense-physiology* of the free-living stages of red scale is still imperfectly known. The crawlers are guided actively by their senses and taxes as long as they crawl. This active stage generally lasts from 2 hours to 1 day, in cooler or very dry weather up to 3 days and even longer. But a great part of the larvae which have not succeeded in fixating themselves within the first 24 hours are too weak to do it later on. A large number of the crawlers are passively transported by the wind, during which time no active orientation takes place.

The dominant taxis of the crawlers is an apparent positive phototaxis towards bright light. In diffused light they generally do not show any reaction towards the source of the light. In the orchard they settle almost invariably on that side of the leaf or fruit which is exposed to the light. Their phototactic indifference in diffused light as well as their ready positive reaction towards bright light strongly suggests that this reaction is not at all a phototaxis, but thermotaxis.

The preference temperature obtained in 136 readings was 29.7° C. The fact that in the Upper Jordan Valley the centre of the trees is more densely settled than in the coastal plain may point in the same direction: The shade temperature in summer is there near the preference temperature, whereas on the periphery it would be exposed to higher temperatures during the day.

Chemotactic movements towards orange- or lemon-rind have been observed in some individuals, but were never very pronounced nor displayed by a great majority. Miss R. Ashbel made a peculiar observation in this respect, which has repeatedly been verified by the writer. A crawler was placed 2 mm from a piece of lemon. It approached the lemon, turned back and again returned. When the lemon was displaced, always at a distance of about 2 mm, the crawlers followed it. At great distances the crawler seemed to be unaware of the lemon. When the distances were less than 2 mm the crawler moved away from the lemon, until the distance of 2 mm was reached; then it again turned towards the citrus.

The conclusion may be drawn that the crawlers are directed to suitable hosts by smell. The phototactic reaction cannot play an important role under natural conditions where the crawler has no choice, having been born on its host. The effect of host-smell on the behaviour of the larvae could be demonstrated also in another way. Citrus leaves were placed in a bottle. Then the air was sucked from the latter through a small smell-chamber into which 10 crawlers of *A. aurantii* were placed. The speed and direction of the larval movements was observed a) before the citrus smell entered the chamber, b) at the moment when it entered, and finally 10—20 minutes later, when the chamber was filled with the smell.

Speed of red scale larvae in the smell chamber.
(mm per 5 minutes).

Experiment No.	before	during	after	before	during	after
	penetration of citrus smell			penetration of pure air		
A	10	15	8	13	13	13
B	25	37.5	17.5	15	15	15
C	12.5	12.5	7.5	12	9	5
D	12.5	13.5	11.5	45	40	0
E	25.5	30	21			
F	7.5	12.5	—			
G	10	15	8			
H	17.5	20	18			
I	25	23	16			
K	—	25	19			
L	5	13	11			
M	10	15	11			
Average (exc.	15.3	19.5	13.0	21.2	19.2	8.2

In 9 cases the activity was increased by 52% on the average (from 8 to 160%) when the citrus smell entered the smell chamber. In one case no change occurred in the speed of activity and locomotion, whereas in another case a reduction of 8% was observed. This excited activity ceased after a few minutes, when the citrus smell had filled the whole smell chamber, so that no directed movement could be effected anymore. A control series wirth pure air excluded the possibility of the wind being the stimulus and not smell. In these series the activity decreased or remained unchanged when the air moved through the smell-chamber with the same speed as in the other experiment.

The settling of the crawlers begins from less than an hour to the next day after they have left the mother's scale. Individuals which remain longer in the crawler stage have, as a rule, little chance of settling later. We dispose of few records only of the percentage of crawlers settling successfully.

In the experiments conducted by Quayle, among 884 crawlers bred during some months 363 or 41.1% settled. He concludes that the seasonal fluctuation was not large and that, owing to the more favorable conditions in the laboratory, this result is higher than that to be espected in the field. In the experiments of Bliss, only 5—10% fo the crawlers died, but only vigorous individuals were selected for the experiments. In similar experiments of Nell 82% of the crawlers settled.

Clark and Friend received the following results in Texas:

month (1930)	IV	V	VI	VII	VIII	X	Total
temperature and humidity	24/76	27/85	27/85	28/78	29/72	24/80	
Crawlers liberated	410	100	90	380	560	340	1980
Crawlers settled	218	58	40	102	349	228	995
% settled	31	58	44	27	62	67	50

Jones obtained in extended series as follows:

month	I	II	IV	V	VII	VIII	XII	Average
on fruits.	90	88	68	34	30	54	64	65
on leaves	72	68	61	38	35	81	47	59

Except in winter (V/VIII) the settling was always higher on the fruit than on the leaves. Percentage of settling as correlated to temperature and humidity is compiled in the following table:

° C	45	50	55	65	75	80	Rel. Hum. %
23.9				85		92	
21.0		52	67	63	40		
18.7		52					
14.5	42						

Temperature seems to be a more important factor than humidity, both showing optimum values at 23—25° C and 65—80 Rel. Hum. Whereas the humidity range is probably valuable at higher temperatures, also, it seems that the full optimum range is at 23—27° C.

Under favorable laboratory conditions between 90 and 100% of the crawlers may easily be induced to settle, especially when vigorous crawlers are selected. Under less favorable conditions in the laboratory or in the insectary 40—60% will settle, as a rule. But such high percentages are rare in nature, where other factor act in addition. The most important amongst them are wind and rains, whereas the activity of predators is negligible in the crawler stage. Both, wind and rain, change not only from day to day, but even from hour to hour. They may blow or wash away about 100% of the crawlers from their substratum and have no influence at all a few hours later. It is impossilbe to evaluate the part of the crawlers thus destroyed even in approximate numbers.

Our own experience in Palestine does not refer to percentage of crawlers settled. We have a few series of counts as to how many crawlers per female succeeded in settling. Klein reports the following numbers from the Upper Jordan Valley:

	on leaves					on lemon fruit	
generation	1	2	3	4	5	4	5
mont of birth	I	IV	VI	VIII	XI	VIII	XI
no. of breedings	2	10	4	10	3	3	4
average settled per female	52	38	38	36	33	96	69
maximum settled per female	61	59	45	59	55	117	81

In the coastal plain, the following series was recorded:

date	10.VIII.28	17.IX.28	22.X.28	24.VI.29
no. of breedings	21	19	4	16
average settled larvae per female	64	70	59	15

The small offspring in June 1929 is the result of the influence and after-influence of the hamsins in May and early June.

These observations made under natural conditions show that during the favorable summer months especially the percentage of settling may be rather high.

BIOLOGY OF NUTRITION.

Hosts and host-relations: The wide range of polyphagy which is shown by the Red Scale is best illustrated by the host list published by Brain (1918) for the Union of S. Africa.

„It is most commonly a serious pest on citrus, rose, apple, pear, oak, mulberry, camphor and privet. It has also been reported on: *Abutilon*, *Acacia spp.*, *Acer*, *Agava*, alder, almond, aloe, *Ampelopsis*, *Aralia*, *Aucuba*, *Araucaria spp.*, asparagus, avocado pear, *Bauhinia*, belhambra (*Phytolacca dioica*), *Benthamia*, *Berberis*, *Bignonia*, *Bougainvillea*, *Bouvardia*, box, broom, *Callistemon*, *Camellia*, carob, castor-oil, *Casuarina*, *Cedrela*, *Cestrum*, chestnut, chilli, *Choisya*, *Chrysophyllum*, *Clerodendron*, *Clematis*, *Clivia*, *Convallaria*, *Coprosma*, *Cordyline*,

Cornus spp., Cryptomeria, Cupressus spp., currant, cycads, *Cyperus, Dahlia, Deutzia, Dombeya, Dorynathes, Duranta, Dracaena, Elaeagnus*, essenwood, *Eucalyptus, Eugenia, Enonymus, Ficus spp., Forsythia*, frangipani, fuchsia, *Ginkgo, Gleditschia*, grape, *Grevillea spp., Greyia*, guava, *Gynura, Hakea*, hawthorn, holly, *Hydrangea, Impatiens*, ivy, *Jacaranda*, jasmine, kaffirboom (*Erythrina sp.*), Kei-apple, *Kennedya, Libonia*, lilac, *Liriodendron*, laurel, *Lagerstroemia, Mackaya*, mango, mangosteen, *Maranta, Melia*, New Zealand flax, oak, oleander, orchids, *Osmanthus*, palms, *Pandanus*, peach, poeny, *Penstemon*, persimmon, phlox, *Pinus spp., Pittosporum*, plane, plum, *Poinsettia*, poplar, quince, *Rhamnus sp., Robinia, Salvia*. sneezewood, *Sophora, Spiraea spp., Statice, Sterculia, Strelitzia, Taxodium, Thuya, Trichilia, Trichocladus, Tristania, Toxicophlaea, Veronica, Vuburnum*, walnut, willow, wistaria, *Yucca* and *Zinnia*."

In the other parts of Africa it is known from citrus, cocospalm, *Aucuba*, agave, *Camilla*, tea, *Cycas*, etc. In Kenya Colony it attacks mainly citrus, roses, apple, plum and risal (Dry).

The host-lists of California and Florida, include beside many of the hosts mentioned from S. Africa (Quayle, Essig, Merril, Chiffin): apple, *Artocarpus*, bamboo briar, *Carissa, Juglans regia, Laurus nobilis*, olive, passion vine, pear, *Pistacia, Podocarpus*, privet. In S. America it is abundant on roses, but rare on citrus (Compere). In Australia it is common on citrus, rose, mulberry trees, willows. In citrus groves it is also met with on pumpkins, etc. (Froggatt). In Queensland it infests also a number of indigenous plants (Summerville).

The Indian hosts include: citrus, rose, aloe, jasmin, agave, *Cocculus, Olax* and thea; the Ducth Indian: *Artocarpus* and djeroek; the Japanese and Chinese; only citrus and palms so far (all other records on *Podocarpus*, etc. have to be referred to *A. citrina* and *A. taxi*); the Formosa records are: *Ardisia, Citrus, Heptapleurum*.

Some of the host-relations within the Mediterranean region have been described above. The following host-genera are recorded from Greece and Egypt:

Greece (Koroneos, Lindinger): *Thuja, Ficus, Eriobotrya, Rosa, Citrus, Pittosporum, Euphorbia, Evonymus, Vitis, Althaea, Olea, Ligustrum, Nerium, Pistacia.*

Egypt: (Hall): *Mangifera, Hedera, Calandula, Pinus, Diospyros, Ricinus, Phyllanthus, Laurus, Bauhinia, Ceratonia, Erythrina, Asparagus, Ruscus, Ficus, Morus, Jasminum, Olea, Rosa, Prunus, Pyrus, Citrus, Salix, Solanum, Vitis*, palms.

The Palestine records are so far:

Anacardiaceae:	* *Pistacia lentiscus* (leaves)
Aralaceae:	* *Hedera helix* (leaves)
Celastaceae:	*Evonymus pulcherrimus* (leaves)
Conifera:	* *Cupressus spp.* (cones), * *Pinus halepensis* (stem)
	* *Thuja orientalis* (cones)
Euphorbiaceae:	*Ricinus communis* (branches)
Lauraceae:	*Laurus nobilis* (leaves)
Leguminosae:	*Acacia spp.* (branches and leaves,)
	Ceratonia siliqua (leaves)
Musaceae:	* *Musa cavendishi* (leaves)
Myrtaceae:	* *Eucalyptus rostrata* (leaves)
Oleaceae:	* *Ligustrum sp.* (leaves, * *Olea europaea* (fruits)
Palmae:	*Phoenix canariensis* (leaves)
Pittosporaceae:	*Pittosporum undulatum* (leaves)
Rosaceae:	*Rosa spp.* (stem and branches), * *Pyrus communis* (branches and leaves)
	* *Maclura aurantiaca* (twig)
Rutaceae:	*Citrus spp.* (fruit, leaves, twigs, stem)
Salicaceae:	*Salix safsaf* (stem and branches)
Tamaricaceae:	* *Tamarix sp.* (stem)
Vitaceae:	* *Vitis vinifera* (leaves)

*) accidental infestation

Either citrus or roses or both are everywhere the main hosts of *A. aurantii*. The first host corroborates the theory that the Oriental region is the home of the species.

Endemic roses occur everywhere and do not furnish us with information in this regard. Other hosts of general importance are *Salix, Morus, Pyrus, Prunus, Ricinus, Cocos, Aloe, Agava, Quercus.* The host-list varies more or less from region to region and occasionally from country to country. In the latter case the date of the introduction is probably a most decisive factor. The different host relations of polyphagous insects in various regions remains yet to be studied. Physical conditions (climate, cultivation, etc.) certainly have a very important influence. Compere hints as much for *A. aurantii*: ,,The widely held belief that the excessive abundance and injuriousness of red scale on domesticated plants is evidence that in a state of nature the species was controlled by parasites is not necessarily correct. Undoubtedly, under primitive conditions, red scale was for less injurious than now. Its injuriousness can now be accounted for by its spread into areas made more favorable for it by man. If citrus trees were not grown under irrigation in arid regions devoid of summer rainfall, and rosebushes were not grown in tropical regions, red scale would not be much of an economic pest. Roses are indigenous to temperate regions and in such regions are usually free from red scale. Citrus is indigenous to tropical or semi-tropical regions with summer rainfall and in such regions is free from injurious attacks of California red scale. The condition under which red scale now becomes so excessively abundant did not exist in primitive times. It is in man-made environment that red scale becomes a pest."

The average monthly number of eggs in the abdomen of 25 females in 1931 was:

Root stock	Leaves				Fruit			
	shamuti	grapefruit	lemon	Average	shamuti	grapefruit	lemon	Average
Sour orange . .	21	21	24	22	36	45	50	44
Sweet lemon . .	20	21	23	21	47	47	51	48
Average	21	21	24		42	46	51	

The wild root-stock had 21 eggs per female on sour orange, 29 on sweet lemon.

Whereas there is no significant difference in the development of the different varieties on both root-stocks, a slight, bur probably significant difference may be observed with regard to the different citrus species.

Fertility was always highest on lemon, which is known to be more exposed to eavy red scale attack than both shamuti and grapefruit.

Some other counts of the same kind may be referred to.

root-stock	on leaves	on fruit
Valencia orange on sour orange	20	47
Citron on sweet lemon	37	44
Citron on citron	20	

On leaves of *Ligustrum* 15 females had 30 intraabdominal eggs on the average on 13.I.1932.

On the effect of manuring citrus trees in regard to the development of Red Scale. Since about 1927, when the sudden rise in red scale in young citrus orchards aroused general attention, the theory was often promoted by old farmers that this increase may possibly be connected with the increased manuring in the citrus orchards. And whereas similar facts were known from literature *), it seemed worth while

*) High potash content of the leaves immunizes the tea-plant against attack by the bug *Helopeltis theivora* (Andrews, 1921), whereas high potash content of the soil renders the coffee-tree less resistant to the attack of the mealy-bug *Pseudococcus coffeae* in Kenya Colony (Kirkpatrick, 1927).

to go into this question. For if such relation should really prove to exist, it would possibly open new ways for the control of the scale, by changing the cultivation methods (especially with regard to manuring) in the orchard. Thanks to the generous help rendered by the Empire Marketing Board in London, the study of this problem was made possible.

One group of citrus plants was given a normal diet of the following composition:

17.5% $MgSO_4$
15.9% KCl
42.5% $Ca(NO_3)_2$
15.9% KH_2PO_4

Other groups were given the same solution, but by leaving out in each case another component. Another group was given the same in addition to a threefold quantity of $Ca(NO_3)_2$. Still another received the same but with only a third of the normal quantity of $Ca(NO_3)_2$. The trees forming the last group grew in clean sand and distilled water only. The salts were given to the trees in the composition described above six times during the year. The drawback of this diet was that whereas N, P and Mg were eliminated or trebled in the experiments, potash still remained, when KCl was eliminated or trebled, in the form of KH_2PO_4. All conclusions drawn for K must therefore consider this fact.

With regard to the influence of the diet on the plant we refer the reader to the preliminary publication (Bodenheimer and Ashbel, 1933). The Jerusalem experiment was carried out in sterile sand and watered with distilled water. The Rehoboth experiments were carried out as water-cultures of young seedlings. Fig. 66 illustrates the method.

Fig. 66.
View on a nutrition experiment in progress. Influence of nutrition on orange-seedlings planted in pure sand, from 1 all salts, nitrates three times the normal dosage to 7 without all salts and 8 with all salts exc. nitrates. (from Bodenheimer and Ashbel).

The processes studied were: the influence of the host on

1) duration of development
2) egg-production
3) number of larvae actually born.

The development started from the day of successful fixation of the crawler to the appearance of the crawlers of the second generation. The number of eggs within the female is an adequate expression of the general fertility of the red scale, as has been pointed out before, provided the number of readings is sufficiently large. With regard to larval production only the Rehoboth record is reliable, as in Jerusalem the experiments were performed in a barrack exposed to wind where many of the larvae were blown away. On the other hand, under laboratory conditions at Rehoboth the number of larvae produced was about twice the normal or even more. However, the relative values of this series are reliable.

The different series of experiments are not all directly comparable as external conditions, season and host conditions were somewhat different. The highest value of every series must therefore be put at 100 and the relative values, as given in the last column of each series, may thus be compared.

The shortest development (the slowest developing of each series = 100) was as follows:

Nutrition	Jerusalem VIII 1932	Rehoboth 1934 VII	Rehoboth 1934 VIII/IX	Rehoboth 1935 VII	Rehoboth 1935 VIII/IX	Rehoboth 1935 IX/XII	Average
1—3 All salts.	100	88	76	82	65	100	85
4. Without $Ca(NO_3)_2$	95	88	78	100	86	97	91
5. 3 times $Ca(NO_3)_2$	95	96	90	94	88	0	93
6. ⅓ of $Ca(NO_3)_2$	95						
7. Without KH_2PO_4	92	86	78	94	79	99	88
8. 3 times KH_2PO_4.		96	97	97	82	100	94
9. Without KCl	92	83	70	94	79	99	86
10. 3 times KCl.		100	100	86	100	0	95
11. Without $MgSO_4$	100						
12. With distilled water . . .	100						

The number of eggs in the abdomen of females was:

Nutrition	A.X.1934, Jerusalem			B. IX.34. Reh.			C. XII.35. Reh.			Rel. values			
	n	M	max	n	M	max	n	M	max	A	B	C	Average
1—3 All salts.	18	13.6	29	39	11.5	31	8	24.4	32	64	93	61	73
4. Without $Ca(NO_3)_2$.	1	(21)	(21)	21	3.5	8	9	22.8	31	100	28	56	61
5. 3 times $Ca(NO_3)_2$; .	26	19.6	35	20	12.4	32	8	40.3	48	94	100	100	98
6. ⅓ of $Ca(NO_3)_2$ . .	27	14.3	34							69			
7. Without KH_2PO_4 .	29	12.4	30	13	6.8	14	10	32.1	44	60	55	80	65
8. 3 times KH_2PO_4 .				3	7.0	12	14	33.4	48		57	84	71
9. Without KCl . . .	30	19.7	45	20	12.4	28	14	23.9	32	95	100	60	85
10. 3 times KCl . . .	28	19.0	35	19	6.5	11	12	25.8	40	91	53	65	70
11. Without $MgSO_4$ . .	16	13.6	21							64			
12. With aqua dest. .	23	15.8	36							76			
13. 2 times KCl . . .	27	19.9	47							96			
14. ⅓ of KCl	28	23.6	45							113			

The number of larvae per female was (single female per leaf):

Nutrition	A. VIII.32. Jerus.			B. VIII/IX.34. Reh.			C. VIII/IX.35. Reh.			Rel. values			
	n	M	max	n	M	max	n	M	max	A	B	C	Average
1—3 All salts.	74	43.5	55		110		6	221	314	70	97	97	80
4. Without $Ca(NO_3)_2$.	17	24.0	30		81		3	131	204	38	71	58	60
5. 3 times $Ca(NO_3)_2$.	118	62.8	90		75		2	185	219	100	66	81	81
6. ⅓ of $Ca(NO_3)_2$ . .	56	53.8	87		—					86			71
7. Without KH_2PO_4 .	173	39.0	64		114		3	165	170	62	100	73	84
8. 3 times KH_2PO_4 .					40		3	229	278		35	100	63
9. Without KCl . . .	86	26.5	44		89		2	201	226	42	78	88	69
10. 3 times KCl. . . .					51		3	175	195		45	77	63
11. Without $MgSO_4$ . .	11	36.0	67							58			56
12. With aqua dest. .	31	26.5	60							42			46

Grouping these data in three classes, we obtain:

	development	eggs per female	larvae per female
Standard (complete nutrition)	85	73	80
Deficiency	88(86—91)	70(61—85)	64(48—84)
Luxuriance	94(93—95)	80(70—98)	69(63—81)

Or for the different salts, we receive the following

relative values:	N	P	K	Standard
Development: Deficiency.	91	88	86	85
Luxuriance	93	94	95	
Eggs per female: Deficiency	61	65	85	73
Luxuriance.	98	71	70	
Larvae per female: Deficiency	60	84	69	80
Luxuriance.	81	63	63	

The development was shortest in nutrition with all salts. (These breedings are most numerous and the chance for a shortest development is therefore highest in this group). The development in the deficiency groups was only slightly longer; this slackening seems to be of no significance. On the other hand, it is almost certain that the luxuriance of any one of the salts lengthens the duration of its development to about 10%. It is important to note that this lengthening occurs in practically every experiment.

As far as egg- and larval-production is concerned, it is rather improbable that low fertility at deficiency plants is significant. The same is true for the luxuriance.

In comparing the relative importance of the salts it is obvious that nitrates are the most important ones, whereas the part played by K cannot yet be regarded as settled for the reason given above.

It may therefore be concluded that under the rather hard conditions of the experiments a prolonged deficiency diet of the host plant does not influence the duration of development, but depresses reproduction very slightly; under luxuriance conditions of the host development is slightly slackened and reproduction increased when nitrates are in excess.

But all these experiment have been performed under extreme conditions of host nutrition, which should never occur in a normal plantation. Most of the plants have been nourished under diet conditions for years or at least months in the laboratory before the beginning of the experiment, so that the plant could not turn back to its reserves on the salts in question. A certain result may be drawn from this study, that extreme overmanuring may inprove reproduction in red scale very slightly, but the normal practice of manuring will have no effect on it.

Another problem arising is, as to how far these results may explain the differential fertility of red scale on fruits and leaves of citrus.

In this connection also the experience gained in the last world war should be mentioned, that cessation of manuring, irrigation and even the cutting back of the trees for some years did not lead to an increase of the red scale populations, but rather to the countrary.

STATISTICAL COUNTS OF RED SCALE ON CITRUS TREES.

The survey of the general trend of citrus scales with progressing age of the orchard have demonstrated that whereas in the first years after planting Red Scale is in predominance, its importance grows smaller in later years. A series of statistical counts have been made in autumn 1935 near Hedera in a heavily infested orange grove which had never undergone any control measures. The results of these counts, part of which have been published before, are as follows:

The first problem was the exact count of all red scales present on orange-trees at different ages. The average of each two trees was as follows:

Age in years	Year of planting	Total scales	Total living scales
9	1926	15.744	5.598
7	1928	13.024	4.980
5	1930	24.855	11.291
3	1932	13.891	6.941
1	1934	2.869	853

There is no doubt that this result (maximum absolute population in the fifth year) is correct and it is supported by the general experience in our gardens. The populations were counted on three other trees, separating the scales living on the leaves, twigs, branches, sten and fruits. The result is as follows (scales per leaves, twigs, etc.):

Age in years	Total scales					Living scales				
	leaves	twigs	branches	stem	fruits	leaves	twigs	branches	stem	fruits
9	1.7	3.6	21.5	403.0	11.6	0.3	0.3	—	1.0	1.7
5	2.5	8.3	41.0	68.0	159.6	0.5	1.1	0.0	0.0	68.2
1	7.2	4.3	9.6	336.0	—	2.2	1.5	2.2	1.0	—

Considering the total population, the result was as follows:

	Total scales	Living scales	Total scales (exc. fruits)	Living scales (exc. fruits)
9	14.886	2.100	13.436	1.898
5	20.125	5.004	13.743	2.277
1	5.213	1.511	5.213	1.511

This seems to indicate that the population maximum is induced mainly by the heavy development of the red-scale on the fruit (159.6 on trees planted in 1930 against 11.6 on trees in 1926). The young tree is especially heavily attacked and with the growth of shade in the tree, the latter develops a process of „self-cleaning".

Fig. 67.
View on five of the trees, the total red scale population of which was counted in autumn 1935.
The trees are 1, 3, 5, 7 and 9 years old.

Counts of complete trees one and five years old have been performed during 1937 at Hedera. The red scale population was:

age		I	II	III	IV	V	VI	VII	VIII	IX	X	XI	XII
1 year	larvae . . .	81	112	86	91	17	14	41	46	128		236	211
	adults . . .	92	93	78	64	42	64	120	122	230		475	398
	total	173	205	164	155	59	78	161	168	358	(535)	711	609
5 years	larvae . . .			241	299	225	210	321	872	989		3475	3017
	adult. . . .	656	621	475	407			510	934	1365		1794	2102
	total	897	920	700	617			831	1806	2354		5269	5119

These results are confirmed by many other, less comprehensive, counts. In one case 70 fruits were collected from both an old (27 years) and a young (7 years) orchard. The average infestation by red scale was 8 scales per fruit in the old, 575 scales per fruit in the young orchard. The decline of red scale is progressive and continuous with age.

In two orchards at Hedera in 1937 the infestation of 20 fruits each with red scale was:

	I	II	III	IV	V	VI	VII	VIII	IX	X	XI	XII
old orchard (30 years)	0	3	13	—	—	1	2	2	8	11	19	7
young orchard (7 years) . . .	378	781	799	771	—	24	42	95	114	163	230	242

This total population trend with age is illustrated by fig. 68 which shows also part of the seasonal population shift.

Further conclusions to be drawn from these counts are as follows (based on the following tables):

1) The vertical distribution of red scale on the trees depends only on the density of the foliage. The percentage of red scale present in the upper, middle and lower third of the tree is equal to the percentage of leaves present in every third.

2) Also the distribution of the red scale from the periphery to the centre of the tree is more even than is generally accepted. Considering the density of foliage it is even slightly higher in the inner third than in the outside one. This is due to the fact that the foliage of the trees under observation was sparse and unequal. The trees were therefore more exposed to the sun than is usual in older ages.

3) The distribution of red scale with regard to the compass is likewise equal and small irregularities show the same deviation as the density of foliage on the trees counted.

4) The distribution of red scale on citrus trees is therefore surprisingly equal, deviations being caused by the difference in the number of leaves and especially of fruits in the different parts of the tree.

Counts of red scale at different heights of citrus trees yielded the following results (each being the average of three trees):

Age in years	Red Scale population on lower	middle third	upper	Leaves on lower	middle third	upper
1	214	2646	757	98	431	161
3	5329	6622	1920	964	1939	377
5	7150	8970	7127	1434	1477	1318
7	4943	4579	3513	1900	1789	1292
9	5056	7115	3297	1233	1487	728
or in percentages:						
1	6	73	21	14	63	23
3	38	48	14	29	59	12
5	31	38	31	34	35	31
7	38	35	27	38	36	26
9	33	46	21	36	43	21
Average	29	48	23	30	47	23

The following are counts with regard to the distribution of red scale from the periphery to the centre of the tree (average of three trees each in percentages only):

Age in years	Red Scale population inner	middle third	outer	Leaves on inner	middle third	outer
1	53	35	12%	44	34	22%
3	28	46	26	14	46	40
5	34	32	33	38	33	29
7	25	28	47	19	32	49
9	26	47	27	23	42	35
Average	33	38	29	28	37	35

The following are counts of red scale populations with regard to the compass (average of three trees each, in percentages):

Age in years	Red Scale N.	E.	S.	W.	Leaves N.	E.	S.	W.
1	21	25	29	25	18	28	30	24
3	34	26	18	22	32	27	16	25
5	39	20	12	29	24	26	22	28
7	24	30	25	21	22	25	26	27
9	28	33	21	18	33	27	20	20
Average	29	27	23	23	26	27	23	25

Klein found red scale on old trees in the Upper Jordan Valley during summer in the centre of the trees only. In October the larvae move from the interior to the external parts of the trees. The scales are to be found on the outside branches during winter and until June, The scale do not move in the summer outward but remain on the old branches and twigs, being thus in the centre of the tree from June to October. i.e. as long as the shade temperature rises above 23° C.

Such shifts occur in the coastal plain also, being often quite as conspicuous as in the Jordan Valley. In 1935 an about 23-years old tree in Petah Tikva showed the following shift of living scales per 100 leaves:

month		I	II	III	IV	V	VI	VII	VIII	IX	X	XI	XII
part of branch	I (outside)	110	67	39	9	18	2	2	4	—	4	—	—
	II	7	—	—	3	—	2	2	7	16	8	—	4
	III	12	6	2	1	—	—	.	11	13	20	12	10
	IV (center)	22	27	13	7	4	10	8	89	99	120	41	42

The centre of red scale development is the periphery in the winter and the centre in the summer.

The red scale is polymerous on citrus, i.e. it attacks almost every part of the plant above soil.

The stem is by far the most unsuitable part. Whereas females are not unusual there, it is, as a rule, almost impossible to find more than a very few living ones

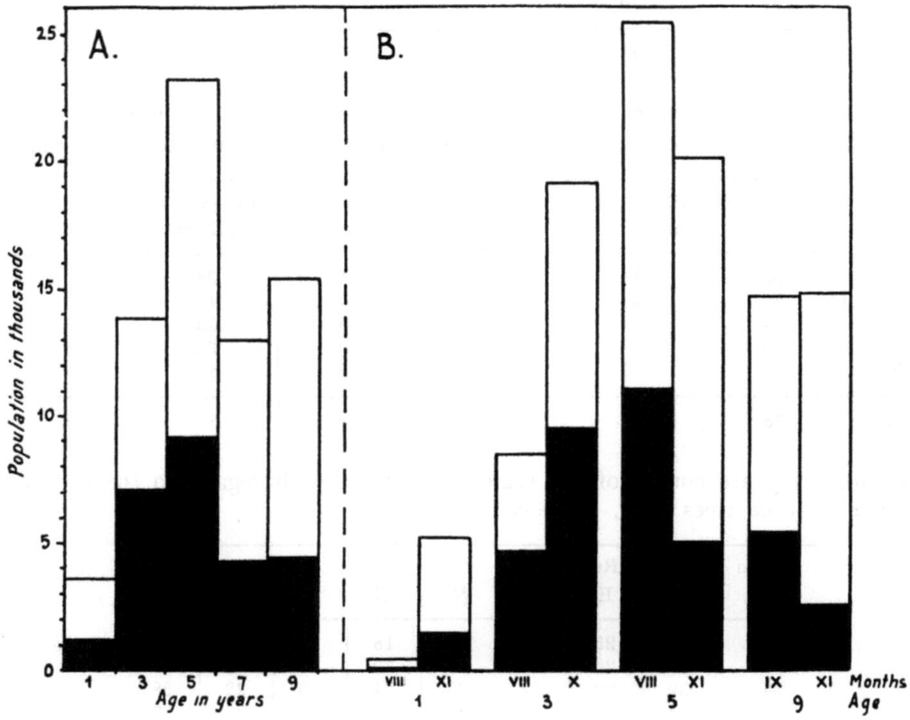

Fig. 68.
Population counts of Red Scale on orange trees of five different ages: A. total averages. B. seasonal shift from August to November. Black: living scales. White: dead scales.

amongst them. The scales of dead individuals stick to the stem for quite a long time. Among 2507 females counted on the stem of shamuti on different occasions only 6 were living (0.2%). The analysis of 300 females on the stem of citron did not reveal any living individual.

The same is true for the branches. As long as their bark is still tender it is exposed to attack by scales, but later on only in a small degree:

Age of tree	Total no. of scales	No. of living scales
1 years	182	41
5	1222	—
9	872	1
	2276	42

Whereas in the young plant still 23% of the scales present are living, the percentage of living scales is almost reduced to zero in higher ages.

The tender twigs remain a suitable host throughout the development of the tree as the new shoots are always being attacked, whereas the older and harder parts grow unfavorable.

Age of tree	Total no. of scales	No. of living scales	% of living scales
1 years	266	94	39
5	1437	193	14
9	2065	142	7
	3768	429	11

That is the reason why the percentage of living scales decreases steadily with age.

Among the leaves the not very tender and not yet hard and old leaves are preferred, but the fruit remains the preferred part of the host plant. But it is a fact that suitably sized fruit (of the size of an apple approximately) is present in the orchard only from July to April and its population is removed in toto from the orchard year after year.

Fig. 69 illustrates the shift of the Red Scale population on the different parts of the plant (November 1935 at Hedera):

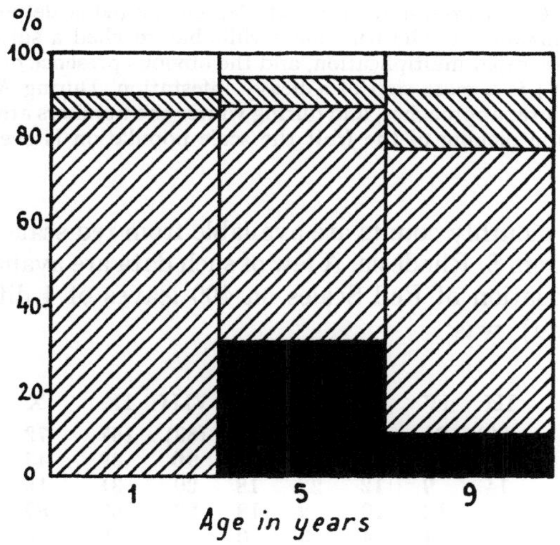

Fig. 69.
Distribution of Red Scale population on fruit (black), leaves (striped towards right), twigs striped towards left), and branches plus stem (white).

Percentage of Red Scales on

Age	stem	branches	twigs	leaves	fruits	Total
1 year	7	3	5	85	0	100%
5	0	6	7	55	32	100
9	3	6	14	67	10	100

SEASONAL FLUCTUATION OF RED SCALE IN ORANGE GROVES.

In California the seasonal fluctuation of red scale is quite as strong as that experienced in Palestine. The writer was privileged to make a trip in the company of R. S. Woglum through the main orange district between Los Angeles and Riverside in July 1931, when only a few individuals of *A. aurantii* could be discovered after thorough search. However, in November of the same year a letter received from Mr. Woglum stated that red scale has, for a long time already, not been as injurious as in that season. Regular control measures in the orchard help to depress the red scale population in California.

In S. Rhodesia the main check of the seasonal increase seems to be effected by the combined effects of winter and fumigation, the extreme values not being as far apart as in Palestine or California:

„In the winter (IV—eVIII) the scale population is at its minimum. A gradual increase is shown in spring, with the temperature rising, but production of crawlers is slow and the production period very short. As the fruit has been removed the scale develops onlt on the branches, stems and leaves. The increase at this stage therefore is not marked and in very little evidence. With the temperature at its highest in October, and November, though the reproduction of scale is not rapid, development is so accelarated that there is a sudden and definite increase in the scale population (XI, XII). With the arrival of the rains the optimum conditions for reproduction are present, and from now up to late March the red scale population increase with extreme rapidity to a maximum. The fruit meanwhile has reached a suitable size for infection, resulting in an even more rapid multiplication, and the obvious presence of scale in large numbers on the fruit (1 III, eIV) is a mere sign of a heavy infestation. During April and May there is a gradual decrease as winter approaches, and during winter the numbers are considerably reduced. With fumigation and the picking of the fruit the scale population is eventually reduced to its minimum." (Jones).

Unfortunately no monthly counts are available from red scale populations neither from California nor from S. Rhodesia. A few of such data are available from Palestine.

The seasonal fluctuation of Red Scales on 400 leaves of 4 different trees in the Atta groves (Hedera) was:

year	I	II	III	IV	V	VI	VII	VIII	IX	X	XI	XII	Total	Max./Min.
1929	(60)	45	(33)	21	**3**	28	10	(83)	157	72	**160**	143	805	53
1930	79	65	56	**21**	55	32	64	**90**	89	47	45	72	715	4
1931	**57**	54	14	**9**	12	22	18	20	31	15	47	34	335	6
1932	28	32	23	12	10	**9**	18	26	67	**87**	78	27	417	10
1933	**44**	36	10	4	4	3	**0**	3	5	10	37	19	175	>44(88)
1934	23	5	(9)	13	8	**3**	4	5	16	10	12	7	115	8
1935	**5**	27	39	14	19	13	55	66	365	225	**371**	337	1536	74
1936	93	83	(57)	33	6	14	10	11	48	26	32	33	446	16
1937	26	29	42	26	12	5	9	21	24	24	38	37	293	8
Average per 100 leaves	12	11	8	4	4	4	6	10	24	15	24	21		

or in percentages:

1929	7.4	5.6	4.1	2.6	0.4	3.5	1.2	10.3	19.5	8.9	19.8	17.8	100.1
1930	11.1	9.1	7.8	2.9	7.7	4.5	8.9	12.5	12.4	6.6	6.3	10.7	100.5
1931	17.0	16.1	4.2	2.7	3.6	6.6	5.4	6.0	9.2	4.5	14.0	10.2	99.6
1932	6.7	7.7	5.5	2.9	2.4	2.2	4.4	6.2	16.0	20.9	18.7	6.4	100.0
1933	25.1	20.6	5.7	2.3	2.3	1.7	0.0	1.7	2.9	5.7	21.1	10.8	100.0
1934	20.0	4.3	7.9	11.3	7.0	2.6	3.4	4.3	13.9	8.7	10.7	6.1	98.5
1935	0.3	1.7	2.5	0.9	1.2	0.8	3.6	4.3	23.8	14.6	24.1	21.3	99.1
1936	20.9	15.6	12.8	7.4	1.3	3.1	2.2	2.2	10.8	5.8	7.2	7.4	99.7
Average	13.6	10.5	6.3	4.1	3.2	3.1	3.6	5.9	13.6	7.2	15.2	11.3	97.6

Additional observations on full branches in different localities yielded the following results:

Full branches *Aonidiella aurantii.*

Year	Locality	Age of orchard	I	II	III	IV	V	VI	VII	VIII	IX	X	XI	XII	Total	Max./Min.
1935	Hedera	8 y	10382	4813	5610	2547	216	26	88	193	483	286	158	138	24620	399
1936	,,		4899				66	103	145	98	112	119	175	156		
1934	,,	7 y			70	36	730	780	1271	3185	8128	13514	12363	13814	(53798)	384
1934	,,	27 y			66	67	38	51	18	67	57	108	77	50	(569)	(6)
1935	,,	28 y	11	21	33	15	6	19	52	59	54	84	108	87	549	18
1936	,,	29 y	37	14	12	6	2	10	8	11	8	5	14	2	129	18
1934	Petah Tikva	10 y	(159)	201	101	53	21	16	19	52	164	120	556	117	1579	35
1935	,,	23 y	111	55	45	26	13	10	8	91	128	169	78	52	766	21

The population ebb always occurs in April/July, and the decisive rise of the scale population in July/September. This is confirmed again by a population count on 5 seedlings of sour orange, which were infested in May 1935 at Hedera (210 leaves—September 1935, 278 leaves in April 1936):

month	First larvae	Second larvae	No. of living Red Scales ♂ pupae + ♂	♀	Total	Dead Red Scales	Ratio of no. of scales of one month to that of previous month
V.1935	13	2	—	—	15	—	—
VI.	2	1	1	2	6	16	0.4
VII.	17	3	2	2	24	6	4.0
VIII.	40	29	8	30	108	89	4.5
IX.	188	82	34	100	404	148	3.4
X.	193	109	51	110	463	103	1.3
XI.	197	99	46	118	460	80	1.0
XII.	207	111	39	121	479	70	1.0
I.36	194	116	39	130	478	53	1.0
II.	176	125	35	133	472	77	1.0
III.	144	101	25	108	380	113	0.8
IV.	49	41	23	97	213	97	0.6
X.	252	235		493	1072	229	(5.0) (420 leaves)

From September to February the population was stable not only with regard to the total number of living scales, but also with regard to the proportion of the different stages.

Similar observations have been made at Dagania in 1933/34 by Klein, where the increase was interrupted through heavy hamsins in September:

month	VI	VII	VIII	IX	X	XI	XII	I	II	III	IV	V	VI
no. of living scales per 4 seedlings	161	59	575	256	724	1470	926	670	224	152	104	51	49
quotient of increase . .	1.0	0.4	9.7	0.4	2.8	2.0	0.6	0.7	0.3	0.7	0.7	0.5	1.0

The numerical increase from the beginning to the end of each generation was:

	month	
I	XII/I	1.2
II	II/III	2.3
III	V/VI	2.0
IV	VII/VIII	24.3
V	IX/X	7.0

These values were obtained in a young grove. In a closely planted orchard the population level as well as the fluctuations are different. The quotient of increase was 5.0 in a 10-years old grove in March/April, whereas only 0.7 in the young orchard. The increase in autumn was also very considerable there.

Some idea as to the seasonal age structure of the (living) scale population may be gained from the following table:

The percentage of adults was as follows:

	I	II	III	IV	V	VI	VII	VIII	IX	X	XI	XII
Branch of 23-years old tree	—	47	33	40	67	32	36	49	37	33	25	26
Branch of 8-years old tree.	23	33	52	48	36	58	37	40	38	43	33	33
Seedling	33	37	36	58	(54)	50	16	36	31	34	35	35
Average	28	39	40	49	52	47	30	42	35	37	31	31

This shows that the highest percentage of adults coincides with the lowest population level and preceeds the population rise.

The seasonal trend of mortality has been studied in various regions. In California 41% of the crawlers succeed to settle (Quayle) under favorable conditions. During further development at a favorable period of low and uniform mortality (25.VII.—29.VIII.1930) the following number of individuals died (Bliss):

	n	Larva I	Larva II	Female (pre-oviposition)	Total
on lemon fruits	501	9.8	7.3	18.9	33%
on citrus seedlings . .	177	5.1	9.1	34.4	43%

Mortality was lower on the fruits than either on leaf or stem; it was highest in the preoviposition stage. In late autumn mortality increased markedly. Of the scales which had settled in September, only 10% survived the end of the month. Practically none of the new scales settling later than October managed to reach maturity. This excessive mortality continued until late winter. However, Bliss believes that this extreme result somewhat exceeds conditions observed in the field, because he was able to collect in the orchards producing females with freshly emerged crawlers throughout the winter.

In the humid part of the Oriental regions mortality is closely correlated to the heavy rains. It approaches 100% during the rains, probably due mainly to washing off, and is reduced in the short dry periods.

The mortality of individuals having settled, at Rhodesia, has been resumed by Jones, as follows:

The percentage mortality among settled females of Red Scale at Mazoe, 1934.

Season	Part of tree	Larva I	Larva II	Preoviposit-ion period	Total mortality
Summer	Fruit	9.3	7.7	18.1	31.4%
,,	Leaves	7.1	13.3	23.1	39.8
Winter	Fruit	25.4	23.3	24.4	56.4
,,	Leaves	48.3	33.3	30.0	72.4

Here, also, mortality is much higher in winter than in summer; also, it is highest in the preoviposition period in summer, whereas in winter it is slightly higher in the early larval period than in the later stages. But it is self-understood, that in all regions the overwhelming mortality really occurs in the first larval stage composed of that of the crawlers as well as of settled first larvae. Mortality is always higher on leaves and stem than on fruit.

In Palestine our observations have now been conducted throughout some years near Hedera. The relative mortality of the different stages from 2 years observations (1956 individual breedings in the orchard under natural conditions) was as follows:

Stage	Larva I (after settling)	Larva II	Male pupa or immature females	Mature adults
% of mortality	49.1	16.8	19.8	14.3%

The percentage of living Red Scales in the same orchard was:

year	I	II	III	IV	V	VI	VII	VIII	IX	X	XI	XII	Average	
1930	25			15	20		13	22	(15)		(27)	25		% living
1931	15	31	33	(18)	2	(17)	84	68	19	60	72	72	41	
1932	66	35	53	53	(7)	59	8	70	54	72	52	63	49	
1933	46	53	60	55	6	31	57	54	67	74	75	71	50	
1934	86	89	65	66	53	46	67	73	75	73	70	54	68	
1935	79	77	59	45	6	(20)	35	45	73	73	64	85	55	
1936	69	76	75	73	61	42	55	49	37	74	75	76	64	
1937	71	61	59	53	36	48	58	55	37					
Average 1931/36 . . .	60	60	57	52	**23**	37	51	60	54	**71**	68	70	55	
Average mortality 1931/36 . . .	40	40	43	48	**77**	63	**44**	70	46	**29**	32	30	45	% dead

A complementary series of observations was obtained by the monthly control counts for spraying experiments. In these counts those individuals are included as dead which died during the last 7—10 days, but not the dried out individuals, which died some time ago:

year	I	II	III	IV	V	VI	VII	VIII	IX	X	XI	XII	Average
1930	24			29	13		22	25				17	(22)%dead
1931	17	21	23		34		2	1	75	6	8	9	20
1932	10	10	14	11		7	13	9	21	7	12	8	11
1933	14	10	9	9	55	4	8	11	5	3	3	3	11
1934	3	7	4	1	49	6	9	12	11	10	7	6	12
1935	6	12	27	ca95	36	4	5	4	2	5	8	6	18
Average 1931/35. . . .	10	12	15	34	44	5	7	7	23	6	8	6	15

The mortality on some full branches was as follows:

year	Locality	Age of tree	I	II	III	IV	V	VI	VII	VIII	IX	X	XI	XII
1934	Petah Tikva	12 years		164	80	113	21	12	19	52	161	120	545	116
				18	10	28	33	17	5	8	18	19	31	6%
1935	Petah Tikva	23 years	111	55	45	26	13	10	8	84	104	126	52	52
			20	2	20	42	60	—	—	1	9	10	23	25%
1936	Hedera	1 year	10	14	23	31		73	20	45	29	18	16	13%

These results are close to those obtained by Klein at Dagania:

	1933				1934							
month	IX	X	XI	XII	I	II	III	IV	V	VI	VII	VIII
Total scales	279	772	1533	954	784	308	159	109	99	64	171	247
% dead	8	6	4	3	15	27	4	5	49	23	5	6%

Mortality is highest after the hamsin winds (May, early June) and during low temperature, high precipitation and humidity in winter. It is highest in both seasons in the first half of the year (21%) and low in its second half (6%) when the seasonal population rise occurs.

Returning to the results of Hedera, we shall now give the values for actually living scales by substituting the values for the actually living scales:

year	I	II	III	IV	V	VI	VII	VIII	IX	X	XI	XII	Average
1929	36	27	18	11	1	10	5	50	85	51	108	100	42
1930	20	39	32	32	11	12	8	20	13	33	12	18	21
1931	9	17	8	2	0	4	15	14	6	9	34	24	12
1932	19	11	12	6	1	6	1	18	36	63	41	17	19
1933	20	19	6	2	0	1	0	2	3	7	28	14	9
1934	20	4	6	9	5	1	3	4	12	7	8	4	7
1935	4	20	23	6	1	3	19	30	266	165	237	286	89
1936	64	63	43	24	4	6	6	5	19	19	24	25	25
Average	24	25	17	12	3	5	7	18	56	46	62	61	28
Average excl. 1929/30	23	22	16	8	2	4	7	9	57	45	60	60	26
per 100 leaves	6	6	4	2	0.5	1	2	2	14	11	15	15	7
%	7.1	7.4	5.0	3.6	0.9	1.5	2.1	5.4	17.2	13.7	18.4	18.2	100.5

But even this table does not offer a real and comprehensive survey: 1) The heavier, explosive scale development on the fruit from July to October is not expressed by this formula, which concerns leaves and twigs only. The real difference between seasonal minimum and maximum is therefore not included.

2) Washing off of scales by the heavy winter rains is often extremely high. Among 240 individuals of all stages which were marked on 12.XI.1928 only 10=4.2 per cent survived until 28.I.1929 and none until 4.II.1929. From among 115 individuals marked on 21.I.29 none was present on 25.III.1929.

One breeding which had started with 4 females under natural conditions on 22.X.1928 gave the following result:

	22.X	29.X	5.XI	12.XI	19.XI	26.XI	3.XII	10	17	24	31	7.I	14	21	28	4.II
generation I adults	4	4	4	4	2	3	5	4	7	7	9	9	9	9	6	—
larvae	—	—	216	236	190	102	86	53	50	23	13	10	6	6	4	—

Such results are due to heavy rains only. "The direct effect of precipitations is relatively unimportant, though rain can and sometimes does wash the surface of the leaf clean from scale." The actual epidemiological importance of this factor depends on the number of days with really heavy rains. In the humid tropics those are undoubtedly the most important of all checks to the increase of red scale populations. But in the interior parts of the tree or on the lower surface of the leaves the chance of a few scales surviving even during heavy rains is quite great.

3) Leaf drop, especially in very young orchards, is quite an important factor. Its extent may be illustrated by some figures:

		I	II	III	IV	V	VI	VII	VIII	IX	X	XI	XII
1935 Hedera 8 yr.													
	old	10	260	203	186	45	25	12	9	4	130	124	117
	young	271	121	106	111	91	119	123	124	124	—	—	—
	total	281	381	309	297	136	144	135	133	128	130	124	117
1936 Hedera 29 yr.													
	old	221	208	167	123	105	114	80	310	308	305	302	291
	young	—	166	178	215	222	239	238					
	total	221	374	345	338	327	353	318	310	308	305	302	291
1935 Hedera 28 yr.													
	old	136	142	562	396	235	187	160	134	1030	1090	960	950
	young	433	458	72	889	821	848	900	919				
	total	569	600	634	1285	1056	1035	1060	1053	1030	1090	960	950
1934 Petah Tikva 10 yr.													
	old	647	649	584	1054	983	968	996	1050	960	909	761	817
	young					622	924	175					
	total	647	649	584	1054	1605	1892	1171	1050	960	909	761	817
1935 Petah Tikva 23 yr.													
	old	255	253	227	187	138	304	288	272	288	281	267	258
	young			192	176	180							
	total	255	253	419	351	318	304	288	272	288	281	267	258

The main leaf-drop occurs in April, coinciding with the heaviest period of new leaves. This may be illustrated by another count on 5 seedlings: On an average 188 leaves were present early November 1935. Until late January 1936, only 7 leaves had dropped whereas 107 had dropped in early April, when the leaf-drop was not yet finished.

The normal number of leaves in autumn on young trees is approximately as follows:

Age	1	3	5	7	9 years
leaves. . . .	689	3279	4221	4981	5550
twigs	62		173		571
branches . .	19		41		41
fruits	—		40		125

This leaf-drop reduces the red scale population enormously as the heavily infested leaves weakened by the attack drop first. This decrease in April—May appears only partially in our monthly counts, as far as the leaves with the heaviest infestations are eliminated from the orchard. On the other hand, we count the remaining leaves only, not considering the number of dropped ones, i.e. 100 leaves in April/May are not comparable to 100 leaves in January or July. The minor leaf-drop in August—September is of no great epidemiological importance.

4) Of major importance is the picking of the fruit. With the fruit a very important part of the Red Scale population (changing at different ages and cases from 30 to 75%) is eliminated from the orchard year after year from November to April. The importance of this fact cannot be overestimated. But as it occurs season after season, it does not influence the annual shifts in Red Scale populations.

5) The destruction of individuals by parasites, predators and diseases is recognisable only when red scale is at its population peak. Even then it is practically certain that in the case of Red Scale no important influence in reducing its population has ever been observed. The effect may be local and occasional, but none over a larger area occurring regularly.

6) The destruction of individuals by control measures (fumigation, spraying) depends on human agency. The reduction occurring just before the beginning of population rise is most important. A good control of 95—99% efficiency at that

season does not influence the quotient of seasonal increase, but reduces the maximum population reached in autumn below the mark of economic damage.

7) The large mortality of the crawlers is not included at all in the counts.

Despite of all those criticisms, it seems that the monthly 100-leaves count offers a fair indication as to the general seasonal trend and the comparative density of red scale in different districts (when the age of the trees is also considered).

ENVIRONMENTAL RESISTANCE.

The environmental resistance in the Atta groves near Hedera since 1929 is calculated in the following table. Four major generations starting in April, July, September and November have been taken into consideration. The formula for the calculation was

$$R_x = \frac{100\,(P_1 - P_2) + R_0 \,.\, P_2}{P_1}$$

where P_1 = mother population P_2 = daughter population R_0 = environmental resistance in case of stable population $P_1 = P_2$. R_0 is calculated as $\frac{100\,(e - 2)}{e}$ where e represents the number of eggs. This formula is in value only when the number of males equals that of females, which is approximately correct in this case.

No. of eggs	100	150	100	50	
R_0	98.00	98.67	98.00	96.00 %	
Year	generation I IV—VII	generation II VII—IX	generation III IX—XI	generation IV XI—IV	Average
1929	99.05	79.12	98.18	99.41	93.94
1930	93.90	98.14	98.38	99.50	97.48
1931	96.00	97.71	97.81	98.79	97.58
1932	97.00	95.05	99.19	99.41	97.66
1933	97.00	98.89	92.40	97.58	96.47
1934	99.38	94.68	99.13	92.00	98.30
1935	92.14	91.17	98.16	99.88	95.34
1936	98.45	93.62	98.62	96.76	96.86
1937	99.31	96.45	96.83	95.89	97.12
Average . . .	96.91	93.87	97.63	97.84	96.75
Range	7.24	19.77	6.73	7.88	4.36

These ciphers would need some correction if, as should be done, only the living scales would be taken into consideration. On the average of 8 years the environmental resistance was as follows:

month	IV—VII	VII—IX	IX—XI	XI—IV
no. of living scales at the beginning.	36	21	172	186 (as pro $^0/_{00}$ of annual total)
eggs per female.	100	150	100	50
R_0.	98.00	98.67	98.00	96.00
R_x.	98.83	89.10	97.86	99.22
Difference R_0—R_x.	—0.83	+9.57	—0.14	—3.22

The destructive effect of the winter is evident as well as the important increase from July to September. The autumn generation allows for a further slight increase, whereas in the spring generation the winter decline continues.

VITALITY OF RED SCALE.

The vitality of red scale varies from year to year and apparently its vitality runs parallel to its average abundage in the respective years. Similar changes in vitality have been described in the chinch bug in Illinois by Shelford. If we compare the average mortality in the monthly spraying experiments (with a well standarized oil, which is used by us as a standard spray) with the average population density during that year, the correlation is quite obvious.

Year	Average living scales per 100 leaves	Average survival after spraying	Relative number of scales p. 100 leaves	Relative survival
1931	13.6	6.3%	65	83
1932	21.0	7.6	100	100
1933	8.5	4.4	41	58

The vitality was increased in the years of high abundance. Further material is needed to decide as to whether these parallel fluctuations are significant. Recent investigations of H. Knight in California confirm the suggestion that density of population itself may influence the degree of mortality induced by fumigation. A summary of his experiments is given below:

No. of scales per lemon	No. of lemons	Count	Alive	Per cent Kill
1—100	213	14506	585	96.0
100—200	98	16016	2630	83.6
200—300	18	5542	1502	72.9

We need not assume that the population density influences the vitality and resistance to control, but it is much more probable that the dense population is the consequence of optimal conditions of increase and development which also raised the scale's vitality.

COMPATIBILITY OF THE DEVELOPMENT OF CERTAIN ARMOURED CITRUS SCALES WITH THE PRESENCE OF OTHER SPECIES.

The problem as to what degree the presence of another species interferes with red scale development can only be touched. No reliable material is available and no clear cut experiments have yet been performed. Certain aspects of the problem have been treated in the chapter on the Egyptian Black Scale.

Comparing the citrus scale situation in the Mediterranean countries, we find almost everywhere one species is *the* predominant one. So *Chrysomphalus dictyospermi* in the W. Mediterranean, *C. aonidum* in Egypt and certain districts in Palestine and Syria, *Aonidiella aurantii* in other parts of Palestine, Syria and S. Turkey. The limits are sometimes well defined, as is the case on the Western shore of the Lake of Galilee. The northern half is the domain of black scale, the southern half that of red scale. The climograms show definitely that in the northern half the climate is drier and hotter during summer, being decidedly beyond the favorable zone for *A. aurantii*. At Dagania, on the other hand, part of the summer (4 months) is within the favorable Red Scale zone, where this species dominates. Yet in the mean time *Ch. aonidum* prevails unto Gesher, whilst at Ein Harod *Aonidiella* is still dominant.

It is decidedly premature to conclude that *C. aonidum* shows a preference for hot and dry weather. It is more probable that its tolerance of drought and heat is greater, whereas the optimal zones are possibly quite similar.

It is probably fair to say that the three mentioned species occupy the same eco-niche on their citrus host. Wherever one species has a decisive advantage over another species of the same group, it will take its place and reduce the eocnomic importance of the less adapted species to a subordinate one. This will occur if the less adapted species has been present before or is just invading. It will occur either if the advantage is a vital optimum which fits better into the climate of the locality or if its range of tolerance is simply wider. Of minor importance is the adhesive power of the scale on the host (especially in connection with heavy rains) and the local host range of the scales. We may draw the conclusion that in any climate of the present orange zones of the world one species will dominate locally either because it is better adapted to the environment or because it has been introduced before and is established by the time the second species invades the area in question.

The epidemiological bonitation of climatic factors at different localities can be calculated only roughly.

We shall employ the following zonation in a preliminary way:

Zone	Temperature	Humidity	Bonitation
Optimal zone:	23.0—27.5° C	70—80% R.H.	2
Favorable zone:	18.0—29.0	60—85	1
Unfavorable zone:	5.0—32.0	45—90	0
Impossible zone:	beyond the limits of the unfavorable zone		0

It is evident that a prolonged season (about 3 months) in the impossible zone makes the permanent occurrence of the species highly improbable. 10—11 months in the unfavorable zone should prevent *A. aurantii* from becoming abundant enough to be injurious. 3—4 months within the favorable or optimal zone are sufficient for a ample increase in number, provided these months are consecutive. 1—2 consecutive months in the optimal or favorable zone within an unfavorable period are, as a rule, insufficient to produce such an increase.

Conditions prevailing at Tel Aviv, Gizeh, Los Angeles, Tampa, Honolulu are favorable, partially even very favorable, for the regular seasonal development of outbreaks.

An important correction must be made in connection with the rains. Heavy rains are decidedly disastrous to Red Scale development. We classify the rains into 4 groups for epidemiological purposes:

0—50 mm	0
51—100	+
101—150	++
150	+++

Rains over 100 mm always classify the climate of the respective month as unfavorable at least, whereas 50—100 mm rains depress the normal bonitation for one class.

A further correction must be made in connection with the host. With regard to citrus the following host-bonitation is advisable for the time being:

1 (lowest)	leaves, twigs and stem
2	fruits (orange, grapefruit)
3 (highest)	fruits (lemon, cederate)

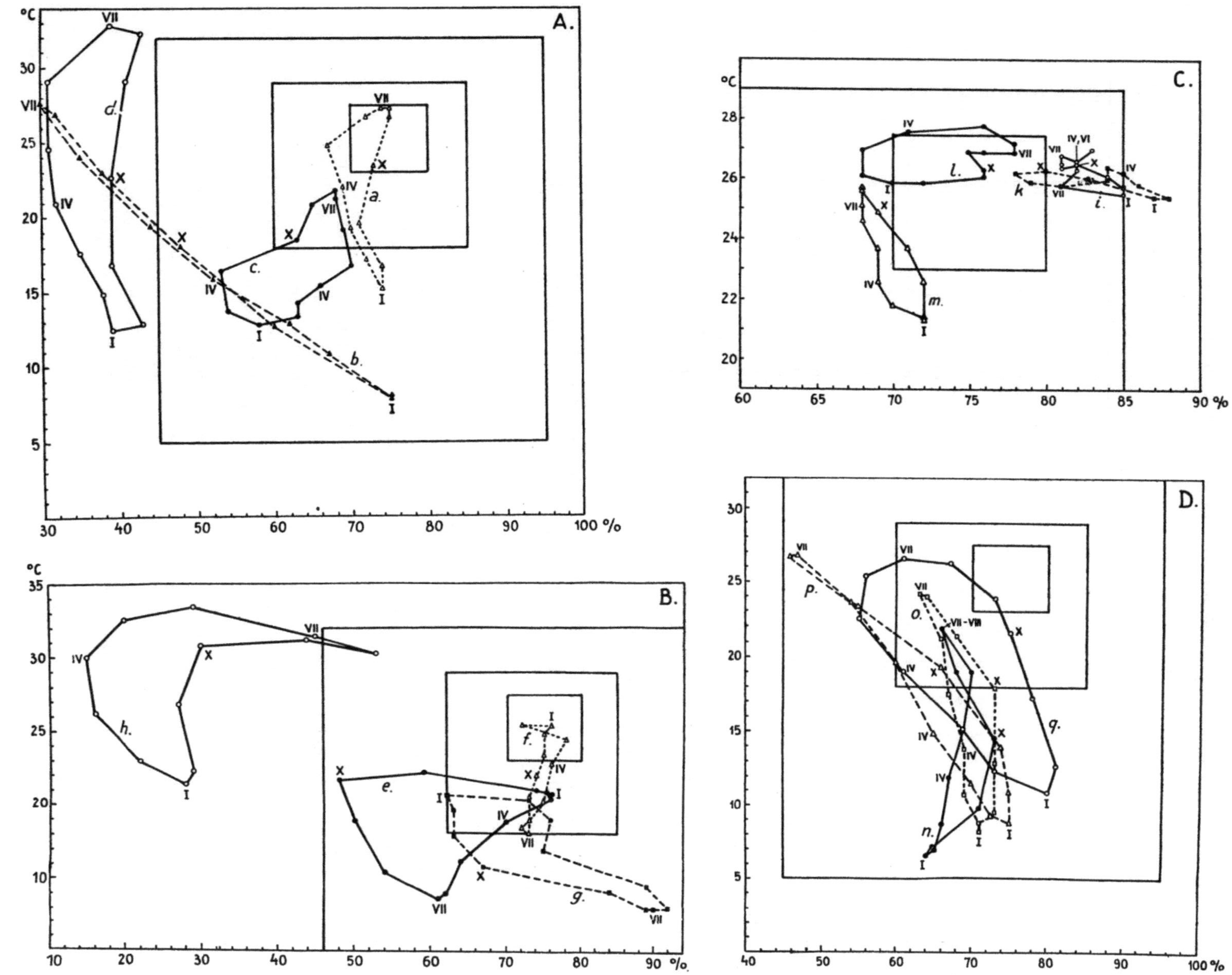

Fig. 70. Ecoclimograms of different localities for *Aonidiella aurantii* Mask. A. U.S.A.: a. Tampa, b. Fresno, c. Los Angeles, d. Yuma. B. Africa: e. Salisbury, f. Lourenco Marques, g. Capetown, h. Khartum, C. Pacific and Indian regions, i. Singapore, k. Batavia, l. Colombo, m. Honolulu, D. Mediterranean regions: n. Nice, o. Naples, p. Athens, q. Giza.

The epidemiological bonitation for various localities is given below:

	Bonitation (with rain correction) Month													Months in zone				No. of annual generations
year	I	II	III	IV	V	VI	VII	VIII	IX	X	XI	XII	Total	O	F	NF	I	
Paris	0	0	0	0	0	0	0	0	0	0	0	0	0	–	–	9	3	0.9
Nice	0	0	0	0	0	1	1	1	0	0	0	0	3	–	3	9	—	1.7
Naples	0	0	0	0	0	1	1	1	0	0	0	0	3	–	3	9	—	2.3
Athens	0	0	0	0	1	0	0	0	0	1	0	0	2	–	2	10	—	3.2
Ankara	0	0	0	0	0	0	0	0	0	0	0	0	0	–	–	6	6	2.0
Tel Aviv	0	0	0	0	1	2	2	2	2	1	0	0	10	4	2	6	—	4.3
Jerusalem	0	0	0	0	0	0	0	0	1	0	0	0	1	–	1	10	1	3.1
Dagania	0	0	0	1	0	0	0	1	1	1	0	0	4	–	4	8	—	5.3
Jericho	0	0	0	0	0	0	0	0	0	0	1	0	1	–	1	10	1	6.2*)
Gizah	0	0	0	1	0	0	1	1	2	1	0	0	6	1	4	7	—	3.8
Karthoum	0	0	0	0	0	0	0	0	0	0	0	0	0	–	–	2	10	9.1
Lourenco Marques	0	0	0	1	1	1	1	1	1	0	1	1	8	–	8	4	—	5.4
Mazoe	1	1	1	1	0	0	0	0	0	0	1	1	6	–	6	6	—	3.4
Salisbury	0	0	0	1	0	0	0	0	0	0	0	0	1	–	1	11	—	3.5
Capetown	1	1	1	0	0	0	0	0	0	0	0	1	4	–	4	8	—	2.3
Singapore	0	0	0	0	0	0	0	0	0	0	0	0	0	–	–	12	—	7.8
Batavia	0	0	0	0	0	0	0	2	1	0	0	0	3	1	1	10	—	7.6
Colombo	0	0	0	0	0	0	0	0	0	0	0	0	0	–	–	12	—	8.1
Honolulu	0	0	0	1	1	1	1	1	1	1	1	0	8	–	8	4	—	6.3
Los Angeles	0	0	0	0	0	1	1	1	1	1	0	0	5	–	5	7	—	2.6
Fresno	0	0	0	0	0	0	0	0	0	0	0	0	0	–	7	5	—	3.2
Yuma	0	0	0	0	0	0	0	0	0	0	0	0	0	–	–	—	12	5.5
Rió Grande	0	1	1	2	0	1	1	0	0	1	1	0	8	1	6	5	—	5.6
Tampa	0	0	0	0	0	0	0	0	0	1	1	0	2	–	2	10	—	5.5

O = Optimal, F = favorable, NT = not favorable, I = impossible. *) not permanently present

This table is illustrated by a series of climograms (fig. 70).

The result of this bonitation describes quite neatly a great number of interesting facts:

1) The lack of economic importance of *A. aurantii* in the Northern Mediterranean regions.
2) The absence of the species from the Euro-Siberian and Irano-Turanian regions.
3) The enormous differences within a small country like Palestine.
4) The dependency of red scale density in the Saharo-Sindian and Sudano-Deccanian regions upon air-humidity.
5) The extremely low bonitation value in the humid parts of the Oriental region, the probable home of citrus as well as of red scale, is caused by the high monthly precipitations. But these are not sufficient to exterminate the scale, for the writer in 1932 saw some specimen in the dryer part of Ceylon at the end of the dry season. In any case, it is of no economic importance in any part of its original home.

This bonitation scheme, however, must be regarded only as a first approach. Regions in which red scale is of great importance, as S. Rhodesia and Cape Colony, show an index highly understating the real conditions. This is probably true to a lesser degree, for the orange district of California. But the absence of sufficient meteorological data prevented the writer from an attempt at a further improvement, for the time being. When the index reaches four, the threshold of economic importance is surpassed.

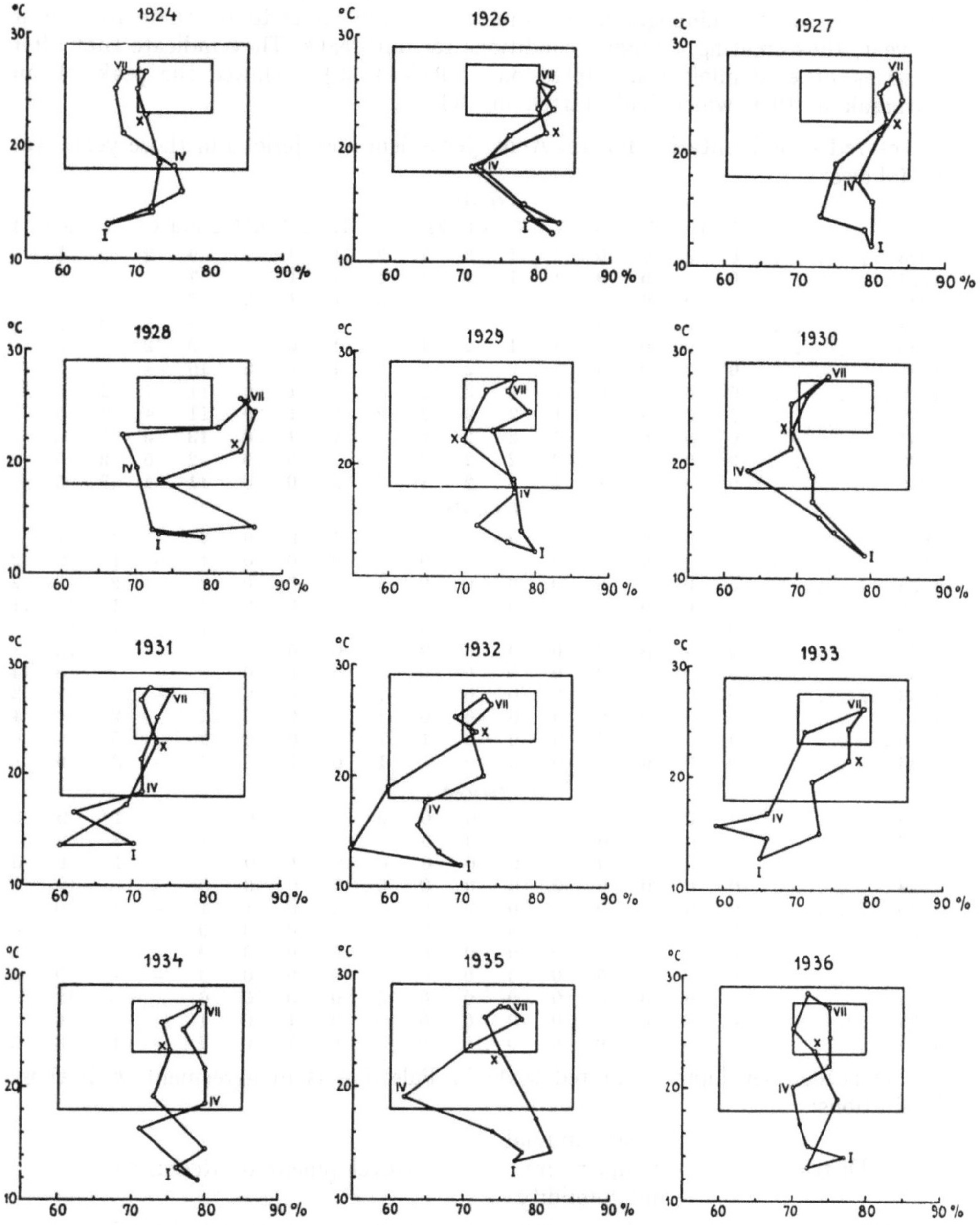

Fig. 71
Eco-climograms of Tel Aviv in relation to Red Scale in 1924—1936.

The climograms given on p. 257, obtained in Tel Aviv in the last decades, illustrate how these average climograms for any locality are liable to fluctuate from year to year, thus creating changing conditions for outbreaks. They indicate that after a series of epidemiological depressions (1927—1930), followed the peak of an outbreak in 1936 which had started in 1931.

The red scale bonitation for Tel Aviv, Jerusalem and Jericho in those years was as follows:

year	I	II	III	IV	V	VI	VII	VIII	IX	X	XI	XII	Total	O	F	NF	I
							Tel Aviv										
1926	0	0	0	0	1	1	2	1	2	1	1	0	9	2	5	5	–
1927	0	0	0	0	1	1	1	1	1	1	1	0	7	–	7	5	–
1928	0	0	0	1	1	1	1	1	0	1	1	0	7	–	7	5	–
1929	0	0	0	0	1	2	2	1	2	1	0	0	9	3	3	6	–
1930	0	0	0	1	1	1	2	1	2	1	0	0	9	2	5	5	–
1931	0	0	0	0	1	2	2	2	2	1	0	0	10	4	2	6	–
1932	0	0	0	0	1	2	2	2	1	2	1	0	11	4	3	5	–
1933	0	0	0	0	1	2	2	2	2	1	1	0	11	4	3	5	–
1934	0	0	0	1	1	2	2	2	2	2	1	0	13	5	3	4	-
1935	0	0	0	1	2	2	2	2	2	1	0	0	12	5	2	5	–
1936	0	0	0	1	1	2	2	1	2	2	0	0	11	4	3	5	–
							Jerusalem										
1926	0	0	0	0	0	0	1	1	1	1	0	0	4	–	4	7	1
1927	0	0	0	0	0	0	0	0	1	0	0	0	1	–	1	9	2
1928	0	0	0	0	0	0	0	0	1	1	0	0	2	–	2	8	2
1929	0	0	0	0	0	0	0	0	1	0	0	0	1	–	1	10	1
1930	0	0	0	0	0	0	0	0	0	1	0	0	1	–	1	11	–
1931	0	0	0	0	0	0	0	0	0	0	0	0	0	–	–	12	–
1932	0	0	0	0	0	0	0	0	0	0	0	0	0	–	–	8	4
1933	0	0	0	0	0	0	0	0	1	0	0	0	1	–	1	9	2
1934	0	0	0	0	0	0	0	0	1	1	0	0	2	–	2	6	4
1935	0	0	0	0	0	0	0	1	1	1	0	0	3	–	3	9	–
1936	0	0	0	0	0	0	0	1	1	0	1	0	3	–	3	9	–
							Jericho										
1926	0	0	1	0	0	0	0	0	0	0	0	0	1	–	1	9	2
1927	0	0	1	0	0	0	0	0	1	0	1	0	3	–	3	9	–
1928	0	0	0	0	0	0	0	0	0	0	1	0	1	–	1	6	5
1929	0	0	0	0	0	0	0	0	0	0	0	0	0	–	–	12	–
1930	0	0	0	0	0	0	0	0	0	0	0	1	1	–	1	10	1
1931	0	0	0	0	0	0	0	0	0	0	0	0	0	–	–	8	4
1932	0	0	0	0	0	0	0	0	0	0	0	0	0	–	–	7	5
1933	0	0	0	0	0	0	0	0	0	0	0	0	0	–	–	9	3
1934	0	0	0	0	0	0	0	0	0	0	0	0	0	–	–	12	–
1935	0	0	0	0	0	0	0	0	0	0	1	0	1	–	1	9	2
1936	0	0	0	0	0	0	0	0	0	0	1	0	1	–	1	8	3

The actual development of red scale in Palestine is in agreement with those bonitations:

District	Average annual temperature and humidity		Development of Red Scale
Coastal plain. . . .	18.4	80%	Strong, also on outside branches
Jerusalem	16.4	62	Never developing heavily on well treated trees
Upper Jordan Valley	20.8	63	Heavy development in winter on the whole tree; in the summer mainly in the centre
Lower Jordan Valley	23.7	55	No development

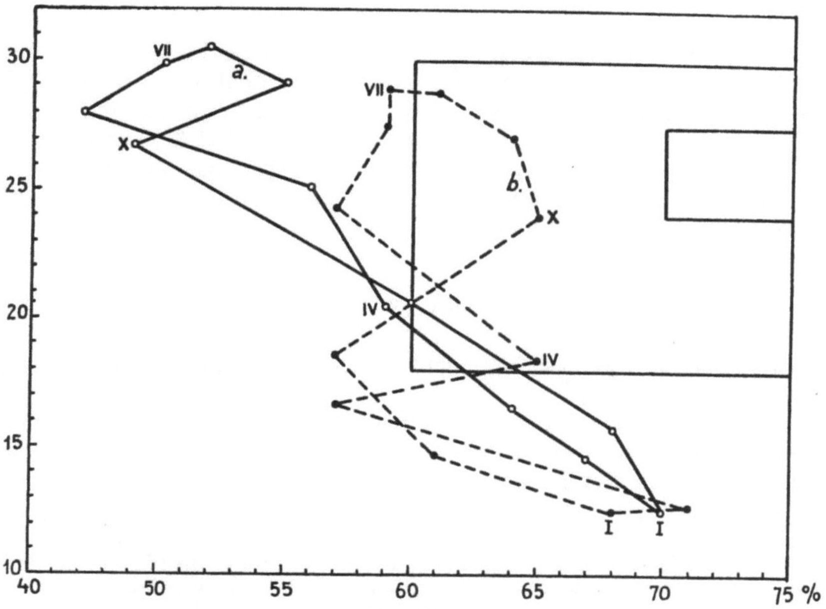

Fig. 72.
Eco-climograms of a) Tiberias and b) Dagania, both on Lake Tiberias.

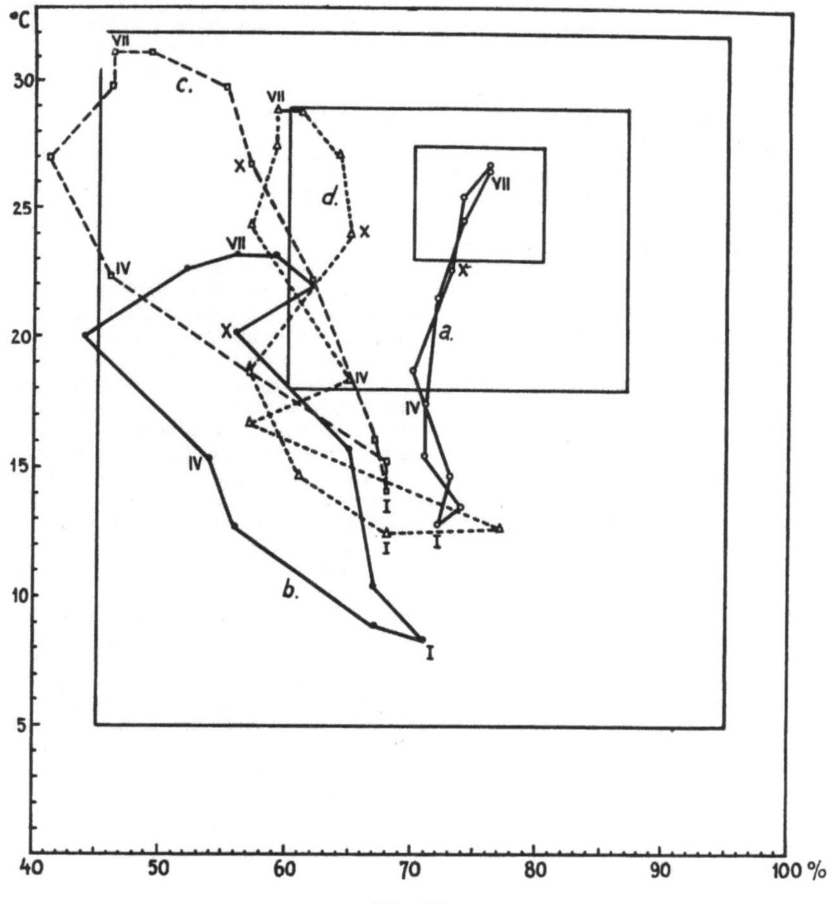

Fig. 73.
Eco-climograms for localities of *Aonidiella aurantii* Mask. in Palestine a. Tel Aviv, b. Jerusalem, c. Jericho, d. Dagania.

Conditions in Jericho are especially interesting. We owe the following information to Mr. Rosenberg, manager of the Government horticultural garden at Jericho; A few years ago, citrus plants infested with red scale were transplanted from Hedera to Jericho. No steps were taken to eradicate the scales, but 3—4 months later not a single scale was present. Despite many careful inspections the writer was never able to discover the scale in any of the older gardens. Its existence is obviously impossible there or almost impossible.

ECONOMIC IMPORTANCE AND CONTROL.

Red scale is one of the most important citrus pests in all semi-arid, subtropical regions.

Quayle questions whether the first rank of importance or the second one should be conceded to *A. aurantii* in California. It certainly ranks first in certain counties, as Riverside and San Bernardino.

,,Practically the entire tree may be killed by Red Scale, sometimes in one or two years' infestation. It infests all parts of the tree, and not only causes a dropping of the leaves, but actually kills large branches. Aside from this permanent and serious injury to the tree, the presence of the scale on the fruit renders it unmarketable. Trees that do not have a severe infestation of the scale, and where the tree itself is not seriously injured, may have its fruit badly infested by the scale . . . The injury by feeding is due to the loss of chlorophyll, toxic effect on the tissues of the plant, and interference with the functions of the stomata when the scale is abundant . . . With the Red Scale on the orange, a heavy infestation on the twigs and branches, as well as on the leaves and fruit, means the destruction of the tree." (Quayle). The same author writes on another occasion: ,,Branches of considerable size are frequently severely injured in a single season's infestation. There may be heavy defoliation. Badly infested fruits are unmarketable. All of this injury is done by the feeding alone, since no honeydew is excreted."

Damages in Southern Texas are described as follows:

,,It was early recognised that red scale was a more serious pest than the other species of scales infesting citrus in this section. Yellowing of foliage, and dropping of fruit caused the growers to realize the necessity for controlling this insect . . . With the increased supply of fruit and with the adoption of grading standards, more critical consideration of the problem of scale control will be necessary.

Red scale injures citrus in several ways. It attacks all parts of the tree, including the limbs, twigs, fruit, and leaves. It not only causes dropping of the fruit and defoliation of the tree, but may kill large branches in cases of severe infestion . . . The scale on the fruit detracts greatly from its marketable value by affecting its appearance. A severe infestation of red scale not only affects the fruit crop during the season in which the damage occurs but may cause a decrease in the crop for several seasons to come. A case is known where an infestation resulted in practically a total loss of fruit for the season. The average yield in this instance was only 21 pounds per tree. Approximately a 75% reduction in yield resulted for two subsequent seasons. Many similar cases of severe injury could be cited; however actual data as to loss of fruit are not available. Thousands of pounds of fruit, which have fallen from the trees on account of injury by red scale alone, have been buried during the last few years. There is no practical way to arrive at a definite measure of the injury caused to the trees by red scale, but it is safe to assume from general observation that the loss in the Valley is very great. This loss is not readily noticed by the average grower, but is reflected in gradually lessened fruit yields." (Clark and Friend).

Summerville reports from Queensland:

,,The red scale reaches its maximum intensity in the drier and hotter parts, becoming more important in abnormally dry periods. The rapid increase in such periods is mainly due to the lower mortality of the young during such periods. Trees with impaired vigour are more susceptible to attack. They suffer more quickly and larger populations are found on them. The Emperor of Canton variety of mandarin is one which when healthy is rarely attacked to any extent by red scale, the pest in most instances, if present at all, being confined to immature fruit. When, however, it loses much vigour from disease or other cause an infestation of red scale frequently quickly follows. It appears that vigorous trees with a free flow of sap offer marked resistance to the pest, and, further, that on any tree the pest is less liable to become established on free-growing parts than elsewhere, for when trees of supple habit of growth are attacked, it is usual to find the insect confined, at least in the first instance and smaller infestations, to the more woody or weaker parts, where there is no great flush of sap.

Varietal susceptibility is in conformity with what might be expected from the foregoing. Thus lemons, which are more woody, are almost invariably infested, whilst the supple-growing mandarins are usually only troubled when other factors are operating strongly. Even in some very dry parts remote from the coast . . . healthy mandarin trees quite free of red scale adjoin lemon trees which are persistently and heavily infested. Also in some of the wetter coastal districts small areas of lemons are sometimes included in an orchard, and though the red scale may be of no significance on the remainder of the trees, artificial control measures have to be applied against the red scale on the lemons. Most varieties of orange may be placed between lemons and manderins in respect to habit of growth and also to the probability of attack by the red scale . . . The foregoing remarks apply essentially to well-grown trees. *With young trees every variety appears to be very susceptible to this pest, particularly trees recently planted out from the nursery.*

Red scale will infest all aerial portions of a citrus tree. When the foliage is heavy by far the largest numbers are generally to be found on more exposed parts — leaves, fruit and twigs. On open and more scraggy trees, however, the limbs and main branches are commonly found to carry large colonies of the pest and it appears that red scale prefers positions exposed to sunlight. It is probably mainly on account of this preference that young trees are so frequently attacked, for in such cases few if any of the trees are effectively shaded.

Red scale is a voracious feeder, and no plant can long sustain the depredations of a large colony. Weakening and killing of leaves and twigs is accomplished rapidly, and young worked trees may be killed back to the union of bud and stock within a few months by colonies which could not be considered abnormally large. Even when actual death does not follow, the damage is often such that the tree never properly recovers. *A tree which has been stunted during its early life by red scale, as many are, is rarely worth keeping.*

On older trees, apart from the damage done directly by the pest, the trees may be so weakened that they are rendered very prone to attack by other insects, such as the borer *Uracanthus cryptophagus* Oll., and particularly by diseases such as melanose, *Phomopsis cirti* Fawcett. Melanose is an ever present menace to weakened trees in Queensland, and it is by the combination of red scale and melanose that a large percentage of older trees in the State are ultimately killed, or at least ruined.

The weakening effect of red scale on trees is overlooked in some instances. It happens frequently that an orchardist finds that some of his trees are heavily infested with red scale, and applies a spray against the pest. A short time later he notices dead wood appearing in the trees, accompanied by a heavy fall of leaves. He may at once condemn the spray material, overlooking the fact that the parts now dead were greatly weakened by the scale, and that almost any spray material would have completed their destruction. In fact, this would not have been long delayed even were no spray used. Whilst spray injury may have similar manifestations, and often does, every year perfectly good spray materials are condemned because orchardists do not take all the circumstances into consideration.

When infesting the fruit, though direct damage is done, in most cases the chieff objection growers have to the scale is that the fruit must be cleaned before it is marketed, and the removal of red scale in large numbers is not very easily accomplished. The fruit must be brushed, and the extra handling, besides costing time and labour, always results in some loss of fruit through injury to the rind. The actual loss directly attributable to the brushing and handling may be small, but, particularly in some seasons and certain districts, the entrance of blue and green mould is greatly facilitated. The direct affects of feeding on the fruit are chiefly arrested development and reduction in size." (Summerville).

A very complete survey on the red scale situation in S. Rhodesia has recently been given by Jones:

,,Red Scale is a major pest of citrus in S. Rhodesia. An annual fumigation campaign is imperative in order to keep it under control. If these measures were relaxed in any manner, the scale population would increase to enormous proportions. Consequently not only would a serious loss to the export crop result, but considerable harm would be done to the trees themselves . . . Thus, in 1931 only 2 per cent of the trees in groves 48—50 were fumigated. The resultant loss through red scale infestation almost equalled in value the total cost of the entire fumigation campaign of the whole estate that year. A more recent example was experienced in 1933 when a block or 3000 trees, some distance from the estate, was not fumigated (through force of circumstances). Some six months later these trees were so heavily infested, that dieback, the symptoms of a tree in an advanced stage of attack, was visible from some distance away. Fortunately it was possible to fumigate the trees just prior to the rains, and although severe burning of the fruit was experienced, no damage to the trees was done and the following crop was practically free from attack.

Various parts of the tree are attacked. After the winter scale is generally present on the branches and leaves. The fresh growth that appears is favoured by the young crawlers and soon becomes infested. Later the developing fruits are attacked.

1. *The branches.* The scales originally present multiply so rapidly during the year that eventually the branches become heavily encrusted with scales. This results in their death, and in severe cases of infestation a tree exhibits severe dieback. The cause of death is the combined effect of the feeding itself and the resulting toxicological condition produced in the branch by the secretionary products of the insect. Bodenheimer is of the opinion that the latter is the sole cause of death. In all probability the ultimate death of the branch is due to the continued poisoning of its cells, but it must be remembered that the persistant feeding of innumerable insects must cause such a drain on the vitality of the plant that its resistance to the toxic secretions is considerably lessened. From this standpoint at least, the drain on the sap must also be regarded as a contributory cause of the dieback that results.

2. *The leaves.* Considerable harm can be done to the leaves, especially when young. The first sign of damage is the appearance of chlorotic spots that sometimes surround the scales. These, as Bodenheimer points out, are caused by the toxic action of the secretionary products on the chlorophyll of the leaves. With a large number of scales present, the vitality of the leaf must be, and is, impaired by the accumulative effect on the chlorophyll. This is eventually shown by the leaf becoming brown and dropping to the ground.

The damage so far discussed is of particular importance in the case of young trees. With dieback and injury to the leaves, especially in a new flush of growth, the development of the young trees is immediately arrested, and its future growth suffers in consequence. *In case of heavy infestation, the young tree may be either killed outright or else permanenlty stunted.*

3. *The fruit.* The damage in this case is generally more indirect than direct. The presence of scale on a fruit precludes it from being packed for export. This a considerable loss to the export crop can result from a severe attack of red scale. The actual damage to the fruit itself is not so great, and though a heavily infested fruit may drop off, the loss thereof is no greater than if it had remained on the tree. A severely attacked branch will not permit the fruit it carries to develop normally, and in bad cases, even though it may not drop off, small and poor quality fruit will result. The colour is also affected, and a badly infested fruit will not develop a full orange colour when mature but presents instead a mottled tint of greenish brown.

4. *Cumulative damage.* The cumulative effect of the damage done to a tree as a whole has been rather overestimated by Quayle... A heavy infestation does not necessarily mean that the death of the tree is assured, but that, on the other hand, there appears to be some correlation between tree vigour and the scale population. *The attack appears to be worst on vigorous and healthy trees, and as the health of such a tree becomes impaired, the degree of infestation appears to decrease* until a stage is reached when the tree seems to be on the point of death. At this stage is practically no infestation or else very slight, and the tree from then on not infrequently shows definite signs of recovery... Two notable examples may be quoted.

An isolated grove of Seville trees on the Mazoe estate has been neglected for many years, receiving neither cultural treatment nor fumigation. Although these trees were very nearly dead as a result of red scale infestation, it was extremely difficult to find any live scale present. None of the trees kept under observation died, and the majority recovered, though remaining thoroughly unthrifty. Receiving no artificial help by cultivation, the recovery of these trees depended solely on their natural vitality. A further illustration was provided by trees in a plot about nine miles from the estate left destitute. When first visited trees in various stages of infestation were to be seen. These were kept under observation during 1933–4–5 and, after a decline, a similar recovery was shown as in the case of the Seville trees. In no instance did death occur, and the trees, though in poor condition, are still very much alive.

Both the above mentioned cases also demonstrated the correlation between tree vigour and scale population. Through lack of cultural treatment the trees were in poor condition. Red scale was present on the majority of the trees, but with few exceptions the population density was comparatively small compared with that of trees on the Mazoe estate, which were in full vigour. The above evidence therefore indicates that the death of a tree is not likely to result in a severe attack of red scale. Nevertheless, the damage done, the slow recovery and thriftless nature of the tree afterwards are almost, if not, as serious and the successful control of red scale infestations in their early stages cannot be over-emphasized." (Jones).

XIV. Damage caused by Red Scale (*Aonidiella aurantii* Mask.) to orange trees and fruit.

After having given these extensive reports on red scale damage to citrus in countries where it is obviously the most important citrus pest, we shall now compare only a few reports from countries in which it is lacking primary economic importance.

In Florida ,,it does not seem to be spreading rapidly nor to be likely to become a serious pest. It has been in the state many years." (Watson).

,,Citrus trees in Egypt never seem to be badly attacked unless they are suffering from water shortage, old age, bad cultivation, or some other factor which lessens their power of resistance." (Hall).

In Ceylon ,,red scale is extremely rare. Anadredapure, the only locality in which a few specimens could be traced is situated in the dry zone. The scarcity of red scale was very significant, as my visit took place just at the end of the dry season, at which period their greatest abundance would have to be expected." (Bodenheimer).

In Palestine red scale damage to citrus is not as extensive as that described from California, Queensland or S. Rhodesia. The following chapter will show that one of the main reasons is the heavy annual mortality induced by spring-hamsins, which follow closely the winter mortality. They reduce the season favorable for increase to a short period from July to September—October. The minor autumn hamsins in September are not of such radical importance but help to stop the further seasonal increase.

The importance of this reducing factor become obvious when conditions in Cilicia (S. Turkey) are compared with those of Palestine. In Cilicia red scale infestation is rather heavy in April and is also reduced by foehn winds, falling from Mt. Taurus into this coastal plain. The maximum is also reached in autumn, but infestation remains heavy the whole year round and is more serious than that occurring normally in Palestine.

Palestinian experience has taught us primarily the paramount importance of planting clean seedlings only. It is greatly regretted that we underestimated the importance of this factor for many years. The heavy infestation from the third to seventh year of the tree depends almost entirely upon the degree of infestation of the young seedlings. Heavily infested seedlings are followed by most heavy attacks during these critical years, whereas in orchards planted with clean seedlings the infestation remains moderate even during this period and often no control measure is called for in such cases until the first fruit is picked. Infestation by wind or spreading from the few individuals present even on very clean nursery stock will occur in any case. But the initial infestation decides whether serious damage will occur during the critical age.

The self-purification of the tree after its fifth year is connected with increased shade and other changes in the host and physical environment. The writer agrees with Summerville that variety differences (heavy attack in citron and lemon, lighter in grapefruit, lighter in orange) depends as well in the first place upon such environmental changes. The moderate distances of the trees (4×6 to 6×6 m) are likewise more unfavorable for Red Scale infestation than the longer distances practised in California and S. Rhodesia.

Brushing the scales off the fruit is practically impossible in Palestine owing to the sensitivity of the rind of our local shamuti orange. The small lesions incurred in such treatment will serve as entrance holes to moulds on the long transport. An infestation of 20—25 red scales on 2 % of the fruit per case is the limit permitted by the fruit inspection service of the government. This limit is only very rarely reached or surpassed in older orchards.

The fact that dieback and other damages which are primarily due to heavy red scale infestation are claimed as spray injury has often been observed by us, especially in the young orchards of the Sharon-district. Summerville's remarks in this respect deserve all attention.

We further agree entirely with Jones that Quayle's earlier statement that heavy red scale infestation kills the tree is overstating the facts. The tree may be reduced to little value from a horticultural point of view, but is not killed. In the orchards in the vicinity of Tel Aviv and Jaffa which have been parcellated for building purposes and have sometimes been neglected for some years, heavy incrustation is sometimes observed. But when these scales are examined, it is often difficult to find one living scale, especially on stem or branches. If left alone, the infestation is reduced, quite in accordance with observation in S. Rhodesia. Healthy trees in young orchards are the preferred host condition for red scale development. These conclusions are entirely supported by the relatively low increase of red scale in the neglected citrus groves of Palestine during the last world war. Klein remarks that on the southern shore of the Lake of Galilee red scale is also prevalent and developing heavily in older closely planted groves. This is in contrast with conditions in the coastal plain, where infestation is light only under such conditions.

The differential susceptibility of citrus species to red scale in Palestine may be illustrated by counts of the scales on 100 leaves through two years. The average monthly infestation was:

year	Host			
	orange (shamuti)	sour orange	grapefruit	sweet lemon
1932	9	4	7	10
1933	8	2	14	80
Average	8.5	3.0	10.3	45

On sour lemon as well as on citron the infestation is as high or higher than on sweet lemon. The fact that lemon is not cultivated extensively in this country has certainly decreased the importance of red scale. It will increase accordingly when the cultivation of lemon will increase in extent, which is a probable trend in the near future for economic reasons.

CONTROL.

Natural control: Fungous diseases, which are detrimental to red scale, as to all armoured scales in a warm, humid climate, are of secondary importance only in semi-arid countries with dry summers.

With regard to parasites and predators, these are of little value as a check of the population trend. The extended explorations of Compere throughout the world failed to produce an enemy of promise in the future. This is the more surprising, as *A. citrinia*, the yellow scale, which is the nearest relative of *A. aurantii*, has a rather effective parasite in *Comperiella bifasciata.*

The most important species are a Hymenopterous parasite *Aphytus chrymsophali* Mercet, which is common in late autumn and early winter, i.e. just at the periods of the highest density of the scale population and the ladybird *Chilocorus bipustulatus* L., which is in mainly active in autumn and spring. Other Hymenoptera and *ladybeetles* as well as *Conwentzia psociformis* (Neur.), the mite *Pediculoides ventricosus* and others are still of minor importance.

All information available hitherto with regard to natural enemies of red scale in Palestine is compiled in a later chapter.

Red scale is the object of very intensive campaigns by spraying and fumigation. Its control will therefore be dealt with extensively in the chapter on pest control.

Technical control and resistant scales: The control work against red scale is almost entirely technical, mainly by spraying or by fumigation. The technique of these will be discusses in detail later.

Another aspect of control of red scale is the development of resistance against fumigation in certain localities. This means that the same dosage consistently gives unsatisfactory results after a certain period of fumigation. This phenomenon is neither restricted to red scale, nor to fumigation. It was observed after continued treatment of San José scale with lime sulphur solutions, of codling moth with lead arsenate and of *Saissetia oleae,* mussel scale a.o. with hydrocyanic acid gas. Fortunately, no signs of resistance have been observed hitherto in Palestine. This does not preclude such possibilities in the future and the problem in itself is interesting enough to be discussed here.

The first observations on red scale developing increased resistance to hydrocyanic acid gas fumigation were made in the Corona District of California in 1914. Commercial fumigators noticed that where previously moderate doses of gas had given very good results; that during the years just previous to and following 1914 they observed that very high concentrations failed to give but moderate control. To this first resistant red scale area other resistant foci have been added and it now covers a considerable part of California's citrus territory. In 1907 - 1909 workers in the Federal Bureau of Entomology conducted much experimental work with hydrocyanic acid gas on red scale and they arrived at a 75% dosage schedule. In 1913 and again in 1924 these same experiments were repeated, and with a 150% dosage not entirely satisfactory results were obtained. Similar experiments were carried out on the black scale with results as with red scale.

Woglum showed that the percentage kill of Red Scale (*Aonidiella aurantii*) and of the purple scale (*Mytilococcus beckii*) was much higher in the lower levels of the tent than higher up.

Red Scale — *Aonidiella aurantii.*

Height	No. of scale	Per cent living
1 ft	1622	1.4
2 ,,	897	3.06
3 ,,	2162	10.29
4 ,,	1346	12.0
5 ,,	727	11.17
6 ,,	219	11.9

This is contrary to what would be expected as it has been shown by Quayle that the highest concentration of gas is in the upper levels of the tree or tent.

Quayle gathered red scale from various districts in California which had reported resistant and non-resistant strains, and fumigated these in the laboratory under controlled conditions The following results were obtained:

Locality	No. of scales fumigated	No. alive	Per cent alive
Orange	6,000	35	.57
Corona	10,170	455	4.47
Orange	1,388	6	.43
La Halva	1,430	280	19.5
Redlands	2,300	6	.26
Highgrove.	2,700	49	1.81
Corona	4,300	173	4.02

Other interesting facts observed from these and other fumigating experiments were: That insects that survive the first fumigation are more resistant to a second fumigation than individuals which had not been previously fumigated. They also found that the greatest resistance is shown by scales on trees that have been fumigated once or twice a year. Differences in the amount of foliage on trees also influenced results: the scales were more difficult to kill on the fruit than on leaves and twigs. Also certain stages of the red scale show more resistance to HCN gas than others. The molting period especially the 2nd molt and the adult are the two most resistant stages.

Boyce found that resistant red scale occurred in spotted areas, and believed that there were other factors that accounted for resistance besides the selection of the more hardy individuals. He carried on fumigation experiments with *Drosophila melanogaster* and showed that there is an indication of a development of the resistance in the population of flies bred from parents that survived the fumigation to which they were subjected. Gray and Kirkpatrick (3) carried on experiments as to the nature of this resistance in the scale. The following is a summary of their work: A group of orange seedlings were infested with black scale from a so-called non-resistant district and another lot with black scale from a resistant district. Twenty-four of these seedlings were fumigated, four at a time, two of resistant and two of non-resistant scales. The counts made one month after fumigation showed that of the 1,092 resistant scales 118 were alive while only 1 out of the 256 of the non-resistant variety remained alive. Later experiments confirmed these results, and proved that individuals were resistant but not immune. Other facts observed:

1. „That the scale surviving the first fumigation were more resistant than those killed off, as a double fumigation showed that the same percentage of kill was not obtained in the 2nd fumigation as in the 1st.
2. Experiments showed that the 2nd molt of the scale may be a natural period of dormancy at which time it is difficult to obtain a satisfactory kill.
3. Scale kills with double fumigation were no better than with single fumigation. With further work the authors found that scale insects are stupefied by a low dosage of HCN gas, and then are much harder to kill. This was termed "protective stupefaction". Gray and Kirkpatrick believed that after the HCN gas is liberated under the tent a short but appreciable time elapses before there is a complete diffusion. It is believed that the scales on certain parts of the tree are exposed to concentrations which are not strong enough to kill them, but instead bring about this inactive or stupefied condition in the insect. In this condition they take no air into their bodies, and therefore can withstand an abnormally high concentration.

Further work was done in the laboratory. It was shown with the aid of a microscope that insects were stupefied within a few minutes, and after remaining in fresh air for a few minutes they recovered. Experiments were carried on with non- resistant Red Scale, resistant Red Scale, non-resistant and resistant *Saissetia*. The insects were exposed to low and then high concentrations, and other insects were exposed to high and then low concentrations.

Results:

	Average per cent survivors			
	Resist. Saissetia	Non-Res. Saissetia	Resist. Aonidiella	Non-Res. Aonidiella
Low first, high second	60.6	20.0	19.3	4.3
High first, low second	22.4	5.7	8.7	1.1

From the above data they concluded that there are resistant and non-resistant varieties of both scales, and that both strains react to stupefaction.

In field experiments Gray and Kirkpatrick came to the following conclusions as to how this „protective stupefaction" is brought about. They found that it was due to either slow diffusion within the tent or by drift gas; that is, the gas diffusing through the canvas of the trees being fumigated, and drifting to adjacent trees in concentrations large enough to stupefy the scale insect.

Moore resumed the problem as follows (1933):

,,The density of the Red Scale population upon a lemon has no striking influence upon the kill obtained. (Working with dense populations only).

The proportion of resistant stages of the Red Scale does not determine the effectiveness of the fumigation but probably determines which insects will survive in an unsuccessful fumigation.

The type of concentration curve has no decided influence upon the kill providing there is a good distribution of gas.

A low concentration built up and maintained and a quick high concentration rapidly leaking away gave similar results.

Under unfavorable conditions, satisfactory kills cannot be obtained by merely raising the concentration or increasing the time of exposure.

High temperature and low relative humidity favor a poor kill of ,,resistant" red scale.

In many cases, the insects are more difficult to kill where the temperature of the lemon is higher than that of the air.

Very low temperatures shortly following the fumigation will tend to reduce the kill.

Exposure of the insects to low concentrations of hydrocyanic acid before the fumigation may materially reduce the kill, have no influence upon it or actually increase the kill.

Under unfavorable conditions, the kill of ,,resistant" red scale is decidedly reduced whereas the kill of ,,non-resistant" red scale is influenced but slightly. Under favorable conditions, there is no significant difference between the kills of ,,resistant" and ,,non-resistant" red scale.

The main difference between ,,resistant" and ,,non-resistant" red scale is the influence, upon the results of a fumigation, of certain environmental conditions."

But all these theories are today only minor contributions to the problem. Under the lead of H. J. Quayle (1938) the true general biological importance and its genetical and selectionistic background have been revealed:

"The greater tolerance or resistance to HCN fumigation that has developed in three of the most important citrus scales in certain localities in California has been brought about, apparently, through natural selection. The areas where the phenomenon first became evident are not necessarily the areas where fumigation has been practiced the longest. It is possible, however, that the areas where resistance first appeared had received as many fumigations as any other area, perhaps even more: that is, fumigations may have been more frequent over a given period. The question of elevation brings in other factors. On account of the warmer winter temperatures on the higher elevations, and also the higher night temperatures in summer, the red scale develops more rapidly under such conditions, as does its favourite host, the lemon. Of the different parts of the tree, the fruit is the one on which the red scale is most difficult to kill. And on the mon fruits are present all round the year. All of these factors, aside from any question of resistance, make the control of the red scale on lemon more difficult. Nevertheless, when red-scale infested lemons are taken from such situations and fumigated under identical conditions with red scale infested fruit from other localities, the difference in resistance is still manifest. And when the scales are transferred to another host, and are grown under identical conditions, this difference still exists . . .

The history of the development of resistance in *Aonidiella* and in *Saissetia* gives some support to the theory that such resistance spread from an original focus. In case of the red scale, it was first observed in the vicinity of Corona, Riverside County, and at about the same time near Orange, Orange County. It gradually spread in these areas and continued La Habra, Orange County; East and North Whittier, Los Angeles County; and finally to Glendora and points also in the Los Angeles County. This, in general, is the higher foothill lemon area . . . The writer believes that for the most part the phenomenon of resistance has developed on the ground, and while some immediate spread would be expected, this factor has not accounted for the present distribution as having come from an original strain in a single locality. The fumigation dosages now used in California even in the nonresistant areas, are much higher than those used many years ago. The schedules have been revised several times and always upward. It is interesting to note also that in Australia, South Africa, and Palestine, countries much younger than California in fumigation practice, the dosage used against the same insect is much lower than in California."

A number of genetical experiments have been performed by Dickson at Riverside and by Yust and coworkers at Whittier. Both series have definitely shown, that HCN resistance is a (completely or) incompletely dominant Mendelian mutation.

The conclusion as about matrocliny, sex-linkage, a.o. still remain under discussion, until more material for analysis is available. Here, we want to insist on certain aspects of population analysis of the phenomenon of HCN resistance, which have not been discussed before.

All facts lead us tot he assumption that the factor R for HCN resistance is a recurrent mutation of probably low population pressure and with probably lowered vitality under normal conditions as compared to the normal r-strain. Under this conception one or more genes which suddenly increase considerably tolerance toward fumigation or even spraying have probably cropped out since many thousands of years regularly. They have always disappeared, because of the low mutation pressure, i.e. the relative rare appearance of the character. And these few mutants were unable to maintain themselves 1. because low mutation pressures have little survival chance in dense populations, 2. because their vitality was lower, or their survival chance under normal conditions lower than that of r. The fact that such resistance has been discovered in red scale, so far, only in a few foci in California, also points to a low mutation pressure. Cross-breeds of both strains show intermediate values, with matroclinous tendency. We shall, however, soon see, that this hybrid component has little importance for the population trend.

This situation was radically changed, when, since 1890, HCN fumigation was made a regular biennual feature in the orange districts of California. Two to three annual generations of *A. aurantii* develop in California. Until 1910 when resistance was first discovered, about 60 to 90, and until 1940, 120 to 180 generations had developed since the early fumigation days. Up to 1910 one fumigation every two years was regarded as satisfactory, later on one every year in the resistant areas. About 15 recurrent catastrophes had been sufficient to produce a highly resistant population pretty speedily. The additional 15 to 25 catastrophes (= fumigations) have in the old areas almost established pure RR-populations, with a very low incidence indeed even of Rr- individuals.The factual difference between survival of resistant to non-resistant strains is as 10 to 20 : 1. The lower vitality may be assumed from 50 to 10%. Under these assumptions the selective value of a catastrophe (fumigation) is 500 to 1800 percent (= 5—18 times the population level) in favour of R. Such enormous selective values are certainly rare in stable natural environments. But experimental transfer in *Drosophila* of known gene-constitution have given similar results, We calculate the dynamic population trend under the following assumptions:

Survival of R = 10%, of r = 1% after each fumigation.
Vitality of R = 10% lower than that of r in normal environment.
Increase factor for both strains = 10 in every generation.
Starting population of R = 20, of r = 1000 individuals.
Roman ciphers: years. $G_{1,\,ff}$ = generations, $C_{1,\,ff}$ = fumigations.
A. One fumigation every year or every three generations,
B. One fumigation every two years or every five generations.

A. Strain	Population immediately after fumigation	Generation 1	2	3	Effect of fumigation	Year
r				1000	10	
R				20	2	
r	10	100	1000	10000	100	I
R	2	18	162	1458	146	
r	10	100	1000	10000	100	II.
R	15	136	1215	10935	1100	
r	1	10	100	1000	10	III.
R	11	99	901	8109	811	
r	1	10	100	1000	10	IV.
R	81	729	6561	59049	5905	

B. Strain population after C		Generation 1	2	3	4	5	Effect of fumigation	Year
r						1000	10	
R						20	2	
r	10	100	1000	10000	100000	1000000	10000	II.
R	2	18	162	1458	10122	91088	9109	
r	10	100	1000	10000	100000	1000000	10000	IV.
R	9	81	729	6561	59049	481441	48144	
r	1	10	100	1000	10000	100000	1000	VI.
R	5	45	405	3645	32805	295245	29525	
r	1	10	100	1000	10000	100000	1	VIII.
R	30	270	8100	243000	7320000	219600000	21960	

The rapidity with which r disappears is amazing. And these calculations are by no means build up upon assumptions of special nature: The difference in survival is experimental, the reproduction factor is within the frame of 50 to 150 eggs per female, equal sex ratio, and the increase quotients of natural red scale populations. The difference in vitality (—10%) is quite considerable. But even if it should be still greater, it would have a small braking effect only on the population trend. The number of recurrent catastrophes and the interval of generations between them are also empirical data.

We have to find out, to which degree hybridization between R and r with its intermediate effects on the survival are liable to influence this population trend. The equilibrium of a mutant in a population of given size depends upon its frequency and recurrency. We do not need assume heavy recurrence of the R-mutation, and have actually neglected this factor, which still would enforce the calculated population trend. This equilibrium is reached at:

$$1 = \frac{1}{r} \cdot \frac{1}{r} + 2 \cdot \frac{1}{r} \cdot \frac{1}{R} + \frac{1}{R} \cdot \frac{1}{R}$$

We now calculate for the populations under A (see above) the population equilibrium for each population immediately after the fumigation. The stable gene distribution would be for:

After fumigation	RR	+	Rr	+	rr	=	Total
I.	0.0004		0.0392		0.9604	=	1.0000
II.	0.3600		0.4800		0.1600	=	1.0000
III.	0.8403		0.0903		0.0694	=	1.0000
IV.	0.97580		0.02405		0.00015	=	1.0000
V.	0.996666		0.033231		0.000003	=	1.0000.

This recurrent calculation shows that hybridisation has a surprisingly small influence for the conservation of the r-gene.

A number of genetical factors remain to be discussed. The best starting point seems to be the fumigation survival series on various crosses performed by Dickson (1941); all at 1 and 2 cc HCN per 100 cu. ft., 40 minutes at 23.9° C:

Cross (f = female, m = male)		Survival percentage 1 cc. HCN	2cc HCN
1. F_1 : RR f × rr m		31.53	13.12
2. rr f × RR m		33.38	12.39
	Average:	32.45	12.75
3. Recrossing: RR × F_1		38.94	17.56
4. rr × F_1		18.26	6.75
5. (RR f × rr m)f × (RR f × rr m)m		38.18	17.77
6. (rr f × RR m)f × (RR f × rr m)m		33.93	18.39
7. (rr f × RR m)f × (rr f × RR m)m		17.25	7.04
8. (RR f × rr m)f × (rr f × RR m)m		17.87	7.48

It should be pointed out, that all fumigation experiments are only performed against adult females. The F_1-females are intermediate, with a very considerable deviation (2/3) of the survival percentage towards the resistant strain (45.43 : 32.45 : 4.06% resp. 22.37 : 12.75 : 0.75% survival). The F_1-females are intermediate in survival between the F_1-females and their parental grandmothers. In these cases the intermediate position is almost complete (38.96, resp. 18.26 in the calculation, only series 3 from 1591 individuals only) showing some deviation.

We begin this discussion with a discussion, if the difference of the two strains is really a mutation. The only alternative is dauermodification, of course. The fact that both combinations of original crosses give almost identical results of incomplete dominance in F_1, shows that the male partner participates in the transmission of resistance (cf. also experiment 7). As the male does not participate in the transmission of the dauermodification characters in metazoa (Jollos) this fact seems to point decidedly in the direction of mutation. The fact that 7 years of interrupted fumigation did not produce a reversal to the original non-resistant population whilst not decisive, strongly points in the same direction. It has hitherto been assumed, that the resistant mutation is a single dominant mutation, as indicated by the symbol RR. Some of the quoted facts on heredity of HCN tolerance suggest matroclinal heredity: The results of crosses of F_2 confirm such an assumption, yet they leave unexplained the incomplete, equal dominance in both types of crosses in F_1 which should be the pure strain of one of both races. Nothing can be decided before before results from F_3 are available.

It is however possible that multiple factors are cooperating in producing resistance. Some alternatives are suggestive: Two or more factors may be more than additive in double occurence, produce an effect stronger than their simple sum or even homozygoty may produce the same effect. In that case the factor appears stronger than in the male's or hybrid's single factor. Not all of such multiple alleles may be of equal value each. Yet then we would have to assume, that by a new catastrophic factor more than one adaptive character has become manifest and selected with enormous speed. It is easier to assume that the new mutant is a modification factor, which acts in the presence of one or more normal factors. If it is not a modification factor, it may still be a factor b, which gains manifestation only in the presence of a, c and d.

It has also been suggested that the mutation R has originated in response to the new stimulus of HCN. Such a theory seems to be extremely dubious in view of our general knowledge of such occurences. Yet in that case the mutation by consistent experimental fumigation should be speedily produced, once the proper procedure of stimulation has been discovered.

But Hartzell and Moore (1937) found that the lilac race of *Mytilococcus ulmi* L. is resistant against coal-tar oils, by which the apple race of the same species is readily controlled. Compton and Kearne (1937) found red mites from roses more resistant to acaricides than those from other hosts. The natural occurence of pre-adapted or accidentally resistant strains in arthropods seems thus clearly indicated. We suppose that this was the initial condition in the various scale and other insects which developed resistant focal populations under the influence of recurrent control measures.

Little is known of the physiological changes connected with the mutation. That such occur, is certain. Some of them have been discovered with regard to survival, fertility, adsorption of HCN, stupefaction and opening of the spiracles, etc. More surprising is the fact, that HCN resistant individuals also are resistant against methyl bromide, ethylene oxide, oil sprays, etc. This is apparently based on the

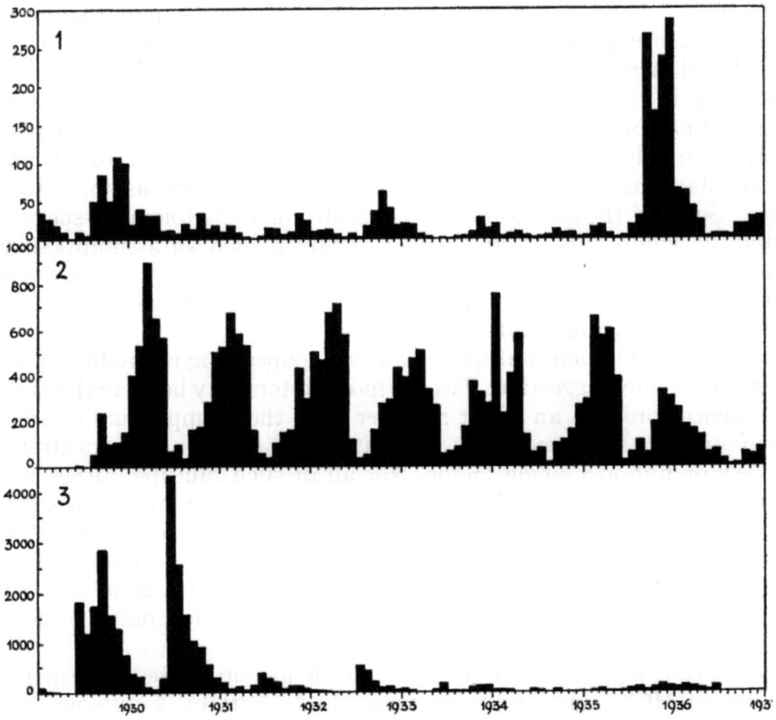

XV. Monthly fluctuations of populations of 1. Red Scale, 2. Chaff Scale and 3. Florida Wax Scale

fact, that all these poisons are neurotoxic, acting progressively on the central nervous system. It may be a general raise of the dehydrogenetic respiration or some similar physiological change which, by chance of course, produces this wide range of resistance.

Dickson's theory that HCN resistance is probably due to one gene of a pair carried by the X-chromosome is not proven in view of the fact that no differential X-chromosomes occur in *Aonidiella* or in other Coccoidea. Whilst the sex can be determined, at least to some degree by factors in autosomes (Winge, in races of *Lymantria*, Bridges, in *Drosophila*), nothing prevents the „less likely possibility", that resistance is due to a group of closely linked genes carried in the X-chromosome.

At Whittier the tolerance was studied in five subsequent generations of the strains and of their hybrids (mature females 36 days old 0.48 mg HCN per 1):

Generation	RR	R female × r male	r female × R male	rr	
F 1	73.6	81.2	82.2	99.8%	mortality
F 2	64.6	68.3	85.9	99.2	
F 2	46.3	49.7	84.1	98.8	
F 3	73.6	83.6	89.3	99.8	
F 4	74.6	70.5	83.8	99.2	
F 5	66.3	64.1	81.0	97.9	
F 1	100	71	68	0.8	The same data, but
F 2	100	89	40	2.3	RR of each line =
F 2	100	94	29	2.1	100.
F 3	100	62	41	0.8	
F 4	100	90	46	2.3	
F 5	100	107	56	6.8	
Average F2/F5	100	88	42	2.9	
Calculated as simple intermediates	100	76	27	3	

Female hybrids of F_1 were intermediate, but nearer to R in both hybrid crossings. This shows that R is incompletely dominant. Male hybrids of F_1 almost equal their mothers. The authors suggest that R be in the sex-chromosome. We think that analysis for matrocliny should be continued.

Males of R females × r males, backcrossed with R females, were not significantly different from R. Males from r females × R males, backcrossed with r females, were not significantly different from r. Here is further evidence that F_1 males from crossings are like the mother parent. Female progeny of the crosses, if permitted to self, were similar to F_2. After F_1 the selfed hybrids r female × R males were always less resistant than R females × r males selfed hybrids. We leave the genetical analysis at this point. Our contribution to the problem is the statement, that in populations which have been not exposed to control for a protracted period, of 10 recurrent catastrophes, for instance, no significant cropping out of r has to be expected. Whilst helping further analysis, the genetical material is still too small to permit definite assumptions.

References.

R. C. Dickson, 1941. Hilgardia **13**(9): 515—22.
Geo P. Gray and A. F. Kirkpatrick. 1929. Calif. Citrog. **14**(8): 308, 336; (9) 364, 380—1.
William Moore, 1936, Jour. Econ. Ento. **29**(1): 65—78.
H. J. Quayle, 1922. Jour. Econ. Ent. **15**(6): 400—4.
H. J. Quayle, 1938a. Insects of citrus and other subtropical fruits. Ithaca N.Y.
H. J. Quayle. 1938b. Hilgardia **11**(5): 183—210.
Edmund W. Sinnot, and L. C. Dunn. 1939. Princ'ples of genetics. Ed. 3, N. York.
R. S. Woglum, 1925. Jour. Econ. Ent. **18**(4): 593—7.
H. R. Yust, R. L. Busbey and L. B. Howard. 1942. Jour. Econ. Ent. **35**(6): 816—20.

H. R. Yust and L. B. Howard, 1942. Journ. Econ. Ent. **36**(6): 821—4.
H. R. Yust, H. D. Nelson, R. L. Busbey, Jour. Econ. Ent. **36.** 1947. no. 5.
C. I. Bliss, B. M. Broadbent and S. A. Watson, Journ. Econ. Entom. **24**, 1931, pp. 1222—1229.
F. S. Bodenheimer, Schaedlingsfauna Palaestinas. Berlin, 1930, pp. 151—155.
F. S. Bodenheimer, Observations on citrus insects and their control in many parts of the world. Tel Aviv, 1932, 105 pp.
F. S. Bodenheimer, Hadar, **7**, 1934, p. 139—148; **9**, 1936, p. 69—70, 74.
F. S. Bodenheimer and R. Ashbel, Hadar **6**, 1933, p. 175—178.
A. M. Boyce, Journ. Econ. Ent. **21**, 1928. pp. 715—720.
C. K. Brain, Bull. Ento. Res. **9**, 1919, pp. 199—200.
J. Carmin and D. Scheinkin, Bull. Soc. Roy. Entom. d'Egypte 1934, pp. 242—274.
S. W. Clark and W. H. Friend, Texas Agric. Exper. Sta. 1932, Bull. 455. 35 pp.
H. Compere, Calif. Citrograph, 1935. p. 371, 382—388.
G. P. Gray and A. F. Kirkpatrick, Calif. Citrograph, 1929. June and July.
W. J. Hall and Ford, Masoe Citrus Exper. Sta. Report for 1932, p.
E. P. Jones, Mazoe Citrus Exper. Sta. Report for 1935, Oxford 1936. pp.11—52, Pl. I.
H. Z. Klein, Hadar **8**, 1935, p. 71—73, 115—116.
H. Knight, Journ. Entom. Zool. Pomona Coll. **24**, 1932, p. 1.
J. Koroneos, Les Coccidae de la Grece. I. Athenes. 1934, p. 22.
V. Lupo, Boll. Labor. Zool. Portici **29**, 1936, pp. 249—261.
H. L. Mc Kenzie, Univ. Calif. Publ. Ent. **6**, 1937 0. 332—336.
W. Moore, Journ. Econ. Ent. **26**, 1933 pp. 1140—1161.
F. Munger, Journ. Econ. Ent. **41**. 1948. p. 422—423.
F. Munger, and A. W. Cresson, Journ. Econ. Ent. **41**. 1948. p. 424—427.
R. G. Nel, Hilgardia **7**, 1933, pp. 415—466.
H. J. Quayle, California Agric. Exper. Sta. 1911, Bull. 222, p. 99—150.
H. J. Quayle, Journ. Econ. Ent. **15**, 1922, p.p. 400—404; **20**, 1927, pp. 667—673.
H. S. Smith, Calif. Citrograph, 1934, p. 263, 280—282.
R. S. Woglum, Journ. Econ. Ent. **18**, 1925, pp. 539—597.

D. THE EGYPTIAN BLACK SCALE (*Chrysomphalus aonidum* (L.))

Synonymy: 1758 *Coccus aonidum* Linnaeus
1880 *Chrysomphalus ficus* Ashmead
1880 *Aspidiotus ficus* Comstock
1900 *Chrysomphalus aonidum* Hempel
Other common names: Florida Red Scale, Citrus Black Scale.

Description: Scale of the adult female circular and moderately convex; brownish black. The margins are occasionally greyish, exuviae in central position. Exuviae reddish yellow, always partially obscured by a layer of secretion which is reddish-brown above the first and pale olivaceous above the second pellicle. Secretionary covering white, circular and nipple-shaped; generally worn off in older individuals. Ventral scale very thin and absent in older individuals.
Diameter of total scale 1.2—2.3 mm; of first larval scale 0.2—0.4 mm; of second larval scale 0.6—0.8 mm.

The male puparium is oval with sides subparallel and ends broadly rounded. Black brown, the posterior extension grey. 0.8—1.2 mm long; 0.7 mm broad.

The sexes of the second stage larvae may be distinguished as follows: in the female the second larval scale is composed of two red circles, whereas in the male only the inner ring is red, the outer black.

The *adult female* is flat, broadly rounded in front, broad, pear shaped and produced behind. Very pale yellow when young, yellow when mature and orange coloured towards the posterior margin. Body shrivels at gestation, the lateral abdominal lobes sometimes reaching the pygidium. Antennal tubercles very small, with a long, fine bristle. Rostrum extending a little beyond the second pair of spiracles. No peristigmatic glands. A spine on each of the metathoracic latero-posterior angle. Long bristles on the lateral margins of the body, 1—2 on each side of every abdominal, 6—7 on every thoracic segment.
0.8—1.4 mm long, 0.7—1.2 mm broad.

Pygidium: Medium-sized, strongly projecting. Three pairs of well developed lobes, subequal in size. The first and second lobes are notched along their outer lateral margin, whereas the third lobe is serrate. Plates rather deeply fringed. The pygidial formula runs as follows:

$$P_1,\ L_1,\ 2P_2,\ L_2,\ 3P_3,\ L_3,\ 3P_4.$$

The margin is finely serrate beyond the last plates and has a single projecting spine.

Dorsal face of pygidium: Anal opening oval, at a distance of about 3 times its own diameter from the median lobes. The latero-marginal tubular glands are numerous, arranged in three rows. The orifices of the first row (2—3 glands) are situated between the first and second lobe, those of the second row (15—30 glands) in double row cephalate between the second and third lobe and the third similar to the second group in arrangement and numbers. No cephalates beyond the third lobe.

Ventral face of pygidium: Vaginal aperture small. Circumgenital glands, generally in 4 groups, as follows:

0—2			0	
1—9	1—9	or on the average	2—6	2—6
2—6	2—6		3—5	3—5

There are 5 pairs of prominent paraphyses, one on each side of the first and second lobe, the fifth on the interior side of the third lobe.

The specific recognition characters, as recently worked out by Mc Kenzie (1939), are: „Prosomatic region at maturity membranous or at times with very slight sclerotization, pygidium short and broad, the apex quite obtuse, perivulvar pores in four groups of three to six pores in each, three pairs of rather small pygidial lobes present, these all of practically the same size and shape, the fourth lobe indicated merely by a minute projection, margin beyond the fourth lobe sclerotized for some distance and twice notched, plates as far as the third lobe all quite finely fimbriate, those beyond the third lobe differently shaped, the first two each with a pair of small, slightly clavate processes, the third very raggedly divided, dorsal ducts abundant, especially in the second and third pygidial furrows, those of the first furrow somewhat shorter and noticably broader than those of the second and third furrows, second and third furrows sclerotized and marked with transverse striations, second abdominal segment with a submarginal dorsal cluster of short ducts, other prepygidial segments with no more than one or two, anal opening relatively large, close to the apices of the median paraphyses, thoracic tubercle small and acute."

Male like the others in the family. Orange coloured with whitish wings, legs, antennae and stylus being brown. 0.8 mm (+ stylus 1.1 m) long.

Egg long oval, lemon yellow. 0.2—0.25 mm long; 1.0—1.5 mm broad.

Neonate larva short oval; yellow. 0.25 mm long, 0.2 mm broad.

Ferris has recently discovered a widely spread tropicopolite species of *Crysomphalus*, *C. bifasciculatus* Ferr., which is not to be distinguished from *C. aonidum* by the naked eye. Microscopical inspection easily permits differentiation of both these species (cf. key on p. 185). We have since especially studied ample material from all Middle East foci of the black scale (excl. Trans-Caucasia). We have never seen anything different from typical *C. aonidum*. We wish to point out, however, that in future determinations of black scale which are not checked under the microscope, are valueless.

Distribution: The original home of *C. aonidum* is probably in the Oriental Kingdom (McKenzie). It was introduced into Florida from the Antilles about 1879. It is known from most tropical and subtropical countries now, extending even into more moderate climates, where it is restricted to greenhouses. In connection with the still open question as to the optimum requirements of this species, some zoogeographical details may be given:

In Cuba 105 hosts are known, which include almost the total host range known from other regions. Density of occurrence and host range are fairly high in Brazil too, but from British Guyana it is recorded as not common and from citrus only. It is known in Mexico, Panama, Argentine, Chili, Florida and California. However, whereas it is recorded from Florida as „of considerable economic importance on citrus, palm, camphor and roses" and from about 75 hosts, its record in California is: „Its attacks in California are almost entirely confined to greenhouses in the central and northern parts and to subtropical ornamental gardens, in the southern part of the State. It is not an orchard pest." (Essig). Quayle does not even mention it in his list of Californian citrus pests. On other islands of Central America it thrives well (Barbados, Jamaica). In Australia it is well known in Queensland, being of occasional occurrence in the other states of the Commonwealth only through introduction.

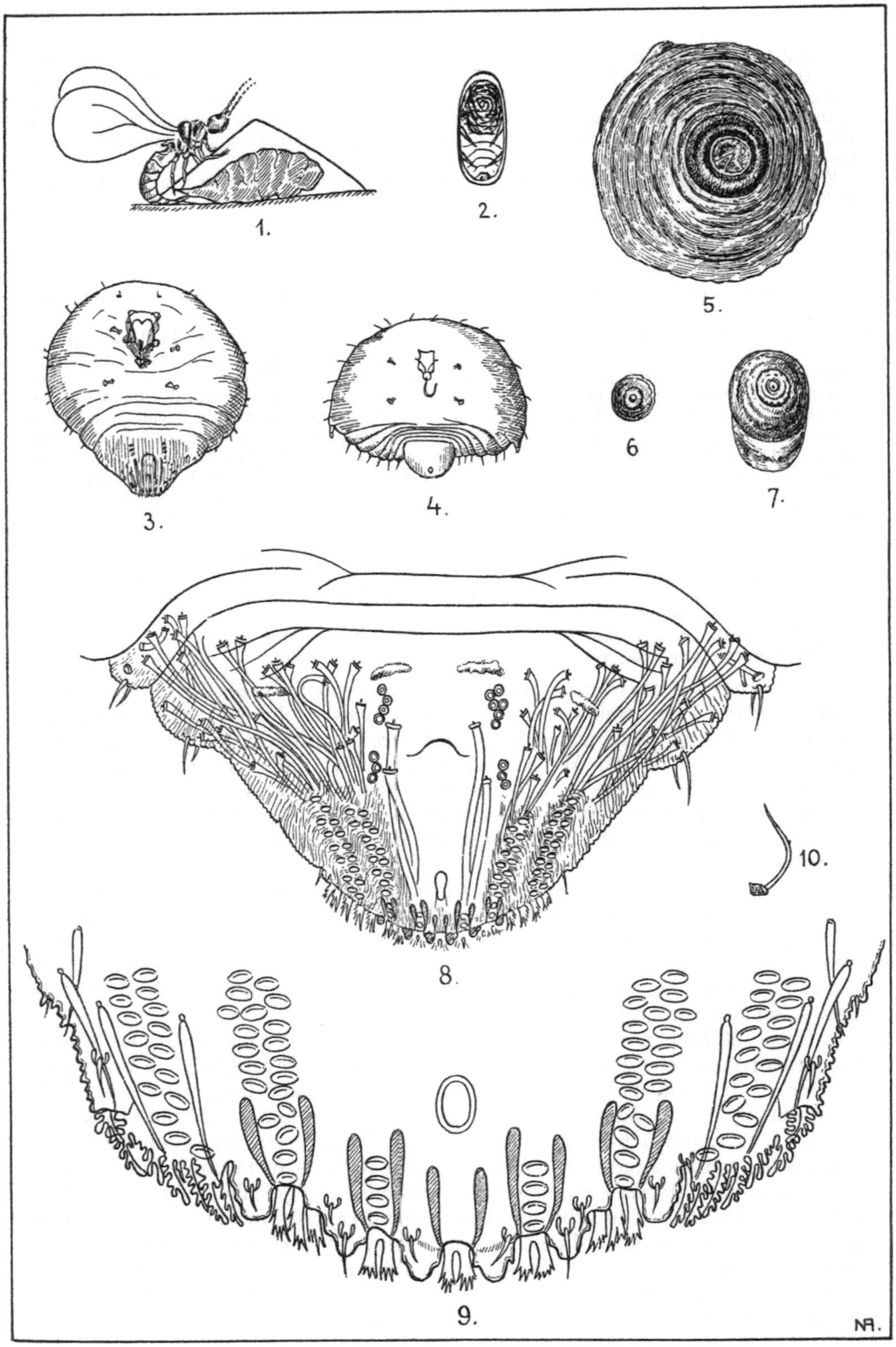

XVI. The Egyptian Black Scale (*Chrysomphalus aonidum* L.) 1) male copulating with female; 2) egg with far developed embryo; 3) young female; 4) old female; 5) scale of adult female; 6) scale of second larval instar; 7) male puparium; 8) pygidium of adult female; 9) posterior margin of adult female.

Summerville (1934) resumes his experience from Queensland: „In so far as citrus is concerned, though the species is distributed throughout the state, it is only in the hotter and drier parts that it can be regarded as a pest of importance. In more humid and milder climates such as at Tamborine Mountain or in the Blackall Range the species is rarely found on tended trees. The species cannot be considered a severe pest on the tree." Dry years are especially dangerous. Infestation on oranges is maintained and supported by the infestation on some native trees as are *Castaneospermum australis, Atlantia buxifolia, Myrtus hillii,* camphor laurel a.o.

At Hawaii, the Egyptian black scale is common everywhere, but the writer did not observe a heavy infestation during his stay there in August 1931.

In the Oriental Region *C. aonidum* is very common, important and polyphagous at Formosa (Takahashi). In S. Japan it is likewise present, but less common and polyphagous. It is no serious citrus pest in the latter country, but certainly so at Formosa. It is not recorded as a pest from Dutch-India. In Ceylon and India it is rare and scarcely a citrus pest. Recorded from China, Philippine Islands, Seychelles, Madagascar, Zanzibar and Mauritius.

The Egyptian black scale is distributed throughout the Ethiopian Region, except the desert tracts. In S. Africa it is common, especially in the coastal area of the Union, on a large variety of plants, including avocado, camellia, citrus, palm, rose, fig and many others (Brain). Hall and Ford (1933) report from S. Rhodesia: Up to the present it has never been found on the company's citrus estates, but it has been taken in widely separate localities in the colony, and citrus trees in Salisbury have been seen heavily infested. *C. aonidum* has been reported from Tropical Western Africa as well as from the East-shore (Dar Es Salam, Lourenco Marques) on plants as *Cycas, Pandanus, Ventia, Codiaeum.* It certainly is not endemic there. On the Atlantic Islands it is present (*Hedera, Ficus, Citrus*), but is evidently not very abundant.

Its distribution within the Mediterranean countries is of great interest: In Morocco an infestation focus was recently discovered near Casablanca, *Strelitzia sp.* being so far the only host known there. It was introduced into Algeria towards the end of the last century. In 1899 it was first discovered in the acclimatisation garden. To-day it has spread all over the town of Algers and its immediate surroundings. In about 30 years it has spread to an area of 15 - 20 klm around the original focus of infestation, but has not left a small area in the maritime zone. In the infested focus considerable damage is done to quite a number of host plants (Balachowsky). In Egypt, the black scale seems to have appeared within the last forty or fifty years, but there is no very definite information on the subject. It is now to be found in almost every garden in Lower Egypt. In Upper Egypt the scale is not very widely or firmly established (Hall) The list of host plants in Egypt is long and varified. The damage done to citrus is very serious, and enforced obligatory fumigation.

In Syria it infests the citrus orchards heavily and many ornamental plants in the coastal plain of the Lebanon. In Turkey it has been reported by Lindinger from a palm. But no other records have since been reported and it is certainly not a pest in the citrus districts of the Mediterranean plains of Anatolia.

In Greece it is often imported with citrus fruits, but the species could not establish itself (Koroneos). In Italy it was collected by Leonardi at Florence on two exotic plants (*Ruppelia, Arthabotris*). Lindinger records it further from *Agave, Citrus, Livingstonea,* and *Phoenix,* but nothing is known hitherto about any focus where *C. aonidum* has become established.

In France it seems to be restricted to greenhouses, not being mentioned by Balachowsky from S. France. In Spain it has once been recorded from citrus, but its importance there is very small. Neither is it included in the pests sent for identification to the local Agricultural Experimental Stations, nor has it been collected by Balachowsky or the writer during recent trips through Spain.

In Palestine, the Egyptian black scale is for a long time established in three main centres. The two more important ones, the coastal plain from Nesle to Ras El Nakura and the Jordan valley from Tabgha to Baisan are connected by a series of smaller foci throughout Galilee. These centres are connected with the Syrian Black Scale area, from which they originated at about 1910. Infestation is very heavy in both centres. About 1925 a third centre was discovered in an old orange grove S.E. of Jaffa. This infestation spreads now slowly to all the neighbouring orange-groves. A fourth centre appeared in 1931 in an old grove at Hedera and was quelled in its beginning by very drastic measures taken immediately by the owner. In 1937 three further centres of the scale were discovered in the Hedera-district. Its further spread all over the coastal plain seems by now unavoidable and in the last 5 years it actually has spread almost all over it.

Host-relations: C. aonidum is an extremely polyphagous species. Its host-list from the humid tropics is a fairly long one, but even in Egypt, where it is known since 40 - 50 years only, it is recorded up to-day from 42 families and 101 species. The following are the most favourite hosts in Egypt (Priesner):

Anacardiaceae	*Mangifera indica*
Apocynaceae	*Nerium oleander, Thevetia nereifolia*
Moraceae	*Ficus spp.*, esp. *F. nitida*
Musaceae	*Musa sapientum*
Myrtaceae	*Eucalyptus spp.*
Oleaceae	*Olea europaea*
Palmaceae	*Latania, Phoenix dactylifera*
Rosaceae	*Rosa centifolia*
Rutaceae	*Citrus spp.*
Verbenaceae	*Duranta plumieri*
Vitaceae	*Vitis vinifera*

The most heavily attacked plants at Algeria are as follows (Balachowsky):

Celastraceae	*Evonymus japonicus*
Rosaceae	*Prunus laurocerasus*
Rutaceae	*Citrus spp., Xanthoxylum americanum*
Saxifragaceae	*Braxia sp.*

Ficus, Nerium, Phoenix are only moderately infested and *Eucalyptus* seems to be not infested at all. Citrus is the only real preference host in every known area of *C. aonidum.*

It is a well known fact that tropical immigrants show a distinct predilection for hosts with tropical relations (Bodenheimer). This fact is fully confirmed upon carefully checking the host lists of *Chrysomphalus* in subtropical countries.

Host plants in Palestine are as follows (1 weak, 3 strong):

Anonaceae	*Anona* (2)
Apocynaceae	*Nerium* (1)
Araliaceae	*Hedera* (2)
Coniferae	*Cupressus* (fruits; 1)
Convolvulaceae	*Convolvulus* (1)
Euphorbiaceae	*Ricinus* (1)
Graminaceae	*Cynodon* (1)
	Sorghum (1)
Leguminosae	*Acacia* (2)
	Spartium (2)
	Ceratonia (1)

Moraceae	*Ficus* (2)
Musaceae	*Musa* (3)
Myrtaceae	*Eucalyptus* (3)
	Myrtus (2)
Oleaceae	*Olea* (1)
Palmae	*Phoenix* (2)
	Washingtonia (2)
Punicaceae	*Punica* (1)
Rhamnaceae	*Zizyphus* (2)
Rutaceae	*Citrus* (3)
Sterculiaceae	*Brachychiton* (1)
Verbenaceae	*Duranta* (1)
Vitaceae	*Vitis* (1)

This list of 20 families and 24 genera of plants attacked corresponds to the slow dispersal towards other hosts of even a polyphagous species of recent introduction in an unfavorable climate. This statement is confirmed by the fact that the overwhelming majority of these hosts are hosts in the Jordan valley only, not having been observed as such in the Acre area. Yet in the parks of Mikveh Israel hosts include avocado, anona, guava, *Eugenia*, mango, *Feijoa, Pytanga*, pecan nut, *Eriobotrya*, a.o. The few that have been observed there (eucalyptus, myrtle, *Vitis, Nerium*) are only moderately infested (Schweig and Gruenberg). A still further confirmation serves the fact that in the latest centre of infestation in the coastal plain, near Jaffa, where about 24 years have passed since the first infestation, no hosts others than citrus are known so far.

All citrus varieties are attacked in Palestine, grapefruit and oranges heavier than mandarines and lemons. Priesner reports from Egypt that lamoon baladi (lime) is less attacked, whereas the Adalia lemon is heavily attacked.

The preferred parts of the plant are leaves and fruit. Twigs and stems are only rarely infested.

The hygienic condition of the host has been claimed to be of primary importance for the intensity of the infestation. But the evidence at hand is still very contradictory:

,,Many observations have shown that healthy plants are never heavily attacked, except, when there is a steady source of infestation, and also in this case severe infections are rare. The plant obvious helps itself against the enemies in a way unknown to us, if it is growing under optimal conditions. Improvement of the conditions under which our citrus cultures are growing, seems to me to be the key to lessening the injury caused by the Black scale . . ." (Priesner).

The richly populated trees of Acacia, Washingtonia and Eucalyptus at Akko Poultry Station are those that are in the pens in which the manure of fowls is lying freely under and round the trees . . . The higher rate of food and water makes the sap of the tree more attractive for the scale. Poorly cultivated citrus (and other) trees incline to establish a poor Black Scale population. Similar observations were made at Migdal. Trees in the vicinity of an irrigation plant showed a bad infestation, though others at some distance were clean. All these trees form a grove and were fumigated on the same day, and there was a good kill of the scale on all of them. The richer food and water supply is to be regarded as one of the chief factors favouring Black Scale infestation. On seedlings at Migdal suffering from lack of water the development of Black Scale stopped after some time. This was confirmed by the high mortality noticed in unwatered or abandoned groves in Acre and Jaffa. Groves that were overwatered and overmanured had a higher population of Black Scale than groves that were neglected." (Schweig and Gruenberg).

It is still premature to arrive at conclusions. The report given above is certainly correct. It may be that a long extended vegetation period with presence of young sprouts is favourable for the settlement of the crawlers. Or it may also be, as the authors assume, that in neglected groves the tree is not always able to replace the loss of water by evaporation to the extent needed to the scales. Systematic observations on the interrelations of vegetation, conditions of host and black scale attack throughout the whole year are needed in order to settle the contradictory evidence on this important problem.

Sporadic infestation of single trees or groups of trees in many orange groves in Egypt and differences of infestation in lemons and limes led Priesner to the theory that plant mutations of special susceptibility may be responsible for those differences. He therefore recommends ,,the possible selecting and growing of scale resisting varieties (mutations), and their possible propagations." However, no immunity or predisposition of citrus against black scale has been observed so far in Palestine. It is difficult to draw conclusions in a country where fumigation is part of the annual routine. It is quite possible though, that the negligence in fumigating some single trees, which easily occurs, is the real reason for such sporadic infestations. And it may also be that more susceptible citrus species, as lemons, may show racial or individual differences in the susceptibility towards black scale. It seems highly improbable that economic results could be gained with regard to grapefruits or oranges by breeding less susceptible varieties.

The Egyptian black scale prefers to settle in the shade during the hot season and on the sunny side of trees during winter. Schweig and Gruenberg found that in the Acre area the autumn-winter generation settles on the sunny side (S.E.) in October and the spring generation (April) remains there. The two summer-generations develop mainly on the shady side of the trees (N.W.). In the Jordan valley the majority of the scales settle from May to October on the S.E. side of the tree, during winter they prefer the N.W. side. The following statistical data illustrate this seasonal shift of the population within one tree:

Distribution of adult females of C. aonidum on trees (in %).

date	Acre				Migdal			
	N	E	S	W	N	E	S	W
II.1934	18	33	33	16	42	18	7	33
III.	13	30	32	25	12	21	19	48
IV.	12	34	34	18	13	32	26	29
V.	15	31	36	16	23	37	22	18
VI.	38	11	15	35	25	27	30	17
VII.	43	12	17	27	29	41	19	11
VIII.	49	13	15	37	26	28	31	14
IX.	32	16	12	40	36	15	25	25
X.	16	31	35	19	27	16	27	30
XI.	15	29	32	23	34	18	27	27
XII.	13	34	29	24	34	20	21	24
I.1935	15	34	26	25	42	21	12	25
Average	23	26	26	25	29	25	22	25

The lower and more central parts of the tree are preferred and more densely infested. The lower sides of the leaves are preferred by the female larvae when the upper surface is strongly exposed to light. The male larvae, on the other hand, are attracted by such exposure. The frontier between exposure and shade séems to be most favorable and in it scales are equally distributed over all points of the compass as well as on upper and lower surfaces. There occurs no extended heliothermy, as a rule, in black scale females and development is therefore equal in all parts of the tree. Schweig and Gruenberg have shown that in compliance with these conditions oviposition is slightly retarded in the shade as compared with individuals in more sunny exposure, but that the peak of oviposition is approximately the same in both groups.

LIFE - HISTORY.

Fertility: The Egyptian black scale is ovoviviparous; i.e. the eggs start to develop within the body of their mother. During the summer the eggs are hatched 1/2 to 2 (-4) days after having been laid. In winter, however, the duration of the egg stage is protracted to 7-10 days. Vivipary has been observed by Balachowsky after fumigation with SO_2.

Balachowsky estimated the average number of eggs per female in Algeria at 60, Priesner in Egypt at 50-80.

Schweig and Gruenberg estimate the average egg-production per female at 150, 211 (Migdal) and 247 (Acre) eggs being the maximum. No significant difference was observed in females exposed to sun or shade. The total egg-production of 9 females at Akko was on orange-leaves as follows:

155, 140, 166, 124, 119, 237, 124, 110, 142 = 146 on the average.

Klein reports the following fertility on grapefruit leaves at Dagania (deposited eggs plus those still in the mother's abdomen):

month	I	II	III	IV	V	VI	VII	VIII	IX	X	XI	XII	Average maximum
average no. of eggs per female	49	32	75	49	49	45	68	44	41	25	31		46
maximum no. of eggs per female.	91	109	120	148	87	87	117	68	98	65	87		148

The maximum figures from November to April show a neat accumulative increase of the eggs, until the hatching of the first larvae in May.

Comparing the results of Klein with those obtained by Schweig and Gruenberg, it is obvious that the latter are to be preferred (owing to the method applied) as they include the result of two months breeding of every female. All females which died prematurely were excluded. Klein's observations did not aim at such comprehensive observations. 150 eggs are to be accepted as the physiological norm, which is by no means identical with the epidemiological average. A very large number of females die before the potential oviposition rate is completed. The average between the results mentioned (46-146) should be accepted for epidemiological purposes, with about 100 eggs per female.

The egg production is higher on fruits than on leaves.

Klein reports for September	on leaves:	41	on fruits:	50 eggs
November	,, ,,	25	,, ,,	66 ,,

Schweig and Gruenberg obtained from 2 females an average of 239 eggs on fruits (per 2 months), against 146 on leaves. In both cases the number of eggs on leaves reached only 60% (57 and 61%) of those on fruits. This agrees with observations on other citrus scales.

The eggs are laid successively at a rate from 1-6 eggs per day. Observations of 19 eggs being laid are certainly very exceptional. The eggs are laid at intervals of from half an hour to one day. In summer, rates as high as 5-6 eggs per day-female were observed at Migdal, in winter of 1-1.5. At Acre no eggs are laid from mid-January to April (Schweig and Gruenberg). The physiological oviposition period must last — in the summer — for at least one month. As many as 25 unhatched eggs were found contemporaneously beneath one scale in the summer, 10 in the winter.

The empty eggshells remain below the scale and may serve, even after the death of the female, for evaluating its total egg-production.

Sex-ratio and reproduction: The sex ratio was computed from the offspring of 11 females to be 58 females to 42 males. Klein counted for every generation the following percentages of males and females on seedlings:

generation	1	2	3	4	average
% males	46	15	28	21	28
% females	54	85	72	79	72

This method, however, is open to criticism, as the female scales remain on the plant much longer than those of the males. The generations 2-4 imply, therefore, an exaggeration of the female percentage. This interpretation is probably more correct than the assumption of a seasonal shift in the sex-ratio. The sex-ratio may be assumed as 6 females to 4 males.

All reproduction is sexual. The male longevity is a few days only, as in all diaspine scales. The copula has been observed by Balachowsky. Parthenogenesis does not occur in this species. Isolated females had normally eggs mature in their bodies, but fell off the leaf or died after a short time, without having oviposited. (Schweig and Gruenberg). The same has been observed by Priesner, whereas Masuda in Japan claims to have observed telytokous parthenogenesis (possibly owing to the presence of a different biological race.)

Priesner discovered that on leaves showing a great difference in the exposure to light of both the upper and lower surfaces, the distribution of the sexes is quite unequal. When the upper surface is directed to the light, the lower averted, almost exclusively males are on the upper surface, while on the lower one almost only females. On one citrus leaf there were upper side: 48 males, 0 females,
31 first larvae
lower sides: 0 males, 42 females.

This preference of the male larvae for the upper side of leaves is also very marked in Palestine.

Development: The crawler settles down, as a rule, during the first day after having hatched; in winter even during the first hour. During this period the larva remains unfed. After 2-5 days it dies, if it has not settled by that time. The crawler stage is greatly decimated by the wind, which blows off large numbers of crawlers, of which only a small fraction is transported to a place suitable for settlement. The crawlers are rather sensitive to drought, high and low temperatures, rains, etc., all these factors reducing their numbers to a minimum. Mortality levies its heaviest toll from this stage and it is fortunate for the scale that the duration of this stage is short. The crawler is positively phototactic, but more so the male than the female. This is a physiological difference between the sexes, which at this stage cannot be distinguished morphologically. The settlement of the crawler is at once recognized by a heavy wax secretion, which forms the ,,white cap". Mortality is often still high at this stage, but it seems to be conditioned mainly by nutritional factors. The ,,white cap" is merely a cover of the yellow chitinous layer below, which is constructed immediately afterwards. 5 to 15 days later the first moult occurs, larval legs and antennae disappear and pygidial structures begin to develop. After 6 to 60 days a thin white membrane appears on the ventral side of the female and the second moult begins. During the following preoviposition period the insect grows and changes its colour from greenish to deep yellow. Eggs mature at this period successively, from 0 to 20, 40, 60, 150 eggs in the abdomen, which lasts about 25-30 days in summer.

The development of the male is much shorter than that of the female, despite the fact that the male passes more moults. Adult longevity is restricted to 2-3 days only.

The total life cycle in summer is less than 40 days, as compared with 90-120 days for the female. In winter it may last as long as 160 days.

LIFE - HISTORY IN PALESTINE.

The life cycle at Akko (Acre) and Migdal has been described in detail by Schweig and Gruenberg:

,,At *Akko* the scale raises 3—4 generations, which are distinctly separated or only very slightly overlap in the later generations. The number, 3 or 4, depends chiefly on the average temperature of the spring and summer months. If the increase of temperature is normal and the summer heat starts in August (as usual in normal years), the spring and summer generations end late, leaving no time for raising a second summer generation, and the generation raised at the beginning of September has sexually mature females in the preoviposition period at the beginning of December or somewhat earlier. Few of these females (of the earlier hatchings) begin to oviposit, but the majority of them find in the second half of December such climatic conditions as to be compelled to pass into a long winter diapause. The progeny of the females which start oviposition at the end of November all perish.

In a year in which the wave of warmth suddenly rises after the first week of spring or early in summer (end May, beginning June), as was the case in 1934 and 1935, the periods of the spring and summer cycles are shorter, and the great heat allows an additional late summer generation.

The spring generation starts with an oviposition of females, descendants of the late summer generation. These females come from two lines of development: a) individuals which entered into the preoviposition period before the beginning of the rainy season (December); and b) adults which were in December in the period after the first moult and passed through the second moult into the preoviposition stage only in the beginning or middle of March. Their numbers, however, are so low, owing to the influence of winter conditions and to the high mortality of settled larvae in March and April, that it is sometimes difficult to fix the exact date of the start of the spring generation. One fails in many cases to get an artificial infestation at this period under field conditions.

The spring generation is very limited and has a high mortality both in settled larvae and nymphs. The summer-brood (July) is stronger and has a very low death rate in all stages. This enables the insect to strengthen its hold on the host plant, to spread more widely, and to supply additional and more copious migrants to fruits which are just forming (the first migration to fruits start with the spring generation in the second decade of April). The fruits have two generations of their own, the late summer and autumn-winter generations, strengthened during all this period by additional migrants from the leaves.

As the low death rate and steady increase in number continues throughout the summer (late summer generation), this part of the yearly life-cycle has to be regarded as the most dangerous to trees and fruits. At the end of summer (beginning of October) there is an abundant population of the black scale on the host plants, and the species can now face the adversities of the winter climate without fear of being totally destroyed. The mothers of the autumn-winter generation continue to oviposit up to (and a bit past) January, but the settling of crawlers has never been observed after 15th December at the earliest, or 1st January at the latest. The mothers usually die after the first half of January. The few surviving up to March have dead, brown eggs in the body and dead eggs and crawlers under the shield.

Owing to the permanent presence at *Migdal* of ovipositing females all the year round, causing overlapping of generations, it becomes very difficult in late summer to distinguish the numerical succession of the generations. This overlapping is due to the fact that the oviposition period of a single female may last from some weeks to some months, which explains why during the summer months all stages of Black Scale are met with. There are five full generations of Black Scale during the year at Migdal,'' which lasted in 1933/34, as follows:

year	1.	2.	3.	4.	5. generation (beginning)
1933	20.IV.	20.VI.	10.VIII.	4.IX.	20.XI.
1934	20.III.	30.V.	20.VII.	10.IX.	25.X.

Klein observed at Dagania that the low temperatures in January/February were responsible for the suppression of the fifth generation, which resulted in the beginning of the respective generations, at the following dates:

generation	1.	2.	3.	4.	
date	early V.	7.VII.	1.IX.	15.XI.	1934

The number of generations (A) was in:

year	1933	1934	1935	1936
no. of generations	4.4	4.3	4.7	5.6

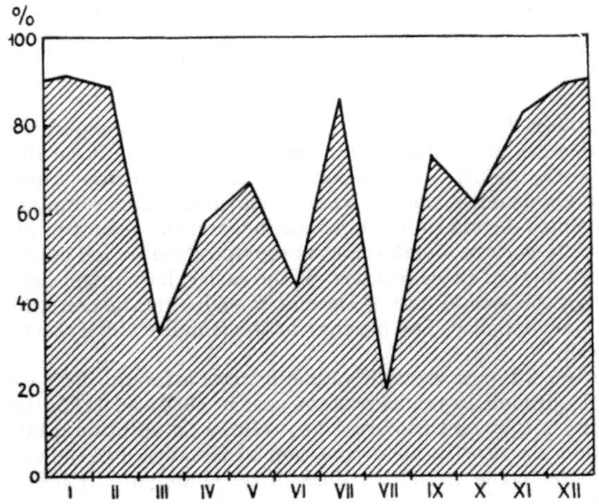

Fig. 74.
Monthly percentage of larvae (white) and adults of the Egyptian Black Scale (*Chrysomphalus aonidum* L.) at Dagania (1933).

It is doubtful whether the late appearance of the first generations in 1934 was really due to retardance of development or, more probably, to extremely heavy mortality of this generation. Whereas the first generations are easily recognisable (fig. 74), the later generations are heavily overlapping.

PHYSICAL ECOLOGY.

1. *Temperature and development:* It is difficult to construe the life-history from a simple hyperbola.

We shall first analyze the following figures obtained from the Jordan valley and Egypt.

1. Dagania	1. generation	26.1° C	65 days	Klein
	2. ,,	27.4	55	
	3. ,,	23.8	70	
	4. ,,	15.3	170	
		c=10.6° C	th.c.=924 days-degrees	
2. Migdal	winter generation	16.0° C	160 days	Schweig & Grunberg
	spring ,,	24.5	60	
		c=10.9	th.c.=816 days-degrees	
3. Cairo	1. generation	19.2° C	97 days	Priesner
	2. ,,	26.1	50	
	3. ,,	25.6	52	
	4. ,,	15.8	160	
		c=11.2° C	th.c.=766 days-degrees.	

These hyperbolae are fairly consistent with one another and give an average of
development threshold (c) = 10.9° C
thermal constant (th.c.) = 835 days-degrees.

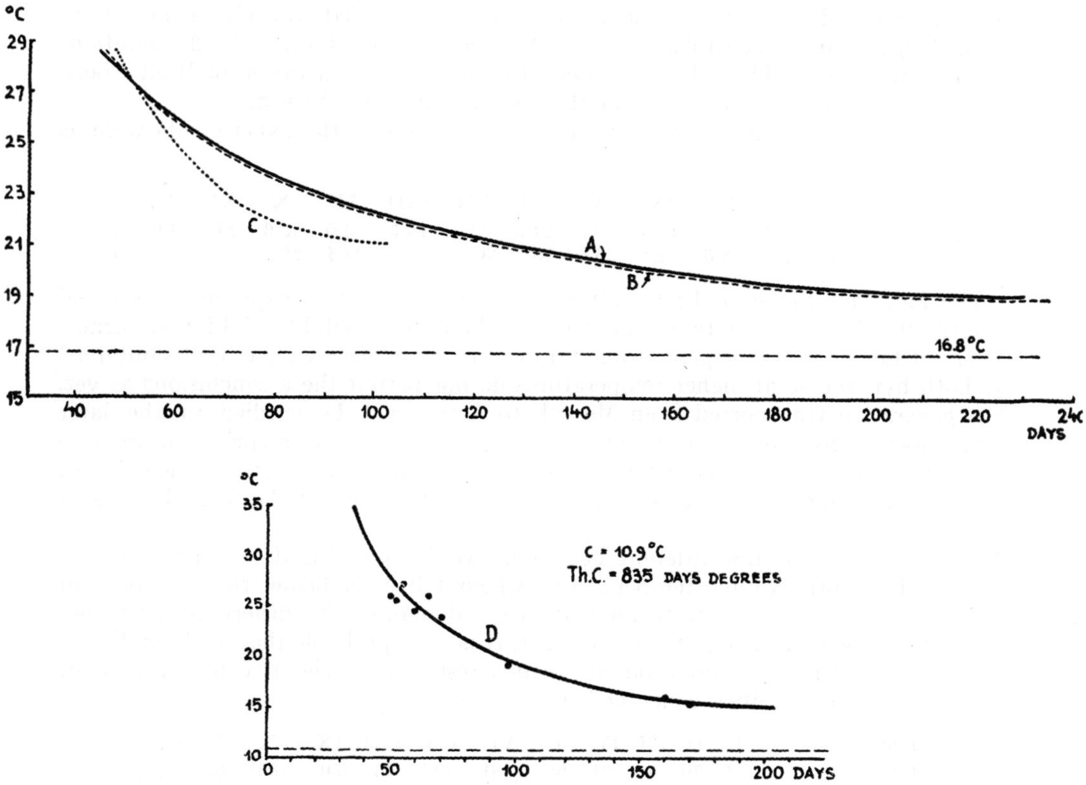

Fig. 75.
Comparison of temperature dependency of shortest development of *Chrysomphalus aonidum* L. at Acre and in the Jordan Valley: A. Empirical data at Acre; B. hyperbola (A) calculated for Acre (c = 16.8°C); C. hyperbola (A) calculated for the Jordan Valley (c = 10.9°C). D. Temperature dependency of shortest development in the Jordan Valley and in Egypt.

The data obtained by Schweig and Gruenberg at Acre (1933/35) deviate decidedly from this hyperbola and follow closely another one, based on the following constants:
development threshold (c) = 16.8° C
thermal constant (th.c.) = 540 days-degrees
The empirical data are as follows:

temperature	empirical data	c=16.8° C	c=10.9° C
17.2° C	190 days	1350 days	133 days
19.1	230	235	102
22.7	92	92	71
23.5	81	81	66
25.2	66	64	58
26.2	57	57	55
27.6	50	50	50
28.7	44	54	47

The deviation is especially marked below 25° C. There is no doubt that this difference is a significant and a real one. Schweig and Gruenberg have pointed out that this behaviour could be interpreted as that of different thermal races. They support this theory by the results of acclimatisation experiments. The development of the first transported generation (Migdal-born crawlers settled on citrus in Acre) lasted longer than the control generation (Acre-born crawlers at Acre), all conditions of the plants being alike. The second generation (second generation of Migdal born crawlers, born at Acre) behaved exactly like the indigenous Acre strain.

The monthly average temperature during the years of the experiments were as follows:

month	I	II	III	IV	V	VI	VII	VIII	IX	X	XI	XII
Acre	12.3	10.5	15.7	17.4	20.5	24.3	24.5	26.5	25.0	22.6	20.0	14.0° C
Migdal	13.3	11.7	18.5	21.8	26.5	29.7	31.0	30.8	28.4	25.3	22.2	15.3° C

The fact that the second generation of the acclimatisation experiment behaved exactly like the normal Acre strain, would exclude the possibility of different strains if correct for the following generations. However, the relatively close results obtained by both hyperbolae at higher temperatures do not permit these conclusions as yet. The generation transported from Migdal to Acre must be studied in the later generations, especially during autumn, winter and the following spring, in order to ascertain, whether this equal development is only induced by the coincidence of both hyperbolas at higher temperatures or by „acclimatisation" of the Migdal-scales at Acre.

In the latter case the different behaviour would be induced by environmental factors: The lower winter temperatures, which fall even below the development threshold at Acre in February could induce a diapause. The differences at higher temperatures would then grow insignificant. It is more probable that air-humidity or other factors such as sunradiation, are mainly responsible. The monthly average air humidity in the localities concerned is as follows:

month	I	II	III	IV	V	VI	VII	VIII	IX	X	XI	XII
Acre (1934)	57	70	54	71	68	(68)	(69)	66	(60)	(66)	56	74%
Dagania (1934)	68	77	57	76	57	59	59	61	56	62	57	79%
Cairo (normal).	80	73	69	61	55	56	61	67	73	75	78	81%

The lower air humidity at Dagania and Cairo during summer act slightly retarding on the development, the higher humidity during winter stimulating. The reverse is true for Acre. The further development of these experiments will be followed with interest.

Still more puzzling are the data published by Balachowsky from Algeria. He evaluates the seasonal development as follows:

1. generation	54 days	development
2. ,,	54 ,,	,,
3. ,,	54 ,,	,,
4. ,,	210 ,,	,,

However, only the second generation has been studied closely and its duration ranged from 44 to 66 days. He remarks that the first and third generation have approximately the same length of development. Should these indications prove to be correct, they could be interpreted as the consequence of heliothermy: As long as the scales are exposed to sun-radiation, their speed of development is equal in spite of the differences in the macroclimate. The just mentioned data from Palestine and Egypt, however, show that this is not the case. It is rather certain that the seasonal history

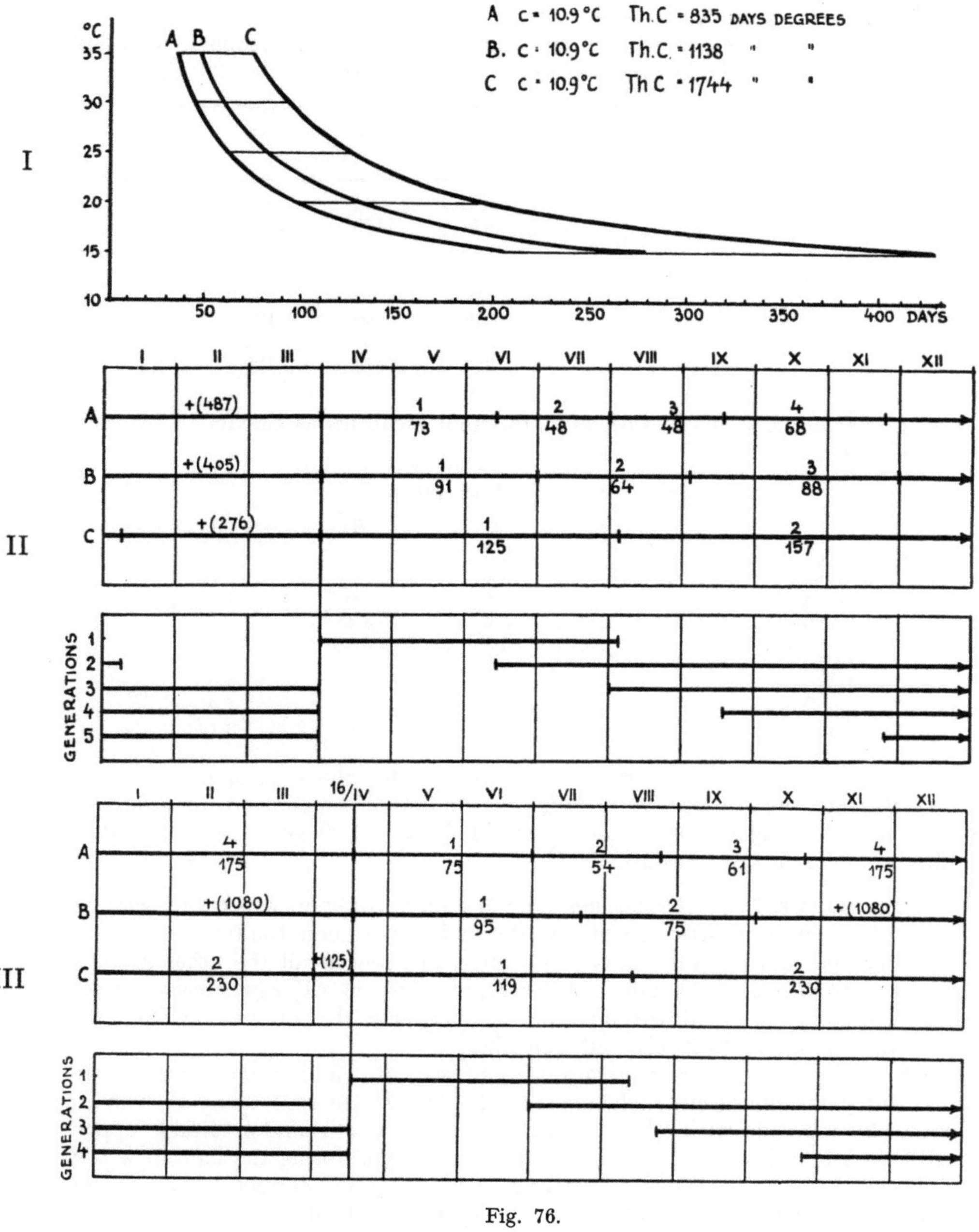

Fig. 76.
The life-history of the Egyptian Black Scale (*Chrysomphalus aonidum* L.) in the Jordan Valley I) the three hyperbolae, II. the life history as calculated from the hyperbolae and the sequence and overlapping of generations. III. The life history, sequence and overlapping of generations of *Chrysomphalus aonidum* L. as calculated for Tel Aviv.

of the Egyptian black scale in Algeria approaches closely that at Tel Aviv (75, 54, 61 days for the first, second, third generations respectively).

The three hyperbolae for the Jordan valley/Egypt scales, assuming an adult ongevity (not including the preoviposition period) of 100 days at 20° C, are:

hyperbola	A	B(⅓)	C
development threshold (c)	10.9	10.9	10.9° C
thermal constant (th.c.)	835	1138	1747 days degrees
duration of development 35° C.	34	47	73
30	44	59	91
at 25	59	80	124
20	92	125	192
15	204	278	420

The annual number of generations at different localities as calculated by these hyperbolae is as follows:

Number of annual generations

	A	B	C
Nice	2.0	1.5	0.9
Naples	2.4	1.8	1.11
Tel Aviv	4.0	3.0	1.8
Dagania.	5.6	3.4	2.1
Alexandria	3.8	2.9	1.7
Algers	3.3	2.4	1.4
Tampa	4.9	3.7	2.2
Los Angeles.	2.7	2.0	1.2
Colombo	6.9	5.2	3.1

It seems that localities which offer opportunity for the development of 3 A, resp. 2 B generations in one year, fulfill the minimum temperature requirements of *C. aonidum,* whereas those localities with 4 A, resp. 3 B generations fulfill the requirements for heavy infestation.

2. *Seasonal mortality and environmental resistance:* Different aspects of seasonal mortality have been recorded repeatedly. We shall now resume the topic on its own merits. Egg mortality is very heavy in winter, but low in all the other seasons. In Egypt up to 87% of eggs are dead in February (Priesner). All eggs deposited in our coastal plain after the end of November perish and even in the Jordan valley only few succeed in developing (Schweig and Gruenberg).

The mortality of the crawlers is heavy in all seasons. The few that hatch in the Jordan valley during winter perish almost all. In the other seasons, many of them are blown off by the wind with relatively little chance to finally find a suitable opportunity for settling down. Climatic mortality is very high during the hamsin winds. And not all survivors succeed in settling. Priesner states correctly that a high percentage (about 60%) die because of meeting with difficulties on the part of the plant (nutritional mortality). This nutritional mortality is not always as high as that. Klein observed that during the May-generation at Dagania, 10 females actually had an offspring surviving the formation of the scale of 20.8 larvae (maximum 54). This average is about 1/5 to 1/7 of the actual egg-production. 14-20% of the eggs laid started development successfully. Fumigation interrupted the observation of the following generations, but Klein assumes that the percentage of successful larvae is higher in the following generations, as no hamsin winds interfere. We shall shortly see that in the orange grove mortality is much higher and a 5 to 7 fold increase rarely occurs.

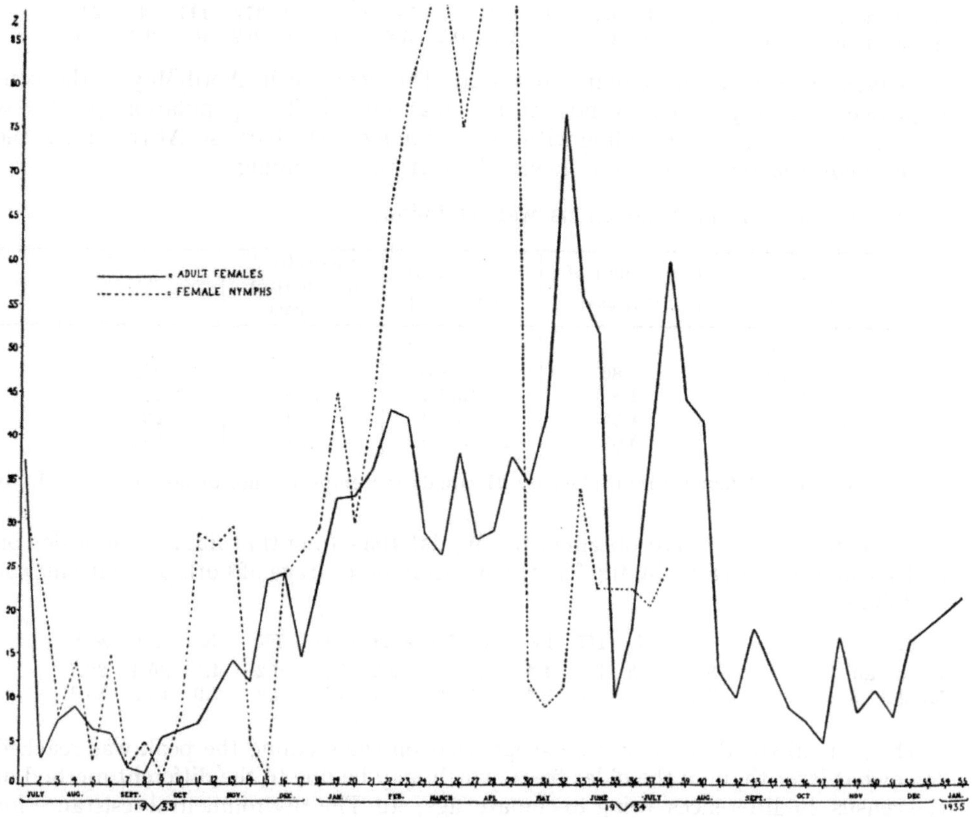

Fig. 77.
Chrysomphalus aonidum, mortality of adult females and female nymphs.
(From Schweig and Gruenberg).

The climatic mortality of nymphs and adult females is rather low (except for the influence of hamsin winds). The accumulation of scales of dead females during summer and their being washed off by the rains in winter tends to suggest a rather high mortality in summer. The natural mortality of nymphs and females at Acre is illustrated in fig. 77. The different peaks are induced: February to April mainly by hamsins and following accumulation of dead scales. In April predators (especially *Chilocorus*) made themselves felt. In May and July 1934 hamsin winds were responsible. The high premature mortality of adult females reduces the average number of eggs laid by the females of that generation rather considerably.

We shall now turn our attention to the actual fluctuation of black scale populations in the Jordan Valley. It may be pointed out beforehand that all these counts are incomplete as they do not consider some or all of the following factors: The annual elimination of the infested fruit from the orchard, the differential fertility of black scale on fruit and leaf, the heavy leaf drop and its parallel elimination of many scales from the orchard, etc.

Klein infested four orange-seedlings with black scale at Dagania, in May 1933. The living scale population throughout one year was as follows:

month	V.1933	VI	VII	VIII	IX	X	XI	XII	I	II	III	IV	V
no. of scales	76	44	65	74	184	135	784	669	413	372	111	41	24
quotient of increase	1.0	0.6	1.5	1.2	2.5	0.7	5.8	0.9	0.6	0.9	0.3	0.4	0.6

Leaves dropped from January to March. The decrease in April/May is the consequence of heavy hamsin-winds during that period. The population peak was reached in November, coinciding with the beginning of the harvest. At the end of the experiment the population was lower than at the beginning.

The increase of the generations was as follows:

generation	actual population	quotient of increase	theoretical increase *)	theoretical quotient of increase	°C	% R.H.
Start	76		76		—	—
1. generation	65	0.86	861	11.3	26.1	58
2. ,,	184	2.83	15.372	17.8	27.4	61
3. ,,	784	4.26	232.428	15.1	23.8	62
4. ,,	24	0.03	3.855.978	16.6	15.3	65

*) empirical percentage of females multiplied by 21 equals the empirical no. of larvae settled by one female.

These results and those obtained for Migdal 1934 (monthly number of scales on 20 leaves) are illustrated by fig. 78. The monthly percentage of both populations was as follows:

month	I	II	III	IV	V	VI	VII	VIII	IX	X	XI	XII
Daganiah	3.9	12.5	3.7	1.4	2.6	1.5	2.2	2.5	6.2	4.5	26.4	22.7
Migdal	8.2	7.8	2.3	2.4	2.1	4.1	5.1	7.7	13.8	17.9	15.2	13.1

The general trend is the same, except, that on the seedling the peak was reached somewhat later than on the older trees, which may be due to the different method of the census, to differences owing to climate, age, etc. The environmental resistance for these generations (e = 150, m = 4, f = 6, Ro = 98, 867; beginning of generation in brackets) is as follows:

generation	I	II	III	IV	V
Daganiah.	99.03(V)	96.78(VII)	95.17(IX)	99.97(XI)	
Migdal 1934	98.06(IV)	97.87(VI)	97.97(VIII)	98.75(IX)	99.82(XI)

The environmental resistance is lowest in summer, highest in winter.

Temperature is undoubtlessly a deciding factor in the mortality of *C. aonidum*. Both moderate cold as well as excessive heat are destructive to the species. The optimum range lies between 25-30° C, perhaps even some still higher degrees included. The high mortality at moderate cold (10-15° C) is remarkable. At temperatures round and below the development threshold (10.9° C) both, growth and reproduction, are interrupted. However, this hibernation is interrupted, immediately when the temperature rises. Regions with a strong daily and seasonal range of temperatures (steppes, deserts) and moderate climates are entirely repulsive to the species.

With regard to humidity opinions are not uniform. Balachowsky regards the species as extremely stenohygr. „The drought arrests its development. It can, therefore, be met only in places, where the air humidity is very high during summer and winter. These conditions are realized in Algeria in some privileged stations of the litoral, where the average summer temperature is 27-28° C, that of the winter 12-13° C and where humidity often reaches saturation."

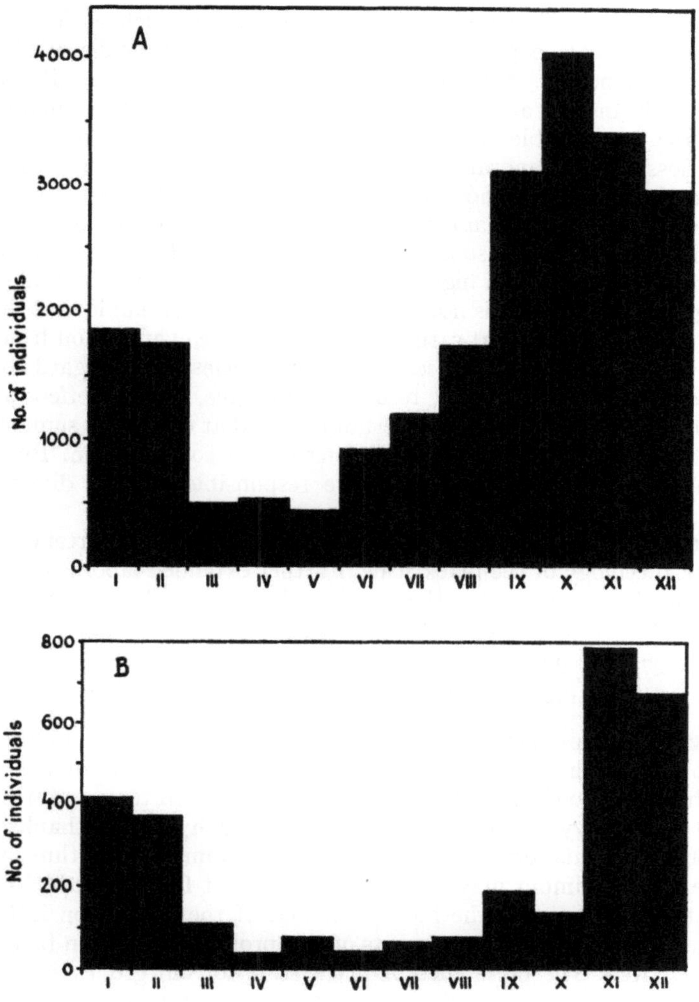

Fig. 78.
The seasonal population trend of *Chrysomphalus aonidum* L. in the Jordan Valley. A. Scales per 20 leaves at Migdal (1934); B. Scales per seedling at Dagania (1933).

Priesner stresses that Upper Egypt has probably not been invaded by the black scale because of its low humidity. Klein regards humidity lower than 60% as unfavorable to the species.

In contradiction to these statements are those of Schweig and Gruenberg: „Atmospheric humidity was regarded as of subordinate importance, as it was thought that beneath the scale there must be always a more or less unchanging amount of humidity higher than that of the environment and uninfluenced by it, except in cases of the hamsin winds. The secondary importance of air humidity is, on the other hand, indicated by the fact that the higher humidity in the coastal plain does not produce an increase in the number of generations there, which are fewer than in the Jordan Valley, where a constant low relative humidity prevails throughout the hot summer months. The area of black scale infestation is not limited to the short radius of Lake Tiberias, as areas beyond its tempering influence, such as Yavneel, have a considerable black scale population besides a very heavy incidence of red scale."

The facts at hand are not sufficient to settle the question. But the general theory that scale insects are not influenced by environmental humidity beneath their scales seems untenable. The fact that plant lice and scale insects are able to compensate losses by evaporation rather quickly by sucking on the host-plant releases them from the influence of drought to a certain degree. However, that is only possible as long as the plant maintains its normal circulation even at drought periods of the day or of the season. The scale certainly contributes, again to a certain degree, to the stabilisation of a higher humidity than the environmental one, as long as the drought is moderate. It is not certain and even not probable that the humidity optimum of the black scale is at saturation. A moderate evaporation from below the scale may even be favorable or necessary to the species. With regard to mortality, a mild winter may counterbalance, to a certain degree, the bad effect of a not too excessive drought in the summer. It is certainly true that the higher summer humidity does not increase the number of generations in the coastal plain. But it is quite probable that the humidity conditions are responsible for the difference in the hyperbolae, with their different development thresholds.

For the present, the facts rather tend to stress the primary importance of humidity for the physical ecology of the black scale. Further extended laboratory experiments at constant humidities can solve the problem definitely. The months of increasing populations have the following average climate:

27.4	23.8	23.8	27.8	26.6° C
61	62	61	60	59 % Rel. Hum.

These low humidities certainly do not indicate that higher humidities were less favorable, but only that they were not available in the areas of observation.

Despite the rather low, but not too low, humidities, high temperatures favour the building up of a heavy black scale population, which is able, thanks to its size, to pull through the unfavorable winter in sufficient numbers and thus maintain the species. The vital optimum may be put tentatively at 65-80% R.H. and 25-32° C.

How far the conditions of the host plant and of the vegetation influence black scale remains to be studied. Some aspects of this problem have been tackled already; others exist as well. The preference of black scale for old groves is very obvious, with the reverse behaviour of red scale. Just which factors are responsible for this preference have not yet been studied (higher humidity, wind protection, shade, etc.).

Dispersal of Black Scale: C. aonidum spreads only through the crawler. In spite of the fact that the crawler may run up to 10 mm per minute at optimal conditions, it tends to settle in the neighbourhood of the mother scale, only very rarely extending its march to 200-500 mm before settling down. Its moderate positive phototaxis leads it to the younger leaves and fruits. The Palestine Government studied factors affecting the spread of black scale in some detail in connection with legislation on interne quarantine.

Seedlings are doubtless the main source of importation of *C. aonidum* into new districts. Even careful inspection before transport is not a full guarantee against its being infested. Scales living on peels, survive for 6-17 days after the fruit has been peeled, whereas those on fruits survived in spring from 50 to 70 days at Acre, in the summer from 20 to 30 days at Migdal. The crawlers first born after the fruit has been picked, often reach maturity and their larvae start to settle down. The picked fruit is dried up or decayed by the time the second generation is fully developed. Infested fruits put under clean trees gave rise to infestation only when the fruit came into direct contact with the tree. „The fruits are little (if at all) responsible for the spread of black scale from one area to another."

Dropped infested leaves may be a more important source of infestation, espec. in nurseries and dense old groves, where the females often survive for 1 to 3 weeks.

In fermenting manure the insects on fruits or leaves are not able to survive for long, the hatching crawlers have no possibility for developing and die. After a period of some weeks, during which no fruits and leaves have been added, the manure seems to be safe (Schweig and Gruenberg).

A tree with moderate infestation remained untreated during a fumigation campaign in the spring of 1934 at Migdal. In December 1934 the 8 surrounding trees were carefully counted, in order to study the manner of spread, with the following result (Schweig and Gruenberg):

No. of tree	direction from source of infestation	No. of fruits	% of infested fruits	remarks
1	E	93	61	
2	S E	157	57	
3	S	178	87	
4	S W	187	27	
5	W	168	69	
6	N W	172	71	
7	N	72	7	(young tree)
8	N E	152	56	

If this one experiment allows a generalization, the conclusion would be justified that reinfestation occurs in concentric circles (not excentrically), despite the strong prevalence of western winds (20-30 hour/miles during 8-10 hours a day) during all summer. But it may easily be that the first infestation was excentric and the interval of 3-4 generations was too long to permit the recognition of these conditions in the immediate neighbourhood. At least 3-5 further concentric circles of trees in the centrifugal direction should have been counted. The importance of wind for the infestation as carrier of crawlers is by no means discarded by this observation.

Priesner points out that infestation spreads easily from tree to tree in dense plantations, where the leaves of one tree touch those of its neighbour and that in plantations with sufficient space infestation is greatly reduced.

Damage: Damage is caused mainly to citrus, which is the favoured host of *C. aonidum.*

The damage is manifold and is by no means restricted to the devaluation of fruit infested with scale insects: defoliation, drying up of young twigs, poor blossom, premature dropping of heavily infested fruits and fruits undersized and lacking juice are common occurrences in heavily infested citrus orchards. In extreme cases the tree is even killed.

Hall describes the damage done in Egypt: „A badly attacked orange garden produces a crop worth approximately one-tenth of the crop produced by a garden of the same age in good condition." In Egypt, it has caused the uprooting of large tracks of orange gardens, before fumigation became obligatory.

Balachowsky reports from Algeria: „From the agricultural point of view *C. aonidum* must be considered as one of the most injurious (if not the most injurious) scale of citriculture, which exist now in Algeria."

From Palestine the following statements may be quoted: „It is no uncommon spectacle to observe the golden yellow colouration of a ripe fruit to be almost totally obscured by the adhering, scab-like incrustations formed by the innumerable, overlapping colonies of the pest. The leaves may be similarly covered." (Bodkin).

„The most prominent of all damage done is defoliation, which was observed twice during the year at both Acre and Migdal (March and October). In some cases the leaf-fall becomes so great as to cause the drying up of the tree. The defoliation for March (Migdal) and for October (Acre) is sometimes so severe that it is impossible to find leaves with a large number of scales upon them. At Migdal the leaf-fall coincides unfortunately with the start of the development of the new scale generation after winter, and soon the new vegetation becomes reinfested by Black Scale migrating from the remaining leaves and from other plants. The October defoliation at Acre is accompanied by an extremely heavy drop of fruit, so that trees are standing entirely bare at the beginning of November. Blossom is very scarce on the defoliated trees, the formation of new leaves absorbing all the creative forces of the tree." (Schweig and Gruenberg).

These statements are certainly not exaggerated and lead to the conclusion that black scale is to be regarded as the most serious citrus pest of this country wherever it has established itself sufficiently. The fruit looses its marketing value as a result of even a light attack, with but a few scales visible.

Priesner describes the damage on various hosts: „Strong attack on *Ficus nitida* causes yellow star-like to net-like markings, the yellow lines then flowing together to spots. On citrus as well the yellow lines are visible, and the whole place is becoming spot-like yellowish if the insects are gathering nest-like around it. On date palm there is little change in the colouring of the leaves owing to the hardness of them. The same is true for *Eucalyptus*, a tree known to be very resistant against insect attacks. On stems of mango I have observed blackish spots and stripes in case of severe attack of *C. aonidum*."

It is noteworthy that whereas the leaves grow chlorotic around the scales, the infested place of the fruit, just beneath the scale-insect, remains green, while the rest of the fruit turns yellow (Balachowsky). All these discolorations are due to the toxic effect of the saliva, injected into the plant-tissue during sucking.

3. *Black scale versus red scale:* Schweig and Gruenberg report interesting observations on the interference of infestation by *A. aurantii* and *C. aonidum*.

„It is of interest to note that Black Scale and Red Scale zones may be situated at a distance of only several kilometres from one another, and in every one of these zones a specific Coccid population remains predominant. Thus, for instance, on the southern part of the Lake of Galilee there is vigorous red scale development and only isolated cases of black scale; on the contrary, the groves of the northern shores of the lake, about 20—25 kilos distant, are badly attacked by black scale and not or rarely by red scale, although it was proved that Red Scale is able to develop there.

A number of citrus seedlings were artificially infested in the insectary at Migdal with red scale in April 1934, and put between seedlings badly infested with black scale. During the first 2—3 months the development and reproduction of red scale was good and a multidude of red scale soon covered leaves and stems of the infested seedlings. Though the developmental cycle of red scale is somewhat shorter than that of black scale, the red scale is less prolific and the mortality of its crawlers and young larvae was higher than that of black scale during the hottest months at Migdal, thus resulting in a more restricted incidence of red scale as compared with black scale. As a consequence of the double attack of red and black scale, leaves of the infested seedlings

Fig. 1.

Fig. 2.

Fig. 3.

Fig. 4.

XVII. Damage caused by *Chrysomphalus aonidum* L. to citrus. 1) Two grapefruits: on the left clean, to the right badly attacked; the fruit is undersized and deformed in a characteristic manner; 2) two lemons; 3) two oranges, both clean and heavily infested; 4) three orange leaves: a) clean, b) and c) infested; the latter being covered with a multitude of young larvae. (from Schweig and Gruenberg).

dropped. Seedlings which developed new leaves after defoliation became immediately infested with black scale. In this way during August—September the black scale succeeded in establishing itself on seedlings previously infested with red scale, the latter disappearing almost entirely from the seedlings, though it remained on the stems and twigs for several months. The second lot of seedlings infested with red scale at end March 1934 and placed in a corner of the insectary not very near to the black scale, suffered less from it, but the same process took place later (February 1935)."

In the citrus nursery and in the experimental plots at Acre red scale did not develop, being rare in the other parts of the stud farm, though no control measures were undertaken.

At Jaffa black scale was certainly able to grow to considerable numbers in some old orange groves which had been previously heavily infested by red scale. It is impossible to judge as to the probable continuation of this process, as fumigation at more or less regular intervals interferes with the natural „struggle" of both species. Both species fill the same ecological niches in the citrus orchard and the ecological requirements of both species must be known a great deal better before one can predict which of the two species will predominate in a given locality, resp. climate.

Klein has made an independent comparison between the damage and danger of both scales in the Jordan valley (Dagania 1935). The number of annual generations and the sex-ratio are nearly equal. *C. aonidum*, however, is about 2 to 3 times as fertile as *A. aurantii*. The intensity of the damage is very different. Whereas red scale leads merely to the exclusion of infested fruit from export, heavy leaf drop, fruit drop and shrinkage of fruit is often caused by black scale, in addition to the export discrimination. Furthermore, black scale is much more resistant to fumigation than red scale, under local conditions. One fumigation always gave satisfactory results with regard to red scale. Trees fumigated against black scale showed reinfestation on fruits and leaves quite commonly through all the years, as early as two months after the treatment. It is quite clear, that, from every point of view, black scale is uncomparably more dangerous and injurious than red scale for the citriculturist.

CONTROL.

1. *Biological control:* Whereas in humid tropics or in subtropical countries with summer-rains (Florida, Japan, Barbados) black scale is heavily infested and perhaps even controlled by fungous diseases, as the Pink scale fungus (*Microcera fujikuroi)*, no such control can be hoped for in our semi-arid climates. The occasional fungi observed in Palestine and Egypt on black scale lack any epidemiological importance (at least during summer). Among the natural enemies, the two most prominent are *Chilocorus bipustulatus* and *Aphytis chrysomphali*, which are dealt with in a separate chapter. Both are practically absent during summer and are useless for preventing the seasonal building up of the scale population. *Aphytis*, in addition, restricts its attacks to male pupae. Some other predators of definitely minor importance are a beetle, feeding on the eggs of black scale, larvae of a *Chrysopidae* and *Coniopterygidae* and undetermined thrips and mites.

The introduction of *Comperiella bifasciata How.* into Palestine is being prepared by the Govt. Entomologist and the results of this experiment will be followed with greatest attention.

Other aspects of biological control (health condition of plant, natural immunity) have been dealt with before.

2. *Technical control:* Fumigation with hydrocyanic acid gas is the only effective control measure known so far in Palestine.

Some series of experiments with regard to spraying did not yield promising results. In the course of the last years spraying with white oils was tried twice, but the results were only a 50-65% mortality of the scales. Klein obtained once a 95% kill at Jaffa in summer 1935, in an experimental spray, but even with this favourable result the tree was fairly heavily reinfested one month later. The high fertility of *C. aonidum* requires a higher kill, surpassing 97%, in order to obtain clean fruit.

Schweig and Gruenberg have published recently similar results, ranging from 60 to 70% kill. They ascribe this low kill mainly to the „frequent occurrence of cobweb on the lower side of the leaves, under which the scale remains sheltered." In our experience the percentage of scale protected in this way is rather small. The fact that black scale is relatively more abundant on the lower side of the leaves than red scale and is therefore less easily accessible to the spray probably plays a more important rôle. However, after all these deductions, black scale must be regarded as better protected against oil-sprays than is red scale. Spraying with white oils (2½%) should be applied, wherever single trees in a fumigated grove show unsatisfactory results. One to two sprayings should be carried out as a supplementary treatment. Experiments with other chemical treatments have been made by Balachowsky.

Experiments conducted by the Shell-Company since 1937 with white oil are rather encouraging to continue with experimentation into spray-materials.

The different system of fumigaton as well as the history of fumigation are treated separately. The pot-method is now more and more replaced by the application of Cyano-fumer and liquid hydrocyanic acid gas.

The experiments were entirely in the hands of the Government Entomologist and the following results are communicated (Schweig and Gruenberg):

Fumigation tests in 1933 at Migdal, in 9 groves:

No.	fruit	Degree of infestation	date	Fumigation method	% of kill	Fruit picked	% of cull due to scale
1	Lemon	Medium	10.VI.	Sulphuric acid (pot method)	100.0	25.IX.	5
2	Grapefruit	,,	31.V.	Cyanofumer	97.7	3.X.	10
3	,,	Medium-Heavy	4.VI.	,,	100.0	3.X.	3
4	,,	,,	4.VI.	,,	100.0	3.X.	1
5	,,	Medium	10.VI.	Sulphuric acid	100.0	8.X.	4
6	,,	,,	25.VI.	,,	98.6	22.X.	11
7	,,	,,	21.V.–VI.	I Cyanofumer } II Bisulphate }	100.0	13.XI.	1 } grove refumigated
8	,,	,,	22.V.–VI.	,,	100.0	10.XI.	2 } grove refumigated
9	,,	Heavy	23.V.	Bisulphate	28.4	14.XI.	27

In 1934, in many groves the fruits were quite clean till October. The cull due to scale insects rarely exceeded 5% of the total yield during the picking season. In regularly fumigated groves the cull should not exceed 5% in the future, as compared with 70-80% in non-fumigated groves.

„Very good results were obtained with winter fumigation both at Migdal and Haifa. At Migdal during the first eight months after fumigation the trees remain practically clean, and at the end of the year they were moderately infested. The mortality of scales in winter fumigation was near 100%, as observed twice (December and January).

In Haifa the effect of winter fumigation was kept under continuous observation during two years. Fumigation was done on 26th December 1932, and was followed by heavy leaf-fall. During the first year after fumigation (1933) the trees employed their sap for the formation of new leaves and had no buds. The leaves were clean, with exception of those on three badly fumigated trees in a corner of the garden. In July 1934 the reinfestation began; this was going on slowly up to May 1935. The fruits of the season 1934 were quite and mandarine trees were observed to be clean even in June 1935."

Fumigation is introduced practically as an annual routine in all infested districts and an attempt to fumigate and spray with white oils in alternate years proved impractible in the Jordan Valley.

The results of fumigation are not yet generally up to the standard set by these experiments on the part of the Government. Our observations in the Jordan Valley (Klein) have shown that a 97% kill of black scale is the normal kill, which cannot be regarded as satisfactory. Two months after fumigation (VI/VII, resp. VIII/IX) quite an important infestation reappears in the fruit bearing groves, especially on the

Fig. 79.
Orange-tree partly killed by *Chrysomphalus aonidum* L.
(Petah Tikva IV. 1944. By courtesy of the Shell Co.).

fruit. A strict and impartial supervision of the commercial treatments must be established and a minimum kill of 98.5 - 99.0% be set as standard. It seems that moulting individuals are somewhat more resistant to fumigation than others.

Experiments aiming at eradication are under way. Schweig and Gruenberg published the following preliminary results, which are now being carried out in the Acre district.

,,In the Jordan Valley two fumigations, one in March and the second in October, could be suggested as an eradication treatment. As the middle of March is the period of blossom and in October the fruit is beginning to turn yellow, this double fumigation can be applied only in the case of non-bearing groves. As to normal groves such treatment would be unreasonable, and the proper time for fumigation is as early in the year as possible, according to the size of the fruit, i.e. at the beginning of May if summer fumigation is practised; otherwise the tendency is towards winter fumigation. Then the percentage of kill will be good (the migration of crawlers to fruit begins in the third decade of April) because the fruits are still clean, and expenses for cleaning infested fruits from killed scales would be saved.

In the Acre area double fumigation applied during the summer can be suggested as a more or less full control of the pest. As the second summer generation is that which established the Black Scale in masses on the leaves and strengthens the population on the fruits, the control has to aim at breaking two links in the yearly cycle at critical points, i.e. in the preoviposition periods of the two summer generations and in the preoviposition period of the autumn generation. Thus two fumigations would have to be carried out in one year, which should leave the trees clean for a period of at least 3—4 years."

In the light of to days experience this conclusion seems to be rather optimistic. Its intrinsic value would be to lower the cost of fumingation treatment to that of spraying. Careful control of fumigation survivors showed that neither negative nor positive deviation is observed in the vitality of these, as compared with normal scales. Observations of fumigated groves demonstrated that the reinfestation of the orchard is due to increase of the survivors and, to a much lesser degree, only, to migration or transport of scales from other gardens and plants.

The recent improvements in spray materials, mainly based on the synergistic principle (combination of another insecticide with the white oil) have given promising results in experiments and control campaigns even against the Egyptian black scale.

The Plant Protection Service of the Palestine Governement has published in 1944 a leaflet describing the results of some years extensive experiments with three such oils, conducted in cooperation with the Shell Co. This commercial company is — unfortunately — the only institution in Palestine which disposes of a highly qualified specialist in insect toxicology, who's activity has much added to the speedy improvement of oil sprays in recent years in this country. Of the three spray materials tested A and B were soluble oils applied in a 1¼% solution, C a white oil emulsion in 2% dilution. The results against *C. aonidum* in 1942, for instance, were at Mikweh Israel (1000 to 7000 fruits counted):

Material	Month of spraying	Month of picking	Clean fruit (0—3 scales)	Slightly infested (4—10 scales)	Culls (over 10 scales per fruit)
A.	July	January	86	12	2
	August	February	82	13	5
	September	February	79	14	7
B.	July	January	77	16	7
	August	February	58	29	13
	September	February	68	21	11
C.	July	January	91	8	1
	August	February	78	17	5
	September	February	85	10	5
Controls.	—	February	1	8	91
	—	February	0	3	97
	—	February	0	3	97

Similar experiments with one of these oils were carried out by my coworker Dr. H. Z. Klein at Kinnereth (Upper Jordan Valley), where other conditions of control exist than in the coastal plain. The experiments were not very promising in 1944, but impoved, probably mainly due to an improved technique, and gave very satisfactory results in 1945. The natural infestation was heavier in 1944 too.

Treatment	Year	clean	slightly infested fruit	culls
1½% emulsion (two sprays: summer and winter	1944	59	31	10
1½% emulsion (one winter spray only	1944	41	38	21
	1945	87	13	3
1½% emulsion (one summer spray only).	1944	5	50	45
1% emulsion (one winter spray only)	1945	80	17	3
Controls	1944	0	0	100
	1945	10	54	36

Whilst the Governement series was on oranges, ours was on grapefruit. A number of interesting points are learned from these experiments. Whilst early summer is clearly the appropriate time for control in the coastal plain, winter control is the appropriate season for the Upper Jordan Valley. The experiments at Mikweh Israel show very definitely in each of the three spraying series the importance of early spraying and of early picking. Both experiments show that spraying has reached already a level where it can compete with fumigation, as soon as the price level will drop, especially in cases of light to medium infestation. No economic damage has been observed by the mentioned materials on the citrus trees.

Attention must be paid to another point. In red scale we were able, especially on the younger trees, to keep down the quantity of the spray material to up to 30 to 40 liter per tree. This standard could not be maintained against either Egyptian black or against mussel scale. These scales infest older trees which are larger and anyhow need larger quantities. They receive a drenching spray of 50 to 70 l. per tree. Another reason why these scales need a drenching spray is the fact that just early in the season of both these scales a considerable percentage settle beneath the sepals and are reached there only by a drenching spray. When these hidden scales are not killed, they will speedily build up large scale populations on the fruit from early summer to early autumn. The higher quantity of the spray certainly causes a better cover of leaves and fruits, especially if sufficient pressure and an otherwise adequate spraying technique is applied. The higher spray quantity does not necessarily mean a much higher oil deposit, considering the large amount of fluid which immediately drops to the soil. Quantitative research in this direction is highly desirable.

Whereas the Government will probably not consider the reintroduction of the internal quarantine line (cancelled in 1936), which prohibited strictly transport of fruits, etc. from the infested northern area to Southern Palestine, it should be well understood that every transport of seedlings from infested foci must be interfered with at all costs. The Plant Protection Ordinance with the supplements actually in force offers a sufficient frame for this purpose. It is the execution which must be enforced with rigid strictness. The public opinion of the citrus growing community must be aroused; its active support and cooperation are needed for the enforcement of these laws and their execution.

The recent introduction of *Comperiella bifasciata* and its result will be followed with great interest. So far, the results are extremely disappointing.

References.

A. Balachowsky, Bull. Soc. d'Hist. Nat. Afrique du Nord. **19**, 1928. pp. 156—180.
F. S. Bodenheimer, Schaedlingsfauna Palaestinas. Berlin, 1930. pp. 155—157.
F. S. Bodenheimer, Hadar **5**, 1932. p. 107.
G. E. Bodkin, Bull. Ent. Res. **16**, 1925, pp. 143—149.
W. J. Hall, Ministry Agric. Egypt., Bull. 45. 1924, 29 pp.
H. Z. Klein, Hadar **10**. 1937, pp. 3—8.
M. Matsuda, Dept. Agric., Research Inst. Formosa, Rept. 39. 1929. 97 pp. (Japanese).
H. L. McKenzie, Mycroentomology **4**, 1939, p. 51-77.
H. Priesner, Ministry Agric. Egypt, Bull. 117, 1931.
C. Schweig and A. Gruenberg, Bull. Ent. Res. **27**, 1936, pp. 677—713.
W. A. T. Summerville, Queensland Dept. of Agric., Div. Entom. Bull. 10. 1934. 99 pp.

E. THE CITRUS MUSSEL SCALE (*Mytilococcus beckii* (Newm.).

Synonymy: *) 1869 *Aspidiotus citricola* Packard
1869 *Coccus Beckii* Newman
1872 *Mytilaspis fulva* Targioni
1881 *M. citricola* Comstock
1902 *Lepidosaphes pinnaeformis* Kirkaldy.

Description: Scale of adult female mussel-shaped. Elongate, usually more ore less curved, narrow in front and broadened behind, with a flattened paler marginal area slightly convex. The scale transversely rugose, with numerous lines of growth. Color generally purplish-brown to pale yellow-brown with exuviae somewhat lighter, red-brown to yellow. The dorsal scale covers part of the lower side of the female, especially the anterior part. Ventral scale white and connected firmly to the substrat, wherefor the scales of dead females stick to the substrat a long time after their death.

The scales are up to 3.0—3.5 mm long and 0.9—1.2 mm wide. The first larval exuvia is 0.5 mm long, the second 1.0 mm.

Scale of male pupa generally straight, narrower and smaller than in the female with sides more or less subparallel. Color usually a little paler than that of the female scale. 1.2—1.5 mm long, 0.4—0.5 mm broad.

The *adult female* is elongate, narrow in front, widest at the first free abdominal segment. Pale yellowish white in color, with the terminal end reddish-fulvous. 1.3—1.5 mm long; 0.7 mm broad. Antennal rudiments consist of a small tubercle with two short, curved hairs. Anterior spiracles surrounded by a group of 6—13 glands, which are absent on the posterior spiracles. The lateral margins of the first four abdominal segments are strongly protracted and pointed. The three penultimate margins bear 4—5 marginal gland spines at the pointed lateral ends. Tubular spinnerets present in all free abdominal segments.

---

*) With regard to the rather heavy changes which the nomenclature of this species has recently undergone, the following excerpt from a letter from Dr. F. Laing of the British Museum, will be of interest:

„Bouché described his *pinnaeformis* from Cymbidium and it has never been satisfactorily established that this is the same as the Citrus Scale; it is possible that *tuberculata*, Leonard, described from orchids may represent the true *pinnaeformis*, and I thought that Green had suggested this but a hunt through his papers today has failed to trace the record.

The first two names available which every worker is prepared to accept are *beckii*, Newman, and *citricola*, Packard, both names published in 1869 but the former has priority by some months. I therefore use *beckii*.

Lindinger has pointed out that *Mytilococcus*, Amerling, has priority over *Lepidosaphes*, with genotype *communis* (= ulmi). I have checked the point and do not see how the adoption of Amerling's name can be avoided, and have decided to adopt it. The reference is Amerling: Lotos (Prag), 8,103, 1858. On p. 29 he cited the genus as *Mytillicoccus*, without giving a genetope, but on the later page he complied with the regulations and altered the spelling to *Mytilococcus*, which should be kept."

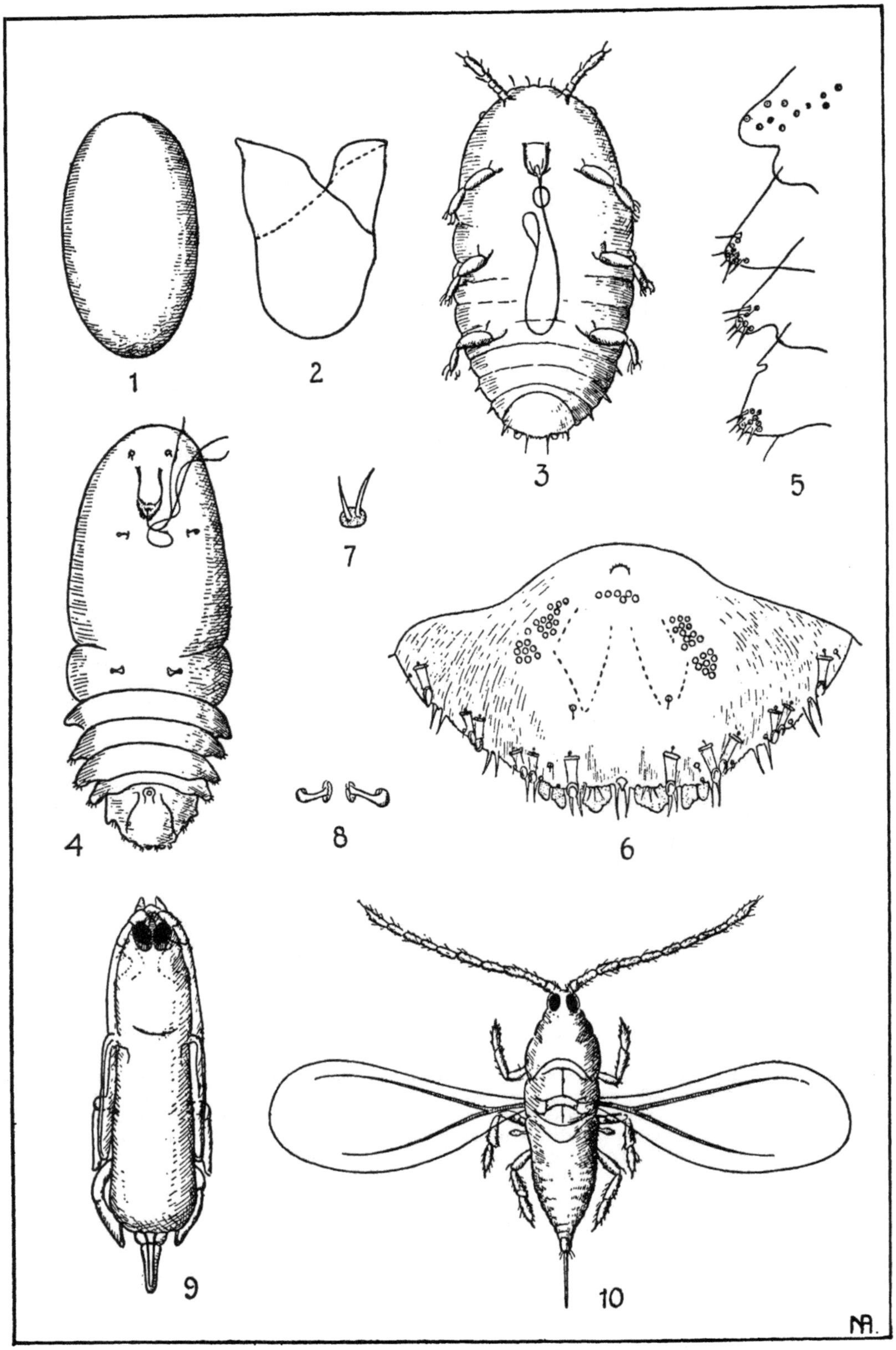

XVIII. The Citrus Mussel Scale (*Mytilococcus beckii* Newm.) I. 1) egg; 2) empty egg-shell; 3) neonate lerva; 4) adult female; 5) lateral margin of the three preanal abdominal segments; 6) pygidium of adult female; 7) antennae and 8) anterior spiracles of adult female; 9) male pupa; 10) adult male. (from Bodenheimer and Steinitz).

Pygidium rather large and broad and not heavily chitinized. The formula for its marginal structure is as follows:

$$2P_1,\ L_1,\ 1\text{—}2P_2,\ 2L_2,\ 2P_3,\ 2P_4,\ 2P_5.$$

The median lobes are large, broad, strongly projecting, crenulate and converging to form a triangular apex. Second lobes less prominent than the median, distinctly cleft to form two lobules, the median one more prominent, both with a more or less distinct lateral notch. The plates, one median and 4 lateral pairs, are simple and spine-like and longer than the lobes. Margin between the second and third pairs of lateral plates distinctly but irregularly serrate. 2 gland spines between the median lobes and 4 further pairs laterally on each side. Marginal gland orifices as follows: 1 between the median and second lobes, 2 between 3rd and 4th, 2 between 4th and 5th plates and 1 beyond the fifth plates.

Anal opening circular and much nearer to the base than the apex of the pygidium. Dorsal gland orifices in 3 irregular longitudinal rows, the inner row being the largest ranging from to 30 in number. Circumgenital glands arranged in five groups, ranging from

$$\begin{array}{cc} \multicolumn{2}{c}{4\text{—}8} \\ \hline 5\text{—}19 & 5\text{—}19 \\ \hline 5\text{—}16 & 5\text{—}16 \end{array}, \text{ in Palestine generally } \begin{array}{cc} \multicolumn{2}{c}{6} \\ \hline 12 & 12. \\ \hline 9 & 9 \end{array}$$

Eggs elongate, pearly white. 0.26 mm long, 0.11 mm broad.

The *neonate larva* elongate oval in shape, more than twice as long than broad. Antennae 6-jointed, the first joint being largest, the last one longest. Two small lobes protrude from the posterior abdominal margin with one spine exceeding their length on each side. Color white to pale yellow. 0.35 mm long, 0.16 mm broad.

The *second larval stage* of the male is still more elongate. The median lobes are large and broad at their base with a serrate margin. The second pair of lobes which is not conspicuous is strongly notched in the middle. Two long tapering plates between the median lobes and also two similar in size and shape between the first and second, and second and third lobes. Two other similar plates or spines found beyond the third pair. General color purple. In the earlier part of this stage the purple pigment is confined largely to the anterior lateral margins. The rest of the body, excepting the posterior tip which is brown, is white in color. There are no distinct eyes at this time, but later the purple granules coalesce into definite areas forming the two pairs of eyes which are conspicuous in the latter part of the stage. The dorsal and ventral eyes are continuous being connected by a narrow neck. They are a very deep purple. 0.8 mm long, 0.25 mm broad. (Quayle).

The second larval stage of the female is very similar to the early second male larva.

In the male *propupa* the antennal, leg and wing sheats are not very conspicuous, lying close to the body. The pygidial characters have disappeared and the posterior end is now truncate with no evidence of the stylus. The ventral eyes are large, very dark purple in color, and lie close together. The dorsal eyes are smaller and farther apart. Color purple. 0.8 mm long; 0.25 mm broad. (Quayle).

In the *male pupa* the antennal, leg and wing sheats are now conspicuous and more or less free from the body. Eyes purple, the ventral pair closer together than the dorsal. Color purple. Length, exclusive of stylus 1.1 mm; width 0.28 mm at widest part, which is near the posterior end. (Quayle).

The *adult male* is exceptionally delicate. Color white to pale yellow with purple pigment scattered about legs, antennae, stylus and thoracic shield yellow-brown. Eyes black. Wings hyaline. The long antennae are 10-jointed with intermediate joints longest. Length of body (stylus

excluded): 0.8—0.9 mm	wing expanse 2.0 mm
antennae 0.7 mm	width of body 0.3 mm
stylus 0.3 mm	

Distribution: The original home of *M. beckii* is tropical America, as will be shown later from its host-relations. However, it has spread from here to practically all countries where citrus is grown and even beyond. It is known from all warmer parts of America and Australia, from S. Africa, from India, Ceylon, Japan, Formosa, Hawaii, etc. In the countries bordering the Mediterranean it is, partially at least, of recent introduction. From Spain and Italy it is known for long, whereas its introduction into Algeria and Tunis, as well as into Palestine, is quite recent.

Into Palestine it was introduced about 1912 with nursery stock of Valencia oranges from Spain, the places of introduction being the Acclimatization Garden of the Ica in Petah Tikva and probably some orange groves near Jaffa. From these sources the infestation spread, but at a rather slow rate, the reason for which will be discussed later.

In Petah Tikva it is distributed to-day over the whole area wherever orange groves sufficiently old are present. In Jaffa quite a series of gardens are heavily infested, centering about the grove belonging to the Russian Church. The dense plantation as well as the surrounding walls detain the wind and sun from the orchard and create optimal conditions for mussel-scale development. The damage is here very heavy. The infestation spreads slowly, mainly in the direction of the prevailing wind. At Mikweh Israel, infestation was first observed in spring 1935.

In 1934 as well as in 1935, some small centres of temporary infestation were observed (caused by nursery stock brought from Petah Tikva), one in the Sharon plain and one in the Upper Jordan Valley, which led to no permanent infestation. In 1937 quite heavy infestations were observed in Hedera and Benjamina. Inland the infestation spread as far as Beth Dagon. The Mussel scale has spread over the whole coastal plain during the last few years.

LIFE - HISTORY.

Sex-ratio: The sex-ratio of *Mytilococcus* is best obtained by breedings. Among 373 individual breedings, 39.4% of the offspring were females, 60.6% males.

In the statistical counts the percentage of living males was on the yearly average 36% on the leaves, 50% on the full branch. This percentage does not agree with the real sex-ratio, as the scales of the adult females remain living and adhere to the tree much longer than the male puparia.

Parthenogenetic breedings confirmed the normal breeding rate. In the offspring of 10 parthenogenetic females from May to July 1935, 60% were males and 40% females. Parthenogenesis occured quite freely, wherever no copulation takes place. But reproduction with copula is the normal way of reproduction. This is confirmed by the occurrence of sperms in the microtome sections of the females. Lindgren and Dickson (1941) however never observed oviposition from isolated females in California. They began oviposition when copulated even 36 days after the normal begin of egg-laying.

None of the sexes show a preference for one side of the leaf. The percentage of males was as follows:

month	1934	III	IV	V	VI	VII	VIII	IX	X	XI	XII	1935	I	II	Average
upper-side of leaves	.	21	2	12	—	—	33	31	45	—	61		40	63	33.1%
lower-side of leaves	.	38	9	—	—	—	32	40	44	—	50		57	35	38.1%

The longevity of the male is as short as that of other armoured scales.

Oviposition, adult longevity, length of development and annual number of generations: The oviposition continues for a number of weeks and still continues when the first larvae begin to hatch. Females which oviposited in April 1934 had an average egg number of:

date		n	M (living only)	n	M (living + dead)
after about	34 days	6	118	6	118
	45	5	136	5	136
	55	5	166	13	144
	65	6	223	11	215
	75	3	221	10	208
	85			3	182

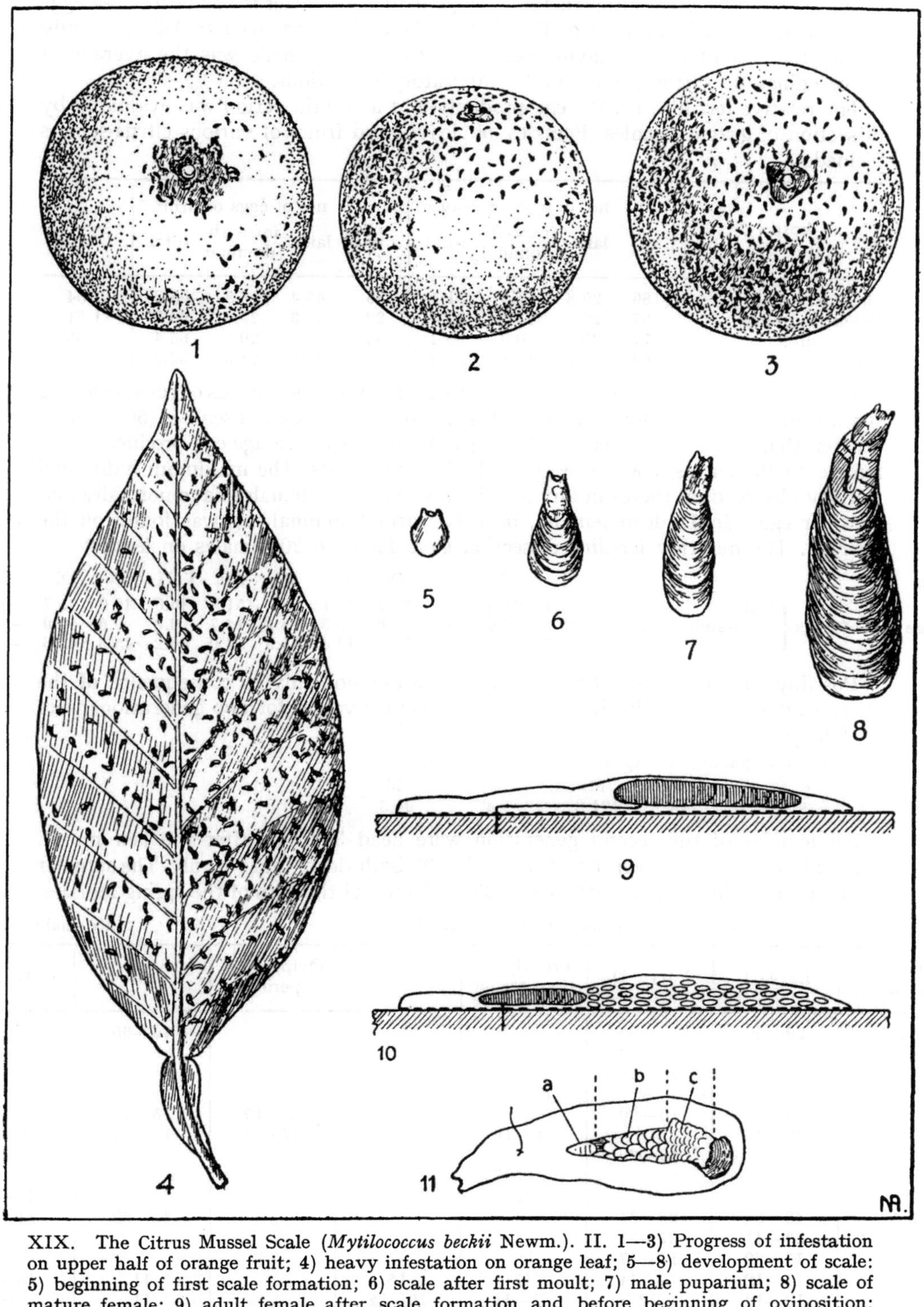

XIX. The Citrus Mussel Scale (*Mytilococcus beckii* Newm.). II. 1—3) Progress of infestation on upper half of orange fruit; 4) heavy infestation on orange leaf; 5—8) development of scale: 5) beginning of first scale formation; 6) scale after first moult; 7) male puparium; 8) scale of mature female; 9) adult female after scale formation and before beginning of oviposition; 10) adult female at end of oviposition period; 11) ventral view of female with eggs: a. posterior end of female body, b. incubated eggs, c. empty egg-shells. (from Bodenheimer and Steinitz).

During summer oviposition therefore stops 9-10 weeks after it has begun. 3-4 eggs are laid in the daily average of the first weeks and are reduced to 1-3 eggs daily during the rest of the egg-laying period. 172 eggs per female was the average of 48 breedings on citrus leaves, under laboratory conditions.

The relative fertility, i.e. the egg-production at a certain period was evaluated by dissection of gravid females. Females on leaves and fruits of various citrus-species yielded:

host	no. of eggs on leaves				no. of eggs on fruit				fruit/leaves
	n	laid	intra-abdominal	total	n	laid	intra-abdominal	total	
Shamuti orange . .	189	29.8	11.3	41.1	86	46.4	16.8	63.2	1.54
Valencia orange . .	67	26.3	11.6	37.9	22	40.3	20.6	60.9	1.61
Grapefruit.	77	29.4	10.0	39.4	63	45.3	20.1	65.4	1.65
Sour lemon	43	18.9	11.1	30.0	16	31.5	21.5	53.0	1.77

It is rather doubtful whether the differences between the various citrus species are significant. The higher fertility on the fruit, known also from red scale is (50 to 70%) greater than that on the leaves. Taking 170 eggs as the average egg-production per female on the leaves, 250-300 eggs are laid on the fruits. The maximum individual fertility observed on leaves in any female was 286. Dead females have normally laid all their eggs. In 13 dead females, only 1.1 intraabdominal egg was found on the average. The monthly fertility dissection gave for each 20 females on leaves:

month		I	II	III	IV	V	VI	VII	VIII	IX	X	XI	XII
average no. of eggs	laid	31.1	21.5	27.1	33.2	34.1	23.0	51.3	36.6	36.0	38.7	—	33.2
	intra-abdominal	13.2	15.0	18.8	6.6	9.3	18.3	12.7	9.6	9.6	10.9	—	12.7
	total	45.3	36.5	46.5	39.8	43.4	41.3	54.0	46.2	45.5	49.6	—	45.9

Possibly that total fertility in summer is lower not only in consequence of the shorter adult longevity in that season. The average values for two generations were as follows:

Generation beginning:	in April		In June	
	n	M	n	M
	48	171.8	4	94.3

All females of the second generation were dead 35 days after the last moult. Even after 25 days the egg number was 63, 78, both dead, and 88, only. The summer generation produces apparently about 30-50% only of the eggs of the spring females.

Development of female stages (in days). dto male

Beginning of hatching	Larva I	Larva II	Preoviposition period	egg	Oviposition period	Larva I	L II + III
XII 1933	29—66					46—65	87—95
I 1934		55—79					62—72
II		16—39		8—22			31—43
III			13—34				
V	15—21	10—20			35—57	15—21	
VI	12—14		8—18		27—41	12—14	19—22
VIII		10—29					24—38
IX			8—17				
X			46				46—77
I 1935	43—54	23—31				43—52	34—42
II	33—41	27—31					
IV	16—20	14—17					
VI	11—12	13—14	13	15		11	15—17
VII	10—12	12	7—11	10—12		11	16

The total quickest development from egg to beginning of oviposition was for the female about 50 days in the summer (44 for the male) and about 110 days in the winter.

The hyperbola (A) was calculated from the data:

48 days at 25.8° C and 64 days at 21.3° C

For calculating hyperbola B, two weeks were added at 25.8 (half of total number of eggs laid) and correspondingly 19 days at 21.3° C.

For calculating hyperbola C, 60 days were added at 25.8° C and correspondingly 80 days at 21.3° C.

Development hyperbola for *M. beckii.*

Hyperbola		A	B	C
development threshold (c)		7.8	8.0	8.8° C
thermal constant (Th.c.)		864	1104	1836
duration of	35° C	31.7	40.9	70.1 days
development	30	38.9	50.2	86.6
at	25	50.2	64.9	113.2
	20	70.8	92.0	163.9
	15	120.0	157.8	296.1

This allows 5.2 generations for the quickest developing individuals of every generation in the coastal plain, 3.9 generations as the epidemiological average and 2.2 for the latest developing individuals.

Consecutive observations in the orange groves confirm that as many four generations occur as a rule at Petah Tikva. The first in April, the second at the end of June, the third in late August/early September and the fourth in the winter from November to January. These observations are in complete agreement with the computations for our hyperbola B.

The local life history as well as the overlapping of the succeeding generations is illustrated by fig. 79.

Host-relations: M. beckii apparently originates from the tropics and subtropics of the New World, where it has an abundance of host. The following hosts are reported from Cuba, Florida, Brasil and British Guyana: Citrus spp.; *Balsamocitrus, Poncirus, Murraya, Croton, Malpighia glabra, Ficus, Elaeagnus, Banksia, Taxus, Cercidiphyllum, Pomaderris, Quercus, Cassia, Hibiscus, Lavanga, Allamanda, Cycas, Viscum, Magnolia,* pecan, palm, mango and many others.

The following hosts are known from

S. Africa: *Citrus,* croton
India and Ceylon: *Citrus, Toddelia,* piper.
Formosa: *Citrus, Murraya.*
Australia: *Citrus,* croton, *Banksia.*

In the Mediterranean region, where *M. beckii* must be of relatively recent introduction (especially in Africa Minor and Palestine) it is reported from the following hosts:

Canary Islands: *Citrus, Codiaeum, Wigandia.*
Spain: *Citrus, Olea.*
Africa Minor and S. France: *Citrus.*
Morocco: *Citrus, Cydonia, Elaeagnus, Prunus.*
Italy: *Citrus, Chamaerops, Elaeagnus, Olea.*
Greece: *Citrus, Pistacia.*

Turkey: *Citrus.*
Syria: *Citrus, Pistacia.*
Palestine: *Citrus,? Ceratonia,? Yucca.*
Egypt: *Citrus.*

The scarcity of hosts, apart from *Citrus*, is surprising, especially considering the polyphagy of the species in its original home.

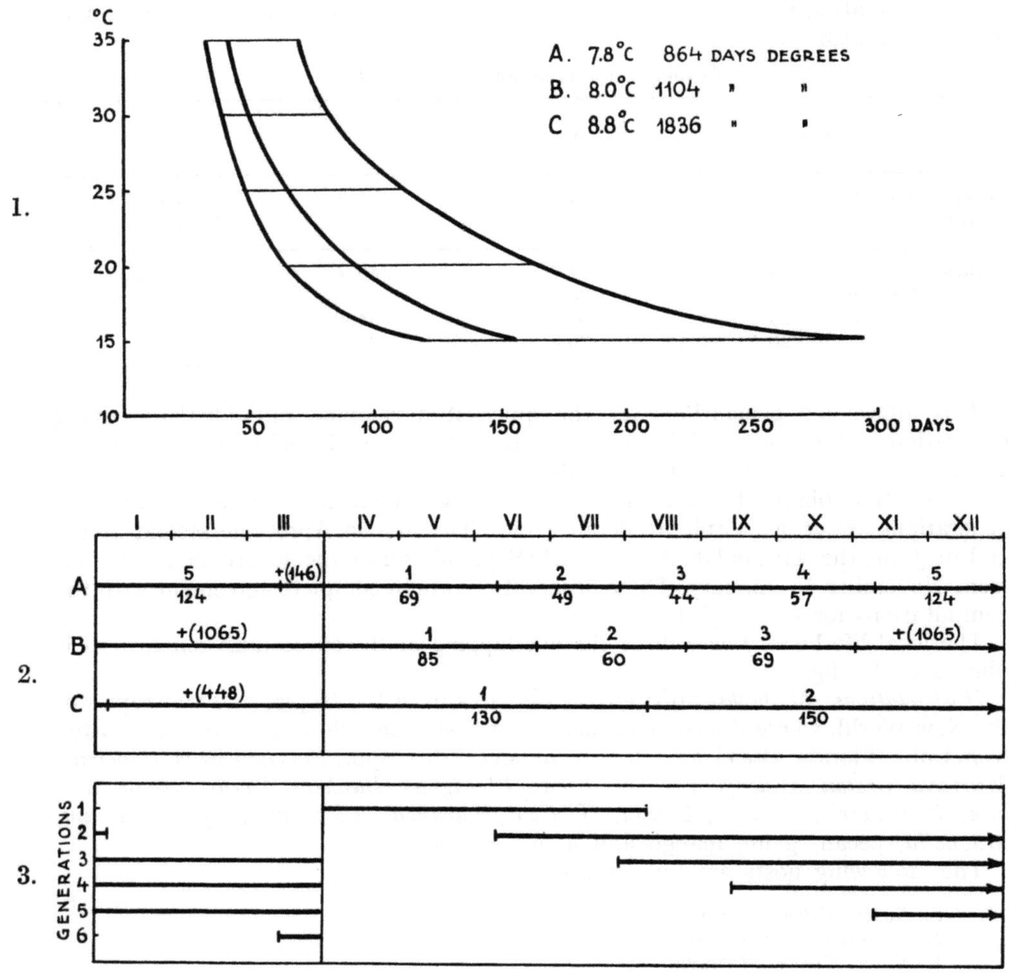

Fig. 80.
The life-history of the Citrus Mussel Scale (*Mytilococcus beckii* Newm.) at Tel Aviv. 1) The three hyperbolae; 2) the life-history as calculated from the hyperbolae; 3) sequence and overlapping of generations.

All citrus species are attacked, and more shady trees are preferred. The Citrus mussel scale is polymeric: leaves and fruits are heavily attacked, whereas twigs and stem only to a very slight degree.

On two (different) full branches, the infestation with living scales was as follows:

year	no. of scales on			maximum	
	branches and twigs	leaves	fruits	no. of leaves	no. of fruits
1934	7	4335	115	1054	5
1935	-	10369	719	419	5

The highest average infestation on fruit was 15 in January 1934, 52 in November 1935; 1 per leaf in December 1934, 9 in January 1935. About 95% of all scales are on leaves, 5% on fruits and practically none on twigs, branches and stem.

Seasonal fluctuation and seasonal mortality: Monthly statistical counts were arranged at Petah Tikva for two consecutive years. It will be noted that neither eggs nor crawlers are included in these counts. Also, part of the first larvae is overlooked, the percentage of which varies with the amount of dust on the leaves.

The first series comprises counts on 100 orange-leaves:

month	I	II	III	IV	V	VI	VII	VIII	IX	X	XI	XII	1934
total scales	—	—	368	117	113	143	22	104	15	156	—	133	
living scales	—	—	159	47	39	42	8	46	3	96	—	94	
% dead	—	—	57	60	66	70	64	56	80	38	—	29	
total scales	99	58	61	37	16	15	13	—	35	76	46	98	1935
living scales	82	35	23	13	5	2	4	—	18	36	17	51	
% dead	18	39	62	64	72	87	72	—	48	53	64	49	

Parallel counts on grapefruit leaves yielded the following results:

month	I	II	III	IV	V	VI	VII	VIII	IX	X	XI	XII	1934
total scales	—	—	260	52	83	48	16	25	36	118	—	194	
living scales	—	—	99	30	20	8	4	17	10	22	—	96	
% dead	—	—	62	43	76	85	77	34	74	81	—	51	
total scales	134	54	25	24	22	7	12	—	46	27	44	57	1935
living scales	103	40	8	11	2	0	6	—	21	15	18	33	
% dead	23	25	68	53	93	100	50	—	54	45	58	57	
Average % dead (all series)	21	32	62	55	77	88	61	45	64	56	61	46	

The population density was lower in 1935 than in the preceding year. It was highest in the winter, combined with the lowest mortality. It was lowest in June/July, following the high mortality brought about by the spring hamsins. High mortality was also caused by the September hamsins. Mortality is lowest in January/February. The counts just mentioned, included all stages, whereas the followings series is restricted to each 200 adults (on shamuti leaves):

month	I	II	III	IV	V	VI	VII	VIII	IX	X	XI	XII
1934	—	22	51	61	91	94	—	60	81	57	—	53% dead
1935	53	45	49	64	93	97	82	—	80	63	47	36

The most important series comprise monthly counts of living scales on whole branches. The results are compiled in the following tables and fig. 81 and 82.

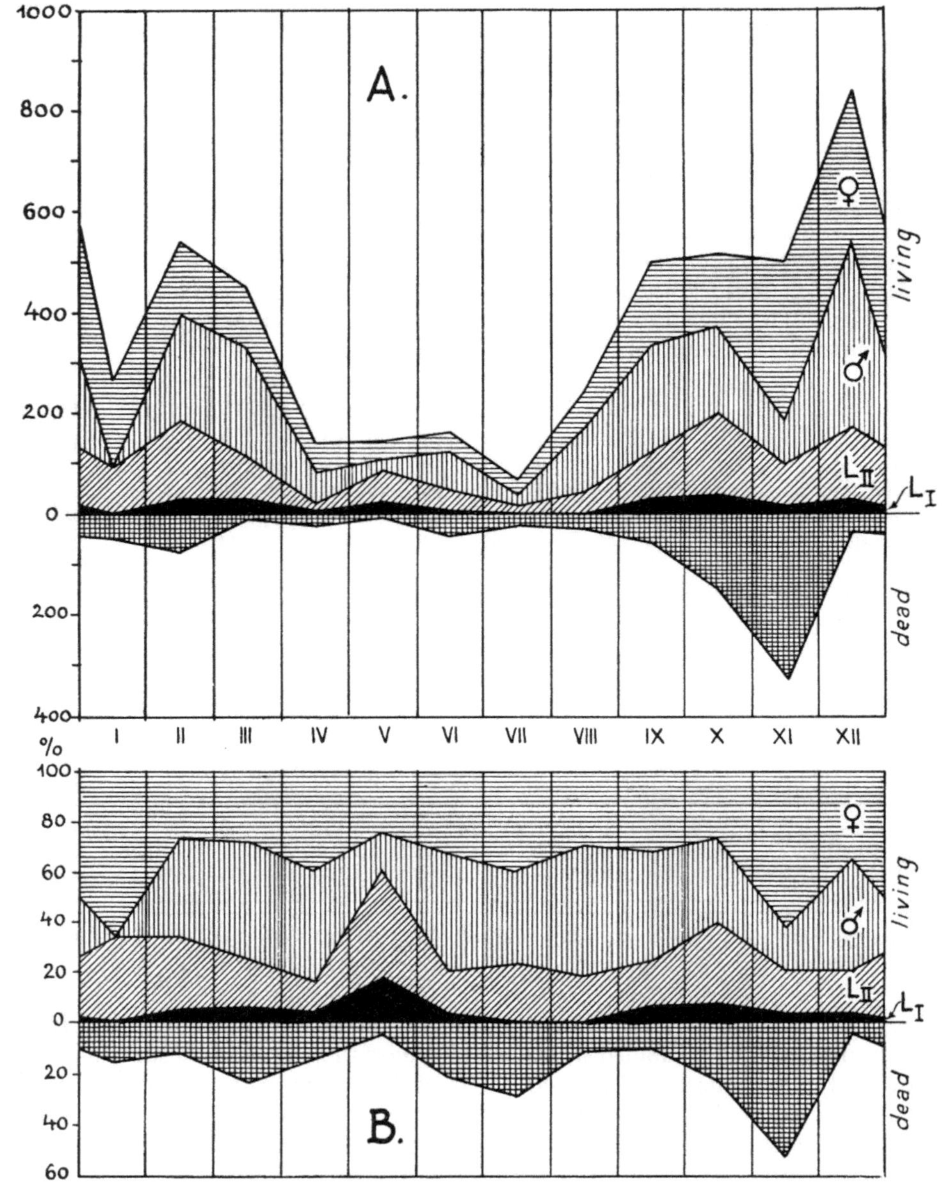

Fig. 81.
Monthly fluctuations of Citrus Mussel Scale population on citrus branch of 12-years old tree (shamuti) at Petah Tikva 1934. A. Absolute numbers. B. Relative numbers.

Monthly count of living mussel scales on full branch of 12 years old tree (absolute values).

Larva Month	I	II	III	IV	V	VI	VII	VIII	IX	X	XI	XII	1934
Larva I	0	30	29	5	24	4	0	0	32	32	16	24	
Larva II	91	156	86	16	61	42	14	43	90	163	82	143	
Larva III ♂+♂. . .	0	213	215	61	21	74	23	124	212	171	85	367	
Adult ♀	178	145	120	55	36	39	25	69	163	142	311	303	
Total living scales .	269	544	450	137	142	159	62	236	497	508	494	837	
Total dead.	51	74	11	25	7	45	25	32	59	148	330	40	

The same in percentages:

Larva I	0	5	6	4	17	3	0	0	6	7	3	3
Larva II	34	29	19	12	73	26	23	18	18	32	17	17
Larva III ♂+♂. . .	0	39	47	44	15	47	37	52	43	34	17	44
Adult ♀	66	27	28	40	25	24	40	29	33	28	63	36
Total living scales .	100	100	100	100	100	100	100	100	100	100	100	100
Total dead.	16	12	24	15	5	22	29	12	11	23	53	5

Dto. of 20—25 years old tree.

Larva I	516	474	364	86	7	—	11	119	33	129	79	99	1935
Larva II	467	386	303	293	13	18	77	196	170	116	90	146	
Larva III ♂+♂. . .	619	636	451	204	22	30	130	313	249	138	79	99	
Adult ♀	572	540	419	194	42	30	88	348	348	246	162	152	
Total living scales .	2234	2036	1542	777	84	78	306	977	800	629	410	496	
Total dead.	163	69	125	410	277	76	119	68	195	321	156	109	

The same in percentages:

Larva I	23	23	24	11	8	0	4	12	4	21	19	20
Larva II	21	19	20	38	16	23	25	20	21	19	22	29
Larva III ♂+♂ . .	30	31	29	26	26	38	42	32	31	22	19	20
Adult ♀	26	27	27	25	50	39	29	36	44	39	40	31
Total living scales .	100	100	100	100	100	100	100	100	100	100	100	100
Total dead.	7	3	8	35	77	49	28	7	20	34	27	18

Here also we find the population minimum in June/July, the maximum in December/January.

In all these cases the average annual fluctuation $\frac{\text{(population max.}}{\text{population min.)}}$ was 21 (13-29).

The quotient of increase $\frac{\text{(population of one month)}}{\text{population of preceding month}}$ for the whole branches is as follows:

month	I	II	III	IV	V	VI	VII	VIII	IX	X	XI	XII
1934	(0.3)	2.0	0.8	**0.3**	1.0	1.1	0.4	**3.8**	2.1	1.0	1.0	1.7
1935	(0.2)	0.9	0.8	0.5	**0.1**	0.9	**3.9**	3.2	0.8	0.8	0.7	1.2

From these quotients we learn that the decisive month for the annual increase are July/August in 1934, May/June in 1935, when the total increase was 9.5 resp. 9 fold. The climate of these months ranged from 25-27° C and from 75-79% relative humidity. These climatic conditions have to be regarded as optimal. During the population peak in winter the quotient of increase is much less favorable and remains below 1.0 on the average. Lindgren and Dickson (1941) report that in breeding rooms with 30% R.H. *M. beckii* did not reproduce satisfactorily, but were reared successfully at 80% R.H.

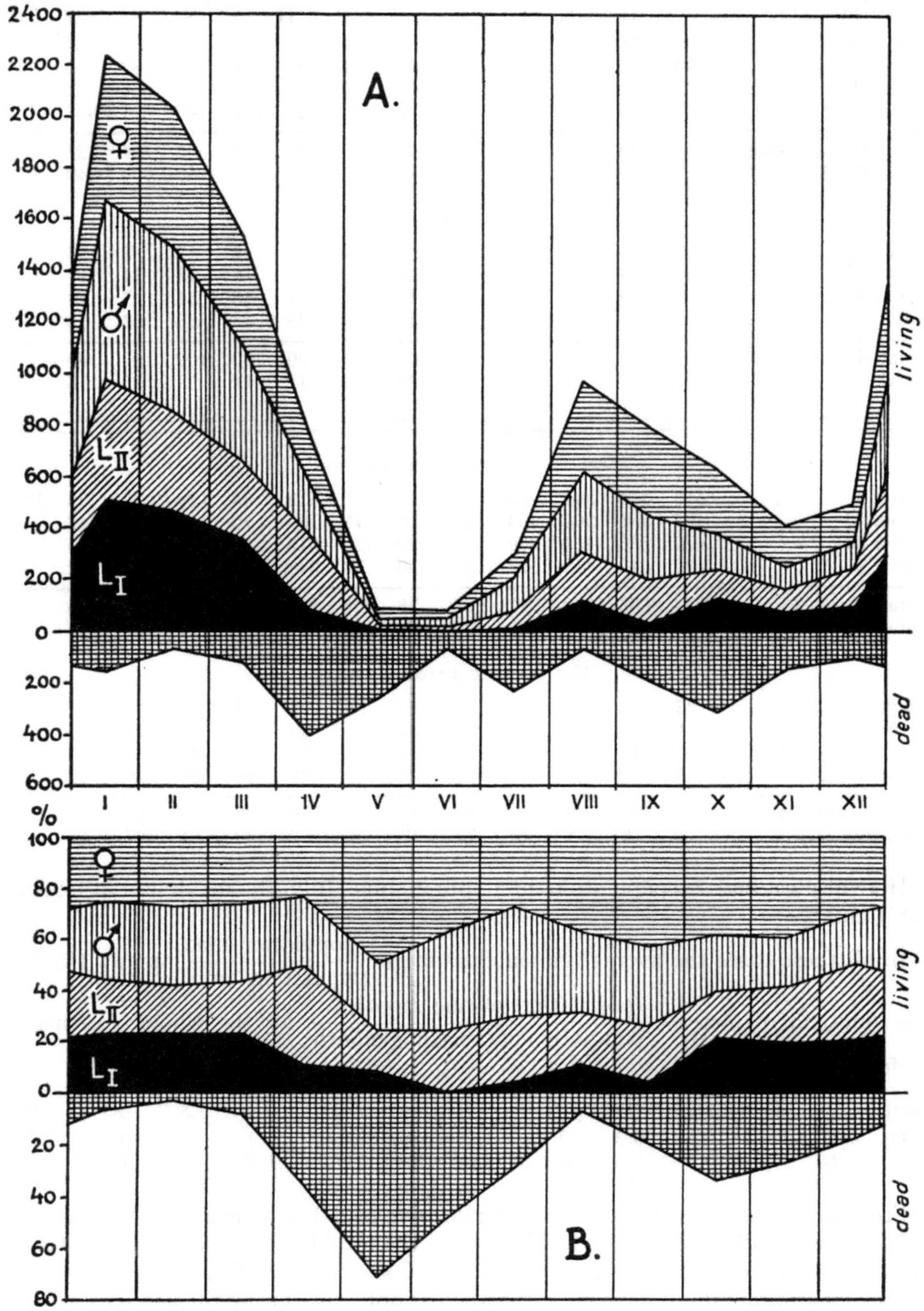

Fig. 82.
Monthly fluctuations of Citrus Mussel Scale population on citrus branch (shamuti) of a 23-years old tree at Petah Tikva 1935. A. absolute numbers, B. relative numbers (from Bodenheimer and Steinitz).

These considerations are in agreement with the results of some other observations. The actual number of settled larvae per female was observed in two series, one running from June to August, the other from November to February:

		summer	winter
no. of breedings		116	33
% of breedings with no larvae settled		22%	49%
no. of larvae per breeding	only positive breedings	28.2	3.4
	all breedings	21.0	1.8
	maximum	97	10

The highest mortality of all stages occurs among the crawlers, which are blown away by the wind. Unfortunately our statistics do not allow the estimation of these losses. The enormous mortality induced by the dry desert winds in spring and in a minor degree in autumn, is sufficiently illustrated by our counts. Other cases of important elimination are the annual collecting of the crop (fruits) and the seasonal dropping of infested leaves. With regard to the leaves it is important, that heavily infested leaves drop prematurely.

Neglecting the overlapping of generations and accepting the populations of April, July, September and December as the final result of development of each generation, we calculate the environmental resistance for each generation, i.e. that fraction of its potential development which has actually been destroyed. The assumption is that the female produces 200 eggs, and that the sex-ratio is 60♂:40♀. The zero resistance, which just about guarantees the maintenance of the population in the coming generation is thus:

$$R_0 = \frac{100 . 200 - \frac{60 + 40}{40}}{200} = 99.9875$$

The actual resistance of every generation is further calculated by the following formula:

$$R_a = \frac{100\ (P_1 - P_2) + R_0 . P_2}{P_1}$$

The actual destruction of each generation is as follows:

Generation	1 (IV—VII)	2 (VII—IX)	3 (IX—XII)	4 (XII—IV)
1934	99,994	99,902	99,993	99,982
1935	99,993	92,448	99,938	99,998

We have thus obtained the percentage of potential offspring which was destroyed. This calculation again confirms the enormous mortality of the spring generation, followed in the summer by the lowest mortality.

The same result is obtained when only adult females are considered:

Generation	1 (IV—VII)	2 (VII—IX)	3 (IX—XII)	4 (XII—IV)
1934	99,993	99,920	99,975	99,937
1935	99,996	99,952	99,993	99,980

Damage, progress of infestation and dispersal: M. beckii decidedly belongs to the toxic group of scale insects. Chlorotic stripes or spots around the infested parts of the leaves are sufficient indication of its toxic effects. The yellow green discoloring finally turns yellow, neighbouring spots unite and the centre of the spot grows necrotic, the leaf falling off in one of these stages. In nature it drops at a much earlier stage of discoloring than in the laboratory, because in the grove the wind accelerates the dropping. Under laboratory conditions heavy incrustation of the twigs takes place as a final stage of the attack which leads to the drying up and death of the plant. Such infestations of twigs and branches may occasionally be observed in the orchard where trees are heavily infested in their centres, twigs being preferred to branches. Whereas in Petah Tikva no such extreme cases of infestation were observed, they were quite common in the dense orange groves of Jaffa. The infestation was here so intense that the general development of the trees was often strongly impaired. Only a few yellow leaves remained on the branches. Twigs and branches were dried up and bare, covered with heavy incrustation of, by now, dead scales.

The crawlers show a general tendency to settle in the immediate neighbourhood of their mother. Only in spring, during the main vegetation period they show a tendency to migrate to a very limited extent, which allows for settling on the tender twigs slowly and later on the almost fully grown leaves. Very young leaves are, as a rule, not infested.

The locomotory power of the crawler is quite considerable, in spite of the fact that it settles close to the mother's scale. Quayle (1912) gives the following values of locomotory activity in two hours:

278 cms at 31.6° C
83 cms at 20.0
32 cms at 17.8

This gives an activity hyperbola

$$1/83\ (20 - x) = 1/278\ (31.6 - x)$$

The activity threshold is 15.3°C for the crawlers. No considerable locomotion will take place below this temperature.

The mussel scale settles mainly on the upper half of the fruit, whereas only a few scattered individuals may be found on the lower half. The infested part of the fruit shows a tendency to remain green longer than the rest of the fruit.

The fruits become infested at a very early stage, from end of April on. This early infestation, however, easily escapes the eye of even a careful observer, because the first larvae settle on the fruit in the shelter of the sepal leaves and cannot be observed without removing the latter. The young larvae do not settle visibly on the surface of the fruit before mid-July. They are part of the crawlers produced by the generation which has settled first in may below the sepal leaves. Counts performed on 6 fruits on 20.VII.1934 illustrate these conditions:

	Larvae I	Larvae II	adult ♀
Free surface of fruit	99%	1%	0%
Below sepal.	80.5	3.5	16.0

However, still 2½ as many larvae of the second generation remain below the sepal, than on the free surface of the fruit. Fruits with females of *M. beckii* below the sepal and those from which these females were removed and which could be resettled only by immigration from parts of the tree beyond the fruit or by wind blown larvae remained nearly clean. The average infestation of fruit was in two series as follows:

Date of experiment	fruits with females below sepal	fruits without females below sepal
27.VIII—25.IX.1934	16.1	1.3
28.X.—19.XII.1934	57.5	4.8

Over 90% of all scales on the fruit originated from the first settlers below the sepal and not from immigrants. With the appearance of these crawlers of the second generation, the colonization of the fruit proceeds at a very rapid rate 1) because most of the offspring remains on the fruit, 2) because of the higher fertility of females on the fruit.

Sheltered places protected by other leaves or fruits, webs of Psocids, etc. are optimal for development since they furnish protection from the wind. The crawlers are positively thigmotactic and often settle along the ribs of the leaves, which is probably one of the main reason why they prefer to settle below the sepal of the fruit. The crawlers are phobic towards exposure to sun-radiation, but do not avoid dispersed light, such as reaches them, for instance, within the tree, under natural conditions. Some experiments, showing that they prefer shady, dispersed light, to darkness, may be recorded:

	n	at end of experiment		%	
		in dark	in light	in dark	in light
Crawlers put in light at beginning . .	13	4	9	31	69
Crawlers put in dark at beginning. . .	29	10	19	34	66

The wind probably plays the most important role in the dispersal and population density of *Mytilococcus*. It removes most or many of the crawlers from their birthplace. It hinders their settling down on strongly exposed places and permits the development of dense populations only on sheltered places. On the other hand, the wind is to a certain degree a means for passive dispersal, for again and again a few of the larvae are blown away to new host-plants and get a chance to settle in a new locality. The number of larvae settled on single leaves, each infested with one adult female was as follows:

	wind-protected	wind-exposed
no. of breedings.	57	23
no. larvae settled	8%	26%
No. of larvae per breeding — average	32.5	14.7
No. of larvae per breeding — maximum	97	55

Wind and light-intensity are the main factors regulating the distribution of the mussel scale in the tree. Shelter from wind and sunexposure lead to the concentration of the scale within the centre of the tree in preference to its periphery. Two illustrations of this concentration in the central parts of the trees may be given:

Monthly counts of a branch, divided in 4 parts of equal length, beginning in the centre, ending at the periphery, showed the following infestation of each part (Petah Tikva, 1935; leaves only):

parts	I (centre)	II	III	IV (periphery)
no. of scales per section	6046	1912	1802	609
% of scales per section	59	19	18	5%
No. of scales per 100 leaves	12052	3904	1564	799
% of scales pet 100 leaves.	68	21	9	4%

165 times as many scales were found in the central quarter than in the distal quarter.

Counts on the average number of mussel scales per fruit showed the following results:

date	fruits in the interior	fruits on the periphery
20.VII.1934	13.4	0.1 scales per fruit
26.IX.1934	53.7	4.4 ,, ,, ,,

The infestation is often quite sporadic, even within the same orchard. Small groups of heavily infested trees may be seen among not or very slightly infested trees. Such cases are probably a transition stage from early to definite and thorough infestation of the orchard.

The estimation of the tree's total loss of vitality induced by heavy infestation is rather difficult. But by heavily infested fruit is excluded from export. Heavily infested old orchards in Petah Tikva showed the following losses in a year of medium intensity of *Mytilococcus:*

date			
11.I.1935	shamuti	12%	
10.III.1935	,,	16	
13.III.1935	,,	19	grapefruit 37%
4.XII.1935	,,	18	
4.I.1936	,,	14	

These data are the percentages of the cull induced by the mussel scale. The total cull (brara) constituted 10-15% of the total crop. The damage may therefore be estimated at 1-3% in shamuti and at 5% in grapefruit. In the gardens infested at Jaffa, the damage must have approached 60-100%.

Grapefruits with their branches hanging heavily down offer shelter from wind and sun-radiation and are therefore infested earlier and are prone to a heavier infestation than young orange orchards. The difference in the infestation of both hosts within the same orchard may be gathered from the above given statistics. In older orchards the difference often disappears.

Infestation of young orchards: The rapid extension of the citrus area in the nearer and farther neighbourhood of Petah-Tikva led to mass-transports of citrus seedlings from nurseries of this locality into other places. Some of these seedlings were slightly infected with mussel scale, especially some older seedlings from densely planted nurseries and from plants inoculated in the nursery. Infestation of the lower leaves of younger seedlings was discovered in one case only.

Those relatively few individuals transported with the nursery stock suffice in the case of red and other scales to lead to a permanent infestation of the newly planted orchard.

Some experimental infestation on seedlings within the orange-grove gave the following interesting result:

date of infestation	infestation					no. of trees still infested					
						I		II		III	
	n	0	+	++	+++	control	result	control	result	control	result
3.IV.35 . . .	22	—	13	6	3	1.V.	7	29.V.	3	VI	0
3.V.	10	10	—	–	–						
25.V.	17	16	1	–	–	5.VI.	–				
5.VI.	6	1	(5)	–	–	10.VI.	–				

Even in April, when many artificial infestations succeeded well, 2½ months later not one single individual was still alive. Later infestations were a failure from the beginning. Lack of shade and full exposure to strong sun-radiation as well as to wind are an insurmountable impediment for the permanent settlement of the scale on seedlings and young trees in the orchard. The average orchard under the normal cultivation usual now in Jewish orchards (spaces, etc.) is not permanently infested, as a rule, before its 9th or 10th year. Although the mussel scale does not apparently decrease in number with age, as does the red scale, exact population counts should still be made in this respect.

Natural enemies are scarce in Palestine. In spite of extended series of breedings no hymenopterous parasite was ever observed. This seems to be the case, so far, throughout the Mediterranean region. The general feeder on armoured scales, the ladybeetle *Chilocorus bipustulatus*, appears also on the mussel scale, on which it feeds quite abundantly from February to April. However, it is never abundant enough to reduce *Mytilococcus*, which is at that season at its population height.

Another occasional predator is the larva of a small Coniopterygid (Neur.) which was observed sucking on the larvae of *M. beckii*. The former larva was occasionally parasitized by the Proctotrypid wasp *Aphanogmus steinitzi* Pr.

Control: The fact that *M. beckii* is extremely well covered by its scale dorsally as well as ventrally makes the penetration of white oils difficult and fumigation is regarded in all countries as the standard treatment against this pest. However, this control measure is believed to be extremely difficult, for the dosis necessary for killing the mussel scale imperils the citrus tree. Gassner (1925) reports from Spain on the relation of the dosis toxica (for the tree) and of the dosis curativa (for the scale)

Temperature	dosis toxica	dosis curativa
below 20° C	0.3 vol. %	0.3—0.35 vol. %
above 20° C	0.1—0.15 vol. %	0.2—0.25 vol. %

This means that in Spain control by fumigation is possible only in winter. In California double fumigation (repetition with highest tolerable dosis in an hour interval) is occasionally applied.

Some preliminary control experiments have been carried out in cooperation with the Imperial Chemical Industries. The first series was carried out with the pot-method on 30.VIII.1934 at Petah Tikva. Each 5 grapefruit-trees were fumigated with 3 different concentrations (C, D, E of the schedule).

Whereas the natural mortality was 74% before fumigation, it was for each 320 individuals, 7 days after fumigation, as follows:

Concentration	C	D	E	Control
mortality.	99.68	99.71	100.0%	74.0%

The first experiment with liquid hydrocyanic acid was carried out at Jaffa on the 9.VIII.1935, the second and third series in the same locality on the 3rd and 4th IX.1935. Each series comprised 5 heavily infested orange trees, which were fumigated with different concentrations. These and climatic conditions during the fumigation-operations, are compiled in the following table:

Date	Method	Schedule	Action of HCN begins first tree	Action of HCN begins last tree	Action of HCN finished	Temperature		Rel. Humid.	
			(a)	(b)	(c)	(a)	(b)	(a)	(b)
30.VIII.34	Pot	C; D; E	18h15	18h25	(ca. 19h30)	24.5	23.0	73%	
9.VIII.35	Liquid	18; 20; 22	18h15	18h24	(ca. 19h30)	26.7	25.6	74%	78%
3.IX.	,,	14; 16; 18; 20	17h23	17h35	(ca. 18h30)	27.2	26.7	73%	74%
4.IX.	,,	14; 16; 18; 20	17h15	17h35	(ca. 18h30)	27.2	26.1	64%	70%

The mortality obtained by these fumigations was (for each 300-500 counts) as follows:

	Schedule	Mortality in control series	Mortality after fumigation				
				3rd day		4th day	5th day
9.VIII.35	18	92.0%		99.4		100.0	100 %
	20			100.0		100.0	100.0%
	22			99.4		100.0	100.0
			2nd day	4th day	6th day	7th day	16th day
3/4.IX.35	14		92.7	98.4	—	100.0	99.8%
	16	69.0%	95.0	100.0	99.9	—	99.4%
	18		97.3	100.0	—	100.0	100.0%
	20		100.0	100.0	—	99.5	100.0%

At first sight, these results appear extremely favorable not only as a control, but even as an eradication campaign. Unfortunately, these figures cannot yet be regarded as final, because they concern only adult and larval mortality. It is reported from Spain that eggs are more resistant to fumigation than the later stages.*) The fact that 16 days after the fumigation no living larvae had settled on the leaves in concentration 18 and 20, seems to indicate that these concentrations might prove satisfactory. A final test with special consideration of the mortality of the egg-stage is still needed. By double fumigation in one year and repetition of double fumigation in all cases, where careful inspection still shows slight infestations in the following years, it might be possible to reach even eradication.

Whereas visible signs of fruit infestation are detected in mid-July, it is desirable to perform the fumigation during June/July. Later operations are still successful, but the scales of the dead females stick to the fruit and spoil its marketing value.

*) Iscander reports from California: This scale is considered to be particuliarly resistant to fumigation owing to the presence of eggs which always occur and are difficult to kill; for this reason a heavy dose is used.

We would like to draw attention to some highly interesting fumigation experiments made by Lindgren and Dickson (1941) in California. The scales were bred at 26.7°C and at 80% R.H. The fumigation was made at 24°C by an exposure of 40 minutes to 0,37 mg. of HCN per liter (the crawlers only to 0.14 mg). The percentage of survival was:

Age	Stage	Percent of survival	Age	Stage	Percent of survival
0–1 day	egg	82.8%	13 days	early first moult	3.0%
1–2	egg	83.5	17	late first moult	31.0
2–3	egg	64.9	21	second instar	3.0
3–4	egg	77.5	25	early second moult	0.3
4–5	egg	35.9	29	late second moult	44.0
5–6	egg	27.4	33	just past second moult	6.9
6–7	egg	29.3	37	one third of last scale formed	2.7
7–8	egg	30.7			
8–9	egg	18.5	41	almost mature (no eggs)	7.3
9	crawler	0.7	45	mature (few eggs laid)	27.4
10	just settled	38.9	49	mature (many eggs laid)	38.4
			53	mature (first eggs hatch)	38.1

This change of susceptibility towards fumigation with age is highly interesting and important. Crawlers and the moulting stages, including the female before it has finished building its scale, are the most sensitive stages. Highly resistent are the eggs, especially during the first half of their development. Resistent are also the just settled crawlers, the late second moults and again the mature female, once oviposition has started. These results show clearly that the age structure of the mussel scale population at the time of fumigation is of decissive influence for the result of the fumigation.

In 1937, similar control experiments were performed in cooperation with the Shell Company with white oils (Shell Oil in 2.5% concentration; Klein 1937). The spraying was done by motor-sprayers. Mortality following one spray was in July 80.1, in August 83.7, in September 75.3%. The lower mortality in September in the consequence of the fact that a very high percentage of *Mytilococcus* are then in the more resistant adult stage.

The population increase was influenced by all spraying experiments, as follows:

Spraying	July Before Spraying	July After Spraying	August Before Spraying	August After Spraying	September Before Spraying	September After Spraying
Check (no spraying)	100		440		2420	
1 spraying (July)	100	19	82		450	
1 spraying (August)	100		440	67	368	
1 spraying (September)	100		440		2420	586
2 sprayings (July, August)	100	19	82	12	450	68
2 sprayings (July, September)	100	10	82		368	109
2 sprayings (August, September)	100		440	67	368	89
3 sprayings (July, August, September)	100	19	82	12	68	17

The population increase in the check-observations was 24-fold, after one spraying 4 to 6-fold. Following two sprayings, the population did not increase until late

September, whereas after 3 sprayings it was considerably reduced to 1/6 of the population in early July.

From 100 females, isolated in test-tubes soon after the spraying the following hatched:

Treatment	Check	After 1 spraying (IX)	After 2 sprayings (VII, IX)	After 3 sprayings (VII, VIII, IX)
Larvae hatched	1144	406	14	8
Egg-mortality	0.0	64.5	98.8	99.3%

One spraying (in August) could therefore be recommended only in the case of very light infestations, whereas two sprayings are needed for commercial control under normal conditions (preferably in July and August). Similarly to fumigation, the results obtained by spraying with white oils are better than in most other countries. The reasons for this greater sensitivity of mussel scale in Palestine towards control media remain to be studied. As a whole the economic importance of the mussel scale after its recent spread over the coastal plain was surprisingly low, so far.

References.

A. Balachowsky & C. Mesnil, Les insectes nuisibles aux plantes cultivées. Paris, 1935, pp. 502—505.
F. S. Bodenheimer, Schaedlingsfauna Palaestinas. Berlin, 1930. p. 158—159.
F. S. Bodenheimer and H. Steinitz, Hadar. **10**. 1937, pp. 153—159.
G. Gassner, Zeitschr. Pfl. Krankheiten u. Gallenkde. 1925, pp. 97—111.
N. Iscander, Min. of Agric., Egypt. Bull. 62. 1925.
H. Z. Klein, Hadar. **10**. 1937. pp. 221—223.
I. Kuwana, Dept. of Finance, Japan, Imp. Plant Quarantine Service. Bull. 2.
D. Lindgren and R. Dickson, Journ. Econ. Entom. 34. 1941. pp. 97—111.
G. Leonardi, Monografia delle cocciniglie italiane. Portici, 1920. pp. 152—157.
H. J. Quayle, Univ. of Calif., Agric. Exper. Sta. Bull. 226, 1912.

F. THE CITRUS LONG SCALE (*Mytilococcus gloveri* Pack.)

Synonymy: 1869 *Coccus gloveri* Packard
1869 *Mytilaspis flava* Targioni
1896 *M. pallida* Green
1902 *Lepidosaphes gloveri* Kirkaldy.

Description: Scale of adult female very long and narrow, straight or curved, the sides subparallel, with a near narrow flattened border. Color brownish-yellow to reddish brown, with pale margins. Exuviae terminal, yellow, occupying the total anterior length of the scale. Ventral scale whitish, well developed.

2.5—3.5 mm long, 0.6 mm broad.

Scale of male pupa. Much smaller and more delicate than the female scale. Generally straight. Color pale yellowish-brown, with white flattened margins. Exuviae yellow.

1.5 mm long, 0.5 mm broad.

The *adult female* elongate with subparallel sides, the abdominal segments slightly wider, but without prominant lateral projections. Thoracic segments much developed in comparison with the abdominal ones. Pale yellowish to pale pink, with wine red tint on the pygidium. Antennal tubercules small with 2 short, curved, spiny hair. 4 anterior parastigmatical glands. Abdominal segments short, with lateral margins only slightly produced, each of the last three with 2—3 marginal gland spines and one sharp chitinous spine-like spur.

Pygidium with three pairs of lobes. Plates dagger-shaped. Formula: P_1, L, $2P_2$, $2L_2$, $2P_3$, $2P_4$, $2P_5$.

Circumgenital glands:

3—5		, generally	4	
6—8	6—8		6—7	6—7
4—7	4—7		5—7	5—7

Anal opening close to the base of the pygidium. Marginal tubular glands large and much developed. Three groups of latero-dorsal gland apertures in groups of 4—5 glands each.

Eggs: white, to pale pink, slightly oblong, deposited in two parallel rows.

Distribution: Quite common in the tropics and subtropics of the world. In Mediterranean regions known to be common in Spain and Morocco scattered records from (Algeria), (S. France), Greece, Italy, Turkey (the eastern Black Sea coast) and Palestine. Its original home is doubtless the tropical part of the Far East.

Hosts: In all Mediterranean regions only recorded from citrus. But in its tropical home it is fairly polyphagous. In Cuba (Ballou 1926) it has been recorded from: *Anona reticulata, Asparagus sprengeri, Calophyllum calaba, Citrus spp., Faramea odoratissima,* and *Murraya exotica.*

In Florida, cabbage palmetto, cherry laural, citrus and magnolia are attacked (Merrill and Chaffin 1923). In other parts of the world *Camellia* is another frequent host.

The *biology* has never been studied. But it is generally assumed that the number of generations is equal to that of *M. beckii.* Two reliable records from Palestine are available, both from 1922/23. P. A. Buxton collected the species once in Petah Tikva on oranges (det. F. Laing - Brit. Mus.) and the writer discovered it about the same time in the Baron's Garden at Petah Tikva in small numbers on Valencia nursery stock. A complete and thorough survey in the Baron's garden and the neighboring orange groves in 1934/35 did not reveal any sign of its further presence. It seems to have died out in the meantime. This agrees with the fact that *M. gloveri* does not seem to flourish anywhere in dry summers. The species which has certainly existed for at least 10 years in small numbers in at least one orange garden at Petah Tikva, must be regarded as extinct by now. And should it even be introduced again by chance it would not seem to be able to develop into a serious pest.

With regard to control measures it is generally reported to be less resistant to all treatments than *M. beckii.*

G. THE CITRUS PARLATORIAS.

a. The Chaff Scale (Parlatoria pergandii Comstock).

Description: Scale of adult female flat, variable in shape. Usually broadly pear-shaped, occasionally circular, broadly ovate or elongate. It is semi-transparent, showing indistinctly the sublaying female. Colour dusky white tinged with pale brown. The dull orange brown exuviae have a marginal position. Secretory covering thin and transparent, opaque white at the margins. Dead specimens become opaque, of a dull ochreous yellow. Ventral scale complete, dusky white.

Total scale	1.2—1.7 mm long	0.7—1.0 mm broad
First exuviae	0.3—0.4	0.2—0.3
Second exuviae	0.4—0.5	0.5—0.6

Male puparium elongate with parallel sides. Colour dusky white. The greenish yellow first exuviae situated on the anterior margin. 0.7—0.9 mm long, 0.2—0.35 mm broad.

Body of the female short oval to subcircular. Bright violet with a yellow pygidium. Young adults frequently quite hyaline. Antennal tubercles flat, with one long, curved spine. Anterior stigmata with only 2—4 parastigmatic glands. Margins of abdominal segments with many short tubular spinnerets. About 0.5—0.8 mm long, 0.3—0.4 mm broad. Pygidium with four groups of circumgenital glands, varying as follows: $\frac{5-10}{3-10}$; generally being $\frac{6-7}{5-7}$ $\frac{6-7}{5-7}$. It is characterized by its regular, rounded appearance, and by its large marginal glands, the openings of which are at right angle to the long axis of the body. Very characteristic are the three pairs of broad trilobate lobes, the fourth lobe being small and punctiform. The plates are mostly broad and fimbriated. The formula of the pygidium runs as follows: P_1, L_1; $2P_2$; L_2; $3P_3$; L_3; $3P_4$; L_4; $3P_5$.

The *egg* is oblong, wine-red. 0.2 mm long, 0.1 mm broad.

The *neonate larva* is oval to short elliptic, wine-red in colour. 0.25—0.3 mm long, 0.15—0.2 mm broad.

The *adult male* is wine-red, with the disk of the thorax nearly colourless: only some wine-red spots and the brown sutures being differentiated. The large eyes are black. Body 0.6 mm long + 0.2 mm stylus; 0.2 mm broad, with extended wings 1.8 mm.

A good technical description of the Chaff-scale, permitting its reliable differentiation from many other closely related forms, has been given by Morrison (1939):

Adult female: Eyespot a small, roughly rounded to oval, inconspicuous fleck, often difficult to observe. Anterior spiracles with 3 to 5 disk pores adjacent; posterior spiracles without any traces of an adjacent dermal pocket but with 2 to several small tubular ducts scattered between opening and body margin. Duct tubercles anteriorly short, flat conical, with bases expanded, somewhat smaller and more elongate conical posteriorly, and varying considerably in size; range in numbers on each half of body, prespiracular 1 to 5, average 2.7; anterior spiracular 3 to 10, average 5.6; intermediate 2 to 7, average 3.6; posterior spiracular 1 to 6, average 3.3; first abdominal 1 to 4 average 1.9; totals 10 to 29, average around 18. Dorsal submarginal tubular macroducts usually rather numerous, but varying considerably; tabulated range, each half of body, 29 to 70, average number around 50 No intermediate or median macroducts and characteristically without, but sometimes with a single microduct above anterior lateral paragenitals, well within dorsal sclerotized area of pygidium, and characteristically with 2 to 5 such microducts on preceding segment, along anterior edge of this sclerotization; no dorsal microducts observed anterior to these. A faint, irregularly rounded to oval cicatrix dorsally, well in from margin, on first abdominal segment, this showing in some well-prepared specimens but not certainly visible in many preparations examined and constancy not established. The 3 pairs of normally developed lobes not conspicuously large as in *cinerea*, or small as in *camelliae*, all similar in shape, but plainly graduated in size from median to outer, the actual size of the lobes and the degree of gradation varying considerably in the material studied, with a corresponding influence on the general appearance of the apex of the pygidium; fourth lobe represented by a definitely sclerotized asymmetrical short conical projection of body margin having usually 1 or 2 minute teeth on inner margin and 2 to 3 on outer; can be described also as a sclerotized projection of body margin having 1 relatively very large and conspicuous projecting tooth just inside its midline and some minute teeth on each side of this; fifth lobe usually present and similar in appearance to fourth, not quite so prominent or so strongly sclerotized. Plates as in most other species, those between the 3 pairs of fully developed lobes rather slender and with roughly parallel sides, strongly fimbriate at apices, but usually with one projection longer and more prominent than the others, the 3 plates between third and fourth lobes broad, especially the intermediate plate, and similarly fimbriate along apical margin; 3 of the 4 plates between fourth and fifth lobes also broad, but the fimbriations fewer, shorter and stouter, the fourth plate more slender; plates continued to posterior corner of first abdominal segment, mostly short, stout at base, rounded, tapering to a single projection of moderate length. Paragenitals relatively few in number, characteristically in 4 groups; tabulated range for anterior paragenitals 5 to 8, average 6.5; for posterior paragenitals 5 to 9, average 6.6; totals, range 23 to 29, average 24.6.

Distribution: The distribution of *P. pergandei*, as established after Morrisons revision, is:

America: Southern U.S.A. (Canada and northern U.S.A. in greenhouses only), Mexico, Central America, the West Indies, Brazil, British Guiana, Ecuador, Argentine.

Asia: Japan, China, Corea, Philippines, Formosa, Ceylon–, Sumatra, Siam, Straits Settlements–India, Palestine, Syria, Anatolia.

Australia, New Zeeland–, Hawaii.

Africa: South Africa, West Africa, Africa Minor, Libya, Egypt, Canary Is.

Europe: Mediterranean countries.

Host-relations: The chaff-scale is a polyphagous species in its native countries, probably within the Indo-Malayan region. In a warm and humid climate the species flourishes well on a number of different host-plants. Many hosts are known from the congenial climate of Florida, as against few from the arid California. A similar

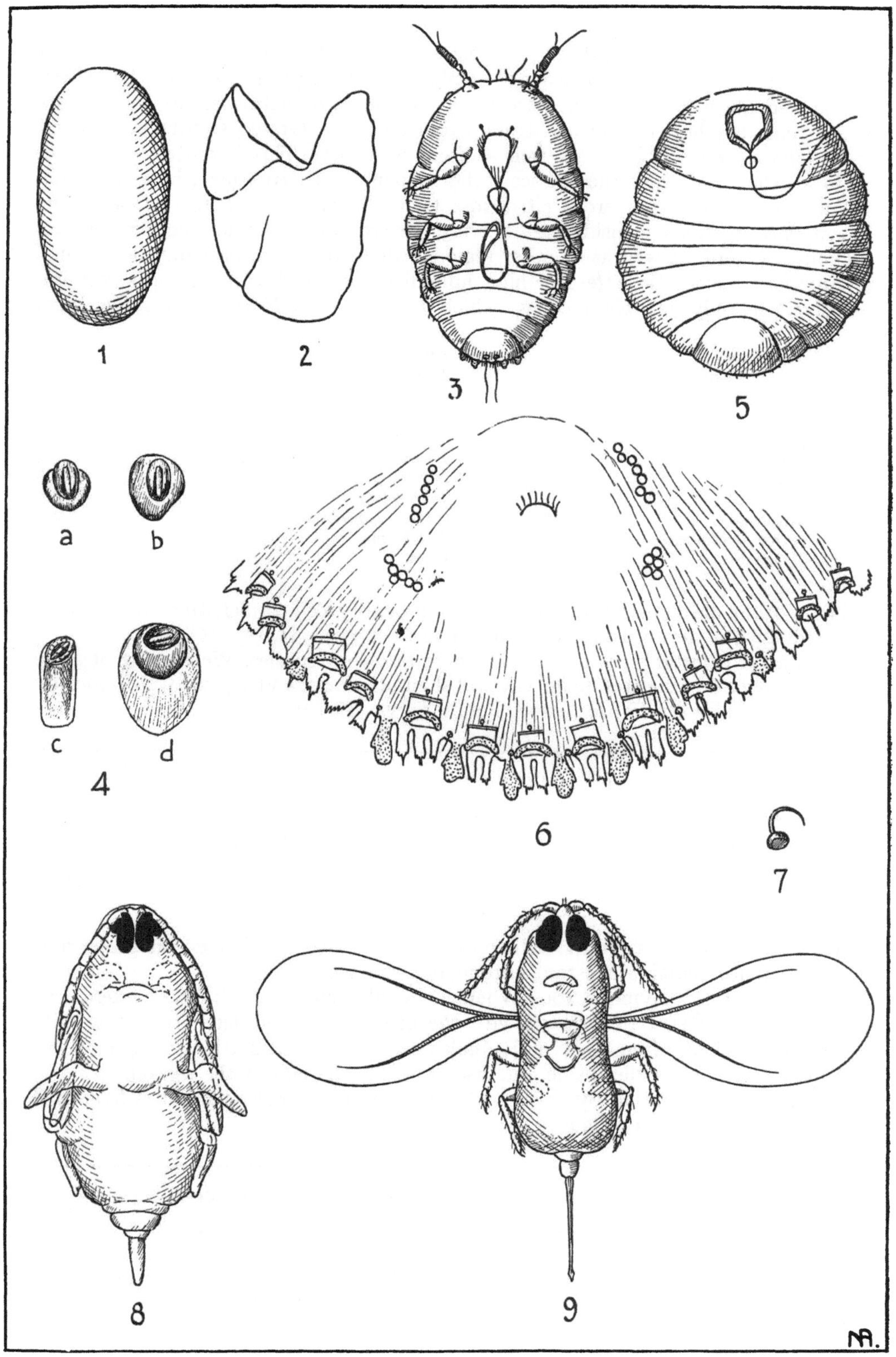

XX. The Chaff Scale (*Parlatoria pergandii* Comst.). 1) Egg; 2) empty egg-skin; 3) neonate larva; 4) scale of: a and b) second larval instar, c) male puparium, d) adult female; 5) adult female; 6) pygidium and 7) antennae of adult female; 8) male pupa; 9) adult male.

difference between the western and eastern half of the Mediterranean region is probably connected mainly with age of immigration. In Algeria it seems to have been introduced quite long ago. In Morocco, where its introduction seems to be quite recent, only one host is known (citrus). From Palestine, Greece, Cyprus, and Egypt it is only known on citrus (as well as from S. Africa). This is quite unexpected, as in all these countries the species rather abounds on citrus (exc. in Egypt) and belongs to the most numerous citrus-insects. In Palestine it cannot belong to the recent introductions, considering its very general and common occurrence. Its monophagy here, as well as in other regions, where there are probably no optimal environmental conditions (espec. high humidity combined with heat), is very impressive.The only reliable records, which remain after Morrisons revision, are:

Buxaceae	*Buxus*
Celastraceae	*Evonymus*
Cornaceae	*Aucuba*
Elaeagnaceae	*Elaeagnus*
Guttiferae	*Garcinia*
Liliaceae	*Yucca*
Myrtaceae	*Myrtus*
Oleaceae	*Jasminum*
Rutaceae	*Citrus, Severinia*
Ternstroemiaceae	*Tutcherma*
Vitaceae	*Vitis*

Many of the older records in literature are probably also correct. But they are here omitted until a new verification. The main host everywhere is *Citrus.*

On Citrus: Infestation is not equally high on every species. We dispose of some data with regard to infestation on leaves of shamuti orange and of grapefruit (monthly average of scales on 100 leaves at Petah Tikva, 1935):

month		I	II	III	IV	V	VI	VII	VIII	IX	X	XI	XII	Sum	Average
no. of scales on	grape-fruit	64	48	31	13	5	6	2	20	26	25	34	49	323	27
	shamuti	94	106	51	33	9	5	10	52	67	40	97	143	707	59

The infestation of shamuti is therefore about twice as high as on grapefruit. This difference remains constant throughout the year.

P. pergandii is polymeric too. Whereas mainly leaves are infested, stem and branches may often be encrusted with the old and dead scales. Fruits are quite normally attacked, often 30-40 individuals on one fruit. Infestation is lower on the outside fruit than on the inside and rises towards the end of the season.

The relative infestation by *living* scales was on one branch at Petah Tikva (tree 23 years old; 1935) as follows:

	no. of	I	II	III	IV	V	VI	VII	VIII	IX	X	XI	XII	Total	
twigs		—	—	—	—	—	—	—	—	—	—	—	—	0	0
leaves	290	1055	929	699	320	44	52	89	233	516	467	288	259	4951	17 per leaf
fruits	5	—	—	—	—	—	—	—	15	44	54	66	—	179	36 per fruit

The number of living scales on a branch of a younger tree (10 years) in the same orange grove was in 1934, as follows:

	no.	sum of living scales	scales per part of tree
twigs	—	9	—
leaves.	1054	6320	6
fruits	4	89	22

In two orange groves at Hedera the infestation of the fruit was lower (20 fruits per month):

month		IX	X	XI	XII	Total
no. of scales in orchard	8 years old	—	—	—	—	—
	27 years old	29	18	26	53	126

The *average* infestation per fruit in November/December fluctuates between 3 and 20 per fruit.

Infestation of citrus trees begins at on age of ten years of age. After the beginning of the infestation, the intensity of attack rapidly gains its maximum. The main reason for this late infestation is the decided shade preference of the chaff scale. This preference expresses itself also in the obvious concentration in the interior of the tree. In an old tree in Hedera lived in the

inner third	medium third	outer third	Total	
3598	1249	915	5762	scales
62	22	16	100	%

Similar results were obtained on an old tree in Petah Tikva (1935). The preference for the interior grows still more distinct, when we relate the number of scales to the number of leaves which are more scarce in the interior. The branch was divided into 4 parts, from the interior to the exterior:

part of branch	sum of scales		percent of scales	
	absolute	per 100 leaves	absolute	per 100 leaves
I (centre)	1850	3755	37	47
II	1357	2463	27	31
III.	1150	1057	25	13
IV (outside)	534	708	11	9

Life-history: Very little is known on the life-history of the chaff-scale. Balachowski (1935) assumes that it has at least 2 generations in the Mediterranean region and deposits 20-30 eggs per female.

Exact observations of the life-cycle are difficult, because length of preoviposition and oviposition period as well as of the egg-stage occur below the scale. Some indications as to the duration of these stages may be obtained only by lifting the scale and thereby interrupting the normal development.

We shall begin with the more obvious length of development (in days) of the larval stages (under laboratory conditions):

Date of infection	Female larva I	Female larva II	Female Larva total	Male larva I	Male larva II	Male Larva total
I.35	52—59	28—35	81—84	59—61	26—28	85—91 days
II.35	41—46	22—35	68—76			
IV.35	19—23	15—18	37—40			
VII.35				12—13	12—13	25
VII.34			30—39			
V.34	10—19	10—16	23—32	10—19	11—13	

These periods last about 50% longer under natural conditions in the orange grove than in the laboratory.

Our data on the preoviposition period are still more scanty. In the two series in V.34 it lasted from 18 to 24 days, in VI.34 from 16 to 21 days.

No data are available for the egg-stage. Of importance is only the fact that under Palestinian conditions no egg diapause occurs and that the incubation of the eggs lasts a few weeks only, at optimal conditions possibly even less.

There are three annual generations in the orange grove in the coastal plain and they overlap and shift seasonally to a certain degree, following the climate of the year. The main larval production takes place April/June, August/September and November/January. This development follows an hyperbola with the following constants (hyperbola B):

development threshold (c) = 7.0° C
thermal constant (th. c.) = 1575 days degrees.

However, more observations are required for this hyperbola. Under laboratory conditions, 4 generations can be easily obtained in the coastal plain. With regard to oviposition, from 20 to 40 females were dissected every month in order to obtain the number of eggs deposited and of mature, intraabdominal eggs.

month	Leaves deposited M	Leaves deposited max.	Leaves intraabdominal M	Leaves intraabdominal max.	Leaves total M	Leaves total max.	fruits deposited M	fruits deposited max.	fruits intraabdominal M	fruits intraabdominal max.	fruits total M	fruits total max.
II.34	10.1	20	4.6	9	14.7	29	12.7	25	11.1	18	23.8	43
III.	8.4	15	8.2	12	16.6	27						
IV.	11.0	19	8.3	15	19.3	34						
V.	12.5	28	6.0	11	18.5	39						
VI.	8.8	17	5.1	13	13.9	30						
VII.	9.4	20	9.3	16	18.7	36						
VIII.	9.0	15	8.2	12	17.2	27						
IX.	9.8	16	8.5	22	18.3	28	13.5	27	11.8	18	25.3	45
X.	16.4	37	8.1	13	24.5	40	20.7	32	9.8	15	30.5	47
XII.	7.7	18	6.9	11	14.6	29	7.3	18	7.8	16	15.1	34
total average	10.6	20.7	7.3	12.9	17.9	33.6	14.0	25.5	10.2	16.8	24.2	42.3
average in the 4 months when fruits were counted	10.8		7.0		17.8	31.5	14.0		10.2		24.2	42.3

The average as well as the maximum egg-production on the fruit exceeds by about 35% that on the leaves. This greater fertility on the fruit is in agreement with observations on other citrus scales.

The numbers of eggs ascertained in these observations do not represent the total egg-production. The oviposition lasts, as a rule, for some months. It lasted 76 days with a female which began to oviposit in early May. On the average, 1-2 eggs are deposited daily. Assuming that the hatching larvae correspond to different days of oviposition, the following table may serve as illustration (infestation late in January in the laboratory, first larvae hatching in early May) the last in mid-July:

Weekly larval production of some females.

	A	B	C	Average daily rate of oviposition
3.V.1935			2	0.1
10.	8	8	8	1.2
17.	11	9	1	1.0
24.	9	6	5	1.0
30.	7	7	3	0.8
6.VI.	4		3	0.5
13.	7		6	0.9
20.			6	0.9
28.			2	0.3
4.VII.			4	0.6
10.			8	1.1
18.			4	0.6
Final control				
living larvae	2	1	2	
dead larvae	19	10	25	
eggs (laid intraabdominal) . . .	5	18	11	
empty egg-shells	13	14	7	
larvae removed.	46	30	52	
Total larval production.	72	59	90	0.75

The dead and living larvae, as well as the eggs discovered in the final control are not included in the average daily count, nor in the weekly list. This means that 42% of the real production was not included in the calculation. The real average daily oviposition rate is therefore 1.3.

In those few cases in which the total oviposition was apparently obtained in laboratory breedings the result was:

age in days	on sour orange	on sweet lemon
20	59	88
26		107
33		128
55	72	
64	57	
78	90	
Average	69.5	107.7 eggs per female

These data seem to indicate that the daily rate is somewhat higher for the first weeks of the oviposition and may approach 3—4 eggs per day in order to fall rapidly later on. It is not improbable that there exists a significant difference in the fertility on orange and lemon leaves. Our data are too scanty to allow any definite conclusion. 87.9 eggs per female is the average of all observations.

No exact sex-ratio have been established. It probably approaches equality. The males show no preference for the upper or lower side of the leaves. The percentage of male puparia was as follows:

year	locality	I	II	III	IV	V	VI	VII	VIII	IX	X	XI	XII	Average (comparative)
1934	upper side	—	17	23	6	10	—	—	24	9	7	—	32	15%
	lower side	—	31	21	6	—	—	—	0	0	14	—	—	12
1935	upper side	14	12	2	10	—	—	—	—	7	15	18	13	11
	lower side	25	31	7	17	—	—	—	—	7	13	21	14	18

The longevity of the adult males is as short as that of armoured scales in general, 1—2 days on the average.

A further analysis of the fluctuations of the chaff scale population is given in the following tables, according to the two branches observed at Petah Tikva:

1934, 12 years old tree.

month	I	II	III	IV	V	VI	VII	VIII	IX	X	XI	XII
L I	—	34	21	2	5	6	0	0	26	1	6	54
L II	—	508	431	63	38	30	3	56	84	93	19	24
♂	—	534	414	23	30	16	0	42	92	87	16	189
♀	—	659	581	190	81	106	66	201	368	350	55	585
Sum living	—	1735	1455	278	154	158	69	299	570	531	96	1073
Sum dead	—	454	117	242	79	19	37	32	83	192	16	40

in %

	I	II	III	IV	V	VI	VII	VIII	IX	X	XI	XII
L I	—	2	1	1	3	4	0	0	4	0	6	5
L II	—	29	30	23	25	19	4	19	15	17	19	23
♂	—	31	28	8	19	10	—	14	16	16	16	18
♀	—	38	41	68	53	67	96	67	65	67	59	54
Sum living	—	100	100	100	100	100	100	100	100	100	100	100
Sum dead	—	21	7	43	34	8	35	10	13	28	14	4

1935, 23 years old tree.

month	I	II	III	IV	V	VI	VII	VIII	IX	X	XI	XII
L I	117	43	44	25	1	0	0	3	18	24	20	13
L II	316	261	196	88	10	18	22	24	75	142	80	81
♂	208	184	117	41	3	5	8	12	32	39	21	39
♀	414	441	334	166	30	29	59	194	391	262	167	126
Sum living	1055	929	691	320	44	52	89	233	516	467	288	259
Sum dead	98	51	78	147	140	29	11	14	54	231	98	83

in %

	I	II	III	IV	V	VI	VII	VIII	IX	X	XI	XII
L I	11	5	6	8	2	0	0	1	3	5	7	5
L II	30	28	28	27	23	35	25	10	14	30	28	31
♂	20	20	17	13	7	10	9	5	6	8	7	15
♀	39	47	49	52	68	55	66	83	77	57	58	49
Sum living	100	100	100	100	100	100	100	100	100	100	100	100
Sum dead	9	5	10	32	76	36	11	6	10	35	25	24

These figures are illustrated by a diagram, which needs some remarks. The relation of the different stages is based on the relative longevity of every stage. When the first larval stage lasts 10 days and the adult female stage 60 days, then the adult females will be six times as abundant in an equibalanced population as the first larvae. This is the case with *P. pergandii*. Whereas the adult females are probably more or less

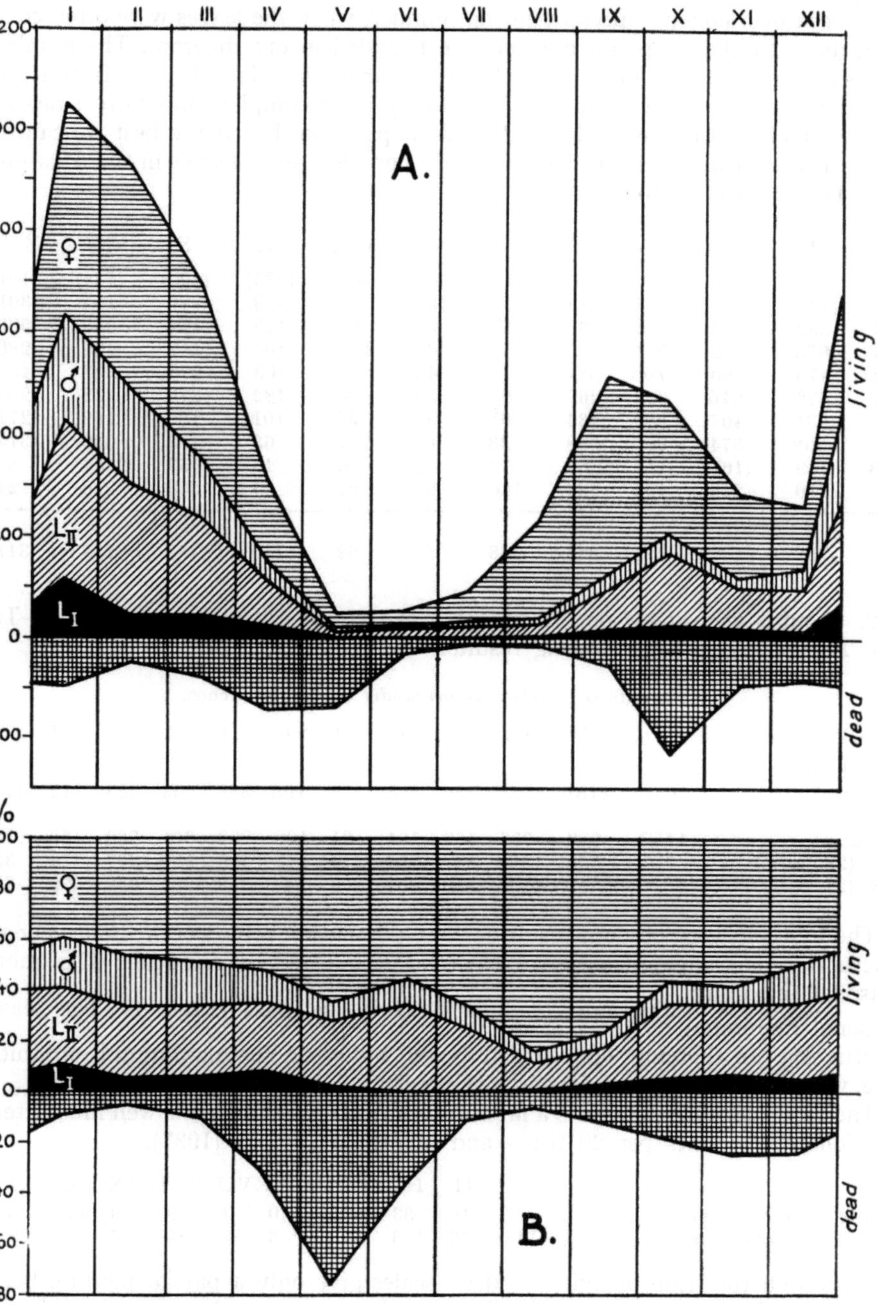

Fig. 83.
Monthly fluctuations of Chaff Scale population (*Parlatoria pergandii* Comst.) of citrus tree (shamuti) at Petah Tikva 1935. A. absolute numbers; B. relative numbers.

completely included in our monthly census, it is certain that many of the first larvae have been overlooked, especially in the summer, when the leaves were often covered with dust. Neither eggs nor crawlers are included in our diagram. The number of males does not refer to adult males, but to the male third larval stage. In conclusion, it may be stated, that the diagram is certainly not a complete and correct picture of the seasonal fluctuation of the chaff scale population, but is the best we can offer.

1. The monthly number of *P. pergandii* per 400 orange-leaves in the Atta-groves near Hedera was as follows:

Year	I	II	III	IV	V	VI	VII	VIII	IX	X	XI	XII	Total	Max./Min.
1928	—	—	—	—	—	(15)	—	(69)	(235)	(109)	(124)	(151)		
1929	—	(121)	—	445	379	52	138	—	459	90	155	301		
1930	406	532	**891**	656	561	87	98	**45**	158	164	288	506	4392	20
1931	523	**673**	588	537	163	55	**44**	106	151	158	439	286	3723	15
1932	502	475	660	**705**	584	112	**42**	59	163	273	297	439	4311	17
1933	392	472	**516**	343	267	**57**	59	84	182	376	322	267	3337	9
1934	**757**	234	(407)	580	134	61	84	**24**	101	120	162	271	2835	32
1925	293	**669**	574	605	398	**28**	67	112	63	198	342	317	3666	24
1936	**258**	189	(167)	144	69	73	32	**6**	15	57	54	88	1152	43
1938	107	280	216	155	225	186	81	92	64	203	216	124	1847	4
Average 1930/36	447	392	**543**	516	311	68	**61**	62	119	192	272	311		9

2. Counts were made on full branches of trees of different ages at Petah-Tikva and Hedera, with the following results:

Counts of *Parlatoria pergandii* on full branches.

Year	Grove and its age	I	II	III	IV	V	VI	VII	VIII	IX	X	XI	XII	Total	Max./Min.
1934	Petah Tikva (12 years)	(1651)	**2189**	1257	520	233	177	**106**	331	953	723	112	1113	9365	21
1935	Petah Tikva (23 years)	**1153**	980	777	467	184	**81**	100	262	614	760	460	342	5980	14
1935	Hedera (28 years)	775	553	**1303**	887	468	230	241	305	**185**	207	279	351	5764	7
1934	Hedera (27 years)			**1011**	513	408	208	**141**	318	232	630	679	791	(4931)	7

The first and very obvious result of these population counts is the relatively small fluctuation as compared with young orchards infested by red scale. The quotient $\frac{\text{seasonal maximum}}{\text{seasonal mimimum}}$ fluctuates in both methods between 7 and 32. The seasonal fluctuation can therefore be regarded as well buffered. The maximum is obtained in late winter, the minimum in early summer.

The fact that *P. pergandii* is a late invader of citrus-orchards is well illustrated by the following counts per 20 fruits and month in Hedera (1937).

	I	II	III	IV	V	VI	VII	VIII	IX	X	XI	XII
Young orchard (7 years)	17	40	33	34	—	0	0	0	9	17	30	42
old orchard (30 years)	148	173	136	—	—	3	29	41	37	50	78	95

Counts on the ratio of living to dead scales give only a partial picture, because neither the climatic influence on fertility, nor the egg- or crawler-mortality is included and all dead scales which have dropped or have been washed off by rains are eliminated. However, the diagram shows the decimating influence of hamsins in spring and autumn very clearly. The havoc caused by the spring hamsins, especially, is well illustrated by the following independent counts in Petah Tikva, in 1935:

month	I	II	III	IV	V	VI	VII	VIII	IX	X	XI	XII
mortality: all stages	32	23	47	53	83	87	67	42	50	57	54	44 %
mortality: adults only	54	57	52	63	93	93	92	—	47	29	34	36 %

It is difficult to evaluate the optimal climogram for *P. pergandii* from the fluctuation in Palestine. We shall compare the climate of the months in which the highest relative increase occurs (August/September as a rule), with that of the maximum population:

Year	temperatures and humidities of highest increase		temperatures and humidities of maximum population	
1930	28.5/27.2° C	74/71 % R.H.	16.4° C	73 % R.H.
1931	28.0/28.6	75/72	14.1	69
1932	26.7/27.5	73/74	18.4	65
1933	26.5/24.5	79/77	15.6	59
1934	25.9/23.3	79/74	13.3	76
1935	26.3/27.1	78/76	14.4	78

Under Palestinian conditions, 24—28° C and 70—80% Rel. Hum. are optimal. The months falling within this zone are responsible for the great summer increase. The lower temperatures and high humidities of winter are still favorable, whereas the

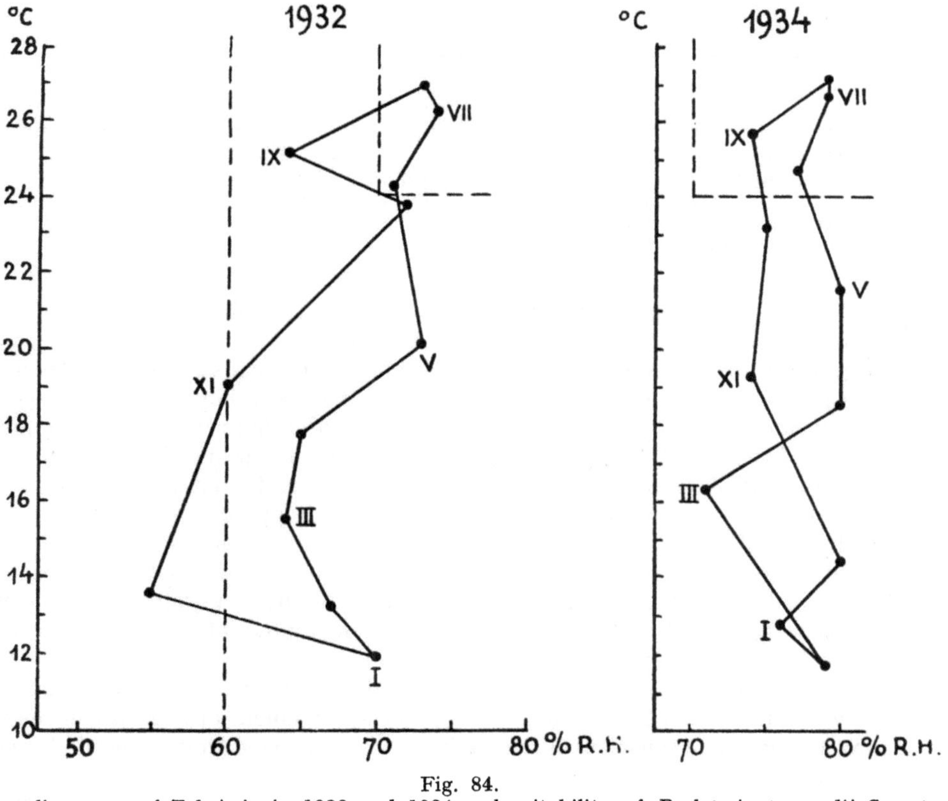

Fig. 84.
Eco-climogram of Tel Aviv in 1932 and 1934 and suitability of *Parlatoria pergandii* Comst.

dry winds of spring are fatal. The optimum regions for *P. pergandii* are undoubtedly regions with humid summers, and probably the humid tropics. The limits of the fair zone may be put tentatively at 10—35° C and 60—100% Rel. Hum.

Natural enemies are few in number. In a few cases *Aphytis maculicornis* Masi (Hym. Encyrt.) was bred from April to August. *Chilocorus bipustulatus* was active from February to April. However, both enemies have no influence on the seasonal fluctuation of the scale, as they are both absent during the main period of increase. Occasional fungal diseases will be discussed later.

A few observations on reproduction in the orange grove are worth-while mentioning The first series is connected with the exposure to wind. Some leaves were marked in a well-protected and in a wind exposed part of the grove:

		wind protected	wind exposed
no. of observations		28	11
observations without settled larvae		14 %	46 %
no. of settled larvae	only positive observations	8.3	9.5
	per total average	7.1	5.2
	maximum	27	14

The influence of wind on chaff-scale is obviously smaller than on other scales. A second series compares the actual reproduction of some females at different seasons:

		summer	winter
no. of observations		53	9
observations without settles larvae		42 %	89 %
no. of settled larvae	only positive observations	8.4	1.0
	per total average	4.9	0.1
	maximum	27	1.0

Rains, storms and lowered reproduction are probably the main reason for the low number of a successfully settled offspring in winter.

Damage and control: In spite of its wide occurrence in the citrus groves of Palestine, and its heavy infestation, the chaff scale is a minor citrus pest only. Two main points deserve consideration: *P. pergandii* does not normally induce any conspicuous intoxication of the host-tissnes, so often connected with red scale. We have only in years of outbreaks observed the chlorotic spots surrounding the later so often on leaf and fruit. We have also never observed desiccation or dropping of leaves connected with chaff scale. Secondly, the scale on the fruit is very unconspicuous. It settles down in the large pores of our shamuti and its greyish colour combined with its small size do not lead to a deterioration of the fruit. As a rule, fruits infested with 20—30 chaff scales pass the fruit-inspection, whereas the same number of red or black Scales are declared unfit for exportation.

In the season 1937/38 for the first time considerable quantities of culls were caused by *Parlatoria*. The scales had often a darker, black-grey colour or a brownish tinge, which made them more conspicuous. In addition, many infested fruits showed a green focus around the scale.

Whereas no control measures were applied before 1938, some experiments were performed in that year by A. Rotman (1939) at Hedera and checked by Klein (1939) The results were (only grade 3 is unfit for export):

Plot	Control measures	no. of graded fruit	grade 1	grade 2	grade 3
A	Spraying with white oil 1¾ nicotine 1 : 1200.	1556	89.3	9.4	0.8
B	dto. (1½ % oil).	1252	92.3	7.0	0.7
C	Fumigation (Cyanogas) . .	1689	99.8	0.2	0.0
D	Control plot	985	77.9	20.9	1.8

Both, fumigation and spraying with white oils, gave satisfactory results.

Even in 1939 Klein concluded, that the economic loss does not justify the expenses for a special spraying or fumigation campain. Every control of other citrus scales will also considerably reduce the grey chaff scale. There were rumours that the chaff scale had considerably increased on the fruit during the transport to the European markets. In order to check this rumour, Klein retained packed boxes of oranges for one month under favourable conditions for development of the scale. During that period the number of scales had not increased. Considering this opinion, which we share, we cannot recommand, under present conditions, the application of any special control measures.

References.

A. Balachowsky et L. Mesnil, Les insectes nuisibles aux plantes cultivées. Paris, 1935. pp. 499—501.
F. S. Bodenheimer, Schaedlingsfauna Palaestinas. Berlin, 1930. P. 160.
H. Z. Klein, Hadar **12**. 1939 p. 179—181.
I. Kuwana, Dept. of Agric. and Commerce, Japan. Insect series No. 14. 1925.
G. Leonardi, Monografia delle Cocciniglie Italiane. Portici, 1920. pp. 145—149.
H. Morrison, U.S. Dept. of Agric. Miscell. Publication No. 344. 1939.
A. Rotman, Hadar **12**. 1939. p. 149.

b. The Tropical Grey Chaff Scale (Parlatoria cinerea Hadden).

Description: The following technical description is given from Morrison:

Adult female: Eyespot normally present, but very obscure and easily overlooked, only slightly differentiated from adjacent derm. Not located on every specimen examined, and perhaps actually wanting in some. Each anterior spiracle with a number of disc pores adjacent, range from 5 toll in material examined; no. dermal pocket observed between posterior spiracle and margin, but a straggling row of small tubular ducts in this area. Duct tubercules varying considerably in shape and size, from relatively short, broad, rounded conical prespiracular 1 to 4, anterior spiracular 3 to 4, intermediate 2 to 4, posterior spiracular 1 to 4, first abdominal 1. Submarginal dorsal tubular macroducts intermediate in number, extending from apex of pygidium onto second abdominal segment; numbers for each side of body ranging from 37 to 50. Intermediate tubular macroducts present, characteristically, on each half of body, with a row of 3 to 9 above of just outside paragenitals and still within sclerotized area of dorsal surface of pygidium, the posterior end of this cluster often joining the adjacent submarginal ducts; no additional intermediate or median dorsal ducts, either macro or micro, observed anterior to pygidium. A circular to oval cicatrix near body margin dorsally on first abdominal segment, this sometimes obscure. Characteristically with 3 small, oval, dorsal, submarginal, sclerotized flecks independent of other pygidial sclerotization on each side, the last located just anterior to the fourth lobe, the other 2 on 2 segments preceding that bearing this lobe. Lobes strongly protruding, median largest, notched once internally and usually twice externally, second and third proportionally reduced, entire along inner margin, usually twice notched along outer

A.

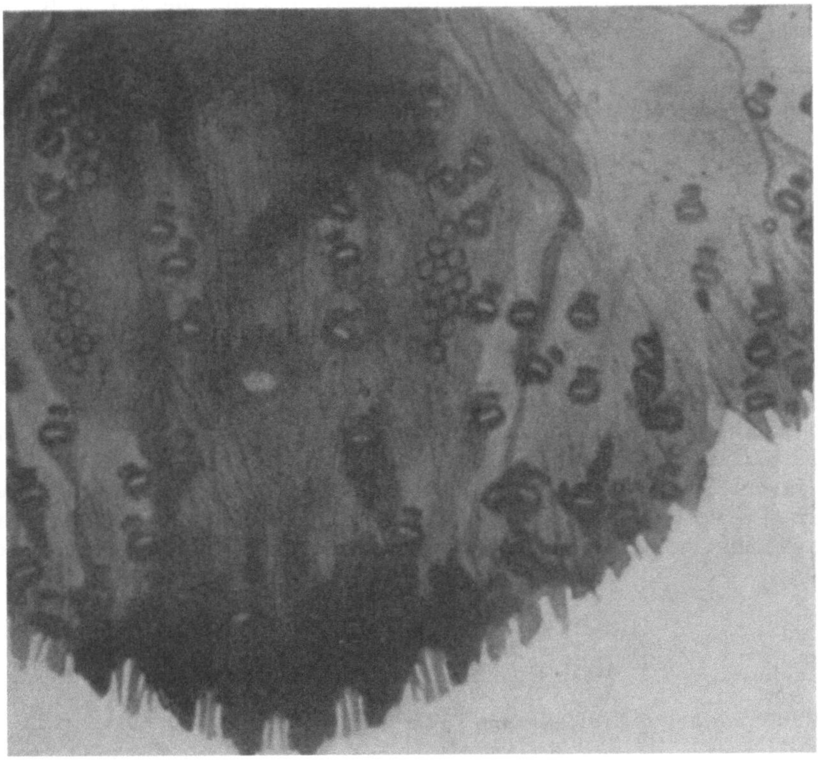

B.

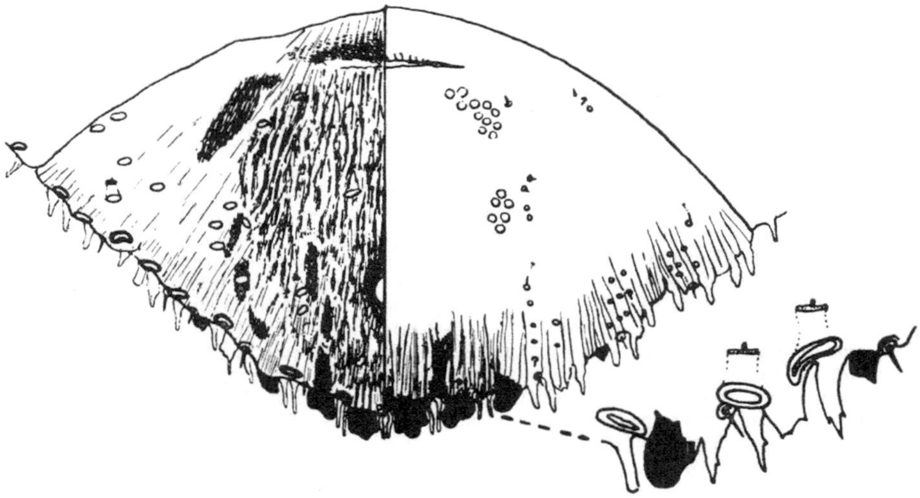

Fig. 85.
Pygidium of: A. *Parlatoria cinerea* Hadd. (from Morrison) B. *P. morrisoni* Bdhmr. (from Bodenheimer).

margin, notches in all lobes sometimes reduced or increased in number; fourth lobes represented by a broad, slight protrusion, not or slightly sclerotized, outer margin paralleling line of pygidial margin, and with or without a few minute teeth. Plates elongate, each narrowing to end in a conspicuous linear elongation, margins toward the apex more or less fimbriate, protrusions short; those plates between the lobes slender, ligulate, those outside the fourth lobes much broader at base and tapering strongly to the slender apex; 3 plates between third and fourth lobes intermediate, broader than those inside third lobes, but elongate and tapering gradually. Paragenitals intermediate in number, characteristically in 4 groups, but sometimes with a single pore representing a median cluster; numbers in specimens tabulated as follows: Anterior median 0 to 1; anterior laterals, range 7 to 16, average 10,66; posterior laterals, range 8 to 14, average 11,85, totals range 33 to 57, average 46.1.

The outstanding differences between *P. cinerea* and *P. pergandii* are:

Character	P. cinerea	P. pergandii
Macroducts between anal-opening and circumgenital glands	3 ducts distinctly present on each side	Absent
Lobes	Twice or thrice notched on outer side	Once only notched on outer side
Plates	Mostly tapering, ending in one elongate extension	Not elongate tapering

Thanks to the excellent analysis of Morrison (1939), the differentiation of both species has now become easy.

Distribution: The home of *P. cinerea* is doubtless the Oriental region, where it is extensively spread. Morrisons records include India, Java, Straits Settlements, Siam, Indo-China, China, Philippines, Formosa.

In America it is now recorded for many countries: Argentine, Brazil, Dutch Guiana, West Indies, Central-America, Mexico, and in greenhouses in U.S.A. In the Pacific it has been found in Samoa, New Caledonia, Society Isl., Marquesas Isl., Tahiti. On one single occasion only the species has been intercepted by the U.S.A. plant-quarantaine from Italy, Spain and South-Africa. One such interception may be based on an error of provenience and is, anyhow, insufficient proof for the establishment of the species in these countries.

The situation is different for Palestine, from where 5 interceptions have been made in different years by the same service, one from a mail shipment.

Hosts: The species is probably rather polyphagous in the Oriental region. Owing to its cryptic existence until rather recently, only the following hosts are safely established, Citrus being by far the most recorded host-plant

Anacardiaceae	*Mangifera*
Caprifoliaceae	*Viburnum*
Nyctaginaceae	*Bougainvillea*
Oleaceae	*Jasminum*
Rosaceae	*Rosa*
Rubiaceae	*Gardenia*
Rutaceae	*Citrus*

Inspite of two special surveys of citrus orchards from all citrus districts of this country, especially on citron fruit, we never saw this easily identifiable species. We must therefore assume, that the species is not present in Palestine (or any of the other regions of the Middle East), and that the patria declarations to the U.S.A. Plant Quarantaine Service were incorrect.

Two further species of *Parlatoria* appear in the Middle East on citrus: *P. morrisoni* Bdhmr. from S. Iran and S. Irak, and *P. zizyphi* Luc. from N. Iran, Trans-Caucasia

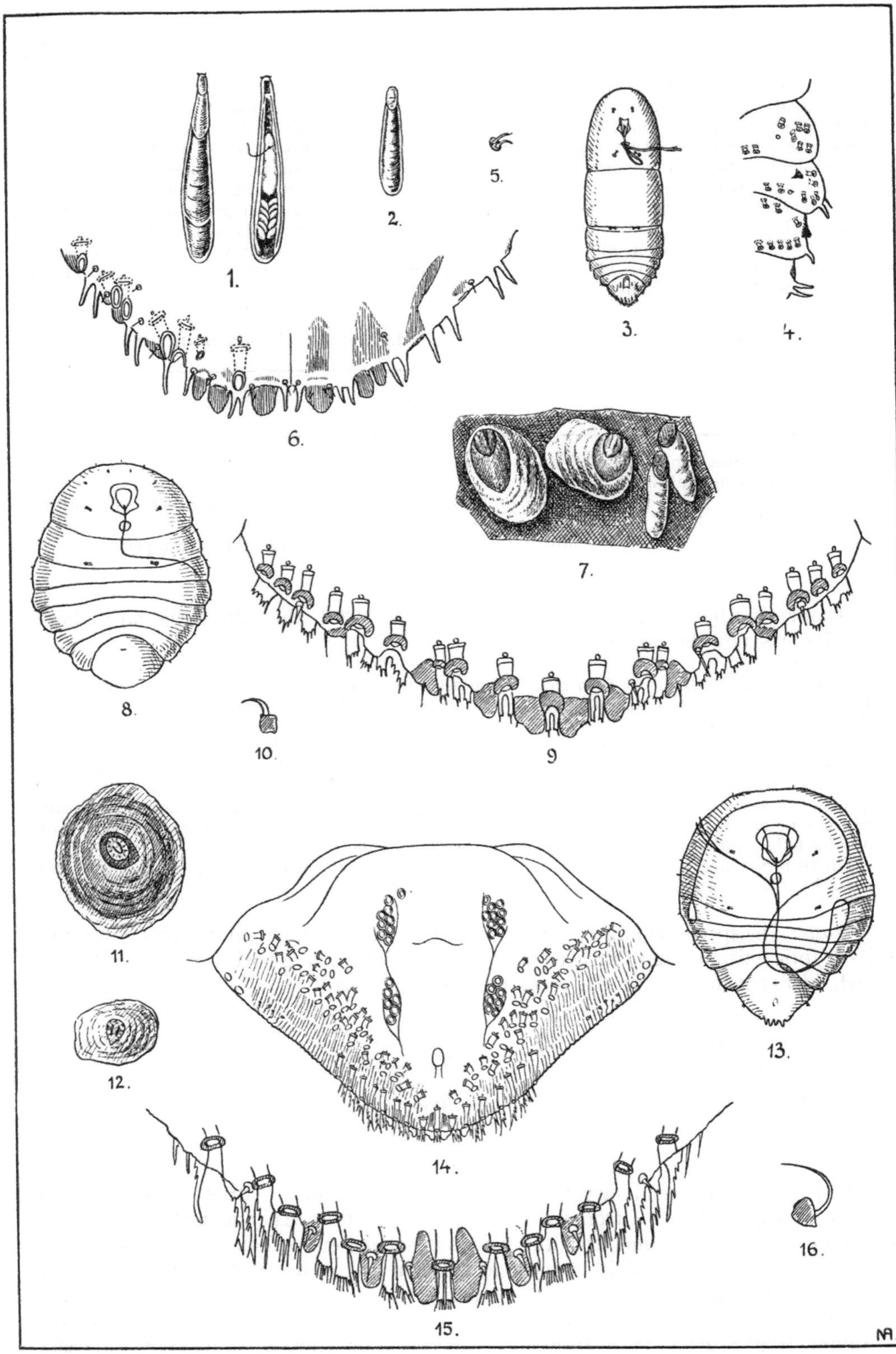

XXI. Minor Citrus Scales. 1—6) *Mytilococcus gloveri* Pack. 1) Female scale from above and from below; 2) male puparium; 3) adult female; 4) lateral margin of the three free abdominal segments of adult female; 5) antenna and 6) pygidial margin of adult female; 7—10) *Parlatoria oleae* Colv.: 7) Scales of adult females and male puparia; 8) adult female; 9) pygidial margin and 10) antennae of adult female; 11—16) *Aspidiotus hederae* Vall. 11) female scale; 12) male puparium; 3) young adult female; 14) pygidium, 15) pygidial margin and 16) antenna of adult female.

and the Lebanon. They are discussed later. Both are easy to identify by the aid of the key on p. 131/2.

C. *The Olive Parlatoria* (*Parlatoria oleae* (*Colv.*)

Synonymy: 1880 *Diaspis oleae* Colvée
1895 *Parlatoria calianthina* Berl. e Leon.
1897 *P. affinis* Newstead

Description: Scale of adult female approximately circular, distinctly convex, dusky white tinged with grey or brown. Exuviae excentric, pale yellow to green brown or almost black with a thin transparent secretionary covering. 1.5—2.5 mm long.

Scale of male pupa subrectangular; very elongate, three times as long as broad. Exuviae terminal. Opaque white, non-carinated, with the exuviae yellow brown. 0.7—1.1 mm long, 0.3 mm broad.

Adult female broad oval, broadest just behind the centre of the body. Segmentation distinct. Antennal tubercle with a short stout bristle. 4—10 anterior peristigmatic glands. 1.0—1.3 mm long. Colour wine-red with brownish pygidium.

Pygidium is broadly round on its free margin. Its formula is: P_1, L_1, $2P_2$, L_2, 3—$4P_3$, L_3, 4—$5P_4$, L_4, 3—$4P_5$. Lobes decrease in size from L_1 to L_4, the latter being short with 2—3 teeth. Plates grow broader from P_1 to P_5. The short tubular, marginal glands well developed. Circumgenital glands as follows:

$$\frac{0\text{—}10}{\frac{12\text{—}31 \qquad 12\text{—}31}{7\text{—}23 \qquad 7\text{—}23}}, \text{ generally } \frac{0\text{—}7}{\frac{14\text{—}26 \qquad 14\text{—}26}{10\text{—}23 \qquad 10\text{—}23}}$$

The dorsal tubular glands smaller than the marginal ones.

Egg: Broad oval, 2.5 mm long, 1.8 mm broad.

Neonate larva oval, yellow with a light wine-red tinge. 0.3 mm long.

Distribution: Common in most Mediterranean countries, especially in Africa Minor, Italy, Greece, Palestine. Occurring also in Crimea, on the western foothills of the Caucasus, in N. Iraq (C. Asia), in the Himalaya. In Egypt it is quite common, but certainly introduced. Two reports from S. India (on Mango at Rajputana and on Michelia at Bombay) need verification. Recently introduced into Arizona. The species is in native of the Irano-Turanian region of Asia, where it abounds on wild rosaceous shrubs.

Hosts: The species is common everywhere on Rosaceae (*Prunus, Pyrus, Eriobotrya, Rosa*, etc.), Oleaceae (*Olea, Jasminum, Ligustrum, Fraxinus*), Apocynaceae (*Nerium*). Occasionally occurring also on *Vitis* and *Citrus*. From *Citrus* it is reported: by Lindinger (1912), from the Canary Islands and from Sicily; by Lepiney and Mimeur (1931) from Morocco and by Bodenheimer (1930) from Palestine. Nichol and Wehrle (1935) report it on this host also from Arizona, but state that no evidence of successful breeding has been obtained. On *Citrus*, it probably always occurs only occasionally, and then mainly on stem and branches, as has been observed by us in Palestine.

Biology: The biology of this species, which is a pest of increasing importance on *Pyrus* and *Prunus* species, is known only for the hill-country. At Kiryath Anavim, in a protected valley 12 klm. west of Jerusalem four annual generations occur:

Generation	I	II	III	IV
Oviposition	12.IV.37	19.VII.37	23.VIII.37	27.XI.37
Second moult of female . . .	22.VI.	18.VIII.	13.X.	
Adult males	27.VI.	22.VIII.	–	
Eggs per female.	55	66	56	29.
Length of generations	99	36	97	136 days

Damage and control: Its damage to *Prunus* and *Pyrus* will be described elsewhere. On citrus it has never been of any economic importance and no control measures were necessary on this host. Spraying with universal winter wash in early spring is quite efficient on other fruit trees, and has the benefit of stimulating early and equal flowering.

References.

A. A. Nichol and L. P. Wehrle, Arizona College Agric. Techn. Bull. 56, 1935.

H. THE NATURAL CONTROL OF ARMOURED SCALES (*Diaspididae*) OF CITRUS.

The armoured scales of citrus do not lack natural enemies, though the number of the latter is much smaller in Palestine than in tropical regions or even in the western Mediterranean basin. The reason for this is not purely a matter of lack of research, but apparently the intensity of attack is actually lower in Palestine than elsewhere. However, even in countries where parasites have been well studied and where the list of parasites and predators of armoured scales on citrus is by no means small, as in the case of Italy, S. France, California, Florida, etc., no economic importance has been attached to the combined effect of all these enemies. The repeated trips of Compere covering almost all continents in search of red scale parasites have yielded quite a series of enemies, yet give little hope for any biological control which would make technical control dispensable. The reason for this scarcity of parasites is not always clear and well understood, and will be discussed for Palestinian conditions for each insect separately. One comes to the general conclusion that the number of these parasites is limited mainly by 1) the scarcity of food from May to July, 2) the great mortality in dry heat. In both cases, which often may cooperate, the parasite population is so reduced at the end of summer that after some months only it reaches a level of marked activity. By this time the scale population has reached or passed its peak. Considering the enormous zero environmental resistance (R_0) of the scales at this season, it is questionable to what degree the combined biological control contributes to the decline of the scale population. Even this small contribution should be appreciated, though it is rather hopeless to increase its efficiency to a degree at which it could replace technical control. Yet in view of the economic interests involved, we are not to neglect even the slightest trace liable to lead to success. The thorough biological study of all parasites, predators and epizootics of the armoured scales should not be abandoned until we are convinced, after the detailed study of ecology and epidemiology of every enemy, that this special factor of natural control cannot be developed into an efficient beneficial one.

The list of such species is small at present. It includes Chalcidid wasps, among which *Aphytis chrysomphali* and *A. maculicornis* are ectoparasites and *Aspidiotiphagus lounsburyi* lives endoparasitic. Among the ladybeetles *Chilocorus bipustulatus* is sufficiently common to be of interest. Some fungous diseases caused by *Aspergillus depauperatus*, *Aspergillus sp.* and *Cladosporium herbarum* appear occasionally in epizootics.

The further programme for biological control of citrus armoured scales in Palestine should include:

1) Further detailed studies of both species of *Aphytis* as well as of *Aspidiotiphagus*.
2) Further detailed studies of their fungous diseases.

3) The biology and economic value of some other enemies neglected, hitherto, as *Coniopterygidae*, *Psocidae*, thrips, bugs and mites should be studied.
4) Attempts to introduce and breed those enemies in Palestine, which promise to be of possible value as *Comperiella bifasciata* How., *Casca chinensis* How. (Compere-Smith) and others.

The intensive campaign against *Aspidiotus destructor* conducted at present in Fiji by means of biological control (Taylor) and the results obtained in California should be followed with greatest interest. The results obtained in Fiji are of doubtful value for Palestine, considering the difference in climatic conditions, whereas those of California would probably be of greatest immediate value for this country, too. *C. bifasciata* has recently been introduced into Palestine and proper experience will be gained within a few years. So far, its acclimatisation has not succeeded.

1. HYMENOPTERA.

Key to the parasites of armoured scales on citrus. (The species not discovered in Palestine yet, but probably present, are included in brackets):

1. Mesonotal scutum not divided by any parapsidal furrows. Median legs with strong and large tibial spur, which is hairy and dentate. 2.
   Mesonotal scutum clearly divided by parapsidal furrows. Tibial spurs on median legs of moderate development only. 3.
2. Antennae 11-jointed (female) or 9-jointed (male). *Aphycus hesperidum* Merc.
   Antennae 6-jointed (apparently only 3-jointed). (*Signiphora merceti* Mal.)
3. Antennae 6-jointed. 5.
   Antennae 8-jointed. 4.
4. Anterior wings almond-shaped. 0.5—0.6 mm long. *Aspidiotiphagus citrinus* Craw.
   Anterior wings with subparallel anterior and posterior margins in the apical half. 0.4—0.5 mm long. *Aspidiotiphagus lounsburyi* B. & P.
5. Antennae, thorax and abdomen uniform „amarillo". *Aphytis chrysomphali* Merc.
   End of antennae, thoracic furrows and sides of the first 5 abdominal segments slightly black. *Aphytis maculicornis* Masi.

a. Aphytis chrysomphali Mercet (*Chalcididae*).

Description of adult female: Body light lemon-yellow, including antennae, legs and wing veins. Eyes olivaceous; apex of mandibles redbrown; central portion of posterior margin of scutellum blackish.

Head as large as thorax; frons hairy; antennae 6-jointed. Pronotum short. Mesonotal shield pentagonal, broader than long, with 10 hairs, arranged as a rule $\frac{\frac{3 \quad 3}{1 \quad 1}}{1 \quad 1}$. Median mesonotal line well visible. Anterior wings broadly rounded at distal end, as long as the body. Posterior wings lanceolate with blunt apex. Abdomen slightly longer than thorax and as broad as the latter. About 0.6 mm long.

The males are rare and always greatly outnumbered by the females. The sex-ratio should be studied in different regions and at different seasons.

The *egg* is elliptical and transparent. The micropylar end is bluntly pointed and bears a short appendage of varied form. The egg is twice as long as broad.

The *larva* is transparent and colourless, except for the gut which is greenish to brown according to the food contents. It shape is roughly oval, the maximal width being a little over two-thirds of its length. Anterior end semi-circular; posterior half tapering rapidly to the blunt anal extremity. The larva is flat, especially in the latter stages of development. Body 14-segmented; 8 pairs of spiracles are present (Taylor). When young 0.18 × 0.16 mm long, full-grown 0.5 × 0.4 mm.

The *pupa* varies considerably in color, usually of a rich light yellow-brown. It is rather flat; the dorsal side turned towards the central scale is quite flat, the ventral side being slightly convex. A dark reddish-brown, longitudinal band is often seen ventrally occupying the median third of the width, being broader and forked at the posterior end. It extends from the first thoracic to the first abdominal segment (Taylor). About 0.6 mm long and 0.3—0.4 mm broad.

Distribution: Mediterranean regions: Spain, S. France, Algeria, Italy, Greece, Palestine, Egypt. Elsewhere: Southern N. America, Brit. Guiana, Pacific Islands (Fiji, Tahiti), Java, Africa.

Host-relations: It has undoubtedly a wide range of hosts among armoured scales, but is possibly restricted to the Aspidiotini tribe.

Life-history: Taylor (1935) who studied the species in Fiji mainly on *Aspidiotus destructor*, reports the following details concerning its life history: „The egg is usually laid under female scales which have already passed the second moult. Scales which have already laid a few of their eggs are preferred, but those which have laid nearly all their eggs are usually avoided. The egg of the parasite is always situated between the scale insect and the leaf. In the host is lifted with a needle the egg remains attached to it (by its stickiness) and can then be seen fairly easily. Superparasitism occurs commonly. It is possible for as many as four parasites to feed on one host simultaneously and to reach maturity, but such individuals are invariably small. The largest number of eggs found on one host was five.

The larva begins to feed as soon as it leaves the egg. It feeds by puncturing the ventral skin of its host with its mandibles and sucking out the body contents, the mouth area being pressed closely against the skin of the host to form a sort of sucker. The activities of the host cease as soon as the parasite begins feeding, but the host does not actually die until about a day later. Although the scale-insect is not affected in any way at the time the egg is laid, and even continues ovipositing, if of a suitable age, with the egg of the parasite upon it, the effect of the larva appears to be immediate, and oviposition, if already commenced, ceases as soon as the egg of the parasite hatches.

The larva grows with great rapidity, and as it grows the body of the host shrinks, until finally there is little left of the host except the shrivelled skin. The space formerly occupied by the scale insect is occupied by the full grown larva of the parasite, and the remains of the host are pushed to one side. In the plate, the two eggs which the scale insect had laid before it was attacked were not devoured, but in these cases where eggs of the parasite are laid on hosts which have already laid most of their eggs, the body of the parent scale insect in each case having shrunk correspondingly, the parasite larvae feed on the eggs instead. They were definitely observed biting the eggs with their mandibles. Being almost as broad as long and entirely without appendages, the larva is almost incapable of locomotion. When feeding upon a scale-insect, rather than on eggs, there is no need for it to move at all. But when feeding upon eggs it is necessary for the larva to advance from time to time and this it does by a series of alternate and very slow contractions and expansions of the body, accompanied by lateral twists. During the contractions the anterior third of the body is almost entirely telescoped into the remainder. The body is very sticky and adheres to the covering scale and the leaf, and this property serves to facilitate locomotion.

No excretion occurs until the larva is ready to pupate. The excrement is in the form of small, shiny black or brown bodies, bluntly pointed at each end and usually 5 or 6 in number, and uniform in shape. They are deposited inside the scale, round the posterior end of the larva. They are very conspicuous.

The pupa is incapable of movement of any kind and changes very little in colour during its development, since the adult also is yellow. It always lies with its back against the ventral scale of the scale-insect.

The adult, when ready to emerge, breaks the pupal skin ventrally, and in the case of harder scales, such as *Chrysomphalus aonidum*, it is impossible for the parasite to make anything in the nature of a flap (as it does in *Aspioditus destructor*), and therefore it bites an irregular and elongate hole. This hole is quite unlike any of the holes made by the internal parasites and can be confused only with the holes made by the predatory Coccinellids. The pupal skin of the parasite and the excrement of its larva remain inside the scale after the adult has emerged.

The adult is very active, crawling over the leaves in an erratic and characteristic manner, but does not fly very often. The ovipositor is inserted through the scale (never under the edge of it) at a point just above the edge of the contained scale insect, then passing underneath the latter, and the egg is laid between it and the leaf.

The feeding habits of the adults are interesting. In captivity it thrives on honey and water only, but there is no doubt that in nature, it frequently feeds on the body juices of some of its hosts (on *Aspidiotus*), it is certain that it never feeds on hard scales (such as *Chrysomphalus*), When about to feed on a scale-insect the parasite first faces away from the centre of the scale and punctures it with its ovipositor. A minute drop of fluid exudes from the puncture, and the parasite then turns round and sucks up the fluid. The parasite never attacked *Ch. aonidum* except for the purpose of oviposition. When the ovipositor is inserted into the scale-covering for the purpose of oviposition no fluid exudes, because the scale-insect itself is not punctured, and it is therefore clear that the puncturing of the scale-insect serves solely for feeding purposes

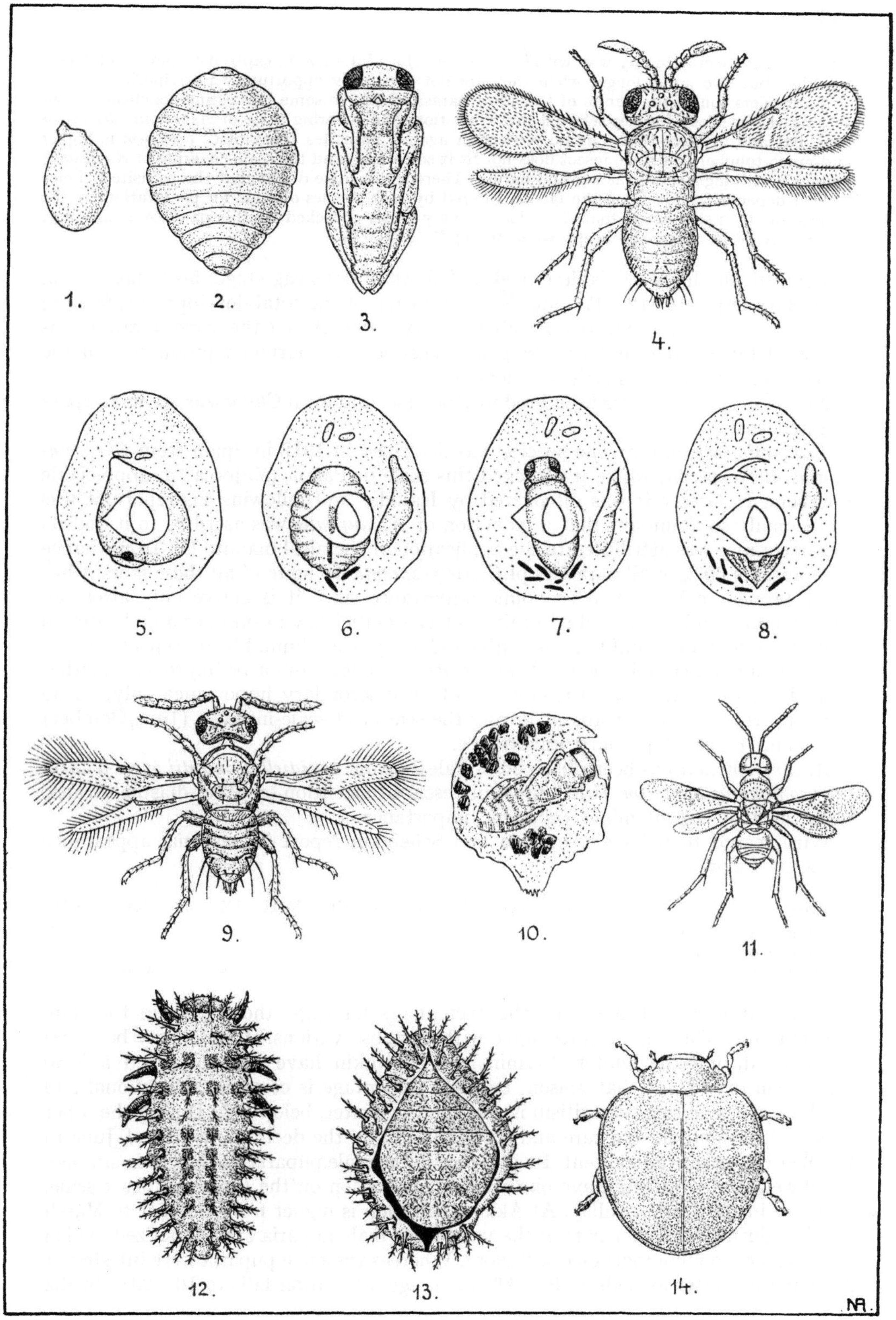

XXII. Parasites and predators of armoured scales in Palestine. 1—8) *Aphytis chrysomphali* Mercet: 1) egg, 2) full-grown larva, 3) pupa, 4) adult female; 5—8) development beneath scale of *Aspidiotus*: 5) young and 6) full grown larva, 7) pupa, 8) empty scale after hatching of the wasp (black: faeces of parasite, 8) the irregular slit in the upper half is the emergency slit of the wasp); 9—10) *Aspidiotiphagus lounsburyi* B. & P. 9) adult female, 10) pupa in the empty exuviae of *Chrysomphalus*. 11) *Homalotylus flaminius* (Dalm.), parasite of *Chilocorus*; 12—14) *Chilocorus bipustulatus* L.: 12) full-grown larva, 13) pupa in larval exuviae, 14) beetle. (Mainly after Taylor Martelli, Malenotti and Balachowski.)

and is not connected in any way with oviposition. The adults live in captivity from 6 to 22 days (in Fiji), but live much longer when they are not given any opportunity to oviposit.

A consideration of the habits of the adult parasite leads to some interesting conclusions concerning the species of scales selected for oviposition. The covering scale of *Aspidiotus destructor* is so flat and delicate that it is no protection against parasites. That of *C. aonidum* is higher and much tougher, the scale-insect does not fit it so closely; and the scale covering of *Aspidiotus palmae* is still higher and thicker and harder. There seems little doubt that the parasite's choice of host depends largely upon the facility offered by each species of scale for penetration by the ovipositor. It is significant that *A. Palmae* is very rarely attacked by this parasite at any age, or by any other parasite except when young." (Taylor).

For Fiji, the life cycle is described as follows: 3 days egg stage, 5—6 days larva, 4—6 days pupa. 14 (12—17) days is the duration of the total development, but the adults grow sexually mature only about 3 days later. About the same duration was observed by Schweig and Grunberg in Palestine. The normal reproduction of the species may be easily parthenogenetical.

No hyperparasites have been bred in Palestine, and even *Chilocorus* seems to spare infested scales.

Economic importance: The species was described in 1912 in Spain from *Chrysomphalus dictyospermi*, where it attacked this scale in Palma (Majorca) in appreciable numbers. In Greece it was discovered by Isakides and following his report, it was occasionally so abundant that a reduction of *A. aurantii* was ascribed to it and its disappearance was attributed to the application of chemical manure (!). In S. France it represents 60% of all parasites of *C. dictyospermi:* In spite of all these facts it has sever proven to be of real economic importance since it is active only after the scale-population has reached its peak. Its lack of efficiency in summer may be due to scarcity of food or to unfitness to withstand a dry heat (humid heat seems to be the optimal climatic combination). However, its efficiency cannot be improved in cither case. It is to be regarded therefore as of very secondary importance only, as no parasite which is rare or absent during the season of scale-increase (June/October) can claim to be of primary importance.

A. chrysomphali has been observed in Palestine on *Aonidiella aurantii* as well as on *Chrysomphalus aonidum*. It has probably escaped attention on other diaspine-scales, being as they are of minor economic importance.

With regard to red scale, Carmin and Scheinkin report its seasonal appearance to be as follows:

months	I	II	III	IV	V	VI	VII	VIII	IX	X	XI	XII
abundance of adults			+		+							+++
abundance of larvae and pupae. . . .	+++	++		++		+			+		++	++

These authors have mistaken the excrements for eggs; the latter are therefore omitted here. The other figures agree with our observations. *Aphytis* may be found regularly during the winter. Carmin and Scheinkin have once observed a local parasitism of 50% at that season, but this percentage is certainly exceptional and local only. The normal parasitism is, even in the winter, below 5%. During the other seasons it is sporadic and rare and especially during the decisive months of June to October it is practically absent. Its presence in the male puparia has not been studied.

Schweig and Grunberg have observed its infestation on the Egyptian black scale. It is low in the Jordan valley. At Akko infestation is higher from October to March and low during the latter part of the year. Only male puparia are parasitized, which limits its economic importance still more. Whereas the male puparia were infested at Akko in the winter as high as 40—60%, average infestation falls to 10—20% in the winter months. The ovipositing females seem to be attracted by the smell of the male puparia. Oviposition lasts for one minute only and may be repeated after 5 minutes. As a rule, only one egg is found in one male puparium and if 2—3 eggs are observed only one larva develops. In both red scale females as well as male puparia of black scale no hole is made in the scale, the parasite leaving through a flap.

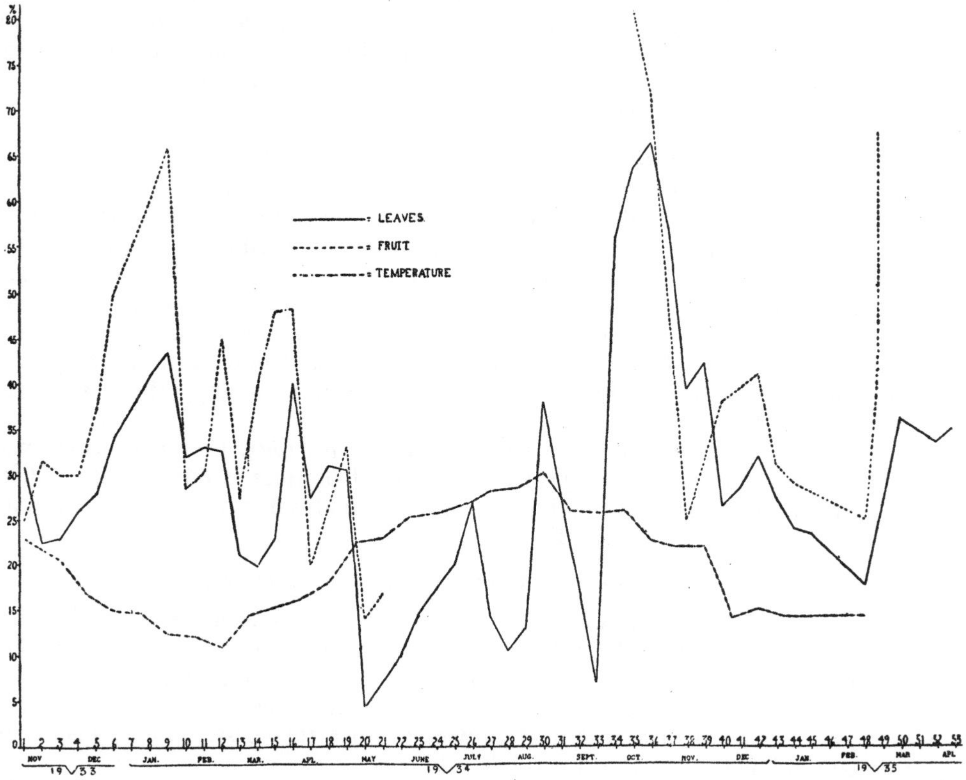

Fig. 86.
Monthly parasitism of male nymphs of Black Scale (*Chrysomphalus ficus*) by *Aphytis chrysomphali* on leaves and fruit in relation to temperature. (From Schweig and Gruenberg).

b. Aspidiotiphagus lounsburyi B. and P. (*Chalcididae*).

(Plate XXXIII, 9—10).

Description of the adult female: Body yellow-brown; antennae, thorax and legs mostly pale yellow; eyes black.

Head as large as thorax; antennae long, 8-jointed. Pronotum short, rounded at the anterior margin. Scutum and scutellum with 4 hairs each. Anterior wing shorter than the body, four times as long as wide. Hyaline with a blackish transversal band below the marginal vein. Marginal hairs on the distal half longer than the width of the wing. The posterior wings hyaline, still narrower than the fore-wings and with marginal hairs in the distal half twice as long and longer than the width of the wing. Abdomen nearly as large as the thorax; ovipositor short.

0.4—0.5 mm long. Anterior wing about 0.35 mm long.

Distribution and host relations: The species has been described in 1916 from Madeira, where it seems to be an important parasite of *Chrysomphalus dictyospermi*. It has been discovered also in Senegal and exists in S. Africa (introduced). Among the Mediterranean countries, it has been observed in Italy on *Diaspis boisduvali* and *Fiorinia fioriniae*, in Algeria on *Chrysomphalus aonidum* and *F. fioriniae*. In Egypt on *Ch. aonidum* and in Palestine on *Aonidiella aurantii* and *Ch. aonidum*. However, in all these countries it is extremely scarce and of no economic importance. Its efficiency seems to be restricted to a warm and humid climate. Although very little is known of its biology, it apparently cannot thrive in our climate. Attention should be paid also to the closely related species *A. citrinus* (Craw.), which is polyphagous on armoured scales and is of wide distribution in tropical and semi-tropical regions. It may be present in Palestine, but has hitherto escaped our attention.

c. *Aphytis maculicornis Masi* (*Chalcididae*).

Description of adult female: Morphology similar to that of *A. chrysomphali*. Easily distinguished by the blackish color of antennal apex, thoracic sutures and side of the first abdominal segments.

Distribution and host relations: A. maculicornis is distributed throughout the Mediterranean regions, but has not yet been recorded beyond them. It is most abundant in winter, when its main host *Parlatoria pergandei* is most abundant. It has also been bred in small quantities from *Chrysomphalus dictyospermi* and *Aspidiotus hederae*. In Palestine it has so far been bred only from *P. pergandii*, but in small percentages only, and we cannot ascribe to it any influence on the numeric fluctuations of its host.

A recently discovered Red Scale parasite, *Habrolepis rouxi* Compere, which was introduced into California a few years ago, does not promise striking results. At its home (Transvaal) red scale is a major pest. As it originates from a dry climate, it may be possible to acclimatize it in Palestine.

d. *Comperiella bifasciata Howard* (*Encyrtidae*).

In December 1945 one of our coworkers brought a few specimens of *C. bifasciata* from the U.S.A. Its establishment is now attempted with the view to add another check on the Egyptian black scale, where any additional limiting factor will be welcome. There is, of course, little hope that the establishment of the parasite will enable us to dispensewith technical control against this pest. So far, inspite of additional sendings, it is not yet established here.

The female oviposits soon after emergence, usually into mature females of the scale-insect, one egg at a time. When two or three eggs are laid into one host, as often occurs under conditions of over-crowding, one larva usually only or none succeed in developing. The fourth larval stage pupates within the host which it leaves by cutting a circular emergence hole through the scale.

C. bifasciata is a normal parasite of *Aonidiella citrina* Coq. in China and Japan. Eggs laid into *A. aurantii* are invariably destroyed within a week. Other hosts in which the parasite is able to develop are *A. orientalis* Newst. and *A. taxus* Leon, *Chrysomphalus aonidum* L. and *C. dictyospermi* Morg. are less succesful hosts according to the experience gained so far. There is, however, no preference for oviposition. The eggs are equally well laid into the not, the little and the very suitable scale species, just mentioned. In its natural home (China, Japan) *C. bifasciata* is heavily hyperparasitised by another wasp (*Marietta carnesi* How) . In California the host, *A. citrina* Coq., has been bred recently in great quantities on pumpkin for mass-breeding of *Comperiella.*

H. W. Simmonds (J. Austral. Inst. Agric. Sci. 10. 1944. p. 38-39) reports that this wasp infests in experiments *Aonidiella aurantii* on lemon much less than on oranges. The reason is still obscure.

2. COLEOPTERA.

The Armoured Scale Lady-beetle (*Chilocorus bipustulatus* L., *Coccinellidae*).

Synonymy: 1758 *Coccinella bipustulata* L.
1903 *Chilocorus minor* J. Sahlb.
1923 *C. meridionalis* Eichl.

Description: Beetle hemispherical, highly arched. Scattered hairs on head and lateral pronotum only. Head retreats to the posterior eyemargin and pronotum farther below. Antennae 8-jointed, hairy. Posterior margin of pronotum distant laterally from basis of elytra. Legs short, the second and third femora not even reaching the lateral margin of elytra. Claws dentate.

Body black to red-brown; head red; antennae yellow red. Ventrally and legs mainly yellow red, base of abdomen darker. Elytra with 3 transversally arranged circular red spots, at the end of the second fifth of the elytra's length. The two median spots often furse into one.

Female 4 mm long, 3—3.5 mm broad; male 3—3.5 mm long, 2.7—3.0 mm broad.

Eggs elongate oval, ochreous or orange in colour. 0.9 × 0.4 mm.

Larva elongate oval, somewhat convex and of dirty yellow colour. Head shiny black, flat, hairy. First thoracic segment with black pigmentation at both sides of the median longitudinal band; with 5 yellow branched and hairy spines 0.5 mm long at each side. Second thoracic segment with 4 spines; third thoracic and all abdominal segments with three pairs of black spines only dorsally. The end of the dorsum of the third thoracic and of all the first abdominal segments covered with a pale transversal spot. Spines of this segment yellow. Legs black. Normally, there exist 4 larval instars in Palestine, but 3 and 5 instars have occasionally been observed. 5—5.5 mm long, 2 mm broad. (Martelli).

Pupa rhomboidal, with a dorsal hunch and 2 small yellow tubercles along the sides of the first abdominal segment. Yellow-brown, with yellow intersegmental lines and a yellow median longitudinal band. Many hairs dorsally. Total length of pupa enclosed within the open larval exuvia.

Distribution: Recorded from the entire Palaearctic domain.

Fig. 87. Ovipositing female of *Comperiella bifasciata* How. (from Compere and Smith).

Host relations: The species is a rather polyphagous predator of many armoured scales, soft scales and Asterolecaniinae. When starved, it may turn to other coccids too, as well as to aphids and honeydew. However, normally it does not seem to feed on aphids, as has been recorded in literature. Armoured scales seem to be its primary and preferred food and amongst them the larger species, as *Aspidiotus hederae*, are apparently its first choice, at least of the adult beetles. Recently (1945) it attacks freely *Matsococcus josephi n. sp.* a Margarodid living on pines, at certain seasons.

The life-history of *Chilocorus* in Palestine has recently been studied by Hecht (1936) in our laboratory, upon whose observations part of the following lines are based.

The eggs are laid singly or in groups of 2 to 6, below the scale of diaspinae, or in the leaf, but not below empty scales. In species with strongly developed ventral scales, as in red scale, the egg cannot be laid below the scale.

The number of adult females of *Aspidiotus hederae* devoured was as follows:

			temperature	observation days	n	eaten scales	per adult/day
beetles:	November	1934	22.5° C	157	10	705	4.3
	August	1934	28.7	34	4	388	11.3
	October	1935	21.0	154	4	428	2.8
larvae:	October	1934	25.0	71	2	241	4.1
(stage 2/4)	November	1935	18.0	61	2	218	3.5
	May	1935	26.0	21	1	87	4.1

These experiments show that at temperatures about 25° C the larva eats about 100 adult females of *Aspodiotus* during its development, at 18° C probably up to 125. The high daily feeding rate of beetles in the summer (11.3 per day) is remarkable. At lower temperatures the daily feeding rate ranged from 3 to 5 scales per days.

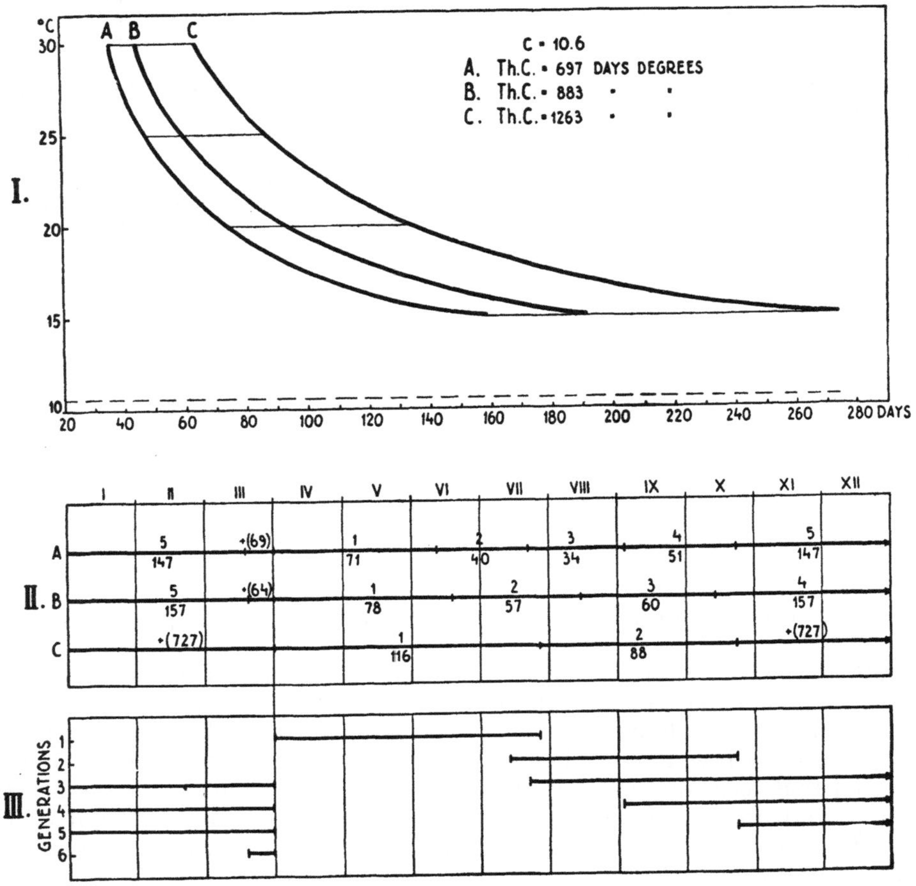

Fig. 88a.
Life-history of *Chilocorus bipustulatus* L. at Tel Aviv. I The three hyperbolae; II the life-cycle as calculated from the hyperbolae; III sequence and overlapping of generations.

It should not be forgotten that only adult females were eaten, whereas in nature *C. bipustulatus* feeds equally well on the much smaller larvae, and that *Aspidiotus* females are much greater than those of *Chrysomphalus*. The actual destruction value of a *Chilocorus* in nature is higher than that obtained by these figures. However, even thus a total destruction value of one pair of *Chilocorus* will be (adult females of *Aspidiotus*):

in winter: 2 larvae (50 x 4.3) + 2 beetles (240 x 2) = 695 females of *Aspidiotus*
in summer: 2 larvae (16 x 4.1) + 2 beetles (50 x 11.3) = 631

Duration of development was calculated from an extensive series of breedings, the extreme data of which were for the

	minimum	maximum
egg	7	28 days
larva	15	81
pupa	4	31
adult life	about 30	about 250
preoviposition period	6	51 (excl. hibernating adults)

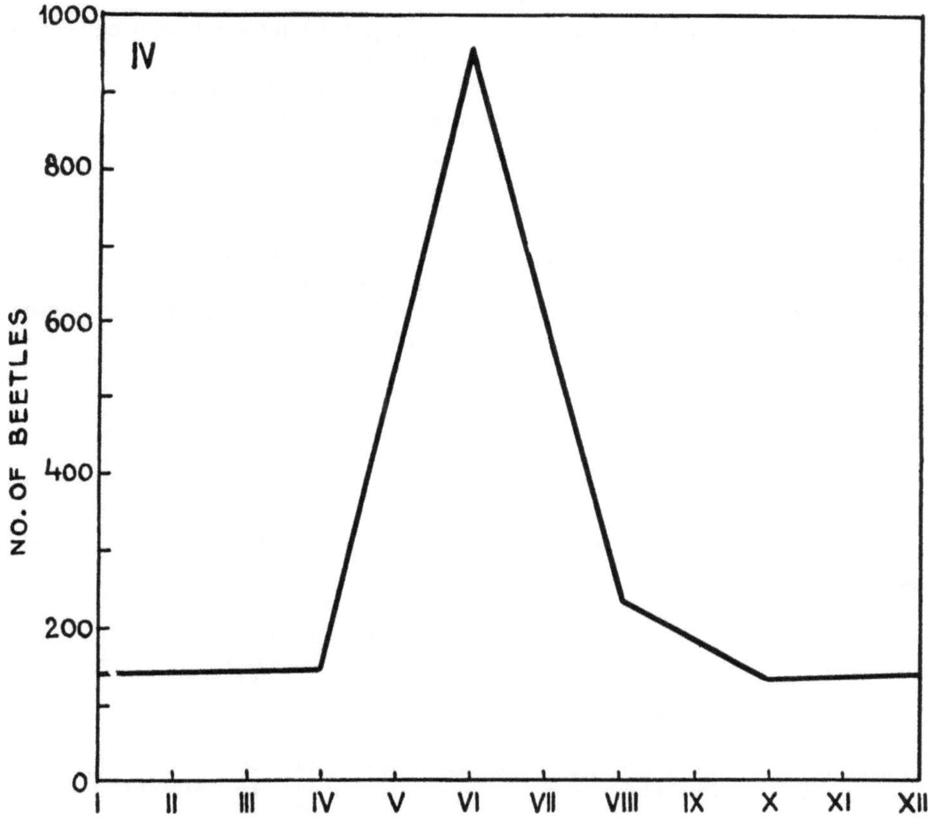

Fig. 88b.
IV Seasonal population trend of *Chilocorus bipustulatus* L. in the coastal plain of Palestine.

The constants for the three hyperbolae (including the preoviposition period) are as follows:

hyperbola	A	B	C
development-threshold (c)	10.6	10.6	10.6: C
thermal constant (th.c.).	697	883	1263
duration of development at 30° C . . .	35.9	45.1	64.4 days
25	48.4	60.5	80.5
20	74.2	93.9	134.4
15	158.8	192.0	214.6

The average monthly mortality of eggs, larvae and pupae was as follows:

month	I	II	III	IV	V	VI	VII	VIII	IX	X	XI	XII
mortality of { eggs		40	(35)	31	47	52	43	45	30	40	44	
mortality of { larvae + pupae. . . .		35	41	52	52	94	95	68	30	55	64	
beetles developing from 100 eggs . .		39	(38)	34	26	3	3	18	49	27	20	

In a total of 298 independent breedings of our own, in all seasons, the mortality was 50 % in the eggs, 73 % in the larvae and 15 % in the pupae. 35 adult beetles appeared.

The adult longevity reaches 6 months (November/April) in the hibernating beetles, whereas in early summer it ranged from 42 to 62 days.

The normal life history of *Chilocorus* in the coastal plain of Palestine is illustrated by fig. 88. The number of annual generations is for hyperbola

A 4.9 generations
B 3.9 ,,
C 2.8 ,,

Up to 5 generations may be reached by the first individuals of every generation, whereas 4 generations are the epidemiological average. The beetles of the third and fourth generation enter the winter and start oviposition in early spring. The number of annual generations for other localities is as follows:

	A	B	
Paris.	1.2	0.8	
Naples	2.9	2.2	(Silvestri: 3 generations)
Jerusalem	3.5	2.7	
Cairo.	4.4	3.6	

The oviposition rate of the generations fluctuates greatly. It is summarized in the following table:

Egg-production of a *Chilocorus*-female.

	n	eggs per female	I	II	III	IV	V	VI	VII	VIII	IX	X	XI	XII
1. Hibernating generation . .	24	60.5	—	65	228	202	91	13	2	—	—	—	4	—
2. dto hibernated af 30° C. .	6	99.7	—	635	345	17	—	—	—	—	—	—	—	—
3. dto hibernated at 22° C .	9	85.8	—	253	320	221	39	24	—	—	—	—	—	—
I. generation	18	11.6	—	—	—	—	18	47	41	10	—	—	—	—
bred at 30° C	11	16.5	—	—	10	148	7	—	—	—	—	—	—	—
bred at 22° C	7	57.7	—	—	—	284	268	24	—	—	—	—	—	—
II. generation.	14	3.9	—	—	—	—	—	—	10	2	13	14	—	—
bred at 30° C	5	2.2	—	—	—	—	22	—	—	—	—	—	—	—
III. generation	9	0.1	—	—	—	—	—	—	—	—	—	—	1	—
IV. generation	3	13.7	—	—	—	—	—	—	—	—	—	26	4	8
Sum of all ovipositions . . .			—	953	903	872	445	108	53	12	13	40	9	8

The winter interruption in egg-laying is conditioned purely by low temperatures (about 15° C and below). The beetles start to oviposit, at any time of the year, shortly after they are brought into a warmer room. A series of independent observations confirms the high oviposition rate in spring and the low one in summer:

month	I	II	III	IV	V	VI	VII	VIII	IX	X	XI	XII
no. of breedings.	—	1	3	2	1	1	4	3	3	4	1	—
eggs per female	—	24	10	3	74	19	4	12	12	6	15	—

In nature, the oviposition rate is not always as low as indicated by Hecht's breedings in the laboratory. 11 females collected in the field in September laid 15.2 eggs on the average, 6 females from October 24.7 eggs. However, this pre-winter oviposition of the third and fourth generation has no epidemiological significance. The beetles of these generations are responsible for the heavy oviposition from February to April. About 30 % of the adult hibernating beetles died before the end of the hibernation.

The normal epidemiology of the four annual generations of *Chilocorus* in our coastal plain, is as follows:

eneration	oviposition	no. of beetles	egg-production		mortality		no. of hatching adults
					eggs	larvae+pupae	
I	1.IV	100	50 × 58	2900	35%	52%	960
II	23.VI	960	480 × 17	8160	43	95	233
III	15.VIII	233	117 × 3	350	45	30	135
IV	11.X	135	58 × 10	680	40	64	147

The environmental resistance of each generation (males=females) is as follows:

generation	I	II	III	IV
R_0	96.55	88.24	33.33	80.00
R_a	66.88	97.14	71.37	78.22

The high environmental resistance in generation II and III (as compared with Ro) explains the low number of individuals during summer.

In nature the mortality may be even higher. From the 135 adults of the third generation 95 would pass the winter successfully, replacing the 100 original beetles. The winter mortality of development stages which was not studied by Hecht is apparently high, so that of the 147 beetles only a few will succeed to develop to adults.

The adults and larvae are common in Palestine from February to April only. They are extremely scare from June to September, and may appear again in October/November in moderate numbers.

Economic importance: The great number of *Chilocorus* beetles feeding intensely on red scale in the spring has been always attracting the attention of the farmer. He therefore claims that it should be possible to breed *Chilocorus* throughout the year in sufficient numbers to suppress its arch-enemy. The following points seem now to be clear:

Eggs and larvae show an enormous mortality, ranging between 95—100 % during June/July and still extremely high in August. Furthermore, the number of eggs per female is reduced in all summer generations to an extremely low level (17 resp. 3 eggs) Adult longevity is very limited during the first two summer generations. Finally, the population density of the host is that low during summer the mortality of

Chilocorus would be enormous owing to lack or scarcity of food, especially in its development stages. Dense host populations are required for any remarkable increase in numbers of *Chilocorus*.

The situation may be summed up as follows: The population peak of *Chilocorus* in spring coincides with the natural decrease of *Aonidiella aurantii, Mytilococcus beckii, Parlatoria pergandii*, etc. It is rather doubtful whether its contribution towards this increase is of any epidemiological importance. From June to September, *Chilocorus* is practically absent in the orchard. During this season the scales build up their dense populations. *Chilocorus* beginning to increase in number in autumn, is soon interrupte in its activity by low winter temperatures.

A biological control is not possible in spite of the fact that *Chilocorus* may be bred throughout the year and at a fairly high oviposition rate in the thermostat at about 22° C. But the same beetles, when transferred into the orchard, immediately display the seasonal low oviposition rate, short longevity and their offspring undergoes the extreme summer mortality. What is more, the scarcity of food in summer would interfere with an increase or even maintenance of *Chilocorus* during that season. The beetle could not be expected to be of practical benefit, as only massive action from June to August would prevent the main population increase.

The only field experiment performed by Hecht at Even Yehuda (11.VII.—9.X.1935) agrees with the above conclusions. Conditions were somewhat artificial, because the confinement in the large field-cage and the dense growth within detained the wind and favoured Red Scale development at least in the same degree as that of *Chilocorus*. The three trees were heavily infested with red scale and the *Chilocorus* population was rather low. They were far more heavily infested with red scale than were the two unscreened control trees.

The question of controlling Red Scale by *Chilocorus* must therefore be answered in the negative. The experienced citrus entomologists of California have instinctively foreseen this result, as they never proposed to control scales by *C. bivulnerus* Muls., a rather common congener of identical habits in California.

Parasites and diseases: The only parasites bred from *C. bipustulatus* is *Homalotylus flaminius* Dalm.

The only other biological control of this lady-beetle is a Laboulbeniaceous fungus (Hecht, 1936). This family is known to be connected with beetles, on which it anchors into the chitin-skeleton by peculiar appendices. Yellow-green coloured Ascomycete-forms were especially conspicuous at the ends of the elytra. It is certain that this fungus does not kill its host and very doubtful whether it interferes at all with its vitality and reproduction. In Palestine its appearance seems to be restricted to the winter season.

The Ladybeetle Parasite Homalotylus flaminius Dalm.

Synonymy: 1830 *Encyrtus flaminius* Dalm.
1844 *E. eytelweenii* Ratz.
1848 *E. apicalis* Ratz.

Description: Head broader than long, broadly rounded frontally. Eggs much more distant ventrally than dorsally. Antennae 9-jointed. Thorax large, rectangular, anterior margin of pronotum only broadly rounded. Abdomen subquandrangular, shorter than the thorax. Oviscapt not visible. Male very similar to female, but antennae somewhat longer, wings somewhat shorter and frontal margin of head more rounded.

Head and mesonotum dark green, antennae black brown with yellow-white final joints. Sides of thorax generally red-yellow; scutellum black, not shiny; abdomen black, shiny. Legs blue-black or yellow-brown, the four last tarsi yellow-brown to black-brown. Wings hyaline, hairy with a large brown median spot.

2 mm long, antennae of female 1.0, of male 1.1 mm; wing expanse 2.3 mm.

Life-history: The adults are rather vivacious at higher temperatures or in sun-light. They feed on sweet substances, as honeydew of scales and aphids. The eggs are laid, one at a time, in half grown larvae of Coccinellids. The maggot feeds in the interior of its host, which, as a rule, fixates its abdominal end as if ready for pupation. But the rest of its organs are eaten instead and the maggot of *Homalotylus* pupates within the empty host skin. Following Martelli (1908) pupation takes place 25 days after oviposition at about 23° C and the adult wasp hatches 4—5 days later, after having made a small circular hole in the dorsal skin of its victim. In S. Italy the wasps hatched quite frequently from *Exochomus* and *Chilocorus* from beginning of July to end of September. During this favorable season about 3 generations of *H. flamineus* may develop. 8 specimens were bred by Hecht in Palestine from the end of October to the end of November. It is of no primary importance for reducing. *C. bipustulatus* in Palestine.

H. flamineus is rather polyphagous in Coccinellid larvae and has been bred once by Walker even from *Galeruca*. It is known from Central and S. Europe, N. Africa, Palestine, Java, Australia.

3. ENTOMOGENOUS FUNGI OF THE RED SCALE AND OF OTHER ARMOURED SCALES.

Entomogenous fungi of red scale may be observed in Palestine every winter. J. Carmin (1936) published a first series of observations on these fungi, contributing thereby very welcome information.

Whereas entomogenous fungi are known to be a decisive limiting factor in the numeric development of scale insects, plant-lice, etc. in tropical countries or in regions where summer-rains occur (Florida, S. E. Black Sea coast), no epidemiological importance has ever been attributed to them in arid or semi-arid regions.

Carmin's observations run as follows:

During January 1931 fungus infestation of citrus scales reached a very high percentage in many orange groves of the Jaffa-district (from Rehoboth to Petah Tikva). Microscopic investigation proved that 2—3 species of fungi were involved in these infestations. The insects shriveled and turned a dirty violet. An actual drop of infested scales began early in February. The fungi were observed the year round, but, as far as our independent, occasional observations showed, heavier infestations occurred only during winter.

The fungi (det. Dr. T. Petch) were:

1. *Aspergillus depauperatus* Petch infested red scale. It looks like protuberant threads from the pygidium of the scale, thickened at their ends. (For a full description cf. Petch, 1932). Petch believes this fungus, which has been met before on *Mytilococcus* in England and on *Aspidiotus* in Ceylon, to be truly parasitic and specific to armoured scales. Carmin records it from Tel Aviv and Ness-Ziona, from September onwards, starting fruiting in early January; only adult females were so far observed to be attacked.

2. *Aspergillus sp.* This second *Aspergillus* has rarely been wet with on red and chaff Scale at Ness-Ziona. Fruiting occurred in June and December. It is not yet fully clear whether it is a true parasite to the scale, but its numerical scarcity renders it unimportant anyhow.

3. A species of *Cladosporium* was the commonest species observed. It appears in two development forms know as *C. herbarum* Lk. and *Dematium pullulans* on red and chaff scale, on the parasitic larva of *Aphytis* and probably on other insects too.

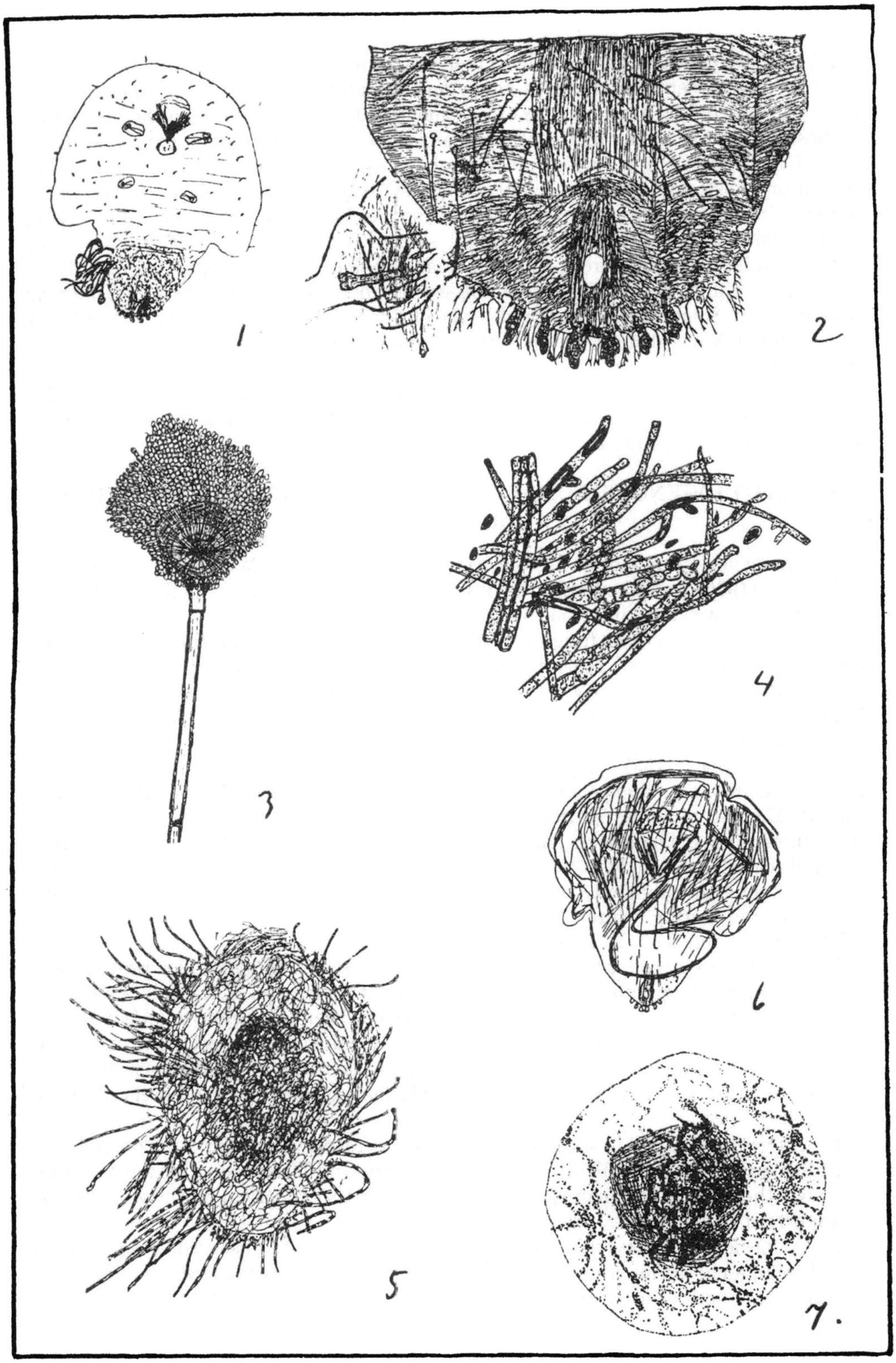

XXIII. Entomogenous fungi of armoured scales in Palestine. 1) and 2) infestation of Red Scale with *Aspergillus depauperatus*; 3) *Aspergullis sp.*; 4—6) *Cladosporium herbarum*: 4) fruiting *C. herbarum*, 5) red scale larva infested with fruiting *C. herbarum*, 6) adult red scale full of hyphae of *C. herbarum*; 7) scale infested with *Dematium pullulans*. (from Carmin).

A variety (*var. citricolum*) of this fungus seems to be connected with the leprosis of citrus trees in Florida. The typical form is polyphagous and probably phytophagous. However, Carmin thinks that it penetrates the body of the living scale and causes their death. He quotes a note of L. I. Klotz from California, stating that *C. herbarum* may be pathogenic and that it develops very well and fruits on the soft body of a young larva of the red scale. Experiments to inoculate red scale with this fungus yielded no results. The spores are distributed by wind as well as by the winged males. The species is to be found in both its forms all the year round. Whole scale populations sometimes seem to be infested by this fungus. But it is still very doubtful whether this fungus really attacks living scale insects to en extent to be of epidemiological importance. Successful artificial inoculation in the laboratory is not a sufficient proof to the contrary.

Earlier records of entomogenous fungi on red scale are scarce. From Guam (1927) a fungus checking red scale is reported. Eveling (1934) reports that in different counties of California an important fungus attacked red scale, causing its death to such an extent that the appearance of crawlers diminished rapidly at Riverside round the first of May, an unusual occurrence in other years. Also, that on the first of August no scales remained alive on the bark of the tree, both young as well as old scales being dead, while on the fruit only some 12 % of the scale population remained alive.

Marchionetto (1934) describes the attack of *Myriangium duriaei* on red scale from Argentine.

Carmin is also most optimistic as to the fungi being a possible control of red scale and ascribes to them a very radical control under certain conditions in nature. He further suggests that older trees are relatively clean of red scale due to the fact that in older groves conditions are more favorable for the development of entomogenous fungi.

Certainly no possibly successful way should remain untried and research should be continued in this direction. However, fungus outbreaks in winter will not help the farmer. Only such fungi can be of practical benefit, which provoke not only individual inoculation, but epidemics under our dry summer conditions, when the annual red scale populations build up. Until such a solution has been found — and the prospects are not good — the judgement of Fawcett (1926) holds: „Under drier climatic conditions, such as those in California, scale insects are not apparently controlled to any appreciable extent by fungus parasites."

References.

A. Balachowsky, Encyclop. Entom. **15**, 1932. 214 pp.
F. S. Bodenheimer, Schaedlingsfauna Palaestinas. Berlin, 1930.
J. Carmin and D. Scheinkin, Bull. Soc. Roy. Ent. d'Egypte, 1934, pp. 242—273.
J. Carmin, Hadar, **9**, 1936, pp. 173—175.
H. Compere and H. S. Smith, Univ. Calif. Publ. Entomol. **4**, 1927, pp. 63—73.
Compere, Flanders and Smith, Calif. Citrograph, **23**. 1937. pp. 58—59.
Eveling, Calif. Citrograph, **19**, 1934, p. 362—363.
H. S. Fawcett, Citrus diseases and their control. New York. 1926, 582 pp.
H. S. Fawcett, Guam Agric. Sta. Rept. 1927.
O. Hecht, Bull. Soc. Roy. Ent. d'Egypte, 1936, pp. 299—326.
J. O. Marchionatto, Rev. Faculd. Agron. y Vet, **7**, 1934, p. 571.
E. Malenotti, Redia **13**, 1918, pp. 17—51.
G. Martelli, Boll. Labor. Zool. Gen. e Agrar. Portici, **2**, 1908. pp. 217—296.
R. G. Mercet, Revista de Fitopat. Madrid. 1924, pp. 1—8.
T. Petch, Transact. Brit. Mycol. Soc. **16**, 1934, part 4.
R. Poutiers, C. R. Acad. d'Agric. de France, 7. V. 1924 (repr. 8 pp.).
C. Schweig and A. Grunberg, Bull. Ent. Res. **27**, 1936, pp. 677—713.
F. Silvestri e G. Grande, Dispense di Entomoligia Agraria. Portici, 1911. p. 325—326.
T. H. C. Taylor, Bull. Ent. Res. **26**, 1936, pp. 22—27.

7. SOFT SCALES, MEALYBUGS AND FLUTED SCALES (RYNCHOTA III).

A. THE BROWN SOFT SCALE (*Coccus hesperidum L.*)

Synonymy: 1758 *Coccus hesperidum* Limneus
1875 *Chermes aurantii* Alfonso
1895 *Lecanium ceratoniae* Gennadios

Description: Adult female bright yellow or greenish-yellow, minutely speckled with red-brown, the specks sometimes agglomerated into transverse bars, especially on the median abdominal region; in other parts tending to form dotted lines radiating from centre to margin. In older examples the ground-colour may be ochreous or pale fulvous; and the maculation may form a broad median fascia. Under surface of older examples with a deep purple-brown or red patch covering the median abdominal area, becoming concave and forming a shelter for the young larvae. Dried specimens straw-coloured and much wrinkled. Form oblong-oval, but generally asymmetric; narrowest in front; more or less concave above according to age (Green).

Body naked; antennae well-developed, slender, 7-jointed as a rule (6—8), third and fourth segments usually the longest; less well developed, without free tibio-tarsal articulation and with merely a transverse septum between the tibia and tarsus and apparantly with no free movement of the joint. Anal cleft relatively short compared to the body length; anal operculum quadrate with cephalo-lateral and caudo-lateral margins equal or nearly so in length and forming rather sharp angles, with one or more small discal setae near the apex of each plate; with no hypopygial setae. Spiracles normal; marginal setae variable, but always more slender than the stigmatic setae; stigmatic setae three, the middle one usually longer than the other two; stigmatic depressions slightly indicated. Tubular ducts apparently lacking; multilocular pores around the ventral anal region; quinquelocular pores in narrow bands in the stigmatic forrows; a few simple, round pores scattered over the dorsum; submarginal tubercles present, 5—7 on each side of the body (Steinweden).

2.5—5.0 mm long (usually on citrus about 3 mm); 1.5—3.0 mm broad.

The *neonata larva* is oval and flat; yellowish-red at birth, but growing pales until the first moult; eyes red. Antennae 6-jointed; legs well developed. Stigmatic setae 3, with the middle one longest; marginal setae short, small; body setae lacking except near the base of each antenna and in the ventral abdominal region; anal plates not protruding beyond the apex of the body and not reticulated. Quinquelocular pores present in stigmatic furrows. Anal ring with 6 hairs (Steinweden).

0.5 mm long; 0.3 mm broad.

It is extremely difficult to observe the moults in *C. hesperidum.* The larval stage is followed by probably two nymphal stages, which are similar in appearance to the adult female. The first of these stages is generally rather elongate, whereas the second approaches the asymmetric oval form. Both stages can be distinguished from the adult by the number of antennal joints, which are still six. The form of the antennae is described by Berlese as less cylindrical and more conical than the larval antenna.

Male: No males have been observed in Palestine.

Distribution: Ubiquist, living in temperate climates in greenhouses.

Life-history: The eggs hatch a few hours after having been laid in the breeding chamber formed by convex elevation of the female's ventral body surface over the plant. In Palestine the birth of mature eggs is the normal and probably only way of birth. The larvae and females remain mobile until the breeding chamber of the mature female is formed. Normally, however, locomotion is mainly restricted to the periods after the moults when they migrate from half an hour to 1—2 days. They remain on the chosen spot when not disturbed. When disturbed, by wilting or drying up of the host plant e.g. they move to a more favorable spot. Migrations up to 1 m have been observed in experiments performed by Dingler during 15—30 minutes. But normally, as in most scales, they do not move very far from their mother, if not transported by wind or stimulated by unsuitable environmental conditions.

The brown pigmentation on the back of the female grows more intense along with the formation of the breeding chamber. The second and third pair of legs are in touch with the substratum. The convex elevation of the breeding chamber is mainly restricted to the central part of the body, whereas individuals showing strong convexity over the whole body are deformed by parasitism.

Coccus hesperidum L. is decidedly positively phototaxic and negatively geotaxic in its mobile stages. It shows distinct positive chemotaxis towards its host plant. Citrus smell entering a vessel stimulates their movement. The rate of activity (mm per min) was:

No.	before	during penetration of smell	after
1	37.5	70	—
2	20	45	0
3	45	55	30
4	45	40	0
5	25	35	25
6	70	75	50
Average	40.6	53.3	35.0

Citrus smell stimulates at the moment of arrival, losing its stimulating effect after the penetration. When the citrus smell was less strong, the brown soft scale sometimes left the leaves and migrated towards the light.

A leaf of *Mirabilis jalapa* infested with *C. hesperidum* was placed in a glass jar at a small distance from a citrus branch. During the night the larvae moved from the *Mirabilis* leaf and settled on the citrus branches and leaves.

When larvae of *C. hesperidum* were left at some distance (about 4 cm) from citrus leaves, they always approached the leaves and settled there.

The larvae are strongly thigmotaxic. They settle mainly along the mid-ribs of the leaves and crawl easily into any microcavern of their environment.

Reproduction and fertility: We have stated before that no males have ever appeared in our breedings of *C. hesperidum*. As these breedings were extended over 4 different localities in Palestine during eight years and close inspections for the presence of males were made at regular intervals, it is certain that the specimens living on citrus in this country belong to the obligatory parthenogenetic race of the brown soft scale.

Only a few of the earlier observers (Bouché, Newstead, Wuenn) have observed males in *C. hesperidum*. All other students failed to observe them. Moniez promoted the theory that the males are living as parasites in the body of the female, producing the spermatozoa and creating thus a "false hermaphroditism". None of the later observers could confirm this statement, Newstead is almost certainly correct saying: „I cannot help thinking that Moniez was mistaken in his identity of the unborn males of *C. hesperidum*, and that what he actually saw were the immature males of one of the Chalcidid parasites which universally infest the females of this species."

Thompson discovered in Danemark two different races of *Coccus*, one showing obligatory parthenogenesis without appearance of any males, the other being bisexual with regular appearance of males, but being able to reproduce parthenogenetically in the absence of males. Near Copenhagen, the obligatory parthenogenetic race lived on laurel trees, which passed the summer in nature, the bisexual race in a greenhouse. The bisexual race is certainly much less common than the unisexual one, being reported up to date from Danemark, Berlin, Alsatia and S. England only. It seems

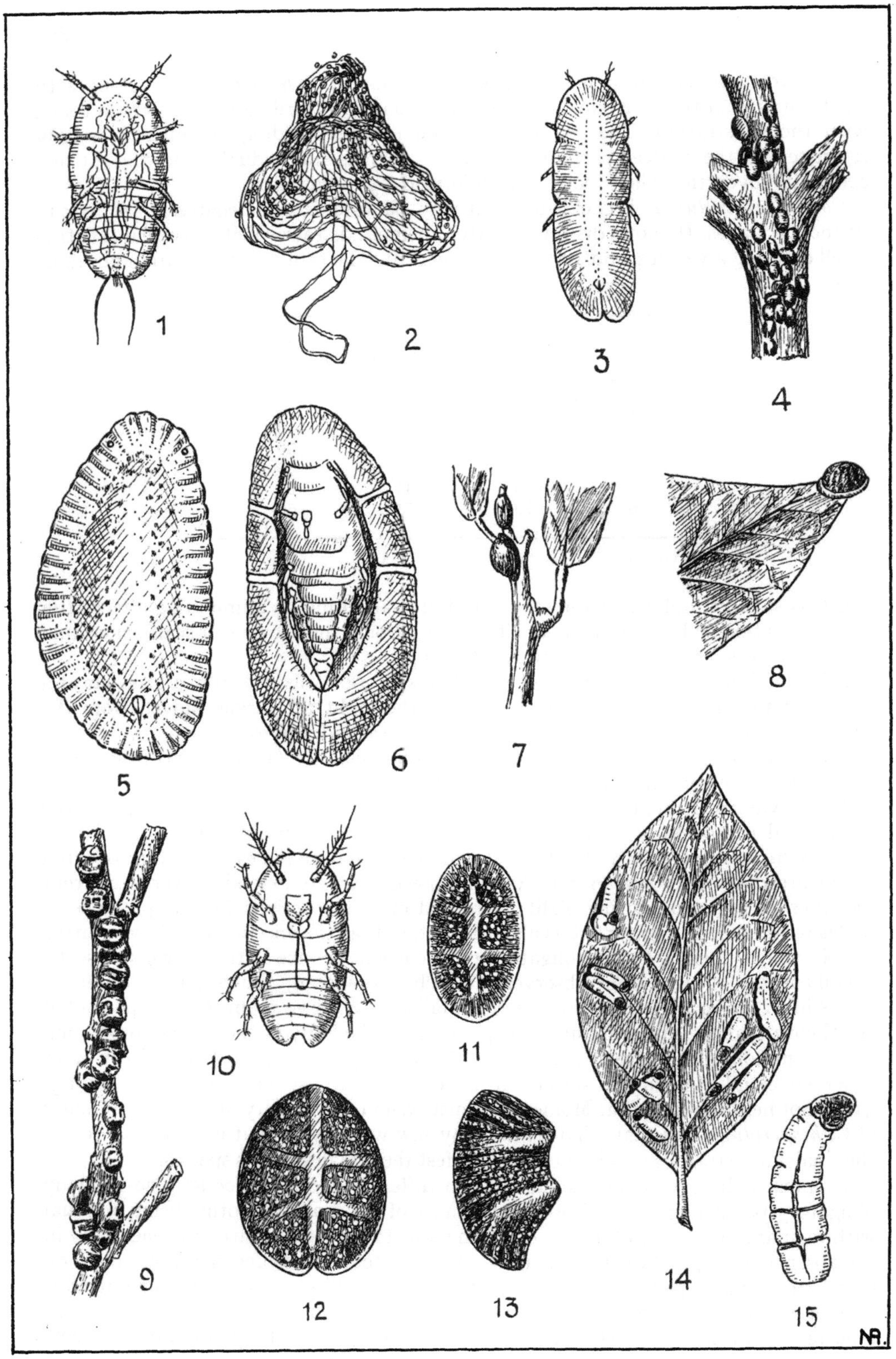

XXIV. Some Soft Scales. 1—6) *Coccus hesperidum* L. 1) neonate lavra; 2) empty egg-shell; 3) nymph; 4) twig with inflated, parasitized specimens of *C. hesperidum*; 5) and 6) female seen from above and from below; 7) and 8) adult females of *Saissetia hemisphaerica* Targ.; 9—13) *Saissetia oleae* Bern.: 9) Adult females on citrus-twig: 10) neonate larva; 11) young female; 12) and 13) adult female from above and in profile; 14) and 15) *Pulvinaria floccifera* Westw. 14) Adult females with ovisac on citrus leaf; 15) Adult female with ovisac (enlarged).

doubtful as to whether it occurs at all in tropical and subtropical regions. The first indication of the presence of males is the occurrence of a number of pale elongate and narrow larvae and nymphs. But Thompson saw 20 males only among many thousands of females. Their number is thus always restricted. During the moults the „scale" of the male often breaks and many male larvae drop to the soil and die, especially amongst those individuals which, as usually, have settled on the lower side of leaves. It seems thus certain that even in the bisexual race most females reproduce parthenogenetically, and if so, always thelytokic (i.e. the giving birth to females only), as is the rule in the other race.

The cytological conditions of both races are described by Thompson as follows:

Cells concerned	Purely parthen. race	Bisexual race in partheno-genesis	Bisexual race in gamogenesis	
	♀	♀	♀	♂
Somatic cells	14x	14x	14x	(7x+7y)
Oocyte I	14x	7		
Oocyte II	14x	7		
Pronucleus resp. gametes		7	(7x+7x)	(7x+7y)
After copulation with second polar nucle		14x		
Offspring	14x	14x	14x	(7x+7y)
Sex of offspring	♀	♀	♀	♂

Like the parthenogenetic eggs of aphids, daphnias, a.o., those of the purely parthenogenetic race of *C. hesperidum* form one polar nucleus only. The chromosome-number in all stages of development is 14 (2 N). No reduction occurs. The one maturation division is an equation-division.

In the bisexual race the normal somatic chromosome number is also 14. The development of the sexual cells shows two reductions and generally follows the normal sheme of gamogenetic reproduction, the number of chromosomes in the mature gametes being reduced to 7 (N). In case of gamogenesis the number of male and female offspring is equal. But this equal sex-ratio is reduced by the small number of adult females (most females developing in facultative parthenogenesis) and the heavy mortality in the immature stages of the male.

The egg-production of *C. hesperidum* during 1934 at Rehoboth (no. of eggs, larvae and empty egg-shells found below the female) was as follows:

month	n	No. of eggs average	No. of eggs maximum	No. of eggs minimum	Remarks
I	11	25.5	82	16	
II	8	53	72	29	
III	6	63	124	34	Many adults with eggs
IV	3	45.5	62	32	Few adults with eggs
V	—	—	—	—	No adults in grove
VI	—	—	—	—	No adults in grove
VII	—	—	—	—	No adults in grove
VIII	12	159	210	112	Few adults with eggs
IX	—	—	—	—	No adults with eggs
X	30	16	20	8	Few adults with eggs
XI	—	—	—	—	All adults parasitized
XII	—	—	—	—	
	70	47	210	8	

The total egg production of 7 generations of *C. hesperidum* at Petah Tikva is compiled in the following table. There exists a distinct correlation between total larval production and duration of larva-producing period:

Length of egg-producing period in days	No. of observations	No. of eggs produced per female average	minimum	maximum
1—9	1	7	—	—
10—19	6	35	10	67
20—29	8	72	35	98
30—39	6	81	34	171
40—49	3	161	117	201
total	24 average	73 eggs per female		

This correlation is stronger than that to climatic conditions. This may be easily understood as the daily egg-production of *Coccus* is small, consisting of a few individuals only. The eggs burst after a short time (four hours) and the larva emerges. The species may therefore be called ovoviviparous. More than 1—2 eggs are met beneath the mother in very exceptional cases only. Empty egg-shells, small waxbodies and a few larvae may be found together with them.

Egg-production begins very shortly after the last moult, even before the breeding chamber beneath the female has been distinctly developed. Whereas the oviposition period lasts in Palestine from 10—40 days, as a rule, in a more moderate climate it may last for 2 months and longer.

C. hesperidum is certainly a species with a short generation cycle.

The three hyperbolae for *Coccus hesperidum* based on these breedings are:

hyperbola		A	B	C
development threshold (c.)		13.0	13.0	13.0° C
thermal constant (th.c.)		420	515	700 day-degrees
duration of development at	30° C	24.7	30.3	41.2 days
	25	35.0	42.9	58.2
	20	60.0	73.6	100.0
	15	210.0	257.5	350.0
number of generations	Paris	1.2	0.9	0.7
	Nice	2.4	2.0	1.4
	Naples	3.4	2.8	2.1
	Athens	4.6	3.8	2.8
	Tel Aviv	6.1	5.0	3.9
	Jerusalem	4.2	3.4	2.5
	Dagania	7.1	5.7	4.2
	Gizeh	6.0	4.8	3.4
	Karthoum	13.0	10.6	7.8
	Honolulu	9.1	7.4	5.4
	Los Angeles	3.5	2.8	2.1
	Riverside	4.0	3.3	2.3

It is necessary to add that above 25° C development grows slower.

Number of annual generations and physical ecology.

Some breedings at Petah Tikva (1936—1938) may serve as representative examples:

Generation	Date of infestation of neonate larvae	Appearance of first larvae of F_1	Length of generation	Appearance of last larvae of F_1	Length of adult-life	No. of larvae
I	13.I.36	23.III.36	69 days	4.V.36	42 days	164
		23.III.	69	25.V.	63	16
		23.III.	69	11.V.	49	39
		29.III.	75	4.V.	36	23
			71		48	
II	23.III	25.V.	63	16.VI.	22	73
		25.V.	63	30.VI.	36	121
		25.V.	63	7.VII.	42	117
		8.VI.	77	23.VI.	75	20
			67		44	
III	25.V.	30.VI.	36	24.VII.	24	82
		30.VI.	36	15.VII.	15	46
		3.VII.	39	20.VII.	17	30
			37		19	
IV	30.VI.	3.VIII.	31	19.VIII.	16	40
		13.VIII.	41	6.IX.	23	61
		19.VIII.	47	21.IX.	32	34
		10.IX.	68	16.IX.	(6)	7
	16.VII.	16.VIII.	31	6.IX.	20	98
			47		25	
V	16.VIII.	16.X.	61	9.XI.	24	81
		26.X.	71	30.XI.	35	37
		27.IX.	41	21.XI.	23	72
		27.IX.	41			
		23.IX.	37			52
			50		27	
VI	27.IX.	15.III.37	169	19.IV.37	35	73
		23.III.	177	3.V.	40	201
			173		38	
VII	15.III.37	20.V.	66	17.VI.	28	75
	(17.III.	8.VI.	83	21.VI.	13	10)
	21.III.	2.VI.	73	23.VI.	21	35
	23.III.	23.V.	61	23.VI.	31	171
			67		27	
VIII	24.V.	12.VII.	49	29.VII.	17	145
	7.VI.	19.VII.	33	29.VII.	10	212
IX	19.VII.	21.IX	64	12.X.	21	96
X	24.IX	5.XI	42	3.XII.	28	131
XI						
XII	7.VI.38	16.VII.38	39	27.VII.	13	35
	15.VI.	27.VII.	44	12.VIII.	14	68
XIII	18.VIII.	21.VIII.	34	4.IX.	13	100
		24.VIII.	37	7.IX	13	26
		16.VIII.	29	11.IX.	25	125
		24.VIII.	37	7.IX.	13	28

The region of Paris is about the northern limit of its occurrance in nature. At Geisenheim, in a xerothermic island in the S. Rhineland, the author met with it on ivy in nature and in the warmer and milder parts of the British Islands it also may be found in nature. This limit closely agrees with the temperature sum for one annual generation (Hyperbola A.) Whenever *C. hesperidum* occurs in such colder climates favorable microclimatic conditions must exist.

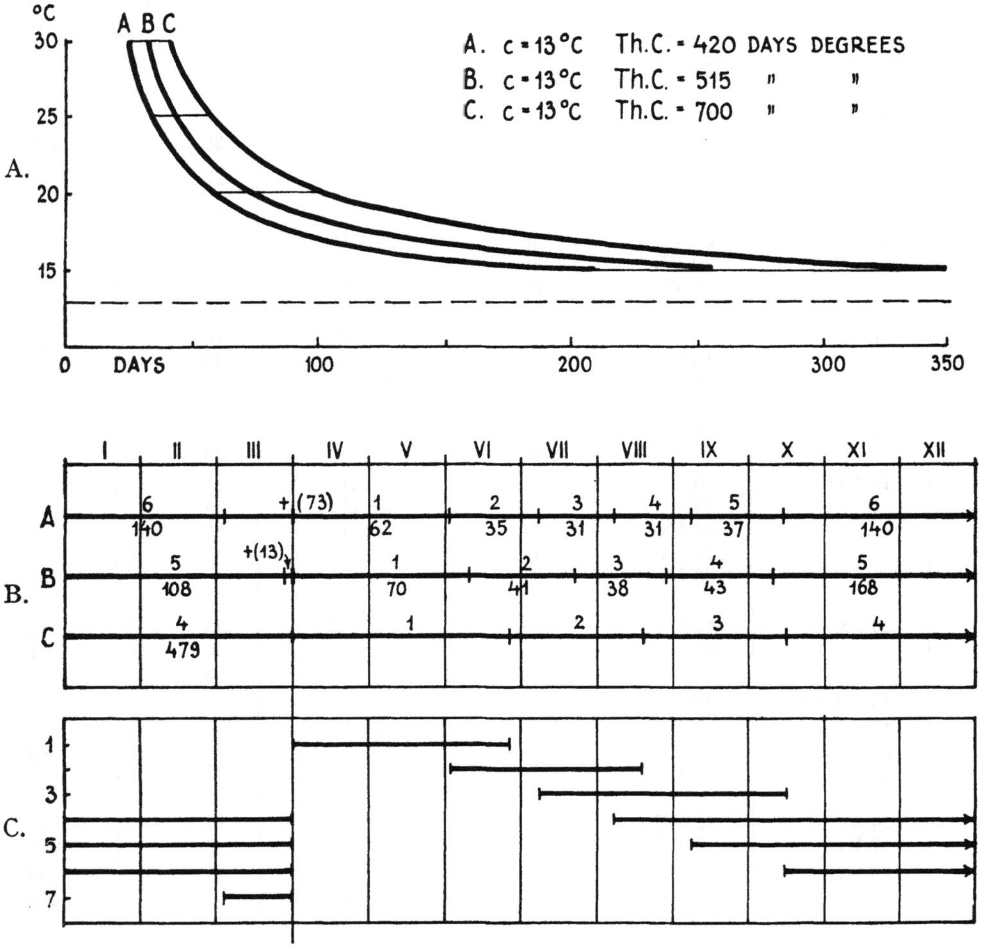

Fig. 89.
Life history of Brown Soft Scale (*Coccus hesperidum* L.) at Tel Aviv. A. The three hyperbolae; B. the life cycle as calculated from the hyperbolae; C. sequence and overlapping of generations.

Seasonal fluctuations in Coccus populations.

The seasonal population trend, as measured per 400 leaves on citrus leaves at Atta, was as follows:

Year	I	II	III	IV	V	VI	VII	VIII	IX	X	XI	XII	Total	Max/Min
1928								1	6	5	1	—		
1929		—		7	2	7	1	—	9	1	3	**13**		
1930	9	**17**	12	13	1	2	3	8	10	11	7	4	97	17.0
1931	**24**	18	19	15	3	6	3	2	8	3	19	8	128	12.0
1932	9	12	7	13	**30**	3	2	2	6	6	11	7	108	15.0
1933	18	**20**	17	16	6	2	10	2	7	5	7	14	124	10.0
1934	11	4	6	7	3	1	**14**	1	—	1	—	1	49	14.0
1935	1	6	**13**	6	3	9	2	7	12	1	3	8	71	13.0
1936	9	4	(4)	5	**13**	—	—	—	4	5	1	—	45	13.
1930—36	12	12	11	11	8	3	5	3	7	5	7	6	89	

or in percentages:

	I	II	III	IV	V	VI	VII	VIII	IX	X	XI	XII	Total
1930	9.3	17.5	12.4	13.4	1.0	2.0	3.1	8.3	10.3	11.4	7.2	4.1	100.0
1931	19.2	14.1	14.8	11.7	2.3	4.7	2.3	1.6	6.3	2.3	14.8	6.3	100.4
1932	8.3	11.1	6.5	12.0	27.8	2.8	1.8	1.8	5.5	5.5	10.2	6.5	99.8
1933	14.5	16.1	13.7	12.9	4.8	1.6	8.1	1.6	5.6	4.0	5.6	11.3	99.8
1934	22.4	8.2	12.3	14.3	6.1	2.0	28.6	2.0	—	2.0	—	2.0	99.9
1935	1.4	8.4	18.3	8.4	4.2	12.5	2.8	9.9	16.9	1.4	4.2	11.3	99.7
1936	20.0	8.9	8.9	11.1	28.9	—	—	—	8.9	11.1	2.2	—	100.0
Average	12.1	12.0	12.4	12.0	10.9	3.7	6.7	3.6	7.6	5.4	6.3	5.9	99.6

The populations on full branches were as follows:

age	year	locality	I	II	III	IV	V	VI	VII	VIII	IX	X	XI	XII	Total
8	1935	Hederah	—	2	2	2	—	1	2	2	—	—	1	1	13
12	1934	Petah Tikva . .	(12)	20	13	4	6	1	—	—	7	9	7	4	83
23	1935	Petah Tikva . .	7	4	2	10	3	2	3	7	3	8	9	5	63
27	1934	Hederah	(4)	(9)	29	11	12	—	1	2	—	2	2	5	(77)
28	1935	Hederah	—	8	14	12	4	1	2	15	6	1	6	5	74
29	1936	Hederah	—	9	9	3	—	—	—	—	—	—	1	—	22
Total			(23)	**52**	**69**	**42**	25	5	8	26	16	20	26	20	

July/August is therefore the season of maximal increase. The range of fluctuation (maximum to minimum population) is relatively low and constant (quotient 10—20). The low level in autumn in the count of hundred leaves is caused by the preference of the half shade in the tree during this season. The population decline in early summer is partly induced by mortality during the hamsin winds, but to a marked extent by great efficiency of parasites. This permanent interference between climatic and biotic factors is the reason why no attempt was made to ascertain the physical optima for *C. hesperidum* from the seasonal population trend.

There are no complete studies with regard to the monthly mortality of *C. hesperidum*. The influence of parasitism is great, especially during the summer months. Observations on full branches at Petah Tikva yielded

month	I	II	III	IV	V	VI	VII	VIII	IX	X	XI	XII
1934 living	—	11	11	3	4	0	0	0	3	2	7	4
dead	—	9	2	1	2	1	0	0	4	7	0	0
1935 living	2	3	2	6	1	1	2	6	2	5	6	4
dead	5	1	0	4	2	1	1	1	1	3	3	1
% dead	71	42	13	36	44	67	33	14	50	59	19	11

Two extended counts at Rehoboth in May 1934 gave the following results:

larvae			adults		
parasitized	living	dry	parasitized	living	dry
94	4	—	203	—	—
21	25	20	75	13	7

Among 462 individuals, 9.1 % were living and not parasitized. Up to 7 parasites were bred from one scale.

The brown soft scale can develop on twigs as well as on fruit of citrus, when artificially infested without any damage. But in nature we have met it only on the leaves of this plant. They prefer half-shade in the parts of the tree between the centre and the periphery. On a full branch at Petah Tikva (1935) they were distributed as follows:

parts of branch (from center to periphery)	I	II	III	IV
no. of *C. hesperidum*.	3	10	20	7
no. of *C. hesperidum* per 100 leaves.	6	18	17	10

Host relations and damage.

C. hesperidum is extremely polyphagous. In Palestine it has been found on:

Amaryllaceae	*Agava americana*
Anacardiaceae	*Schinus molle*
	Mangifera indica
Apocynaceae	*Nerium oleander*
Araceae	*Arum italicum*
Aralaceae	*Hedera helix*
Asclepiadaceae	*Calotropis procera*
Begoniaceae	*Begonia sp.*
Cactaceae	*Echinocactus sp.*
Caprifoliaceae	*Lonicera sp.*
Labiatae	*Coleus sp.*
Lauraceae	*Laurus nobilis*
Leguminosae	*Acacia longifolia, Robinia pseudacacia*
Lythraceae	*Lawsonia alba*
Malvaceae	*Abutilon indicum*
Moraceae	*Ficus carica, F. sycomorus, Morus alba.*
Myrtaceae	*Psidium guayava, Myrtus communis*
Nyctagineae	*Bougainvillea sp.*
Rosaceae	*Rosa spp., Cydonia vulgaris*
	Prunus armeniaca, P. persica, P. communis
	Pyrus communis
Rutaceae	*Citrus spp.*
Salicaceae	*Populus alba*
Solanaceae	*Solanum nigrum*
Vitaceae	*Vitis vinifera*

also on mango, *Gazania* and *Adianthum.*

It is rather curious that *C. hesperidum* did not settle on potato sprouts, whereas it develops on *Solanum nigrum.*

It prefers to settle on young leaves and young twigs of citrus plants. On the leaves they are common along the mid- and less frequent along the side ribs. They are met with occasionally on the fruit, where they develop normally.

Ornamental plants are often encrusted with them and it is easy to get rid of them by brushing or spraying the plant with an emulsion of soap or of white oil.

On citrus we have never seen it to be present in sufficient numbers to cause damage by sucking. The damage is done by heavy honeydew secretion, which spoils the fruit by sooty mould development. In the larval stages, especially, the fluid excrement is sprayed by pressure through the anal aperture throughout the day. During this excretion the anal opercula are lifted. The permanence as well as the force of this honeydew excretion can easily be demonstrated when a tent out of microscopic coverglasses is erected around the animal. The drops stick visibly to the glasses and their continuous increase can be shown by changing the glasses. The total honeydew-excretion of the larvae is considerable.

Notorious damage has been observed in citrus orchards only very locally, especially on very young trees. Heavy sooty mould develops on the areas sprayed with honeydew. These areas are generally well circumscribed and limited in extent, but intensely black. The relative scarcity of the species on the trees prevents damage on older trees. Rivnay (1944) reports also occasional local damage by June drop of fruits caused by larvae settling beneath the sepals.

Special control measures have never been found advisable. On old trees it never was necessary, on young trees heavy parasitism had occurred at the time of sooty mould development.

Heavy parasitism, mainly through *Coccidophaga scitula* and *Eucomys lecaniorum,* but also through *Chilocorus,* thrips, predacious bugs; etc. is a very characteristic appearance of *C. hesperidum.* There is certainly no other citrus scale which is checked to such extent by parasites as the common brown soft scale. 77—99 % of parasitized individuals are not uncommon in late spring and this is certainly one of the main reasons for the summer and autumn depression of this species.

Ants are also frequent visitors of the honeydew excreted by soft scales. *Crematogaster inermis, C. i. armatula, Camponotus compressus sancta* and *Monomorium subopacum phoenicia* have been especially collected.

During summer occasional groups of *C. hesperidum* may be found unmolested by predators and only slightly parasitised. These develop quite well and the protection given by the ants interfered with earlier parasitism. Only in mid-summer when the ants have an extended rest at noon, the scales are parasitized during the hours of rest of their protectors. A similar observation has been made by the writer in the Tulure district of California.

The symbionts of *C. hesperidum,* as well as of all other soft scales, live singly or as short germinating chains within the haemolympha or in cells of the fat-body. They reproduce by germination only and belong to the Pyrenomycetes (order: Sphaericales; lower fructification form *Dematium pullulans*). It was impossible to cultivate them hitherto. Suppressed chain-formation leads to the formation of long multicellular bags. The suggestion of Buchner that the symbionts of *C. hesperidum* are different on different hosts, could not be confirmed by Schwartz. The egg-infection with symbionts occurs through the connection between egg-and nutritive cell.

References.

F. S. Bodenheimer, Schaedlingsfauna Palaestinas. Berlin, 1930, pp. 149—150.

P. Buchner, Tier und Pflanze in intrazellular Symbiose. Berlin, 1921, pp. 228—235.

M. Dingler, Zeitschr. angew. Entom. **9**, 1923. pp. 191—246.

R. Moniez, C. R. Acad. sc. Paris **14**, 1887. 4 pp.

R. Newstead, Monograph of Coccidae of British Isles, London, Vol. 2.

E. Rivnay, Journ. Ent. Soc. S. Africa 1. 1944. p. 73—81.

W. Schwarts, Arch. f. Mikrobiol. **6**, 1935, pp. 369—456.

J. B. Steinweden, Ann. Ent. Soc. Amince **22**. 1922. pp. 197—243.

M. Thomson, Zeitschr. Zellforsch. u. mikrosk. Anatom. **5**, 1927, pp. 1—116.

M. Thomson, IV. Internat. Congress Entom. Ithaca (1928), Vol. **2**, pp. 18—24.

B. THE FLORIDA WAX-SCALE (*Ceroplastes floridensis* Comst.)

Description: Adult female. The Florida wax-scale is a soft-scale, heavily covered with a thick wax-layer. This layer is broad-oval and strongly convex. It is of an opaque white colour, often shaded pinkish along the dorsal portion, the margins remaining white. The dorsal shield is very large, the eight marginal shields small. The shields are divided by shallow furrows only and the division of the different shields is never as distinct and clear as in *C. rusci*, for example. In older specimens the distinction may disappear altogether and the wax-layer becomes an homogeneous shield. The shield is from 3 to 4 mm long, from 2,3 to 3,3 mm wide and from 1,3 to 1,8 mm high. The bare body is reddish-brown in colour. The antennae are 6- of 7-jointed; the mouth-parts small, the legs well-developed. The margin of the body beneath the spiracles is furnished with characteristic stigmatic spines. A band of multilocular pores extends from each spiracle to the corresponding group of spines. The dorsally situated anal-opening lies on the apex of a small cone-like protuberance. The heavily chitinized plates almost form a circle. The anal ring is small and nearly circular. The vaginal opening is ventrally situated.

The egg is elongate-oval in form, purplish-red or yellow in color. Its surface is smooth and shiny. The average length and width is 0,3 mm and 0,15 mm respectively. The egg-stage lasts for less than a week, from spring to autumn.

The first stage larva is flat and oval. The 6-segmented antennae bear a number of long hair at the terminal joint. The eyes are dark red-brown, the mouth parts and legs well developed. The larva is light red-brown, with slightly paler legs and antennae. It is 0,35 mm long, 0,20 mm wide. Within three days, or generally during the first day, this active larva settles down after a few hours. Once having settled, no further locomotion takes place. It prefers to settle on the twigs and the nerve-ridges of the leaves, but always on that surface of the leaf which is exposed to light and sun, i.e. as a rule, on the upper surface. The larva produces waxy protuberances, which give it a star-like appearance, soon after its fixation.

The second stage larva has an oval and strongly convex layer. Similar to the first stage, we observe eight short, pointed, white, waxy processes along the sides; one rather slender process at the front and a short one along each of this; four rather short processes along the posterior part. It is about 1,4 mm long, and 0,8 mm broad. The star-like appearance of the first larval stage is preserved, but this larva is larger and the shields are easily recognized.

The insect itself is also oval, strongly convex and dark red-brown. It is about 1,2 mm long 0,6 mm wide. Its antennae are 6-jointed, the mouth parts small and the legs much reduced in size

The third stage larva has a broad-oval, slightly narrow posteriorly and convex layer. It is white. During the transition period between second and third stage larvae the longwaxy protuberances are some what smaller. The ventral and the marginal shields grow in size and merge. The scale does not look star-like any longer, but is similar to a nipple. It is 1,8 mm long and 1,5 mm wide

The body of the bare insect is broad-oval, convex and dark red-brown. The antennae and legs are greatly reduced in size as compared with the growth of the body.

The development of a small group of the Florida wax-scale is described in the following lines:

27.V.34	50 active larvae have been transferred to a young citrus tree.
28.V.34	Most of the larvae have settled.
29.V.34	All the larvae have settled. The central was shield has started to secrete.
31.V.34	The dorsal wax-shield has grown considerably and the shields cover the entire body.
2.VI.34	The lateral protuberances appear in some of the larvae.
3.VI.34	The lateral protuberances appear in most of the larvae. 36 larvae still live.
6.VI.34	Most of the larvae develop normally.
13.VI.34	The larvae are 0,75 mm long, 0,5 mm wide and 0,25 mm high.
22.VI.34 – **6.VII.34**	No change.
13.VII.34	The larvae are 1,5 mm long, 1,0 mm wide and 0,8 mm high.
21.VII.34	No change.
28.VII.34	Rapid development.
5.VIII.34	The larvae are 2,2 mm long, 2 mm wide, 1 mm high.
12.VIII.34 – **20.VIII.34**	Increase in height.
28.VIII.34	An increase of height from 1 to 1,25 mm. The first oviposition has taken place. 16 scales still live.
30.VIII.34	Most of the individuals are inflated and have begun to oviposit.
7.IX.34	Many crawlers (newly hatched larvae) appear and they leave the females until 22.IX.34.

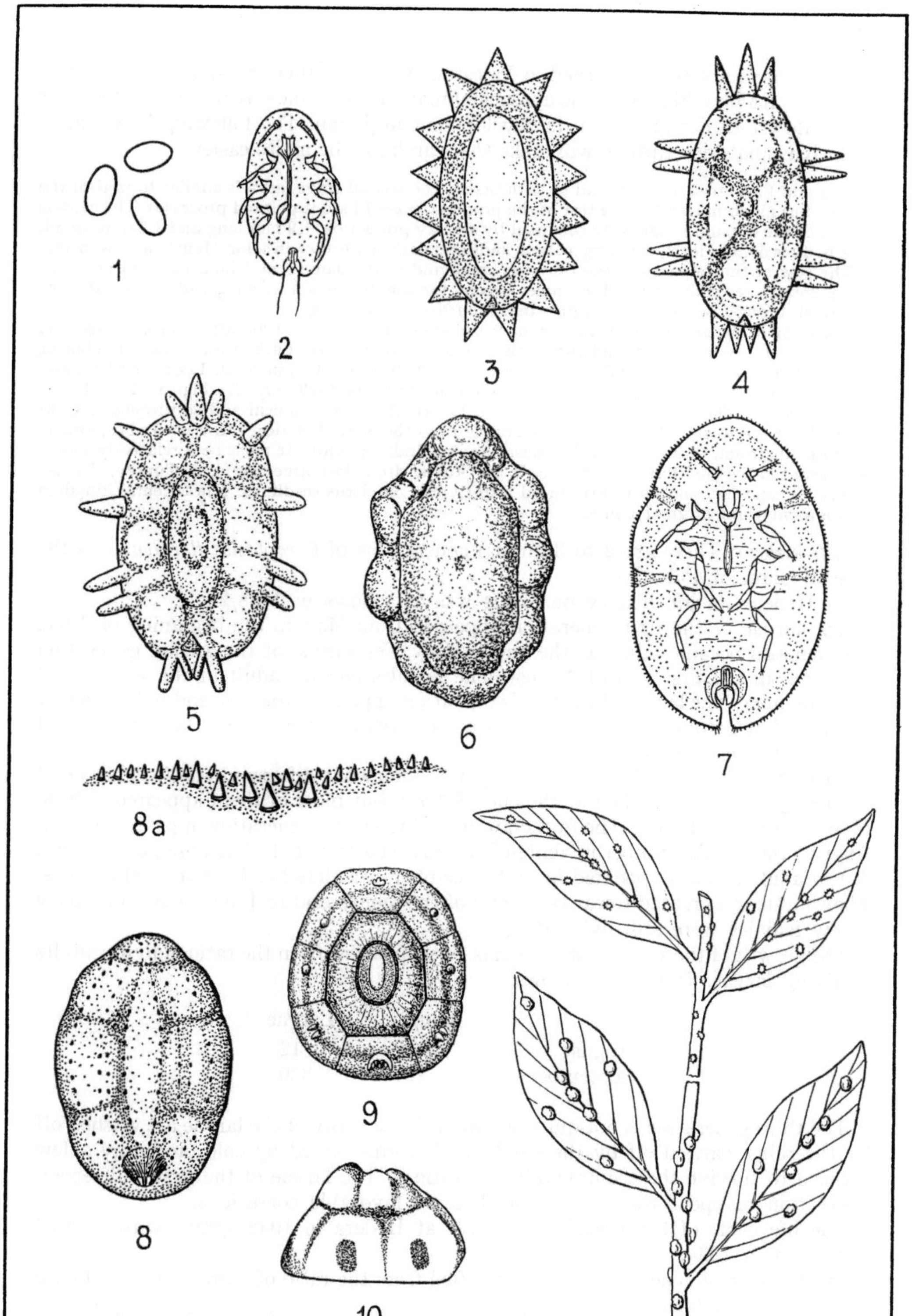

XXV. The Wax Scales. 1—8) *Ceroplastes floridensis* Mask. 1) eggs; 2) neonate larva; 3—5) larval instars; 6) adult female; 7) and 8) naked adult female from below and above; 8a) paraspiracular marginal spines of adult female; 9) *Ceroplastes rusci* L. (seen from above) 10) *Ceroplastes sinensis* Del Guercio (profile); 11) infestation of *C. floridensis* on orange tree: above third stage larvae, below adult females. (from Bodenheimer)

Males have not been observed in Palestine as yet, neither in our breedings nor in nature. Balachowski reports the absence of males in S. France. Kuwana, on the other hand, describes the male and its development in Japan. The following lines, taken from Kuwana's description, will make the search for the males easier:

Egg and first larva are identical in both sexes. The second stage larva is similar to that of the female but much narrower. The test of the propupa bears 13 stout pointed processes. The central region of the dorsum is narrowly occupied by a waxy projection. 1,9 mm long and 1,3 mm broad. The body is oval in form, tapering at both ends and dark red-brown in color. Mouthparts wanting.

The test of the pupa as in the previous stage and of the same size. The general form of the pupa greatly resembles the adult male, except for the possession of wing-pads. General color yellowish-brown. Antennae 10-segmented. Mouth-parts wanting.

The body of the adult male is rather short and stout, with the head round; in general yellowish brown in color with the legs and antennae pale, while the eyes are dark. 1,3 mm long (including stylus), 0,5 mm broad and 2,5 mm wing expanse. Antennae 10-segmented. Legs very long and slender. (Illustrations of these stages are given in Kuwana 1923, Pl. IX, Fig. 8—11, 19, 20).

Distribution: The Florida Wax-Scale is widely distributed in tropical and subtropical regions throughout the world. It is known for example from the S. E. America, the West Indies, Brazil, Hawaii, Australia, China, Japan, Formosa, Assam, India, Ceylon. It must be a relatively recent immigrant in the Mediterranean region. It is known from Palestine, Syria, S. Turkey, Egypt, S. France, and has just been introduced into Cyprus on citrus seedlings. The Oriental kingdom is apparently its original home.

Life-history: There are 2 to 3 annual generations of *Ceroplastes floridensis* in the coastal plain of Palestine.

In Hedera its life history on citrus was as follows in 1928/29:

The larvae of the first generation appeared from May to the beginning of June. The females became adult at the end of July. The larvae of the second generation appeared in the first half of August. The females became adult in the second half of September. The larvae of the third generation appeared from the end of September to the end of October. These grew adult until December, yet the first larvae appeared only by the next spring.

In Rehoboth, in 1934, the life history was somewhat different. The first larvae of the first generation hatched at the end of April, but most of them appeared only in May and until mid-June. The first crawlers of the second generation appeared in mid-August, their main number in September and still a few at the beginning of October. A few adult females appeared in mid-December, crawlers hatching from their eggs. Their number increased towards the end of the month and in January and February a few females were still ovipositing.

The growth of the second generation is very apparent from the ratio of living adults to living larvae. It was as follows:

	Adults	Larvae
August	388	12
September	16	330

The third generation is not quite as simple. Very many of the larvae are washed off by the rains, carried off by the winds, and storms, killed by cold, and only a few succeed to survive these unfavorable conditions. The larvae of the first and second generation reappear there under much more favorable conditions.

The life cycle of the Florida wax scale at Hedera in 1932 agreed with that of Rehoboth in 1934.

The number of generations can be verified from the ratio of living adults to living larvae $\left(\frac{\text{living adults}}{\text{living larvae}}\right)$ in two different orange-groves in Rehoboth (1934):

Month	I	II	III	IV	V	VI	VII	VIII	IX	X	XI	XII
Orange grove A . .	0,08	0,19	0,44	**4,69**	**5,44**	0,13	0,33	**3,00**	0,02	0,02	**0,05**	**0,10**
Orange grove B . .	0.11	0.15		**3,28**		0,01	0,34	**4,98**	0,02	0,00	**0,00**	**0,04**

Only the third generation cannot be verified in this way, but the higher ratio of December as compared with that of October is readily seen.

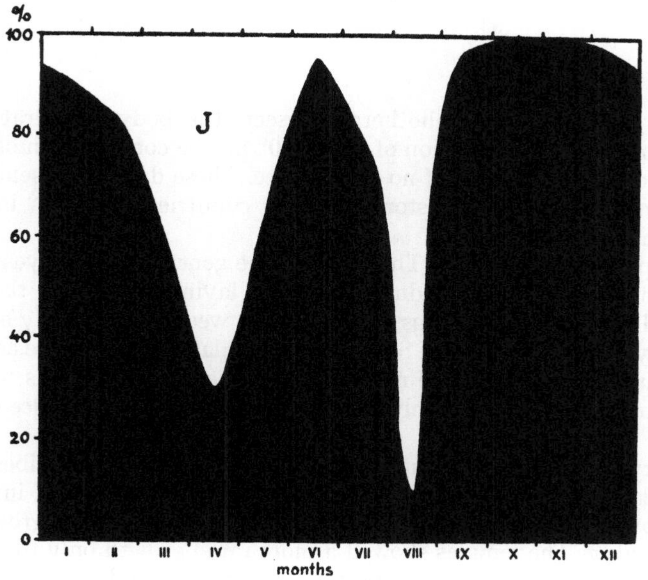

Fig. 90.
Monthly ratio of adults to larvae of *Ceroplastes floridensis* Comst. at Rehoboth (1934). White: adults; black: larvae.

The number of living adults and larvae in the same groves presents the same picture:

Month	Orange grove A			Orange grove B		
	Adults	Larvae	Total	Adults	Larvae	Total
I	33%	76%	69%	21%	56%	**48%**
II	38	70	59	26	54	**47**
III	74	79	78			
IV	**92**	48	80	**86**	43	70
V	**87**	76	83			
VI	14	82	51	21	87	**84**
VII	29	45	40	**86**	61	**66**
VIII	**66**	43	58	**79**	69	**75**
IX	17	84	80	15	75	**69**
X	8	83	70	0	84	**80**
XI	26	72	67	0	73	**73**
XII	**59**	77	75	**16**	72	**64**

The three annual generations are even more conspicuous than in the preceeding table.

The great majority of the *Ceroplastes* sit on the upper surface of the leaves and are exposed to sun-radiation for a longer or shorter period of the day. During that period its body temperature is much higher than that of the environment. Consequently, the body temperature is subject to a great deal of variation.

A 20 minutes observation in Rehoboth (7.IX.33; Black bulb 50,2° C; White bulb 40,7° C; shade temperature of the air 30,5° C) yielded the following results:

Time	Body temperature	Natural shade among the leaves
1 : 28 P.M.	34,8° C	+++
1 : 3	34,1	+ +
1 : 8	33,1	+
1 : 23	35,0	+
1 : 28	40,7	–

Ceroplastes floridensis is a heliothermic insect, the body temperature of which rises when exposed to sun-radiation of above 40° C. The computation of the development threshold was practically of no importance. These data are essential, however, for the explanation of its life history in other countries, which is in parts very different from that in Palestine.

Kuwana reports from Japan: "There is but one generation each year. The female passes the winter as a fertilized adult and begins laying eggs about the first part of June in the following year. The eggs hatch within a week. These newly hatched larvae reach maturity by the first part of September. Copulation takes place as soon as both sexes are mature, after which the male dies and the female remains to produce the next generation." (Japan is the only country from which the presence of the male is known.).

Balachowsky has likewise observed one generation only at Antibes (S. France). The neonate larvae hatched on the 25th of June, 1930, developed into the second stage larvae by mid-September, and from that time on the larval growth slackened down condiderably. The females showed a more rapid growth only by the beginning of April. The first oviposition was observed about the 22nd of April, but the first neonate larvae only on the 3rd of June. This proves that active development takes place only from June to September. 1½ month incubation of the eggs is an abnormally long period and illustrates the degree to which the development has slackened down.

Watson writes from Florida: "Three or four months are required for growth and there are three principal broods of crawlers which appear during April and May, July and August, and October and November. The last is a smaller, less important brood." These conditions agree fairly well with conditions in Palestine.

The following table illustrates the monthly temperatures of the four different countries, for which reports are present:

Locality	I	II	III	IV	V	VI	VII	VIII	IX	X	XI	XII
Tokio.	2,9	3,5	6,7	12,4	16,4	20,5	23,9	25,4	22,0	15,9	10,2	5,2° C
Antibes. . . .	8,3	7,8	11,3	13,2	16,2	21,9	24,0	23,9	23,2	16,5	12,8	9,4
Tel Aviv . . .	12,8	13,5	15,5	17,9	21,5	24,5	26,4	26,7	25,5	22,6	18,7	14,7
Jacksonville. .	12,2	13,8	16,6	19,8	23,4	26,1	27,2	26,7	25,2	20,9	16,3	12,8

It is obvious that active development takes place in all regions only, when the air shade temperature surpasses about 20° C: From June to September in Tokio and Antibes, from May to October in Palestine and Florida. Development continues even at lower temperatures, but at a considerably slower pace.

The annual effective temperature sum above 20° C is for the four mentioned places, at:

Tokio	324	days degrees
Antibes	393	,, ,,
Tel Aviv.	747	,, ,,
Jacksonville	885	,, ,,

However, the absolute temperatures are not decisive in heliothermal animals. The geographical latidude, the humidity of the atmosphere, altitude, etc. influence the intensity as well as the duration of exposure to the sun-rays. As to how much the sun-radiation influences the speed of development can easily be deduced from the fact that in a bright laboratory fully protected from direct sun-radiation, at room-temperature, not one specimen survived 8 months' time. This condition was so unfavorable, that none of our breedings lasted until oviposition, all individuals dying before that from various reasons. The difference in the life-history of *C. floridensis* in these different countries is now easily explained.

Life cycle in Palestine: The all-year-round development of *C. floridensis* in an orange grove in Rehoboth (1934) may serve as an illustration of the normal life history of the Florida Wax-Scale in Palestine.

January: Several adult females oviposit. The large larvae and immature females produce heavy honey-dew deposits. The rains wash off most of the old sooty mould from the leaves (not from the fruit), whereas most of the sooty mould, which developed on fresh honeydew, remains. The development of the scales is slow. Many scales are washed off by the rains and many die.

February: The very heavy rains wash off most of the wax-scales from the trees. A light frost brings on an additional mortality. No honey-dewis observed, no change in the size of the scales.

March: Several females oviposit, most of the scales which still secrete some honey-dew approach sexual maturity. Sooty mould develops rapidly on the honey-dew.

April: Gravid females are abundant. The first larvae of the spring generation appear towards the end of this month. 1—2 wax-scales per leaf may be found. The number of parasites is low. The older larvae and the young females exude a moderate amount of honey-dew.

May: There are numerous gravid females and scrawling first stage larvae. The number of eggs per female is very large. The mothers of the spring generation begin to dry out. The larvae develop rapidly under assumingly favorable climatic conditions. Honey-dew excretion and development of sooty mould are scarcely observed, as the adults are too old and the larvae too young. Parasites and predators are found in moderate numbers.

June: Very many young larvae after fixation and the last larvae of the spring generation settle down. They develop quickly. Most adult females are empty and dead. The small Hymenoptera are welldeveloped parasites. Honey-dew and sooty mould are not conspicuous, as most larvae have not reached the suitable age as yet.

July: Most of the wax-scales approach sexual maturity and excrete large amounts of honey-dew, on which sooty mould develops freely. Parasites and predators are present in moderate numbers only, but the climatic mortality is rather high.

August: A considerable honey-dew excretion and a rapid development of sooty mould still observed. They cease, however, with oviposition, which takes place in this month. The first crawling larvae appear towards the end of the month. There are numerous parasites and predators and the climatic mortality is still high.

September: The crawlers of the second generation appear and settle down. Living adult females become scarce, as is also the case with the honey-dew production. However, sooty mould continues to develop on the honey-dew secreted during August. There are many parasites and predators.

October: The settled larvae, which begin to produce honey-dew towards the end of the month, develop quickly. Sooty mould develops only moderately.

November: The larvae approach sexual maturity. A large amount of honey-dew is secreted, but sooty mould spreads only moderately. Parasites and predators are scarce. The development slackens down and the beginning of the cold weather may be responsible for the high mortality of the larvae.

December: Part of the females of the second generation begin to oviposit and the first crawlers appear, but not in great numbers. The majority of the individuals of the second generation are still immature. Those continue to produce some honey-dew. Sooty mould develops only little. The heavy rains wash off the old sooty mould from the leaves, but not from the fruit. During winter neither parasites nor predators are seen.

Fertility and Oviposition: The number of eggs per female varies from season to season and from year to year. The completest count has been made in Rehoboth during a mass outbreak of the Florida wax scale, in 1934:

Month	n	Number of eggs per female Average	Minimum	Maximum	Remarks
I	11	200	45	282	Only a few females with eggs.
II	8	203	63	406	Only a few females with eggs.
III	8	334	109	620	Only a few females with eggs.
IV	26	460	134	953	Many females with eggs.
V	16	515	298	804	Many females with eggs.
VI	17	280	51	1278	Much less females with eggs.
VII	—	—	—	—	No female with eggs.
VIII	24	418	52	1125	Numerous females with eggs.
IX	20	722	196	1013	Numerous females at the beginning, very few at the end of the month.
X	3	156	119	212	Only a few females with eggs.
XI	6	289	93	610	Only a few females with eggs.
XII	10	372	107	594	Only a few females with eggs. beginning of winter generation
Total resp.	149		45	1278	
Total average		415			

Other counts have been made in Hedera in earlier years, when the population level of the wax scale was low:

Date	n	Number of eggs per female Average	Maximum
V.1930	11	465	590
VIII	9	220	594
IX	10	307	375
V.1932	9	240	400
VI	3	75	85
VIII	10	220	251
VI.1933	2	50	54
VIII	5	207	364
IX	10	153	299
X	10	209	327
XI	10	168	319
XII	6	140	218
	95	233	594

The number of eggs is thus extremely variable and dependant upon environmental conditions. The range of fertility known at present is as follows:

in Japan	90— 800	eggs per female	(Kuwana)
in Florida	75— 100	,, ,, ,,	(Watson)
in Palestine	45—1278	,, ,, ,,	(Bodenheimer)

A warm temperature of not less than 23° C or even above 25° C seems to be optimal for the maturation of the eggs, but the same temperatures and the low humidity prevailing at that temperature affect the vitality of the larvae decidedly. The same is true of low temperature and much rainfall, with regard to both fertility and larvae, the effect being an unfavorable one.

There exists a correlation between the size of the adult female and its fertility, in as much as tiny specimens produce only few, large specimens many eggs. During the oviposition period the female becomes inflated.

The eggs are laid one after another. During the first half of oviposition, the eggs are covered thickly with a whitish powder, probably a product of some of the wax-glands. These eggs pass the vagina easily and drop to the surface of the plant. The eggs are thus protected — as is the case in most soft scales — by the body of the mother, shield-like. The eggs are usually purple and occasionally yellow. Their surface is smooth and shiny. In the second half of oviposition, the white powder mentioned above gradually becomes scarcer. The eggs are then viscous and stick to the vaginal opening until expelled by the following egg. During that period the eggs often stick to one another. In this manner, small chains of eggs are often produced. The female shrinks with the progress of oviposition.

The exact extent of the oviposition period per female has not been ascertained. The following observations are based on the assumption that the period of hatching of the crawling larvae coincides with the duration of the oviposition period.

No.	Date of first crawler	Total number of eggs	Total number of hatching period
1	17. V.34	443	10 days
2	22. V.34	307	8 ,,
3	19. V.34	311	10 ,,
4	25. V.34	229	9 ,,
5	22. V.34	167	7 ,,
6	22. V.34	113	8 ,,
7	2.VI.34	512	10 ,,
8	9.VI.34	418	10 ,,

The oviposition period is thus relatively short and lasts from spring to autumn about 1—2 weeks, or even somewhat less.

The daily number of hatching larvae — which is supposed to be approximately the same as the rate of daily oviposition of the mother — was observed in May in two cases, as follows:

<table>
<tr><th>Day</th><th></th><th>1</th><th>2</th><th>3</th><th>4</th><th>5</th><th>6</th><th>7</th><th>8</th><th>9</th><th>10</th><th>Total</th></tr>
<tr><td>No. of larvae hatched</td><td>A. . . . .</td><td>43</td><td>78</td><td colspan="4">268</td><td>21</td><td>13</td><td>14</td><td>6</td><td>443</td></tr>
<tr><td></td><td>B. . . . .</td><td>52</td><td>68</td><td>54</td><td>61</td><td>39</td><td>17</td><td>10</td><td>6</td><td></td><td></td><td>307</td></tr>
</table>

As a rule, oviposition is strongest in its first half, gradually declining towards the end. The larvae hatch from the first laid eggs when the female still continues to oviposit. They leave the empty shells of their eggs below the scale. The female dies, as a rule, a few days after having finished to oviposit. But the dead ones cannot generally be recognized as such by the naked eye. They are dried out, and this can be best observed when the animal removed from the plant is examined under the binocular microscope. The mother's body shrinks when its ventral side approaches the dorsum. The hole which contains the eggs is thus progressively enlarged.

About 80 eggs per 24 hour period was the maximal daily oviposition rate that we observed. Kuwana reports from Japan up to 159 eggs per female and day and it is certain that the more fertile females in Palestine must approach or even occasionally surpass that daily oviposition rate. In one specimen, which produced 26 eggs per 24 hours, 5 eggs were expelled at intervals of 15, 15, 27, 17 minutes. This rate would allow for the birth of 77 egs per day, supposing oviposition continues day and night at an equal rate. There is no doubt, that the intervals can be reduced to at least a quarter of the above quoted ones.

Host-plants: C. floridensis is a polyphagous species. The following host-plants have been recorded in Palestine:

Anacardiaceae	*Pistacia lentiscus, Schinus molle*
	Schinus terebinthifolius
Anonaceae	*Anona squamosa*
Apocynaceae	*Nerium oleander*
Celastraceae	*Evonymus japonica*
Compositae	*Chrysanthemum hortorum*
Convolvulaceae	*Convolvulus sp.*
Lauraceae	*Laurus nobilis*
Leguminosae	*Ceratonia siliqua*
	Acacia farnesiana
Meliaceae	*Melia azederach*
Moraceae	*Ficus carica, F. infectoria, F. sycomorus*
	Morus alba
Musaceae	*Musa cavendishi*
Myrtaceae	*Myrtus communis*
Palmaceae	*Washingtonia sp.*
Polygonaceae	*Polygonum sp.*
Punicaceae	*Punica granatum*
Rosaceae	*Cydonia vulgaris, Prunus persica,*
	Prunus domestica, Pyrus malus,
	Pyrus communis, Eriobotrya japonica
Rutaceae	*Citrus spp.*

Honey dew-secretion and damage: The honey dew secretion of the Florida wax scale is by no means a solved problem. It is at present certain that *C. floridensis* may occasionally secrete a rather considerable amount of honey-dew. But it is an equally well-established fact, that in „normal" years, when the absolute number of the scales is quite high, no honey-dew is ever observed. Climatic conditions probably influence the quantity of honey-dew secreted and its variation from year to year and from season to season.

The honey-dew leaves the body through the anal-opening at the dorsal side of the posterior end of the body. Two triangular anal valves and several bristles surround it. Towards the end of the first and in the second larval stage (1—2 month old) a few very tiny drops are excreted in the immediate neighbourhood of the anal opening of the wax scale. Flies, ants etc. are often immediately at the spot to consume the drops speedily. The third stage larva and the young female are responsible for the most important honey-dew production. The role of the second larval stage is still unknown, but the production of honey-dew is probably still low at this stage.

The size of the honey-dew drops, the intensity of secretion and the power, with which the drop is expelled from the body, increase as the insects grow older. The older larvae never expel the drops in the neighbourhood of the producing insect.

On 23.VIII.33 the production of several large larvae was as follows:

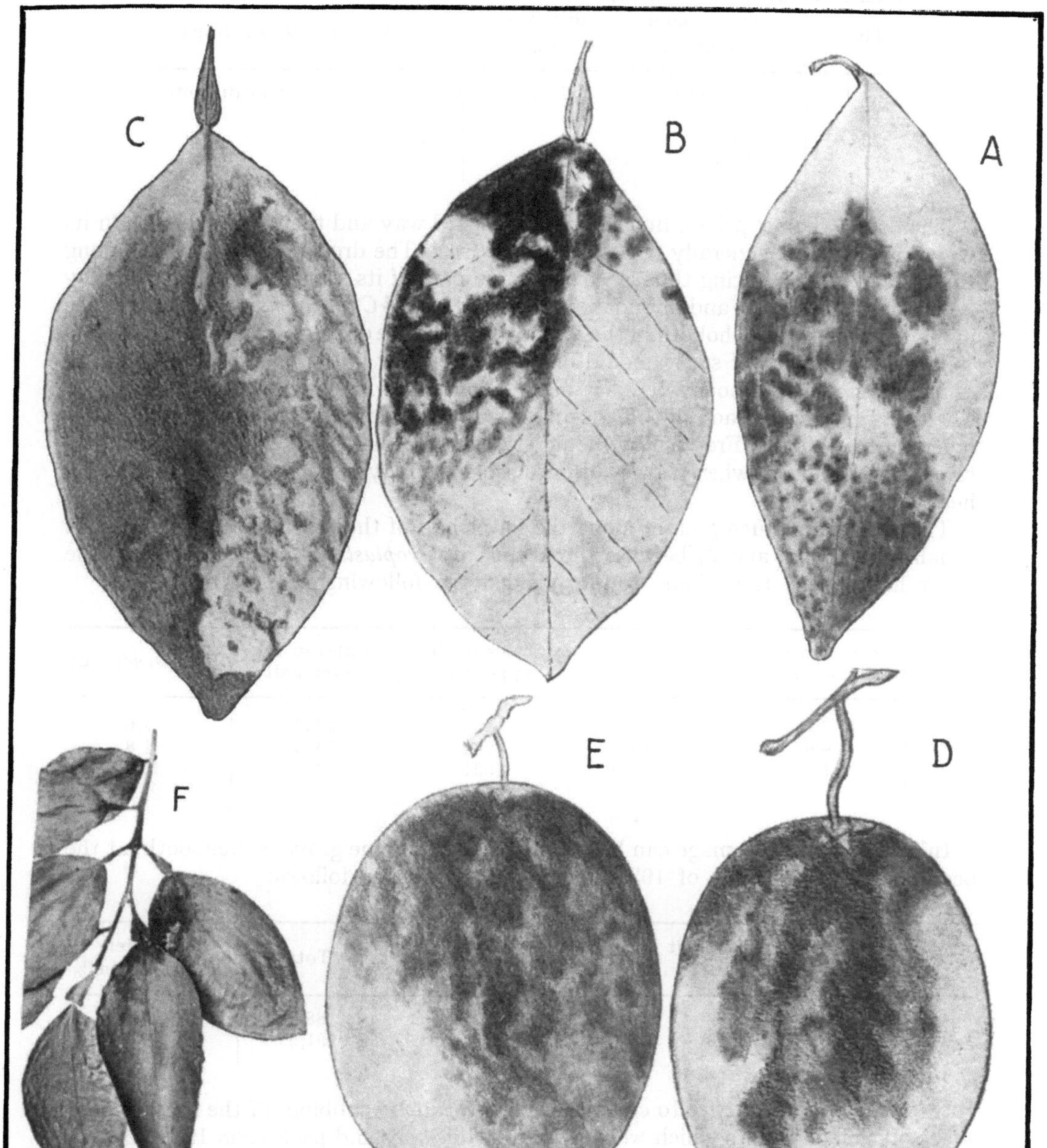

XXVI. Sooty mould development on citrus caused by *Ceroplastes floridensis* Mask. A—C: successive stages of sooty mould development on leaves; D—E: sooty mould on orange fruits; F. The sooty mould on the lower leaves is caused by honeydew exudation of the scale above. (from Bodenheimer)

Time	Drop found at a distance horizontally	vertically	No. and size of drops	
7 h	10,5 mm	7 mm	1	1 mm diameter
10	17,0	10	1	4
23	15,0	10	1	2
11	12,0	5		

The drop, when expelled, first follows an arched way and then drops down. On its way downwards it generally falls on a leaf or fruit. The drop then slowly rolls along the fruit or leaf, covering thus a considerable part of its surface. In years of heavy honey-dew production and abundance of individuals of *C. floridensis*, as was the case in many orchards in Rehoboth in the autumn of 1934, the damage may be considerable.

The development of sooty mould on the honey-dew is the cause of the damage. A few days after the honey-dew has been excreted the first grey spots appear on it. The spots grow into blackish threads and finally the whole area is covered with black layers of this fungus. Fresh sooty mould sticks to the plant and is not washed off by rains. Only later on, when it dries up into a blackish layer, is it easily washed off by heavy rains.

The damage is much greater along the lower half of the tree 1) because the drops usually drip down and 2) because the attack of *Ceroplastes* is much heavier on the lower half of the tree. Some counts yielded the following results:

No.	No. of leaves per each half of tree	Fruit	Scales in the upper half	Scales in the lower half	% in lower half
1	700	Shamuti	1492	2292	61,0
2	900	Shamuti	964	1513	61,8
3	1100	Grapefruit	4621	5998	56,5

In a bad year the damage can be extremely high. In one grove in Rehoboth, at the beginning of the winter of 1934/35, the attack was as follows:

Fruit	No. of fruit counted	% of fruit attacked heavily	lightly	Total	Absolutely clean fruit in %
Shamuti	2000	42	46	88	12
Grapefruit	2000	32,5	59	91,5	8,5

It is the general practice to clean the spoiled fruit by rubbing off the sooty mould with soft linen. Fruits which were treated in this way and packed on 19.II.35 were examined on 1.IV.35.The percentage of rotten fruit was in

the treated cases (rubbed): 28,3 %
in the not treated cases (not rubbed): 8,7 %.

It is evident, therefore, that dry rubbing should be avoided and further experiments with desinfecting liquids as sublimat, borax, etc. should be arranged for. This kind of treatment can never replace, however, the control of the insects which produce the honey-dew. The damage should be prevented by using these control measures. *C. floridensis* excretes most of the honey-dew in July, in a slighter degree again in

September/October. This is explained by the trend of the fluctuation in seasonal abundancy, which will be treated later on.

Mortality and seasonal population trend of Ceroplastes.

The distinctions between the main natural sources of death are as follows:

1) Physiological death. The body of the female is shriveled up and dry. Below it are found the empty egg-shells, the larvae of which have left the mother's body. As a rule, physiological death shortly follows oviposition.

2) Death caused by parasites or predators, which are sometimes rather numerous. These soft scales are easily recognized by the exit hole, through which the parasite leaves the scale. When the ventral side of the scale's body is opened, the larva or pupa of the parasite, if still in the insect's body, is easily noticed. The presence of predators is indicated by a generally very conspicuous feeding hole.

3) Death caused by fungi. When honey-dew accumulates in proximity of the scale, the latter may become covered with sooty mould, which apparently leads up to the scale's death.

4) Death caused by climatic conditions is especially common among the young scales. Drought has a most unfavorable effect. The specimens effected dry out immediately and may be thus recognized.

The real total mortality can scarcely be ever fully classified. When we consider only 250 eggs per female on the average, then the potential increase from generation to generation is 250-fold. As a matter of fact, a 210-fold increase was observed once at Hedera in May/June, 1929. But the number of eggs per female averaged then about 500. As a rule, the increase from generation to generation is only 10-50 fold, or quite commonly even less. Most of this mortality is due to climatic conditions in the very early larval stage and to the blowing winds.

The importance of the wind was established by the following experiences:

1) On a quiet morning, 60 crawlers were observed running along one citrus leaf. The leaf was painted with oil at its base, so that no larva could actively leave the leaf. After five days only one larva, which had settled down, was still present. The rest were dispersed by the wind.

2) Most of the crawlers, once blown away by the wind, will never find a suitable place for further development. However, a fraction will reach leaves of citrus or other suitable hosts and settle down. Some days later, a small number of settled larvae of *Ceroplastes* was observed on bare leaves, which were previously isolated at their base.

3) Ovipositing females were attached to leaves of twigs, which were immediately transported to places almost completely protected from the wind. After a few days, the hatching larvae have settled on the neighbouring twigs and leaves in large numbers and only very exceptionally on spots at a greater distance.

4) Wind must influence the distribution of the scales over the different points of the compass. The side exposed to the prevailing winds is necessarily poorer, the opposed and wind-protected side richer in individuals. In one case, 0,7 individuals per leaf (and twigs respectively) were counted on the lee side (WSW) and 1,4 on the luve-side (E N E). Trees more exposed to the wind must show even larger differences.

The very large number of larvae which never find a suitable place for development and which are certainly the overwhelming majority of all larvae under normal conditions, can never be exactly counted. The same is true for the mortality observed among the very young larvae, caused by climatic conditions. This type of mortality is also very considerable, especially during the period of hamsin prevalence, when the environmental conditions are not favorable. Young larvae which die due to climatic reasons fall off the leaves easily. We should, therefore, not forget, that all counts as to the causes of mortality are performed only on those individuals which have survived these critical stages. When the increase from generation to generation is tenfold and 250 eggs are the average egg-production per female, then only 4 % of the offspring are included in our counts, 96 having died before and escaped observation. But when the next generation decreases tenfold, then only 0,4 % are accessible to us and 99,6 % have been previously destroyed.

A few of such partial counts follow:

Rehoboth 1934

Month	Adults				Larvae				Total
	Living +	Fungi +	Parasites +	Climate	Living +	Fungi +	Parasites +	Climate	
I	2	2	—	2	25	4	1	3	39
II	17	14	3	20	92	3	2	37	184
III	29	—	—	10	65	6	—	11	121
IV	85	4	—	3	18	9	—	10	129
V	87	3	5	6	16	4	—	3	124
VI	10	8	21	35	78	3	7	11	173
VII	8	5	7	8	26	9	4	17	84
VIII	27	5	7	2	9	4	—	8	62
IX	3	3	—	11	142	4	—	18	181
X	2	6	4	13	97	7	3	10	142
XI	5	5	3	6	103	10	7	23	162
XII	13	3	2	4	137	8	3	31	201
	288	58	52	120	808	71	27	182	1606
	518				1088				

68 % out of the 1606 specimens studied were living, 32 % dead: 19 % due to climatic causes, 8 % due to fungi and 5 % due to parasites and predators.

In Hedera some counts were performed in 1932 and 1933.

Month	1932			1933		
	Living +	Climate +	Parasites	Living +	Climate +	Parasites
I				69%	31%	0%
II				72	28	0
III	96%	3%	1%	84	16	0
IV	100	0	0	83	17	0
V	95	4	1	58	38	4
VI	82	8	10	73	27	0
VII	94	6	0	67	33	0
VIII	82	8	10	58	32	10
IX	16	56	28	21	63	16
X	55	38	7	17	68	15
XI	73	23	4	37	46	17
XII	72	23	5	42	53	5
Absolute	765	169	66	681	452	67
Percent	76,5	16,9	6,6	56,7	37,7	5,6

67 % out of 2200 specimens studied were living, 33 % dead: 27 % due to climatic factors, 6 % due to parasites and predators. It is probable that in this count the mortality due to fungi was classified under the climatic conditioned mortality.

The average monthly mortality in these counts was:

Month	I	II	III	IV	V	VI	VII	VIII	IX	X	XI	XII	Average
Rehoboth 1934 A. . .	31	41	22	20	17	49	60	42	20	30	33	25	=33%
Rehoboth 1934 B. . .	52	53		30		16	34	25	31	20	27	36	=32
Hedera 1932			4	0	5	18	6	18	84	45	27	28	=24
Hedera 1933	31	28	16	17	42	27	33	42	79	83	63	58	=33
Average mortality. . .	38	41	14	**17**	21	28	43	32	**54**	45	38	37	=33

When we wish to know the approximate number of living scales, these respective percentages should be deduced from the actual population count in nature.

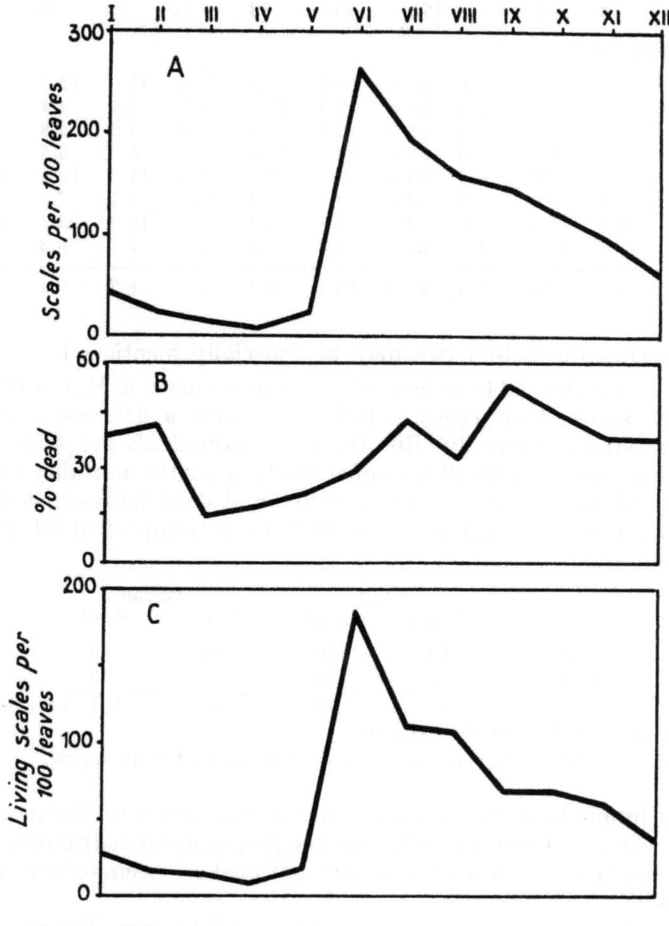

Fig. 91.
Seasonal fluctuation of *Ceroplastes floridensis* Comst. The abscissa represents the months. A. Scales per month and 100 leaves; B. percentage of dead scales per month and 100 leaves; C. living scales per month and 100 leaves (from Bodenheimer)

The number of *Ceroplastes floridensis* per 400 leaves of orange-trees at Atta was as follows:

	I	II	III	IV	V	VI	VII	VIII	IX	X	XI	XII	Total	Max./Min.
1928	—	—	—	—	—	245	(188)	132	100	164	336	196	—	—
1929	(118)	39	(26)	12	**9**	1891	;1298	(1783)	**2248**	1568	1365	865	11.222	270
1930	413	375	190	**119**	325	**4420**	;2546	1516	1069	941	565	318	12.801	37
1931	239	83	110	**72**	165	**495**	355	243	74	158	106	**93**	2.195	7
1932	55	34	20	**14**	15	**516**	433	209	51	80	103	58	1.588	37
1933	75	46	**18**	30	58	**206**	27	37	33	100	105	109	844	11
1934	61	42	(24)	**6**	8	**68**	35	24	53	28	28	20	397	11
1935	**15**	30	50	22	18	22	76	93	53	105	177	132	792	12
1936	100	125	(104)	84	33	**149**	16	**7**	16	19	16	15	686	21
Average per 100 leaves 1929/36	34	24	17	**11**	19	**243**	149	122	112	94	77	57	959	22
dito living scales only	21	14	15	**9**	15	**175**	85	83	52	52	48	36	605	19
or in percentages of the animal fluctuation:														
1929	1.1	0.3	0.2	0.1	0.1	16.8	11.6	15.9	20.0	13.9	12.2	7.7	99.8	
1930	3.2	2.9	1.5	0.9	2.5	35.3	19.9	11.9	8.5	7.4	4.4	2.5	100.9	
1931	10.9	3.8	5.0	3.3	7.5	22.5	16.2	11.1	3.4	7.2	4.8	4.2	99.9	
1932	3.5	2.1	1.2	0.9	1.0	32.1	27.3	13.2	3.2	5.0	6.5	3.7	99.7	
1933	8.9	5.4	2.1	3.5	6.9	24.4	3.1	4.2	3.9	11.8	12.4	12.8	99.6	
1934	15.4	10.6	6.0	1.5	2.0	17.1	8.8	6.0	13.4	7.1	7.1	5.0	100.0	
1935	1.9	3.8	6.3	2.8	2.3	2.8	9.6	11.7	6.7	13.3	22.7	16.7	100.6	
1936	14.6	18.2	15.2	12.2	4.8	21.7	2.3	1.0	2.3	2.7	2.3	2.2	99.5	
Average	7.4	5.9	4.7	**3.2**	3.4	**21.6**	12.4	9.4	7.7	8.7	9.1	6.9	100.4	7

Among the climatic factors two may be especially mentioned:

1) Strong and prolonged hamsin-winds, such as occurred in May, 1935, reduce the number of the scales, their egg-production and have a deleterious effect on the younger larval stages especially. But the dead individuals are soon replaced by neonates at that season. After 10 days of heavy hamsins the mortality was extremely high, especially when considering that many dry and dead individuals did fall from their hosts and the settlement of the crawlers was interrupted at all. The mortality in two orchards was:

	adult		larvae		
	living	dead	living	dead	
grove A	13	50	43	58	
grove B	11	24	40	43	
	24	78	83	101	i.e. 63 % dead as

compared with 21 % in May in the average.

But the dead individuals are soon replaced by new ones at that season.

2) The air humidity must have a considerable influence on the mortality of the Florida wax-scale, as it seems to influence the geographical distribution of the scale. In semi-arid regions the scale — or at least its outbreaks — seem to be restricted to the coastal plains.

Neither are citrus fruits attacked, nor stem nor branches. The young twigs are infested, but never highly. The leaves are the normal habitat of the Florida wax scale and they prefer the habitat along the ribs of the leaves. In two annual counts of full branches the percentage on twigs and leaves was:

	1934	1935	1936
on leaves	100	96	91%
on twigs	—	4	9%

Young leaves are preferred:

month	I	II	III	IV	V	VI	VII	VIII	IX	X	XI	XII
no. of scales on young leaves. .	41	20	—	—	—	42	630	762	} 228	293	406	470
on old leaves	33	7	34	7	3	16	127	177				

As far as active locomotion is concerned, not choice of food, but taxic movements lead the larvae to their definite habitat, which is never very far from their mother. Unfortunately, no analysis on the taxes of *C. floridensis* has been made yet.

On the whole, the moderately exposed parts of the host are preferred:

		Part (from center to periphery)					per 100 leaves of part				
		1	2	3	4	5	1	2	3	4	5
Branch, Peta Tikva	1934						134	139	123	91	165
Branch, Peta Tikva	1935	68	86	274	162		136	159	239	227	
Branch, Hedera . .	1935	542	934	1586			175	234	159		
Branch, Hedera . .	1936	210	120	75			140	240	75		

But, if the infestation is calculated per 100 leaves, the highest infestation is half way between centre and periphery. The infestation does normally start about the tenth year of the tree. But it is less dependent on age than shade-and wind-conditions. Single trees replanted in an old and shady orchard are infected earlier than those in a spacy planted young orchard. Some counts may illustrate these conditions:

The monthly infestation in an old (about 27 years old) and young grove (about 10 years) at Hedera was per 100 leaves:

		I	II	III	IV	V	VI	VII	VIII	IX	X	XI	XII	Total
old orchard .	1935	12	16	18	5	45	87	124	67	27	56	63	77	607
	1936	41	50	27	38	44	53	8	12	5	6	3	4	291
young orchard	1935	—	—	—	—	—	—	—	—	—	—	1	—	1
	1936	—	—	—	—	—	—	1	—	—	—	—	1	2

The monthly infestation on full branches was:

Age	locality	year	I	II	III	IV	V	VI	VII	VIII	IX	X	XI	XII		Max./Min.
7	y. Hedera .	1934	—	—	—	—	—	—	—	—	—	—	—	—	0	
8		1935	—	—	—	—	—	—	—	—	—	—	—	—	0	
		1936	—	—	—	—	—	27	17	8	7	20	9	5	93	
12	Petah Tikva	1934	(125)	105	51	23	217	132	88	51	141	217	127	146	1428	9
23		1935	21	15	14	11	8	202	105	77	43	36	45	42	619	28
27	Hedera . .	1934			35	18	182	526	103	81	270	204	169	90	(1683)	29
28		1935	75	28	36	10	10	63	772	956	245	312	432	501	3440	96
29		1936	52	29	26	17	99	99	53	5	—	—	4	—	384	

These counts may be completed by others on full branches at Petah Tikva:

year and age	I	II	III	IV	V	VI	VII	VIII	IX	X	XI	XII
1934 (12 years) living . . .		95	49	22	206	122	84	49	125	208	97	143
dead		10	2	1	11	10	4	2	16	9	26	3
% mortality		10	4	4	5	8	5	4	11	4	22	2
1935 (23 years) living . . .	19	15	14	11	6	198	100	68	36	36	44	41
dead	1	—	—	—	2	4	4	9	7	—	1	1
% mortality	5	—	—	—	25	2	4	12	16	—	2	2

The environmental resistance can be computed for 25 generations (three annual generations V: 240 eggs, VIII: 220; X: 250 eggs per female, for which actual counts are now available). The respective environmental resistances are:

year	generation I	generation II	generation III
1928		99,435	99,976
1929	53,928	99,600	99,906
1930	98,996	99,748	99,943
1931	99,928	99,745	99,962
1932	94,194	99,815	99,653
1933	99,734	98,771	99,680
1934	98,906	99,469	99,714
1935	97,847	99,487	99,876
1936	99,909	98,714	
	92,555		
excl. I.29	98,073	99,420	99,839

Wo (250 eggs)=99,600; (200 eggs)=99,500; (500 eggs)=99,800.

It is obvious that the spring generation is in most years the generation of increase, but in 1931, 1933 and 1936 the environmental resistance was rather high. The summer generation remains a little below R_0, as a rule, and shows an increase. The winter generation is responsible for the seasonal decrease, the destruction being always higher than R_0.

Control.

Control experiments, as spraying with white oils, were performed. Volck concentrated was used in our experiments The concentration advisable for other brands of white oils corresponds to that published annually by the Citrus department of the Jewish Farmers Association.

The results of our spraying experiments are as follows;

Spraying with white oils as a measure against the Florida wax-scale.

Month	Counts before spraying: Living scales (adults/larvae)	Dead scales (adults/larvae)	Percentage Living	Percentage Dead	Counts after spraying with Volck conc. 2,5%: Living scales (adults/larvae)	Dead scales (adults/larvae)	Percentage Living	Percentage Dead	Percent dead Volck 2%	Percent dead Volck 1,5%
I.1935	223(23/200)	278(91/187)	44,5	55,5	103(36/67)	402(132/270)	20	80	70	49
II	260(45/215)	246(94/189)	51,5	48,5	90(18/72)	394(188/236)	18,5	81,5	67	57,5
III	394(106/287)	114(16/98)	77	23	147(121/26)	330(109/221)	31	69	51	37,5
VI	443(3/440)	64(8/53)	87	13	20(2/18)	490(21/469)	4	96	57,5	37,5
VII	374(63/318)	218(18/200)	65,5	34,5	27(14/13)	528(15/513)	5	95	89	31,5
VIII	400(12/388)	120(62/58)	77	23	77(68/9)	413(295/118)	17,5	82,5	66	27
IX	346(16/330)	173(69/104)	66,5	33,5	24(—/24)	490(160/330)	5	95	96	28,5
X	385(—/385)	129(46/83)	78	25	26(2/24)	480(40/440)	5	95	93	19,5
XI	356(—/356)	174(52/122)	67	33	42(—/42)	464(28/436)	8,5	91,5	88	31,5
XII	306(12/294)	244(100/144)	58,5	44,5	127(11/110)	373(12/361)	25,5	74,5	66,5	27

Spraying is a most efficient measure against larvae, whereas it has a relatively small influence on vigorous adults. The three concentrations of white oil, tried in our experiments, gave the following larval mortality:

Concentration	Month I	II	III	VI	VII	VIII	IX	X	XI	XII	Average mortality
Volck 2½%	80	77	89	96	98	93	94	95	91	76	89%
Volck 2%	59	59	81	88	90	95	94	93	88	63	81
Volck 1½%	39	50	68	13	38	21	25	15	29	22	32

It is therefore advisable to begin with the spraying only after the young generation has almost finished hatching. Only at such periods can full success be expected. From this view-point, the second half of June or the beginning of July should be selected for spraying, provided the conditions are those which prevailed in 1934. As this season is not unfavorable for fighting Red-Scale as well, both control measures should be carried out at that season. The use of white oil concentration, equivalent to the insecticidal value of Volck concentrated 2½ %, is recommended, even if Wax-Scale control alone is aimed at.

The impression gained from the tables just given is misleading; the results are not as good, even with the larvae. Part of the larvae dead after the spray, died previously from natural factors (climate, parasites, etc.). For a concentration of 2½ % the percentage of larvae killed by spraying is as follows:

Month	I	II	III	VI	VII	VIII	IX	X	XI	XII
% killed by spraying. . . .	42	40	66	85	60	81	71	78	68	51

However, that percentage is lower than the real one, for the following reasons: 1) Because all crawling larvae are killed by the spray, yet do not appear in our counts. 2) Because a very large, but also very variable percentage of the dead scales is washed off from the tree by spraying. The real control effect is probably somewhat lower than the figure given by our count and considerably higher than the figure obtained from the just related computation.

References.

A. Balachowski, Ve Congres Internat. d'Entom. Paris (1932) 1933 pp. 79—87.
F. S. Bodenheimer, Die Schaedlingsfauna Palaestinas. Berlin. 1930, pp. 147—148.
F. S. Bodenheimer, Hadar 8. 1935., pp. 187—191, 223—228.
I. Kuwana, Dept. of Agric. and Comm. Imp. Plant Quarant. Sta. Yokohama, Bul. 3. 1923, pp. 34—42. Pl. IX.
J. R. Watson, Florida Agric. Exper. Sta. Gainesville, Bul. 183. 1926, pp. 322—324.

C. THE OLIVE SOFT-SCALE (*Saissetia oleae* Bern.)

Description. — *The female* is dark to black-brown, broadly oval, almost circular; highly convex, with distinct ridges on the back, which form the letter H. Surface somewhat shiny and regular and faintly ridged towards the margin. 4.0—6.0 mm long, 3.0—5.5 mm broad, 2.5—4.0 mm high. The slender antennae (0.3 mm long) are 8-jointed, the third joint being the longest. Legs large and long (3rd pair 0.7 mm) with two long, slender tarsal digitules and two shorter, very stout digitules, which originate from the claw. The anal plate is almost twice as long as broad, with three apical setae. Anal ring with eight large setae. Abdominal and thoracic segmentation marked in mature females. Spiracles normal with three stigmatic spines. Dermal cells elongate, enclosed within an irregular polygonal tesselation. Ventral dermis with numerous small, tubular spinnerets and a few scattered, short spines.

The immature female is flat, yet distinctly marked with the dorsal H-ridges. It is light brown colored and darkens with age. After death it has a greyish tint.

The eggs are laid in a cavity formed by the ventral side of the female and the surface of the host. With progressing ovipisition, the ventral surface of the scale approaches the dorsal surface. The cavity which contains the eggs is enlarged in this way. Simultaneously with oviposition, the female secretes small wax-scales, which cover the eggs powder-like and prevent their sticking together.

The *eggs* are oval and pinkish.

The pinkish *larva* is very similar to that of *C. hesperidum*, from which it can be distinguished by minor details in the six-jointed antennae only. It is 0.4 mm long and 0.2 mm broad.

The *nymphe* or second stage larva has also six-jointed antennae, which lack the long terminal setae, characteristic of the antennae of the first larval stage. The long setae of the anal plates of the first stage larva likewise disappear in the nymphal stage. The flat body shows the first indications of the dorsal H-shaped ridges. The body is pale brown in color.

Males, or any development stage of the male, have not yet been observed in Palestine.

Distribution. — Distributed over the tropical and subtropical regions of the world; further north in greenhouses. It originates, most probably, from the Sudano-Deccanian region of Africa. The endemic fauna of parasites is richest there.

Life-history: The typical development of the olive soft-scale on shamuti in 1934/35, at Rehoboth, was as follows:

On 28.III.34. 50 crawlers were put on a tree. The development was very slow until the end of April. From then on until November 1934 no development took place. The development was interrupted and the young larvae were in diapause. In mid-November, the larvae began to develop rapidly. In December, moderate honeydew-production appeared. In January, 1935 the larvae grew in size and the young females, which became darker, showed the H-shaped elevation on their backs very distinctly. The honeydew-production was considerable. In February, all the larvae became females. The honeydew-production was heavy in the first half of the month and decreased in its second half. The first eggs were deposited at the beginning of March and oviposition continued throughout the month. The honeydew-production stopped with the beginning of oviposition. On 22.III.35, the first crawlers were observed. Many crawlers appeared at the end of the month and in mid-April no eggs were seen anymore.

In contrast to these laboratory breedings, the development in nature was as follows:

The first few females with eggs appeared in January and February. In March, the number of ovipositing females increased and towards the end of this month the first crawling larvae appeared; the number of both increased in April and in May the crawlers were most abundant. At the beginning of June many crawling larvae were still seen, but towards the end of this month had all settled or died. At that period, the mothers were empty, dried out, shrinkled and dead.

The crawlers are present during two months and longer, whereas the individual larvae settle down during the first two days, only in very exceptional cases somewhat later. They prefer to settle on the young and tender twigs, but may be also observed on the leaves, along the central nerve. After settling, the larvae develop only very little and then their development is interrupted from May to the middle of November, on the average. In mid-November, rapid growth begins and also a honeydew-production, the increase of which is parallel to the development of the scale and stops only close before oviposition. When mature, the scale is entirely black. The color changes to grey after oviposition is over.

The olive soft scale has, therefore, only one annual generation in Palestine, including a diapause, from May to November. The number of generations is different in other countries. In the drier interior parts of California, it has one generation only. But in some of the southern coastal areas of the same state, we find *S. oleae* to hatch „unevenly". Development is not absolutely interrupted here during summer, but accelerated and retarded development takes place contemporaneously, in between overlapping generations. For some localities, two annual generations are assured in this uneven hatching area.

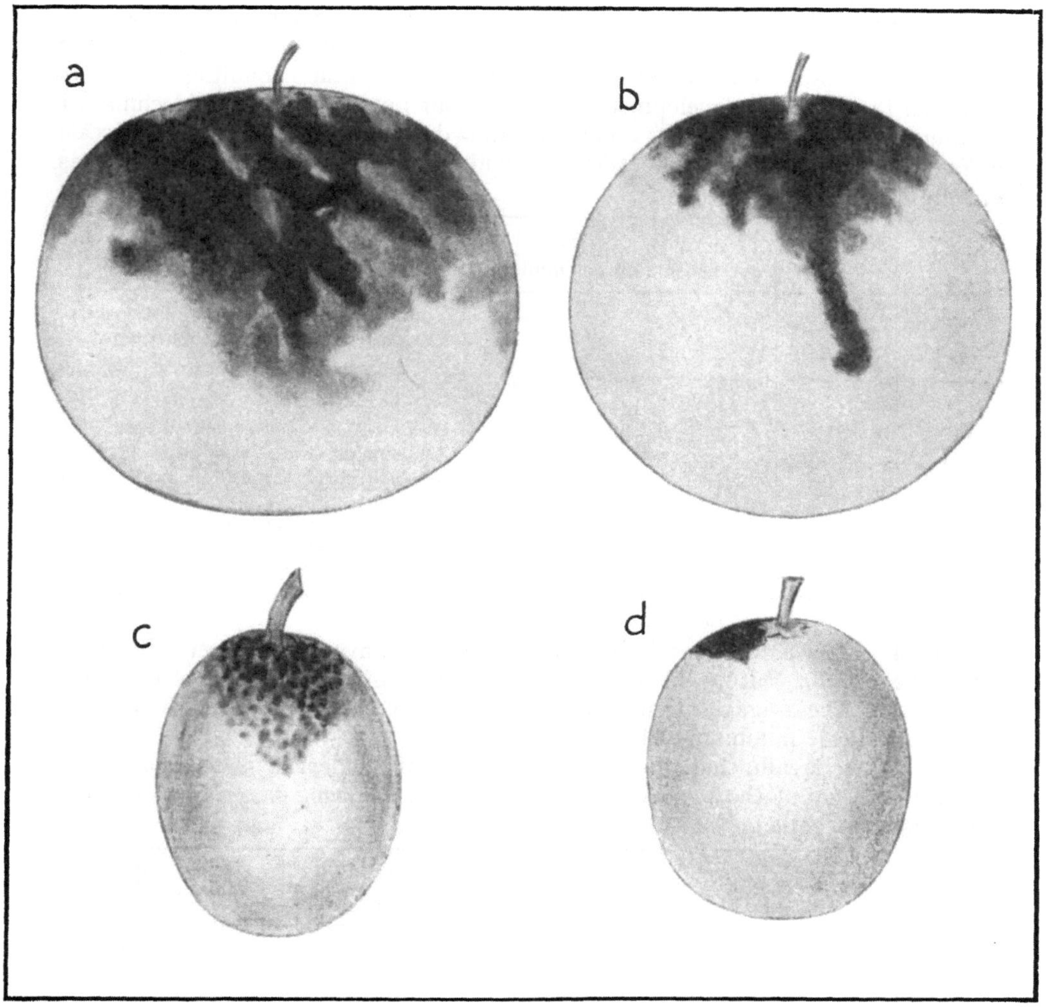

XXVII. Sooty mould caused by various soft scales: a) and b) by *Saissetia oleae* Bern. on grapefruit; c) and d) by *Coccus hesperidum* L. on orange fruit.

Two annual generations are the rule in S. Italy. The first generation lasts from May to August, and the second from August to May. But this is not invariably the case. Silvestri reported, that in Portici only one generation developed on oleander under natural conditions from the beginning of June, 1906 to the end of May, 1907. The specimens which develop in two generations do not show an „even" hatch in Italy. Martelli reports that the larvae which hatched on 30.V.07 became adults from 25.VII until 15.IX; and larvae which had hatched on the 27.VI. became adults from the beginning of October to the end of November.

It seems that every unfavourable environmental influence induces a diapause in the first larval stage. This diapause is caused by dry heat, by cold and probably also by unfavorable host-conditions. Unfavorable environmental conditions likewise influence the speed of development. A warm (but not hot) and humid climate is apparently optimal and permits a homodynamic development of the olive soft scale.

Fertility and oviposition: The all-year-round egg-production (at Rehoboth) was, as follows

Month	No. of observations	No. of eggs average	maximum	minimum	Remarks
I.34	—	—	—	—	Young females, still without eggs
II	—	—	—	—	Young females, still without eggs
III	—	—	—	—	Young females, still without eggs
IV	8	631	1105	305	Many females, with eggs
V	14	935	1416	504	Many females, with eggs
VI	21	725	992	317	Many females empty and dead
VII	—	—	—	—	All females empty and dead
VIII	—	—	—	—	All females empty and dead
IX	—	—	—	—	All adults disappeared
X	—	—	—	—	All adults disappeared
XI	—	—	—	—	All adults disappeared
XII	—	—	—	—	All adults disappeared
		776			

In 1934, eggs were present from April to June, the average number of eggs per female being 776. But in 1935, the period of oviposition fell somewhat shorter (April/May.) The average number of eggs per female, out of 20 females, was 756 (maximum 1341, minimum 249).

If we assume, again, that the period of hatching of the larvae is — identical in lenght with that of the oviposition, we receive the following data for the length of the oviposition period:

No.	No. of eggs	Date of appearance of first crawler	of last crawler	Length of oviposition period
1	713	4.V.34	22.V.34	18 days
2	616	6.V.	24.V.	18
3	518	30.IV.	15.V.	16
4	893	20.IV.	15.V.	25
5	467	25.IV.	9.V.	15
6	581	25.IV.	9.V.	15
7	694	2.V.	15.V.	20
8	677	15.V.	28.V.	16
9	942	20.IV	8.V.	19
10	851	25.IV.	12.V.	18
11	753	30.IV.	15.V.	16
				18

The oviposition period thus, lasts two to three weeks. The length of the egg-stage in S. Italy is from 10—20 days.

The environmental resistance (R_0) which is almost entirely conditioned by climatic mortality, ranges between 99.841 to 99.893.

Host-plants.

The number of hosts of this polyphagous scale, hitherto observed in Palestine, is rather small:

Apocynaceae	*Nerium oleander*
Cycadaceae	*Cycas revoluta*
Lauraceae	*Laurus nobilis*
Oleaceae	*Olea europaea*
Rosaceae	*Prunus persica, P. armeniaca.*
Rutaceae	*Citrus spp.*
Solanaceae	*Solanum nigrum*

In other regions, economic injury seems to be restricted to *Olea* and *Citrus*. In Italy, it may be regarded as an olive-pest, in S. California as a very important citrus-pest. A wild *Olea* is no doubt the original host of *S. oleae*. It is certain, that in eveiy region the population level of the scale is higher on *Olea* than on *Citrus*. Fertility is probably also highest on the former host.

Mortality. As there is only one annual generation of *S. oleae* in Palestine, no seasonal fluctuations in abundance are observed, except for the strong decimation of young crawlers and larvae due to wind and climate. The further decrease of the successfully settled larvae is relatively small. As the attack of *S. oleae* is only very local, no exact population counts could be made. The monthly count of 100 scales gave the following results:

Month	% mature females	% larvae and immature fem.	living adults	living immatures	Causes of death of adults: dried	Causes of death of adults: parasites predators	Causes of death of immatures: dried	Causes of death of immatures: fungus	Causes of death of immatures: predators
I.34	(5)	95	—	80	5	—	15	—	—
II.	22	78	16	68	4	2	10	—	—
III.	32	68	24	54	8	—	12	2	—
IV.	100	crawlers and first stage larvae excluded —	100	crawlers and first stage larvae excluded —	—	—	—	—	—
V.	100	—	100	—	—	—	—	—	—
VI.	100	—	—	—	—	—	—	—	—
VII.	100	—	—	—	88	12	—	—	—
VIII.	22	78	—	66	18	4	12	—	—
IX.	19	81	—	71	13	6	10	—	—
X.	7	93	—	82	7	—	11	—	—
XI.	5	95	—	87	5	—	8	—	—
XII.	8	92	—	86	8	—	6	—	—

The dried out females have probably all died a ± physiological death, wheieas the dried out larger larvae have died due to climatical and other factors. The role of parasites and predators (24 out of 1200 scales) is surprisingly small. There is little doubt that the decisive mortality occurs through wind and dry heat during the crawling stage or the first few days after settling. At the present, little is known of the influence of these factors and of parasites and predators as well on the local fluctuations of numbers in different years.

The combined effect of environmental conditions (resistance) on the population level of the olive soft scale is clearly illustrated by the status of this scale in California: In some of the hottest and driest valleys in the interior, as e.g. the Imperial Valley, the extreme heat and drought of spring and summer prevent any increase of the scale and reduce it to some scattered individuals. In other, less unfavorable valleys, *S oleae* develops in slight numbers and may, in especially favourable years increase considerably on the preferred host, but never on a less favourable host, as *Citrus*. In other parts of the „even hatch" area, e.g. Tulare Country or Riverside, it is able to increase more or less regularly, even on *Citrus* and to cause serious damage. The parasites have very small value in this area, because either they have more generations than the host or the supply of suitable ages of the host is inadequate. The controlling factors of *S. oleae* on Citrus in California may be tabulated, as follows:

Locality	Population level of S. oleae	Effective factors of environmental resistance	Technical control
Imperial Valley	very low	climate	unnecessary
Other interior valleys	lower on Citrus	climate + host resistance	unnecessary on Citrus
Other 'even hatch' area	over damage mark	fumigation + climate + host resistance + parasites	necessary
Uneven hatch area	over damage mark	fumigation + parasites + climate + host resistance	necessary

In the uneven hatch area of some coastal belts the climatic resistance of *S. oleae* is much lower than in any of the other belts, but the parasites find here better possibilities for an all-year-round development and reduce the population level of *S. oleae* even below that observed in Riverside, etc.

Honeydew-production and damage. The main injury of the scale is done by honeydew production and the following infestation by sooty mould. Sooty mould appears from mid-December to the end of February and March in an ever increasing amount. As the production is highest in the later part of this period and most of the fruit is picked before, the damage to the citrus-crop is limited. But here, again, we deal with a pest whose importance would rise if later ripening fruits (Valencia late e.g.) would be grown in numbers. The honeydew-production of this scale is higher than that of all the other soft-scales. Branches, which they infest, are frequently seen in the early spring covered entirely by the black sooty mould: twigs, leaves and fruits. In S. *oleae*, the honeydew is also actively expelled from the body, in large drops. Sooty mould formation occurs in large patches and does not start from special centres as in the case of *C. floridensis*.

The farmer generally observes the damage long after the honeydew has been excreted. We received in a typical year, e.g., inquiries from farmers:

month	IV	V	VI	VII
no. of questions	4	3	4	2

Damage on Citrus, mostly to a very limited and localised extent is reported from: Rehoboth, Ness Ziona, Rishon Lezion, Nahlath Yizchak, Jaffa, Schechonath Borochov, Petah Tikva, Hedera, Gan Shmuel a.o. It can be regarded to-day in Palestine as only a minor pest of Citrus.

No control experiments were performed until now, but the control of the Florida Wax-Scale can certainly be applied to *S. oleae*. With the only difference, that the spraying should not be done during the diapause (May/November), during which time the species is very resistant to all environmental influences, but either somewhat earlier or later.

References.

F. S. Bodenheimer, Schaedlingsfauna Palaestinas. Berlin, 1930, p. 150.
G. Leonardi, Monografia delle Cocciniglie Italiane. Portici. 1920. p. 278.
G. Martelli, Boll. Ent. Agr. Portici. 2 1908 p. 217.
R. Newstaed, Monograph of the Coccidae of the British Isles. Vol.2 .London 1903. p. 126, Pl. LVII.
H. S. Smith and H. Compere, Univ. of Calif. Publ. in Entom. **4**, 1928. pp. 231—334.
J. B. Steinweden, Mo. Bull. Calif. State Dep. Agric. **19**, 1930, p. 565.

D. OTHER SOFT SCALES.

a. The Hemispherical Scale. (*Saissetia hemisphaerica* Targ.)

Description: The red-brown to dark-brown adult female has a more or less hemispherical profile and is short oval in contour. The surface is smooth and shiny. The immature, flat females, which vary from pale yellowish to light brown, possess three carniae which form the letter H. This carination disappears almost entirely in the adult female. The adult female is 2—3 mm long and 1.5—2 mm broad.

This species is easily distinguished from the other Soft Scales living on Citrus by its pronounced convexity and the lack of the H-shaped dorsal ridges.

Distribution: Ubiquitous in tropical and subtropical regions.

This polyphagous species has, so faı, been met with in Palestine in and near Tel Aviv only, on *Eriobotrya japonica* (Rosaceae), *Asparagus plumosus* (Liliaceae), *Cineraria* sp. (Compositae) and *Agave americana* (Amaryllaceae). It was discovered on Citrus only once, on a single tangerine tree at Mikveh Israel (13.IX.33). Infestation spread from the neighbourhood. No damage whatsoever is known up to today from *S. hemisphaerica*. It has apparently only one annual generation, with oviposition in June/July.

b. The China Wax-Scale (*Ceroplastes sinensis* Del Guer.).

Description: The adult female is covered with a thick, grayish white wax-layer. A dark red grove is in the centre of each plate, in the midst of which a small white point is recognizable. The body-color is red to black-brown. The regular wax-plates are very distinctly divided: a large, flat central plate and six trapezoidal lateral plates, which are not much smaller than the central plate.

It is up to 7 mm long, 5 mm broad and 4 mm high.

This species differs from *C. floridensis* by its district wax-plates, from *C. rusci* by the number of the lateral plates (6 as against 8 in *C. rusci*).

Distribution: China, introduced to Italy, S. France and Turkey.

This polyphagous species has been introduced into Palestine from Italy on ornamental and fruit-trees, on at least two different occasions. As far as we know, it has not acclimated in Palestine yet. It should be especially looked for on the Plant-Quarantine stations, as well as in our orchards. It has one annual generation in Italy, where oviposition takes place in summer and at least 2000 eggs per female are produced. Males are common.

Lindinger has recently attempted to synonymize *C. sinensis* with *C. floridensis*. This is obviously a mistake.

D. Spille (New Zealand J. Sci. and Techn. 25 A. p. 129-130. 1943) gives some data on this species on *Vitex lucens*. This recent introduction has in New Zealand also one annual generation with oviposition in January/February.

c. *The Fig Wax-Scale* (*Ceroplastes rusci* L.).

Description: The test of the adult female is very strongly convex, waxy white in colour, often with a pink tint. It is divided into a domeshaped central portion and eight distinct quadratic marginal-plates. In the centre of the dome is a slightly depressed, reddish brown, broadly oval area, in the centre of which is a small elongated raised white pad. The shallow depressed lines which subdivide the test are also of a reddish brown colour and in the centre of each plate is a more or less similarly coloured, circular depression, with one or more opaque raised white points. The stigmatic bands of opaque white wax are well marked. In old adult females the test becomes globular, smooth surfaced and almost devoid of all signs of the characteristic plates of the earlier stages. (Hall). It is 3—6 mm long, 2—3 mm broad and 2—3 mm high.

Distribution: Widely distributed over the subtropical and tropical regions of the world. The eastern Mediterranean or the Sudano-Deccanian region are the probable original home of the species.

Hosts: C. rusci is a common pest of figs in the hills. It has been found on the following hosts in Palestine:

Anacardiaceae	*Schinus terebinthifolius*
Apocynaceae	*Nerium oleander*
Aralaceae	*Hedera helix*
Moraceae	*Ficus carica, F. infectoria, F. sycomorus.*
Rosaceae	*Crataegus azaolus*
Rutaceae	*Citrus spp.*

Life-history: The species is very common in the mountains, relatively scarce in the coastal plain and absent from the Jordan valley. Only a few individuals have been observed on citrus at Hedera. Should it become a common parasite of citrus, its menace would be great in years of mass-increase, as its honeydew-production surpasses that of *C. floridensis* and other Soft scales living here on citrus considerably.

It has two generations in the mountains: one from the beginning of May to mid-July (70—80 days), the second hibernates as adult females.

Carmin and Scheinkin maintain that in the coastal plain three generations occur on sycamore: at the beginning of November, at the end of January and at the beginning of October respectively. Such a life-history would not be impossible if a summer diapause occurred. But it does not seem very probable (full development in 35 days, at 22,1° C!). As the conclusions of the just mentioned authors were based on monthly visits and not on breedings, it seems more probable that the „generations", which are supposed to begin in October and November are specimens of the same scattered generation. The life-history of *C. rusci* in the coast plain of Palestine remains to be studied.

„At first the young larvae are exceedingly mobile and wander over the twigs and leaves in search of a suitable feeding spot. They are dull red in colour, without wax and by no means conspicuous. In a weeks time they have settled down and become more conspicuous owing to the production, laterally, of 15 glistening-white, waxen processes. At this stage they prefer leaves shaded from the direct rays of the sun. The upper surfaces of the leaves are invariably chosen, usually along the mid-rib or branch ribs. After a month they again migrate, this time to the stems of the leaves or to the upper, more tender portions of the young shoots." (Bodkin).

The number of eggs varies from 800 to 1500 per female. Males have not been observed here, but they have not been properly searched for. The mortality is very great at the mobile larval stage. Cold, parasites and the falling off of infested leaves are the main causes for later mortality.

If control should become necessary, spraying with white oils might be attempted during the mobile or young larval stages, as well as fumigation at the same season and in the winter.

d. *Pulvinaria floccifera* West.

Description: The adult female is distinctly cordate, but becomes greatly wrinkled after gestation. Its color is yellowish brown and mottled; after parturition the color usually changes to dusky ochreous. After treatment with potash the form is more ovate. Antennae 8-jointed. Legs about one third longer than the antennae, very strong; digitules to claw very long, broad and dilated at apex; those of the tarsus scarcely longer and faintly clubbed. Stigmatic spines arranged in a triangle. Dermis above with numerous irregular ovate pores or glands; spinnarets on ventral dermis long and tabular. The ovisac of adult female very elongate, straight or curved; overcrowded females form irregular masses of white flocculent matter occasionally (Newstead). Female is 2.0—3.5 mm long; 1.5—2.5 mm broad. Ovisac 5—11 mm long and 2 mm broad. The immature female is pale yellow with red-brown anal-lobes.

Distribution: In the subtropical countries of the world.

This polyphagous species has been intercepted by the plant quarantine service of Palestine from Egypt on *Ficus*. It is a minor pest on citrus in S. France, Italy and Greece, where it has two annual generations. This species might be introduced into Palestine and should by looked for, even if no serious damage is to be expected from its part.

References.

F. S. Bodenheimer, Schaedlingsfauna Palaestinas. Berlin, 1930, pp. 269—270.
G. E. Bodkin, Bull. Ent. Res. **17**, 1927. pp. 259—263.
J. Carmin and D. Scheinkin, Bull. Soc. Roy. Entom. d'Egypte, 1931, p. 181.
W. J. Hall, Techn. Bull. 22 Ministry of Agric. Egypt, 1922, p. 17.

E. PARASITES AND PREDATORS OF THE SOFT-SCALES IN PALESTINE.

The number of parasites, belonging mostly to the Chalcidoidea, which develop in soft-scales, is rather large and will increase with further research. The known species are as follows:

Pteromalidae	*Scutellista cyanea* Motsch.
	Pachyneuron coccorum L. (Sec. parasite).
	Enargopelta obscura Foerst.
Aphelinidae	*Coccophagus scutellaris* Dalm., C. spp.
Encyrtidae	*Eucomys lecaniorum* Mayr
	Blastothrix erythrostethus Walk.
	Trichomasthus cyanifrons Dalm.
	T. cyaneus Dalm., *Diversinervis elegans* Silv.
	Euaphycus flavus How.
	Microterys consobrinus Merc., *M. frontatus* Merc., *Encyrtus sp.*
Noctuidae	*Coccidophaga scitula* Rbr.
Coccinellidae	*Chilocorus bipustulatus* L.

The combined effect of all these parasites and predators on the different soft scales varies greatly. Only *Coccus hesperidum* is heavily parasitized throughout the year and shows sometimes locally a very high parasitation, especially in dense colonies. In two cases the percentage of specimens killed by parasites or predators was as follows:

	Larvae		Adults		Total %	
	living	parasitized	living	parasitized	living	parasitized
1.	4	94	—	203	98.5	1.5
2.	25	21	13	75	71.5	28.5

The parasitism on other soft-scales is much lower. *Ceroplastes floridensis* showed in three years the following average loss per year:

Locality	Year	% killed by parasites and predators
Atta	1932	6.6
Atta	1933	5.6
Rehoboth.	1934	4.9

The average parasitism of *Saissetia oleae* is even lower, a phenomenon which is well known from the „even hatch" area of this scale in California. The restriction of its larval stage, the most suitable stage for parasitic infestation, to one short period throughout the year, lessens the effectiveness of the parasites, which all show *some* annual generations. Only the quick developing *C. hesperidum*, which always has all stages of development in each season, is heavily attacked by all parasites.

By far the most important parasites *Scutellista cyanea,Eucomys lecaniorum*, *Coccidophaga scitula* and *Chilocorus bipustulatus* are sometimes quite numerous, but they are of no decisive importance in the general epidemiology of the soft-scales. They are abundant wherever and whenever soft-scales are abundant, but neither do they prevent the increase of the latter, nor prove to be the primary factor in their decrease.

Short notes on these parasites will follow. *Chilocorus* has been treated before.

a. *Scutellista cyanea Motsch.* (*Pteromalidae*).

Description: The *adult* is thick and short, somewhat depressed. Head larger thorax, triangular in the front. Basis of the antennae low. The eight joints increase distally in breadth. The scutellum is very largely developed and nearly covers the anterior-dorsal region of the abdomen. The color is metallic black-blue, the antennae, tibiae and most tarsi light brown. The male is 1.3—1.4 mm, the female 1.6—1.9 mm long.

The white *egg* is bottle-shaped, the collar-like appendix of the bottle being as long as the body of the egg. It is 0.6 mm long.

The young *larva* is approximately fusiform (broadest near the cephalic end), of a pinkish white color and bent. Besides the head, it consists of 13 body-segments. It has 4 pairs of spiracles, is 0.3 mm long and 0.1 mm wide. The mature larva is broader, is 2—5 mm long, 1.3—2 mm wide. It has 9 pairs of spiracles.

The dark grey *pupa* is 1.5—3 mm long, and about 1.1 mm wide.

Distribution: Throughout the tropical and subtropical parts of the world, probably originating from the old world.

Life-history: The biology of the species has been worked out by Silvestri, Quayle, Smith, Compere, and others. It is a very common parasite of soft-scales everywhere and has been bred in Palestine from *Saissetia oleae*, *S. hemispherica*, *Ceroplastes floridensis*, *C. rusci*, and occasionally from *Coccus hesperidum*. The large circular hatching holes in the empty body-walls of the dead females are mainly the work of this species. The adult wasp cuts this exit hole after several hours of work. The wasps remain in the interior of the host from one to three days, before leaving it. They are poor flyers and generally walk actively among the scales and jump or fly short distances only when disturbed. They are greatly attracted by the honeydew excreted by the scales, which constitutes their nourishment. The copulation lasts only a few minutes.

The females may reproduce parthenogenetically, although the males generally equal or even exceed the females in numbers (Smith and Compere). The normal longevity of the wasps in nature is probably about 1—2 weeks in the summer, under favorable laboratory conditions up to about two months.

The female oviposits through the anal cleft of the soft scale, after inspecting the scale with its antennae, beneath the body of the host. It is reported from California, that they do not always show good judgement in the selection of their hosts: eggs have been deposited under old scales, from which all crawlers had hatched, etc. Oviposition starts about one day after hatching.

The egg of *S. cyanea* can be easily distinguished from its hosts eggs, among which it is laid, by its long tail-like appendix. As a rule, one female lays only one egg per host, at a time. Up to six, seven, in one case even ten larvae developed from one scale, but this is the result of repeated oviposition of different females.

The hatching larva sucks the eggs of its host empty.

,,When full grown, the larva, preliminary to pupation, casts the meconium and hollows out a cell among the old eggshells and debris, which it mates together with a small amount of silk. Strands of silk are frequently spun from the inner edge of the scale to the twig to prevent the scale from becoming detached during the pupal life of the parasites . . . . Scales with exit holes are therefore more likely to accumulare . . . . than are those which have not been inhabited by *Scutellista*. The pupa inhabits the old scale." (Smith and Compère).

The shortest period of development observed (from oviposition until the adult leaves its host) was 14 days at about 27° C. Comparing this value with the seasonal history, known from Italy and California, we arrive at the following approximate constants for the hyperbola of *S. cyanea:*

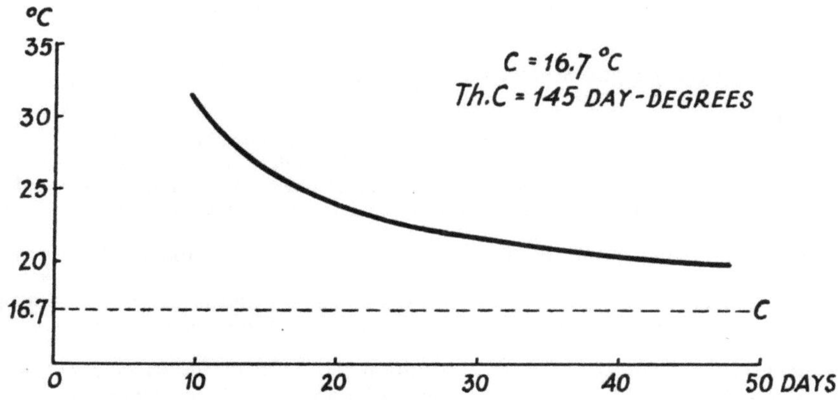

Fig. 92.
Temperature dependency of shortest development of *Scutellista cyanea* Motsch.

	Temperature	length of development
development threshold (c): 16.7° C	30° C	(10.9) days
	25	17.5
thermal constant (Th.C.): 145 days-degrees . . .	20	43.9

Following this hyperbola, 5—6 annual generations have to be expected at Naples, 9—10 at Tel Aviv. *Scutellista* may enter a diapause when the environmental conditions are unfavorable or when the host enters a diapause. Silvestri reports from Italy: ,,The duration of the larval stage is 9—10 days in summer, or 9 months (from September to June of the following year). In this latter case, which has been observed in *Ceroplastes*, the larva remains throughout this period small and inactive under the host's body, until the later starts to oviposit."

b. *Pachyneuron coccorum L.* (*Pteromalidae*).

This species is a common secondary parasite of soft-scales.

Description: Body dark green, often of a metallic shine. Antennae brown, wings hyaline with brown veins. Legs scraw-coloured. Up to 2.5 mm long.

The *egg* is 0.3 mm long, 0.1 mm broad. The mature *larva* is elongate-oval, convex, with small antennae. Nine pairs of stigmata, 1.5—1.7 mm long, 0.7 mm broad. The *pupa* is brownish; 1.5 mm long.

Distribution: over the whole of Europe, Palestine; U.S.A. (probably introduced).

Life-history: In Palestine it is bred repeatedly from *Coccus hesperidum*. Rondani and Silvestri have shown, that it does not live on the soft-scales, from which it hatches, as a parasite, but as a hyper-parasite mainly of various *Encyrtidae*.

The female inserts its ovipositor into the body of the scale. When the oviposition hits upon a parasitic larva in the host's body, it paralyses the larva by stinging it, and leaves an egg on its body surface. The larva, harching after a few days, sucks on the parasite larva, to which the egg has been attached. During one to two weeks the body of the prey is reduced to a small amorphic mass. The *Pachyneuron* larva excretes its meconia and pupates. The whole development lasts from 3—4 weeks, in winter longer. From 1—10 specimens may hatch from one soft-scale, each having destroyed one Chalcidid larva in its development. Silvestri observed one case, in which 2/3 of the primary parasites were destroyed by *P. coccorum*.

c. *Coccophagus scutellaris Dalm.* (*Aphelinidae*).

Syn.: C. lunulatus How.

Description of female: Head and body black, with scutellum more or less extensively yellow or orange yellow at the apex. The extreme apex of the scutellum may be spotted with black. Legs yellow with middle, hind coxae and hind femora blackish. Wings faintly and uniformly infumated. 1.1 mm long. The male resembles the female, except that its scutellum is entirely black. (Compere).

The *egg* is 0.15 mm long and 0.035 mm wide. The neonate larva is about 0.2 mm long and 0.035 mm wide.

Distribution: Original home unknown. Has followed its soft scale hosts all over the subtropical and tropical regions of the world: America, Australia, Africa, Mediterranean regions, Europe, etc.

Life-history: C. scutellaris Dalm. is a common parasite of *Coccus hesperidum* and *Ceroplastes floridensis* in Palestine. In other regions it is reported from a number of other soft-scales.

Second stage larvae (1—2 mm long) are preferred for oviposition. The female runs over the body of the scale, the antennae permanently moving and inserts its ovipositor generally into the mid-dorsal region of the host, leaving the scale after having laid one egg only. The oviposition itself rarely lasts longer than one minute, generally much less. The egg floats in the body fluid of its host. Its first nourishment is the body-fluid and the fat-body of the soft-scale. Finally all organs are attacked. But pupation takes place before the host is dry.

The last larval skin acts as a protecting shroud and becomes filled with air. The meconia do not float at random in the tissue of the host, but are enclosed in a shroud, close to the abdomen of the pupa.

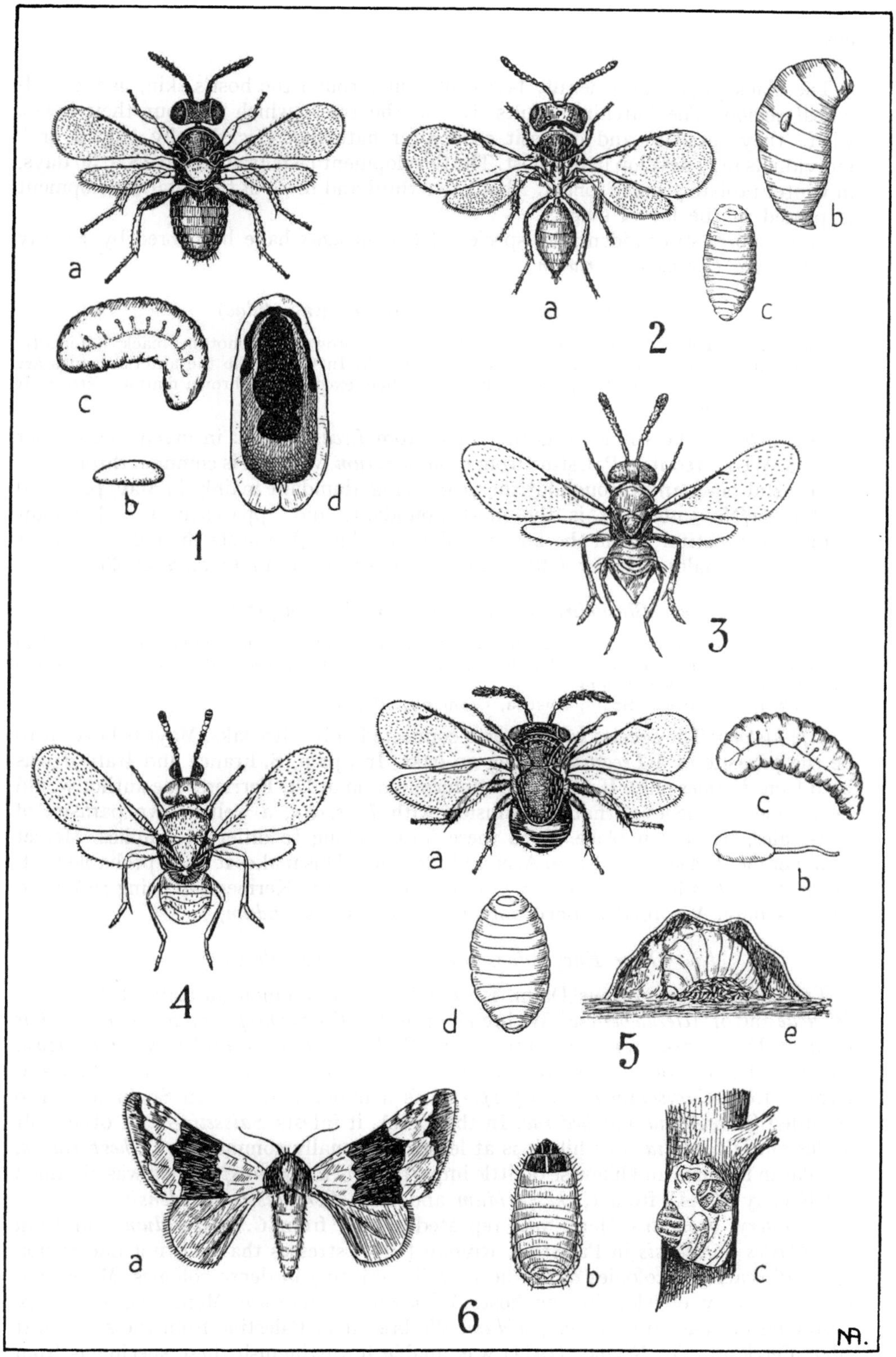

XXVIII. Parasites and predators of Soft Scales in Palestine. 1) *Coccophagus scutellaris* Dalm.: a) female, b) egg, c) full-grown larva, d) pupa and larval feaces in empty host exuvia. 2) *Pachyneuron coccorum* L. superparasite. a) female, b) egg on parasite larva, c) full-grown larva from below. 3) *Trichomasthus cyanifrons* Dalm, 4) *Blastothrix erythrostethus* Walk, 5) *Scutellista cyanea* Motsch, a) male, b) egg, c) young and d) full-frown larva, e) cross-section through *Saissetia oleae* Bern. infested with a full-grown larva of *S. cyanea.* 6) *Coccidophaga scitula* Rbr. a) moth, b) caterpillar, c) pupal cocoon. (After Silvestri, Smith, Compere, Martelli and Mercet).

The black pupa can generally be easily seen through the host's skin, in the mid-dorsal region. The hatching adults live on the trees, which harbour their hosts, where they copulate and oviposit soon after hatching. Occasionally only 2 or 3 individuals may develop in one host. The development lasts from about 18 to 30 days, in winter probably much longer. About two third and more of the total development is passed in the larval stage.

Three more, still unidentified species of *Coccophagus* have been bred by Rivnay (1944) from *Coccus hesperidum.*

d. *Eucomys lecaniorum Mayr* (*Encyrtidae*)

Description: Yellow-brown; antennae and abdomen brown. Mesonotum black anteriorly, covered with yellowish-white hairs at its posterior half. In the female the anterior wings are fumated at the distal half; the proximal half is hyaline, except for a brown marginal stripe. It is 1.5—2 mm long.

Life-history: The species was first bred from *Lecanium* sp. in greenhouses, near Vienna. We reared it in Palestine from *C. hesperidum,* where it is common throughout the coastal plain and throughout the year, being abundant mainly in May/June and September/November. It is the most abundant, and apparently the dominant parasite competitor, when the host-population is high. It infests the older larvae or the young females. Rivnay (1944) bred the entire cycle in 18 days at 25° C.

e. *Blastothrix erythrostethus Walk,* (*Encyrtidae*)

Description: Head, back of thorax and abdomen dark golden-green. The lateral and ventral parts of thorax and abdomen red-yellow; scapulae and legs yellow-white. Male 1.4—1.5 mm long, female 1.7—1.9 mm long.
Distribution: Europe (Spain, Austria, Germany), Palestine.

Life-history: The species was hitherto connected only with oaks. Mayr believes it to be the parasite of a *Lecanium* living on oaks. In Spain, S. France and Italy it was bred from *Kermes* only. If not for the determination of Dr. Ferriere, the author would be inclined to believe, that a confusion with *B. serica,* a well known parasite of soft-scales, has taken place. Both these hosts belong to different families. Mercet however met the species also on Acer and Platanus. This makes it quite probable that. *B. erythrostethus* lives in Spain on Coccidae as well, the Kermesidae being restricted to oaks only. We bred it once only from *Coccus hesperidum,*

f-i. Other Encyrtidae, parasites of citrus soft scales.

Trichomasthus cyanifrons Dalm. is in Palestine a common parasite of *Eriopeltis festucae* and of *Aclerda berlesei*. We bred it once from *Coccus hesperidum. Trichomasthus cyaneus* Dalm. also was once bred from *C. hesperidum. Euaphycus flavus How.* was bred by us from *C. hesperidum,* but it is rare. It is known in Spain, Italy and Algeria from *Chrysomphalus dictyospermi* as a minor parasite. In Spain it is also recorded from *Aspidiotus hederae.* In the U.S.A. it infests *Saissetia oleae,* other soft scales and *Tachardia.* In Chile it is at least occasionally common on *C. hesperidum.* Similar in features and likewise of little importance is *Encyrtus sp.* which was obtained by Rivnay (1944) from *C. hesperidum* and from *Ceroplastes floridensis.*

Microterys frontatus Merc. was repeatedly bred from *C. hesperidum* and from *Ceroplastes floridensis* in Palestine. Rivnay (1944) stresses that it is not uncommon in small scattered colonies of the host, whilst wanting in dense colonies. More than one wasp may develop in one host. *Microterys consobrinus* Merc. was originally described in Spain from *C. hesperidum.* We bred it in Palestine from the same host (det. Ferriere). Yet Rivnay (1944) was unable to distinguish these specimens from what he believes to be the preceding species. *Diversinervus elegans* Silv. or a closely related species was bred by us from *C. hesperidum*, by Rivnay from *Ceroplastes floridensis*. It has not been bred yet in Palestine, where it is rare, from the typical host *Saissetia oleae* (Eritrea, Queensland; in Natal on *S. persica*).

The fungus *Cephalosporium lecaniorum* Zimm. can easily be observed, especially during the wet season. Conditions are however unfavourable for epidemics as the summer is dry and warm, the winter cold and wet, whilst warm humidity is the favourable condition for *Cephalosporium*-plagues. The fungus is extremely polyphagous on soft scales (Coccidae).

k. Coccidophaga scitula Rbr. (*Noctuidae*).

Description: The *moth* has whitish anterior wings with a broad, undulated brown median transversal band. The apical tip of this wing is brown with pale red-brown and one black dot. Posterior wings grey with a border of whitish fringes. Head and prothorax pale red-brown, the rest of the body silvery-white. Wing expansion 11—20 mm.

The hemispherical *egg* is about 0.4 mm long. Tubercles surround the pale concentrically. When recently laid, it is of a whitish color.

The mature *larva* is conical, enlarged at the posterior end, with the anterior end retracted. The head is dark-brown. The second joint of the three-jointed antennae bears a long bristle; four ocelli are on each side of the head. The body is whitish. Two blackish areas are on the pronotum and a conical, black protuberance on the last abdominal segment. 3—6 mm long, 2—3 mm broad.

The light-brown *pupa* has a smooth surface and is provided with six small hooks at its posterior end. 4—5.5 mm long, 2—3 mm broad. It is enclosed in an elliptical, dark cocoon. The latter has 3 to 5 wrinkles at its anterior half and 1 to 4 „scales" of *Coccus*, *Saissetia*, etc. on its back. 4—8 mm long, 2—5 mm broad (Silvestri).

Distribution: W. Africa, S. India, Mediterranean countries.

Life-history: The moths hide during the day among the leaves of trees. Copulation, oviposition and nutrition take place only during the night. The copulation act lasts for some hours. The eggs are deposited on leaves in the immediate neighbourhood of larvae or immature females of a soft-scale, only one egg per leaf. The female lays 3—12 eggs in one night. At 24° C, the incubation of the egg lasts about one week.

It is certainly a mistake to assume that the caterpillars of *C. scitula* feed only on the eggs. Even that they live mainly on them is also not correct. The caterpillars enter below the body of the host, pierce its soft ventral skin and devour the tissues of the host. In this way, 20—30 soft-scales may be destroyed by one caterpillar during the latter's development. Eggs may be also occasionally devoured. The caterpillar webs together some empty scales, which he carries along for protection. The scales remain on the grey to brown pupal cocoon. In corners, where twigs branch off, two or even five may be seen together. They often pupate on the root-collar of the tree and the cocoon is then covered with a layer of soil.

Silvesti observed three annual generations in S. Italy, Rouzaud five in S. France. The whole biology of the moth needs further investigation. It is certainly not a very effective control agent of soft scales in nature. In Palestine, it is common on *Coccus* and *Saissetia*, especially in autumn and spring.

References.

F. S. Bodenheimer, Die Schaedlingsfauna Palaestinas. Berlin, 1930, p. 389—390.

H. Compere, Proceed. U.S. Nat. Mus. **68**, 7: 1931, p. 99.

G. Martelli, Boll. Labor. Zool. Agrar. Portici. pp. 217—296.

G. Mayr, Verh. Zool. Bot. Ges. Wien, 1875, pp. 675—778.

R. G. Mercet, Fauna iberica: Encirtidos. Madrid, 1921, 732 pp.

E. Rivnay, Journ. Ent. Soc. S. Africa 1. 1944. pp. 73—81.

H. Rouzaud, Bibl. Progrès agricole. Montpellier, 1893.

F. Silvestri, Boll. Labor. Zool. Agrar. Portici. **13**, 1919. pp. 110—115.

H. S. Smith and H. Compere, Univ. of Calif. Public. Entem. **4**, 1928, pp. 231—334.
H. Timberlake, Journ. Econ. Ent. **6**, 1913, p. 299—300.

F. THE CITRUS MEALYBUG (*Pseudococcus citri Risso*).

Synonymy: 1813 *Dorthesia citri* Risso
1867 *Coccus citri* Boisduval
1875 *Dactylopius citri* Signoret
1881 *D. brevispinus* Targioni
1881 *D. destructor* Comstock
D. vitis Niedl. etc., etc.

Description: Adult female slightly elongate ovate, covered with white mealy waxy secretion, except at the articulation of the segments, where the yellow colour of the body shows through. The 17 pairs of wax-appendages are short, equal in length and equidistant; the outer caudal pair of these appendages is often much longer (up to 10 times) than the others; the inner caudal pair is very short and lies close to each other. Antennae 8-jointed. Legs long; tarsus about ½ as long as tibia; digitules to the strong claw, and tarsi simple. The anal lobes bear a very long and a much shorter seta. The surface of the last segment is dotted with orifices of wax-glands. Six setae (115 μ) ½ as long as those of the long setae of the anal lobe (225) surround the anal ring.

On the anal lobe 2 (to 3) strong spines surrounded by 15 to 20 small trilocular pores. This area is neither strongly chitinized nor well delimited. A few short setae (about 5) are present (on this segment only). Similar arrangements of 2 smaller spines surrounded by 4 to 8 small trilocular pores on the lateral margins of the preceeding segments.

2.5—5.0 mm long; 2.0—3.0 mm broad.

The *eggs* are long ellipsoidal and light straw yellow. They measure 0.33—0.35 × 0.18—0.20 mm.

The *first instar larva* is of a brown yellow colour. It is long ovate, slightly narrower in front than behind. The legs and the 6-jointed antennae are well developed. The loops of the rostral filament nearly reach the anal orifice. Tarsi longer then the tibiae. On hatching they measure 0.38—0.40 × 0.17—0.18 mm. The last antennal joint is ⅔ of the total length of the feeler.

The *second instar larva* is more distinctly brownish coloured. It is very similar to the first stage larva. The antennae are still 6-jointed, but the last joint measures only ⅓ of the total antenna. 0.5 mm long.

The *third instar larva* is very similar to the adult female, but smaller and with 7-jointed antennae. 0.8—1.5 mm long, 0.3—0.7 mm broad.

The larval stages, especially the first and third ones, are described in detail by Berlese and Leonardi.

Male yellowish to reddish-brown, legs and antennae paler; eyes and ocelli black; the hyaline wings are blue irridescent and longer than the body. Caudal filaments long and stout. Antennae 10-jointed and hairy. Two halterae on the metathorax. Abdomen cylindrical. Genital armature like the stern of a boat, with very long caudal setae.

0.9—1.0 mm long, 0.2—0.3 mm broad.

The *egg* and the *first larval stage* of the male are indistinguishable from those of the female. In the second larval stage the colour is of a darker brown than in the female and the wax filaments are more loose and cotton-like (Vos). Towards the end of this stage the male larva produces a cottony cocoon. Within this cocoon it moults the second, third and forth moult. The *male prepupa* is pale reddish brown, wing-sheds generally not yed developed. The male pupa is pale yellowish brown with well developed wing-sheats, which are pale and translucent like those of the legs. The two last mentioned stages remain normally below the wax cover prepared by the second instar. When disturbed, they move rather clumsily.

Distribution: Ps. citri is of world-wide distribution. Especially from the subtropics of the old and new world it is recorded every-where, living also in the tropics and common in the greenhouses of more moderate climates. It has been recorded on ivy from England even; however, it certainly does not live there the year-round in the open.

Host-relations: The citrus mealybug is very polyphagous, but not every plant on which oviposition is observed is a host. The female may have migrated here just before egg-laying. It may occasionally be found even on the dry wood of posts or boxes. Most of the following hosts recorded from Palestine are true hosts. Preferred hosts have a star. This list is decidedly incomplete and could easily be lengthened.

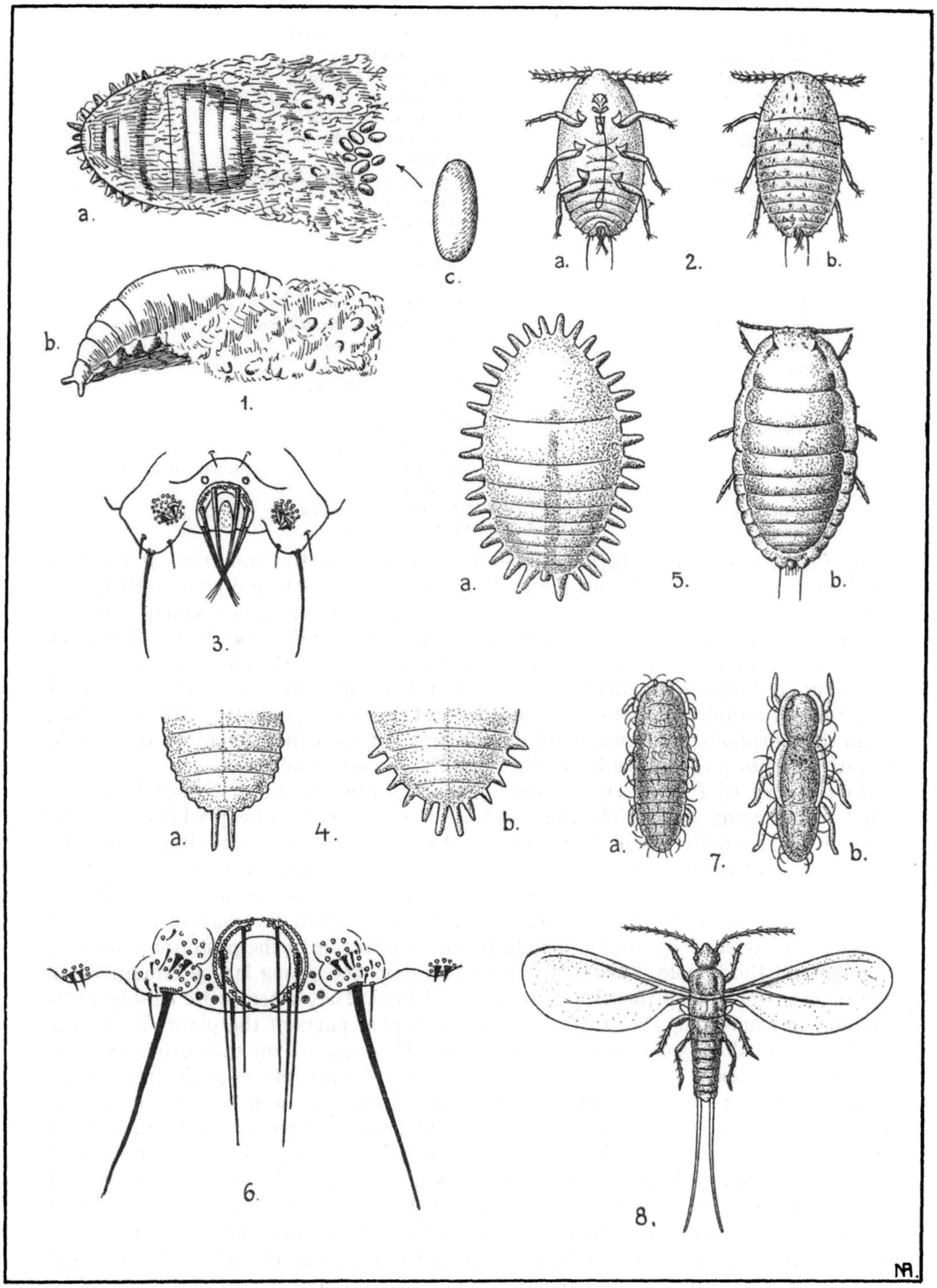

XXIX. The Citrus Mealybug (*Pseudococcus citri* Risso). **1**) ovipositing female, ovisac and egg; **2**) neonate larva; **3**) anal segment of neonate larva from below; **4**a and b) posterior body-end of female larvae of the second and third instar; **5**) adult female covered with wax-secretion and bare; **6**) Pygidial margin of adult female seen from below; **7**) male prepupa and pupa; **8**) adult male.

Host-plants of Pseudococcus citri in Palestine.

Aizoaceae	*Mesembryanthemum sp.*
Apocynaceae	*Nerium oleander*
Capparidaceae	*Capparis spinosa*
Cucurbitaceae	**Cucumis melo*
	Citrullus vulgaris
Gramineae	*Sorghum halepense*
Labiatae	**Coleus sp.*
Leguminosae	*Ceratonia siliqua, Acacia farnesiana*
Moraceae	**Ficus carica*
	F. sycomorus
	Morus alba
Musaceae	*Musa cavendishi*
Oleaceae	*Olea europaea*
Palmae	Ornamental palm
Polygonaceae	*Polygonum equisetiforme*
Punicaceae	*Punica granatum*
Rosaceae	*Cydonia vulgaris*
	Eriobotrya japonica
	Crataegus azarolus
Rutaceae	**Citrus ssp.*
Salicaceae	*Salix babylonica*
Solanaceae	*Solanum tuberosum* (artificial infestation),
	S. nigrum
Thymelaeaceae	**Thymelaea hirsuta*
Vitaceae	**Vitis vinifera*

In *Pseudococcus*, as well, fertility is dependent upon the condition of the host. In *P. brevipes* on pineapple, mealybugs living on the peduncle and green fruit tend to produce more young than those feeding on leaves. This is in agreement with our observations on the fertility of *Aonidiella aurantii, Mytilococcus beckii, Parlatoria pergandii*, etc. on leaves and fruit of citrus. The decrease of *Pseudococcus* populations on citrus seedlings with older leaves agrees in this respect too. The differentiation is due to unfavorable physiological conditions of nutrition (=undernutrition) arising from the morphological structure of the substrat, for on such different hosts as tender potato sprouts and orange-fruits the fertility is very much alike.

Life history of P. citri in Palestine. The eggs are deposited within the ovisac, the female resting on top. The eggs generally form a compact heap which is covered with a dense woolly tissue. Occasionally the eggs are loosely distributed and the tissue of wax-threads is looser. This occurs especially in small egg deposits or when the female was disturbed during oviposition. The form of the ovisac is determined by the distribution of the multilocular pores, which secrete the wax-threads of the ovisac. The latter are restricted mainly to the ventral side of the posterior abdominal segments of the female. For about one or two days the young larvae remain within the ovisac, which they then leave in search of food. They often settle closely to the mother, but may migrate, to some extent, to higher parts of the plant. They often settle in rows along some nerves of the leaves. They may cover each other like tiles, but may also leave and migrate to another spot. Shady places with great opportunities for contact are preferred. The young larvae, as well as uncovered eggs, are very sensitive to dry heat. The male and female larvae are undistinguishable until the end of the second larval stage, when the male grows browner in color, becomes covered with woolly wax-secretions and the first beginning of wing-sheds becomes visible. The female larvae begin with short wax-secretions, after 8—12 days in summer, but their body colour remains visible all the time. Some individuals start to oviposit early (even at a length of 2.5 mm) and these lay relatively few eggs. Other individuals, offspring of the same mother, grow to a length of 4 mm and lay many eggs. Unsuitable host, overcrowding, etc. further early oviposition.

The seasonal history is as follows in citrus groves: In years of increase the white ovisacs are often seen, from the second half of June, covering young fruits densely. This heavy attack lasts until October. At that period natural enemies (*Dicrodiplosis, Sympherobius, Scymnus, Hyperaspis, Leptomastidea*) appear in numbers and help to clean the orchards. The first rains also wash many of them to the ground. Large quantities of mealybugs drop to the ground with the fruits during autumn. Here they often infest the roots of some weeds, as *Polygonum equisetiforme, Thymelaea hirsuta,* etc. In the breedings, females show, at the same season, a tendency to oviposit on the soil, not on the host-plant. But the infestation of these roots certainly occurs by dropped fruit. No active migration from the citrus trees to the soil in winter, up the trees in spring, does exist. During winter development is very slow, the number of eggs per female extremely small and the loss of individuals by storms and rains rather heavy. The larvae of the hibernating females (development requires at least 3½ months in winter) migrate towards the young fruit, where they are commonly found in May and June between the fruit stem and the navel.

The counts per 100 leaves do not give a very satisfactory illustration as mostly larvae are on the search for food. But if the picture is unsatisfactory for the individual year, the sum of the 8 years count seems to give a more or less suitable illustration of the seasonal population trend:

Monthly counts of *P. citri* on 400 orange-leaves at Atta.

year	I	II	III	IV	V	VI	VII	VIII	IX	X	XI	XII	Total
1929	—	—	—	—	—	—	3	—	20	—	—	1	24
1930	1	—	—	—	1	4	5	2	—	2	—	2	17
1931	1	—	1	—	2	7	20	—	3	—	2	1	37
1932	28	32	23	12	10	9	18	26	67	87	78	27	417
1933	—	1	—	—	—	1	1	—	—	—	—	—	3
1934	6	—	—	—	—	—	—	—	—	—	—	—	6
1935	—	—	1	2	1	6	3	8	2	1	—	6	31
1936	—	1	—	2	2	20	3	18	—	—	2	3	49
Sum	36	34	25	14	16	47	53	54	92	99	82	35	

Some breedings may be compiled here. All were made on young potato-sprouts. The most rapid seasonal increase is in early May/June.

No.	Begin. of oviposition mother	daugther	Average temp.	Interval between begin of oviposition of mother and first oviposition of daughters	No. of daughters not yet ovipositing
1	4.XI.26	3.I.27	14.9°C	60, 66, 86, 88, 88, 99, 99, 104, 106, 109	7
2	2.XI.	3.I.	14.9	62, 68, 82, 92, 92, 94, 96, 101, 101, 101, 107, 112, 112	9
3	9.XI.	13.I.	14.9	65, 68, 75, 81, 83, 88, 88, 88, 89, 90, 91, 92, 92, 92, 92, 97, 97, 97, 97, 99, 99, 99	25
4	8.II.27	13.V.	16.3	94, 96, 96, 97, 97, 97, 99, 99, 99, 99	8
5	22.II.	5.V.	16.6	72, 72, 74, 76, 79, 79, 79, 82, 85, 85, 85	20
6	20.III.	15.V.	18.4	56, 56, 57, 57, 57, 59, 59, 59, 59, 59, 59, 59	32
7	20.III.	16.V.	18.4	57, 59, 59	25
8	20.III.	22.V.	18.9	63. 63, 65, 65	6
9	14.IV.	2.VI.	20.5	49, 52, 54, 54, 55, 55, 55, 55	40
10	21.VI.	20.VII.	27.0	29, 31, 31, 33, 33, 33, 33, 35, 35, 35, 35, 35, 35, 35, 35	7
11	1.VII.	22.VII.		21, 21, 23, 23, 25, 25, 25, 25	12
12	1.VII.	7.VIII.	27.1	36, 38, 41, 43, 43, 45, 45, 45, 45, 45	2
13	29.VII.	23.VIII.	26.6	25, 30, 31, 34, 34, 34, 34, 34, 34,. 34, 34, 34, 34, 34, 34, 34, 34, 34, 34, 34	14
14	7.VIII.	3.IX.	26.4	27, 29, 30, 32, 32, 32	13
15	8.IX.	16.X.	24.5	38, 43, 46, 46, 48, 48, 50, 50, 50, 52, 52, 52, 52, 52, 52	5
16	4.XI.	29.XII.	16.7	55, 58, 58, 58, 59, 59, 59, 59	20
17	12.XII.	23.IV.28	15.1	133, 133, 133, 133, 133, 133, 133, 133	36

The Citrus Mealybug has 8 annual generations in the coastal plain of Palestine. The development on potato-sprouts lasted as follows:

Temperature	n	Duration of development (egg to oviposition) average	minimum	maximum
15.5° C	7	76	72	79 days
16.5	14	94	82	95
17.5	7	45	37	49
18.5	28	60	34	75
20.5	12	50	39	55
22.5	92	45	26	58
23.5	116	45	34	59
24.5	4	37	34	38
25.5	4	29	25	30
26.5	100	35	20	42
27.5	21	33	21	45

The corresponding hyperbola has the following constants:

Development threshold (c) = 8.4° C
Thermal constant (th.c.) = 525 days degrees.

This hyperbola represents the average epidemiological, not the quickest development (including the preoviposition period).

The development of individuals starting development in every month lasted:

Month	I	II	III	IV	V	VI	VII	VIII	IX	X	XI	XII
Average development	—	88	59	47	35	38	32	47	41	—	88	133 days
Shortest development	—	72	37	34	30	29	21	20	26	—	58	— ,,
Longest development	—	98	75	55	38	42	45	59	52	—	112	— ,,
n	—	21	32	15	12	81	62	112	140	—	110	8

The development is not interrupted throughout the winter, but lasts longer. It takes 2—3 months in winter, but only about 1—1½ months from spring to autumn.

Fig. 94 illustrates a general observation: The dispersal of hatching becomes the larger the longer the average development of a generation.

The development under average conditions follows the following hyperbolae:

hyperbola		A	B	C
development threshold (c)		8.4	8.4	8.4° C
thermal constant (th. c.).		525	698	884 day-degrees
development at	30° C	24.3	32.3	40.9 days
	25	31.7	42.0	53.2
	20	45.8	60.2	76.2
	15	79.5	105.7	133.9
	12	145.8	193.9	245.6
number of annual generations at	Tel Aviv	8.0	6.0	4.7 generations
	Jerusalem	5.8	4.4	3.5
	Daganiah	8.8	6.6	5.2

Under extremely favorable conditions of nutrition, etc., the earliest development may still be shorter and the thermal constant lowered up to 338 days degrees (30° C = 15.7 days; 20° C = 29.1 days). But such speeds probably occur very rarely

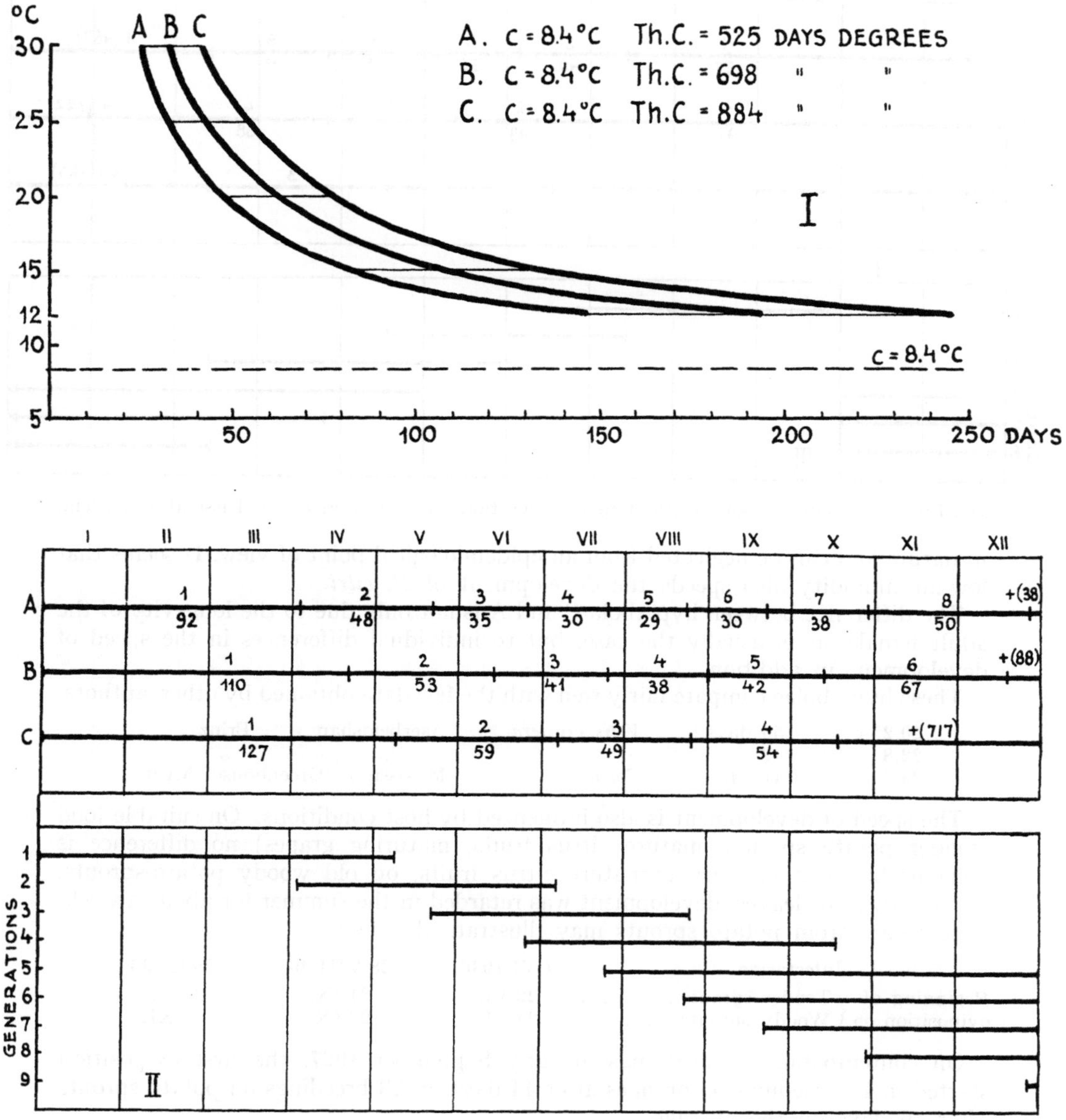

Fig. 93.
Life-history of the Citrus Mealybug (*Pseudococcus citri* Risso) at Tel Aviv. I. The three hyperbolae; II. above: the life-cycle, as calculated from the hyperbolae; below: sequence and overlapping of generations.

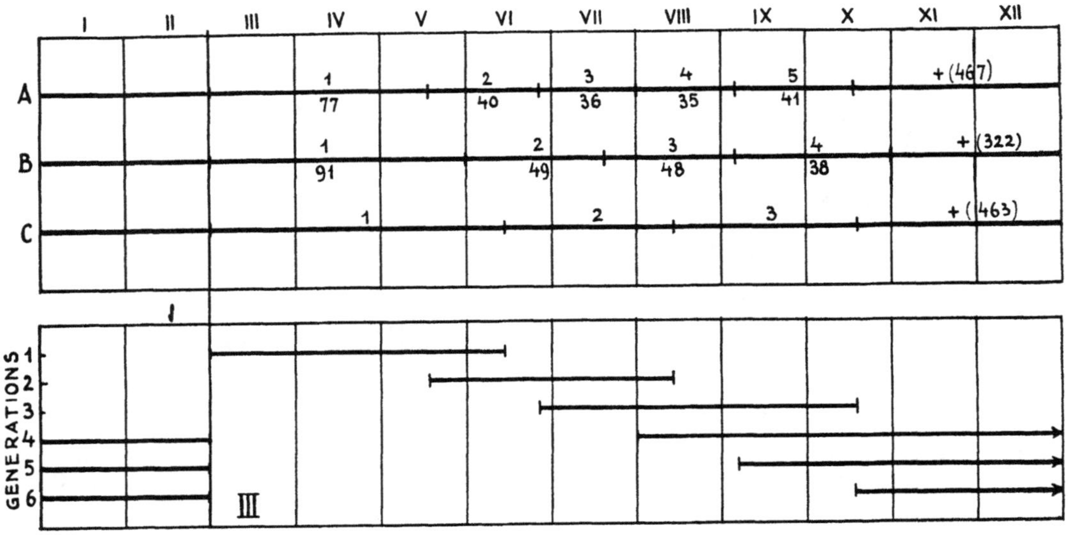

III. Life cycle, sequence and overlapping of generations of *Pseudococcus citri* Risso at Jerusalem.

in nature and can be neglected from an epidemiological point of view. It seems that low air humidity also speeds the development of *P. citri*.

The difference between hyperbola C and A is not only due to the longevity of the adult female, as is usually the case, but to individual differences in the speed of development in addition.

These hyperbolae compare fairly well with the few data obtained by other authors:

20.2° C	50 days	*Vitis vinifera*	Aserbeidshan	Prinz
22.8	42	,,	,,	,,
23.9	41—45	*Coleus*	Mississippi (Greenhouse)	Myers

The speed of development is also influenced by host conditions. On suitable food (tender potato sprouts, mature citrus fruits, maturing grapes) no difference is recognisable. But on very immature citrus fruits, on old woody potato-sprouts, on vine-twigs or leaves, development was retarded in the summer for about a week. A few data from potato sprouts may illustrate this fact:

Infestation		1.VI.1936	28.VIII.36	28.IX.36
Beginning of oviposition on	Tender sprouts	22.VI.	21.IX.	30.X.
	Woody sprouts	28.VI.	28.IX.	4.XI.

On contemporaneous breedings in early September 1927, the first oviposition started in 14-breedings on oranges after 44 days, in 23 breedings on potato sprouts after 46 days on the average.

Oviposition and fertility. Oviposition was observed repeatedly. A female collected on 7.I.25 began to oviposit the following day. Other females which probably had started with egg-laying still in the orchard, hid during the transport in crevices and immediately started with oviposition. The first mentioned female started oviposition in the room at 20° C on 7.I. The ovisac extended on 9.I. from the last third of the female and beyong, extending to about half of the length of the body. The ovisac was as broad as the female, but twice as thick. The eggs were accumulated mainly under the body of the mother. The ovisac is very scanty on the lower side, where the eggs generally lie in the substratum. Two days later the ovisac was still larger and the first larvae crawled within. Convulsive movements of the female are probably connected with oviposition. From the 11.I. on, the size of the ovisac did not show a further increase. The female was dead and dry, many larvae crawled within the ovisac and on the leaf beyond it.

Oviposition is much extended at low temperatures (20—30 days and more), whereas in summer it is much shorter (3 to 11 days only). The interval between the hatching of the first and last larvae is 30 days and more in winter, 6—12 days in summer.

The number of eggs depends on environmental conditions:

Number of eggs of *Pseudococcus citri*.

Country	Season	Host	eggs per ♀ (min./max.)	Author
S. California	summer	Citrus fruits	482 (300—587)	Clausen
S. Spain.	,,	,,	200—250	Clements
Florida	,,	,,	350—400	Watson
N. Italy.	,,	,,	250	Berlese
Mississipi (glasshouse)	year	Coleus	307 (95—592)	Myers
Cambridge (glasshouse)	,,	Coleus, Amaryllus	239 (43—404)	James

High fluctuation of eggs-number deposited under different conditions has been observed by Prinz on *Vitis* in Transcaucasia:

Oviposition	n	M
stem	31	19 eggs per female
leaves.	45	49 ,, ,, ,,
grapes.	52	115 ,, ,, ,, (third generation at 25°C)

But temperature has also a great influence. At low temperatures the number of eggs is very low:

Temperature	Rel. Hum.	Localisation	n	No. of eggs per ♀
13°C	78%	stem	31	19
17	65	grapes	9	12
21	63	,,	12	180
25	67	,,	59	164

Our Palestinian experience confirms this depressing influence of low temperatures. On citrus fruits the following have been counted:

date	No. of ovisacs	average no. of eggs per ovisac (minimum and maximum)
29.VI.1925	3	400 (330—448)
23.IX.	6	176 (110—240)
4.X.	3	418 (345—480)
31.I.1927	10	107 (34—163)
24.II.	6	91 (39—172)
4.IX.	6	95 (32—212)
13.X.	6	136 (34—201)
6.XII.	8	68 (25—150)
30.I.1928	25	44 (24—90)
25.VII.1929	6	210 (107—300)
17.VIII.	9	174 (93—290)
13.IX.	8	106 (37—198)

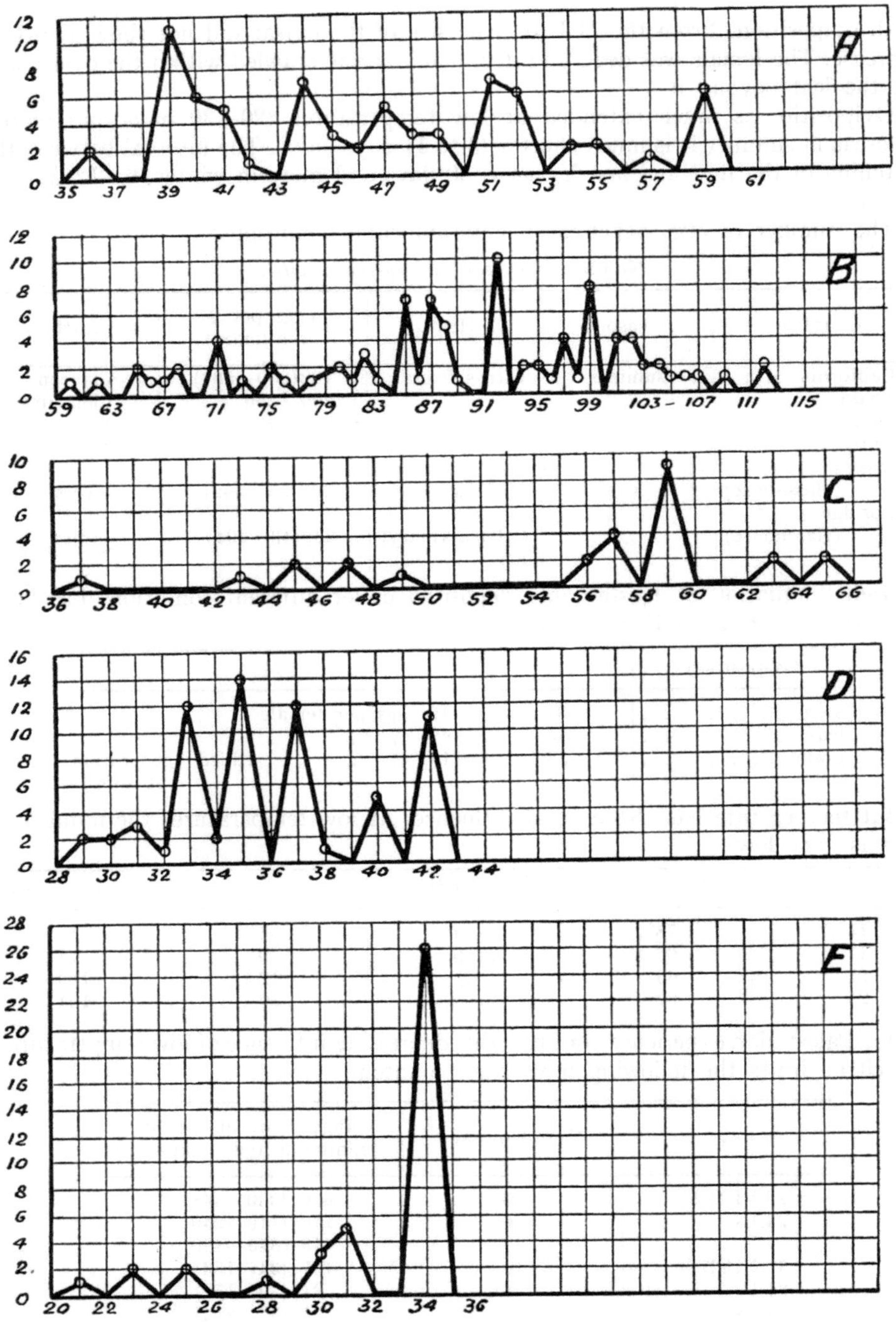

Fig. 94.

Duration of development in different seasons and generations of *Pseudococcus citri* Risso. The abscissa counts the days from larval appearance to oviposition, the ordinate the number of females beginning oviposition. Dates: A. 25—31. VIII. 1926; B. 4—9. XI. 1926 (abscissa in half the scale as in the other diagrams); C. 20—25. III. 1927; D. 21—26. VI. 27; E. 17—29. VII. 29. (from Bodenheimer and Guttfeld).

The monthly average on citrus fruit was:

month	I	II	III	IV	V	VI	VII	VIII	IX	X	XI	XII	Average
Average no. of eggs per female	61	91				400	210	174	123	194		68	119
maximum no. of eggs per female	163	172				448	300	290	240	480		150	
n	35	6				3	6	9	20	9		8	96

The monthly average on potato sprouts was:

month	I	II	III	IV	V	VI	VII	VIII	IX	X	XI	XII	Average
average no. of eggs per female	112	92		116		198	114	153	89	99	152	56	134
n	9	6		21		12	11	8	4	16	8	5	85

Whereas the average egg-production of every female was 171 (n=25) on long, tender potato sprouts, it was 119 (n=12) on short but tender sprouts where the animals developed under crowded conditions and 67 (n=15) on old, woody and short sprouts only. This shows that population density as well as quality of food and temperature influence the egg-production considerably. It may be of interest to mention that on young orange seedlings egg-production remained extremely low (below 100 eggs per female). They cannot apparently develop for some successive generations on citrus leaves and twigs only.

Balachowski (1933) counts 100—150 eggs per ovisac, but maintains that one female lays 300—600 eggs. In order to do so the female should deposit 3—5 successive ovisacs. During the many years in which we have been breeding *P. citri* in our laboratory, not one case of a second ovisac formation was observed. The female dies over its first ovisac, having performed the parturition, or — less commonly — it retires into some crevice, dying soon afterwards.

The limiting factors of the Citrus Mealy-bug are manifold in Palestine. Parasites and predators abound at certain seasons and localities. Limiting factors of the host cooperate inasmuch as the host and its condition influence the fertility as well as the larval mortality. The nutritive larval mortality has unfortunately not been sufficiently studied. Overcrowding favours early and small oviposition. The increase of the number of males in summer, its decrease in winter should be mentioned. Among climatic influences the dry desert-winds induce high egg and larval mortality. The unfavorable influence of winter rains and storms has been mentioned before.

A series of breedings was set up to study climatic mortality. But these were arranged in the laboratory in lamp-cylinders, not much exposed to direct climatic influences and certainly living in a more humid microclimate than in the open. (Fig. 95).

In some breedings of Vos on *Coleus* the larval mortality fluctuated between 44 and 80 %. The egg-mortality was not considered. Spring and autumn are the most favorable seasons for development in Palestine. The climatic mortality (not considering rains, etc.) ranges between 10 and 32 %. Temperatures above 25—26° C, especially when connected with low humidity, grow decidedly unfavorable. We are still unable to say as to how far the favorable conditions in spring and fall are changed by free exposure to the hamsin winds in May and September. As in most other Coccids, the wind transports quite considerable numbers of young larvae from their host, of which a fraction only may succeed in settling on a suitable host. Large quantities of mealy-bugs are taken annually from the orange grove with the crop resp. by dropping with the fruit to the ground. All these individuals are practically lost to the mealy-bug population on the trees. When the climate in late spring and early summer is highly humid, conditions are favorable for a summer outbreak.

Sex-ratio, parthenogenesis and sex-determination. Very little is known on the sex-ratio of *Pseudococcus*. Vos states that in eight breedings of 416 individuals only 84, i.e. 20 %, were males. Schrader obtained among 267 individuals 125=47 %

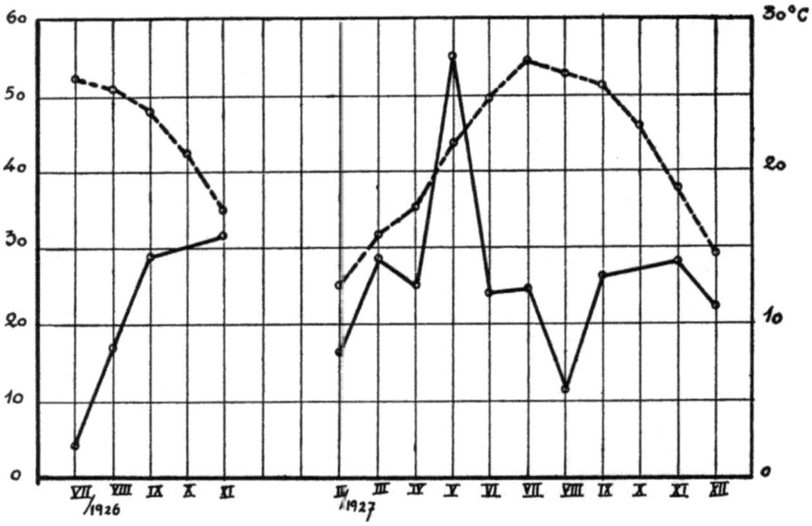

Fig. 95.
Average number of female offspring per female of *Pseuducoccus citri* Risso which succeeded to develop to oviposition (— —) as related to the average monthly temperature (full line; from Bodenheimer and Guttfeld).

males. Newstead (1902 p. 166) states that he has never met with the males, except in the hottest weather. But the species described by him as *Dactylopius citri* probably is another species. Kirkpatrick (1927) writes with regards to *P. kenyae* on coffee in Kenya Colony:

„It is certain that the proportion of males to females which arrive at maturity is larger on caged potted plants in the laboratory than in the field. It appears also that in a plantation more males are produced under shade than in the open. It is therefore possible that a damper atmosphere, more even temperature and absence of wind tend to produce a larger proportion of males. But it is conceivable that the proportion produced is always the same, but that the mortality among the immature males in exposed situations, probably at the time when they migrate downwards to change to the pro-pupal stage, is comparatively greater.

It is also very probable that food conditions affect the production of males. Observation tends to show that the males are comparatively much more scarce at the times when there is an abundace of flower buds and young cherries on the coffee trees, i.e. at the times when the females are multiplying at a greater rate than at any other; while later on in the year, from July to September, when the females are breeding more slowly consequent upon there being only leaves and the more mature cherries available, the proportion of males increases.

It is therefore suggested that parthenogenetic reproduction is the rule, when there is an abundance of the most nutritious kind of food, and that when this fails and the race tends to become less vigorous, males are produced in order to strengthen it again."

The following sex-ratios were obtained in breedings through one annual cycle:

generation	n	series A	n	series B	n	series C
1	901	58% ♀	333	71% ♀	131	60% ♀
2	746	40	121	48	714	41
3	230	64	657	57	1381	15
4	623	52	296	72	220	25
5	329	51	514	67	888	48
6	295	61	299	45	557	41
7	681	60	159	63	559	54
8	423	77	640	77	106	92
9	278	66	100	73		
10	61	79	233	72		
	56	58	65	63	31	45

The percentage of females from these and other series is as follows when distributed over the months:

month	I	II	III	IV	V	VI	VII	VIII	IX	X	XI	XII	Average
% of ♀	81	—	—	64		46	64	60	51	53	62	—	60

The increase of males in June and September is possibly connected with the seasonal occurrence of the hamsin-winds. The decrease in the number of males during winter is certainly significant, as it was observed in all breedings.

W. Emeis believed in the normal occurrence of parthenogenetic development since he neither obtained males nor met with spermatozoa within the recepticula seminis. But F. Schrader did not succeed in obtaining oviposition from isolated, unfertilized females, in which the ovaries degenerated and the body shriveled. After copulation two hitherto isolated females started to oviposit a few days later. He concludes that no parthenogenesis is possible in *Pseudococcus*.

Vos supposes that in *P. citri* the normal propagation is parthenogenetic, because the percentage of males is too low to permit the fertilization of females. She observed normal oviposition in fertilized as well as unfertilized females, but she has not made it clear if copula was observed and excluded or not. Myers isolated second stage female larvae and obtained normal development in four successive parthenogenetic generations. Nothing is reported on the sex-ratio of their offspring. This seems to indicate that males and females appeared, because the appearance of females only would certainly have been notified.

Preliminary experiments seemed to show that early isolated females oviposit readily and that males as well as females appear in the offspring. Breedings were therefore arranged through nine consecutive generations which all apparently reproduced parthenogenetically, when the daughters were isolated at the end of the first larval stage. We were therefore certain that facultative parthenogenesis of *P. citri* may occur in Palestine, and that the offspring of these females includes males and females in the normal sex-ratio of the general seasonal trend. The cytological basis of this observation did not seem to be easily explained and the cylinders of the last three generations were therefore carefully searched. In all 15 cylinders, which seemed to be quite sufficiently isolated by mull and cotton, dead males were discovered. In an additional series the active intrusion of living males was actually observed. The chemotaxic attraction of the mealy-bug males towards the females must be very high, indeed.

In five comprehensive series of breedings in which every possible precaution against the intrusion of males was taken not one female oviposited. The initial erroneous results were recorded here with the purpose to show the importance to avoid a faulty techniqne in experiments on parthenogenesis.

The cytological basis of sex-determination is therefore the same for our strain as for that studied by Schrader:

	♀	♂
Parents	10 x	5 x + 5 y
Gametes	5 x 5 x	5 x 5 y
Offspring	10 x	5 x + 5 y
	♀	♂

No parthenogenesis has yet been definitely demonstrated in *P. citri*.

These conclusions are confirmed by a recent, very comprehensive study of H. C. James (1937). In many thousands of breedings of virgin females of *P. citri* (both, aerial and root forms) as well as those of four other species of *Pseudococcus*, the females were entirely incapable of producing offspring. Only fertilized eggs developed.

In breedings made under uniform conditions of food and climatic environment the females showed a wide range in variability with regard to fertility and sex-ratio. Whereas the offspring of some females produced only from 30 to 50 eggs, the offspring of others produced from 350 to 400 eggs. And whereas the total average sex-ratio was 101.62 ± 1.54 ♂♂ per 100 ♀♀, different families showed a fairly wide range of sex-ratio, independent of the total fertility. In 12 families with a fertility of 250—300 eggs this ratio was: 120, 113, 109, 109, 106, 104, 95, 95, 92, 92, 92, 88 ♂♂ per 100 ♀♀. Delayed fertilisation in *P. citri* lead to an increased proportion of males, which was proportional to the period of delay before fertilisation:

delay in weeks	n	sex-ratio per 100 ♀♀
—	47	101.62 ± 1.54 ♂♂
6	36	180.89 ± 6.20
8	15	327.40 ± 11.63
10	10	991.09 ± 460.50

On a purely genetical basis an equal sex-ratio would have to be expected. James explains the shift of the sex-ratio by an abnormality in the reduction division of the male. Whereas one of the two fiver groups of chromosomes are massed close together and appear to act together in many respects like a unit, the other five chromosomes may be called the loose group. ,,Random segregation does not occur at reduction and the massed group invariably goes to one pole and the loose group to the other. The spindle fibres attached to the loose group are always late in appearing and are imperfectly formed and the chromosomes of this group exhibit a lack of coordination in travelling to their pole. It is suggested that this imperfection in the spindle fibres of the loose group of chromosomes may be indicative of an incipient degeneracy in one set of gametes, which is not evident in the fully formed sperms, but which is expressed in a variable vitality that causes a variability in their functional efficiency.

In a most interesting series James tested the fecundating capacity of 13 individual males, which proved to be: 7, 3, 17, 12, 0, 23, 14, 10, 2, 1, 17, 0, 2 = 9.07 females with a maximum of 23. It follows that even at the lowest sex-ratio observed, there are always ample males produced to ensure the fertilisation of the females.

In a later study James (1938) showed a significant, but not very substantial influence of air humidity on the sex-ratio of *P. citri*. The number of males obtained per 100 females was:

Series	n	air-humidity	age of *Pseudococcus*	no. of males
A.	35	dry air	during maturation	153.04 ± 10.51
B.	37	dry air	at ageing of over (42 days)	164.17 ± 10.38
C.	36	saturated air	during maturation	180.89 ± 6.20

The symbionts of Pseudococcus citri. Beneath the intestinal tract of *Pseudococcus citri* we find an unpaired yellowish voluminous organ which is richly supplied with tracheae. This mycetome extends over one third of the length of the female's body, being smaller in the male. The large mycetocyts are enveloped in a strongly pigmented epithelium. They possess large nuclei which abound in chromatin and 10—12 round to elongate parcels, including the symbionts, which are within the cytoplasma. Whereas the younger of these parcels are enveloped in a mucous substance, a membrane is excreted around the older ones. The great number of chromosomes (up to 200) within the nucleus of these cells as compared with 10 chromosomes of the normal diploid phase indicates that degenerative changes occur within these cells. The intimate cytological study of the origin of these mycetocyts is of no concern to us here, but it is of interest to note that they offer a very close parallel to the cytological phenomena observed by Silvestri in the polyembryony of certain parasitic hymenoptera. They have been studied in detail by Pierantoni, Buchner, Schrader, Walczuch and others.

The symbionts of the citrus mealy-bug are bacteria. Their shape is rather variable, from roundish to oval and elongate, sausageformed. The plasma is comb-like in structure containing many granulae. Sausage-shaped forms prevail almost entirely in the embryo and larva, whereas they are replaced later on by the round and oval forms. As soon as the ovarial eggs reach maturity, some of the symbionts lose the original plasmotic structure. The round and oval forms become small and curved: the infestation stage. The formation of special infestation stages is observed only in those scale-insects in which the symbionts undergo a distinct degeneration during their development. Degenerated or intact mycetocyts filled with these infestation stages leave the mycetom. Having left the mycetom they degenerate and fall to pieces. The symbiont parcels are thus liberated and eventually transported to the eggs. The infestation occurs on the anterior pole of the egg. Not single symbionts, but a complete parcel, always enters the egg plasma during the latest stages of its maturation. Here they rest and only after the invagination of the germ stripe on the posterior pole has much progressed, do some cells migrate towards the symbionts and inclose them, giving later rise to the mycetom. The nuclei of these cells are derivatives of the „Richtungskoerperchen".

The physiological role of these symbionts, which seem to occur in all Coccidae, is still unknown. The symbionts are not always bacteria, as in the case of *P. citri*. Some species have two or more species of regular symbionts. But the presence of such an elaborated system of egg-infestation as is met with in the Citrus Mealy-bug is very exceptional in this group.

The recent interesting studies of Carter concerning *P. brevipes* in Hawaii turn the attention of the applied entomologist to these forms. Not only is one of the symbionts concerned in the production of phytotoxic secretions by the insect, but the change of the host plant changes the symbiotic flora and therewith the toxicity of *Pseudococcus*. And just in *Pseudococcus* where ants often help in transporting individuals from one plant to another, the knowledge of such potentialities is desirable.

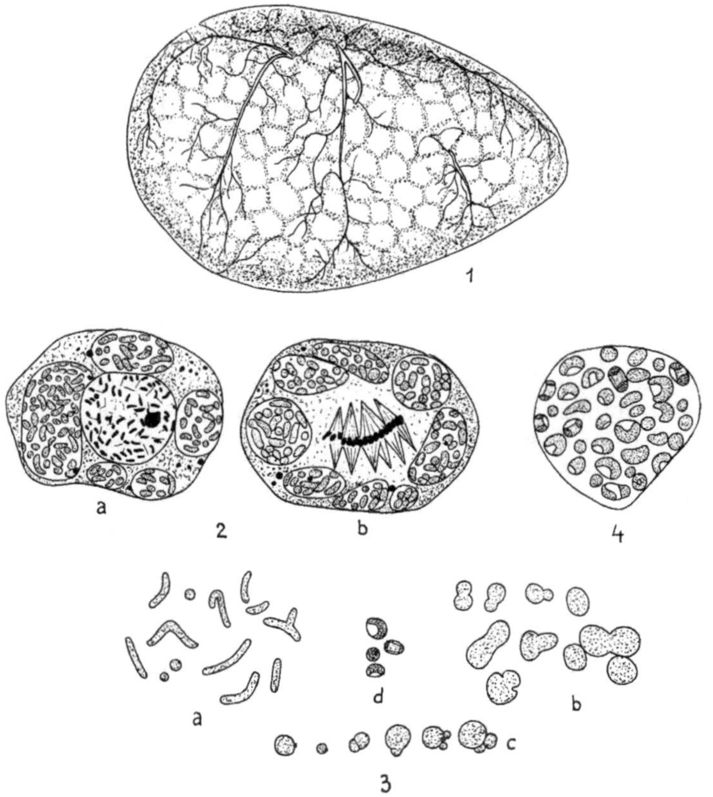

Fig. 96.
Intracellular symbionts in *Pseudococcus citri* Risso. 1 Mycetom, total view from the side; 2. Mycetocyts: a) before b) in cell-division; 3) different forms of the symbiont: a—c from the mycetome, d infectious stage; 4. infection-parcel of symbionts within a mucous substance (from Buchner).

Sense-ecology. All stages of *Pseudococcus citri* show a negative phototaxis, except the ovipositing females, which are not very sensitive to any external stimulus. This negative phototaxis is very pronounced in strong light, its effect is small in diffused light and it has very little influence in semi-darkness. It is highly probable that this negative phototaxis is a main factor regulating the diurnal activity. In the early morning or late evening the larvae and young females may be seen crawling quite actively over fruits, leaves, etc. and as soon as full daylight has set in, they disappear, hiding themselves in crevices, in the space between fruit stem and fruit, between the contact surface of two neighbouring fruits etc., They leave these refugia readily in darkness and find them again with the help of a very strong positive thigmotaxis.

When exposed to light or in the dark room, they use every opportunity to hide in crevices. They are also negatively geotaxic. In the dark room *Pseudococcus* moves up the shortest way, but takes the route furthest away from the source or direction of the light. We have never observed positive geotaxis in satiated individuals, as did Bayer in *P. adonidum*, but young larvae as well as full grown females often fail to give any reaction to gravity.

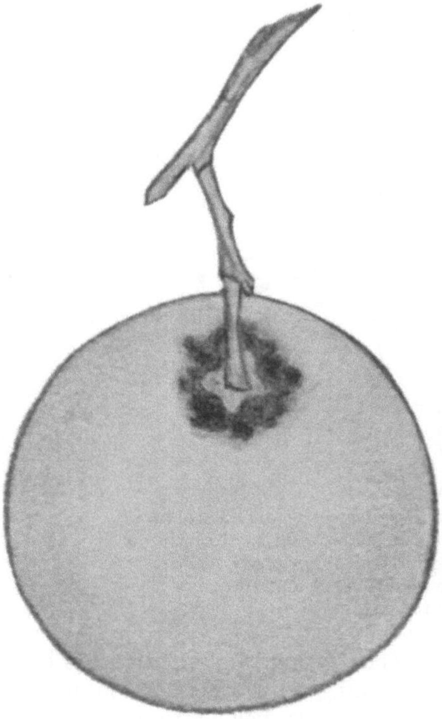

Fig. 97.
Sooty mould on orange fruit caused by *Pseudococcus citri* Risso.

Chemotaxic reactions to citrus are positive only when the distance from the animal does not surpass 2 to 3 cm. In closed petridishes no orientation occurs, probably because no orientation with regard to the source of the smell is possible. In open petri-dishes the large majority moves towards the source of the stimulus and settles on it. Fully grown females do not react to smell at all.

In the absence of fruit, the mealy-bugs prefer to settle on the shadowy, lower side of the leaves, near the median vein or in angles or crevices of the branches. This tendency is much stronger in the females than in the larvae. When disturbed by full light, the mealy-bugs move away from the light. But whereas females remained and settled in angles of twigs or leaves, the larvae moved up again. Whereas the larvae are more mobile than the young females, they seem to react somewhat less to stimuli of light and touch them later than the latter.

The preference temperature of *P. citri* was 21.5° C (n=106). Its scale of activity is as follows:

scale of activity	1	2	3	4	5	6	7	8
about	0.0	10.2	13.0		30.0	35.4	40.3	42.7° C

Damage and injury. The citrus mealybug, when living on citrus, does not belong to those plant lice which excrete an abundance of honeydew. They excrete it in large drops, which often are attended by ants. These drops are not sprayed to a distance nor do they disperse over a large area, so that they do not lead to an extensive formation of sooty mould. Locally restricted development of sooty mould on individual fruits, as illustrated in fig. 97 may be observed,though this damage as well has never gained any importance as far as our experience goes. This lack of honeydew excretion is connected with the prevalence of intracellular feeding habit of *Pseudococcus*. The cell-content of the plant tissues which are a highly nutritive and concentrated food is sucked out. Only when sap-vessels of the plant are met with by chance, does honeydew excretion occur on citrus.

We have repeatedly observed that the sooty mould which develops from *P. citri* and other mealybugs dries out rather quickly and then often falls off the fruit. In September, 1932, special attention was paid to this process.

Whenever it feeds on saps, as on the fruit of grapes, it excretes large, viscous drops in abundance.

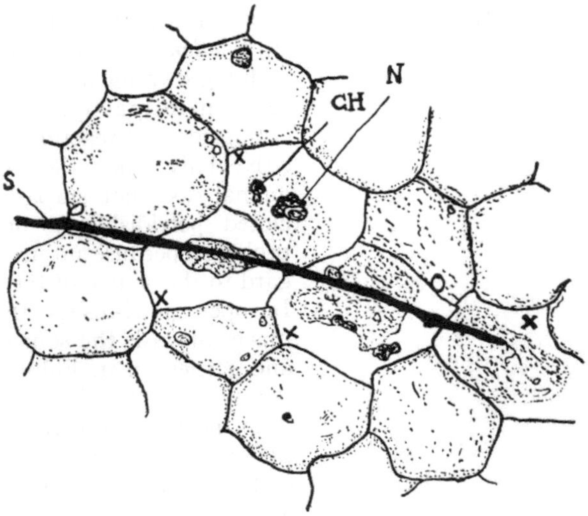

Fig. 98.
Anatomy of sucking of larva of *Pseudococcus citri* Risso in parenchym of stalk of *Impatiens*. s = sucking channel of the mealybug, N = nucleus, Ch = grains of chlorophyll. In the cells marked by an x, the plasma is shreveled and the chlorophyll grains degenerated (from Vos).

The connection with ants is a primitive one. The visiting ants are attracted by the sweet exudation which they are in the habit of licking up. No further contact takes place between mealybugs and ants. The visiting ants belong to many species, among which the first mentioned is the most numerous. The following species were observed:

Crematogaster inermis Mayr.
C. i. var. armatula Em.
Camponotus compressus thoracicus var. sanctoides For.
C. c. var. fellah Em.
Tetramorium punicum var. lucidula Em.
Paratrechina jaegerskioeldi Mayr.
Plagiolepis pallescens var. ancyrensis Sants.
Tapinoma simrothi var. phoenicia For.
Monomorium subopacum var. phoenicia For.

These observations were made mainly during the summer-months:

month	I	II	III	IV	V	VI	VII	VIII	IX	X	XI	XII
no. of collections . . .	2	1	—	—	—	1	2	10	3	1	2	—

The visits of ants are by no means restricted to mealybugs living on citrus. Root inhabiting colonies of *P. citri* are often heavily visited, as was the case with a colony of *P. citri* on the roots of *Thymelaea hirsuta* (Sargonia II.1926).

The „June-drop" is a normal phenomenon in almost all orange groves in Palestine. Fruits up to the size of walnuts drop during the second half of May and early June, sometimes up to 200 fruits per tree. This excessive shedding of young fruit is a symptom only, the causes of which are manifold and complex. In Palestine, the main cause certainly are disturbances in the water-metabolism of the plant, which are correlated with the hot desert winds in the spring. Nutrition seems to be another important factor. Antebi describes the effect of these factors as follows: ,,The sudden gusts of heat caused by the ,,hamsin" winds result in quick and sudden transpiration and cause a disturbance in the water balance and movement in the plant. Since citrus trees are very reluctant in dropping their leaves, they rid themselves of a certain percentage of the newly formed fruit until the balance of the water metabolism is regained. Now, if the tree finds in the soil a sufficient supply of water at this stage, it would certainly suffer less and the balance would be regained quickly. Similarly if the tree is given an adequate supply of nitrogen just before blossoming, the young fruit is helped to develop quickly and vigorously and withstands the sudden change better."

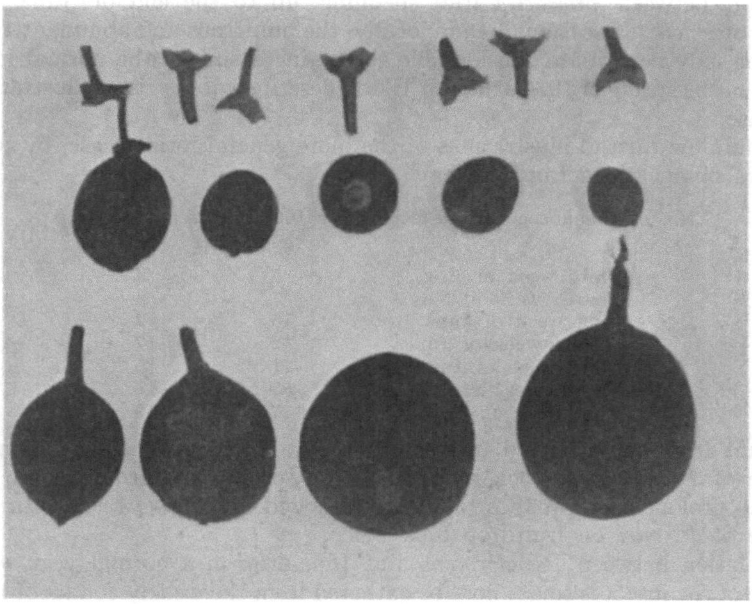

Fig. 99.
Orange fruits of different ages dropped after infestation by *Pseudococcus citri* Risso.

The young fruit is obviously the most sensitive part of the plant at that season and any injury being either of general influence on the health of the plant or of special influence on the fruit will lead to shedding of the fruit. There is no use to discuss ,,the" cause of the June-drop, in itself a symptom only which may be produced by a large variety of factors. Even the fact that some fruits drop together with the pistil, others without it, is of no importance. Pea-sized fruit tends to drop without, walnut-sized fruit with the pistil. The presence or absence of the pistil on the shed fruit does not serve as an indication of the lacation of the injury. Fruits drop, whatever the source of injury, on this most sensitive part, which shifts from the naval in young fruits to the base of the fruit-stem in older ones.

Having thus made clear that the dominating causes of the normal ,,June-drop" are disturbances in the water-metabolism, nutrition, etc., we must now turn our attention to a secondary agent, the citrus mealybug. It is quite probable that unless the preliminary weakening of tree and fruit by both mentioned primary causes, no damage would be inflicted to young citrus fruits by the normal sucking of a few mealybug larvae between stem and fruit. But the weakened fruit is more sensitive. It has recently again been demonstrated that the „normal" June-drop has no connection with mealy-bug infestation (Carmin 1938).

The first experience of damage by *Pseudococcus* was gained in spring 1924 at Petah Tikva. In early June up to 150 yellowish, chlorotic fruits, up to the size of a walnut, were seen beneath every tree. In dropped fruit, as well as in yellow fruit on the tree from 87 to 92 % were infested with young larvae of *Pseudococcus* between the fruit and the pistil. Among normal, healthy fruit on the tree, from 3 to 5 % only were infested at the same time.

Experience of many years has demonstrated clearly that often there exists a correlation between *protracted* fruit shedding, up to the end of June, and the presence of *P. citri*. Occasionally and locally, the numerous and abundant *Pseudococcus* larvae may contribute considerable to the increasing of the normal fruit drop. The case of the spring of 1924 at Petah Tikva, described above, is an illustration of the latter case.

We shall now turn to illustrations of the more general former case, by comparing two years observations from Hedera:

No. of Pseudococcus infestations per 100 fruits ready to drop.

	1930	1934
Third week of May. . . .	5	3
Fourth week of May . . .	7	8
First week of June. . . .	15	12
Second week of June. . .	42	17
Third week of June . . .	71	15
Fourth week of June. . .	82	11
First week of July. . . .	87	13

The fruit drop was prolonged in 1930 until the end of June. Whereas no considerable influence of *Pseudococcus* took place until the second to third week of June, its abnormal prolongation was certainly conditioned by *Pseudococcus*. In 1934 no influence of *P. citri* on fruitdrop took place.

The relation between *Pseudococcus* and June-drop in a normal year, where the influence of the insect is small, may be gathered from the weekly counts of 100 fruits, ready to drop, in the spring of 1935 at Petah Tikva (3 groves: A heavily, B moderately C slightly infested by *P. citri*):

	A		B		C	
	no. of fruit dropped	% of infested fruit	no. of fruit dropped	% of infested fruit	no. of fruit dropped	% of infested fruit
Third week of May . . .	30	8	17	9	15	3
Fourth week of May . .	18	20	25	19	8	12
First week of June . . .	12	35	10	25	2	6
Second week of June . .	12	43	10	15	5	1
Third week of June. . .	7	35	5	18	8	7
Fourth week of June . .	10	30	8	27	12	5
First week of July . . .	5	37	9	25	2	15
Second week of July . .	0	40	0	35	0	10
Third week of July. . .	0	25	0	25	1	15
	94	31	84	23	53	8

The increase of *Pseudococcus*-infestation from late June on has no further influence on the shedding of the fruit. The connection of the fruit with the tree has grown sufficiently strong to resist the disturbances caused by the sucking of a few mealybugs.

Woglum (1926) has shown that a heavy infestation of young fruit at the same season in spring by old and ovipositing mealybugs leads to the same heavy fruit drop as does the sucking of larvae beneath the sepal.

In normal years the fruit drop connected with older fruits is small. The following observations were made in Petah Tikva (a = number of dropped fruit per tree; b = number of infested fruit per 100 fruits):

Grove	A (heavy infestation)		B (mediocre infestation)		C (slight infestation	
1936	a	b	a	b	a	b
Second week of July	9	30	25	50	4	51
Third week of July	27	21	12	8	5	16
Fourth week of July . . .	2	75	18	5	0.5	27
First week of August. . . .	4	43	—	—	—	12
Second week of August. . .	1	41	—	—	3	32
Third week of August . . .	0.5	59	—	—	—	10
Fourth week of August. . .	—	50	—	—	—	—
First week of September . .	0.5	41	—	—	—	—
Second week of September .	—	12	—	—	—	—
Third week of September. .	—	3	—	—	—	—
Fourth of week September .	—	—	—	—	—	—
Total dropped fruit (per tree)	44		55		12.5	
Average no. of infested fruit per tree.		7		1		3

Locally, some fruit-fall occurs on certain trees or certain restricted areas within older orange groves, where 20 to 50 fruits may drop after a heavy *Pseudococcus* infestation. This damage is increased because heavy mealybug infestation prepares the fruit for heavy dropping in September caused by *Cryptoblabes gnidiella*. Later in the season, also, heavy mealybug infestation causes fruit-shedding. In November 1935 observations were made on 4 trees in centres of heavy *Pseudococcus* infestation (a = no. of dropped fruit; b = number of heavily infested fruit among a).

Tree No.:	I		II		III		IV	
	a	b	a	b	a	b	a	b
First week of November	26	7	19	7	35	14	17	5
Second week of November	23	9	15	8	40	8	31	11
Third week of November	11	—	10	2	3	1	—	—
Fourth week of November	7	2	9	—	—	—	—	—
	67	18	53	17	78	23	48	16

The total fruit drop caused in centres of infestation is in some years quite considerable locally. Such local damage has in the meantime been observed in all old centres of orange-cultivation, from Rehoboth to Hedera.

However, W. Carter (1936) shows in his recent studies on the heavy damage inflicted by *Pseudococcus brevipes* in pine-apple-fields in Hawaii that, besides the purely physical damage described by Vos, toxic effects are certainly produced. Before going into a further discussion the clear resumé of Carter on the fundamental differences between phytopathogenous toxin and virus may be reproduced:

Toxin	Virus
,,No evidence for multiplication. New tissue arising after bugs are removed is symptom-free.	Multiplication of virus within the host plant typical and without exception.
Recovery usual; depends on dosage and length of time dose operates.	Recovery extremely rare.
No symptoms ever observed on any other host.	Wide range of host plants usual.
Length of colony's feeding time correlated with amount of wilt produced.	Feeding time necessary for virus transmittal is a matter of minutes.
Wilt is determined by number of bugs and length of time they feed i.e. a mass-action phenomenon.	Mass-action not a factor.
Sub-wilting populations commonly encountered invariably produce wilt if left to develop.	Non-infectious colony produces no virus symptoms however long it operates or however large it becomes.
Capacity to produce wilt is inherent in the insect.	Insect acquires virus by feeding on diseased plant.
Vegetative reproductions normal except where mother plant reserves have suffered on account of long continued mealybug feeding.	Vegetative reproductions all carry virus.
Control is effected by control of established insect colonies.	No control effected by removal if insect vector.''

The pineapple mealybug may have a series of different effects on the host: Mass-action of mealybugs on the base of a pineapple fruit leads to cracking between the fruit eyes followed by the ingress of Nitidulid beetles and *Drosophila* and subsequent souring of the fruit. Relatively few individuals of *P. brevipes* may cause a wilt of one leaf as well as of the whole plant. Others may produce a green-spot disease. In both latter cases only certain individuals or strains of the mealybugs were toxic, whereas in all other individuals damage was restricted to the destruction or deformation of cells on which they had sucked; just in the way described by Vos. Carter succeeded in showing that in all the toxic individuals one of the symbiontic organisms had undergone a morphological change and that this change is correlated with the toxicity of the mealybugs. This case is more fully treated in the chapter on the symbiontic organisms of *Pseudococcus*. But as this type of toxicity may appear wherever *Pseudococcus* infest cultivated plants, it was necessary to mention this possibility.

The sucking of *Pseudococcus* has a decidedly injurious effect on the host. The growth of the entire plant is handicaped. This has been proven by Vos: The increase of length of Coleus was:

Time after infection	n	on infested plants	on uninfested plants
1 month.	24	69%	118%
2 months	17	185	347
3 months	12	305	441

On tomatoes the difference was 304: 500 %. This damage does not decrease with time. The monthly increase on Coleus was:

	1. month	2. month	3. month	Total increases in 3 months
infested plants. . .	368	321	164	853%
uninfested plants. .	643	409	487	1539

The number of leaves is much inferior on infested plants:

	increase of length (1 month)	number of leaves (1 month)
infested plants. . .	147 cm	165
uninfested plants. .	348	424

The drop of leaves was decidedly heavier on the infested plants.

Similar results were obtained by Van der Goot with *P. crotonis* on cacao.

Following Vos the influence of the injury on the growth of the host plant is, in the case of *P. citri*, not due to the influence of toxic saliva, because only those cells in the plant tissue show degenerative signs (loss of plasma, shriveling and discoloring of chlorophyll and nucleus) in which a sucking channel has been observed. No plasmolysis has been observed, but only the direct influence of sucking. The intracellular type of sucking of our species is illustrated by fig. 98.

Our own experiments on the effect of different sized *Pseudococcus* populations on young citrus seedlings just suffice to demonstrate the toxic influence of the mealybug on the plant. They are not conclusive with regard to quantity effect, because the population which develops is by no means proportional to the initial effect.

The general series of observations lasted from 24.VI.1935 to 19.I.1936. The results were as follows:

Initial mealybug population	0	6	12	24
Change in original hight of seedling. . . .	+9.	+2.	+2.	—1.7%
Change in number of leaves	—2.5	—25.0	—63.0	—30.0

24. VI. 1935.

28. VII. 1935.

13. VIII. 1935.

13. IX. 1935.

4. X. 1935.

3. XI. 1935.

Fig. 100.
Influence of populations of *Pseudococcus citri* Risso in different densities on the growth of young citrus plants. Number of mealybugs at the beginning of experiment: 1 = 0, 2 = 6, 3 = 12, 4 = 24.

Out of the three seedlings with 6 mealybugs at the beginning, one died rather quickly (13.VIII.). In the two others the plants showed partial drying at the beginning, followed by self-cleansing of *Pseudococcus*. The injury seemed to be highest at a density of 12 mealybugs. One seedling was dead on 13.VIII. The two others were fairly damaged, but later the mealybugs were much reduced and at the end of the observation these plants were still living, but in a bad condition. The injury was smaller at an initial population density of 24 mealybugs per plant. This was, however, only apparent, as the seedlings die rather quickly at this initial density, before any remarkable increase in abundance of mealybugs occurs. The seedlings infested on 24.VI.35 showed beginnings of heavy chlorosis on the 9.VII. and died on the 13.VIII., i.e. in about 6 weeks. A second series was arranged on 13.IX.35. Drying began in early October and was complete towards mid-January, 1936. One plant cleaned itself completely and was in fair conditions at the end of the experiment.

Pseudococcus citri in vineyards.

A certain serial disease of vine, in which *Pseudococcus citri* is an important joint, was described by Mangin and Vialla as phthiriosis. A vineyard heavily suffering from phthiriosis gives a very similar impression as that attacked by the phylloxera. At first one or more plants die, preceded by signs of progressive degeneration. Death expands to all sides from this centre, the more centrally situated plants always suffering more than the more peripheral ones. 3 to 5 years at least pass from the first sign of degeneration until death.

The diagnosis is easily made as soon as the roots are laid bare. They are enveloped by a thick greyish white felt, formed by fungous hyphae. Cutting through this hyphal layer mealybugs are met, often in large numbers, between the felt and the roots. Cutting through further, the root galleries abounding with mites are seen, at least in the later stages of the infestation. The relation between these three partners and their relative importance is by no means definitely settled.

The fungus has been fully described by Mangin and Vialla as *Bornetina corium*. But a higher form of fructification, a grey mushroom, appears towards the end of winter in the vineyards. It originates from the felt-layer of *B. corium* and has been determined by Dr. I. Reichert as *Polyporus corium* M. and V. The injurious stage of the fungus is doubtlessly the *Bornetina*-stage. It is also by no means parasitical, and its hyphae do not penetrate into the roots, but feed on the honeydew excretion of the mealybugs. The felt is often 30—50 mm thick and one cylinder may weigh up to 1 kg. The spores are transported readily by cultivation, by ants and germinate on honeydew only.

This means that through the honeydew excretion the presence of the mealybug is the necessary condition for the development of the fungus. Whereas the mealybug was believed, until some years ago, to be a separate species (*P. vitis* Niedl.), it is now commonly accepted as being identical with *P. citri*.

Amongst the mites, the most important is the Tyroglyphid *Rhizoglyphus echinopus* Clap., which is small, white, up to 1 mm long. It seems to be pretty certain by now that this polyphagous mite, which is mainly saprophagous, attacks the roots only, after they have suffered beforehand from *Pseudococcus* and *Bornetina*. This is also true with regard to graps attacked by phylloxera. The plant dies when the galleries of the mite have reached the pith of the root. The last degeneration of the plant, the small size of the leaves, the small number and low quality of fruit, seem to be mainly the work of these mites. Whereas it is certain that healthy roots are not attacked, it is still doubtful whether they ever attack healthy plant tissues. But even should they

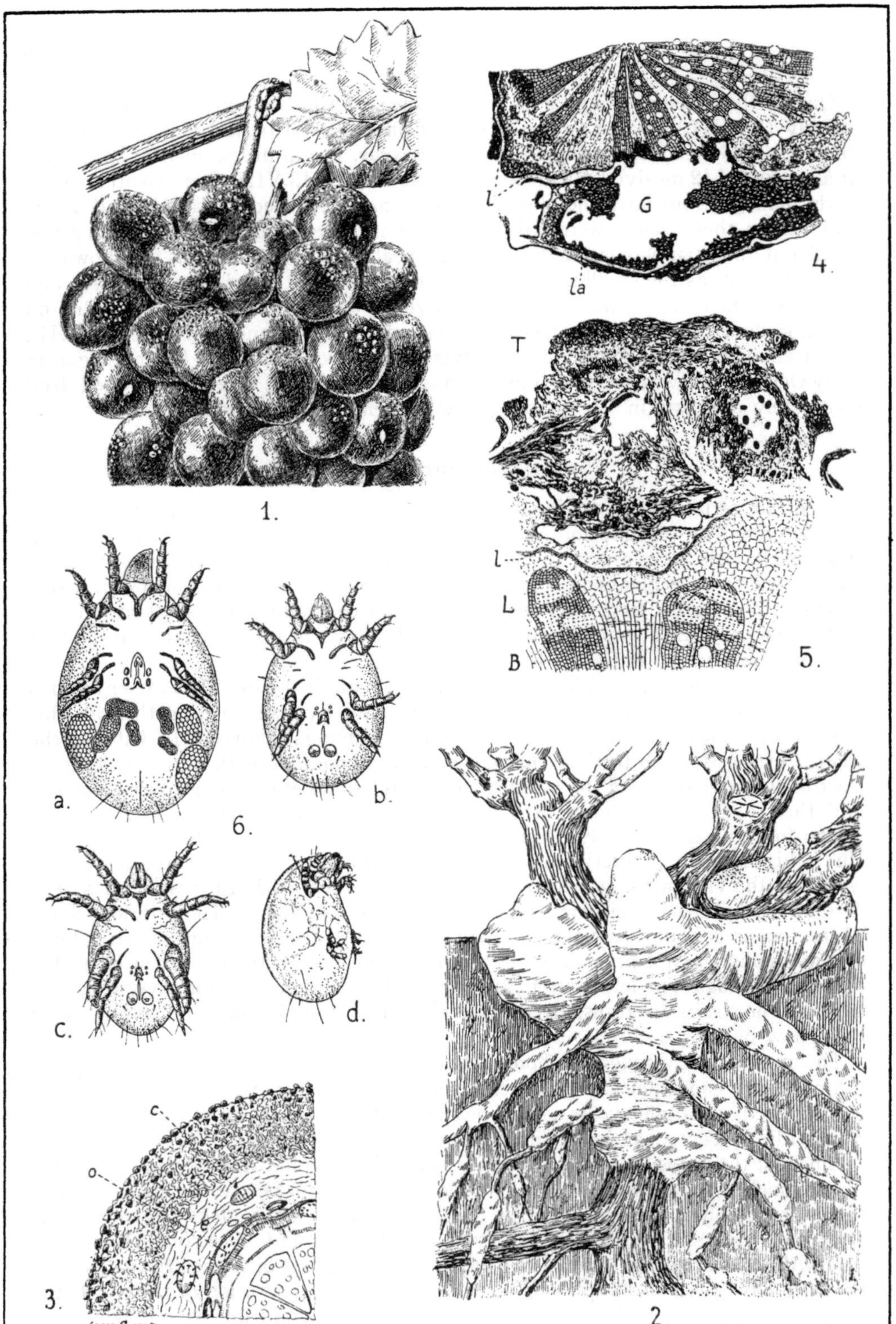

XXX. The Citrus Mealybug as pest of vineyards in Palestine. 1) Bunch of grapes covered with honeydew. The white ovals are adult females of *Pseudococcus citri*; 2) Vine-roots infested and covered with the felt of *Bornetina corium* (phthiriosis); 3) schematic cross-section through an infested vine-root in the later stage of phthiriosis (g = felt layer of Bornetina, c = mealybugs, o = galleries of mites); 4) and 5) cross-section through roots infested with phthiriosis and mites. (G = galleries of mites, L = pith, B = wood, l and la = cork, T = tuberosity); 6) *Rhizoglyphus echinopus* Clap. a) adult, b) homoeomorphous male, c) heteromorphous male, d) first nymphal instar. (Partially after Mangin, Viala and Foa.)

do so, their boring activity accelerates the appearance of lesions and decay in the neighbouring tissues, thus preparing them for future infestation. Quite a number of other saprophagous mites are found together with *R. echinopus* in the decaying roots.

The damage done by the sucking of the mealybugs is probably negligible. Its honeydew excretion provides a source of *Bornetina*-development. By cutting off the roots from water and gaseous metabolism, it weakens them, opening the way to the attack by mites. We are now under the impression that the slight primary damage caused by the *Pseudococcus-Bornetina* association is only preliminary to the more serious final damage inflicted by *Rhizoglyphus*. But no definite opinion can yet be given.

Phthiriosis occurs in Palestine mainly in the southern coastal plain (Mikveh Israel, Rishon Lezion, Rehoboth, Ekron). Not all varieties suffer to an equal degree. American root-stock is believed to be immune, as are some native varieties from the hills (Hebroni, Es Salti).

Reviewing the complete reports of Mangin and Vialla about the damages caused by phthiriosis to the viticulture of the Judean plain some 30—40 years ago, and comparing these reports with the author's own experience during the last 15 years, it becomes very obvious that the importance and intensity of phthiriosis have been reduced during all this period. At the present moment it has reached an economic insignificance which is quite independent of the lower interest in viticulture, in recent decades. 15 years ago, it was difficult to find acute foci of phthiriosis, and to-day this has grown even more difficult. The author has never seen dense felt covers 30—50 mm thick. Only very rarely, some 5—10 mm thick. Extended and progressive foci have never been met with in this period. This breaking down of phthiriosis is possibly connected with the extensive use of American root-stock, but probably more so with the secular rain-cycles, which have just reached the bottom of the dry period, whereas the last peak was 50 years ago. Sequences of rainy winter and — probably more important — springs favoured mass outbreaks of mealybugs in the vine-roots, possibly increasing also the relative amount of honeydew. The ample honeydew-excretion favoured luxuriant growth of *Bornetina*. When the felt was well established it probably helped to create a moist environment around the roots, thus permitting the mealybugs to propagate themselves. throughout the dry season on the roots, thus again stimulating the growth of *Bornetina*, etc.

It has been pointed out be Mangin and Vialla that phthiriosis is an adaptation to an arid climate. This seems to be justified only as far as in semi-arid regions only heavy attacks on roots are made. But this development on the roots is normally restricted to the rainy winter-season, being interrupted by some kind of estivation during the summer. Only after the association with *Bornetina* active development may proceed in summer. The root-infesting habit is very general in *P. citri* and on the roots of many plants during winter. Roots of *Polygonum*, *Thymelaea*, *Tamarix*, palms, etc. have been found heavily infested here, often attended by ants. In Kenya *P. coffeae* is even popularly known as the „Coffee Root Mealybug". Nothing peculiar can therefore be discovered in the root-infesting habit of *P. citri*

In the hilly region around Jerusalem and elsewhere heavy infestation of the grapes in summer has been observed in recent years. The mealybugs hibernate under the scaly bark of the trunk and the root collar. Here they oviposit in early spring. Later generations migrate to the green twigs and leaves. But mortality of the first two or three generations must be enormous, as they are never observed until mid summer. By then the leaves and grapes offer ample shade and a highly humid microclimate is probably found between the grapes. Here, and on the luxuriant food offered by the grapes, they soon multiply from late July to early September, and often cause heavy damage. The white wax-secretion of the mealybugs and the many sticky honeydew drops covering the grape, make the infested grapes unfit for table grapes. Damage to sprouts and leaves has not yet been observed. From the coastal plain only one attack of this type is reported by Mangin and Vialla from 1903. The main damage occurred in May and June. The preceeding winter and spring was especially rainy and the development of leaves and fruit abundant.

There are 5 to 6 annual generations in the hills. Of these only one tends to increase and to become injurious, the late summer generation. This is valid for Kiryath Anavim, where the climate is intermediate between that of Jerusalem and that of the coastal plain. Whereas all other generations suffer from low temperatures (winter), lack of shade (spring) or lack of suitable food (autumn), the late summer generation develops in heavy shade on optimal food and in an especially humid microclimate in the interior of the plant and of the grape.

We had entirely negative results with different winter controls. The parcels treated did not differ in intensity of infestation from the untreated check-parcels. Despite these negative results, it has become a routine work to paint the stem in early winter with a mixture of lime and white oil. This mixture does not enter either the crevices, or sufficiently under the scaly bark and certainly does not come in touch with the toots, where many individuals hibernate.

We have no indication that anything else than the microclimate of the plant, when grown on spalier or highly erected, creates the condition for mealybug increase. On vine which was grown spread over the soil, as was usual until some years ago, no *Pseudococcus* damage of some extent was observed. The damage appeared with the introduction of modern methods of cultivation.

It is worthwhile to attempt summer control only early in the summer before infestation has developed to any extent. Experiments are going on to study this question. Early harvest helps to depress the extent of damage, which develops mainly during the harvest season.

Conditions of microclimate and food govern the mass-fluctuations of the species. Enemies and diseases of mealybugs are very rare in vineyards.

THE CONTROL OF PSEUDOCOCCUS CITRI.

Technical control. Efficient control of *P. citri* by chemicals is rather difficult in citrus groves.

Fumigation with hydrocyanic-acid gas remains rather unsatisfactory, even when repeatedly applied. The killing concentration for the mealybug is higher than the tolerance of the tree. This general experience (Clausen) is contradicted by Balachowsky who regards fumigation as rather effective.

Strong pressure spraying with water has often been reported as satisfactory. But application of oil-sprays is certainly more satisfactory. Carter applied Diesel-oil in 1 % concentration with success. White oils (1 %—2 % concentrations are also useful (light oils preferable). Other sprays recommended are (Watson):

Water 100 gallons; lime sulphur solution 7 quarts; Nicotine sulphur 1 pint.
Water up to 50 gallons; soap 1½ pounds; Kerosene 3 gallons.
Water 99 gallons; crude carbolic acid 1 gallon; etc.

No experience with spraying exists so far in Palestine. Outbreaks of sufficient intensity to justify the expenditure are rather local. Both spraying and fumigation have not been tested yet on an economic scale. Some recent progress in control by spraying, applied especially against the mealybugs on grapes, will be discussed under *Pseudococcus comstocki.*

It seems that citrus mealybug presents one of the few cases where our present experience offers cheaper and safer means of biological control.

References.

S. Antebi, Hadar **1**, 1928, Nr. 7, p. 11.
C. G. F. H. Bayer, Over de Biologie van Pseudococcus adonidum (L.). Leiden. 1929. (Thesis).
A. Berlese, Rivista di Patol. Veget. **2**, 1893, pp. 70—109, 129—193.
F. S. Bodenheimer, Hadar **1**, 1928. Nr. 5 en Nr. 6.
F. S. Bodenheimer and M. Guttfeld, Zeitschr. angew. Entom. in 1929, p. 67—136.
F. S. Bodenheimer, Schaedlingsfauna Palaestinas. Berlin, 1930, pp. 130—141.
P. Buchner, Tier & Pflanze in intrazellulaerer Symbiose. Berlin, 19121, p. 240—250.
J. Carmin, Hadar. **11**. 1938. p. 131—138.
W. Carter, Phytopathology **23**, 1933, pp. 207—259.
W. Carter, Ann. Entom. Soc. America **28**, 1935, pp. 60—71.
W. Carter, Proceed. Hawaii Entom. Soc. **9**, No. 2. 1936, pp. 159—170.
W. Carter, Phytopathology **26**, 1936, pp. 176—183.
C. P. Clausen, Agric. Exper. Sta., Berkeley (Calif.). 1915. Bull. 258.
W. Emeis, Zool. Jahrb., Anat. **39**, 1915.
S. M. Fedorow, Plant Protection. Leningrad, 1935 fasc. 7. pp. 73—82. (Russian)
H. Cl James, Bull. Ent Res. **28**, 1937, pp. 429—461.
H. C. James, Proc. R. Ent. Soc. London A. **13**. 1938. p. 73—79.
E. Malenotti, Anz. Schaedlingskde **6**, 1930. pp. 79—83, 91—94.
L. E. Myers, Journ. Econ. Entom. **25**, 1932, pp. 891—896.
J. Prinz, Beitrage zur Biologie & Bekaempfung der Rebschaedlinge Tiflis, 1925, pp. 62—80.
J. Prinz, Anz. Schaedlingskunde, 1932, pp. 145—148.
F. Schrader, Arch. f. Zellforsch. **17**, 1923.
F. Schrader, Zeitschr. indukt. Abstamm. & Vererbungslehre, **30**, 1923.
Fl Schrader, Biol. Bull. Woods Hole, **45**, 1923.
H. C. C. A. A. Vos, De Invloed van Pseudococcus citri op de Plant, Baarn, 1930, (Thesis).
A. Walczuch, Zeitschr. Morph. Oekol, Tiere **25**, 1932, pp. 623—729.
Woglum, Calif. Citrograph. 1926, p. 317.
(Anonymous), Agric. Suppl. 11; Pal. Gazette 1936, p. 245—248.

The biological control of the citrus Mealybug.

Before discussing the various actual and potential insect enemies of mealybugs in Palestine, we may just mention the efficient seasonal control of mealybugs in tropical regions and in subtropical ones as well, whenever summer rains prevail. In moist heat entomogenous fungi develop rapidly and sweep out whole and dense mealybug populations in almost a few weeks. Speare reports on the effect of *Entomophthora fumosa* Sp. against *P. citri* in orange-groves in Florida (Winter Haven, 1921) as follows:

date	13.VI.	22.VI.	29.VI.	6.VII.	14.VII.	22.VII.	8.VIII.
% mealybugs killed by *E. fumosa*	10	19	63	79	80	87	94%

No entomogenous fungi have hitherto been observed on mealybugs in Palestine. Some species are probably present, but none is known to be able to produce an epidemic amongst the mealybugs under our conditions.

The more important local insect enemies are:

The Brown Mealybug-Lacewing *Sympherobius sanctus* Tjed.
Some Lady beetles: *Scymnus quadrimaculatus* Hbst.
S. apetzi Muls.
S. suturalis Thnbg., also *S. fenestratus* Schlbg. and *S. includens* Kirsch
Hyperaspis polita Wse.
Oxynychis marmottani Fairm. (= *H. pumila* Muls).

The Mealybug Gallmidge *Dicrodiplosis* sp.

The wing-banded Mealybug Parasite *Leptomastidea abnormis* Gir. and *Anagyrus pseudococci* Gir.

The Mealybug thrips *Karnyothrips flavipes* Jones and the predacious mite *Pediculoides ventriculosus* Newst.

An unsuccessful attempt has been made to introduce the ladybeetle *Cryptolaemus montrouzieri*. *Sympherobius* and *Scymnus quadrimaculatus* were bred before the war in special parasite breeding stations maintained by the Jewish Farmers Association at Petah Tikva, Hedera, etc.

a. *The Brown Mealybug-Lacewing* (*Sympherobius* (*Nefasitus*) *sanctus* *Tjeder*)

Syn.: Sympherobius amicus Navas. (ptm.).

During the summer of 1926 a brown lacewing appeared in numbers in our breedings. It was determined by K. Morton-Edinburgh, who wrote on 23.X.26:

„I am disposed to consider this insect as *Sympherobius amicus* Navas, which was taken on Olea chrysophylla, in Eritrea (Nefasit). The description of the wings fits exactly; any slight differences in the colour of the body being probably due to different degrees of maturity of the specimens. Perhaps these of yours are bred examples and a little less strongly marked and paler than old mature wild specimens would naturally be. Later, Navas, created for the species, a new genus wich he named Nefasitus. It differs from *Sympherobius* only in some trifling details.

In my collection I have a species from Funchal, Madeira, which at present I can hardly distinguish from your Palestine insect. I had determined it as *Sympherobius fallax* Navas, described from the Canary Islands. This would also fall into his genus Nefasitus and I am not at all satisfied that it is really different from *amicus*, but these small species of Navas are often difficult to determine in the absence of typical material as he neither describes nor figures the very important appendices of the male. If the two are the same *fallax* is the older name. It may be better, at least provisionally, to use *amicus*, looking at the widely separated geographical localities. However it must be kept in view that some of the Hemerobiids, more or less attached to certain trees of wide distribution, have an equally wide range, and moreover they are very likely liable to transportation by human agency on fruit trees and other plants." B. Tjeder, in 1939, separated the Palestine specimens as *S. sanctus* from the Erythraean species.

Description: The *adult* is very similar to the Common Green Lacewing (*Chrysopa vulgaris*) from which it can be easily distinguished by its colour. The body is of a bright yellow-brown, only eyes and antennae being dark brown. The wings, also, are not as slender as in the first named species. The wings are hyaline with darkbrown venation. They are covered with long bright yellow hair. The wings are roof-like folded during rest over the body of the insect.

Body length 5 mm, spread of wings 9 mm.

The recently laid *egg* is white and becomes darker with time. It is short elliptical, somewhat broader at its posterior end. The egg is not stalked and is pasted to the surface of a leaf, glass, wood, etc. They are laid singly or in very small groups.

Length: 0.3—0.5 mm, maximal breadth 0.1—0.2 mm, breader at posterior and.

The white, tender, hatching *larva* has the form of an elongate spindel. The colour becomes pink and red later, but the head with the long mandibulae and the legs become yellow-brown. When full-grown they are 6—8 mm long.

The white, later bright yellow-brown coloured *pupa* is slightly curved and 4—5 mm long. It rests in a long oval cocoon of white silk, which is somewhat longer than the pupa and is composed of two layers. The inner one is thin and tight, the outer one loose.

Some typical life-histories of this species from Palestine are recorded in the following table:

Typical life-histories of *Sympherobius sanctus* in Palestine.

No.	Date of oviposition	Egg-stage in days	Larval stage in days	Pupal stage in days	Date of hatching	Total development
1	6.I.27	25	45	26	12.IV.27	96
2	6.III.27	14	23	20	2.V.27	57
3	24.IV.27	10	9	12	25.V.27	31
4	1.V.27	7	7	16	31.V.27	30
5	1.V.27	7	12	13	2.VI.27	32
6	30.V.27	6	9	12	26.VI.27	27
7	13.VI.27	5	10	10	3.VII.27	25
8	23.VI.27	5	8	10	16.VII.27	23
9	7.VII.27	5	6	13	31.VII.27	24
10	20.VII.27	4	7	11	12.VIII.27	23
11	5.VIII.27	5	7	10	27.VIII.27	22
12	15.VIII.27	5	10	11	10.IX.27	26
13	25.VIII.27	5	8	12	19.IX.27	25
14	4.IX.27	5	7	13	29.IX.27	25
15	16.IX.27	6	10	11	13.X.27	27
16	23.IX.27	6	10	12	21.X.27	28
17	4.X.27	6	11	13	3.XI.27	30
18	17.X.27	7	12	16	21.XI.27	35
19	28.X.27	6	52	38	1.II.28	96
20	8.XI.27	7	20	40	3.II.28	87
21	18.XI.27	10	67	54	28.III.28	123
22	12.XII.27	18	43	47	28.III.28	108
23	25.XII.27	18	41	55	17.IV.28	114

The eggs are laid singly or in very small groups on citrus twigs in the immediate neighbourhood of *Pseudococcus*. The incubation period fluctuates between 3 days in summer to 25 days in winter.

The young larvae are very active and find their food even at relatively long distances. They moult twice. The shortest larval stage lasting 6 days is met with in summer, the longest lasting 67 days in winter.

Pupation takes place in crevices of the bark or in the angles of branchings. It lasts from 7 to 116 days.

The average total development of all those individuals which started their development in the respective month was in 1927:

Month	I	II	III	IV	V	VI	VII	VIII	IX	X	XI	XII
Average duration of development	83		52	33	28	25	24	25	27	51	106	118

Duration of development and early mortality, as observed in our Petah Tikva laboratory in 1937 were:

Development and early mortality of *Sympherobius sanctus*.

month	duration of development in days				from 100 eggs develop into adults in				
	egg	larva	pupa	total dev.	eggs	larvae	adults	♂	♀
XI.36	8 days	77	27	112	100	89	63	37	26
XII.	25	46	26	97	100	93	16	9	7
III.37	16	14	21	41	100	56	37	24	13
IV.	9	11	15	35	100	78	59	29	30
V.	7	13	20	40	100	67	55	37	18
VI.	5	6	11	22	100	92	48	36	13
VII.	5	11	7	23	100	86	63	40	23
VIII.	4	10	9	23	100	83	61	33	28
IX.	4	9	10	23	100	88	41	22	19
X.	4	9	11	24	100	64	46	26	20
XI.	5	30			100	86	43	27	16
XII.	9				100	89	57	37	20
I.38					100	78	34	15	19
II.					100	68	35	17	18
III.					100	74	48	29	19

The average longevity of the adults was 26 (12—70) days for the females, 37 (10—101) days for the males in 1928:

month	♀ longevity	♀ food *)	♀ n	♂ longevity	♂ food *)	♂ n
IV	23 days	82	2	22 days	18	4
V	25	121	4	—	—	—
VI	26	244	2	—	—	—
VII	12	68	1	—	—	—
VIII	21	62	2	37	70	3
IX	35	108	3	60	74	1
X	22	82	5	52	36	4
XI	45	36	2	40	36	6

*) i.e. number of *Pseudococcus*-females devoured.

Some recent figures on egg-production and adult longevity of *Sympherobius* are compiled in the following table (Petah Tikva 1937):

Egg-production and adult longevity of *Sympherobius sanctus*.

n	hatching of adults	no. of eggs per ♀		adult longevity ♂		adult longevity ♀	
		min/max	average	min/max	average	min/max	average
2	25—27.X.36	144—223	183	35—46	41	53—70	62
1	1.XI.	504	504	140	140	99	99
10	15.—17.XI.	51—398	229	62—140	111	53—107	73
1	1.XII.	481	481	88	88	88	88
1	23.II.1937	297	297	63	63	78	78
8	1.—9.III.	19—424	238	28—53	43	28—68	49
10	6.—7.IV.	23—520	235	16—74	46	14—61	40
10	14.—16.V.	326—603	429	35—57	49	31—57	42
10	15.VI.	82—349	233	27—40	34	14—43	28
10	15.VII.	5—314	108	21—46	33	10—38	27
10	15.VIII.	20—306	108	21—53	38	26—49	38
9	14.—16.IX.	11—207	77	40—55	48	17—59	42
9	17.—21.X.	48—313	241	41—123	76	21—80	58
7	26.—30.XI.	12—431	168	91—128	112	68—119	101
3	5.—6.XII.	242—272	257	95—127	113	73—121	105
2	23.—24.I.38	6—143	75	48—67	58	27—67	47
8	1.—15.II.	22—198	112	56—107	78	16—116	63
10	6—24.III.	1—174	89	14—64	45	31—70	53

No true hibernation is observed in *Sympherobius*. Oviposition generally stops in mid-December (with only 5 exceptions) and starts again in the beginning of March. It is thus interrupted for 2½ months. The hatching of the eggs continues uninterruptedly, until mid-January. The larval development continues at a very slow rate only, but without interruptions or a diapause. Pupation likewise continues uninterruptedly. The hatching of adults, however, stops in mid-November (7 exceptions only) and begins again in the last days of February. It becomes more frequent from the beginning of March on. The winter is thus passed mainly in the larval and pupal stage, which last from December to February, 50, 48 and 77 days respectively for the larval and 71, 49 and 40 days respectively for the pupal stage. Whilst all processes from the hatching of eggs to pupation continue permanently, hatching of adults and oviposition are stopped during the winter season.

Sympherobius sanctus has 8—9 annual generations in the coastal plain, i.e. 2 generations more than *P. citri*. The length of its winter generation (94 days) is almost identical with that of the Citrus Mealybug (93 days). Its total development in the summer lasts only 3—4 weeks. The computation of the hyperbola is based on:

34 days development at 20.8° C
22 ,, ,, ,, 26.4° C

development threshold (c) = 10.5° C
thermal constant (th.c.) = 350 days-degrees.

This hyperbola coincides remarkably well with the actual breeding data:

Temperature °C	Total development in days			n
	Average	Maximum	Minimum	
13.3	109	125	97	40
14.4	121	160	66	26
16.5	57	73	52	8
17.3	54	60	50	8
18.7	50	57	40	22
19.5	49	55	37	4
20.6	35	40	28	24
21.6	31	40	22	40
22.1	30	32	28	32
23.7	28	32	24	9
24.8	26	33	24	31
25.8	27	31	22	52
26.7	24	27	21	66

This hyperbola is computed for the epidemiological average of each generation, not for the shortest development. An interval has to be intercalated for the average preoviposition period, which we put at 5 days, despite its shortest duration of only 1—2 days. The band of three hyperbolae runs, as follows:

Hyperbola		A	B	C
Development-threshold (c.)		10.5	10.5	10.5° C
Thermal-constant (th.c.)		350	463	567 day-degrees
Duration of development at	30° C	17.9	23.7	29.1 days
	25	24.2	31.9	39.1
	20	36.9	48.7	59.7
	15	77.8	102.9	126.0
number of annual generations at Tel Aviv		9.8	7.4	6.0

It seems that of our whole local enemy series of *Pseudococcus S. sanctus* is best adapted to our climate. It has further no important enemies, which reduce the value of its near relatives in California for biological control. Cannibalism occasionally occurs among the larvae in breedings and feeding on the eggs by adults. But these phenomena are probably rare under natural conditions, in absence of crowding. No local parasite of *Sympherobius* is yet known.

Of among all insects analyzed in this chapter, *S. sanctus* is the only one which best survives the hot summer months. It is sensitive, especially in the larval and pupal stage, to the drying hamsin winds. All development stages die soon after having been exposed to direct sun-rays. In the average of our breedings, 45 % of the eggs developed into adults, during no season less than about 20 %. Careless handling, cannibalism in

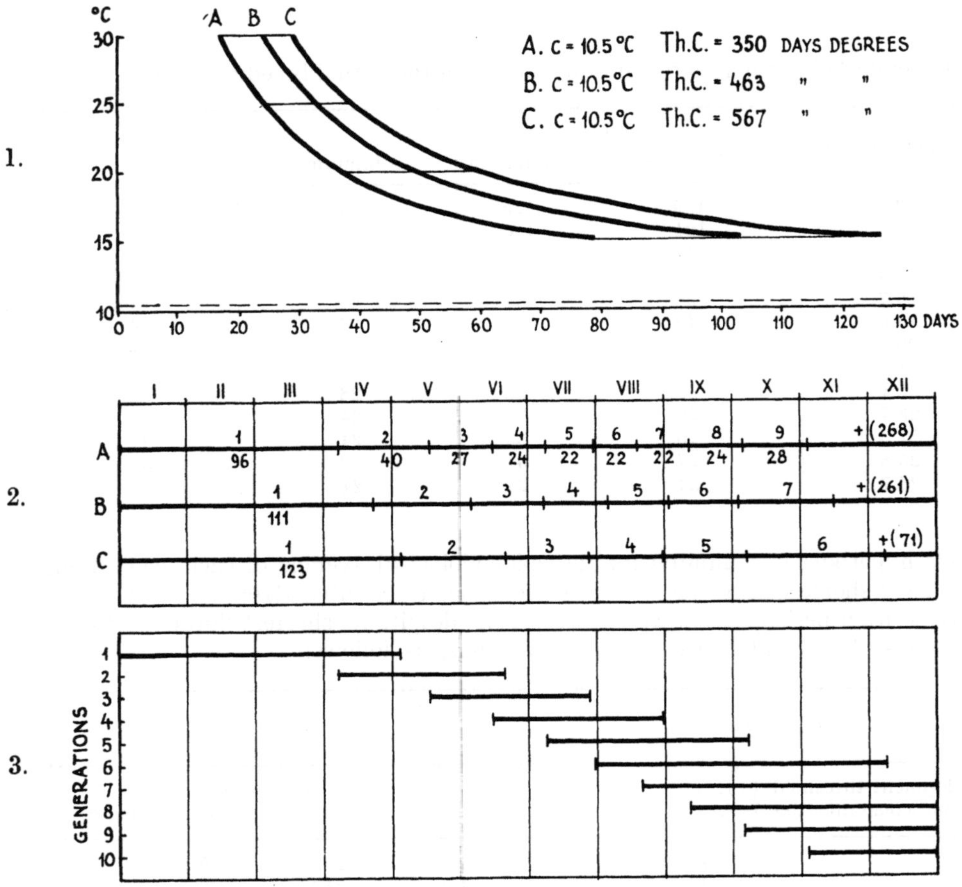

Fig. 101.

Life history of the Brown Lacewing (*Sympherobius sanctus* Tjed.) at Tel Aviv. 1) The three hyperbolae; 2) the life history as calculated from the three hyperbolae; 3) sequence and overlapping of generations.

crowded conditions, etc. have certainly increased the natural mortality in our breedings. The winter is also fairly well passed in our climate. The average of successfully developing eggs is as follows:

month	I	II	III	IV	V	VI	VII	VIII	IX	X	XI	XII
% of successful eggs . . .	60	—	35	**23**	39	52	44	48	**64**	43	34	40

The mortality in every stage is:

	From 100 eggs still survive	Mortality in every stage
egg	100.0	10.5
larva.	89.5	**45.7**
pupa.	48.6	15.1
adult.	41.0	—

Fig. 102 illustrates again that *Sympherobius* is even better adopted to our climate than its host.

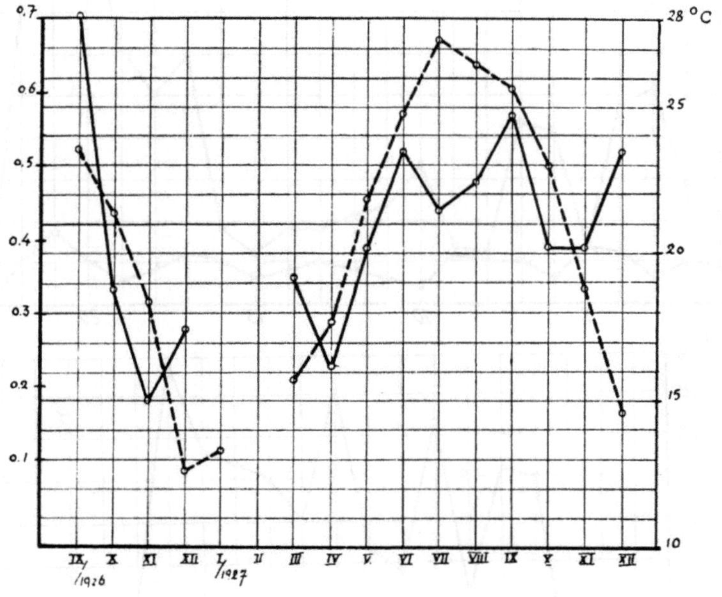

Fig. 102.
Temperature (---) and seasonal mortality of *Sympherobius sanctus* Tjed. (from Bodenheimer and Guttfeld).

In all experiments which aimed to determine the quantity of food of larvae of adults of *Sympherobius* we provided only adult females of *Pseudococcus*. But under natural conditions both of these stages, and especially the larvae, devour also eggs and young larvae of mealybugs, and those in much greater numbers. Their true biological utility is therefore higher than could be concluded from these figures.

The larvae crawl very actively on their search for food, spying with their head to all sides and finally rushing to the discovered prey. They prefer ovisacs to adult mealybugs. They enter the ovisac and remain there until the eggs are sucked empty more or less radically. Larvae and adult mealybugs are attacked from the ventral side and slowly sucked empty. Sometimes 3—5 larvae of the brown lacewing were observed beneath one large female of *Pseudococcus*, which was sucked empty by their final attack until only the empty skin remained.

The table on the following page gives some information as to the total consumption of the adult *Sympherobius*. The average daily food was 3.9 for the females, 1.0 for the

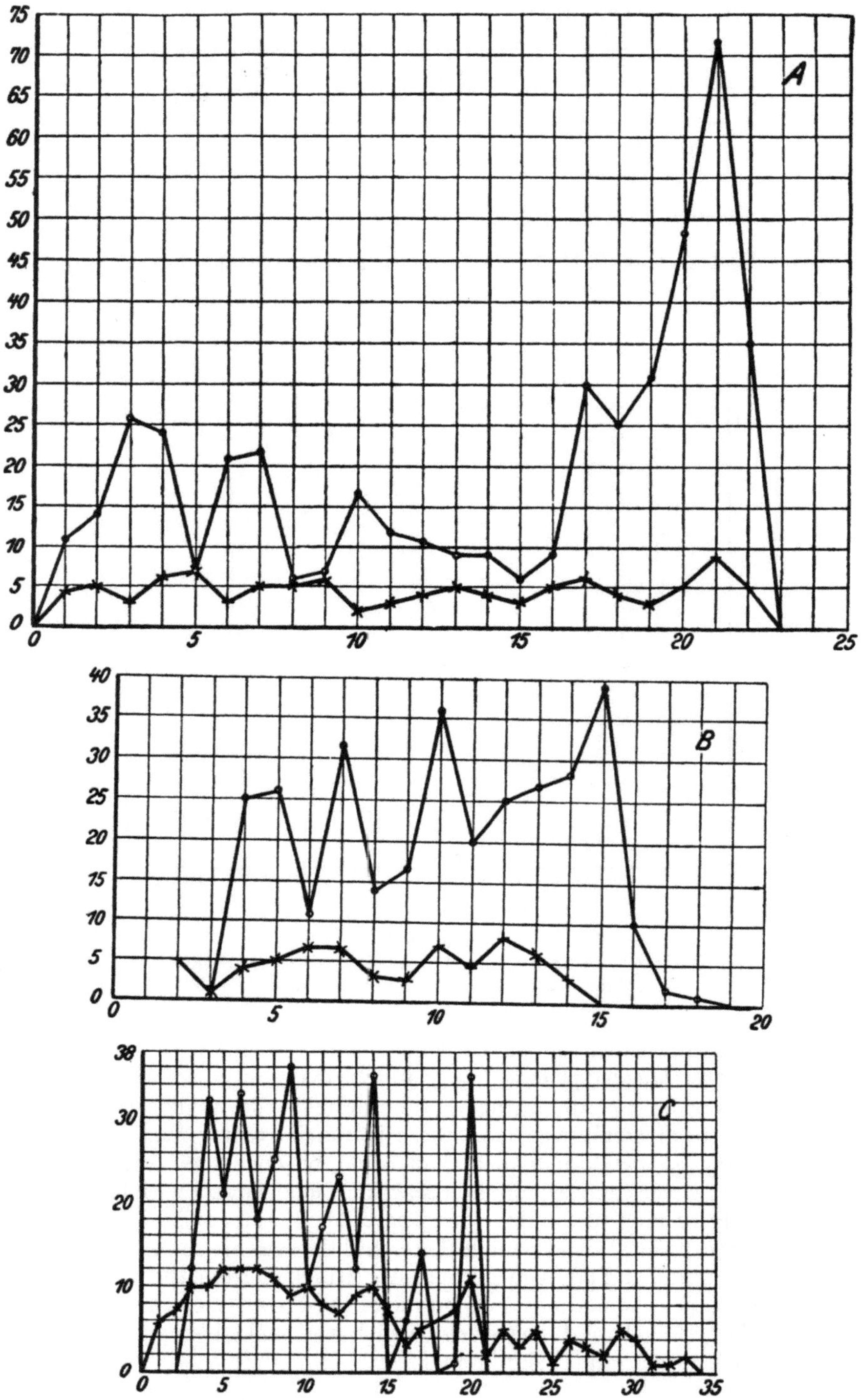

Fig. 103.
Egg-production and food quantity of three females of *Sympherobius sanctus* Tjed. The abscissa designates the adult longevity of the females, o——o the number of eggs deposited every day, x——x the number of *Pseudococci* devoured every day. (from Bodenheimer and Guttfeld).

males. During the total longevity of the adults, this makes 111 mealybugs per female or 37 per male.

There exists a direct correlation between length of development and quantity of food. In summer and autumn about 25 mealybugs are the ratio per total development, in winter up to 67. 43 mealybugs are the average of 22 series of observations.

The consumption of the larvae is more or less equal to that of the adult male, but that of the females is decidedly higher.

The adults easily endure some days of hunger. From 12 freshly hatched adults left without food (VIII.27) the following died:

day	1	2	3	4	5	6	7	8	9
still living	12	12	12	9	9	4	2	1	–

Copula takes place during the night, in the evening or early morning. Oviposition generally starts during the first or second day after the hatching of the female. But occasionally it may be delayed to the fifth day (in one case it even occurred only on the 16th day). Parthenogenetic reproduction has never been observed in many isolated females. Not a single egg has been laid under such conditions. But when a male was introduced — even after 10 days isolation — oviposition started immediately on the next day.

Details on the oviposition of 22 females are compiled in the following table. No. 22 is anormal, as at that season normally no oviposition takes place anymore.

Mealybugs are always the preferred food of *Sympherobius*. Aphids are devoured (4 specimen per day), when no other food is available, also *Icerya* to some degree. Neither *Ceroplastes*, nor *Aonidiella* or *Aspidiotus* have been touched. On the aversion against *P. comstocki* later!

No. 1	Living 2	Longevity in days 3	Oviposition period in days * 4	Effective oviposition days 5	No. of eggs 6	No. of eaten Pseudococcus 7	Oviposition per day 8	eggs consumption 9
1	21.X.—13.XI.26	23	18	17	254	66	14	3.9
2	21.X.—14.XI.26	23	17	14	314	58	18	6.4
3	21.X.—17.XI.26	27	22	19	365	100	17	3.7
4	21.X.—13.XI.26	21	13	8	262	62	20	4.9
5	22.IV—17.V.27	25	25	22	243	105	18	4.2
6	22.IV.—11.V.27	20	10	16	144	58	8	2.5
7	12.V.—24.V.27	12	11	10	176	46	16	3.7
8	11.V.—24.V.27	13	13	11	181	52	14	3.5
9	26.V.—30.VI.27	35	21	16	192	183	9	1.4
10	26.V.—4.VII.27	39	20	16	338	205	17	2.1
11	9.VI.—11.VII.27	32	29	25	589	282	20	2.1
12	11.VI.—2.VII.27	21	20	18	522	200	26	2.6
13	19.VII.—31.VII.27	12	11	11	198	68	18	2.9
14	20.VIII—15.IX.27	26	20	15	260	104	13	2.5
15	20.VIII.—4.IX.27	15	11	8	144	60	13	2.4
16	12.IX.—15.X.27	32	29	20	190	152	7	1.2
17	13.IX.—21.X.27	37	12	11	69	119	4	0.8
18	29.IX.—3.XI.27	35	27	16	208	125	8	1.6
19	12.X.—27.X.27	15	11	7	132	52	12	2.5
20	2.XI.—21.XI.27	19	17	11	177	54	10	3.3
21	3.XI.—22.XI.27	19	17	11	177	53	10	3.3
22	22.XI.—31.I.28	70	28	12	42	78	2	0.5

* Longevity minus preoviposition and postoviposition period.

This gives 254 eggs per female and roughly 10 eggs per day of adult life.
Seasonal correlations of egg-productions are:

Months	Longevity	Effective oviposition	No. of eggs	Mealybugs eaten	eggs/food
X/XI.26	24 days	15 days	299	72	4.2
IV/VII.27	25	16	319	148	2.5
XI.27/II.28	26	12	171	92	1.9

The highest egg-production is in late spring and early summer. The most economic use of food for egg-production is made in the fall.

Of the four combinations between food consumption and egg-production, all four were met in our observations:

I. The egg-production lasts until death
 A. Feeding continues until death 9 times
 B. Feeding stops earlier 3 times
II. The egg-production stops before death
 A. Feeding continues until death 8 times
 B. Feeding stops earlier 2 times

The tendency is to continue feeding, independently of egg-production. Some of these correlations are illustrated in Fig. 103.

During the months July to September 1927 we liberated 4500 adults of *Sympherobius sanctus* in six orange groves in Petah Tikva.

At the beginning of August strong foci of mealybugs started again in the grove Ga. The attack was not smaller than that in the previous year. 800 adults were liberated and started immediately a very intense activity. As soon as one small infested spot was cleaned they fled to other fruits and branches. Their activity was occasionally disturbed by ants, though the latter could not interfere with the feeding of Sympherobius in a radical manner. But a heavy attendance of mealybugs by ants is not desirable. During the autumn of 1927, no fruit-fall whatsoever was observed in this grove and a thorough inspection in November showed that all foci of *Pseudococcus* were empty and cleaned and only some scattered individuals were living. As the activity of the Mealybug Gallmidge was very low in this orchard, this radical success must be attributed entirely to *Sympherobius.*

In a second orchard Ja the success was somewhat obscured by heavy fruit drop, caused by the citrus fruit moth (*Cryptoblabes gnidiella*). 600 lacewings were liberated in August. The number of 20—30 fruits which have dropped in the previous year was greatly reduced this year. And every fallen orange showed the symptoms of *Cryptoblabes.* Only very few scattered mealybugs were still present in November.

In two other groves which showed only a mediocre infestation, every focus was cleaned 4—6 weeks after the liberation of 400 lacewings in each in August.

A very heavy attack was reported to us only at the end of August. It extended over an area of more than 10 dunam in the grove Po. 1600 adults were liberated here during the following month. Despite their small number they cleaned the orchard very effectively. The dropping of oranges ceased pretty soon. In November a small mealybug population was still scattered over the orange grove, but any serious damage — as that experienced in previous years — has been effectively prevented. These few samples are a good illustration of the easy and thorough control and prevention which the biological control by *Sympherobius* offers in late summer and autumn. They help to diminish the damage of the mealybug in the following spring by reducing strongly the population level of *Pseudococcus* in these seasons.

The most handy recepticle for mass breeding of *Sympherobius* are test-tubes. Jam jars or larger vessels make the handling of the young larvae more difficult and when opened, the adults may easily escape. Small cardboard boxes are likewise unfit, as the young larvae leave them immediately without any difficulty. The test tubes are closed with a stopper of cotton, wrapped in gauze. If the gauze is left out, many of the young larvae are caught in the cotton and die there. If the test tube is shaken before handling, all larvae easily fall down from the gauze of the stopper.

The breeding starts by putting one or a few impregnated females into the test-tube. A wooden splinter (2—5 cm long, 1 m thick) is entered, because *Sympherobius* prefers ovipositing there and the eggs can easily be handled on this substrat. Some mealybugs must always be present as food. When lacking other food, the females devour their own eggs.

Eggs from the same patch hatch in summer more or less contemporaneously. Some mealybugs should be entered into the test-tube one day preceeding the expected hatching of the young larvae, in order to prevent cannibalism. It is advisable not to open the tube during the first two days after the hatching of the larvae, in order to avoid unnecessary losses of the young larvae, which are not yet easily seen. After the first feeding, the larvae change from white to pink and red. New food, in sufficient quantity, should be added every one or two days in the warm, every 4 to 6 days in the cooler season. The *Pseudococcus* is introduced into the test-tube with a brush or on a needle.

Pupation generally occurs between the wooden splinter and the glass or between the stopper and the glass. Many larvae often pupate together on the same spot. During winter pupae lacking the pupal cocoon have occasionally been observed.

The test tubes need no moistening. High humidity, as that induced for example by parts of living plants which are included in the tubes, is fatal to the lacewing larvae. Exposure to direct sun-rays is likewise fatal. On hamsin days, which are highly unfavorable for all stages, the tubes may be put into a light cardboard box wrapped in moist gauze.

Mass hatching is easily obtained in the large breeding cages. From May on, 20—60 adults may be collected in each cage every day. They are caught easiest in test tubes in the evening between 7 and 11 o' clock. when they sit quietly on the gauze of the breeding cages. 20—30 adults may be caught in one test tube, into which some mealybugs have been previously entered. The few remaining adult lacewings are easily caught the next morning. These adults are ready for liberation in the orchard, but may also be kept for 2—3 days in captivity, if desirable.

It is very important to collect from the cages as many adults as possible, every day. In this way only a very small fraction of the eggs is laid on the mealybugs in the cage and a regular development of *Sympherobius* is therefore secured for some months. If too much oviposition is permitted in the cage, the whole reserve of mealybugs is quickly destroyed. One obtains a mass-appearance of *Sympherobius* in a period when there is possibly no use for them and must therefore build up the infection of the cage from the beginning.

It is very troublesome to have *Sympherobius* populating the room intended for the breeding of the mealybug trays. It may then eventually interfere radically with the preparation of sufficient food for the maintenance of *Sympherobius* or other beneficial insects. During a heavy infestion of our preparatory trays in July/August 1926 and 1927 no infestation of the potato sprouts with *Pseudococcus* was possible. Every artificial breeding was quelled at its beginning.

Under such circumstances it seems necessary to breed the preparatory trays in cupboards with glass doors. The latter are besmeared with oil, and during the night a burning lamp should be put before the cupboard and the adult *Sympherobius* to be disturbed by knocking on the cupboard. They are then disturbed, fly towards the light and are caught by the oil on the glass. This oil-paint should be renewed every 4—6 days. Before starting this manipulation, the cupboard may be searched for adults in the regular way and this crop is ready for liberation in the orchard.

It seems to be impossible to have permanent mass-breedings of *Sympherobius* on open trays under the modest room conditions of our laboratories.

We can be brief in discussing a paper of Rivnay (1943) on *Sympherobius*. Inspite of a series of sweeping statements it is mainly based on bad technique. As our breeding experiences of 1926/28 are fully confirmed by those of 1936/38 we are able to make unhesitatingly the following statements:

1. R. points out that in two groves both which were heavily infested by *P. comstocki* (sic!) he collected during 1939/41 the following numbers of enemies of the mealybug:

Enemies	Grove A.	Grove B.
Various parasitic *Hymenoptera*	616	114
Scymnus spp.	668	84
Sympherobius	37	5

In grove A over 1000 individuals of *Sympherobius* were liberated in 1938. „Both in the laboratory and in the field it was found to feed unhesitatingly on *P. comstocki*. However the outcome of the attempt to use it for the biological control of P. *comstocki* was not satisfactory."

This statement has to be compared with that of Klein and Perzelan (1940): The brown lacewing (*Sympherobius*)„devours *P. comstocki* only when deprived of all other food and even then only in small numbers." What importance has under these circumstances the small number of *Sympherobius* in a garden which was heavily infested by *P. comstocki* only, and where all typical predators and parasites of the latter species swarmed in correspondence with the abundance of their host? It just confirms completely the statement of Klein and Perzelan.

2. Another sweeping statement of R. „Availability of food is not the primary reducing factor of *Sympherobius*, as *P. comstocki* (sic!) was present in abundance" falls dawn by the same argument.

3. When we compare the averages of R.'s breeding tables about adult longevity of 8 to 33 days, egg-production per female 15—165, larval and pupal mortality of 70—95 % plus an egg-mortality of 25—80 %, etc., with our own results the conclusion it unavoidable that the breeding technique was extremely poor. That R. obtained an 80 % egg-mortality after a 5 hours exposure to 42° C (sic!) and high larval mortality under similar conditions will astonish nobody. The contrary would be more astonishing: His experiments created conditions more similar to those of a concentration camp than to those suitable for a study of natural environment. Equally inadmittable is the attempt of R. to add all monthly hours of high temperatures and low humidities into one additive diagram, because intermittent high temperatures of 30 to 40° C spread over a number of days are very different in their effect from a massive exposure during 20 to 40 hours.

That mortality is raised during the spring hamsins is not new. It was clearly stated in our first publication (1929) and the extent of this mortality is clearly demonstrated in our tables.

4. R. was never able to collect *S. sanctus* from citrus mealybug on vine in various parts of the country. We actually concur with him and no such occurrence has ever been claimed by us. To the contrary, the tropical affinity of this lacewing characterises it as originating from savanna with trees with shady crowns. And such is the typical biotop in citrus crowns in Palestine. The bare, shadeless vineyard is a biotope where no experienced ecologist would ever look for it, except perhaps in a much more humid climate.

5. We have seen that neither parasites nor climate are the major limiting factor for *S. sanctus* in Palestine. The truth is that — in spite of R.'s inadequate experience on *P. comstocki* — the host abundance, i.e. the abundance of *P. citri*, is the main factor determining seasonal and annual and local abundance of the lacewing. In periods and foci of scarcity of the host the lacewing grows still rarer. Yet it quickly builds up high populations when the host abundance suddenly increases. It is therefore definitely worthwhile to breed *Sympherobius* permanently when one wishes to have it ready for outbreaks, even if R. could seem no use in this for *P. comstocki*, which is not its proper prey and for which its utility never has been claimed by anybody.

We have selected only a few points among the many of that paper which are equally provoking criticism, but which are even less substantiated than those remarked on.

b. The Australian Mealybug Ladybird (*Cryptolaemus montrouzieri Muls.*)

The excellent results which were reported on the biological control of mealybugs by *Cryptolaemus montrouzieri* in citrus groves in California encouraged us to attempt the introduction of this ladybird into Palestine. It had been established with little difficulty some years previously in the French Riviera and large scale attempts were undertaken in the same period to establish it in Egypt.

We received the first consignment from Mentone in the winter of 1924/25 by the courtesy of Dr. Poutiers. But none of about 30 beetles survived the transport during that unfavorable season. Prof. Efflatoun-Bey-Cairo was kind enough to arrange for the transmittal of 30 beetles and larvae from the breeding-station of the Egyptian Ministry of Agriculture at Gizah. This consignment arrived in excellent condition and all our breedings descended from these individuals.

The breeding was performed according to the method described by Smith and Armitage, which will be described in detail later on. The work was started at Tel Aviv, but was soon transferred to Petah Tikva. The artificial breeding made no difficulty, except during 1927, when special conditions prevailed.

The species is a native of Australia, where it feeds on *Pseudococcus aurilanatus* Mask. and on other mealybugs.

Description: The long, ellipsoid adults have the general form of ladybirds, but the abdomen is decidedly pointed posteriorly, when compared with other species. Head, pronotum, the end of the elytra (except a narrow black margin) and abdomen are cinnamon red to orange-coloured. Elytra, which are covered with pubescent hair, body and legs are black, the first pair of legs being testacous. Our specimens were 4.9—5.3 mm long, 3.2—4.0 mm broad. They were larger than those measured by Poutiers at Mentone (4.0 × 2.5 mm). But some dwarfish specimens also occurred in our breedings (ex. g. 3.8 × 2.8 mm).

The *eggs* are oval and lemon yellow; 1.0 × 0.3 mm. They are deposited singly among the eggs of the mealybugs, where they are distinguished easily from those of its host by size and colour.

The true, yellowish green colour of the larva is very soon covered with heavy, white wax-filaments. These excretions make it sometimes difficult to distinguish the young larvae from mealybugs, at first sight. In its later stages, the size of the *Cryptolaemus* larva is much larger than that of an adult mealybug. It remains there until the first moult (5 mm length), when it has more or less emptied the ovisac and may start for another prey until the second moult (8 mm).

For *pupation* it retires into crevices of the bark, etc. The pupa has the normal shape of a Coccinellid pupa and remains in the split skin of the last larval instar.

The following table compiles the shortest development in 18 of our breedings.

The eggs are laid singly among or within the egg-masses of *Pseudococcus,* where the hatching larva finds plenty of food. The shortest incubation period fluctuates from 6 days in summer to 24 days in early spring. The prolongation of the egg-stage in the extraordinary hot summer of 1927 is remarkable.

The young larvae are relatively lazy and probably perish in nature, when they do not meet food in their immediate neighbourhood. They moult twice. Their development lasts 2—3 weeks on an average, fluctuating between 31 days in April and 9 days in July.

Pupation always takes place only on the shady and dark parts of the cage. The pupal stage generally lasts 1—2 weeks, from 39 days in winter 1925/26 to 7 days in October 1925.

Hibernation occurs only in the adult stage. Oviposition stops from beginning of December on. Large larvae and pupae still continue to develop and the hatching of beetles has been observed until the beginning of January. The pupae which have not hatched until that time, generally rot before the spring. Also young larvae, etc. die during the winter. The oviposition of the beetles starts again in March. The first larvae appear in the beginning of April. The mortality of the hibernating beetles is very high.

Life-histories of *Cryptolaemus montrouzieri* in Palestine.
(Shortest development per breeding).

No.	Date of oviposition	Egg-stage	Larval stage	Pupal stage	Total development
1	2.X.25	14 days	15 days	8 days	37 days
2	22.IV.26	23	28	4	55
3	25.V.26	24	16	8	48
4	20.VIII.26	9	21	14	44
5	23.VIII.26	6	27	9	42
6	3.X.26	16	16	33	65
7	7.X.26	8	20	22	50
8	19.X.26	12	18	35	65
9	2.XI.26	16	30	39	85
10	3.V.27	12	15	13	40
11	3.VI.27	9	14	14	37
12	3.VI.27	9	25	17	51
13	23.VI.27	11	10	13	34
14	30.VI.27	13	17	7	37
15	10.VII.27	10	9	10	29
16	21.VIII.27	9	23	17	49
17	15.IX.27	13	16	10	39
18	20.IX.27	19	12	24	55

The average longevity of the adult beetles in different months was as follows:

month of hatching	average longevity	n	Maximal longevity
VI	45 days	28	94 days
VII	30	12	61
VIII	32	17	41
IX	53	16	79
X	57	18	76
XI	96	12	174
XII	106	30	168

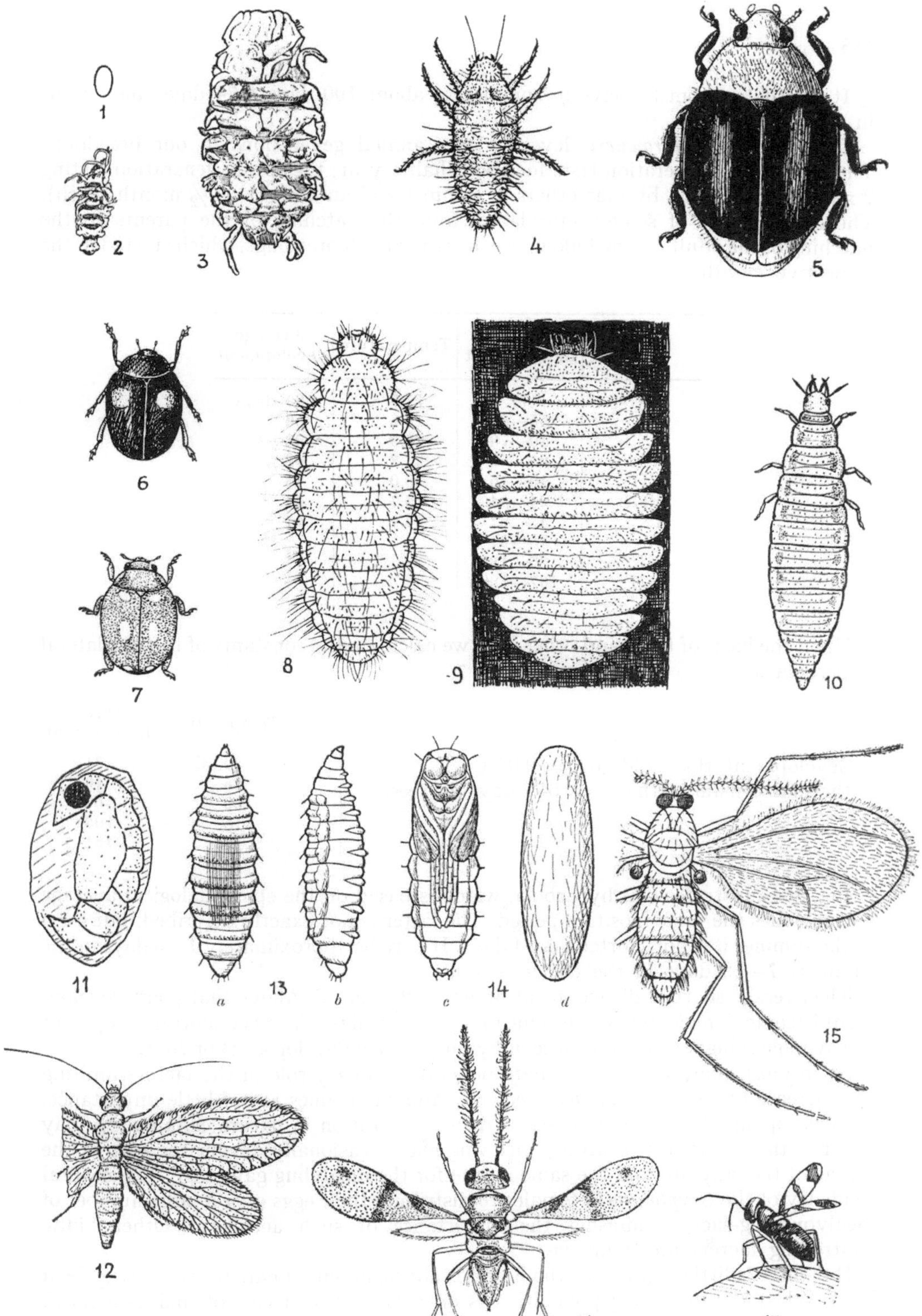

XXXI. Parasites and predators of the Citrus Mealybug in Palestine. 1—6) *Cryptolaemus montrouzieri* Muls. 1) egg; 2) neonate larva; 3—4) larva III—I; 5) adult; 6) *Scymnus quadrimaculatus* Hbst.: beetle; 7—9) *Hyperaspis pumila* Muls. 7) and 8) larva bare and covered with wax-secretions; 9) beetle; 10—12) *Sympherobius sanctus* Tjed. 10) larva; 11) pupa in cocoon; 12) adult. 13—15) *Dicrodiplosis sp.* 13) larva; 14) pupa; 15) adult; 16) and 17) *Leptomastidea abnormis* Gir. 16) adult female; 17) ovipositing female. (Mainly after Bodenheimer and Guttfeld).

It fluctuates from 30 days in summer to about 100 days (174 days maximum) in winter.

Cryptolaemus montrouzieri developed six annual generations in our breedings: The hibernating generation lasts for about half a year; the spring generation lasting 2 months is followed by four others in summer and autumn (1—1½ months each). The development of some beetles lasted from the hatching of the parents to the hatching of the adults of the following generation in all breedings, which started in the respective month:

Month	Days development	Temperature	Average development
III	77	17.9° C	85 days
IV	55—66	19.6	56
V	40—79	20.0	53
VI	29—52	24.3	42
VII	40—55	25.0	48
VIII	24—54	25.7	46
IX	32—62	26.7	44
X	38—97		
XI	85—184		
XII	167—189		

Upon the basis of these and other data we calculate the constants of the equilateral hyperbola as:

development threshold (c) = 9.0° C
thermal constant (th.c.) = 645 days-degrees.

Temperature	Days development
30° C	31
25	40
20	57
15	108
13	161

Fig. 104 illustrates this hyperbola, which is based on the epidemiological average and includes the preoviposition period. The latter is not exactly described, but even in the summer it is not shorter than 4 days. It may be approximately 4—6 days in the summer, 7—12 days in the cooler seasons.

The average shortest development occurs at 25—26° C. Above that point development becomes longer and even if some individuals still reach a very short development the average is higher than for those individuals which develop at 24 or 26° C.

As in most analyzed cases, climate plays the primary role in the factors limiting the increase of *Cryptolaemus montrouzieri*. Animal enemies are of little importance. In overpopulated cages or in cases of dense oviposition in nature cannibalism may occur in the larvae, the mealybug lacewing may occasionally devour the eggs or the larvae of the ladybird and the same is true for the mealybug gallmidge. The natural tendency of the *Cryptolaemus* females to distribute their eggs over a large number of mealybug-egg-sacs, diminishes the importance of such accidents. Other biotic controlling factors are unknown.

But there is little doubt that the beetle is not fit in our climate for the ,,struggle of life." When the mealybug population is on a low level, it cannot maintain itself. In considerable centres of infestation it increases in autumn satisfactorily and disappears entirely during the winter. The young larvae perish, when mealybugs are not in their immediate neighbourhood and the adult also is very lazy in its search for food, even during the summer. They die, instead of starting a survey flight in the

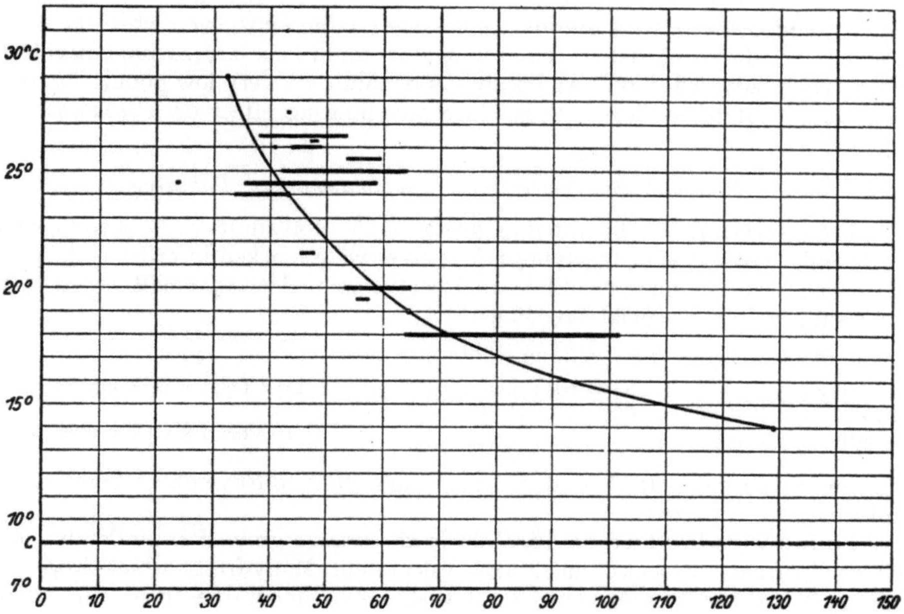

Fig. 104.
Temperature dependency of duration of development of *Cryptolaemus montrouzieri* Muls. (from Bodenheimer and Guttfeld).

neighbourhood. This agrees with the experience in California. The rate of actual increase is also very low, even in spring and autumn:

Month	offspring / parents	temperature
V	1.0	21.6° C
VI	1.5	24.3
VII	**0.6**	26.7
VIII	1.1	26.0
IX	1.5	26.1
X	**1.9**	22.4
XI	1.2	18.6
XII	1.2	18.4

During the extraordinarily hot summer of 1927, the decimation was especially heavy. In July 1927 (27.3° C) the relation of increase was only 0.2, but in the normal July 1926 (26.0° C) it rose to 1.1,

If we consider the number of eggs per female as 100 in *Cryptolaemus* the mortality is very strong in our climate. It must occur mainly in the first days of the larval stage. Similar decimation occurs in other subtropical ladybirds. The Brazilian *Curinus zonatus* Muls. has three annual generations at Porte Allegre with a relative abundance of: 1.0 : 2.5 : 1.8 *

*) Wille, Zeitschr. ang. Ent. 1926, p. 372.

Cryptolaemus is thus cramped between two seasons of heavy mortality: a lethal winter and a very unfavorable summer. The favorable seasons, spring and autumn, are too short for a sufficient recovery. And finally the generally low population level of Pseudococcus citri is another depressing factor. Thus, even when the winter would not be absolutely lethal, environmental conditions are so unfavorable for the mealybug ladybird that it would always remain on a very low population level, which would not be sufficient to produce any limiting effect on *Pseudococcus* in Palestine.

c. Practical experiences with Cryptolaemus in Palestine.

The beetles were always liberated in groups of 25—100 adults to 1—3 trees, which were heavily infested with mealybugs. The trees were marked in order to facilitate later inspections. The number of liberated beetles was relatively small, owing to the poor results in our artificial breedings:

	1926	1926
Petah Tikva	535 beetles in 6 groves	1182 beetles in 9 groves
Hedera.	60 ,, in 2 ,,	
Rehoboth.		100 beetles in 1 grove

This makes a total of 1877 beetles in two years (June to November). Some typical results may be reported:

110 beetles were liberated on 16.VI.26 in the orange grove Pa. at Petah Tikva, in which a permanent focus of mealybugs was present. Many fruits, which dropped in consequence of *Pseudococcus* activity, lay scattered under the trees. The similar heavily attacked grove Ba. was furnished with 50 *Cryptolaemus* beetles on one focus a week later.

Much heavier was the attack in the orchard G., where thick wax filaments of mealybug conglomerations were conspicuous. 125 beetles and 50 larvae were liberated.

The heaviest infestation was in the grove Ja. 265 adults and 30 larvae were exposed.

In none of these cases a radical change was observed. Even the few trees on which the *Cryptolaemus* were concentrated, did not become clean in a reasonable time. Only in the orchard Ga, where 175 *Cryptolaemus* were liberated in July/August, a visible effect was observed. An infested focus of 15—20 trees was cleaned before the beginning of winter and *Cryptolaemus* persisted up to that period. How far other beneficial insects were concerned in this success, was not studied. But it was probably much higher than the effect of our lady-beetle.

The number of liberated individuals is relatively limited. But twice this number was sufficient to acclimatize the beetle in five different localities at the French Riviera.

It was impossible in any case to trace any instar of *Cryptolaemus* in the orchards, where this species was colonized, in the following year. Its acclimatisation in Palestine ended with failure. Even if our attempts had been intensified and if the acclimatisation of a few individuals had proved succesful, there is no doubt that *Cryptolaemus* would never have gained any epidemiological importance. The climate of our coast-plain is decidedly unfavorable to its existence.

d. Experiences with Cryptolaemus in other countries.

The attempts to acclimate *Cryptolaemus montrouzieri* in Egypt ended with a similar failure. The experiences started earlier in that region and are maintained up to the present day. The expenses are enormous and some hundred thousands of beetles have been colonized. During the first years of the work in Egypt were exposed (Hall):

Month	1923	1924	1925	1926	Total	Average monthly temp. at Gizah
VII	—	—	—	—	15000	26.6° C
VIII	—	1750	10000	9750	21500	26.3
IX	400	1050	13500	25750	40700	23.9
X	600	10200	8500	38500	**57800**	21.6
XI	1800	8500	8000	41000	**59300**	17.3
XII	2000	6000	3000	4800	15800	12.7
I	—	11200	500	—	11700	10.9
	4800	38700	43500	134800	221800	

From these 221800 liberated beetles it was possible in one case only to prove the successful survival of a few individuals throughout the winter. The winter season is just as fatal in Egypt as it is in Palestine. The staff engaged in breeding *Cryptolaemus* at the Gizah laboratory estimated the winter mortality of the larval stage alone in the insectary to be higher than 90 % as compared with an average mortality of 70—80 % of the larvae during the other seasons. The summer temperatures are just as unfavorable in Egypt as in Palestine.

In recent years successful hibernation of a few individuals has been repeatedly observed. A real acclimatisation is not yet proven. The number of survivals is too small, compared to the total population liberated and bred during autumn on the spot. But it is absolutely certain that even in the optimal case the population level of *Cryptolaemus* will remain extremely low and remain without any effect on the population density of the local mealybugs. Concentrated artificial liberations may more or less clean a group of infested trees from mealybugs. But no permanent focus and no centre of dispersal has been created therewith. The writer's prognosis in 1928 has been fully confirmed by the later development in Egypt, where the attempts for its establishmont are now abandoned.

In California it was introduced in 1892 from Australia by Koebele.

"From the importations made at that time the insect was firmly established in the orchards and was thoroughly distributed to all parts of the stage where mealybugs were known to exist. It did not, at that time, seem to be able to adapt itself to the varied climatic conditions in southern California, and very largely disappeared excepting in San Diego and Santa Barbara counties, where it has persisted ever since the original introduction. In these counties, according to reports, it has often without artificial aid increased to such numbers as to clean up the mealybug infestaition.

At the time of the severe outbreak of the common mealybug in Ventura County in 1909 it was reintroduced in considerable numbers and did excellent work in the control of the pest. In general, however, where it occurred at all in infested orchards outside of San Diego and Santa Barbara counties, it was relatively unimportant so far as mealybug control was concerned, and it failed entirely to persist in the interior sections where there is a relatively low humidity and greater extremes of temperatures.

These facts are of great interest since, compared with the present position of *Cryptolaemus* in the control of mealybugs, they indicate very clearly what has been accomplished by the propagation and distribution of the beneficial insects in large numbers. At the present time, although the mealybugs occur over a far greater area than formerly, and under even more varied conditions as regards temperature and humidity, biological control is almost the sole method in use, and has given excellent results in the interior as well as along the coast." (Smith and Armitage 1931).

Ever since the installation of many costly insectaries for *Cryptolaemus* breeding throughout the Californian Citrus belt, field outbreaks of *P. citri* are practically unknown. But the control was not so satisfactory with the more injurious-citrophilus mealybug (*P. gahani* Green), a species of Australian origin with heavy honeydew exudation. The recent introduction of two Australian hymenoptera: *Tetracnemus pretiosus* Timb. and *Coccophagus gurneyi* Comp. proved to be so successful that all the insectaries for *Cryptolaemus* breeding are closed down since the summer of 1931.

It is quite obvious that *Cryptolaemus* established itself on its own only in a very limited district near the sea-shore. Everywhere else it died out. But artificial annual colonization at a rate of 10 beetles per tree in any infested area proved to be a very cheap and satisfactory method of control. The average number of annual generations at Los Angeles may be calculated from our hyperbola to be four: 95 days early summer 59 days late summer-, 80 days autumn- and the hibernating generation. It is a great pity that the biological and climatological data of the rich experience in California have never been published.

On the French Riviera a small introduction was made in 1918. 1585 beetles were liberated at 9 different localities along the seacoast during 1919 and 1920, of which five colonies became well established and the remainder died out probably owing to scarcity of food. The writer has witnessed in 1926 the establishment of the species at Mentone, feeding on *P. citri* on lemons, on *Ps. adonidum* on oleander, on *Phenacoccus aceris* on plane trees. The latter especially, were seen in large masses in October on the trees. *Cryptolaemus* was doubtless established everywhere in the foci of infestation, but no real control could be realized. This means that under the conditions of the French Riviera artificial *Cryptolaemus* mass-breeding is necessary, when good results of biological control are to be obtained. But it seems to have acclimatized for good in this district.

Three annual generations are the mean for the French Riviera. Some breeding results may be compared with calculations from our hyperbola:

Date of oviposition	Average development (Poutiers)	Average development (hyperbola)
20.IV.	72 days	69 days
12.V.	49	56
10.VI.	40	46
17.VII.	39	41

This agreement is fairly good, especially considering that the average climate, not that of the special breeding year, has served as basis.

Repeated attempts to establish *Cryptolaemus* in Algeria have all ended in failure (1912—17, 1925—26).

The real reason for the failures in the introduction of the Australian mealybug-ladybird into Palestine, Egypt and Algeria, is the fact that two fatal seasons have to be tolerated there: a humid and cool winter and a dry, very hot summer. The favorable seasons, spring and autumn, are too short to enable a real recovery in the number of individuals of the ladybird. Average temperatures of 18—23° C during 3—5 months are decisive for a good result in biological control with this species.

The climate of Los Angeles and Nice is therefore much more favorable to the mass-increase of *Cryptolaemus*. 2—3 efficient generations with relatively low mortality develop from spring to autumn, and are interrupted only *once* by the winter generation, which develops under less favorable conditions. The hibernating beetles are used for the beginning of breedings, the adults of which may be liberated in the summer.

Tentatively the optimal climatic zone for *C. montrouzieri* may be assumed to lie between 18—23 C° and 86—90 % relative humidity.

e. *Native Coccinellidae feeding on P. citri in Palestine.*

Scymus quadrimaculatus and *Hyperaspis polita* are fairly regularly connected with citrus mealybug in the coastal plain. *S. apetzi* and *S. suturalis* are not rare, but not so often met with on *P. citri*. Their other hosts are still unknown.

The first two species are of auxiliary importance for the biological control on the citrus mealybug. They were bred in some quantities in our Parasite-Breeding Laboratory at Petah Tikva.

Key for determination of the lady-beetles connected with *P. citri* in Palestine.

1.	Upper side and eyes covered with pubescence. Elytra never longitudinally punctuated.	2.
	Upper side and eyes not covered with pubescence.	6.
2.	Elytra without spots.	3.
	Elytra with spots.	4.
3.	Elytra red to red brown with black marginal sutures.	*Scymnus suturalis.*
	Elytra straw coloured and with black margin.	*Scymnus fenestratus.*
4.	Elytra with oblong brown spot in centre surrounded by irregular straw-coloured patch.	*Scymnus includens.*
	Elytra selfcoloured beside the maculae.	5.
5.	Elytra with one, rarely with two, small round spots each.	*Scymnus apezti.*
	Elytza with two, rarely with one, broad oblique bands.	*Scymnus quadrimaculatus.*
6.	Colour a shiny black, 2.2 mm long.	*Hyperaspis polita.*
	Colour a faint black, 2.8—3.0 mm. long.	*Oxynychus marmottani.*

a) *Scymnus (Nephus) quadrimaculatus Herbst*

Synonymy: 1783 *Scymnus quadrimaculatus* Herbst
1797 *Sc. pulchellus* Herbst

Description: Adult oval, moderately convex; black with grey pilosity. Two orange or yellow coloured spots on each elytra. The anterior and larger spot runs obliquely backwards in a median direction, the posterior spot transversal. The form of the elytral spots is variable. The posterior spot may even be absent. Elytra fine and densely punctuated and with a red distal border. Antennae, mouthparts and legs (all or partially) and end of abdomen reddish-yellow; base of hind-femora in the female often blackish. Young specimens are wholly red-yellow; dots not clear, whitish.

1.5—2.1 mm long.

Eggs round, bright yellow, very small.

Distribution: Almost over the whole Palaearctic Kingdom, being more common in its warmer parts.

This common predator of *P. citri* is less eager to feed on *P. comstocki* (Klein and Perzelan **1940**).

b) *Scymnus apetzi Mulsant*

Description: Adult broad oval, moderately convex; black. Elytra bearing one yellow-red spot in the anterior half, rarely a second spot present in the posterior half. Labrum of female reddish; in the male labrum, anterior part of head and often a border in the anterior angles of the pronotum reddish. Legs red-brown, hind- and mid-femora black.

2—3 mm long.

Distribution: C. Europe, S. Siberia, Mediterranean Region.

Both these species have been bred by us from *P. citri.*

Scymnus fenestratus Sahlb.

Similar to *S. suturalis*, but elytra straw-coloured instead of red.

Scymnus includens Kirsch.

Also similar to *S. suturalis.* Beetle dark to light brown. On each elytron one oblong brown spot surrounded by an oblong irregular straw-coloured patch (Rivnay and Perzelan **1943**).

c) *Scymnus (Pullus) suturalis Thunberg*

Synonymy: 1795 *Scymnus suturalis* Thunberg
1798 *Sc. discoideus* Paykull
1834 *Sc. atriceps* Stephan

Description: Oval, only slightly convex; of faint black colour, with fine gray pilosity. Elytra brown yellow, yellow-red or yellow, and heavily punctuated; all margins, or more commonly, only the median and basal margin blackish, rarely without any black margin. Body colour black, antennae and palpi yellow-brown, legs dark with black femor; tibia and tarsus brown red. Young specimens are unicoloured red-yellow, head and pronotum often darker.
1.5—2.2 mm long.
Distribution: Europe, Siberia, Mediterranean Region, Caucasus.

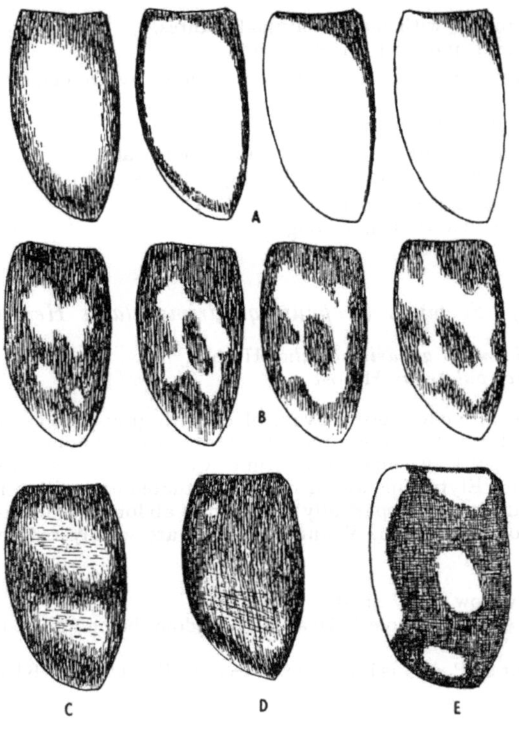

Fig. 105.
Elytra of some lady-beetles feeding on *Pseudococcus*: A. *Scymnus suturalis*, B. *S. includens*, C. *S. quadrimaculatus*, D. *S. sp.*, E. *Oxynychus marmottani* (from Rivnay).

Life-history of S. quadrimaculatus: The life-history of this most common lady-beetle of *Pseudococcus* in Palestine stands as an example for the very similar one of all three species.

The life-cycle of 12 generations at Petah Tikva was:

No. of generation	adults hatched	first larva observed	days from copula to larva	larval stage in days	pupal stage in days	preoviposition egg period in days	date of death of parents
I	12—14.VII.36	16-23.VIII.36	5—38	5—8	5—6	10—40	15.IX.36-7.II.37
II	28-31.VIII.36	9-15.IX	3	10—12	9	9—12	15.VI.36-2.VI.37
III	16-21.X.36	6.XI-13.IV.37	21—26	10—39	11—28	21—179	24.V.37-15.VIII
IV	2-6.V.37	23-28.V.	14—21	5—10	7—9	19—26	20.VI.-13.IX.
V	9-15.VI.37	20-24.VI.	7—11	6—8	6—9	9—13	23.VII-19.XII.
VI	4-17.VII.37	9-25.VII.	5—18	5—11	5—10	5—18	29.VII.37-3.I.38
VII	29.VII-1.VIII.37	9-17.VIII	8—19	6—10	5—7	11—19	1.IX.-15.III.38
VIII	24-27.VIII.37	3-8.IX.	7—12	8—10	5—6	10—29	15.IX.-6.VII.38
IX	23-30.IX.37	8-19.X.	9—20	8—15	6—15	12—48	5.XII.-18.VII.38
X	1.XI.37	22-28.XI.	13—19	29—35	6—8	22—31	28.III.38-4.VIII.38
XI	27.V.-6.VI.38	9-22.VI.38	9—16	8—13	5—10	12—21	8.VII.-6.IX.38
XII	27.VI-1.VII	29.VI-30.VII	6—23	5—7	4—9	9—25	22.VII.-1.XI.38

The adult longevity and the production of adults per female (equals total egg-production minus mortality of developments stages) was as follows:

generation	adult longevity			total adult production per female		
	n	min/max	average	n	min/max	average
I	18	23—327	102	8	5—46	20
II	13	42—317	191	6	5—71	22
III	16	203—315	255	8	9—176	73
IV	19	47—131	84	8	4—154	102
V	20	44—187	101	8	26—130	77
VI	20	25—183	99	9	3—69	37
VII	18	15—229	101	6	18—73	47
VIII	20	21—313	194	5	1—81	36
IX	20	13—309	214	8	1—56	18
X	20	89—282	243	8	2—57	24
XI	19	37—106	78	10	11—72	42
XII	19	40—119	69	10	1—75	27

With regard to length of development from first larva to first larva we separate between development without and with imaginal diapause:

Generation	n	development in days without inhibition	average	development in days with inhibition	average
I	4	20 20 24 38	26		
II	5	42 55 58	52	192 202	197
III	9	40 46 47	44	189 200 200 204 217 265	213
IV	8	38 45 45 47 49 50 51 58	48		
V	8	24 24 24 26 26 26 26 29	26		
VI	10	19 21 23 25 25 25 26 26 28 35	25		
VII	6	24 24 24 26 27 32	26		
VIII	7	25 26 26 27 27 32	27	65	65
IX	8	30 31 32 33 33 36 37	33		
X	8	38 45 45 48 48 48 56 57	48		
XI	10	37 39 40 40 40 42 43 43 44 49	42		
XII	10	27 27 27 29 29 29 32 37 37 45	32		

In winter no oviposition was observed.

S. quadrimaculatus is feeding almost exclusively on mealybugs. Aphids are devoured very little, if no other food is available. Soft scales (*Ceroplastes*) and other scales. (*Aonidiella, Aspidiotus, Icerya*) lave not been touched even in hunger.

The number of mealybugs devoured by beetles of *S. quadrimaculatus* per diem, is:

Month	I	II	III	IV	V	VI	VII	VIII	IX	X	XI	XII
n.	10	10	11	17	15	12	16	11	10	7	10	10
eaten per beetle/day . .	2.3	2.8	4.4	12.1	15.1	27.8	31.7	43.4	34.6	28.0	15.4	4.7

Scymnus deposits its eggs among those of *P. citri*. It feeds on *Ps. citri* and its eggs, in summer on about 30—50 eggs daily, in winter on 2—15. The exact number of eggs deposited is unknown, but 50—100 larvae hatch per female and 20—70 of these develop into beetles.

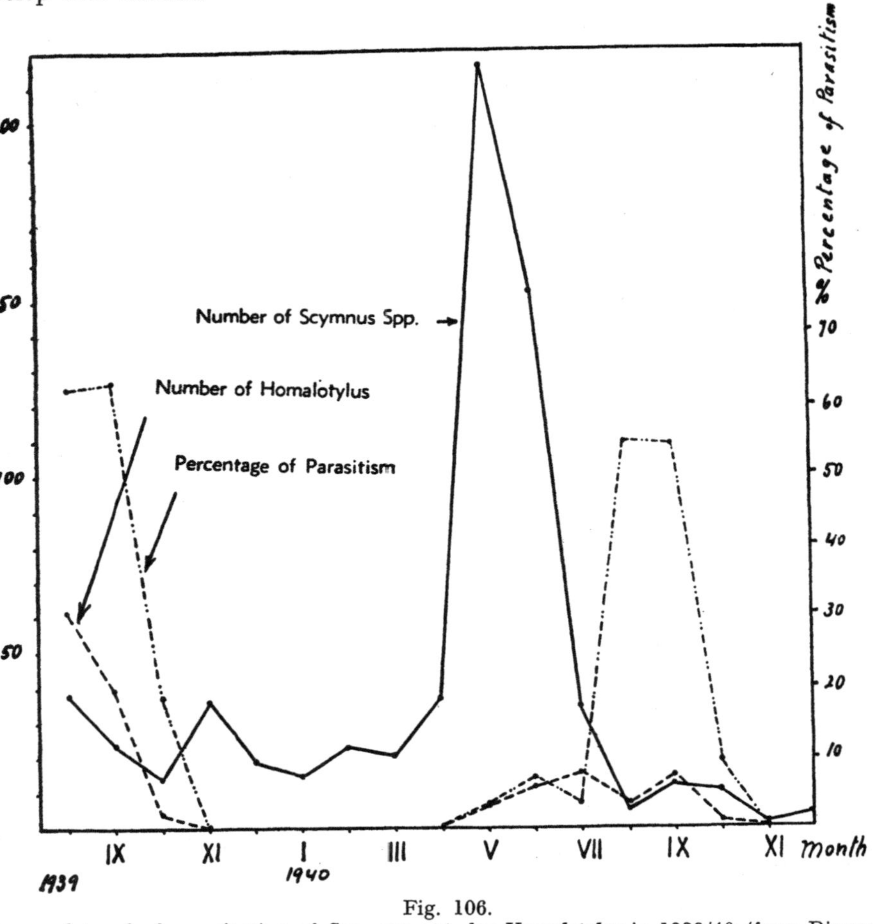

Fig. 106.
Seasonal trend of parasitation of *Scymnus spp.* by *Homalotylus* in 1939/40 (from Rivnay).

The larva also feeds on *Ps. citri* and its eggs. In summer the larval stages last 7 days, in autumn 14—38 days. During this stage the larva devours 2 ovisacs of the mealybug. The pupal stage lasts from 5 to 10 days during most generations.

The beetles copulate on the day of their emergence or a few days later and start to deposit their eggs. From 5 to 14 days after hatching of the adults the first larvae appear in summer, after 1 month and later in winter. Adult longevity is 3—4 months, but up to 8—10 months in the hibernating individuals, one of which lived almost one complete year.

The *Scymnus*, especially *S. quadrimaculatus*, prove to be very efficient enemies of the mealybugs during summer and autumn. Their distribution together with *Sympherobius* or alone proved to be decidedly beneficial.

S. suturalis and *S. includens*, especially the former, were common predators of *Pseudococcus comstocki*. The beetles hide during winter and become active in early spring. Rivnay and Perzelan (1943) say that they reach their peak in May/June. In mid-summer their numbers are rapidly declining due to lack of food, sensitivity towards the high temperature, and especially due to parasitation by *Homalotylus quaylei*, the development of which lasted 16 days at 29° C., and which attacks larvae of all ages. These wasps were observed from May to October, with the peak in August. Whilst the percentage of *Scymnus*-parasitation is low in May, it reaches 40 % in August/September. Thus the main activity of *Scymnus* against *Pseudococcus* in the orchard is restricted from March to early June.

d) *Hyperaspis polita Weise.*

Description: Adult oval, strongly convex, shiny black. Elytra with very fine punctuation, looking as if polished; with 3 spots on each elytron: The first one is semi-circular on the basis, near the scutellum; the intermediate is oval and situated in the centre; the third, at the end of the elytra, is oblique-oval. Epipleura yellow. Lower side and legs reddish-yellow-brown; middle and hind sternit and first abdominal sternit tar-brown.

2.8—3.0 mm long.

Egg: oval, greenish, 0.-8 mm long, 0.5 mm broad, flat. It is laid among the eggs of *Pseudococcus* or near the ovisac.

Distribution: Egypt, Palestine, Syria, Asia Minor.

Life-history in Palestine: Larvae of *Hyperaspis* were first collected on 6.VIII.1936 in orange groves at Petah Tikva, which were heavily infested with mealybugs. They pupated on 16.VIII. and the beetles appeared on 24.VIII.1936. The latter started oviposition in early April 1937 i.e. after about 225 days of imaginal life, when temperature had just passed 19° C.

Four to five annual generations develop and adults of all them, except the first, hibernate: very few of the second, but a good many of the third and fourth generation. The adults of some generations appeared.

No. of generation	hatching of beetles	death of last beetles	n	adult longevity in days	average
I	24.VIII.36		10	♀ 223 224 224 224 224 225 226 226 224	227
II	4.-9.VI.37	12.VII.37-1.IV.38	20	♀ 89 152 166 172 192 192 202 246 261; ♂ 38—300	♂186 ♂173
III	25-30.VII.37	14.III.38	12	♀ 196 213 215 232; ♂ 60—239	♀214 ♂171
IV	25-29.VIII.	21.III.38	12	♀ 168 189 238; ♂189—204	♀ 98 ♂197
V	21-26.IX.37	20.II.38	11	♀ 154 161 183 215 222; ♂117—169	♀187 ♂141

The duration of development and mortality of early stages is compiled in the following table:

Date of oviposition	duration in days of stage				preoviposition period	number of survivors		
	egg	larva	pupa	total		eggs	larvae	adults
6.VI.1937	9	17	16	42	2—3	100	76	23
9.VI.	6	9	15	26	5—16	100	72	55
12.VII.	7	12	8	27	8—24	100	28	18
13.VIII.	5	13	11	29	6—8	100	61	25
15.IX.	7	13	17	37	10—15	100	18	11
21.X.	7	22	26	55	—	100	31	4

The productivity of the different generations is:

Generation	n	No. of eggs per female: during the same year min/max.	average	during the next year n	min/max	average	Total average
I	7	243—477	391	—	———	—	391
II	8	47—862	424	—	———	—	424
III	5	111—542	262				
IV	7	28—196	91				
V	6	5—102	37				

e) *Oxynychus marmottani Fairm.*

Synonymy: Hyperaspis pumila Mulsant.

Description: Oval, above shiny black. Elytra with 3 yellow maculae each. The first one covering $^1/_5$—$^3/_5$ of the width, prolonged to $^1/_5$ of the length; the second, central spot oblong, covering $^1/_4$—$^1/_5$ of the width; the posterior spot oblique oval. Tarsi red-yellow.
2.2 mm long, 1.3 mm broad.

Distribution: This is the local species and not *Hyperaspis pumila*, which is widely distributed in Tropical Africa from Senegal and N. Nigeria to Lake Rudolf, Sudan and Abyssinia. Capra*) regards the records from S. Arabia, Egypt, Algeria and Canary Islands as doubtful and belonging to the closely related *O. marmottani* Fairm.

Life-history: The little known on the life-history of this beetle was observed from 1930 to 1932 in the field-station at Hedera.

On 7.IX.30 the first beetle was discovered. Feeding experiments showed that it feeds on *P. citri*, but not on soft or announced scales. In breedings eggs were seen first on 5.X.30, larvae on 15.X.30. During December 6 beetles hatched, but these beetles as well as the larvae still present died in January 1931.

In 1931 the first beetles were observed in late August. Larvae appeared in breedings one month later, but all died.

In 1932, when special attention was paid to this species, the first beetles were seen on 10.VI. More beetles and larvae were found on 1.VII in the orange groves. The beetles which developed from these larvae started oviposition on 31.VIII. and continued to lay eggs until October.

5 beetles, which had hatched that day, were put into one glass-jar on 31.X.32. On 15.XI.32 one egg and two larvae were present. The beetles were then isolated, and one female deposited 10 eggs until the end of October, which all dried and died. In a second parallel series of breedings one female deposited 44 eggs during November and December, all of which dried out and died except 3 which hatched, but died after a few days. The parent beetles of all these and other groups died in March and April 1933, without further oviposition. These beetles survived the winter, all other beetles died at the beginning of the „cold" season. This sensitivity to cold is in agreement with the tropical affinities of the species, and explains its low abundance in Palestine. The life-cycle seems to be similar to that of *H. polita*.

---

*) Capra, Bull. Soc. Ent. It. **61**, 1929, p. 107.

D. *The Mealybug Gallmidge* (*Dicrodiplosis n.sp.*)

In the autumn of 1925 a species of gallmidges appeared in our breedings, which occasionally disturbed even the mass breeding of *Pseudococcus*. This same gallmidge was common in the orchards in autumn. Its breeding showed that its development is considerably shorter than that of *Pseudococcus citri*. Some attention was therefore paid to this insect.

Description: The adult is 2.0—2.5 mm long. The head of the tiny midge is black, the 25-jointed antennae brownish. The thorax is redbrown. The hyaline wings have dark hair, the veins are colourless. The halterae are dark-brown, with hyaline shaft. The legs are hyaline and colourless. The first two abdominal segments are dark-brown, the rest of the abdomen vine- to orange-red.

The full-grown *maggot* is 2.2 mm long. It is yellowish with a large brick-red spot on its back or flesh-coloured with two bright yellow longitudinal stripes on the back at both sides of the blackish shining intestinal tract. On the fifth to tenth segments a white lateral stripe is present. On the ventral side it bears 8 pairs of skin-cones.

The *pupa* is enclosed in a thin transparent, but very tight cocoon. It is red-brown, the margin of the abdomen being somewhat brighter. 2 mm long.

It was impossible to get a determination even by specialists. Some remarks have only been made concerning the figure which was published by us some years ago.

Oviposition must take place on the adult female of *Pseudococcus*, before the latter begins with the formation of its ovisac. In 105 experiments with mealybugs with

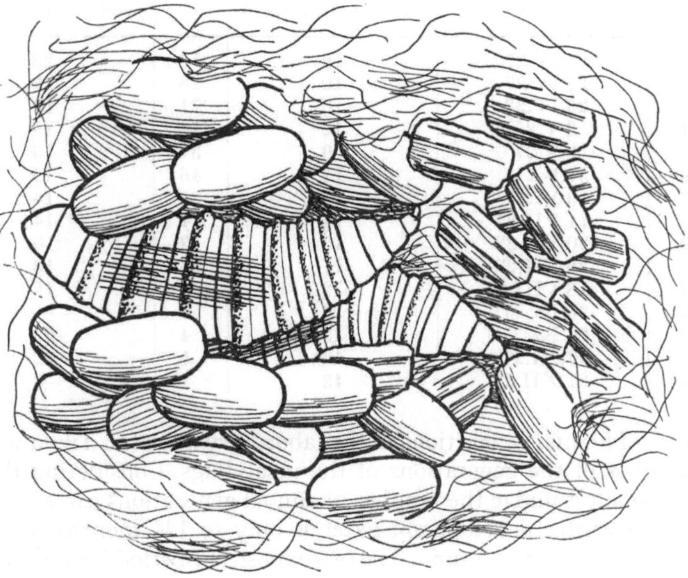

Fig. 107.
Maggots of *Dicrodiplosis sp.* feeding in ovisac of *Pseudococcus citri*.
(from Bodenheimer and Guttfeld).

ovisacs, oviposition or normal development were not observed in any case. But in 19 out of 25 cases of breeding on *Pseudococcus*-females, which had not yet started oviposition, normal oviposition and development of the gallmidge occurred. The eggs are laid on the mealybugs in small groups, mostly singly (1—3). The egg-stage must be very short. The larvae enter the base of their hosts ovisac, where they devour the eggs eagerly. From an infested ovisac with 1—2 larvae a few *Pseudococcus* larvae still hatch occasionally. The infested ovisac shrinks and eventually collapses. When no sufficient eggs are present, the larva crawls under the mealybug-female and sucks it empty, leaving only the bare skin. About 100 eggs are destroyed by one of these maggots. The whole larval stage lasts only for 3—5 days in summer and autumn.

The full-grown maggots spin a thin, but strong coccon, in the interior of which they remain for two days (in autumn) immovable, before they change into the long red-brown pupa.

The adult has a very limited longevity. Among 41 adult midges, the following lived (from June to October):

longevity	1	2	3	4	5 days
No. of individuals	12	20	8	1	0

The average longevity is therefore two days. The adults are decidedly nocturnal in their activities. At these hours copula and oviposition probably take place, but have never been actually observed. In the early morning, the midges are often seen buzzing in a peculiar, restless, up and down dancing flight over the mealybugs.

Some typical life-histories are given as an illustration:

No.	Date of oviposition	Shortest total development in days	n	Average total development in days
1	21.IV.27	19	13	22
2	22.IV.27	19	28	22
3	26.IV.27	16	71	19
4	3.V.27	12	14	15
5	9.V.27	9	34	13
6	31.V.27	12	46	15
7	9.VI.27	13	28	14
8	11.VII.27	14	4	14
9	20.VII.27	10	3	10
10	26.VII.27	11	26	13
11	18.VIII.27	12	6	12
12	25.IX.27	16	1	—
13	24.X.27	19	4	20
14	17.XI.27	40	5	42
15	21.XII.27	43	1	—

The number of annual generations in the laboratory at Petah Tikva was 16—17 i.e. twice the number of the generations of the mealybugs. Considering the very short longevity of the adult midge the number of annual generations can scarcely be less in nature. Even in winter development extends only to 1½ months, whilst it is only 10—15 days from spring to autumn. The preoviposition-period is very short; probably only a few hours long.

The data: development in 40 days at 15.9° C
„ „ 10 days at 27.3° C

permit the computation of the hyperbola with the constants

development zero (c) = 12.1° C
thermal constant (Th.C.) = 152 days-degrees.

This hyperbola agrees fairly well with our empirical data (see table above).

Biotic limiting factors of this gallmidge have remained unknown. Cannibalism and mportant decimation by *Sympherobius* are counteracted by the habit of the mother-midge to lay only single eggs in the ovisacs of *Pseudococcus*.

However, the number of eggs per female remains restricted and these are very sensitive to climatic factors. The quotient of increase $\frac{\text{(No. of offspring)}}{\text{(no. of parents)}}$ is very low during summer and winter and not too high in spring and autumn:

Month in 1927	IV	V	VI	VII	VIII	IX	X	XI	XII
Quotient of increase	**2.16**	2.09	1.27	0.97	0.44	**0.25**	1.33	0.71	0.33
Temperature in ° C	17.8	21.9	24.9	27.3	26.5	25.7	23.0	18.9	14.6

Besides the climate, the quantity of food has certainly some influence, which probably explains why this Cecidomyid is more abundant in autumn than in spring.

Specimens of this gallmidge were sent on request to the Parasite Breeding Laboratory in Riverside (California) and arrived in good condition, but died out after some months. This confirms the climatical sensitivity of the species.

Its climatic sensitivity as well as other difficulties in its behaviour make it difficult to breed this gallmidge in large masses. But in the period in which it is common in nature, it is easy to infest with it large masses of mealybugs on potato-sprouts. By putting them into centres of heavy infestation, surprisingly quick and good results can locally be obtained. But this can be done only in September and October.

The breeding of the species is done in test-tubes, where 4—5 just hatched midges and 4—8 mealybugs, which are ready to start with oviposition, are kept together. If heavily infested *Pseudococcus* are present in the breedings, these are also taken in test-tubes for breeding purposes.

When some cages are heavily infested with this gallmidge, corrugated paper is put between the potato-sprouts. Many ovipositing mealybugs settle there and their ovisacs are infested with the midges. This corrugated paper is then put into the orchard and fixed near heavy centres of infestation. This handling may be practised from May to November.

When many midges hatch in the test-tubes, they may likewise be liberated in foci of infestation. No reliable method of permanent breeding could yet be worked out for this insect. If found, it would probably be the easiest and quickest method of biological control.

It becomes quite troublesome occasionally by interfering with breedings of other beneficial insects. When it should invade the mealybugs on the potato-sprouts or in the breeding cages, it will become necessary to fight it. The glass-windows of the breeding cages are besmeared with oil. Those midges which hatch during the night, are caught in the early morning by trapping with the hand, besmeared likewise with oil. The latter method is also useful for catching the midges in the open trays of potato-sprouts.

E. The wing-banded Mealybug Parasite (Leptomastidea abnormis Gir.).

The complicated synonymy of the species was cleared by Mercet (1924):
Leptomastix sp. Viereck, 1915.
Paraleptomastix abnormis Girault 1915. H. S. Smith 1917.
Leptomastidea aurantiaca Mercet 1916, 1921.
Tanaomastix abnormis Timberlake ab. 1920.

Description: The adult parasite is yellow with some grayish drawings. The antennae are grey and white, the legs whitish.

The long antennae are covered sparsely with long hair in the male, but are bare in the female. The pronotum is very short. The anterior wings are relatively long and conspicuous by three characteristic oblique bands. The posterior wings are small and hyaline. They bear bristles on the posterior margin. The triangular abdomen does not surpass the thorax by its length. The first abdominal segments are very short, the last one long. The ovipositor of the female is very short. The male is much smaller than the female:

Sex	Length of body	Length of anterior wing	Length of posterior wing
male	0.67	0.77	0.45 mm
female	1.04	1.06	0.61 mm

The species was originally discovered in Sicily, later in Spain, Japan and Palestine. It was imported into the southern U.S.A., Hawaii and Australia as a mealybug parasite. It is polyphagous on *Pseudococcus*. It seems to prefer *P. citri*, but infests freely *P. comstocki*.

Oviposition takes place as soon as the adult has emerged and its wings have hardened. The first and second larval stages are preferred as hosts, oviposition in mature females never observed. In the process of oviposition there is very little preliminary work, the female simply examining the host casually with her antennae, then turning, inserting the ovipositor into the host's body, and depositing the eggs in a comparatively short period of time. The mealybug resents the attack by a considerable amount of squirming which is not sufficient, however, to deter the parasite from carrying out its purpose. In nature she apparently places but one egg in a host, or at least only one reaches maturity, but when forced to do so she will deposit as many as 100 eggs in a single mealybug. The time required for development from egg to adult insect is 25—45 days, depending upon the temperature. (Smith and Armitage).

From the breedings (18 days at 24° C, 30 days at 18° C, both excluding the preoviposition period) the following hyperbola is calculated:

threshold of development (c.) = 9.0° C
thermal constant (th.c.) = 270 day-degrees.

Duration of development at:		
	30° C	12.9 days
	25	16.9
	20	24.5
	15	45.0
	13	67.5

Unfertilised eggs develop into males only. The reproductive rate is low. 57 wasps at most were bred from one pair of parents. Rivnay and Perzelan (1943) observed a parasitation of *P. comstocki* in winter of 1—2 %, up to 6 % in summer.

This wasp is distributed throughout our coast-plain. It is not readily seen in nature, but large breedings of mealybugs are rarely made without its appearance. In masses we obtained it only in autumn. *Leptomastidea* seems to be least dependent on the population density of the host among all specific enemies of *Pseudococcus*. But its activity is less obvious to the farmer. In the orchard it may fly and spread to considerable distances.

A careful rotation of potato-sprouts infested with first and second instar larvae of mealybugs gives a continuous emergence of adults for liberation in the ordinary type of breeding cages. After some failures at the beginning, we can now confirm this Californian experience. The hatched adults may be collected from the cages and liberated. Having obtained satisfactory results with *Sympherobius*, we did not attempt to breed this wasp for biological control. In times of need, mass-breeding may be attempted.

F. *Anagyrus pseudococci Girault* (*Encyrtidae*).

Synonymy: A. kivuensis Comp.

This species was bred in 1938 from Rehoboth and Rishon Lezion (Rivnay and Perzelan 1943). The female is dark brown, with hyaline wings, 1.5 to 2.0 mm long, the male metallic dark blue, 1 mm long.

P. citri and *P. comstocki* are both infested, but the former is preferred. The third larva to the immature female are infested. The hyperbola constants are:

Threshold of development (c.) = 1377° C
Thermal constant (th.c.) = 215 day-degrees.

At 30° C	13.2 days duration of development (exclusive pre-oviposition period).
25	19.0
20	34.0
15	165.0

No eggs are laid below 15° C, above 17° C about five eggs daily. Rivnay and Perzelan obtained little reproduction in outdoor breedings. The wasp is rare in winter and again in summer, its main abundance being confined to the periods of May/July and September/November. It never is really abundant.

G. *Karnyothrips flavipes* (*Jones*).

Synonymy: K. longisetis Bagn.

Description: Blackish brown, tibiae and tarsi yellow. Fore-legs incrassate. Last abdominal segment elongated into a tube. 2 mm long. Larvae pinkish, antennae and legs greyish.

Life-history: The species is well known as predacious on coccids. Its common prey are soft scales, Astrolecaninae and mealybugs. Rivnay found it in large numbers in Hedera between the sepals and the fruit, where it probably searched for eggs and larvae of *Pseudococcus citri*. It is common in the coast plain from December to February.

It is widely distributed in Africa and America.

H. *Predatory mites.*

A mite, probably *Pediculoides ventriculosus* Newst,. was repeatedly observed feeding on eggs of *Pseudococcus* within the ovisac. In many cases not one egg of the mealybug survived, in others only a few. We know still very little on the importance of this and other predaceous mites for the naturel control of mealybugs and other Coccidae.

References.

A. Balachowsky and L. Mesnil, Les Insectes Nuisibles. Paris, 1935, p. 518.
F. S. Bodenheimer and M. Guttfeld, Zeitschr. ang. Ent. 1929. pp. 67—136.
F. S. Bodenheimer, Palestine Citrograph **1**. 1938. No. 5—16.
F. S. Bodenheimer, Schaedlingsfauna Palästinas. Berlin. 1930.
A. D. Borden, U. S. Dept. Agric. Farmers Bull. 1309. 1923.
W. Carter, Journ. Econ. Entom. **24,** 1931, pp. 1233—1242.
W. Carter, ibidem. **28**, 1935, pp. 268—284.

F. G. Clemente, El Cryptolaemus montrouzieri Muls. Muls. Parasito del Pseudococcus citri Risso. Valencia. 1932.
W. J. Hall, Bull. Entom. Res. **17**, 1927, pp. 385—392.
T. W. Kirkpatrick, The Common Coffee Mealy Bug in Kenya Colony. Nairobi. 1927.
P. Marchal, Ann. des Epiphyties **7**, 1922, p. 1—2.
R. Pontiers, Ann. des Epiphyties. **7**, 1922, pp. 3—18.
E. Rivnay and Perzelan, Journ. Ent. Soc. S. Africa **6**. 1943.
H. S. Smith and H. M. Armitage, Monthly Bull. California State Dept. Agric. **9**, 1920, pp. 103—158.
A. T. Speare, U.S. Dept. Agric. Bull. 1117. 1922.
J. R. Watson, Florida Agric. Exper. Sta. Press Bull. 392. 1930.

G. THE GROUP OF COMSTOCK'S MEALYBUG (*Pseudococcus comstocki Kuw.* 1902, *etc.* *) (G.))

Synonymy: 1902 *Dactylopius comstocki* Kuw.

Description: Adult female broad oval, light greysih pink to reddish brown; covered by a fairly thick coating of white, powdery wax-secretion. Of the 17 pairs of conspicuous thick lateral wax-filaments 15 are short and robust (0.65 mm. long). The caudal pair reaches one to two thirds of the body length, the penultimate pair being somewhat or considerably shorter. Body 2.3 × 1.3 mm.

Antennae 8-jointed, the last joint about 1½ times as long as the subequal joints II and III (10, 11, 11, 7.5, 9, 7, 7, 18 μ long), of the antennal formula:

VIII, (II, III), I, V, (VI, VII).

The rostral setae reaching to second coxae. Legs robust with subequal femora and tibiae. The tibiae slender, narrowing at the base, but broad at the tarsal joint The length of the tarsus is one third that of the tibia. Scattered translucent pores on the hind legs. Anal ring with two bands of small, apparently simple pores, with six anal setae, subequal in length, between them. Of similar length are also the long caudal setae.

Cerarii with 2 to 5 stout, short cones, a few hairs and a varying number of trilocular pores. Cerarii I to XIII (excepted II) with two cones each, cerarii XIV, XV and XVII with three, and cerarius II (the paraocular one) with five cones. The last ceraris on a rather weekly chitinised area only, and with the trilocular pores well spaced, not massed in a cluster as in P. *adonidum* L.

Derm with many trilocular disc pores on both surfaces. On the dorsal surface with six big tubular duct openings. The first of these, in the cephalic area, isshowninplate XXXII. The first sub-median pair in the cephalic area, the second sublateral pair on the metanotum, the third sub-lateral pair on apparently the first or second abdominal tergites. Fine hairs, longer and shorter ones, are scattered over very segment on both surfaces.

Apparently three moults occur, which may be separated:
Larva I with only caudal wax-filaments. Antennae 6-jointed.
Larva II and all following stages with 17 pairs of wax filaments. Antennae 7-jointed.
Larva III: Antennae 7-jointed.

Adult male: of the typical structure of *Pseudococcus* males. Reddish-brown to purple. Antennae 10-jointed, with 3 pairs of eyes and with two long caudal filaments. The first and second male larvae are almost identical with those of the female, followed by prepupa and pupa, both covered by a small white cottony cocoon.

Identity and distribution. We have purposely abstained from giving the synonymy of the species. The truth is that the present stage of mealybug taxonomy does not yet permit a clear answer. The facts shall be given here in extenso.

A *Pseudococcus comstocki* Kuwana was originally described from Japan, where it is a considerable pest on mulberry and pears. In Niigata county it has three annual generations, the crawlers appearing mainly in middle of May, middle of July and in late August. The winter is passed in the egg-stage. Each female lays about 300 to 400 eggs in sheltered places on the tree (Takahashi). It is apparently rare in Formosa. In China it is the most common mealybug on citrus in the southern and central regions, and from the coast inland to Changsha. But real damage was only once observed on citrus there by Silvestri on a pomelo tree, where the mealybugs

*) For reasons to be described below, in every place the name *P. comstocki* Kuw. should be read: *P. citriculus* Green.

2

3

1

9

5. 6

7

8

4

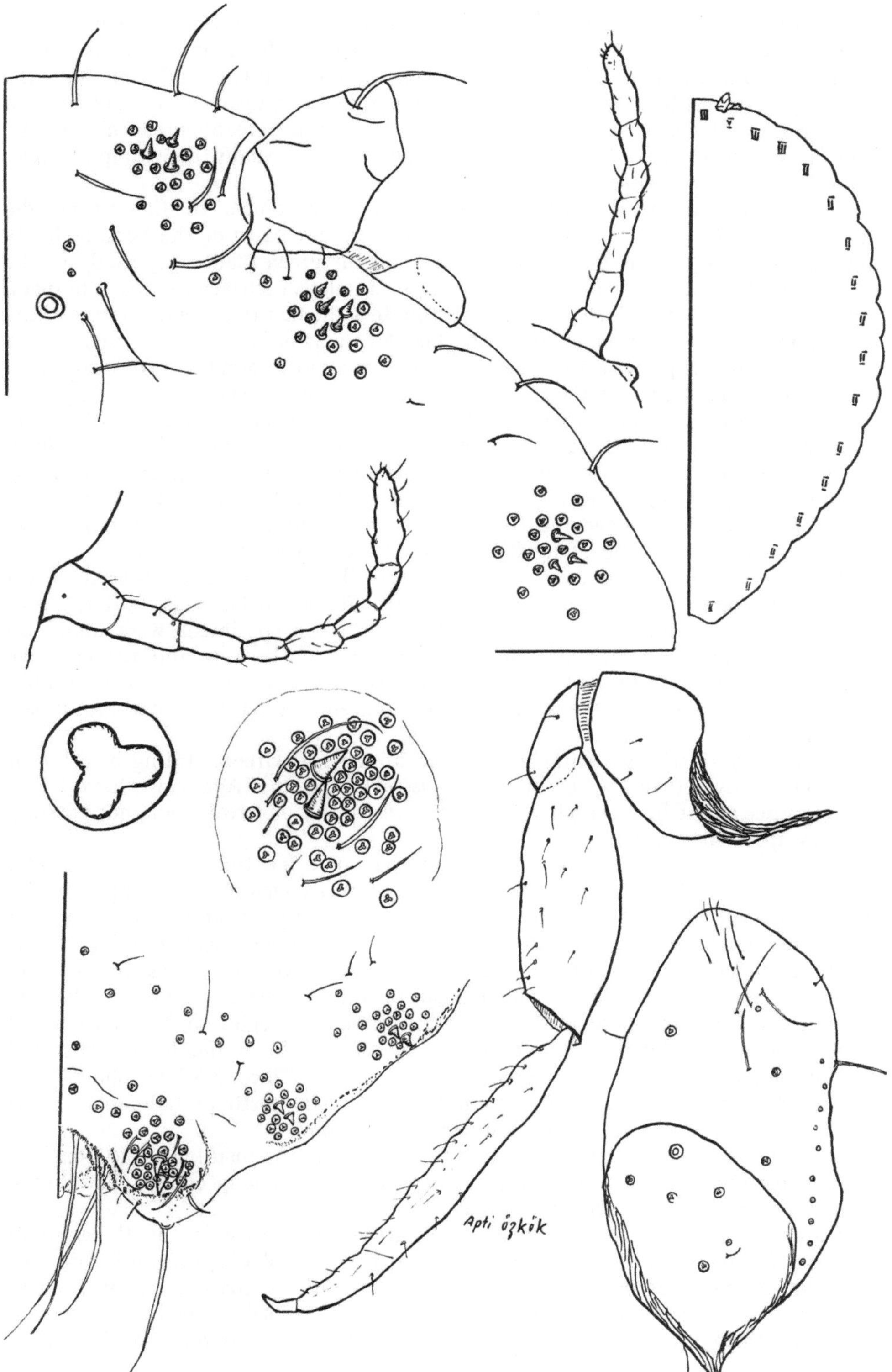

XXXII. Adult female of *Pseudococcus citriculus Green.* 1) Distribution of cerarii, 2) antenna, 3) para-ocular area, 4) last abdominal segment and three last cerarii, 5) last cerarius, 6) trilocular pore, 7) hindleg, 8) hind-coxa, 9) antenna of second larva.

were protected by ants against natural ennemies. Other records include Ceylon, Samoa, New Zealand, Australia, Puerto Rico, Brazil, Paraguay and Argentine.

Ferris recorded *P. comstocki* from a pine(!) in California in 1918 and from mulberry and maple trees in New York. Since then it has spread, and is a considerable pest in the eastern and southern states of the U.S.A., even in greenhouses in Canada. The N. American mealybug is rather polyphagous, especially on catalpa, apple, privet, ivy, wiegelea, *Photinia*, and many others.

P. comstocki is recorded as a major pest of bananas in some of the Canary Islands. Mc Dougall describes it: „On all of the above ground parts of the banana fruits and partly hidden in the bunch, on the axis, on the green blades, especially in the grooved part on each side of the central rib; on the lower surface in great numbers, especially along the midrib. Specially unfortunate is the fact that great colonies of the mealybug have as their favourite place for their existence the inner sides of the leaf-sheats. In this position the mealybugs receive almost complete protection from treatment directed against them." Badly infested banana leaves suffer and poorer fruit bunches result. This damage is kept small as the colonies of *P. comstocki* are permanently crushed with rags by gangs of children going over the plantations. On the fruits they cause — beginning at the bottom of the bunch — „going off" of the „fingers" and premature ripening. Considerable labour is also wanted for cleaning the fruit before the sale. The generations are overlapping and no true hibernation occurs in the Canary Islands Mc Dougall counted about 125 eggs per ovisac.

Of special interest to us is the record from Egypt (Hall 1923): „It is a little doubtful whether this species is an inhabitant of Egypt. Found on bananas, said to be of Egyptian origin, bought in a Cairo shop. Since, some specimens were found on *Agava* at Alexandria, which are doubtfully *P. comstocki* because and although the translucent pores were present on the hind limbs, they were much fewer in number than in typical *comstocki*." We have heard nothing, since about the occurrence of the scale in Egypt.

In the early thirties we obtained a letter from E. E. Green, stating in a list of scales he discovered in Cyprus also *P. comstocki* on citrus. Again, we have heard nothing since about this mealybug in Cyprus, where any outbreak could not possibly have escaped observation.

In Palestine a new citrus pest was first observed on a few trees in one of the older orange groves at Mikweh Israel. Aggregations of mealybugs formed compact white clusters, mainly along the midribs of leaves, on their upper- and lower sides, but always in the shade, very often between two leaves woven together by *Psocidae*, which abounded around the mealybugs, or in curled leaves. The trees were heavily covered by sooty mould. The initial outbreak took a very menacing form in the first initial foci, others were found at Petah Tikva, Rishon and Rehoboth. The trees looked white as snow on a black background of sooty mould. Heavy defoliation, fruit fall was followed by partial or complete death of the trees. Plate XXXIV shows such victims. Its area at present reaches from Hedera in the north to Rehoboth in the south, and from Jaffa inland to Ramleh, and beyond.

As the taxonomic literature available to the writer was insufficient to obtain a reliable identification, we sent specimens to various competent specialists. The first to answer was F. Laing of the British Museum:

„I am placing the citrus mealybug which arrived the other day as *P. comstocki* Kuwana. I don't seem to have had it on citrus from the Mediterranean before, but I see that Ferris, Jour. Econ. Ent. 12, 294, 1919 mentions citrus as one of the host plants and again in Bull. Ent. Res. XII, 211, 1921, records it again on citrus in Formosa. When I read your letter I mentally set the species down as *P. gahani*

Green, but it is not that species. The chief characters lie in the 17 pairs of cerarii, the anterior of which have 3—4 spines, the scarcely chitinised ultimate cerarius (except the underside), and the sensoria on the hind legs."

Upon examination of further material F. Laing found beside the species designated as *P. comstocki* Kuw. two more mealybugs with long caudal cerarii from citrus in Palestine. One of these was *P. adonidum* L. (=*P. longispinus* Targ.), the other a species of *Pseudococcus* without translucent pores in the hindlegs, which Laing could not identify with any known species. The former species was found by the writer 15 years ago in orange groves near Beirut and on *Sparmannia* in Jerusalem. It is always rare. We have never seen the second species, as since our return from Turkey in 1941 the mealybugs had all grown rather rare.

Other authorities have given divergent opinions. A young student, G. J. Rou from New York, says: „I checked the specimens with our catalpa mealybug and find them to be identical." Yet perhaps the leading and most competent authority, Dr. H. Morrison of Washington, comes to a different conclusion. He is certain that the Palestine species is not identical with the N. American one and concludes: „While variation plainly exists within the species in the mealybugs, I have no background that would indicate that it might extend so far as to include within one species specimens showing the differences enumerated above. It seems to me, that it would be much more practical and probably more nearly correct, to regard the Palestine specimens as representing a species thus far unplaced that is quite close to but definitely distinct from *comstocki*. In the absence of examples from Kuwana's type material, any question as to the correctness of Professor Ferris' placing of our material as this species will, of course, have to remain in abeyance."

Without feeling competent to decide the synonymy of the *P. comstocki*-group at present, we share the opinion of C. P. Clausen, that the *Pseudococcus* attacking citrus in Palestine be identical with the form attacking citrus in Japan, which apparently is not the same species as that found on apples, catalpa, etc. in the eastern states of the U.S.A. The latter is a close form of the same group. For the time, we identify the species which invaded Palestine in 1937, as the true *P. comstocki* Kuw. from Japan and China. We assume, that the catalpa mealybug of the eastern U.S.A. is a different, but related species. With regard to the banana mealybug from the Canary Islands no opinion can be given, before specimens have been authentically restudied.

This seems also the opinion of H. Compere: „I question whether the mealybug in the United States known as *P. comstocki* is the same as the species which is now known in Brazil as well as in Asia under the same name. The tendency to form compact groups and the characteristic marginal filaments are rather characteristic for this form. *P. comstocki* Kuw. is the common citrus mealybug in Brazil, but only rarely abundant enough to be rated as injurious. Scattered specimens on citrus in Misiones (Argentine) and small, but exceedingly injurious infestations on ornamental citrus trees at Asuncion (Paraguay). The individuals on citrus tend to congregate in small compact colonies which are sheltered between adhering leaves, within the folds of curled leaves, and between fruits that are touching each other."

Silvestri, however, still lumps all forms from all over the world. With the help of almost all competent specialists of the world we can at present apply the name *P. comstocki* Kuw. to our species. Yet we are unable to establish the authentical synonymy of the group to which this mealybug belongs. In addition to *P. comstocki* Kuw. and *P. citri* Risso, at least two additional species have been observed during these last years on citrus in this country: *P. adonidum* L. and another species (either *P. gahani* Green, *P. maritimus* Ehrh. or a new species), both of which are, so far,

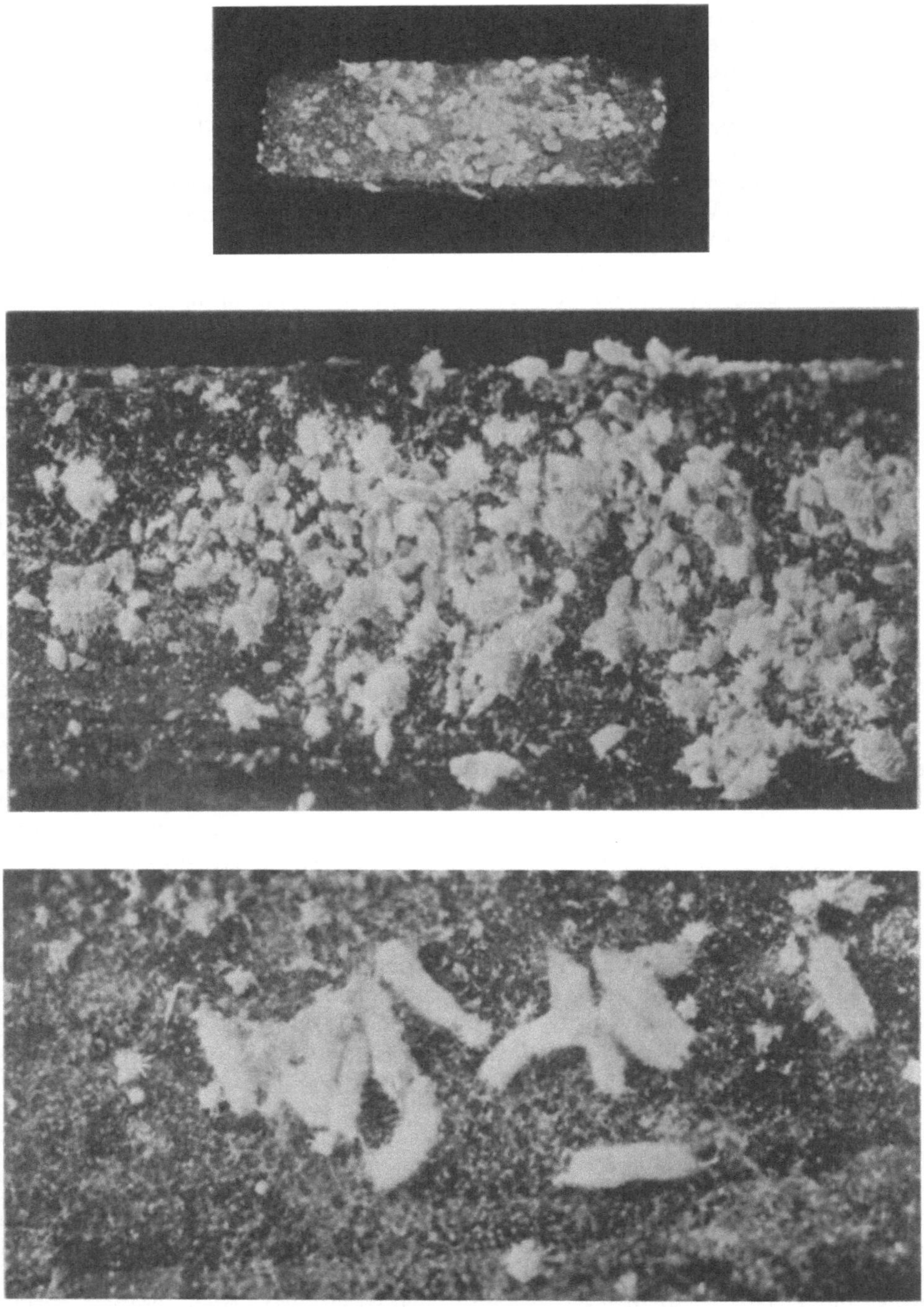

XXXIII. *Pseudococcus comstocki* Kuw. on bananas in Canary Islands. Adult female: A. Natural size, B. 4 times enlarged, C. Male puparia (from Mac Dougall)

extremely rare in Palestine. The origin of the species and the history of its introduction into Palestine remain obscure.

Quite recently N. S. Borkhsenius (1948) has utilised new characters for the identification of the *Pseudococcus comstocki*-group. His key to the adult females runs, as follows:

1. With 2—3 discoidal pores near eyes. Body-setae along posterior margin of the anterior dorsal ostioles up to 0.022 mm. long, those along the median dorsal line of the thorax up to 0.021 mm. long. Hind-coxae without translucent pores. Mushroom-shaped ducts usually with one discoidal micropore and without body-setae. Without irregular discoidal pores. *P. maritimus (Ehrh.)*

Without discoidal pores near eyes. Body-setae of posterior margin of anterior dorsal ostiole either 0.038 or 0.056 mm. long; those along mediodorsal line of thorax 0.055 or 0.084 mm. long. Hind-coxae with translucent pores. **2.**

2. Without a multilocular pore near thoracic spiracles. Longest body-setae on the posterior margin of the anterior dorsal ostiole 0.032 to 0.038, those of the median dorsal thoracic line 0.032 to 0.054 mm. long. *P. comstocki (Kuw.)*

With one multilocular pore near the posterior thoracic spiracles, and often one near the anterior spiracles. Longest body-setae on posterior margin of anterior dorsal ostiole 0.050 to 0.056, those od the median dorsal thoracic line 0.072 tox0.084 mm. long. *P. citriculus Green.*

The study of our material leads to the conclusion, that most local observations refer to *P. citriculus* Green, a few mealybugs intermixed with them point to *P. comstocki* (Kuw.). Borkhsenius also reports *P. maritimus* (Ehrh.) from Palestine. We have, so far, not yet found that species in this country.

The most complete description of *Pseudococcus citriculus* Green has so far been given by E. C. Zimmerman (Insects of Hawaii. Vol. 5. Honolulu. 1948. p. 210 ff, fig. 117 (2)—119). This species was first described by E. E. Green from Ceylon on Citrus. In Hawaii also on Citrus.

„Recognition characters:" 3 mm. long with 17 pairs of well developed cerarii. Anal lobe cerarii with two stout cones surrounded by a moderately crowded group of trilocular pores and a few slender setae of varying lengths, the whole borne in an elongate oval sclerotized area. Other cerarii each with two smaller cones, but in the cephalic region with 3 to 4 cones surrounded by a dense cluster of pores, and 3 or 4 slender setae. Dorsum sparsely beset with trilocular pores and a very few tubular ducts. These ducts are confined to a submarginal row, are as broad as a trilocular pore, and have a pronounced oral collar. Dorsal body setae quite numerous, of various lengths, mostly very long and slender.

Ventrad with distinct sclerotization on each anal lobe. Multilocular pores in small numbers in the midregion of segments 10 to 5. Also a few tubular ducts, but trilocular pores common.

Circulus present. Antennae 8-jointed, quite slender. Legs not distinctive. The especially distinctive features are the unusually long and slender dorsal setae combined with the sclerotized area about each anal lobe cerarius."

Gruenberg (1950) has made a number of counts of the characters of Borkhsenius and Ferris on material from various parts of Israel. They lead to the conclusion, that the characters of Borkhsenius are of doubtful importance. In most specimens a sclerotized area on the anal cerarii is present, which very often is pyriform. But multilocular pores near the posterior spiracle occur in less of half of the specimens. No correlation between the shape of the sclerotized area and the presence or absence of these multilocular pores exists. Neither is there any difference in the range of length of the longest dorsal setae or the dorso-ostial setae and the presence or absence of multilocular pores near the posterior spiracle. The greatest length of the dorsal setae ranges from 55 to 110 with the overwhelming majority between 70 and 85.

The number and the diameter of the trilocular pores on the area of the anal cerarii are both much greater than in the drawing of *P. citriculus* by Ferris.

The local species may be new. Also the habitual absence of coxal pores points in this direction. But in the present confusion we prefer not to name it until a comprehensive revision of the group is made. We shall call it here *P. citriculus* Green without claiming that it actually belongs to this species.

Host-plants: The *P. comstocki*-group is rather polyphagous. It has been reported from citrus: in China, Formosa, probably in Japan, S. America, Palestine and Cyprus,
from bananas: in the Canary Islands, in the West Indies and in Egypt(?).
from apple and pear trees: in Japan, New Zealand and the U.S.A.

Other hosts in Japan are: *Morus, Acer, Vitis, Evonymus,* persimmon, roots of soya-bean, water melons, etc. At Ceylon Green records it from *Loranthus, Trema* and *Cinnamomum.* In the U.S.A. the local form is especially polyphagous, being mentioned, among others, fromavocade, *Ampelopsis,* ivy, boxwood, privet, catalpa, gooseberry, hollymaple, pine, yew, etc.

In Palestine Comstock's mealybug has remained, so far, rather monophagous. Whilst infesting all citrus species, apparently without discrimination, it has been found once only in nature on any other host, on *Aralia sp.* (*Araliaceae*). Mr. A. Gruenberg has performed long series of breedings on various hosts, and has kindly put at our disposal the following unpublished results:

1. Plants on which apparently permanent breeding of *P. comstocki* is possible and which may suffer damage: *Citrus spp.* (*Rutaceae*), apples, pears and quinces (*Rosaceae*), potatoes (*Solanaceae*), *Arelia sp.* (*Areliaceae*), *Eugenia, Pytanga,* guava (*Myrtaceae*).

2. Plants on which *P. comstocki* may develop for one or two generations, but disappears afterwards: *Catalpa, Musa, Olea, Asparagus, Mesembryanthemum, Carica, Ceratonia, Begonia,* plums, peaches and avocado.

3. Plants on which no crawler could develop to maturity: *Opuntia, Agave, Ficus, Pittosporum, Eriobotrya, Acacia.*

None of the hosts of group 1 or 2 was infested in nature, inspite of their growing often in close neighbourhood of infested citrus trees, except citrus and *Aralia.* The other important fact is that the potential host range of the Palestine *P. comstocki* agrees well with that known from the entire *P. comstocki*-group.

Life-cycle: The breedings in our laboratory at Petah Tikva showed that 6 to 7 annual generations of *P. comstocki* can breed in our coastal plain. The generations, maintained on citrus nurslings, developed from beginning of oviposition (A) to the beginning of oviposition of the descendant generation (B) as follows:

No. of generation	Begin of oviposition A	B	Duration of development (days)	Mean temperature
1	1.VIII.1938	7.XI.	37	28.5° C
2	7.IX.	12.X.	35	25.8
3	12.X.	12.I.1939.	92	18.3
4	12.I.1939	27.IV.	105	16.2
5	27.IV.	9.VI.	43	23.9
6	9.VI.	12.VII.	33	26.3
7	12.VII.	14.VIII.	33	28.0

Breedings which began in November developed in 120 to 140 days. The development of the males lasted from 2 to 14 days less than of the female. The males hatch at the begin of the pre- oviposition period of the females. About six annual generations are thus the normal in the coastal plain.

The epidemiological hyperbola calculated from our breeding data, is:

Threshold of development (c.): 11.6° C,
Thermal constant (th.c.): 492 day-degrees.

Duration of generation at:	38° C	21.0 days,
	30	26.7
	25	36.7
	20	58.6
	15	144.7.

This hyperbola gives 6.2 annual generations in the climate of Tel Aviv. Two annual cycles were calculated, and are in very good agreements with all breedings:

Beginning of generation and days development			Beginning of generation and days development		
1.	10.I.	114 days	1.	1.XI.	144 days
2.	12.V.	43	2.	25.III	65
3.	24.VI.	35	3.	29.V.	39
4.	29.VII.	33	4.	6.VII.	33
5.	31.VIII.	36	5.	8.VIII.	34
6.	6.X.	66	6.	11.IX.	42
7.	2.XII.	(remain 93 day-degrees)	7.	23.X.	(remain 99 day-degrees).

As development is often slightly retarded in nature, for epidemiological purposes 5 to 6 annual generations will be correct. In any case, there is no winter diapause and all generations are heavily overlapping.

Adult longevity is very short in the males, which die soon after copulation. That of the female is rather long, as it deposits its eggs over a long period. It lasted in winter 88 (54 to 112) days, in spring 54 (45 to 76) and in early summer 36 (30 to 42) days. Of this period oviposition lasted about 60 days in late winter, 35 days in spring, about 27 days in early summer.

When not paired, the female begins to secrete the ovisac, but never deposits eggs. No parthenogenesis was ever observed in some series of experiments. Unpaired females lived longer than mated ones: 104 days in winter, 75 days in spring, and 65 days in early summer. Females which were late mated only, also had a longer longevity, but laid a much reduced number of eggs: 120 days (166 crawlers) in late winter, 97 (69) and 68 (53) days and crawlers in spring and early summer.

Males are always abundant, apparently approaching an equal sex-ratio. The dances of the male swarms around the heavily infested citrus trees in the morning and evening are an interesting mass-phenomenon, ressembling the well known dances of midges. Mr. P. Lichtenstein kindly observed this activity at Mikweh Israel during one complete October day, from sunrise to sunset. From 6 to 8h in the morning (temperatures below 20° C on thermometer exposed to the sun) only few males were on flight. With rising temperatures their number increased and huge swarms were dancing at 27° C and more. With the onset of the western breeze shortly before noon, the males disappeared. The swarms were always concentrated on the sunny side, towards south-east. From time to time they settled on leaves or fruits in the shade, but inside the crown very few males could be seen. Large numbers of males found their death sticking to the honeydew or caught in spider webs. Whilst on that day no evening dances were observed, we have seen repeatedly evening swarming of the males on windstill days.

Active migrations of the females of *P. comstocki* are rather conspicuous during mass outbreaks. Such migrations have not been observed to such extent in *P. citri*.

When a storm has blown the mealybugs by the thousands to the ground, they can be seen crawling upwards on the trees. Klein and Perzelan (1940), in order to determine the seasonal trend and extent of these migrations, put, from January to June, two burlap bandages on the stems of nine trees at the heights of 10—20 and of 50—60 cms. above the soil, each bandage being 15 cms. broad. The fortnightly counts revealed (per average tree):

Date	Upper bandage females	+ ovisacs	Lower bandage females	+ ovisacs
20.I.	1		1	
3.II.	1		2	
19.II.	1		1	
9.III.	5	1	3	
26.III.	42	16	5	2
12.IV.	33	17	9	4
25.IV.	88	27	30	9
14.V.	88	47	26	10
2.VI.	28	12	14	3
16.VI.	13	2	6	1
29.VI.	6	1	5	1

The mass migration in spring is probably connected with pre-oviposition migrations of the mature females, in search for sheltered places in crevices, between leaves or fruits, and such like. The peak coincides with the main period of hamsin winds. Yet more important still is the fact that more females were caught in the upper than in the lower bandage. This shows clearly that active migrations of such females which normally lived on the tree are involved and that they are by no means restricted to females deplaced from the trees by storms or other accidents. In spring there is a definite downward migration. This migrations lead, however, not to heavy root infestations on herbs, such as are observed quite frequently in *P. citri*. Gruenberg, in long quests, observed a very few mealybugs only on roots of citrus seedlings (up to 40 cm. deep) and once on roots of *Aralia*. The females are frequently associated with ants. These are the same species which we recorded for the citrus mealybug.

A very characteristic habit of the ovipositing females of *P. comstocki* is the aggregation of the females and their ovisacs to solid lumps. We have never seen similar aggregations in any other mealybug. Some dozens up to some hundreds of females are forming such a solid patch, where even two to three layers of females, one above the other may be observed.

We have mentioned before that oviposition is rather prolonged in Comstock's mealybug. This leads to the possibility that a disturbed egg-laying female may move a short distance away and begins to form a new ovisac. This may be repeated several times. This is, however, not the normal, as Rivnay (1939) assumes, but an exception. Under normal conditions the female remains on the selected spot, and forms one ovisac only.

Ovisacs were opened and their eggs counted, when most eggs were laid. This leaves the possibility of higher egg-numbers. These counts were checked by counting the number of crawlers leaving isolated ovisacs of similar age (Klein and Perzelan 1940):

Month	Av. eggs in laboratory	in grove	Av. crawlers in laboratory	in grove
March	207(167—238)			
April	319(169—494)	294(242—328)	136(79—282)	
May	295(114—475)	307(214—363)	216(90—421)	219(215—243)
June		324(294—372)	220(90—371)	193(93—338)
July			182(44—291)	

During the months of the counts about 300 eggs per female were laid, with the maximum at 500. Only 200 crawlers left in the average these 300 eggs with the maximum at 400 crawlers. About one third of the eggs die in conditions, where they are protected to be blown off by wind as well as before feeding by predators or parasitation. In April this climatic mortality was highest, reaching over 60 %. This is the influence of the devastating hamsin winds.

The scattered oviposition is expressed by scattered hatching of the crawlers. A few examples may illustrate this point (in days after appearance of first crawler):

Month	Percent of crawlers hatched			Death of female
	50	80	100%	after
March	31	46	58 days	62 days
April	12	21	32	–
May	11	22	40	40
June	4	10	35	30
July	3	6	13	22

The oviposition period ranges from two months in early spring to two weeks in summer. Up to 25 eggs per day have been laid.

The crawlers soon leave the ovisac and scatter around the maternal area, settling down at an early opportunity. They are bare, without wax-coating or filaments, and hiding in crevices, between two leaves etc. We have mentioned before, that during a heavy outbreak, very often an abundance of citrus leaves were woven together by *Psocidae,* which found ideal conditions of food on the dried honeydew, sooty mould, wax-secretions, etc. of the mealybugs. The relative duration of the various stages of the female have been compiled by Gruenberg, from whom we borrow the following illustrative data on two generations (Begin of):

Oviposition	Larva I	Larva II	Larva III	Adult	oviposition	cycle
4.IV.1940.	12.IV.	1.V.	17.V.	26.V.	3.VI.	60 days
23.VIII.1940.	17.VIII.	30.VIII.	8.IX.	24.IX.	26.IX.	33 days.

The eggs as well as the crawler are very sensitive to low temperatures. None of these stages survived a 24 hours stay at -2° C, and very few survived a 6 hours stay at that temperature. Later stages are much more resistent towards cold, but about +5° C was tolerated and survived by very few individuals only after an exposure for 3 to 5 weeks. (Gruenberg). Such lethal cold conditions do not occur in our coastal plain, but these results clearly indicate that cold is an important limiting geographical and epidemiological factor for this species in general, whilst still without primary importance in Palestine. All stages are able to survive 42—46° C during the hot hamsin periods, says Gruenberg, if properly sheltered and not exposed to sun and wind. As a matter of fact, climatic mortality is high during hamsin winds. How far this mortality depends on high temperatures or on low humidity we are yet unable to decide. Obviously Comstock's mealybug develops better and prefers higher humidities, the upper limit of which we were unable, so far, to establish. The females, and apparently all larvae, are restructed to shade. We had, owing to our absence in Turkey during the outbreak, no opportunity to study the physiological base of this taxis. The females are also very sensitive to wind, even to light breezes. They always seek shelter, where they are not exposed to them. Optimal conditions for the sense-physiological equilibrium of the female is a dark, rigidly sheltered (before wind and sun exposure), hide, where possibly the high thigmotaxic stimulation is another positive factor. The characteristic oviposition-aggregations of the females, when not in hidden places, obviously is the outcome of the creation of these favourable sence-physiological conditions by chance encounters.

The big initial outbreak, which took place within quite normal climatic extremes, as well as the following experience shows clearly, that climate alone is a heavy limiting factor, but by itself unable to explain the great population regression of Comstock's mealybug since 1940.

Damage: The very serious damage caused by the initial menace in 1937/38 has been roughly sketched out. Klein describes the damage in more details: In cases of heavy attacks the honey-dew can be seen dripping from leaves and fruits. Sooty mould covers enormous surfaces on those trees. In thick layers covering the fair layer of dried honey-dew, it covers thickly all parts of the plant. The crevices between this crust and the plant offer ideal hides for development, oviposition and male pupation.

With the appearance of the young growth in early spring the mealybugs migrate to the new growth, and in cases of heavy infestations may soon arrest the development of the young buds. The old crust of honeydew/sooty mould dries and peels off as a result of the hot winds blowing. Then the mealybugs migrate in quest of new crevices and hides. In early summer other mealybugs migrate to the young fruit where they settle beneath the sepals. Infested fruit in this way or on the contact area of two neighbouring fruits begins to drop from August onward increasingly. Klein counted over hundred fruits dropped in that way under one single tree in September. The drop of the fruit continues, often until half of the crops has been lost, accompanied by severe shedding of leaves and drying of twigs. In the autumn the natural colour orgy of the orange grove: green and golden, has given way to another one in black and white only: white by the wax-secretion of the aggregated ovisacs, black by sooty mould. The heavy winter rains wash off a large part of these white and black features. But in 1939 there were not a few cases where there was no, or very limited, blooming only. The sooty mould sticks heavily to the fruit. After having been cleaned, the fruit was excluded from grade A, being partly packed as grade B and C and partly rejected altogether. The abundant honeydew attracts of course a great number of other insects, amongst which we would like to mention only one minor citrus pest, the rindboring orange moth (*Cryptoblabes gnidiella*), which contributes on its part to the autumn drop of the fruit.

But the most important feature of the introduction and the outbreak of Comstock's mealybug into Palestine is the almost just as dramatical regression of the outbreak, as was the menace and terror connected with its first invasion. We have stated before, that the early history of the introduction of *P. comstocki* into this country is obscure. But information obtained by Dr. Klein tends to show, that at 1937, or before, the pest was established and well known in some groves at Jaffa. The troubled state of the country had prevented earlier information on this point. The actual area, within the limits indicated above, comproses, following the census of the Dept. of Agriculture, as follows:

Degree of damage	1939	1940
Heavy	3590	790 dunam
medium	5680	4630
light	16640	29020
Total	25900	34440

Since then the actual area infested has slowly grown, but all heavy and medium damage has disappeared, and even only a very small part of the infested area only shows even light damage. We have thus a very violent focal outbreak from summer 1937 which spread into other foci and reached its peak in 1939. The damage was much reduced in the older foci in 1940 and even in the new foci never reached by far

XXXIV. Damage of *Pseudococcus comstocki* Kuw. in Palestine. At right: Trees partly or entirely defoliated during the early invasion of the mealybug. Since 16 years no such damage is anymore seen. At left: small clusters of Comstock's mealybug, above: on grapefruit surrounded by heavy sooty mould; below: on orange leaves: 1. on upper, 2. on lower surface; 3. early stage of infestation. The single females are conspicuous along the midrib. In centre: adult female Cherd downwards).

the intensity of the oldest initial foci. Since then only light damage is observed in a very few foci. Even in new foci the damage remains neglegible. Apart of this major trend of the outbreak of *P. comstocki* there exists a seasonal population trend. The highest population density occurs from March to June. The number falls rapidly from June to a minimum in September. A slow increase begins in November, which continues throughout winter to reach its peak in the following spring. It is without

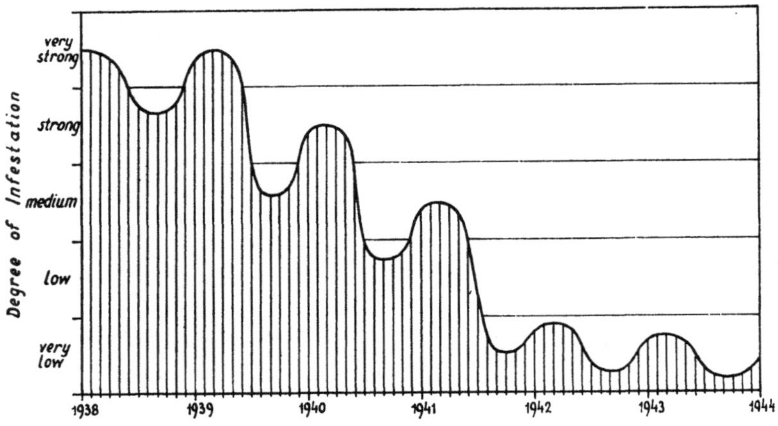

Fig. 108.
Schema of fluctuations of abundance of *Pseudococcus comstocki* Kuw. in Palestine following its introduction.

doubt that the main factors responsible for this summer decrease are natural enemies. In one experiment of Gruenberg, where the mealybugs were caged in an insectary free from any ennemies the mealybug population remained high throughout the summer. Climatic mortality is highest in the spring months. It is most severe against eggs and crawlers.

In 1949 the area infestation covers about 19.000 dunams of orange groves. According to Gruenberg (1950) it extends in the north beyond Hedera, in the east to Ein Hachoresh and Tulkarem (15 klm. from the coast), in the south to Majdel. The areas on the verge of infestation are: Gan Shmuel, Pardess Hanna and Benyamina. No infestation is, so far, known from the interior plains (Emek Yesreel, Jordan Valley).

The newly invaded areas may show foci of heavy damage for one year, which disappears in the second year. The infestation then drops to a very low level. So far, no damage has reappared in any area after the first years of infestation. The central citrus area, the first to be infested, is at present free from any damage.

CONTROL OF COMSTOCK'S MEALYBUG IN PALESTINE.

Technical control. As all attempts to introduce spraying or fumigation as control measures against *P. comstocki* in this country have since long been abandonned, we do not wish to insist upon critisisms of experiments reported and formulas recommanded. In fumigation the dosis toxica was close to or higher than the dosis curativa. In spraying with many substances added to mineral oils or similar insecticides none of these substances survived the tests either because of danger for the plant, or because of its unsatisfactory kill, or because some temporary claims never became substantiated in due time. The fact that together with highly qualified materials spraying of water in a strong jet and banding with cleaning the trunks have seriously been recommanded (Rivnay 1930), best illustrates the base for the claims of technical control. The progress of biological control in the following year made all these experiments and discussions superfluous.

Natural equilibrium and biological control: The following natural enemies and parasites are recorded by Silvestri (1920), Ishii (1928) and by Compere (1933) from China and Japan:

Hymenoptera.	*Anagyrus subalbipes* Ishii
	Cheiloneurus nagasakiensis Ishii
	Clausenia purpurea Ishii
	Coccophagus pseudococci Compere
	Leptomastix citri Ishii
Cecidomyidae.	*Diplosis sp.*
Coleoptera.	*Scymnus bipunctatus* Kug.

We remenber Silvestri's remark, repeated by other observers, that in its Oriental home *P. comstocki* inspite of being abundant, is very rarely injurious, and than only because ants, such as *Polyrhachis* or *Iridomyrmex*, protected the mealybug colonies by their permanent visits to the honeydew from their natural enemies. The entomophagous *Aspergillus perniciosus* was found on dead specimens of Comstock mealybugs, but artificial infestation even under conditions usually favouring *Aspergillus*-development, failed.

In Palestine it was the first aim to study, to what extent the endemic parasites of the citrus mealybug are attacking the newly arrived mealybug, inhabiting the same host and to a high degree the same ecological niche. Sufficient chance certainly was given for the biotic component of *P. citri* to meët with individuals of *P. comstocki*. The breedings of Klein, Rivnay and Perzelan showed that the following local mealybug parasites and enemies appeared on *P. comstocki* a few years after its introduction:

Encyrtidae:	*Leptomastix flavus* Merc. (parasitised by *Thysanus sp.*)
	Leptomastidea abnormis Gir.
	Anagyrus pseudococci Gir.
Coccinellidae:	*Scymnus suturalis*
	S. fenestratus
	S. includens (parasitised by *Homalotylus quaylei*).
	S. quadrimaculatus
	S. apetzi
	Oxynychus mormottani
Cecidomyidae:	*Diplosis sp.*
Neuroptera:	*Sympherobius sanctus* Tjed.
	Chrysopa sp. (parasitised by *Perilampus sp.*)

Most of these enemies and parasites we have discussed under *P. citri*. We wish to add here only a few words on their passing over to *P. comstocki*. It seems almost certain, that the original material of Comstock's mealybug, when introduced into this country, was free of parasites and predators. The first settlement must have occurred in groves where very few citrus mealybugs were present. In early 1939, this means only a few months after the outbreak of local foci, but in an area where citrus mealybug was always present, all three mentioned *Encyrtidae* were bred from *P. comstocki* and were increasing in importance towards the end of that year. The most important two are *L. abnormis* with fair activity in autumn and winter, *L. flavus* with fair activity during summer, but hempered by a hyperparasite (*Thysanus sp.*).

The Cecidomyid was common but sporadic, the *Chrysopa* rare and heavily parasitised. And with regard to *Sympherobius sanctus* experiments of Klein and Perzelan

(1940) revealed, „that the brown lacewing devours Comstock's mealybug only when deprived of all other food, and even then only in small numbers." Also among the *Scymnus*-beetles the sequence of preference was not the same for both mealybugs. Thus, *S. apetzi* and *S. quadrimaculatus* were much more active on citrus than on Comstock's mealybug. Yet the *Scymnus*-group as a whole was extremely valuable in the biological control of *P. comstocki*. But they show little activity in winter, and as fig. 106 shows, they are heavily parasitised by *Homalotylus* in summer. This summer parasitation, in addition to scarcity of food and perhaps the dry summer climate leads every year to a rigorous reduction of the lady-beetles from July onwards.

The conclusion is that most enemies of *P. citri* did pass fairly quick by over to the massive food offered by the outbreak foci of *P. comstocki*. Two years after the appearance of the first foci at Mikweh Israel all those local enemies, which could be expected to attack the new pest, had begun to be thoroughly active. Some others, such as *Sympherobius* and some *Scymnus*-species, did not participate in this transition, but the eagerness of other species and the excellent food conditions amply counterbalanced this deficiency. The main activity of the local enemies was concentrated on spring and autumn, but —as a whole — very low in summer and winter.

In November 1939 Dr. J. Carmon brought from a horticultural mission to Japan living breedings of *Clausenia purpurea* Ishii, one of the important native parasites of *P. comstocki* in Japan.

Comstock's mealybug parasite (Clausenia purpurea Ishii, 1923 *Japan).*

Description: Adult wasp metallic blue and green. Female 1.3—2.0, male 1.3 mm. long. Egg elongate. The originally rhomboid larva grows by two moults into an elongate sausage which pupates within the host.

Host: Develops in young, more common in older larvae and young females of *P. comstocki*, but not in those of *P. citri*.

Life-cycle: The adults feed eagerly on the honeydew of Comstock's mealybug. Males appeared in the breedings in small numbers only. Oviposition is shown in fig. 108. A number of eggs may be laid into one host, but one egg only develops to maturity. The number of offspring (not of eggs laid!) was 55 at 25, 29 at 30° C and still lower at 20° C. 20 to 27° C are the optimal range for reproduction with the peak at 25° C (Rivnay 1942).

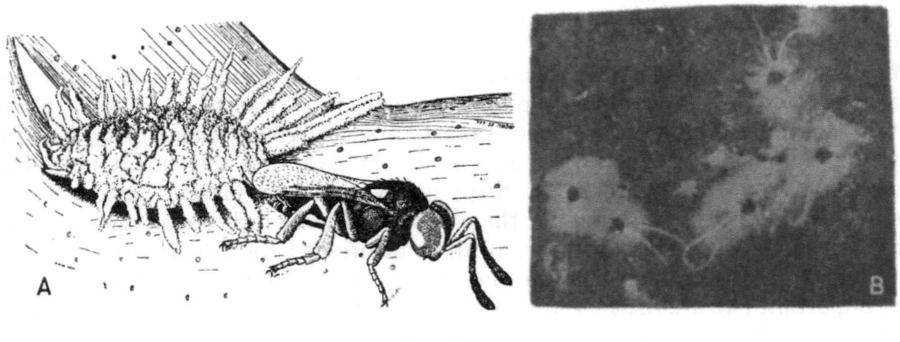

Fig. 109.
Clausenia purpurea Ishii: A. female ovipositing in Comstock's mealybug, B. mealybug with exit holes of parasite (from Rivnay).

The development lasts longer than in the other Encyrtid wasps. This means that the number of its annual generations is much lower. The following hyperbola was calculated:

Threshold of development (c.): 9.0° C,
Thermal constant (th.c.): 5.40 day-degrees.

Development at	34° C	21.6 days
	29	27.0
	24	36.0
	19	54.0
	14	108.0
	12	180.0

The number of annual generations is about eight, with a duration of 90 to 120 days in the wintering generation, the development of which is never really interrupted in the coastal plain. Adult longevity lasted 10 to 20 days at 20 to 30° C, 25 days at 15 to 20° C. The lower threshold for slow activity of the wasps is at 16° C, for normal activity at about 20°. 16° C are also the lower threshold No offspring developed at constant temperatures of 33° C to maturity, whilst oviposition still continues at 35° C. At 4 to 7° C in the refrigerator high pupal mortality begins after three weeks (Rivnay 1942).

No oviposition takes place in nature from December to March, when development lasts three months, as just stated. The winter mortality of the larvae and pupae of *Clausenia* is high and the survivors have a low reproductive rate. The summer heat apparently also reduces the reproductive potential by killing of larvae. The actual life history with 7 to 8 annual generations in the coastal plain is: Two generations of 45 days development each from March to May, four (30 days each) from June to September, one of 60 days in October/November, and one of 90 days from December to February. *Thysanus sp.* is a hyperparasite of *Clausenia* and other Encyrtids, with a low reproductive rate.

In all groves, where *Clausenia* was liberated, a prolonged low initial percentage of infestation was followed — depending upon season and upon host density — by high parasitation (60 to 70 % have been observed and higher ones may occur) and by a breakdown of the host, in the second year usually. Once established it sticks to the grove as long as a reasonable, even scarce, host population is maintained and spreads rather speedily into neighbouring groves.

Leptomastix flavus Mercet.

Female black or yellow, 2 mm, male 1 mm. long.

Development similar to *Leptomastidea abnormis.* It breeds in *P. comstocki,* but Rivnay and Perzelan (1943) could not breed it in *P. citri.* We also never bred it from the latter host. The normal host of this species in Palestine is still unknown. Also Mercet (1921) in his original description of the species from Madrid gives no host.

L. flavus is absent in winter, whilst its summer abundance is not affected by the heat, but depends upon host abundance. When *P. comstocki* was abundant, up to 35 % parasitism was obtained in summer, in years of less abundance of the mealybug less than 10 % were observed, and by now it has become very rare. The species of *Thysanus,* mentioned as a hyperparasite of *Clausenia,* also infested full grown larvae and perhaps pupae of *L. flavus.* This hyperparasite seems to have little influence on the abundance of its hosts.

Rivnay and Perzelan (1943) also mention a Proctotrypid wasp (*Bothriocaera sp.*) of uncertain status, which was once bred. Also *Isodromus sp.*, obtained once by Klein and Perzelan (1940) was not found again. They also mention a Pteromalid wasp (*Pachyneuron coccorum*), which almost certainly developed in *Coccus hesperidum* but not in the mealybug.

There is little doubt that is mainly due to the biotic complex, that the initial intensity of the outbreak of *P. comstocki* was broken. This is not an isolated event. Some mealybugs which invaded Egypt after the first world war, especially *Phenacoccus hirsutus* Green and *Pseudococcus filamentosus* Ckll., have lost after some years their initial impetus and are no more regarded as major pests. And the introduction of *P. comstocki* into Cyprus was not accompanied by any major outbreak at all. We have not the slightest doubt that a similar sequence of events would have been observed in Palestine, even if *Clausenia* had not been introduced. This assumption is strengthened by actual observations. Wherever new groves are now infested by Comstock's mealybug we observe a quick parasitation in spring by the local parasites, which later are deplaced by the *Clausenia*. The details of this process are now under study. This statement should not be misunderstood as a blaim for the introduction of *Clausenia*. Considering the very menacing initial situation, it was imperative to bring parasites from the natural home of *P. comstocki* into this country (Bodenheimer 1938). It is a retrospective consideration on the trend of events, if *Clausenia* had not been introduced. The final result would have been rather similar. Perhaps the regression of the mealybug would not have been as quick, as it actually was, perhaps it would not have been quite as radical as it is today, when Comstock's mealybug is not anymore even a minor pest. But the menace would have disappeared anyhow to little economic importance.

There is little doubt, that enemies and parasites are the most important single factor to reduce the outbreaks of this mealybug. Gruenberg (1950) comes to the following conclusions:

The first important phase in the diminution of the mealybug populations is the activity of *Scymnus spp.* (*Scymnus*-phase). The invasion of local parasites is a more slowly process, and it shows a great range of variation. In four groves under observation at Majdel, Hedera, Kfar Yavetz and Tulkarem different parasites played a dominant role. At a distance of 20 klm. one grove was cleared by *Leptomastidea abnormis*, the other by *Clausenia*. *L. abnormis* is in some cases doubtless responsible for the elimination of the pest. Yet it is still absent, both on *P. citri* and *P. comstocki*, in the Gaza area, where *Scymnus spp.* were helped by *Leptomastix flavus*. *Anagyrus* sp. which is active only in late summer, was of little importance, as the host population decreases before August. In other groves, such as that in Kfar Yavetz, *Clausenia* definitly was the checking agent.

In a grove close to Tulkarem a heavy infestation occurred in spring 1947, with all the symptoms of a heavy primary infestation. In the middle of June the damage was still very heavy. But in early August no living mealybugs could be found, and a few perforated exuviae only were seen. As no climatic extremes were observed in the period of observation, this rapid reduction seems to suggest the possibility of an epidemic. However this be, the rapid spread of this radical mortality over the entire area unto Hedera, makes it certain that other factors than entomophagous insects were responsible. In the biocomplex of *P. comstocki* none of the enemies (*Scymnus spp.*, *Clausenia*, *Leptomastidea*, *Leptomastix*, *Anagyrus*) has gained dominance. This is true, when the infestation is left to itself. The massive introduction of *Clausenia purpurea* at an early stage of infestation will always be useful. Its introduction at a later stage may just interfer with the work begun by the other parasites (Rivnay 1946).

We wish to add, that Gruenberg recommands a spray of summer-oils at the end of the second season of an infestation, in order to clean the trees from the crusts of sooty mould, which else may remain for a long time on the trees.

References.

F. S. Bodenheimer, Hadar 11. 1938. no. 7.
N. S. Borkhsenius, Bull. Ent. Res. 39, 1945, pp. 417—422.
H. Compere, Canad. Entom. 65. 1933. pp. 243—247.
E. E. Green, The Coccidae of Ceylon. London. Vol. V. 1922. p. 386, fig. 3.
A. Gruenberg, Pseudococcus citriculus Green in Israel. Thesis Jerusalem, 1950.
W. J. Hall, Min. Agric. Egypt, Techn. Bull. 36. 1923. p. 4.
W. S. Hough, Virginia Agric. Exper. Sta. Techn. Bull. 29. 1925.
T. Ishii, Dept. Agr. and Comm. Japan. Imp. Plant Quar. Sta. Bull. 3. 1923.
T. Ishii, Bull. Imp. Agric. Exper. Sta, Japan. III, 2. 1928.
H. Z. Klein and J. Perzelan, Hadar 13. 1940. no. 4.
S. S. Kuwana, Proceed. Calif. Acad. Sci. 3. 1902. p. 52 (Or. Descr.)
R. S. Mac Dougall, Bull. Ent. Res. 17. 1926. pp. 85—90.
E. Rivnay, Hadar 12. 1939. no. 7.; Bull. Soc. Fouad Ier Entom. 26. 1942. pp. 1—19, ibidem, 30, 1946. pp. 11—19.
E. Rivnay and J. Perzelan, Journ. Ent. Soc. S. Africa 6. 1943. pp. 9—25.
F. Silvestri, Transact. IVth Internat. Congr. Entom., Ithaca., Compendio Vol. I. Portici 1939 p. 670.
N. Takahashi, Niigata Agric. Exper. Sta. Rept. 23. 1927; Rept. 36. 1936.
N. Turner, Proceed. 7th Nat. Shade Tree Confer. New York. 1931. pp 131—132.
A. M. Woodside, Journ. Econ. Entom. 29. 1936. pp. 544—546.

H. PSEUDOCOCCUS ADONIDUM (LINNÉ).

Synonymy: Coccus adonidum Linné 1676.
Dactylopius longispinus Targioni 1869.

Description of adult female: Long oval with 17 pairs of wax-filaments (lateral $^1/_5$ to $^1/_7$ of body-length; two very long subcaudal ones (½ to twice the bodies length; the two preceding filaments half as long). Colour pale yellow with brownish median longitudinal stripe on back and legs of the same tinge. Body covered by white mealy secretion.

2,5 to 4,0 mm. long, 1,5 to 2,0 mm. broad.

Antennae 8-jointed, longer than in *P. citri*, the last joint being the longest, the subequal fourth and sixth joints the shortest. Legs also longer than in *P. citri*.

On most segments 2 (to 3) spines, 12 to 20 small glands and 1 to 3 hairs on each side. The later are absent in *P. citri*. Characteristic is the structure of the anal lobe. Two groups of stout spines with a group of small, dense, trilocular pores within a strongly chitinized and well delimited area. A similar arrangement, as mentioned above, but smaller, on the margin of the preceeding segments. The apical setae of the anal lobes about as long as the six anal setae.

Eggs: yellowish.

Distribution: Ubiquistic in tropical and subtropical countries.

Hosts: Polyphagous. In the Mediterranean region mainly on ornamental plants in gardens, parks and greenhouses (*Ficus, Strelitzia, Myoporum, Dracaena, Musa* e.a.) Common in many Mediterranean and other countries on citrus, but has never been observed in large numbers or as a major pest. The writer has seen it some years ago near Beirut on oranges. It has lately been discovered also in Palestine on this host. In this country it was first collected in 1924 on an ornamental *Sparmannia africana* in Jerusalem. *P. adonidum* is almost certainly a relatively recent introduction into Palestine on ornamental plants. It has also been lately found here in small numbers on citrus.

References.

A. Balachowsky et L. Mesnil, Les insectes nuisibles aux plantes cultivées. Paris. 1936. pp. 1606—1609.

I. THE CITRUS FLUTED SCALE (*icerya purchasi maskell*).

Description: The *adult* (called female in earlier papers, but really being hermaphrodite) is broadly oval, reddish brown. Ventral surface flat; back relatively flat in the immature adult, growing more and more markedly convex. Lead-red marginal and dorsal spots appear on the dark red ground colour of the mature individuals, lateral wax bristles up to 46 in number. Body of ten covered with a thin, powdery, white wax-secretion. Antennae and legs black, ventral surface light orange. All these colour indications are subject to rather large variations. Body segmentation indistinct.

Ovisac white, as broad as the adult body. Dorsally convex and distinctly fluted. The adult is raised, often in a right angle, the anterior end only still attached to the host. The number of ridges which form the ovisac is 14 to 16. Its back is of firm texture, growing less compact and thicker ventrally. At the posterior end, ventrally, an oval opening for the hatching of the larvae. The doughy stringy consistence of the ovisac changes toward death into a more fibrous structure.

Size of body 3.0 to 4.5 × 2.0 to 3.3 mm. (in Japan 3.3 to 8.2 × 2.0 to 7.3 mm). Length of ovisac 2.0 to 6.0 mm (up to 10 mm in Japan).

Microscopic characters: Antennae short and slender, 11- (10-)jointed, the last joint being by far the longest, moderately covered with short bristles. The prominent, subhemispherical eyes are placed behind the antennae. Mouthparts small, mentum 2-jointed, the rostral loop very weak, scarcely reaching the second coxae. Legs short, robust, heavily chitinized and hairy; the tibia being longest, the femur slightly and the tarsus much shorter; claw slender, curved, tarsal digitules slender and hairlike, those of the claw knobbed and extenting beyond the tip. The two pairs of thoracical spiracles conspicuous, located behind the forelegs; the second pair about the hight of the hind coxae. A groove extends from each thoracical spiracle to the margin. The two pairs of abdominal spiracles are on the back, near the margin of the seventh resp. eight segment. They are much smaller and not very conspicuous. Anal opening large, membranous, oval and hairless. Its location is dorsally near the posterior end of the abdomen and surrounded by many long, brown hairs.

Numerous black hairs, raising from small cones, of varying size (up to 0.5 mm) on the dorsum, forming dense clusters on the median portion of meso- and metathorax as well as on the margin. Much finer, brownish hairs on the ventral surface, densily in the thoracic region, much more dispersed on the abdomen. Derm pores of two general types. Large, open centre pores with a marginal band of 8 loculi among the marginal hairtufts, which produce the long, straight, glassy filaments. Smaller pores with solid, uni- or bilocular centre and a marginal band of 9 or 11 loculi are densely spread all over the derm. A conspicuous abdominal band of closely set pores on the ventral surface, extending across the abdomen near the posterior legs, curving back and running parallel and close to the body margin. This ovisac band secretes the ovisac. The thin filaments of adjoining pores melt together, thus forming the ridges of the ovisac.

Eggs elongate oval, smooth of pink colour. Size 0.6 to 0.8 × 0.3 mm. The empty egg-shells have a very characteristic shape (Dingler).

Larva I elongate oval, flat below and slightly convex above, reddish in colour with blackish antennae, eyes and legs. Size 0.6 to 0.7 × 0.3 to 0.35 mm. Antennae long and slender, 6-jointed, the last joint being by far the longest and bearing 4 long bristles, surpassing the length of the whole antenna. The prominent eyes are on the margin of the head, behind the antennae. Mouthparts well developed, rostral loop surpassing the hind coxae. Apical margin of back with spatulate hairs. Legs long and slender, the tibia being longest, the slightly shorter femur and tarsus being of subequal length; claw long, slender, slightly curved and with a tiny denticle near the tip; digitules of claw knobbed and extending beyond the tip. Two pairs of thoracic and the same number of abdominal spiracles. Body segmentation indistinct. Caudal end of abdomen with 6 thin hairs, as long as the entire body. Anal opening round and hairless.

Shortly after fixation on the host, the body becomes covered with white wax plates. It is not possible to accept Dingler's viewpoint, who regards both, the bare egg-larve and the *wax-covered* larva before the first moult as different larval stages. Derm with scattered hairs and pores *of the* smaller type.

Larva II reddish brown, soon covered with wax-plates, antennae and legs black, ventrally light orange. Body thicker, legs and antennae much more robust than in the first stage larva. Size 2.2 × 1.3 mm.

Antennae still 6-jointed, but whereas joint II—V were subequal in the first larva, the third joint is now almost twice as long as the more or less sub-equal joints II, IV and V. Legs proportionally shorter, but more robust, with similar proportions. The six caudal hairs much shorter than in the first larva. Both types of derm pores appear, and the marginal open centre pores secrete long conspicuous white wax bristles of 2 to 3 mm length.

Larva III broadly oval, reddish brown, soon covered by wax plates. Size 3.0 × 1.6 mm. The stout antennae are now 9-(8-) jointed, the last joint being by far the longest, the other joints being of subequal size. Legs stout. The relative proportions approaching those of the adult, especially the relative size of the tarsus much reduced. Dermal hairs stout and black, much more abundant and grouping in tufts, especially along the margin. Both types of dermal pores much more numerous than in the previous stages. The lateral wax-bristles grow longer and their number increases. A lead-red dorsal protuberance is conspicuous, but soon covered by wax secretions.

Development stages of the male (after Kuwana, Menozzi and Leonardi). The first two larval stages are identical for both sexes. Toward the end of the larva II the male is much more active, larger and slender as compared with the female larva (2.5 × 1.5 mm, Kuwana). The *prepupa* is orange coloured. The elongate oval, flat body is restricted between thorax and abdomen. Antennae stout, 9-jointed. Mouthparts wanting or rudimentary. Legs short and stout, claws short, curved and digitules as fine hairs. Size 3.5 × 1.5 mm.

The *pupa* is elliptic, orange-red with dark purple eyes, yellowish-brown wingpads and legs. Body appendices free. Abdominal segments distinct. Derm with the smaller type of pores only. With 4 pairs of spiracles. Cocoon irregular elongate, white 6 to 10 × 2 to 2.5 mm.

The beautiful *male* is reddish purple with shining metallic blue wings. 3 mm long, 6 mm wing expanse. Head triangular. Antennae long, feathery, dark brown, 10-jointed, all joints being subequal, except the two basal ones, which are much shorter. Two ocelli besides the compound eyes. Mouthparts absent. Legs slender, the tibia being longest, the femur shorter and the tarsus very small; claw slender, simple, slightly curved with the digitules as fine hairs. Wings with costa, subcosta and media and with two white longitudinal lines. Halteres foliated with 2 to 3 hooks at the tip. Four pairs of spiracles on the usual location. Two short fleshy appendages with long bristles at their caudal end. No stylus.

Distribution: Ubiquitous in tropical and subtropical. regions.

Life-history: It is difficult to obtain the individual life-cycle of *Icerya* under natural conditions. Out of 196 individual breedings in 1928, 130 were prematurely interrupted by *Novius*, 65 other individuals died before reaching maturity, (probably as a result of *Novius* or of being washed off by rains), and only one individual survived to oviposit. Similar results were obtained in the following years. Only very few breedings yielded exact data on the length of all stages and these may be compared with similar ones from Japan.

Month of oviposition	Duration of					
	Egg	Larva I	Larva II	Larva III	Preoviposition period	Total
VII 1930	35	17	18	11	15	96 days
VII	35	12	16	24	11	88 ,,
VII	35	14	16	16	17	98 ,,
VI 1931	16	19	14	18	12	79 ,,
VII	16	18	17	23	15	88 ,,
Japan (Kuwana)	21—27	14—21	14—21	21—50		70—119 ,,

It is easy, however, to follow the development of different generations: Ovisacs were generally formed at the end of December 1928 to early January 1929. The females of this first generation matured in the second half of May. Oviposition of this second generation took place in the second half of August. The ovisacs of the third generation appeared at the end of October to early November.

In February of 1932 ovisacs were generally formed, from which the larvae hatched at the end of March. In May and August, the first and second generations matured, in November the third.

3½ generations per year are the rule. In exceptional years, either 3 or 4 generations may occur. As a rule seven generations are to be expected in two years.

We calcule the hyperbola from the average number of 2½ generations per annum in Shizuoka (*Kuwana*), of 3½ generations at Hedera (*Bodenheimer*).

The total temperature sum above 0° C is for two years
at Shizuoka 10.332 days degrees : 5=2.066 day-degrees per generation
at Hedera 14.568 days degrees : 7=2.081 day-degrees per generation

The identity of the quotient per generation proves, that the development-threshold is at about 0° C. Starting the calculation with these constants (c=0° C, Th.C.=2074 day-degrees), we obtain the hyperbola, illustrated by fig. 110. The development will last

at 35° C	59 days	15° C	138 days
30	69	10	207
25	83	5	415
20	104		

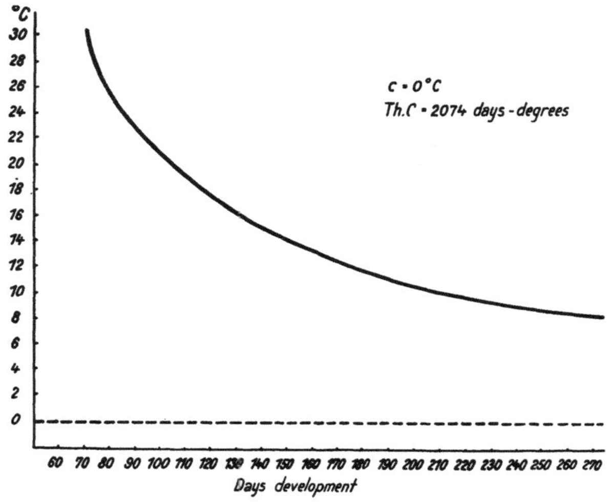

Fig. 110.
Temperature dependency of shortest development of *Icerya purchasi* Mask.

The preoviposition period is included in the above computation and the average duration of development of each generation has been chosen as the starting point of the period. The results may, therefore, be applied to epidemiological problems.

Balachowsky has arrived at another formula for the hyperbola. At Antibes (S. France) two breedings of generations from larva to larva lasted as follows:

19.V.30—22.VIII.30 95 days at 21° C average temperature
28.VIII.30—8.V.31 258 days at 13° C average temperature
c = 8,5° C Th. C. = 1188 days degrees.

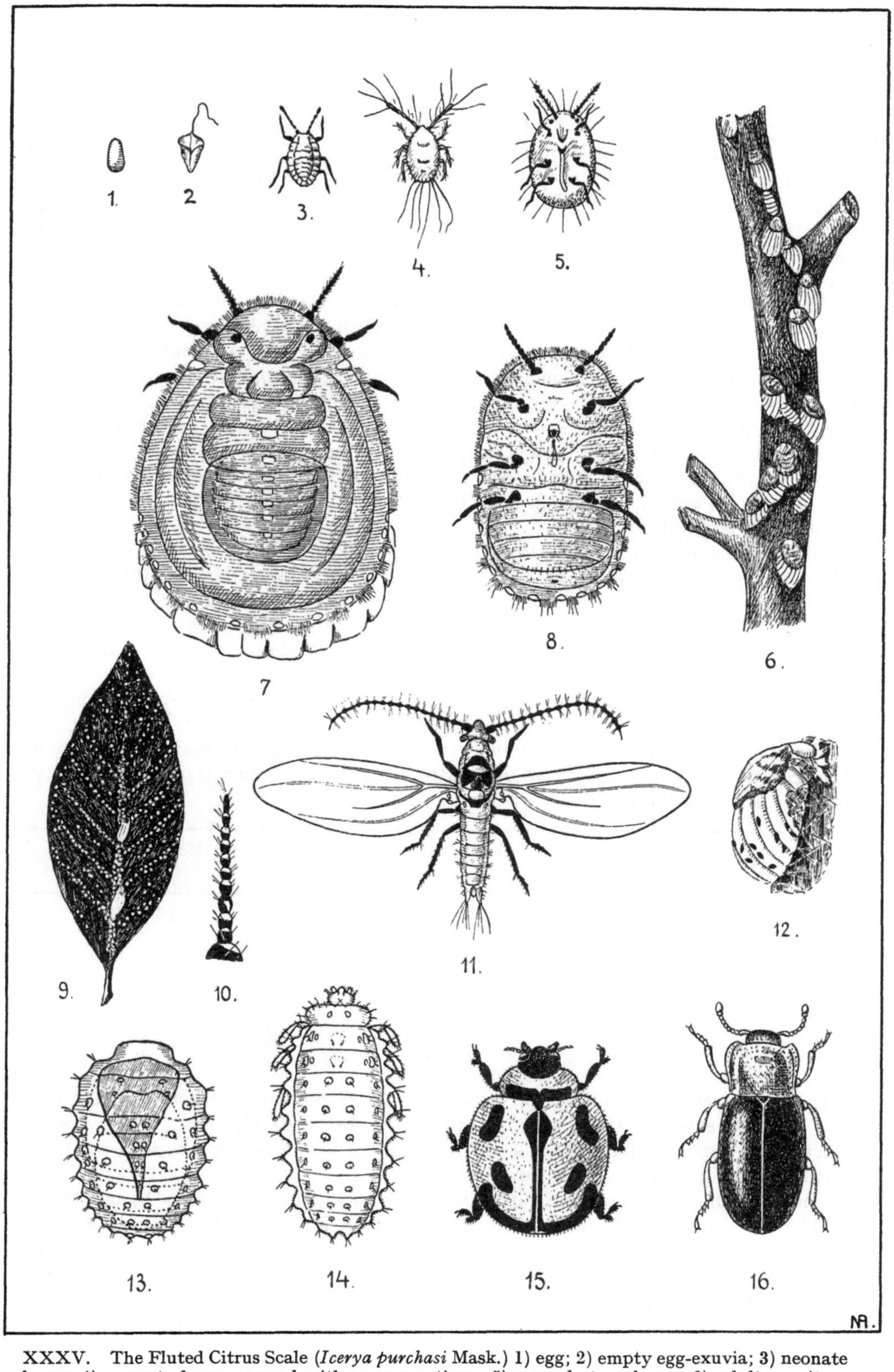

XXXV. The Fluted Citrus Scale (*Icerya purchasi* Mask.) 1) egg; 2) empty egg-exuvia; 3) neonate larva; 4) neonate larva covered with wax-secretions; 5) second stage larva; 6) adult on citrus brqach; 7) adult at the beginning of ovisac formation; 8) adult seen from below; 9) larvae settled along the veins of a citrus leaf; 10) antenna of adult; 11) adult male; 12) adult with ovisac on which *Novius*-eggs are pasted; 13—15) *Novius cardinalis* Muls.; pupa, larva and beetle; 16) *Lithophilus marginatus* Rtt.: beetle.

The corresponding values from our hyperbola for the average temperatures quoted, are 99 and 159 days respectively. The first of these values agrees fairly well with our hyperbola, (deviation of 4 %). The second value shows such a deviation, that an explanation must be looked for.

Fortunately, this discrepancy can be explained. Low average temperatures in winter (8,0 — 9,9° C from XII.30 — IV.31), parallel hibernation and leaf-drop of most trees, much lower minimal temperatures, create a rather unfavorable environment for *Icerya*. And *Icerya* seems to respond to these unfavorable conditions with a very usual response of animals: It enters a period of lowered activity, a kind of torpor or diapause, an interruption of development. Balachowski himself (1930) agrees that such a period lasts three months (December to February) at Antibes. No generation, however, with an interrupted development can safely be made the starting point for calculating the hyperbola. Balachowsky's observation seems, therefore, an important proof, that *Icerya* enters a true diapause on its northern boundaries, in the zone of regular winter-frosts and leaf-drop. This diapause limits the number of annual generations to two, instead of the calculated 2.6. The effect is again the synchronization of the two generations in nature. The first larvae of the third generation would meet, in winter, with such unfavorable conditions, that a complete mortality would be the result.

The number of annual generations for some localities is compiled on page (498): The most important result of this hyperbola is the explanation of the hitherto unexplained fact, that *Icerya* does not develop more than four annual generations, even in its tropical home. Accordingly, *Icerya* can be considered as a new type of regulation in abundance of populations, which has not been previously analyzed. The extremely low development-threshold, together with the high value of the thermal-constant, cause a conspicuous stability in the development-speed at higher temperatures.

Limiting factors for the development of *Icerya* and its geographical distribution. The scale of activity for *Icerya purchasi* introduces us to the normal range of activity, as well as to its limits.

Scale of activity	Young, first stage larvae ($n = 30$)	Young adults ($n = 29$)
1) Beginning of cold torpor. . . .	19,0° C	16,7° C
2)		
3) Crawling with interruptions. . .	23.2° C	23,8° C
4) Normal activity.	23,4° C (22,5—27,5)	24,2° C (23—27)
5) Strong activity	28,6° C	27,7° C
6) High irritation	33,0° C	35,6° C
7) Beginning of heat paralysis. . .	—	41,3° C
8) Instantaneous heat-death. . . .	38,9° C	42,2° C

In the temperature gradient the insects remain immovable at their original place. The preference temperature was, therefore, not observed. The limit of the phobic reaction to heat was for young larvae at 33,0° C. This value agrees well with the corresponding result in the scale of activity.

The low value for heat death explains the detrimental effect of a hamsin wind of 38—42° C and of several days duration on the *Icerya* population. The spring-hamsins influence the then increasing population trend decidedly; their absence aids the increase, their presence interrupts it.

This low temperature for heat death as well as the optimal condition of humid warm weather characterize a true tropical, in contrast to a subtropical animal.

The latter is fit to live in the subtropics, provided extreme temperatures are avoided and a certain minimal humidity is maintained. This physiological behavior fits in very well with the climate of tropical islands, like Mauritius and Bourbon (to the east of Madagascar), as well as with that of S. China. The first was mentioned by Riley as the home of *Icerya*, the last was recently suggested by Silvestri.

The theory prevailing in textbooks that Australia is the original home of *Icerya*, is based on the fact that *Novius cardinalis* and a number of other parasites have been first discovered there. If it could be definitely shown that *Novius cardinalis* is really an Australian native, this theory would gain in strength. But an assumption that *Novius* has been imported into Australia together with its host is even more probable.

The scale was first introduced in California in 1868, probably from Australia. It reached Florida (via California) only in 1893. In S. Australia Koebele found it in 1888, in the southern coast district only. It was first discovered in Formosa in 1905 and established there definitely as early as 1907. It has been introduced into Japan from California and Formosa in 1908 and 1910, with citrus stock. To-day, it is widely distributed on the main island.

On the French Riviera it was first observed in 1912, at St. Jean Cap Ferret. Its gradual extension is described by Balachowsky and Molinari. Up to date, it has not surpassed in France any point of the isotherm of 12,5° C. In Algeria it appeared in a nursery, as late as 1920, and it rapidly spread from there over the Mediterranean coast plain. In New Zealand, where it was first described by Maskell, in Australia and in Formosa, it is associated mainly with Acacia. If this plant should prove to be the original host of *Icerya*, the theory of Australia (its N. Eastern territory perhaps) being the original home of *Icerya*, would gain in strength. Should, however, citrus, with which *Icerya* is associated as well, prove to be the original host, the Oriental origin of *Icerya* would be more probable. No definite criterion, which would solve this question satisfactorily, exists at present.

In Palestine, it has been repeatedly introduced, probably since the beginning of this century, or somewhat earlier. In 1911 and 1912, it showed a large increase in the orange groves of Petah-Tikva, which is the first reliable record of this species in Palestine. It is now widely spread over the coast plain. In gardens of Jerusalem it is by no means uncommon, occurs at Tiberias, but is not as yet discovered in the lower Jordan Valley. In the mountains its distribution is not continuous, but is restricted mainly to ornamental plants, on which it has been directly or indirectly introduced from nurseries in the coastal plain.

Identical factors govern the annual fluctuations in the *Icerya* population and its geographical distribution. Frost-temperatures during the winter are of great importance in the moderate climate. Balachowsky and Molinari report, that at the French Riviera, in the extremely severe winter of 1928/29 (—12° C being the minimum) many foci of *Icerya* infestation died out. The severe winters of both the massives of the Alps in France and the high plateaus in Algiers, are insurmountible obstacles to its spread. *Icerya* develops fairly well during summer as far to the north as Paris and may even be able to maintain itself there during mild winters. But on the whole it perishes in free nature, in the first somewhat severe winter.

With regard to the northern limit of *Icerya* distribution Balachowsky has misunderstood the present author, thinking that the development-threshold and its isotherm are the determining factors. The fact, that in *Gueriniella serratulae* *) this isotherm and its northern limit of distribution coincide, is only accidental. In *Icerya*, as in many other species, the area is much earlier limited than at this isotherm. Other species may

---

*) Bodenheimer, Bull. Soc. Ent. France. 1927. pp. 195—198.

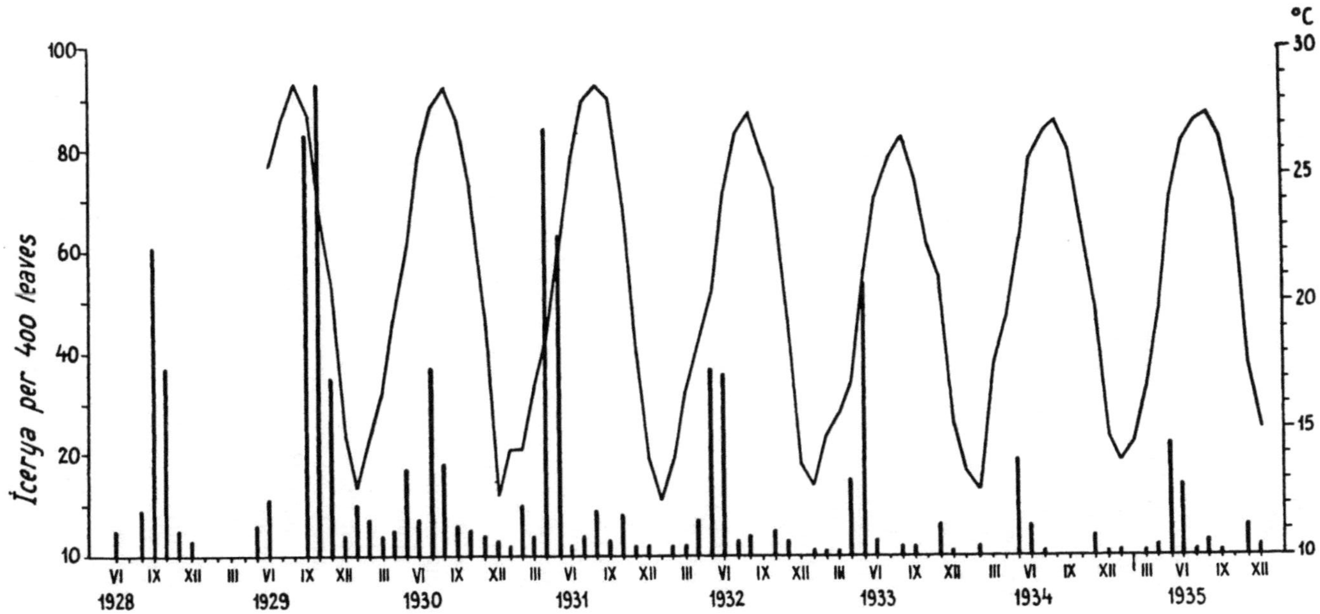

Fig. 111.
Fluctuation of *Icerya purchasi* population in the Atta groves (near Hedera; per 400 leaves) and that of temperature from 1929 to 1935).

be able to develop even in an area where two years are required for the development. There is not reason whatsoever to expect a future spread of this tropical insect up to the 0° C isotherm.

The low thermal death point (38,9° C for young larvae, 42,2° C for young adults) and a certain sensibility to dry atmosphere, are both limiting factors in hot climates. These limitations are not as severe as those of frost. *Icerya* may be found in the Saharo-Sindian territory, where its population level is always very low.

The mortality of the eggs is low. Kuwana observed a 20 % mortality in Japan. In Palestine it proved to be in general even less, but unfortunately we failed to make observations after hamsins. The main mortality probably occurs in the first stage larvae, Dingler observed that about 14 larvae, out of 115, died as a result of failure to settle on the host. The host (*Laurus*) is probably not a favorite of *Icerya*. Besides this host-conditioned mortality, the climatic mortality under unfavorable climatic conditions as hamsins, dry heat, frost, heavy rains, etc., is very high. Cold and high dry heat are the main limiting factors. A warm, humid climate, as that of greenhouses, is favorable for a mass increase and is probably its vital optimum.

Host plants and food-ecology: I. purchasi is considered to be extremely polyphagous, as the number of recorded hosts is very large. The number of hosts, however, on which the whole development is performed successfully, is practically much more limited.

In the literature we find some indications of *Icerya* as a pest of potatoes. This induced the author to attempt a mass-breeding of *Icerya* on potato-sprouts (in connection with the artificial mass-breeding of *Novius*). Adult females commenced with the formation of the ovisacs without any difficulty. But not a single larva settled on this host. All the larvae left the potato-sprouts and died. The mothers, which migrate to some extent before oviposition — as do many Monophlebinae — and in all likelihood do not take any food beginning with this migratory period, would have oviposited even on a wooden box. Many records of hosts are based only on the presence of adults with ovisacs on those plants. When no further development stages are found on the plant, the observation of an adult bearing an ovisac is not sufficient evidence to establish that plant as a definite host.

Only very few larvae develop successfully on other plants, Acacia farnesiana for example. Many fail to suck at all and may die during early development. Such plants are true, second grade hosts. Among such may be included certain other plants on which older larvae may develop, younger ones die. Only such plants as Citrus, Spartium, etc., on which the whole development takes place successfully and on which under suitable environmental conditions the population is rapidly increasing are true, primary hosts.

The host may influence the fertility, the insect's inclination to suck on it, the speed of development and the larval mortality. We want more data on these points. According to our observations, fertility is decidedly affected by the host, the speed of development only in to a minor degree, if at all. Both are insufficient, though, to explain the great difference in the susceptibility of the insects towards the various hosts. It is neither very probably that the larval mortality can be the decisive factor; but it may well be that the taste, especially that of the young larva, can serve as such. In that case, the whole foodecology of *Icerya* would be based primarily on the inherited taste.

Among citrus species, sour lemon is most heavily attacked. On young inoculated trees those grafted on sweet lemon are dicidedly preferred to those grafted on bitter orange. The reason for this difference in host relationships (mortality, taste, etc.) is still unknown.

All stages may endure hunger for some days. Kuwana gives the following average longevity without food:

Larva I	stage (recently hatched)	6,9 days
	(older)	11,3 ,,
Larva II	,,	14,0 ,,
Larva III	,,	15,5 ,,
Adult female	(recently hatched)	10,0 ,,
	(older)	14,2 ,,

This great tolerance of hunger is certainly an important ecological fact, which helps towards the maintenance of the species in an unfavorable environment.

List of host plants in Palestine.

Anacardiaceae:	*Schinus molle*
	Schinus terebinthifolius
Casuarinaceae:	*Casuarina tenuissima*
Compositae:	*Artemisia monosperma*
	Xanthium strumosum
Euphorbiaceae:	*Ricinus communis*
Geraniaceae:	*Pelargonium sp.*
Juncaceae:	*Juncus*
Lauraceae:	*Laurus nobilis*
Leguminosae:	*Acacia farnesiana*
	Acacia sp.
	Calycotome villosa
	Spartium junceum
	Gleditschia sp.
	Caesalpinia sp.
	Cassia
Marantaceae:	*Maranta bicolor*
Pittosporaceae:	*Pittosporum*
Polygonaceae:	*Polygonum equisetiforme*
Punicaceae:	*Punica granatum*
Rosaceae:	*Rosa spp.*
	Fragaria
Rutaceae:	*Citrus spp.*

The following are the main host plants in Japan (out of more than a 100 species, from over 40 families, after Kuwana):

Rutaceae:	*Citrus spp.*
Rosaceae:	*Pirus malus, Pirus imensis*
Moraceae:	*Morus alba*
Berberidaceae:	*Nandina domestica*
Leguminosae:	*Lespedosa sp.*
	Indigofera tinctoria
	Glycina soja
	Cytisus scoparius

The most susceptible host plants of Egypt (after Hall) are the following:

Bignoniaceae:	*Jacaranda mimosaefolia*
Euphorbiaceae:	*Acalypha sp.*
Pittosporaceae:	*Pittosporum sp.*
Rosaceae:	*Rosa sp.*
Rutaceae:	*Citrus spp.*

Distribution of Icerya on the host: Four citrus trees of medium infestation were counted carefully, in order to study the total *Icerya*-population per tree, its distribution in the height of the tree and towards the direction of the compass. The number of adults is fairly correct, but many larvae have been surely overlooked, though equally in all directions.

Distribution of *Icerya purchasi* on four orange trees in September 1931 (larvae in brackets).

Height	North	East	South	West	Total ad. (l.) total	Novius
0—1,0 mm	53(12)	44(23)	37(15)	64(15)	198(15) 213	10
1—0,5	47(10)	55(13)	46(15)	44(11)	192(49) 241	4
1,5—2,0	46(18)	48(28)	34(17)	38(10)	168(73) 241	14
2,0—2,5	11(10)	15(—)	36(2)	18(3)	90(15) 105	8
Total.	**207**	**226**	**202**	**203**	**648** (197) L.	**36**

The distribution is therefore more or less equal with regard to the height and the directions of the compass.

The average number of 162 adults + 49 larvae per tree is much higher than could have been estimated even by a careful observer. It explains easily the large honeydew production per tree, especially if we add the probably considerable number of larvae overlooked.

The number of *Novius* (beetles only) related to the number of adult *Icerya* amounts to 4,3 %.

The early larval stages migrate all towards the outer parts of the trees, where there is much light and settle along the mid-ribs of the leaves. The migration of the young adults leads invariably back into the shady interior or to the trunk of the tree, where they settle on stem, branches and twigs as well as on leaves. This cyclic migration is of importance for the understanding of the distribution of the different stages on the tree.

Sense ecology, locomotion and dispersion. Sense reactions. The larvae of *Icerya purchasi* Mask. are very positively phototaxic, only that they react at noon much slower than in the morning hours. They are negatively geotaxic. In a dark room, *Icerya* travels directly upwards. When a few individuals are in the same experiment, they climb upwards in parallel ways to the top margin. In the case of interference of photo- and geotaxis, the response generally follows the parallelogram of forces. But they may also travel first to the side nearest the light, then up to the highest point. The response to geotaxis is much slower in the dark than in the light.

Icerya is decidedly stimulated by chemical irritations, but the chemotaxis is always outweighed by photo- or geotaxis. Any real orientation towards the source of chemical irritations (odour) could not be proven, though it is probably present (Exp.1). The preference of the larvae, especially before moulting, to settle on the central vein of the leaf, is due to positive thigmotaxis.

Experiment No. 1: Icerya larvae, when placed in a cylindrical tube of a smell chamber, migrated directly toward the upper part of the tube, nearest the direct light, and settled there. When the water was withdrawn from flask no. 2, the *Icerya* immediately headed toward the end of the cylindrical tube through which the ditrated air came in. Only for a moment did they turn and then scattered in various directions. When the withdrawal of the water was stopped, the *Icerya* again made their way toward the upper part of the tube nearest the direct light. This experiment was repeated four times with the same result.

Experiment No. 2: Three closed petri dishes were placed in the centre of the table facing the direct light of the window. In petri dish no. 1 a citrus leaf was placed at the end nearest the light. Petri dish no. 2 was empty. In petri dish no. 3 a citrus leaf was placed at the end farthest from the light. In the centre of each petri dish were placed larvae of *Icerya*. In each case the larvae migrated directly toward the light. In petri dish no. 1 they stopped at the citrus leaf for a moment, then continued their way toward the light, settling at the top of the dish. Only occasionally did one larva migrate toward the leaf in petri dish no. 3 and settle there.

Experiment No. 3: One citrus leaf and several *Icerya* were placed in a petri dish. The dish was left over night in the dark, at room temperature (22—25 C). The following morning the *Icerya* were found gathered mainly on the central vein of the leaf.

Experiment No. 4: Two citrus leaves were placed in a petri dish, one at the side nearest the direct light of the window, the other at the opposite side. *Icerya* were placed on each leaf. Those placed on the farthest leaf made their way directly toward the light. Those on the leaf nearest the light travelled over the leaf and settled on the side of the plate. Only an occasional *Icerya* remained on the leaf.

Experiment No. 5: A fresh citrus branch bearing two leaves was fixed in a petri dish containing *Icerya,* in such a manner that one leaf faced the light, the other leaf was just opposite. The *Icerya* which were at the farthest end of the dish travelled across to the light, and after a time, crawled up towards the leaf nearest the light and settled there. When the branch was reversed, the *Icerya* thus being on the leaf farthest from the light, they slowly returned to the leaf nearest the light.

Experiment No. 6: A small fork of glass was placed so that the intensity of light was equal on the two branches of the fork. *Icerya* were placed at the beginning of the tube. On one branch citrus leaves were put, the other remained empty. The *Icerya* immediately went up and when it reached the fork, it went into both branches of the fork. In three experiments of the same nature, it was found that in the citrated as well as in the aeriated branches there was an equal number of *Icerya.*

Experiment No. 7: A tube with a citrus leaf at the top was placed in a vertical position on the table. *Icerya* were placed at the bottom of the tube. They travelled toward the top of the tube, over the leaf and to the side. When the tube was reversed, so that both the *Icerya* and the citrus-leaf were at the bottom again, the *Icerya* again travelled toward the top of the tube.

Experiment No. 8: A tube with a citrus leaf at the bottom and *Icerya* at the top was placed vertically. The upper part was covered with black paper. The *Icerya* travelled down to the bottom as far as the leaf, turned and travelled up again, settling at the lower margin of the black covering.

Experiment No. 9: A tube was placed on the table horizontally with the closed end toward the direct light. Several females bearing ovisacs were put at the opening of the tube. The larvae travelled to the light, gathering at the closed end. The tube was then set up vertically and a branch bearing several citrus leaves was inserted. The larvae travelled up to the top of the branch and settled on the leaves. In about three hours it was noted that over 100 larvae travelled from the tube to the citrus branches. The large majority of the *Icerya* settled on the leaves nearest the light, only few going to the other leaves.

Experiment No. 10: Two small flasks were closed with cork stoppers in the centre of which was a small opening. In one flask, a twig bearing two citrus leaves was inserted, in the other flask, a thin piece of wood was inserted. The latter served as a control. In both flasks *Icerya* larvae were placed at the bottom and left over night in a dark room. The following morning it was found that in flask no. 1 most of the *Icerya* had travelled up the twig, some remaining there and some settling on the middle vein of the leaves. In flask no. 2 one *Icerya* was found on the piece of wood, the others remained in the upper part of the flask.

Locomotion and dispersion: The neonate larvae of *Icerya* are comparatively mobile. Vayssiere (1926) obtained a locomotion of:

at 25° C	1 m in 10,5 minutes
at 15° C	1 m in 35,5 ,, .

The corresponding hyperbola of activity has the following constants:

Activity threshold:	6,8° C
Thermal constant:	191 minute-degrees.

Some of the values are:

35° C	6,8 minutes per meter
30	8,2
25	10,5
20	14,5
15	35,5
10	59,7

But it would be wrong to assume that many meters are really normally covered by the migrating larvae. The above values are maximal values, which have been gained by utilising the strong positive phototaxis of the larvae. In nature, the conditions of the tree's crown stop these tactic migrations very quickly. After the larvae have arrived at a sunny place on the periphery of the tree and a high place on its branches, (to satisfy their negative geotaxis), they settle down and begin to suck. Similarly, the long migration of the young adults is soon ended on the branches or the stem of the interior. Each active migration probably never exceeds 1—3 meters for any stage.

The natural spread, which extends a few meters in the tree or in the nearest neighbourhood, occurs either through passive transport of the young larvae by the wind, or by mechanical transport (on clothes), by carrying the adults or ovisacs on wood, other host etc. into a new environment. Under natural conditions the wind is probably by far the most important of these factors.

The importance of the wind for its spread is easily recognised in Palestine. In 1924, for example, *Icerya* were discovered on some *Juncus* shrubs on the dunes between Haifa and Akko, at least 4 km from the nearest habitation. Similar observations of isolated occurrence, for which only the wind can be responsible, have been made repeatedly.

Wilcott and Sein report on the detrimental effect of a hurricane in Puerto Rico, which destroyed *Icerya* almost completely in part of the island.

Reproduction and fertility. Earlier writers (Kuwana, Pierantoni) have shown quite definitely, that *Icerya purchasi* develops normally without copulation, during some generations. Parthenogenesis seemed to be the prevailing method of reproduction. The great and regular scarcity of males, observed by all authors, seemed to prove that an originally bisexual species is now successfully transformed into a parthenogenetic one. Hughes-Schrader has recently shown definitely that the so-called females of *Icerya* are really hermaphrodites, which, as a rule, self-fertilise their own eggs with their own sperm.

„The hermaphroditic gonad is basically an ovary . . . . In the majority of Coccids the ovaries are hollow structures with the eggs forming in pouches or ovarioles from their walls. In *Icerya purchasi*, however, the hermaphroditic gonad is solid during the early instars. In the first instar certain of the centrally lying cells become reduced to the haploid condition, and haploid cells, proliferating rapidly, come to form a solid central core in the gonad. The outer cells remain diploid and from them the ovarioles are all derived. The haploid cells give rise to sperm . . . . The gonad is actually hollowed out by the process of sperm formation. From the outer walls of the gonad, meanwhile, normal ovarioles are developing in which the eggs come to maturity. Fertilised eggs always become diploid and develop into females. Unfertilised eggs develop parthenogenetically into haploid males. 2x+2x=♀; 2x=♂. The sperm of the same individual matures long before the eggs." (Hughes-Schrader 1930).

This type of hermaphroditism seems to be general and not restricted to any geographical location of *I. purchasi*.

Hughes-Schrader got from her breedings:

from 31 mated hermaphrodites:	2548 ♀	and	39 ♂	=	1,5 %
from 10 unmated „ :	1616	and	5	=	0,3 %.

In nature, however, the percentage of males seems to be subjected to large fluctuations. In most years much less than 1 % of the offspring are males. Many authors confess to have never seen the male. We have only by chance seen two males during all these years of mass-collecting and mass-breeding. But some observers have seen real mass-appearances of males as did Chapin in the U.S.A. (fide Riley) and Menozzi (N. Italy). The latter observed a few males only in 1924, but 1925 could be called a „year of males."

Copulation is not necessary for propagation and it is questionable whether the male sperm is really fertile. „Copulation occurs in nature usually during the late afternoon. The males evince a weak mating instinct as compared with the females. The latter react at once to the near approach of the males by raising the posterior tip of the abdomen. When copulation occurs the male mounts the up-tilted body of the female and clings to its posterior margin. The abdomen of the male is then bent under that of the female and the penis inserted into the genital aperture of the female. The female retains the position with abdomen upturned during copulation thus holding the male up from the surface of the host plant. Copulation lasts from 2—5 minutes. Females will mate frequently and with many different males; the males rarely mate more than once, and it is not unusual for a male entirely to ignore the presence of females even though the latter may be reacting vigorously at the time." (Hughes-Schrader, 1930).

The adults are ready for copulation immediately after the third moult, until the formation of the ovisac begins, which covers the genital aperture. The formation of the waxy egg-sac as well as the oviposition generally take from 2—3 weeks. The ovisac is formed by mated as well as by unmated females. The first and perhaps also the latter are self-fertilised. And it should be well understood that speaking of females, we mean the hermaphrodites, which were derived from the true females of an earlier period.

Egg-production. The egg-production of *I. purchasi* is very variable. Textbooks generally talk of 500 and more eggs per female. Reliable averages are published only by Kuwana, who counted the following at Shizuoka (Japan):

Generation	Month	No. of observations	Average eggs per ovisac
I	May	30	411
I	June	7	1592
II	September	24	220

About 10 % are eggs which remained in the body of the dead adult.
Five years observations in Palestine gave much lower values:

Eggproduction of Icerya on Citrus from 1928—1933

Month	1928				1929				1930				1931				1932			1933			Total	
	n	Average	Maximum	Length of ovisac	n	Average	Maximum	Length of ovisac	n	Average	Maximum	Length of ovisac	n	Average	Maximum	Length of ovisac	n	Average	Maximum	n	Average	Maximum	n	Average
I	—	—	—	—	—	—	—	—	—	—	—	—	—	—	—	—	9	61	143	5	74	109	14	66
II	—	—	—	—	—	—	—	—	—	—	—	—	10	199	345	1.7	9	166	210	10	189	250	29	185
III	—	—	—	—	—	—	—	—	—	—	—	—	10	374	602	2.4	27	149	406	30	213	310	67	211
IV	—	—	—	—	10	167	266	3.2	10	216	376	3.1	10	375	525	2.5	20	185	307	30	215	470	80	219
V	—	—	—	—	10	354	544	4.4	17	146	270	2.4	—	—	—	—	30	169	313	30	205	301	87	198
VI	10	312	700	—	—	—	—	—	—	—	—	—	—	—	—	—	20	132	223	—	—	—	30	192
VII	4	355	400	4.1	10	260	415	3.2	20	149	340	1.6	10	126	180	1.6	18	110	197	—	—	—	62	164
VIII	7	274	830	5.2	10	104	225	2.3	9	38	56	1.2	8	127	194	1.5	19	94	213	—	—	—	53	115
IX	—	—	—	—	10	115	231	2.7	10	56	80	—	4	104	122	1.5	10	83	128	—	—	—	34	87
X	9	130	215	4.0	10	76	127	2.3	—	—	—	—	7	109	151	1.7	8	85	134	10	59	72	44	90
XI	—	—	—	—	5	102	172	2.1	9	73	130	1.3	10	75	119	1.2	5	81	123	—	—	—	29	80
XII	—	—	—	—	—	—	—	—	10	70	145	1.5	10	141	308	1.7	4	38	72	—	—	—	24	94
Total	30	221			65	173			85	127			79	176			179	126		115	187		553	171

The Palestinian average amounts to 171 eggs per family only, or to 188 if we add 10 % of eggs still in the adult.

Comparative breedings on different hosts showed:

Comparison of egg-productions on different host plants in 1931/33.

Month	on Citrus		on Spartium		on Xanthium	
	n	eggs per ♀	n	eggs per ♀	n	eggs per ♀
I	14	66	22	138		
II	29	185	30	148		
III	67	211	30	140		
IV	60	228	20	190		
V	60	187	20	165		
VI	20	132	8	93		
VII	28	116	27	123	4	44
VIII	27	104	20	103	1	163
IX	14	89	20	92	10	230
X	25	81	29	64	10	164
XI	15	82	30	74		
XII	14	110	30	65		
Total.	373	169	286	115	25	171

The average ration of the egg-production on *Citrus* to that on *Spartium* is 100:71,9. The data for *Xanthium* are too small and restricted to one season to allow a final conclusion. But since *Spartium junceum* is one of the host plants on which *Icerya* shows a fairly good development, it is certain that the difference in fertility on more unfavorable host plants will be still more pronounced. Greenhouses, with more stable temperature, would be very suitable for further experiments in this direction.

However, this important role played by the host does not explain the difference in fertility between Japan and Palestine. Both observations were made on citrus. It is therefore of importance to know whether *Icerya* has a different egg-production on various species or varieties of citrus. Only those months were chosen for the experiment, in which 10 observations on all three citrus varieties have been performed.

Egg-production of *Icerya* on different Citrus species.

Citrus species	1932 (III.V.VIII)	1933 (III.VI.V)	Average
Shamuti Orange. . . .	139	206	172
Sour Lemon	168	199	179
Grapefruit	132	227	177

There is, therefore practically no difference in the fertility of *Icerya* on different citrus species.

Other factors, which influence the numbers of eggs, may be mentioned:

The temperature influence is strongest in the period just preceeding oviposition. The data for breedings on citrus are:

Temperature ranges	No. of observations	Average eggs per ♀
10—15° C	70	96
15—20	141	200
20—25	153	161
25—30	174	151

The spring peak of fertility may be influenced by favorable temperature conditions.

Humidity alone also seems to have only little influence, but humid and hot weather seems to be optimal. Palestine appears to differ from Japan only in her dry summer heat, in contrast with Japan's humid-warm summer.

It is possible that the climatic differences are sufficient to explain the difference in fertility in both countries. It may well be, however, that their influence is not entirely direct, but more indirect perhaps through their influence upon the conditions of the host plant.

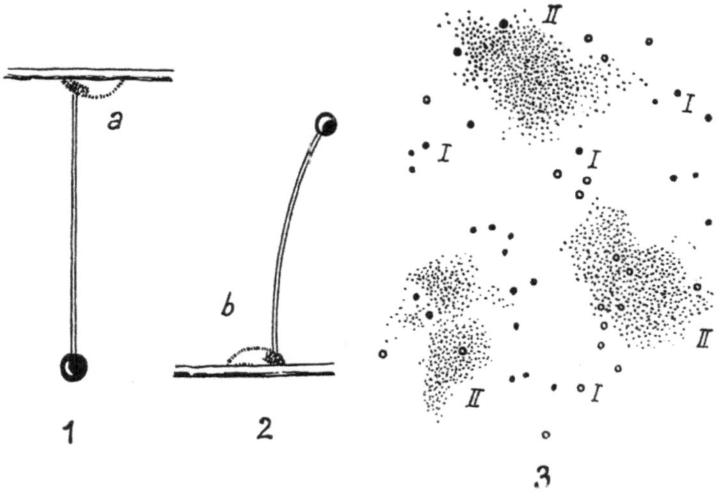

Fig. 112.
Honeydew excretion of *Icerya purchasi* Mask. 1) first larva with excrement on lower side of leaf, 2) idem, on upper side, 3) drops of honeydew of young larvae (after Dingler).

Honeydew, Ants and Damage.

Honeydew. An abundant excretion of honeydew is very characteristic for *Icerya.* Immediately after the fixation of the first larval stages, simultaneously with the development of a wax-cover, a wax-thread is excreted from the anal opening, which is present on the dorsum, near the posterior end of the abdomen. At the end of this wax-filament, a semi-liquid, hyaline drop of honeydew is seen which may reach half the volume of the larval body. Generally, long before it reaches that size, the filament breaks near its base and another wax-thread is excreted.

The whole phenomenon has been first described by Dingler. He observed that in absolutely quiet air those wax-threads may reach a length of 50 mm. In their natural environment they are scarcely 5 mm long, before the filament breaks and drops down. The wax-thread is probably excreted immediately after the exudation of the honeydew and remains at its distal top with the latter's increase. In those animals which settle on the upper side of the leaf, the drop falls occasionally on the larva itself, which in turn becomes spoiled by this viscous mass and generally dies.

In the following two larval stages no such transport of honeydew by wax-filaments is to be observed. The excretion is squirted from the rectum with force and drips down in small drops similar to those of a drizzling rain. Fig. 112 shows very well the difference in the size of drops between the first and the second larva stages.

The most copious excretion occurs in the young female, just before oviposition. The leaves on which the excretion drips down, are not sprinkled with single dots, but are covered with large continuous areas of honeydew. Most of the damage is done in this stage. In our dry summer climate the wet parts dry out quickly and are infested by the black-fungus or sooty mould.

Fig. 113.
Sooty mould following honeydew excretion of *Icerya purchasi* Mask. Note the black girdle underneath the insect.

In more humid regions, the copious liquid at that stage discolours the trunk and drips continuously from the leaves, as Kuwana reports from Japan. The excretion stops with the beginning of oviposition, which indicates a simultaneous interruption in feeding.

Icerya and ants. Icerya purchasi is not among those coccids, which are regularly frequented by ants. But one observation made in a recently planted citrus grove at Rehoboth (30.XI.1927) has been mentioned before.

Damage. The author has not observed *Icerya* in Palestine in such numbers, that whole branches of citrus trees were ,,covered with snow.'' No destruction of branches or trees has ever come to our knowledge. The damage done is produced by the heavy exudation of honeydew, which covers the leaves and fruit of trees. The latter show only a mediocre infestation. A heavy layer of Sooty Mould is the result. As our fruit undergoes no brushing process, as that of California, the spoiled black fruit is more or less unmarketable. And it is not likely that a mechanical brushing process will be introduced into Palestine, for the rind of the Jaffa orange is very susceptible.*) The demand for efficient control will therefore be directed toward a reduction of the population level of *Icerya* below the damage mark.

A heavy damage in the Hedera district, since 1926 was the immediate incentive for a more intensive stüdy. The damage resulting from the deterioriation of the fruit alone in that district's small groups of orchards was estimated at 5.000 LP. in the winter of 1927. Damage of such value has not been repeated since and not yet observed in other districts, but serious outbreaks in the future are very likely to occur. At Gaza, as well as in the Tulkarem district, neglected, densely-planted Arab orange groves have been repeatedly observed to be very heavily infested.

---

*) In the meantime such a brushing machine has been built in the country.

K. THE BIOLOGICAL CONTROL OF THE CITRUS FLUTED SCALE.

a. The Icerya-ladybird (Novius cardinalis Muls.).

Life-history and ecology: Icerya purchasi and the ladybeetle *Novius cardinalis* are the classical example of a successful biological control. Since the first shipments of *Novius* by Koebele from Australia and New-Zealand to California in 1888 and the removal of the *Icerya* menace in California since 1889, this ladybeetle has been introduced into many regions with equal success. Into Palestine it has been introduced about 1912 from Portici (Prof. Silvestri) and Beirut mainly into the orange groves of Petah Tikva. To-day it is present in Palestine, whereever *Icerya* occurs in the coastal plain. Its definite establishment in Jerusalem is not yet proven, despite repeated introductions. But in general it is not very abundant, except when and where *Icerya* is on the increase.

The economic importance of this ladybeetle justifies a more extensive study of its ecology and epidemiology.

Description. The *adult beetles* are broad oval and hemispherical in shape. The ground-colour is blood-red, but the head, base of prothorax and a very characteristic drawing on the elytres, which is best studied on Plate XXX, are black. The antennae are 8-jointed. Red predominates in the females, and black in the males. But the colour difference is not always reliable. The dorsum is covered with very minute brown hair. 3—4 mm long, 2—3 mm broad.

The *egg* is long oval and of a pink colour. 0,8 × 0,3 mm.

The *larva* has four stages, which vary from 1,3—5 mm in length. They are elongate, tapering towards the caudal end. Their coloration is dark pinkish, with a dark brown head and some black spots. The legs are orange red with dark hair. The body of the fourth stage larva becomes covered with a light gray powder.

The *pupa* is partially enclosed in the old larval skin, which is of a whitish colour marked with black dots; the skin is rent dorsally from the front edge of the first segment to the middle of the eighth. The exposed part is mottled light and brownish red, the first segment is marked with two dorsal black dots, or is entirely black as are the second and third segments. The abdomen has an interrupted polished black dorsal line. The length of the body is about 4 mm.

Life-history: From the many life-histories which we worked out in our observations, 15 typical cycles of development are chosen from all seasons in order to illustrate the normal life-history.

No.	Preoviposition period in days	Date	egg-stage days	Larval-stage in days	Hatching of adults	Pupal stage in days	Total development in days
1	4	13.XI.28	5	34	29.I.29	35	78
2	4	11.III.29	6	21	9.IV.29	12	39
3	7	14.V.29	2	10	3.VI.29	8	20
4	5	4.VI.29	5	9	25.VI.29	7	21
5	6	17.VII.29	8	10	13.VIII.29	9	27
6	3	14.XI.29	8	13	8.I.30	4	55
7	4	15.XI.29	7	9	5.I.30	35	51
8	11	2.III.30	9	16	14.IV.30	18	43
9	5	16.V.30	4	9	8.VI.30	10	23
10	—	29.VII.30	2	8	14.VIII.30	6	16
11	—	5.X.30	5	11	31.X.30	10	26
12	8	8.XII.30	6	30	9.II.31	27	63
13	3	23.VIII.31	4	7	14.IX.31	11	22
14	—	18.IX.31	4	5	11.X.31	14	26
15	5	1.XI.31	5	16	13.XII.31	21	42

The breedings were made in a well aired stone barrack. The empirical data for the lowest development permit the evaluation of the hyperbola with the following constants: Dev.zero (c)= 9,0°. Thermal constant (Th.C.)= 371 days-degrees

The empirical and the calculated values may be compared from the following table:

The relation of the development of *Novius cardinalis* (egg to adult) to the average environmental temperature.

°C	n	Average development empir. in days	calcul. in days	Deviation in %	Shortest development empir. in days	calcul. in days
13,8	1	86,0	77,3	+11,2	86	78
14,3	7	74,9	70,0	+ 7,0	68	68
15,5	3	65,3	57,1	+14,4	58	54
16,4	6	50,2	50,1	+ 0,2	39	46
17,5	5	42,6	43,6	— 2,3	40	40
18,9	3	36,7	38,2	— 3,9	33	35
19,5	6	33,8	35,3	— 4,2	30	32
20,5	11	31,8	32,2	— 1,2	28	29
21,5	7	29,2	29,8	— 2,0	26	26
22,5	6	28,3	27,5	+ 2,9	22	24
23,4	11	25,7	25,6	+ 0,4	20	22
24,5	7	24,3	23,9	+ 1,7	21	21
25,5	7	22,6	22,4	+ 0,9	19	19
26,6	11	24,6	21,1	+16,6	21	18
27,6	18	21,8	19,9	+ 9,5	18	17
28,2	23	20,9	19,3	+ 8,3	16	17

Fig. **114** illustrates the three hyperbolae. The shortest development occurs at 26° C. Above that temperature the deviation from the hyperbola becomes significant. From 16—25° C the average deviation of the actual from the calculated value is only 0,8 %. Below 16° C the actual development is longer than the calculated one, a quite usual phenomenon.

Copulation takes place immediately after hatching. The preoviposition period lasts for 1—4 days in summer and for 1—3 weeks in winter. For the quickest developing individuals of every generation 12 annual generations may therefore be expected in the coastal plain of Palestine. The combination of the average development time of each generation with the shortest preoviposition period gives 9 annual generations, with the average preoviposition period about 6.

No seasonal interruption of development, no diapause, is ever observed in our coastal plain. In winter the development is only slackened down, but not interrupted. The shortest development observed was 16 days in the summer, the average development at the same season being 21 days. The longest development observed was 110 days in the winter, the average development at the same season being 70—77 days

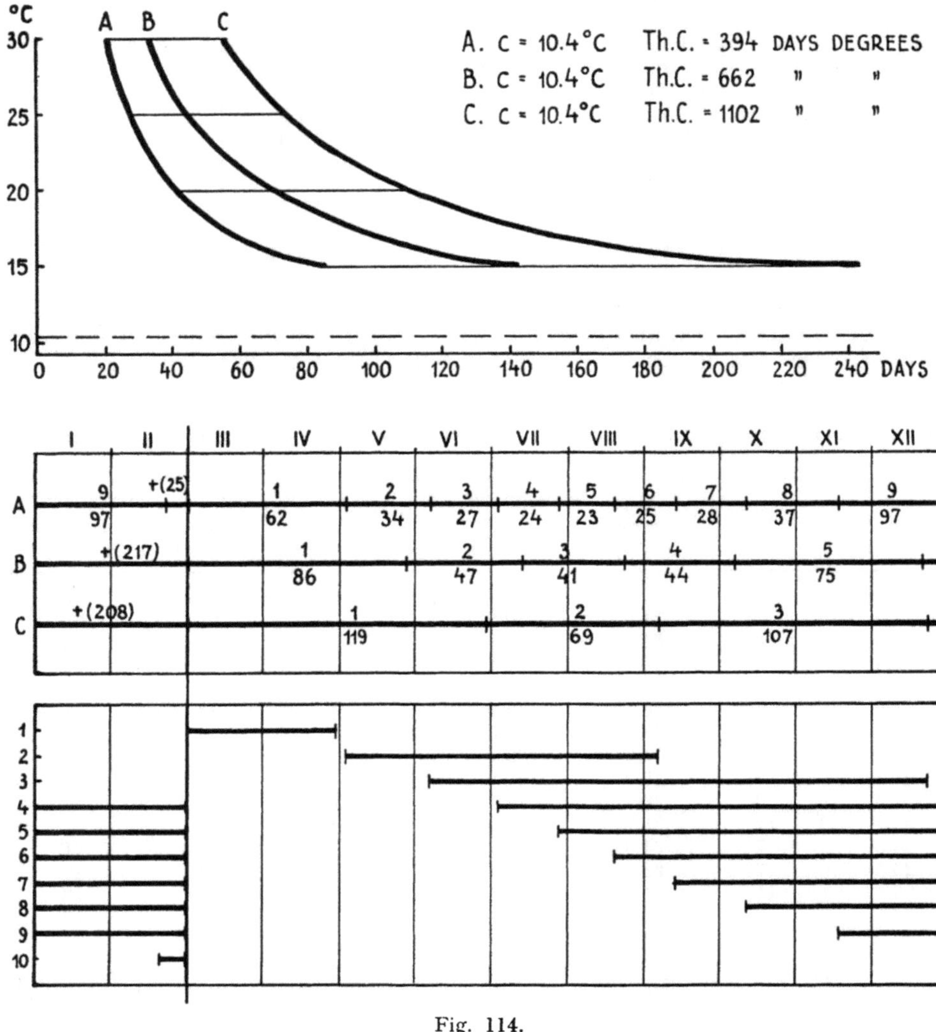

Fig. 114.
Life-history of *Novius cardinalis* Muls. at Tel Aviv. A) The three hyperbolae; B) the life-history as calculated from the hyperbolae; 3) sequence and overlapping of generations.

Oviposition takes place on or near the ovisacs of *Icerya,* mostly only one size egg. Its oviposition takes 20—30 seconds. The newly hatched larva is very active and burrows immediately into the ovisac, where it feeds on the eggs and newly hatched larvae of Icerya. The older larvae are able to attack any stage of its host, usually attacking the underside of the abdomen.

The longevity of the male is generally longer than that of the female:

	Longevity in days		
	average	longest	shortest
Male	81	216	20
Female . . .	70	178	16

The female oviposits during the major part of its life. The average number of eggs per female was **307**, **1037** the maximal number and **35** the minimal one. There exists a correlation between the length of oviposition and the longevity. The average oviposition period was **57** days, the longest **170**, the shortest **11**.

Longevity and oviposition of Novius cardinalis.

No.	Beginning of breeding	Longevity ♀	Longevity ♂	No. of eggs	Oviposition period of date	Oviposition period of in days	No. of eggs per day
1.	9/XI/28	178	24	193	13/XI/28—2/5/29	170	1,14
2.	22/XI/28	138	167	35	26/XI—24/XII/28	29	1,21
3.	7/III/29	157	57	300	11/III/29—2/V/29	53	5,64
4.	8/V/29	28	47	136	14/VI—3/VII/29	20	6,30
5.	29/V/29	30	22	82	4—17/VI/29	13	6,31
6.	10/VI/29	17	20	106	16—27/VI/29	11	9.64
7.	13/VI/29	21	21	47	16—27/VI/29	11	4,27
8.	30/VI/29	16	22	44	3—14/VII/29	11	4,00
9.	11/VII/29	44	—	70	17/VII—11/VIII/29	25	2,80
10.	14/VII/29	41	30	250	11/VII—23/VIII/29	38	6,58
11.	20/IX/29	51	128	151	23/IX/3/XI/29	41	3,68
12.	8/XI/29	107	109	190	14/XI—31/X/29	79	2,40
13.	8/XI/29	63	216	128	15/XI—3/I/30	48	2,66
14.	19/II/30	109	132	300	2/III/30—6/V/30	66	4,54
15.	11/V/30	76	80	621	16/V—23/VII/30	68	9,13
16.	25/VII/30	88	80	383	29/VII—26/IX/30	59	6,67
17.	3/X/30	131	95	274	5/X/30—1/II/31	119	2,30
18.	30/XI/30	175	191	607	8/XII/30—16/V/31	160	3,78
19.	13/V/31	36	—	73	18/V—14/VI/31	28	2,68
20.	12/VI/31	60	—	591	14/VI—10/VIII/31	58	11,92
21.	12/VI/31	72	—	1037	16/VI—22/VIII/31	58	15,25
22.	12/VI/31	83	—	605	16/VI—2/IX/31	79	7,66
23.	18/VI/31	90	—	502	21/VI—3/IX/31	74	6,79
24.	18/VI/31	73	81	501	19/VI—18/VIII/31	61	8,21
25.	20/VIII/31	27	43	244	23/VIII—10/IX/31	19	12,85
26.	3/IX/31	89	—	559	6/IX—26/XI/31	81	6,89
27.	3/IX/31	112	—	365	6/IX—13/XII/31	98	3,77
28	10/IX/31	50	54	402	18/IX—24/X/31	37	10,81
29.	18/IX/31	136	76	366	18/IX/31—5/I/32	112	3,27
30.	4/X/31	70	—	119	11/X—19/XI/31	39	3,05
31.	27/X/31	127	177	502	1/XI/31—19/II/32	110	4,56
32.	13/XI/31	127	143	269	17/XI/31—15/III/32	119	2,24
33.	14/XI/31	175	147	311	24/XI/31—6/V/32	164	1,90
		81	90				

The natural mortality: Our laboratory breedings are the basic material for the discussion of this question. The mortality of a group of eggs is followed up to the hatching of the adult beetles. Every group is reported under the month in which the oviposition of the eggs took place.

Monthly mortality of Novius laboratory-breedings.

Month		I	II	III	IV	V	VI	VII	VIII	IX	X	XI	XII
	No. of eggs	—	—	—	—	—	101	72	35	3	28	60	20
	No. of larvae. . . .	—	—	—	—	—	66	41	21	3	22	45	18
1929	No. of pupae. . . .	—	—	—	—	—	34	35	9	3	20	43	15
	No. of beetles . . .	—	—	—	—	—	31	28	4	3	18	40	15
	% of mortality. . .	—	—	—	—	—	69,4	61,1	88,6	0	35,8	33,3	25,0
	No. of eggs	—	—	26	11	16	106	116	116	123	121	120	120
	No. of larvae. . . .	—	—	19	5	15	95	88	60	94	97	100	103
1930	No. of pupae. . . .	—	—	10	5	15	81	64	41	59	64	83	99
	No. of beetles . . .	—	—	7	5	15	81	46	28	48	58	81	97
	% of mortality. . .	—	—	73,1	54,6	6,3	23,6	69,4	75,9	61,0	52,1	32,5	17,5
	No. of eggs	126	120	125	120	120	120	100	100	100	100	100	100
	No. of larvae. . . .	109	89	97	118	101	99	57	92	74	76	84	53
1931	No. of pupae. . . .	86	82	91	104	98	89	5	71	53	60	74	33
	No. of beetles . . .	84	79	91	79	93	86	5	25	49	58	72	31
	% of mortality. . .	33,3	34,2	27,2	34,2	22,5	28,4	25,0	75,0	51	42,0	28,0	69,0
	No. of eggs	100	100	100	100	100	100	100	100	—	—	—	—
	No. of larvae. . . .	46	47	63	84	96	92	57	57	—	—	—	—
1932	No. of pupae. . . .	39	37	62	79	93	88	12	44	—	—	—	—
	No. of beetles . . .	37	36	60	78	93	77	9	42	—	—	—	—
	% of mortality. . .	63,0	64,0	40,0	22,0	7,0	23,0	91,0	58,0	—	—	—	—
Average mortality % . . .		46,5	47,7	39,5	29,9	12,2	35,6	77,3	77,5	56	43,3	31,4	37,2

The lowest mortality was in May (12,2 %), the highest in July and August (77,4 %) Of 100 eggs laid survived into the following stage:

Year	egg	larvae	pupae	beetles
1929	100	67,7	49,8	**43,6**
1930	100	77,2	59,4	**53,4**
1931	100	74,7	63,7	**56,2**
1932	100	80,2	56,7	**53,1**
Average. . .	100	76,1	59,6	53,2

The mortality for the different stages was:

Mortality	egg	larva	pupa
	%	%	%
absolute.	23,9	21,7	10,8
relative to original no. of eggs . . .	23,9	16,5	6,4

All this mortality is influenced mainly by climatic conditions, as food, etc. were optimal. One gains the impression, that relative humidity is of relatively small influence. The low mortality in May, during the hamsin-period, corroborates this impression. That the macro-humidity has little importance for mortality during the development is probably mainly due to the protected microclimate in the ovisac itself. The environmental relationship maintained in the interior of the ovisac between insect and host plant is much more stable with respect to the daily fluctuations of humidity, than is the macroclimate. Those adults, living under out door conditions suffer from a considerable mortality during the hamsin winds.

The total mortality during the development is on the whole rather small. Only about half of the eggs laid, die due to climatic reasons. This high resistance of the species to a considerable range of climatic conditions is certainly one of the promising factors for the efficiency of this ladybird in a large variety of regions.

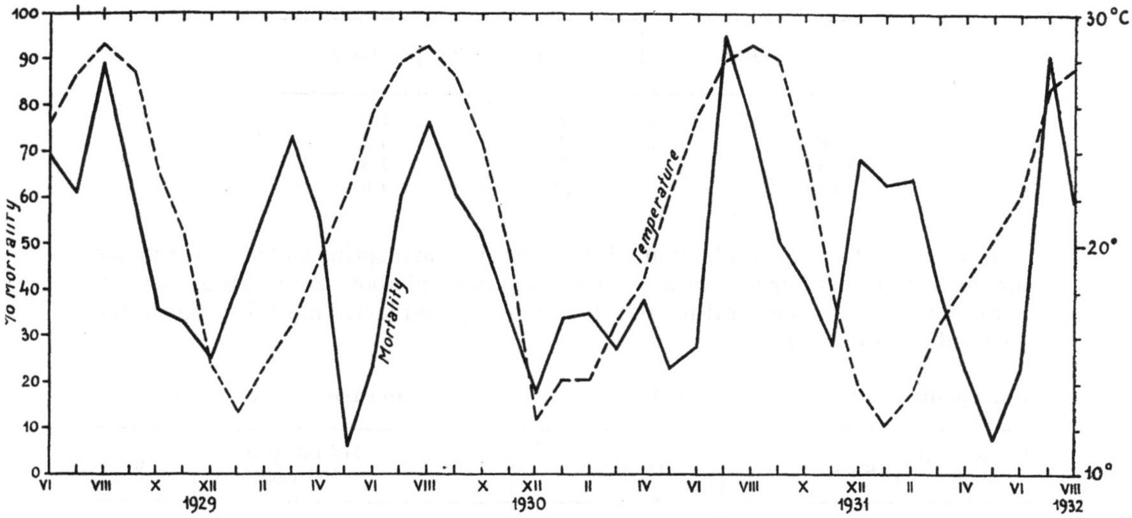

Fig. 115.
Mortality of *Novius cardinalis* Muls. and monthly average temperature (------).

Fig. 115 illustrates the parallel course of the mortality and the temperature curves in the summer. The peak of both is in July/August; between 23 and 26° C both curves fall, the bottom of the mortality curve is reached at 18—22° C. The bottom of both curves does not coincide, but that of the development is reached either sooner or later than that of the temperature-curve.

The temperatures scale of activity for *Novius* is:

		Beetles (n 25)	Larvae (n 25)
1.	Beginning of cold torpor.	11,9 C	16,9 C
2.	Only slight movements of eggs or antennae	13,3	—
3.	Crawling with interruptions	10—28	—
4.	Normal activity.	15,5—38,5	23,5
5.	Strong activity	23—36,5	26,8
6.	High irritation	35,5	32,5
7.	Beginning of heat-paralysis.	41,0	—
8.	Instantaneous heat-death.	43,0	41,3

The young larvae seem to be fairly well protected in the microclimate of the *Icerya* ovisac against the host's dry hamsin winds. But adults, older larvae and even many individuals of the most protected stages are decimated by every protracted hamsin. After such an occurrence *Novius* seemed to have disappeared from the orchard for weeks at a time. Dry summer heat is equally unfavorable. Both hamsin and dry summer heat are the causal factors of the large population drop which generally in summer occurs.

A heavy hamsin wind in about 14.IX.31, which mounted up to 47,5° C, caused an almost complete disappearance of *Novius* in the orchards and a new colonization was everywhere necessary.

The wet and moderately cold winter is much more favorable. Spring and autumn are the optimal seasons in Palestine.

The fertility of the female may also be correlated to season and temperature:

Season	eggs per ♀	eggs per day and ♀
Winter	198	2,2
Spring	320	4,2
Summer.	335	7,8
Autumn.	272	4,9

The maximal total eggproduction takes place in the early spring and in mid-summer. But the fertility per female and day reaches a maximum above 28° C. Since the total maximum is at that temperature also, the high egg- and larval mortality is counterbalanced to some degree.

The oviposition of *Novius cardinalis* and its mortality in relation to environmental temperatures.

Temperature in °C	Eggs per day	Total eggs	n	Mortality of 100 eggs	n
12					
14				69,0	1
16	2,40	(190)	1	30,7	3
18	2,39	164 254	4	50,0	2
		(15—20° C)			
20	3,91	**394**	3	37,2	4
22	3,05	(119)	1	31,3	3
24	3,71	258 187	2	21,7	3
		(20—25° C)			
26	7,71	171	5	45,2	4
28	7,55	317 414	5	60,4	3
		(25—30° C)			
30	7,90	4,75	8	77,2	5

On the quantity of food, eaten by Novius cardinalis in its different stages and on Icerya from different hosts. It was important to gain serviceable data on the food ecology of this ladybird.

Fig. 116 illustrates very well, that the curve for the total food-quantity during development runs antibat to that of temperature. At the beginning of winter one *Novius*-larva eats five times as much larvae of *Icerya* during its development than in July. The consumption of one generation is therefore higher in winter than in summer. But the lenght of one generation is only 3—4 times as long. The feeding per day shows quite a different picture, but even here the distinct depression of the daily food quantity in mid-summer is of high interest. Since in winter the longevity of the females (adult beetles) is 6—7 times longer than in mid-summer, the beetles of the winter-generation devour in absolute numbers more larvae of *Icerya* than those of the summer-generations. In all these observations the larvae of *Icerya* used as food belonged to the second and third larval stages. Another series of observations was made with regard to the quantity of *Icerya*-eggs devoured by a *Novius*-larva during its development:

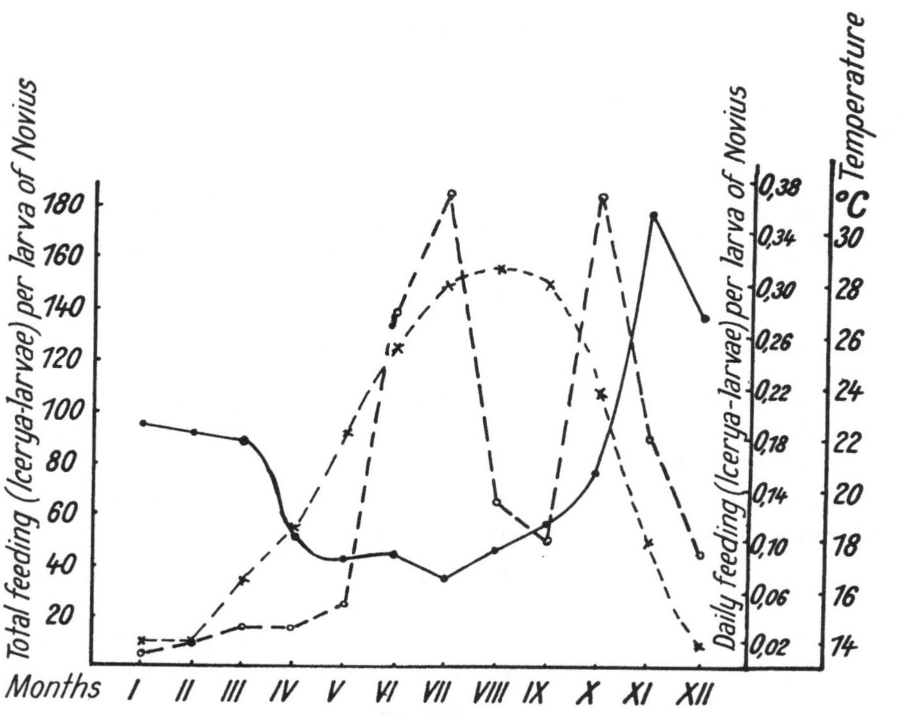

Fig. 116.
Total number of *Icerya* larvae devoured by *Novius* larvae. A. during total development (—); B. daily per annum (o----o) and monthly average temperature (X----X).

Total feeding of Novius-larvae on the larvae of Icerya.

Month	I	II	III	IV	V	VI	VII	VIII	IX	X	XI	XII
					On Spartium 1931							
n:	7	7	8	7	—	—	—	—	—	3	—	8
Total feeding days	246	69	217	156	—	—	—	—	—	44	—	227
Total devoured	27	9	44	31	—	—	—	—	—	14	—	31
Feeding per Novius-larva	3,8	1,3	5,5	4,4	—	—	—	—	—	4,7	—	3,9
Development of Novius-larvae in days	35,1	9,8	27,1	22,3	—	—	—	—	—	14,7	—	28,4
					On Citrus 1931							
n	7	9	8	8	7	2	3	4	7	3	3	3
Total feeding days	240	213	165	124	83	18	18	37	64	25	55	69
Total devoured	20	35	35	27	26	10	18	20	45	28	29	18
Feeding per Novius-larva	2,8	3,9	4,3	3,3	3,7	5,0	6,0	5,0	6,4	9,3	9,7	6,0
Development of Novius-larvae in days	34,3	23,6	20,6	15,5	11,8	9,0	6,0	9,2	9,1	8,3	18,3	23,0
					On Citrus 1930							
n	—	—	3	3	2	2	1	4	8	8	7	8
Feeding per Novius-larva	—	—	4,0	4,0	3,5	6,0	5,0	4,5	5,6	6,6	11,0	6,8
Dev. of Novius-larvae in days	—	—	16	12	14,5	12,5	10,0	9,2	8,3	9,8	15,8	32,2

Total feeding of Novius-beetles on the larvae of Icerya.

Month	I	II	III	IV	V	VI	VII	VIII	IX	X	XI	XII	I
			On Spartium 1931										1932
n	3	3	3	3	—	—	—	—	3	3	3	3	—
Total feeding days	60	60	57	60	—	—	—	—	60	59	60	60	—
Total of Icerya-larvae devoured	16	8	16	20	—	—	—	—	75	38	55	16	—
Average per beetle and 20 days	5,3	2,7	5,6	6,7	—	—	—	—	25,0	12,7	18,3	5,3	—
			On Citrus 1931										
n	3	3	3	2	3	3	3	3	2	3	3	3	3
Total feeding days	60	60	66	20	60	39	60	60	40	60	60	60	60
Total of Icerya-larvae devoured	18	9	25	15	34	27	46	55	56	100	57	20	8
Average per beetle and 20 days	6,0	3,0	8,3	15,0	11,3	13,8	15,3	18,3	28	**33,3**	19	6,7	2,7
			On Citrus 1930										
n	—	—	1	3	2	1	1	1	3	3	3	3	
Average per beetle and 20 days	15,0	8,3	10,5	14,0	8,0	13,0	23,6	21	17,3	11,3	—	—	

These values are very much below those observed in Japan by Kuwana, who reports 210 second stage larvae as the average food during the larval period. This big difference remains unexplained.

Months in 1931	II	III	IV	V	VI	VII	VIII	IX	X	XI
Eggs eaten per Novius-larva. .	35	74	89	142	22	24	181	198	215	355
Average feeding days per larva	21	22	26	20	12	12	12	10	12	44
Eggs eaten per day.	0,2	0,4	0,4	1,3	3,0	9,8	3,7	6,4	5,8	8,1

This series is of a higher ecological importance than the previous one. In nature *Novius*-larvae feed almost exclusively on the eggs or on the newly hatched larvae of *Icerya*.

Of more theoretical interest are some data on the food-quantity of *Novius*-beetles, fed only with *Icerya*-eggs:

No.	Sex	Date	Longevity in days	No. of eggs devoured	No. of eggs produced	No. of oviposition days
1.	male	13/V—28/VII/31	76	2700	—	—
2.	female	13/V—18/VI/31	36	1430	73	31
3.	female	18/VI—10/IX/31	84	3750	502	74
4.	female	4/X—13/XII/31	39	1258	119	39

Poutiers and Borg have stated in 1930 for S. France and Malta respectively that *Novius* does not attack the *Icerya* living on *Spartium junceum*. In breeding cages *Icerya* have been eaten eagerly from *Pittosporum* and oviposition of *Novius* was normal. In another cage *Icerya* were not touched on Spartium and all females died of hunger without oviposition. Poutiers concluded, that *Icerya* absorbed plant alcaloids from the food plant and when those are fresh in its fat body they have a deterrent effect on *Novius*. This ,,protective effect'' was not found to be hereditary.

Our own experiments show, that the deterrent effect is by no means as strong as Poutiers has observed. The following table proves, that *Icerya* is readily eaten on *Spartium* in Palestine. *Novius* is able to undergo on this food its full developmental cycle as well as oviposition and normal longevity of the adult beetles, even when any other food is excluded. Our experiments have been repeated so many times, that the results are sufficiently based. They show, however, the following differences in development:

	Icerya purchasi on Spartium	Citrus	Remarks
1. Average food of beetles in 1931 per 20 days.	10,6	16,1	During the comparable 6 months
2. Average food of larvae during total development in 1931 . . .	3,9	4,9	During the comparable 6 months
3. Duration of development per Novius-larva in 1931	25,5	14,3	During the comparable 5 months
4. Duration of development per Novius-larva in 1932	33,2	26,2	During the comparable 5 months

A significant difference in development-speed and food-quantity is present. On *Spartium*, *Icerya* is still a normal food of *Novius*, but decidedly not an optimal one.

This fact is of importance for the biological control. We always have to face the question how to breed the injurious insect in sufficient enough numbers to provide food for the breeding of the parasite. Silvestri has been breeding since decades *Icerya* and *Novius* at Portici on open shrubs of *Spartium*. This combined breeding of both insects is always successful. Individuals of both species are always present in moderate numbers and some *Novius* beetles may always be taken for colonization into a new *Icerya*-focus. On optimal food plants as *Citrus* or *Pittosporum* such balance between the two species is never reached. There *Icerya* is devoured entirely and consequently *Novius* dies out.

Some surviving larvae of *Icerya* may occasionally give rise to a new attack on the tree. When some individuals of *Novius* survive until the complete revival of *Icerya*, they may be able to destroy the *Icerya*-population over again. But on isolated trees the result is not a balanced equilibrium between *Icerya* and *Novius* but either the extinction of both species or an undisturbed development of *Icerya* after the dying out of *Novius*.

This nutritional resistance of *Icerya* developing on *Spartium* to *Novius* is of economic importance. Foci of *Icerya* on *Spartium* guarantee the permanent existence of *Novius* in their surroundings. They must therefore be regarded as useful. An artificial creation of such foci is therefore advisable, whereever *Icerya* tends to be a permanent pest even of minor importance.

Novius cardinalis is often regarded as monophagous on *Icerya purchasi*. But it feeds eagerly on eggs and larvae of other Monophlebinae too, (*Waricoccus, Monophlebus, Gueriniella*) even in nature, Yet it is unable, at least in the Mediterranean environment, to maintain itself throughout the year on any other species except *Icerya purchasi*. In Egypt *I. aegyptiaca* may play a similarly important role.

Further details on the physiology of the feeding rate of *Novius* on *Icerya* are reported by Cressman and Dumestra (1930).

The fluctuations of Icerya and Novius in the orange grove: The ecological constants of both species. The orange groves everywhere are artificial biotops, in which only few host plants of *Icerya*, besides citrus, are met with. *Novius* is the main limiting biological factor. The few other species of predators are relatively unimportant.

The citrus plant, the basic joint of this food-chain, is present in abundance throughout the year. *Novius* is restricted in this biotop to *Icerya purchasi* as the only host of importance, while *Guerriniella serratulae* appear there for a short season only, in very limited numbers. The biotic relations are as a whole very simple. The climatic constants for the development of both species are:

	Development threshold	Thermal-constant
Icerya	0° C	2074 days-degrees
Novius	9° C	371 ,, ,,

These constants permit the calculation of the number of generations for both insects in some localities of extensive distribution.

Calculated number of generations of Icerya and Novius per annum.

Place	No. of generations of				
	Icerya		*Novius*		
	average	empirical	average	empirical	maximal
Paris	1,8		3		4
Antibes (S. France)	2,6	2	5	6	6—7
Naples (S. Italy)	2,7	3	6	5	8—9
Tel Aviv (Palestine)	3,5	3,5	9	9	11—12
Shizuoka (Japan)	2,5	2,5	6	8	8
Koshum (Formosa)	4,2	3 <	14	13	15
Adelaide (S. Australia)	3,0		7		8—9
Honolulu (Hawaii)	4,0		13		14
St. Mauritius	3,7		12		13
Los Angeles (California)	2,5	2	5		6—7

The higher the temperature, the greater the increase in the number of generations of *Novius* with regard to *Icerya*. Where the sum of temperatures per annum is equal, more *Novius* generations will develop in the more equally warm climate than in that with stronger seasonal contrasts (cf. e.g. Mauritius and Tel Aviv). *Novius* seems to favor a humid warm climate. The high summer-mortality in Palestine would then be due more to the dry, than to the hot component of our climate.

Some climograms of the actual distribution area of both species are given in fig. 117. Some of these localities belong to the Mediterranean type of climate (Antibes, Sarona, Los Angeles, Adelaide), others to the moderately humid tropics (Honolulu, Mauritius). Japan (Shizuoka) has, at least during summer, a climate similar to that of the latter islands and in winter the temperature is low enough, to stop the activity of both species. In Paris we find a locality to the north of the area of permanent distribution. The winter-cold of Paris does not permit a permanent acclimatization of *Icerya*. In the Sahara-Sindian or in the Irano-Turanian region both species may be encountered, but *Icerya* exhibits a low population level, which has no chances for outbreaks. In Egypt, for example, it has no economic importance.

The large tolerance of *Icerya* is proven by its occurrence in very different types of climate. Its hermaphroditic propagation helps this high resistance to environmental influences. *Novius* hibernates in regions of the cool northern Mediterranean climate, thanks to its high development threshold (not in Palestine); *Icerya* may develop in those regions for some months, unmolested by *Novius*. Balachowsky and Molinari estimate that period for Antibes to four months. In addition to those four months there are, as was shown before, three winter months in which the development of *Icerya* is interrupted by diapause.

It is obvious that *Novius* suffers more than *Icerya* from dry summer-heat and from cold in winter. It is a great deal less tolerant. But the vital optima for the most sensitive stages of both species seem to be between 18—23° C. The corresponding relative humidity is probably above 70 %.

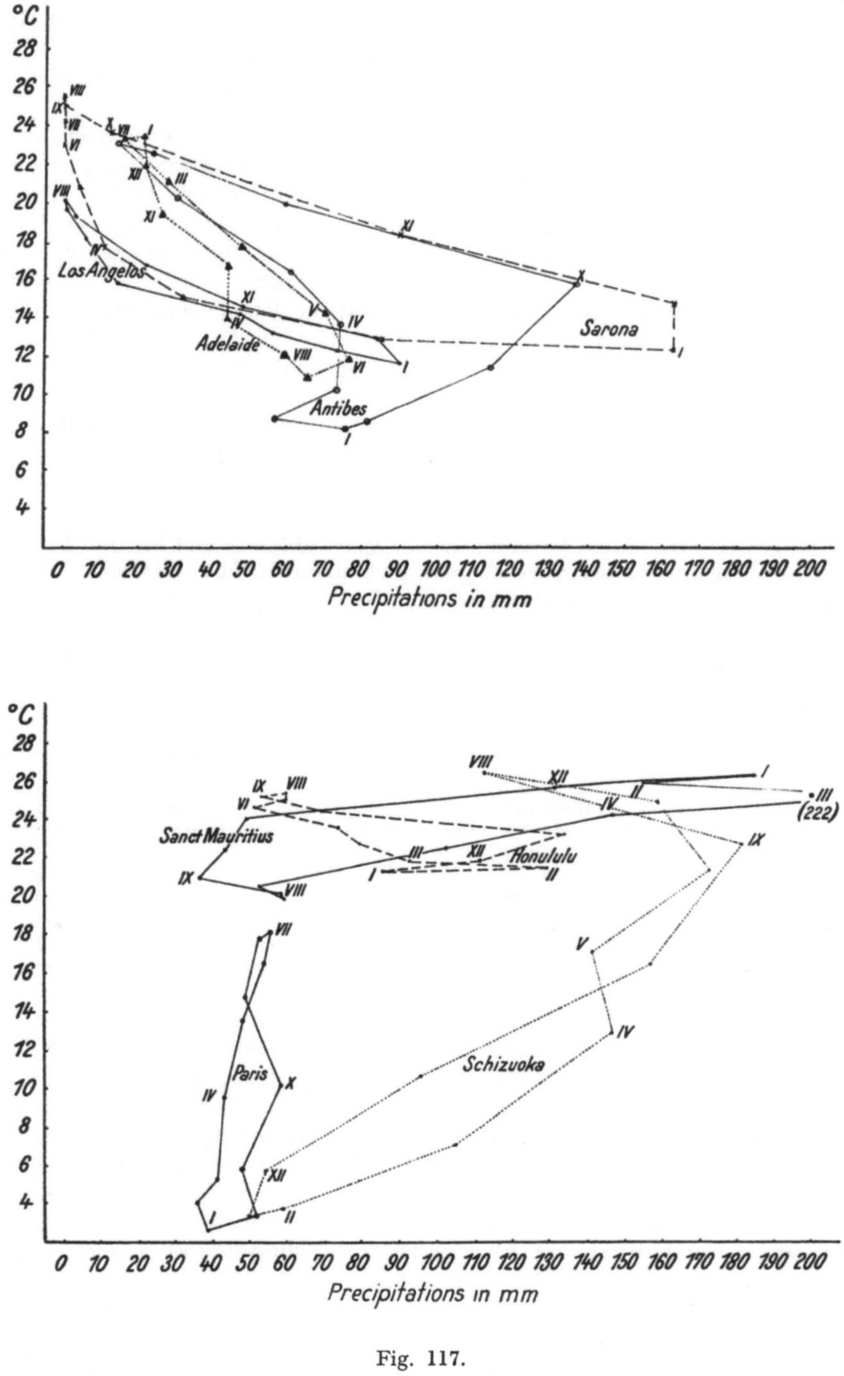

Fig. 117.

Climograms from the area of *Icerya* and *Novius*. A. Mediterranean climates (Antibes, Sarona, Adelaide, Los Angeles). B. Tropical climates (Honolulu, Mauritius, Shizuoka (partim) and climate beyond the northern border of distribution (Paris). The Roman numbers refer to months.

It is evident from many experiences that a mass development of *Novius* is possible in the Mediterranean climate as well as in the moderately humid tropics of Hawaii. Balachowsky and Molinari believe: „Les exigences biologiques d'*Icerya purchasi* et de *Novius cardinalis* étaient les mêmes, le prédateur d'étant toujours acclimaté dans les localités oû *Icerya purchasi* a pû lui même se maintenir."

Their conclusion is correct, not so their premise. *Novius* can become acclimatized in every region where *Icerya* outbreaks develop. But that does not necessarily mean, that both belong to the same ecological type and have the same ecological requirements. Those ecological constants, which have been evaluated up to date, enable us to perceive a large difference between both species. *Novius*, however, is able to allay outbreaks of *Icerya* completely in every climate, which

1) has not a range extreme to that degree that the *Novius* population is reduced almost to zero, once or twice yearly,
2) does not deviate from the just described optimal range for a certain minimal sequence of months.

Only at such climatic conditions does the high biotic potential of *Novius* develop its constant efficiency. In a tropical climate the density of the *Icerya* population determines primarily the fluctuations of the population of *Novius*. In the more humid tropics the rainy season has possibly another adverse influence.

Theoretical considerations of the mutual fluctuations of Icerya and Novius: About four generations of *Icerya* and thirteen of *Novius* develop under the conditions of Honolulu or Mauritius. We estimate the increase potential of an *Icerya*-hermaphrodite as 200 eggs per individual. For *Novius* we assume 300 eggs per female and an equal sex-ratio. In a starting population one pair of *Novius* may be present for every 100 of *Icerya*. If only the natural increase would determine the development of our population, we would get:

Icerya		*Novius*	
Generation	No. of individuals	Generation	No. of individuals
P	100	P	2
1	20.000	1	300
		2	45000
2	4.000.000	3	6750000
		4	1012500000
3	800.000.000	5	151973000000
4	160.000.000.000		

In its third generation *Novius* would have already reached the second one of *Icerya*. But *Novius* does not destroy one but many hosts during its development and adult life. If we select for computation the data from summer-months in Palestine, we find that one *Novius*-larva destroys 126 *Icerya*-eggs; and one adult beetle 56 second stage larvae in 80 feeding days. One *Novius*-beetle, which dies a normal death, destroys more than 180 hosts.

The destruction would be even much higher, when one takes into consideration that Kuwana estimated the total average larval consumption of *Novius* to be 210 second stage larvae of *Icerya*.

Under the Palestinian conditions the computation would therefore be:

	Icerya	*Novius*	Consumption of the *Novius* population
Starting population	100	2	
1. Generation	20.000	300	54.000 *Icerya*

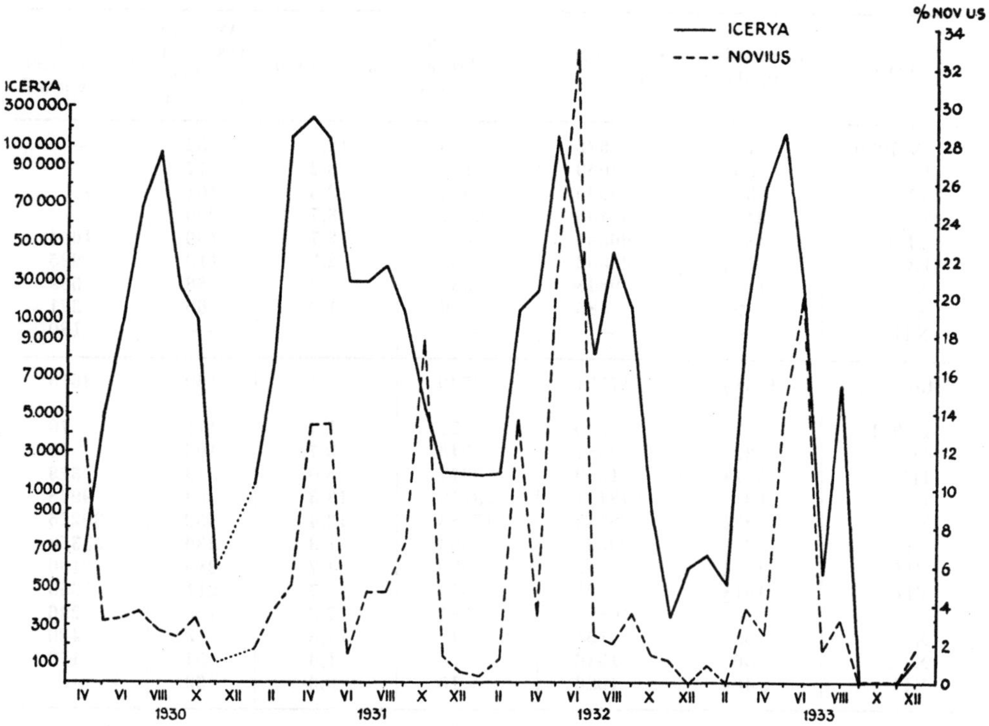

Fig. 118.
Fluctuation of *Icerya* and *Novius* populations in the orange groves at Atta (1930—33): percentage of *Novius* infestation (----), number of *Icerya* calculated as product of collecting days multiplied by the average day-worker result per month (——) (from Bodenheimer).

Even after considering that many eggs may be replaced by much fewer larvae, it is obvious that *Novius* should destroy its host entirely in the course of one generation, whenever the 1—2 % ratio between predator and host is reached.

When we add 1) that *Novius* is extremely active in its search for food and for suitable places for oviposition, 2) that a double or greater oviposition per *Icerya*-ovisac is relatively rare, 3) that *Novius* may migrate, probably supported by wind, for some kilometers, the merit of Novius is even greater. A climatic mortality of only 30—70 %, as observed in Palestine, is small compared with these advantages, especially as a certain climatic mortality occurs also in the host.

The epidemiology of Novius and Icerya in Palestine: The main facts are based on regular counts in the Atta-group of orange groves (40 hectars).

Fluctuations in the Icerya- and Novius-population in the Atta Groves (Hedera) in 1930—1934.

Date	No. of collecting days	No. of Icerya collected	No. of eggs and larvae of Novius	% of Novius	Average result of a collecting day	No. of Novius beetles exposed
IV.1930	13	672	86	12,8	52	—
V	65½	5059	162	3,2	77	—
VI	96½	9685	324	3,3	101	—
VII	347½	69495	2591	3,7	200	—
VIII	487	96865	2628	2,7	199	1654
IX	231½	25552	602	2,3	110	825
X	111½	9848	331	3,3	88	589
XI	9	585	6	1,0	65	371
XII	—	—	—	—	—	180
Total . . .	1361½	217761	6730	3,1	160	3619
I.1931	7	1200	21	1,7	171	70
II	39	7587	346	3,7	196	90
III	581½	141369	7377	5,0	243	378
IV	631	248454	23158	13,3	393	9978
V	283½	128143	17183	13,4	452	39225
VI	123½	29201	399	1,3	236	53350
VII	102½	29202	1390	4,7	285	18100
VIII	181½	38321	1819	4,7	211	1655
IX	70	10827	787	7,2	155	296
X	62	5154	919	17,8	83	420
XI	36	1946	28	1,4	54	173
XII	46½	1826	15	0,6	39	—
Total . . .	2163	643230	63422			123635
I.1932	35	1716	7	0,3	50	—
II	28½	1868	21	1,1	65	215
III	129½	13650	1887	13,7	105	110
IV	139½	23187	790	3,4	165	2255
V	412	121946	25496	21,2	300	7880
VI	237½	55673	18437	33,0	235	37550
VII	40	8069	120	2,3	200	4025
VIII	130	31320	655	2,0	340	1300
IX	131	13353	459	3,7	100	1175
X	15	897	13	1,4	60	—
XI	5½	346	4	1,1	60	—
XII	7	588	—	—	85	—
Total . . .	1310	272613	47887		208	54510
I.1933	9½	664	6	0,9	70	—
II	4	515	—	—	129	—
III	35½	10527	412	3,9	197	287
IV	231	77093	1926	2,5	334	7860
V	382½	161638	22579	13,9	423	59300
VI	99½	25475	9091	20,3	256	16650
VII	4½	572	—	1,6	127	—
VIII	31	5464	177	3,2	176	300
IX	—	—	—	—	—	—
X	—	—	—	—	—	—
XI	—	—	—	—	—	—
XII	2	109	2	1,8	55	—
Total . . .	799	282057	31193		353	84397

These figures give the ratio between the absolute number of *Icerya*-ovisacs collected and the number of larvae and eggs of *Novius* contained in them. As a matter of fact it is obvious, that from an epidemiological point of view, 5 larvae in one ovisac must have another influence than a single one. In order to gain a survey on the influence of this factor, special counts were arranged for in 1933:

Number of eggs of Novius cardinalis per Icerya-ovisac

Date	Total counted	Average	Novius eggs per Icerya ovisac																					
			1	2	3	4	5	6	7	8	9	10	11	12	13	14	15	16	17	18	19	20	22	25
I.1933			No eggs observed																					
II			No eggs observed																					
III	59	4,21	4	1	3	1			2	1	2													
IV	255	4,05	3	19	7	8	13	4	4	3		2												
V	2229	5,48	17	46	74	62	74	21	39	17	21	7	5	7	3	3	4			4	1	2	1	2
VI	495	4,85	19	7	18	10	24	2	4	9		3	1	1				1	1					2
VII	3	(1,0)	3																					
VIII	6	(1,0)	6																					
	3047	5,10	52	73	102	81	111	27	49	30	23	12	6	8	3	3	4	1	1	4	1	2	1	4

Number of larvae of Novius cardinalis per Icerya-ovisac.

Date	Total counted	Average	Novius larvae per Icerya ovisac																			
			1	2	3	4	5	6	7	8	9	10	12	13	14	15	16	17	18	21	28	30
I.1933	6	1,00	6																			
II	—	—	—																			
III	237	1,16	180	14	8		1															
IV	919	1,18	740	19	25	3	1		7													
V	10.039	1,18	7641	491	149	76	32	22	7	5	7	9	1	1	1		3	1	1	1		
VI	1.860	1,56	987	99	55	25	14	3	8	6	5	4	1			3			1		1	1
VII	6	1,00	6																			
VIII	130	3,61	14	1	7	3	2	2	1	2	4											
13.197		1,23	9574	624	244	107	50	27	23	13	16	13	1	1	1	3	3	1	2	1	1	1

The most interesting result of these observations is the difference between the number of eggs and larvae of the ladybird per *Icerya*-ovisac. The number of 5,10 eggs per ovisac is reduced to 1,23 larvae per ovisac. There is no doubt, that the overwhelming majority of the difference is due to the cannibalism of *Novius*. The *Novius*-egg is as readily eaten as that of *Icerya* and a young conspecific larva is devoured by an older one as well as an *Icerya*. As a rule only one *Novius*-larva will develop successfully in any one ovisac of the host. There are certainly not many exceptions. Among 10706 ovisacs 9574 contained only one larva and only in 1132 more than one *Novius* larva was present. In the latter case most larvae were probably very young and would as a rule have disappeared in the later development. Only 24,1 % of the eggs developed into larvae, 75,9 % were devoured by older larvae.

We may now study the influence of double and higher infestations of the same ovisac on the intensity of predatism. As a basis for the computation we will use

1) the actually observed relation of the number of eggs and larvae;
2) the monthly reduction factor, e.g. in March 1,16 for the larvae and 4,21 for the eggs.

The number of successfully developing *Novius* and consequently the percentage of *Novius* attack change as follows:

Date	No. of *Novius* absolute	No. of *Novius* corrected	% of *Novius* attack absolute	% of *Novius* attack corrected	difference in %
I.1933	6	6	0,9	0,9	0
II	—	—	—	—	—
III	412	303	3,9	2,5	—36
IV	1926	1379	2,5	1,8	—28
V	22579	16353	13,9	10,1	—28
VI	6091	3285	20,3	12,9	—36
VII	9	9	1,6	1,6	0
VIII	177	54	3,2	1,0	—70
					—28

About —28 % is the deviation factor for all those series, which were large enough to present a reliable average. The epidemiological value of predatism has to be assumed therefore about one third below the value given in the general table. The general epidemiological trend is not changed by this correction.

We compare the average *Icerya* collecting-day per worker in the orchard (as index of population density of *Icerya*) with the percentage of *Novius* infestation:

% of *Novius*	n	Average collecting-day per worker
0—2	13	93 Icerya
2—5	14	185
5—10	4	224
10—20	6	251
20—30	2	278
30—40	1	235

This means that a parasitism percentage, which is higher than 5—10 % has only very little influence on the population density of *Icerya*. The maximal efficiency is reached at about 10 % of *Novius* parasitism. Any further increase adds only very little to the further biological control of *Icerya*. And a low population density of *Icerya* is always accompanied by a low percentage of *Novius*. Fig. **118** shows again the mutual fluctuations of *Icerya* and *Novius* in the Atta orange-groves. The result of average collecting-days per worker, multiplied by the number of collecting-days per month, is taken as an index of the *Icerya* population. The increase in the *Novius*-predatism follows the increase of *Icerya* after 1—2 months, and as a rule its decline starts one month later than that of the host. An phenomenon of fundamental importance is the steep descent following each peak of *Novius*-infestation. The highest value for the year is always followed by a minimal one. This shows that not only the optimum of efficiency is surpassed, but that, in the epidemiological sense, every peak creates conditions which lead to its own breaking down. The optimal density of *Novius* in a dense *Icerya* population is reached at a 5—10 % of parasitism. Any further increase in this percentage thins out the *Icerya* population below the optimal density. Of more importance are the over-oviposition on *Icerya* and the greater scarcity of food.

Fig. **118** shows clearly two peaks for the *Icerya* population: a major one in the spring, a minor one in late summer. It is very probable that this summer cessation is the effect of the *Novius*-activity. The year 1930, in which our control measures were hardly efficient, shows no summer-cessation. It shows a different type. Predatism was in former years high in winter and spring, low in summer and early autumn. From May to October the activity of *Novius* was much depressed and not very conspicuous to the observer in the orchard. Since the beginning of the control this depression occurred as a rule, only in July and was of much shorter duration.

In order to get a critical comparison as to how far our competitions give a fair cross-section of general importance for the Atta groves, we compare now the number of *Icerya* in our monthly counts on 400 leaves (multiplied by 1000) with the *Novius* percentage:

Comparison of Density of *Icerya*- and *Novius*-populations.

Ionth	1930			1931			1932			1933			Average		
	Icerya × 1000	Novius %	Novius absolute	Icerya × 1000	Novius %	Novius absolute	Icerya × 1000	Novius %	Novius absolute	Icerya × 1000	Novius %	Novius absolute	Icerya × 1000	Novius %	Novius absolute
I	—	—	—	2000	1,7	34	—	0,3	7	1000	0,9	6	1500	1,3	20
II	—	—	—	10000	3,7	370	2000	1,1	21	1000	—	—	6000	2,4	195
III	—	—	—	4000	5,0	200	2000	13,7	1887	1000	3,9	412	2330	7,5	832
IV	5000	12,8	640	87000	13,3	11571	8000	3,4	790	15000	2,5	1926	28500	8,0	3307
V	17000	3,2	544	63000	13,4	8442	37000	21,2	25496	54000	13,9	22579	42750	12,9	14265
VI	7000	3,3	231	2000	1,3	26	36000	33,0	18437	3000	20,3	6091	12000	14,5	6196
VII	37000	3,7	1369	4000	4,7	188	3000	2,3	120	—	1,6	9	13700	3,6	626
VIII	18000	2,7	162	9000	4,7	423	4000	2,0	655	2000	3,2	177	8250	4,2	429
IX	6000	2,3	115	9000	4,2	648	2000	3,7	459	3000	—	—	5700	3,4	407
X	5000	3.1	124	8000	17,8	1424	5000	1,4	13	—	—	—	6000	7,4	520
XI	4000	1,0	90	2000	1,4	28	3000	1,1	4	6000	—	—	3000	1,2	21
XII	3000	—	—	2000	0,6	12	—	—	—	—	1,8	2	2000	0,6	12

This table confirms all the conclusions which we draw from the monthly counts. If we divide the number of 400 leaves (multiplied by 1000) by the corresponding number of *Novius* collected, we are again confirmed that our counts give a fair picture of the mutual fluctuations of host and parasite:

Novius %	n	Ratio of *Icerya* to *Novius* average	range
0—1	2	167	167
1—2	7	78	39—133
2—4	14	27	4—111
4—6	4	19	14—21
10—20	6	5	1—8
20—30	3	1,3	0,5—2

It is now obvious that the theoretical computations on the epidemiological efficiency of *Novius*, with which we started this chapter, are in strong disagreement with the facts. Considering the very low climatic mortality of *Novius* and that, as far as could be ascertained, no other factor of major importance was overlooked, we conclude that host density must be mainly responsible for this disagreement. And the influence of this factor is twofold: Low population density of *Icerya* is always accompanied by a minimum of *Novius*. The reason for that is obvious: If the density of *Icerya* surpasses the mark of optimal density, then the density of the predator becomes a limiting factor. Cannibalism, double and triple oviposition on the same ovisac, etc., limit the further increase in *Novius*, and condition a limitation of its epidemiological efficiency. In our observations, the optimal conditions were about 10 % of *Novius* parasitism at a high population density of *Icerya*. But that percentage certainly varies with the density of the host's population, and is certainly not of absolute value. Population density of host and predator is in any case of primary importance,which can never be neglected in studying the mutual fluctuations.

The following table shows the seasonal and local differences in the *Novius* infestation of the various gardens in the Atta group:

	A	B	C	D	E	F	G	H	I	K	L	M	N
1930													
IV.										4,1	50,8		
V.		2,6								1,1	6,2		
VI.		3,6	1,3					2,2		4,1	3,6		
VII.	5,3		6,1	4,6		1,3	3,3		2,9	5,5	3,2	3,4	4,6
VIII.	3,2			0,8	2,5	2,2	3,0		3,1	2,3	2,6	2,0	2,0
IX.	8,3				4,4		2,6		1,0	1,2	3,1	1,5	
X.	8,3						1,8		3,1	4,7	6,6	5,1	
XI.												1,0	
XII.	—	—	—	—	—	—	—	—	—	—	—	—	—
1931													
I.					3.5						0,7	1,6	
II.		9,0			5,2		2,7	0,3	2,6	2,6	2,5	5,0	
III.	5,4	3,0	3,8	3,1	5,8	3,5	4,7	6,0	5,0	4,4	4,5	5,0	
IV.	14,0	7,1	7,9	20,5	23,0	8,2	15,1	16,2	7,4	3,9	7,2	4,1	6,4
V.		21,0	16,1	10,8	—	7,0	23,0	31,8	3,5	6,4	10,0	4,9	16,3
VI.				6,5	8,0	2,3				1,8			
VII.		1,1			3,5	3,3	7,7		3,8	3,6	4,6	4,4	
VIII.			6,5		5,9	4,5	5,0	2,8	4,6	4,0	4,6	4,2	3,9
IX.					4,7				7,1	7,4	10,3	6,9	
X.		0,9					15,0		18,6	10,4	11,9	25,6	14,4
XI.				1,1	0,6							2,6	
XII.				0,5	0,5						0,5		

The list of the single groves of the Atta-district illustrated the changing intensity of *Novius*-infestation in the different gardens according to time and space. This changing intensity is of decisive importance for the mutual fluctuations. Simultaneously with the reduction of *Icerya* to a minimum in an old focus, new growing foci of *Icerya* are formed, which are the centers of future local outbreaks and which attract the beetles, which leave the old focus. This mutual correlation of always alternating centres of *Icerya* and *Novius* is typical for avery large *Icerya*-outbreak. In contrast to many other pests, no permanent focus is formed in *Icerya* in consequence of the activity of *Novius*. Another table may illustrate these changes:

Maximal Icerya population per 100 leaves in Atta.

	grove			
Year	A	J	K_1	K_2
1928	12	2	34	—
1929	14	9	**70**	**32**
1930	14	3	12	10
1931	**59**	5	9	20
1932	16	**10**	10	10
1933	12	9	5	28

Apart from these observations made during outbreaks, we may study an isolated infestation in Jerusalem. One lemon tree on the campus of the Hebrew University was infested in May 1930 by some 20 ovisacs of *Icerya.* In July the tree was covered with the ovisacs of the following generation, as if covered with snow. Ten beetles of *Novius* (presumably five couples) were put on the tree at the beginning of August. One month later not one living *Icerya* could anymore be discovered. In October/November some scattered ovisacs were present, which originated from the few *Icerya*-larvae which had been overlooked by *Novius.* In the following year a new *Icerya*-outbreak occurred, which was decimated by *Novius.* It is not clear, whether these originated from those of the previous year, or were unintentionally introduced by us. This time the extinction of both species was radical. In the winter 1931/32 no ovisac has been discovered again. The tree, however, was not entirely isolated. Scattered *Icerya* were found on Spartium and Fragaria in the University gardens. It is easy to guess what would have occurred on an entirely isolated tree. Either 5—10 eggs or young larvae of *Icerya* would have escaped extinction by *Novius.* This number would have been sufficient to build up a new *Icerya*-outbreak, unmolested by a *Novius*-attack. And then either the latter would have died from hunger in the mean time, or both species would have died out. The first probability is more likely to occur in fully isolated trees. The shift of *Icerya*-foci in citrus groves prevents any such event in nature.

b. Other parasites and predators of Icerya in Palestine.

The real importance of the high biotic potential of *Novius cardinalis* is very obvious when we compare it with that of another predacious beetle of *Icerya* with *Lithophilus marginatus Rtt.* (Coccinellidae.)

Description: The adult ladybird is relatively broad, of short oval form and not much convex. It is yellow-red, except its black elytra, which have only a narrow red margin. 4 mm long.

Eyes large and round. Wings lacking. The body is covered with a velvet-like pubescence. Prothorax, with elevated margins, much retracted in the middle of the posterior margin. Its lateral margins form an obtuse angle in their middle. The elytra are larger than the thorax. The shoulders are distinctly angular.

Distribution: Originally described from Beirut; has been met by us at Hedera and Petah-Tikva. Apparently occurring all over the coastal plain of Palestine.

Only 16 % of the eggs develop to maturity. The mortality is especially high in the hamsin-month of May and during the hot summer months (July/August).

The potential development of this species is very low as compared with *Novius* and when taking into consideration the slow development of this species (one generation per annum only). We have to consider now its feeding activity. The total consumption of *Icerya* larvae (II stage) was in 1933:

Month	n	Average development of larvae	Average consumption of Icerya-larvae
III	4	72 days	26 larvae
IV	1	64	16
V	7	31	14
VI	11	23	19
VII	22	15	14
VIII	1	24	27
IX	46	23	27

During the summer and autumn, the main period of development, about one larva is devoured per day or about 20 larvae per development. During the spring the consumption is relatively less owing to the colder weather conditions.

The adult beetles devoured:

On Citrus				On Spartium		
No.	Period of observation	No. of feeding days	No. of Icerya devoured	No.	No. of feeding days	No. of Icerya devoured
1	19/VII/32—25/V/33	311	288	1	453	384
2	19/VII/32—18/XII/33	518	489	2	520	462
3	21/VII/32—18/XII/33	516	461	3	520	453
4	21/VII/32—12/XI/33	480	449	4	520	498
5	24/VII/32—18/XII/33	513	486	5	518	463
6	24/VII/32—18/XII/33	513	519			
7	24/VII/32—26/XI/33	491	481			
8	24/VII/32—18/XII/33	513	494			
9	24/VII/32—18/XII/33	513	508			
Average consumption per beetle			464			452

The average duration of the different stages from May to September 1932 was in 100 breedings:

Egg-stage	7	days
Larval-stage	19,5	,,
Pupal-stage	14	,,
Total average	40,5	days (Minimum 32 days, Maximum 48 days)

The longevity of some adult females which hatched in July/August 1932 was:

No.	Longevity	Period of oviposition	No. of eggs
1	425 days	163 days	170
2	467	117	135
3	551	189	229
4	549	188	181
5	549	187	219
6	546	179	206
7	539	21	15
8	539	155	200
9	535	202	373
10	535	163	310
11	512	246	161
12	527	174	318
Average. . .	523	169	227

The longevity of the adult female lasts from the spring to the autumn of its hatching into the winter of the second following year. Its oviposition takes place from February to October of the first following year, mainly from June to September. The eggs are laid singly, on the average 1—2 eggs per day of the oviposition period. No oviposition was observed during the second winter and spring, but under optimal conditions it may occasionally occur in nature. The longevity of the males is approximately the same or a few weeks longer.

The natural climatic mortality of the development stages is much higher than in *Novius:*

Month	Eggs	Larvae	Pupae	Beetles
IV	100	47	41	35
V	100	21	9	7
VI	100	35	21	14
VII	100	29	21	8
VIII	100	24	8	8
IX	100	28	28	21
Average. . .	100	31	21	16

The average consumptions per beetle (464 resp. 452) or their average daily consumption (0,96 resp. 0,89) is identical for both plants. Somewhat less than one larva per day is the average food. About 480 larvae are devoured during the total life-period.

The high natural mortality, the low number of generation and the low feeding index, make *Lithophilus* a relatively unimportant predator of *Icerya.* It is not very common.

No other natural enemies of any importance are known from Palestine. The fly *Cryptochaetum grandicorne* Rond. has been mentioned by mistake. It is restricted as a host to *Gueriniella serratulae.* But the caterpillar of the Noctuid *Coccidophaga scitula* Rbr., which feeds normally on soft scales, occasionally also attacks *I. purchasi.* Small mites and some neuropteroid insects are also suspected of occasionally preying on *Icerya,* but they certainly have no importance for its natural control.

Technical control: Since *Icerya purchasi* develops in numbers in citrus-seed-beds. nurseries, or in ornamental gardens, a technical control seemed to be advisable, The thick wax-layer of the ovisac is an important impediment, as mosts sprays, etc. do not moisten the wax but flow off, without having touched the eggs. The chemical used must therefore possess a high wax-dissolving power on one hand, without damaging the plant on the other hand.

Satisfactory results were gained with painting oils of the carbolineum type, with a broad brush on the *Icerya* females and ovisacs. Painting with a 10 and 20 % solution of the Dutch Carbolineum Carbokrimp (Utrecht) yielded: (control on 17.IX.26)

Concentration	1st Inspection (2 days after control)	2nd Inspection (8 days after control)
10%	young larvae dead, ovisac white, ♀ dying	♀ dead, all eggs hatched in the meantine
20%	young larvae dead, ovisac white, ♀ dying	♀ dead, ovisac shrivelled, all eggs and larvae dead in its interior.

No damage to foliage was observed. Carbolineum 20 % gives satisfactory results. Owing to the difference in the different brands of carbolineum, the concentration should always be tested beforehand, to see whether it is not injurious to the plant. In any case, it should be applied, as far as possible, to the scales and ovisacs only, and not to the plant itself.

In some orchards lime was dusted on the trees. This method proved to be entirely valueless. In two series of observations even the first larval stage remained entirely undamaged. The later development of sooty mould over the lime layer was not prevented.

Resumé of control recommendations: Besides the technical control in nurseries just mentioned, the following method proved to be most satisfactory in orchards, where heavy damages are liable to be caused by *Icerya* and Sooty Mould.

In every centre of *Icerya*-increase the conspicuous ovisacs are collected by special workers, mainly girls. All collected ovisacs are delivered every evening into a central breeding station, where they are counted (quantity per daily worker) and put into special breeding cages. The hatching of many millions of eggs is thus easily prevented, but all *Novius* eggs or larvae, which may be contained in these ovisacs, develop and are eventually collected as adults and brought into such parts of the orchards, where new centres of *Icerya* are in the process of formation. This combination of collecting *Icerya* and breeding *Novius* proved to be entirely satisfactory.

This method was the more desirable, as all our attempts to breed *Icerya* quickly in large quantities for artificial *Novius*-breeding were in vain.

The practical results were striking: no economic loss was ever noted, where this system was functioning, and this despite a very high spring development in at least two years. Under normal conditions this development would have led to a very high peak in the autumn and extensive Sooty Mould formation, which were actually observed in Arab orchards of the same district.

Counts of Icerya on 400 leaves

	Month												
Year	I	II	III	IV	V	VI	VII	VIII	IX	X	XI	XII	Total
1928						16		28	**61**	37	5	3	(300)
1929					6	22			87	**93**	35	4	(488)
1930	9	7	4	5	17	7	**37**	18	6	5	4	3	122
1931	2	10	4	**87**	63	2	4	9	—	8	2	2	196
1932	—	2	2	7	**37**	36	3	4	2	5	3	—	99
1933	1	1	1	15	**54**	3	—	2	3	—	6	—	85
1934	—	1	—	—	19	6	1	—	—	—	4	1	33
1935	1	2	1	2	22	14	1	3	1	—	6	2	53
Average 1930/35	2	4	2	19	**35**	12	8	6	2	3	4	**1**	98

These counts demonstrate two phenomena very clearly: The first one is the influence of intensive control measures. Since the beginning of the regular campaign a peak is reached in April or May, which is definitely cut in the second half of the year. In earlier years another important peak appeared in the autumn (IX/X), which caused the main damage. Since the beginning of control measures no further peak in autumn has appeared, despite the fact that the development of the *Icerya*-population was rather menacing in the springs of two out of four years. In another orchard, where no *Novius* was brought in artificially, the fluctuation on 400 leaves was even in 1932:

Month	V	VI	VII	VIII	IX	X	XI	XII	Total
No. of Icerya	**58**	1	2	14	**26**	1	5	4	(166)

The second phenomenon is that the annual rate of increase under the present conditions is larger than 1:>43; the maximum being in May, the minimum in December.

The costs of control vary in different years proportionally with the intensity of the menace. When only a few *Icerya* are present a few collectors only will be sent into the orchard and their number may be increased daily according to the number of *Icerya* present. This is a very desirable feature of a control method, since its expensee depend on the intensity of the potential menace. In spraying or fumigation the costs remain entirely independent of the annual intensity of the menace. Some figures from a large orchard in Atta may describe the expenses of collecting.

A thoroughly clean fruit was the product of this collecting. The total cost in **1931**, a year of a very heavy attack, was LP. ½ per dunam which is very low compared with the cost of spraying or fumigating. In those parts of Samaria where *Icerya* has developed into a permanent menace, it is performed every year.

Collecting has to be started, whenever an increase of *Icerya* is observed, especially during July/August, when it has to replace the lowered activity of *Novius*.

References.

A. Balachowsky et L. Molinari, Annales des Epiphyties. **16**, 1930, pp. 1—24.
A. Balachowsky, Encyclopedie Entom. Vol. **15**, Paris, 1932, pp. 124—128.
F. S. Bodenheimer, Die Schaedlingsfauna Palaestinas. Berlin 1930, p. 141—147.
F. S. Bodenheimer, Zeitschr. ang. Ent. **19**, 1932, pp. 514—543.
F. S. Bodenheimer, and Tenenbaum, Hadar. 1933, pp. 32—34.
P. Borg, Entomological Notes. Malta Dept. Agric. typescript. 1930.
A. W. Cressman and J. O. Dumestre, Journ. Agric. Res. **41**, 1930, pp. 197—203.
M. Delassus, Revue Agricole Afrique du Nord. 1923.
M. Dingler, Biol. Zentralbl. **50**, 1930, pp.
S. Hughes—Schrader, Zeitschr. Zellforsch. und Mikrosk. Anatomie **2**, 1925; **6**, 1927. Science **63**, 1926. Ann. Ent. Soc. Amer. **23**, 1930.
C. E. Johnston, Ann. Ent. Soc. Amer. **5**, 1912, p. 383—388, Pl. 28.
I. Kuwana, Dept. Agric. and Comm. Imper. Plant Quarantine Sta. Yokohama, Bull. 2. 1922.
C. Menozzi, Boll. Soc. Ent. Ital. **16**, 1927, pp. 105—109.
U. Pierantoni, Monit. Zool. Ital. **21**, 1910.
U. Pierantoni, Archivo di Zool. Ital. **5**, 1912; **7**, 1914.
R. Poutiers, C. R. Soc. Biol. **103**, 1930. pp. 1023—1025.
C. Ribaga, Riv. Patologia Vegetale **10**, 1902.
C. V. Riley, U.S. Dept. Agric. Div. of Entom. Bull. 15, 1887.
G. O. Shinji, Entom. News **28**, 1917.
F. Schrader and S. Hughes—Schrader, Zeitschr. wiss. Zool. **128**, 1926.
P. Vayssiere, Ann des Epiphyties **12**, 1926,
G. Wolcott & F. Sein, Journ. Dept. Agric. Puerto Rico **17**, 1933.

8. THRIPS AND TERMITES (CORRODENTIA).

1. THYSANOPTERA.

Of the 17 species knows so far from citrus in Palestine, only *Heliothrips haemorrhoidalis* can be regarded as a major pest. This species is the only one among the 17 species which is apparently not native in this country.

Key to the Thysanoptera living on citrus in Palestine.

1. Female with a saw-like ovipositor composed of two pairs of gonapophyses arising from the 8th and 9th abdominal segments. Last abdominal segment rarely tubular, usually conical in the female, and bluntly rounded in the male, never tubular. Wings furnished with microscopic hair and bristles. Fore-wings with a marginal, and at least one longitudinal, vein reaching to the tip of the wing. (Subordo *Terebrantia*) 2.
   Female without ovipositor. The last abdominal segment closed all round usually tubular. Wings without microscopic hair and with only three basal bristles on their anterior pair, at the base of it the rudiment of a vein. (Subordo *Tubulifera*) 11.
2. Ovipositor turned upward; wings broad and rounded at the ends, body not depressed; antennae 9-jointed. (Fam. *Aeolothripidae*) 3.
   Ovipositor turned downward; wings narrower, usually pointed apically; body more or less depressed; antennae 6- to 8-jointed. (Fam. *Thripidae*) 4.
3. A smaller, blackish insect. Fore-wings with cross veins. *Aeolothrips deserticola* Pr.
   A larger, more vividly coloured insect. Fore-wings without distinct cross veins. *Franklinothrips myrmicaeformis* Zan.
4. 10th abdominal segment with two spines; head somewhat produced in front of the eyes. Male wingless, female with wings. *Limothrips cerealium* Hal.
   10th abdominal segment without a pair of spines; head not or little produced in front of the eyes. 5.
5. Body surface with distinct polygonal areas. Tip of antennae needle-shaped. *Heliothrips haemorrhoidalis Bouché.*
   Body surface with more or less confluent cross wrinkles, at the most. 6.
6. Fore-tibia with one or two very small curved teeth apically within; antennae 8-jointed, maxillary palpus 3-jointed. *Odontothrips karnyi rivnayi* Pr.
   Fore-tibia unarmed. 7.
7. Anterior angles of prothorax with a long strong bristle on each side. Blackish insect. Male with a horn-like projection on either side of abdominal segment 8. *Kakothrips robustus* Uz. (var.)
   Anterior angles of prothorax without a strong bristle. 8.
8. Antennae 7-jointed (style 1-jointed). 9.
   Antennae 8-jointed (style 2-jointed). 10.
9. Distal portion of upper vein of fore-wing with usually 4 scattered bristles; ocellar pigment yellow; comb of fine hair at hind margin of tergit 8 complete. *Thrips tabaci Lind.*
   Distal portion of upper vein of fore-wings with usually 3 bristles; ocellar pigment reddish; comb at hind margin of tergit 8 wanting in the middle. *Thrips major* Uz.
10. Large dark species; joints 4 to 8 of the antennae dark. *Taeniothrips meridionalis* Pr.
    Small yellowish to light brown species; joints 3 to 6 of the antennae yellowish at least at base. *Taeniothrips discolor* Karny.
11. Cheeks with some small bristle-bearing warts. Wings not narrowed in the middle. *Haplandrothrips* sp.
    Cheeks without bristle-bearing warts. Wings narrowed in the middle, more or less distinctly sole-shaped. 12.
12. Anal hair at tip of abdomen moderately long; fore-tarsal tooth directed inward or wanting; fore-femora enlarged in the male only. *Haplothrips spp.*
    Anal hair three times as long as the tubular terminal abdominal segment; fore-tarsal tooth directed inward and forward; fore-femora always enlarged. *Karnyothrips flavipes* Jones.

A. THE GREENHOUSE THRIPS *(Heliothrips haemorrhoidalis Bché.)*

Synonymy: 1833 *Thrips haemorrhoidalis* Bché.
1836 *Heliothrips adonidum* Hal.
1836 *Heliothrips haemorrhoidalis* Bché.

Description: The adult female is small, elongate and flat. Colour mainly dark black-brown; the last abdominal segments tinged with brown; antennae, legs, wings and ocelli yellow; eyes black.

Head: Anterior margin between the eyes forming 2 sides of a triangle; cheeks constricted posteriorly. Ocelli on the dorsal surface, between the eyes. Antennae strong, 8-jointed. The last joints grow narrow progressively; the last joint is much longer than the preceding one. Maxillar palpi 2-jointed; labial palpi 3-jointed. The length (resp. breadth) of the single antennal joints is as follows: 19 (32), 46 (35), 81 (22), 51 (23), 45 (23), 39 (18), 16 (8), (68 (5) μ. The prothorax widens gradually to the postlateral angles, with a small constriction in the middle. Wings long and straight, pointed terminally. Fringe of bristles on margin of wings sparse. The entire body heavily reticulated. This reticulation becomes less distinct in the central dorsal part of the abdomen and on its last segments. The recently hatched adult is pale grey to pale yellow, but it darkens soon. 1.3—1.7 mm long, 0,3—0.4 mm broad.

Males unknown.

The *eggs* are short banana-formed to long reniform, colourless. 0.25—0.35 mm long, about 0.1 mm broad.

The *first larval stage* is whitish with small pink eyes, when hatching. The colour changes later to yellowish. The last antennal joint in this and the following stage is needle-shaped. 0.25 mm long when hatching; antennae 0.16 mm long.

The *second larval stage* is yellowish. The body is relatively broader than in the first stage. 0.75 mm long when hatching, but reaches later up to 1.0 mm and slightly more. The abdomen is shrivelled after hatching in both larval stages, extending considerably after feeding.

The *prepupa* is whitish yellow with pink eyes. Antennae slightly broader and shorter than in the adult larva with indistinct joint limits. Mouth-parts inactive in this as well as in the following stage. Short wing-pads, reaching the second abdominal segment. 1.05 mm long.

The *pupa* is slightly larger than the prepupa (1.15—1.25 mm long). Whitish-yellowish, growing darker with age. The eyes are much larger than in the preceding stages; the ocelli appear in this stage. The antennae which were straight before are now bent backward over the head, reaching up to the middle of the pronotum. Wing-pads reaching the fifth abdominal segment.

Distribution: H. haemorrhoidalis is an ubiquistic species. It develops in nature in tropical and subtropical countries, espec. of the New World, were its home seems to be as long as they are not too dry for its existence. It is restricted to greenhouses in the cooler climates of Northern N. America and of C. and N. Europe.

In Palestine it is restricted mainly to the citrus belt in the coastal plain. But it has been found also at Ben Shemen and at Nazareth, far from citrus cultivation. In the younger citrus belt between Hedera and Haifa it seems to be still absent, whereas in the old groves between Haifa and Akko, as well as in those near Gaza no survey has yet been made. In the hills it is scarce, if present at all. In the Jordan Valley it seems to be absent. It has been recorded first from Palestine in 1913 by Morgan („on orange").

A similar type of distribution, restricted to a more humid coastal belt, has been observed in Egypt where the species occurs at Alexandria, but neither at Cairo nor in California, where it is recorded from the coastal belt only. It is not based on recent introduction, but on the physical ecology of the species.

Life-history: The eggs are deposited normally singly at the rate of 0.5 to 2 per day by the female which inserts its ovipositor into the parenchym of the host plant. As a rule, a shady place is chosen for oviposition. When the egg is laid, some fatty substance is excreted at its upper end by special glands at the base of the ovipositor. The latter is withdrawn and, as a rule, a large protective drop of excrement is given forth over the aperture of the oviposition gallery. The egg is not always deposited into the parenchym; occasionally between epidermis and parenchym, oftener with

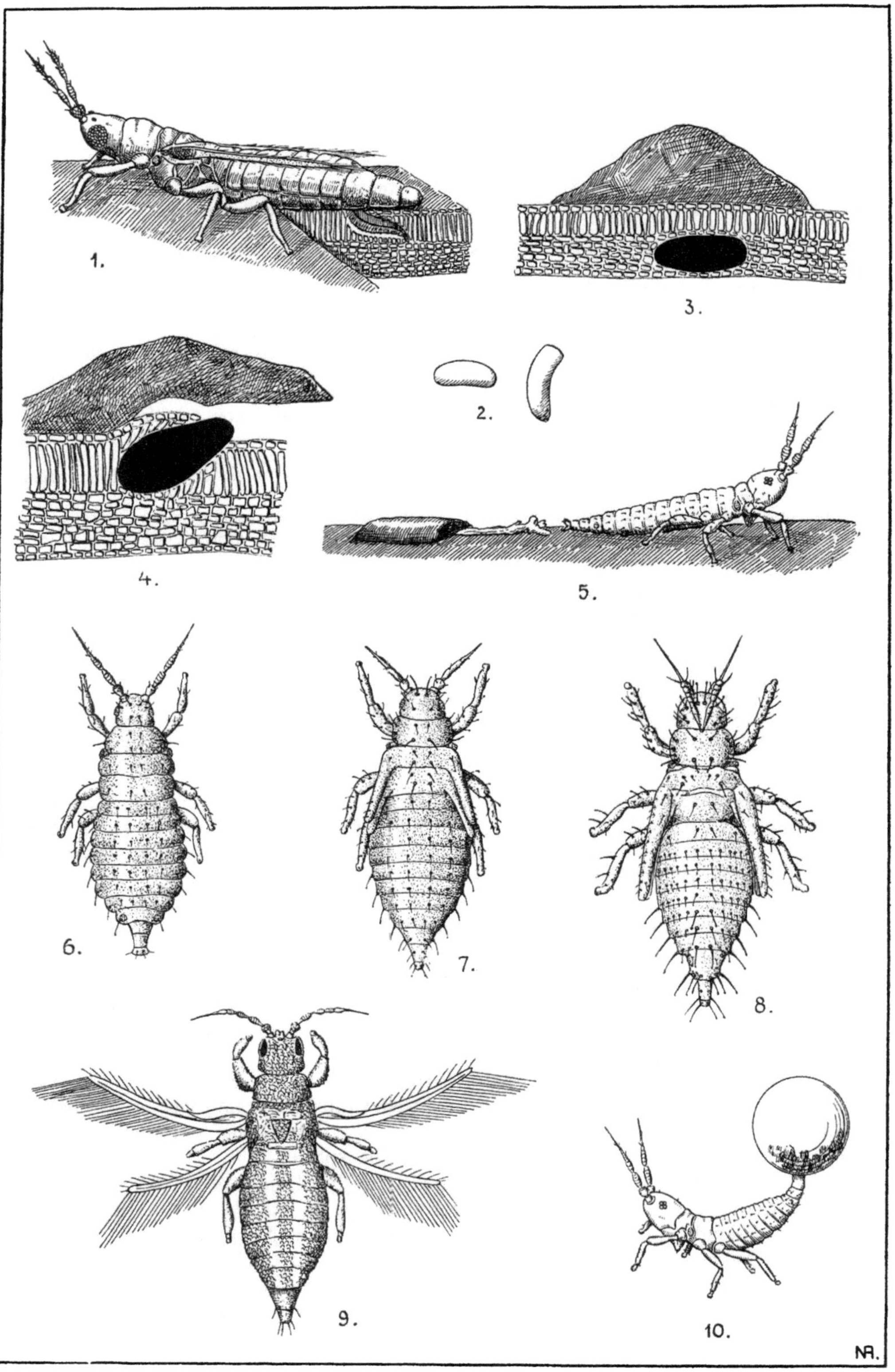

XXXVI. The Greenhouse Thrips (*Heliothrips haemorrhoidalis* Bché.) 1) ovipositing female; 2) eggs; 3) and 4) eggs (black) in leaf, covered with excrements; 5) neonate larva leaving the egg, after having shed the amnion; 6) full-grown larva; 7) first nymph; 8) second nymph; 9) adult; 10) larva carrying a large drop of excrement. (after Buffa).

one end lying in the epidermis and occasionally even partially protruding from the leaf. On citrus the egg causes the leaf to bulge, which can often be easily detected with a magnifying lens. After oviposition the plant produces a cork-layer, separating the egg from the plant tissue (Rivnay). The process of oviposition has been carefully described by Buffa (1911). The water-condition of the plant has an influence on the egg during its development (Rivnay, 1935). When water is scanty and the leaves wither the eggs are injured and, depending upon the intensity of withering, only 10—50 % of the eggs hatch. Such conditions are not rarely created in Palestine in under-irrigated orange groves during or following a strong hamsin-wind. Similar mortality occurs when water-metabolism and other conditions are optimal, leading to a strong proliferation of cork-cells. The eggs of *Heliothrips* may then be crushed and killed, or the epidermis may burst and the egg pushed out of the leaf. Good irrigation followed or accompanied by warm weather would stimulate such accidents. These phenomena are probably of very common occurrence in endophytic eggs of insects and more attention should be paid to this type of mortality, which is the more important as endophytic eggs are not exposed to the influence of air humidity.

The length of the egg-stage has been indicated to vary considerably. Whereas Pomeyrol (1928) has observed 5—6 days under laboratory conditions and Russel (1909) 8 days, Buffa (1911) indicates it to last about 10—11 days in *Viburnum*-leaves during summer. In the observations of Rivnay (1935) on citrus in Palestine it fluctuated from 2 to 6 weeks, according to temperature. It is difficult to explain these great differences, owing to the fact that the Palestine series of observations covers the whole normal temperature range. It would seem as if the host and its conditions possibly greatly influence the duration of the egg-stage.

Both larval stages last from 16 to 20 days (Russel), about 15 days (Balachowsky). In Palestine these stages ranged on citrus from 9 to 30 days, following the environmental temperature, about 10—12 days in summer being the normal (Rivnay).

Neither prepupa nor pupa feed. They usually retire into some hidden place, being often reported to hide under vegetable mould or in the soil, but in Palestine they are mostly met on the lower surface of the leaf, often closely to the mid-rib. The whole life-cycle is thus completed on one leaf. The entire content of the intestinal tract is emptied before pupation. Both stages do not move, as a rule, when undisturbed. The mouth-parts of the pupa are still fit for sucking, but more developed than those of the prepupa.

The prepupal stage lasts less than one day only. The total pupal stage ranges from 4 to 5 days (Russel) and from 3 to 14 days in Palestine (Rivnay).

The two larval stages are the only feeding stages during development. Their feeding habits will be discussed later on. Very soon after hatching an excremental drop appears beyond the anal opening, which is held by the six hairs of the last segment and which may exceed the larva in length. Its colour is generally greenish, but changes with the food. When this drop is sufficiently large, it falls down, covering sometimes the thrips which leaves this „bath" immediately. Oftener, the drop falls near the animal. Buffa thinks that it serves a protective purpose against enemies. When in danger, the larva places these excrements between itself and the enemy. This method of slow permanent defecation of the thrips larva is remarkable. The steady growth of the drops and their immediate replacements by new ones shows the intensity of feeding in these species. While feeding the larva is motionless and only the head jerks downward every now and then.

The moulting has been described by Rivnay (1934) as follows: „Before moulting the insect retires to a hidden place and remains quiet and motionless for some time. The moulting begins from the head. The skin along the epicranial suture splits and shrinks sidewards, thereby exposing the soft, new skin of the head. When a greater part of the cranium is freed, the antennae are drawn out one by one, thus enabling the freedom of the ventral surface of the head. The shrinking of the old skin extends posteriorly to the pronotum. In the meantime, the insect bends forward, attaches its front to the surface of the object upon which it stands, and lifts its legs so that the entire body is obliquely supported by the head and antennae. The legs, being free, can be drawn out from their old sheath, beginning with the front pair, and as the old skin shrinks further posteriorly, the second and third pair (and in the pupa the wings from their pads) are also freed. This entire process is accompanied by shivering-like movements, which apparently help to free the organs. Immediately after the moulting, the legs are soft and folded together. But they soon harden and begin to move, first the front pair, then the others. As they get hold of the surface, the head and antennae are detached and the insect stands upon its feet. The abdomen now performs movements sideways and upwards and the skin gradually slips off."

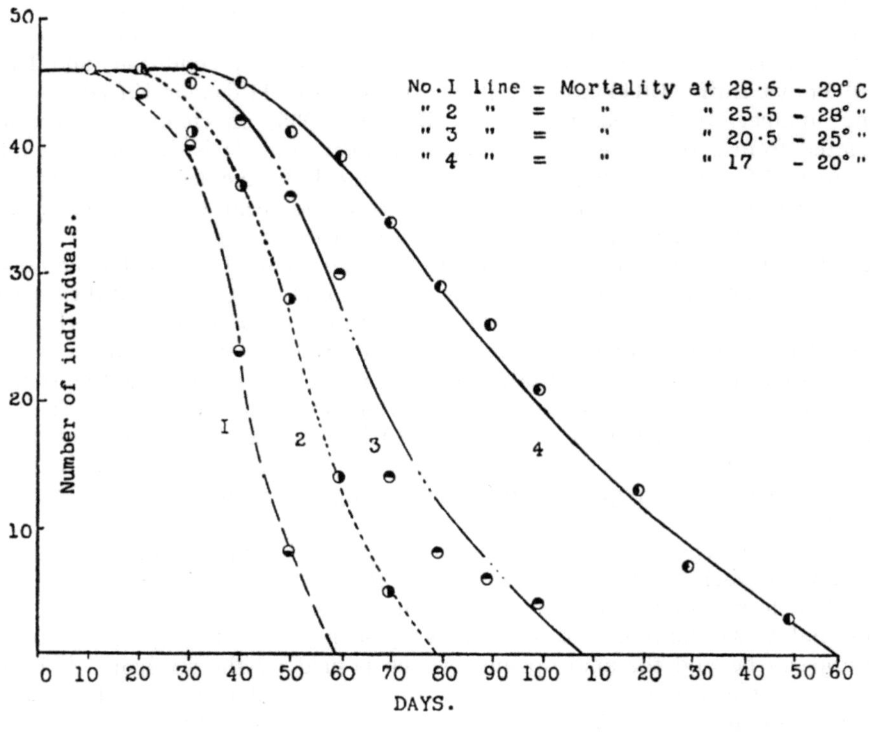

Fig. 119.
Mortality curve of adults of *Heliothrips haemorrhoidalis* Bché. at different temperatures (after Rivnay).

Crawling is the normal locomotion of the adult female. The female's occasional active flights are only jumps of a few centimeters and these occur at warm weather only. The passive transport of all stages by wind is of greater importance as it carries them from one host to another. Many individuals perish in these trips. The distance thus covered must sometimes be considerable. Berland (1935) captured five species of Thysanoptera in the aeroplankton at an altitude of between 500 and 2000 m. The permanent mass transport in the upper air strata during the optimal season is necessarily of great importance for the dispersal of the species. The still limited area found in Palestine is due to unfavorable environments in the limitroph regions on which the dry summer-heat as well as the short vegetation period annually kill the arriving individuals, even if they manage to develop during a short period in winter or spring.

The longevity of the adults depends, normal host conditions given, on environmental temperature and humidity. It was in some experiments (Rivnay, 1935) as follows:

Temperature	Reared in test tube	Reared on plant
29—30° C	40 days	24 days
25—27	55	35
18—20	110	48

The difference in longevity between individuals reared on plants and in test tubes is due to the difference in air humidity, which was above 85 % in the test tubes and only 60—70 % in the other series.

Adults survived without food at:

	65%	75%	87%	95% Rel. Humidity
28° C	1	1	2.5	3.1 days
26	1	2	10.2	9 days
13	1	3.6	13.2	17 days

The optimal longevity conditions are above 85 % rel. humidity and below 29° C. Every deviation from this optimum depresses adult longevity. A mortality curve of adults at different temperatures is reproduced in fig. **119**.

Reproduction: H. haemorrhoidalis is a purely parthenogenetic species, producing only female offspring (thelyotokous parthenogenesis). No one, except Heeger, has ever seen a male and Heeger has probably seen the male of another species.

In Palestine, the preoviposition period ranges from 4 to 19 days. The eggs are laid singly and with intervals, which range from 1 egg every 6 days at 14—15° C to 2 eggs daily at 29° C.

The influence of temperature on reproduction is computed from data obtained by Rivnay (1935):

Temperature	n	average no. of eggs per female	daily oviposition rate: egg per female-day	Computed length of oviposition period
15.5—20° C	63	25	0.5	48 days
20.5—25	28	47.5	1.0	48
25.5—28	49	46.6	1.5	31
28.5—29	34	37	2.0	19

The maximal individual egg production observed was 62. The optimal temperature range is from 20 to 28° C. Air humidity influences the reproduction only slightly, for loss by evaporation is, as a rule, replaced by the water content of the food.

Oviposition increases with the age of the individual, decreasing soon after the climax has been reached (fig. 120).

The effect of temperature on the rate of development has been studied in detail by E. Rivnay (1935). The results are compiled in the following table:

Temperature	development in days				Combined total	Calculated total
	egg	larval stages	prepupa + pupa	pre-oviposition period		
29° C	15	10.3	3	5	33.3	28
28	14.5	9.3	3	4	30.8	29
27	14	10.6	3	5	32.6	32
26	14.7	10	3.5	5.5	33.7	34
25	16	10	4	6	36	36
23	19	12	5	6	42	42
21	22.3	15.5	6	10	53.8	50.4
19	25	19	7	12	63	63
16	35	28.5	12	17	92.5	101
15	44	30	14	19	107	126

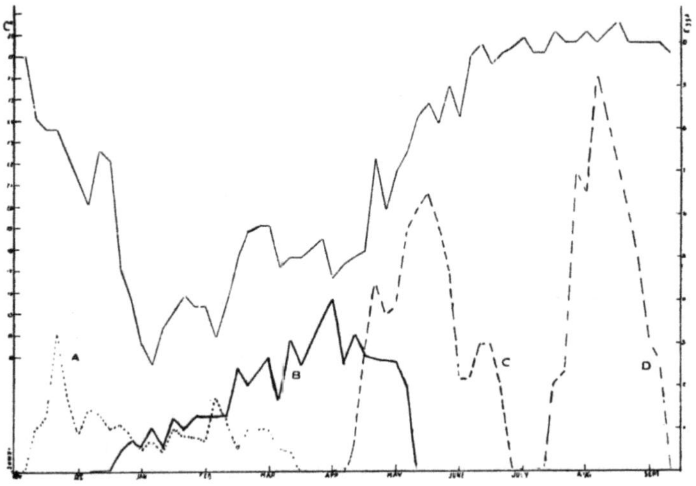

Fig. 120.
Oviposition curve of *Heliothrips haemorrhoidalis* Bché. at five days interval (after Rivnay).

Within the optimal range of 16—26° C these data follow rather closely the hyperbola with the constants

Development threshold (c) = 11.0° C
Thermal constant (Th.C.) = 504 days-degrees.

Rivnay points out that a parabola suits the total range slightly better. From 27° C on, unfavorable influence of heat is observed, first in the larval stage, then in the egg and at 29° C the total development is prolonged.

Air humidity influences especially the pupal stage. Below 75 % no complete development took place. At lower temperatures development is somewhat retarded at a humidity below 87 %.

The scale of activity for adults and larvae was studied by Miss Ben-Ami in our laboratory:

A.

B.

C.

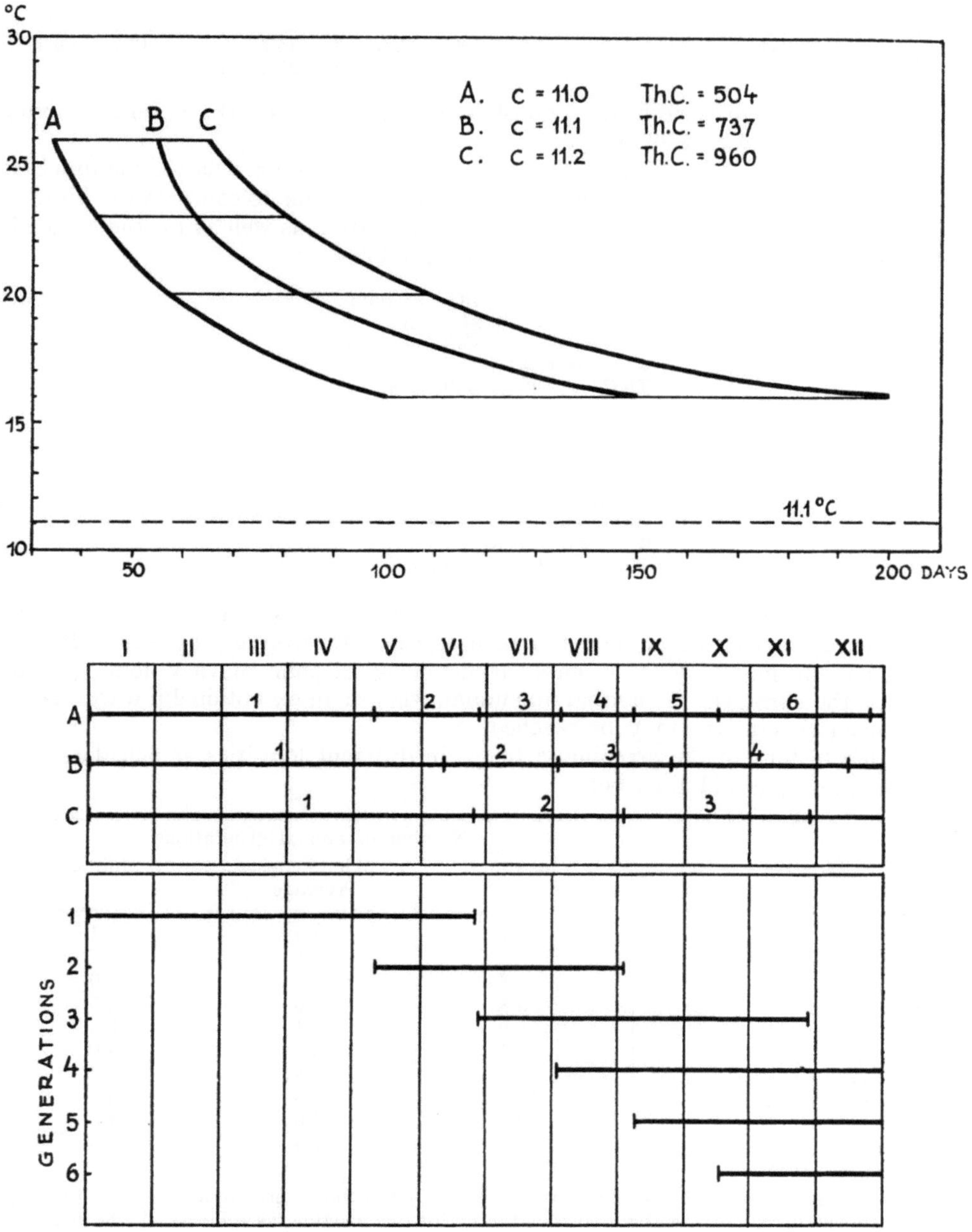

Fig. 121.
Life-history of *Heliothrips haemorrhoidalis* Bché. in Palestine: A. the 3 hyperbolae. B. life-history at Tel Aviv calculated from hyperbola A—C. C. sequence and overlapping of generations at Tel Aviv.

	month	n	1	2	3	4	5	6	7	8
			Scale of activity							
Larvae	VIII	20	2.2	4.5	7.8	20.6/33.3	35.6	—	43.7	46.9° C
Adults	VIII	20	3.6	—	7.5	21.2/30.0	33.4	38.5	44.0	44.2° C

The larvae are more resistant against both heat and cold. The optimal activity occurs between 20 and 30° C.

Whereas the above mentioned hyperbola represents the formula for the quickest developing individuals of every generation, we are now able to compute the formula for the slowest developing individuals of each generation as well as for the epidemiological average. The former is computed as follows:

21° C = 50.4 + 48 = 98 days
26° C = 34 + 31 = 65

c = 11.2° C
Th.C. = 960 days-degrees.

The epidemiological average is calculated as follows:

21° C = 50.4 + 24 = 74.4 days
26° C = 34 + 15.5 = 49.5

c = 11.1° C
Th.C. = 737 days-degrees.

Fig. 121 illustrates the life history of *H. haemorrhoidalis* in the coastal plain of Palestine, as calculated from these formulae, as well as the overlapping of the different generations during the year. From an epidemiological point of view four generations are the norm, six the normal maximum, whereas in exceptionally warm years even seven generations may be reached.

The life history of the greenhouse thrips in different localities as aclculated by these formulae, would thus be:

Locality	Number of annual generations maximal	epidemiological average	minimal
Paris	2.0	1.4	1.1
Nice	3.3	2.3	1.7
Naples	3.8	2.6	2.0
Jerusalem	3.8	2.6	2.0
Tel Aviv	6.0	4.1	3.2
Tiberias	7.7	5.2	4.0
Alexandria	6.7	4.6	3.5
Sierra Leone	10.6	7.3	5.6
Honolulu	8.8	6.0	4.6
Los Angeles	3.4	2.3	1.8

Epidemiology in Palestine: "Upon a survey, it was found that there are at least five limiting conditions which control the abundance of the thrips in Palestine, as follows: 1) The removal of the fruit from the grove; 2) the physiological state of the leaf (proliferation or cell turgor); 3) low relative humidity; 4) high temperature; 5) predators.

Of these factors the fifth is of least importance in reducing the number of thrips. The removal of the fruit is a more important factor, while the extent of the influence of the physiological state of the leaf cannot be established. Low relative humidity and high temperature are the most important factors in checking the population of the insect.

As mentioned above, the most favourable relative humidity is 85 per cent.; for at lower humidities many eggs dry up. Furthermore, a humidity below 60 per cent. is detrimental to the pupal stage, the pupa dying before maturing. It is safe to conclude, therefore, that a hot desert wind, which lasts a few days, with a relative humidity of 30—60 per cent., is certainly detrimental to the pupae and probably also to the eggs. Such winds, and even drier ones, are quite common in Palestine during spring and autumn.

A temperature above 33.5 ° C is detrimental to the adult insect in general. It is true that the insect does not die immediately at this temperature, but its life is shortened and reproduction reduced. The pupa, however, is certainly harmed, especially when this temperature is coupled with a low relative humidity of below 70 per cent., which is a common occurrence." (Rivnay, 1935).

Counts on four branches of orange trees (3 old, 3 young ones) in years of low population level of *H. haemorrhoidalis* yielded the following results:

month	I	II	III	IV	V	VI	VII	VIII	IX	X	XI	XII	Total
3 old trees.	—	—	1	—	5	3	2	1	—	19	4	3	38
3 young trees	—	—	—	—	—	1	—	—	—	—	—	—	1

Statistical counts on 400 leaves at Atta yielded the following monthly results:

	month												
year	I	II	III	IV	V	VI	VII	VIII	IX	X	XI	XII	Total
1931	3	6	8	42	23	77	51	4	2	—	—	—	216
1932	1	1	—	3	3	1	1	1	—	1	1	—	13
1933	3	3	8	6	1	—	1	—	—	—	2	—	24
1934	1	—	—	1	—	—	—	—	—	—	—	—	2
1935	—	—	—	—	1	2	—	—	—	5	—	—	8
1936	2	—	8	27	—	5	—	—	—	—	2	—	44
Average	2	2	4	13	6	14	9	1	—	1	1	—	

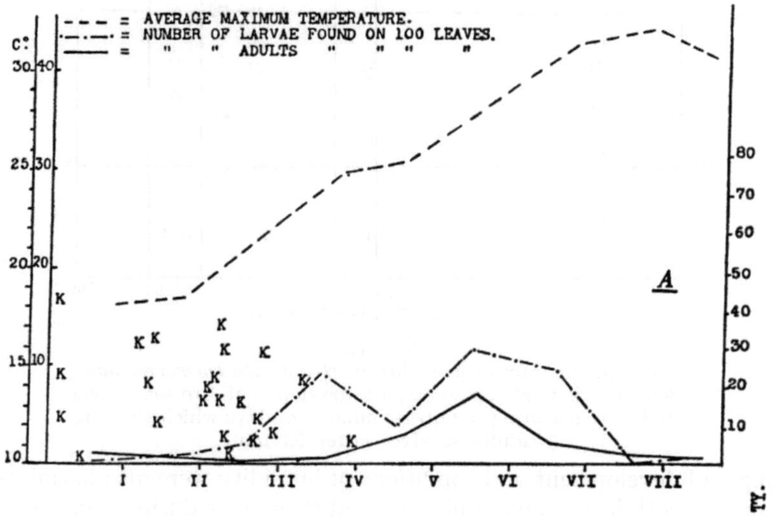

Fig. 122.
Population trend of *Heliothrips haemorrhoidalis* Bché. near Hedera in 1931 (k lowest relative humidity on hamsin days). (after Rivnay).

Parallel to these observations runs another series performed by Rivnay (1935) in 1931 in the same neighbourhood by means of the same methods, only differentiating between larvae, pupae and adults. (Fig. 122). His explanatory text is, as follows: „It is to be noted that with the rise of temperature towards the end of February, many eggs that were probably laid in November/January, begin to hatch, the cold temperature during these months having prevented hatching before the warmer days of February set in. However, the number of adults in the following month did not increase accordingly, but on the contrary even decreased during March and also in early April. This was due to the fact that the hamsin-days, with their low humidity, which could not affect the larvae, killed many of the pupae, so that few of them matured. The reduced number of adults during April caused a drop in the number of eggs and consequently in the number of larvae during May. During the latter part of May and early June the temperature was very favourable for

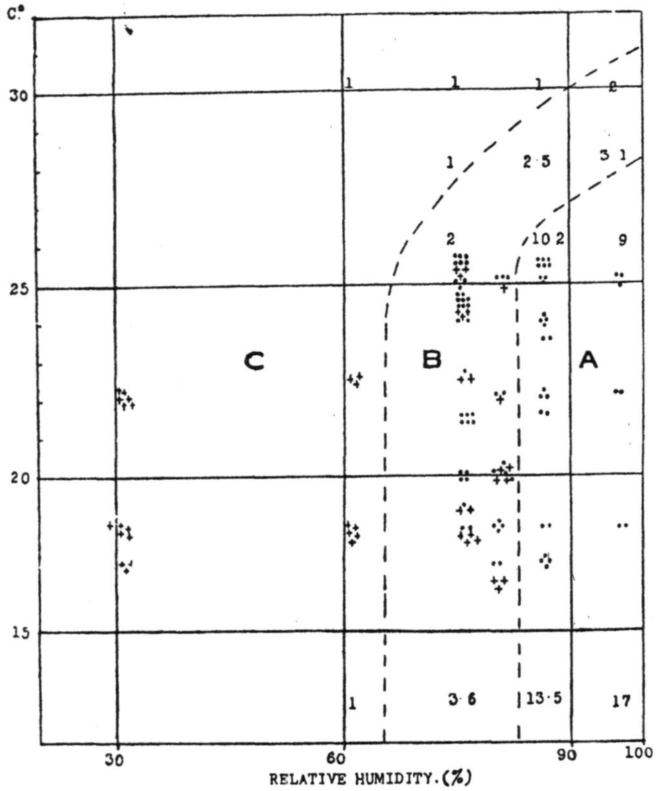

Fig. 123.
Thermohygrogram of mortality of *Heliothrips haemorrhoidalis* Bché. Black dots: pupae complete development. Crosses: pupae die before maturing; numbers: number of days which starving adults survive. (after Rivnay)

egg-laying and development and conditions of humidity were also harmless, so that the number of both larvae and adults reached their peak during June. Towards the end of this month, however, temperature reached beyond the optimum limits, and we note a drop in the population, espec. in the adults. The drop was still greater in August. It should be remembered that the thrips in June began to migrate to the fruit, and, therefore, the record is not altogether dependable."

Physical ecology of H. haemorrhoidalis: From the normal epidemiology of the greenhouse thrips in Palestine, we may now turn to the general epidemiology of this species, as far as it is expressed by its general distribution and by it outbreaks.

The basis of this analysis forms the physiological study of Rivnay (1935), as expressed in fig. 123. The normal range of temperature-humidity correlations is divided into three zones: In zone A adults without food survive over 9 days and

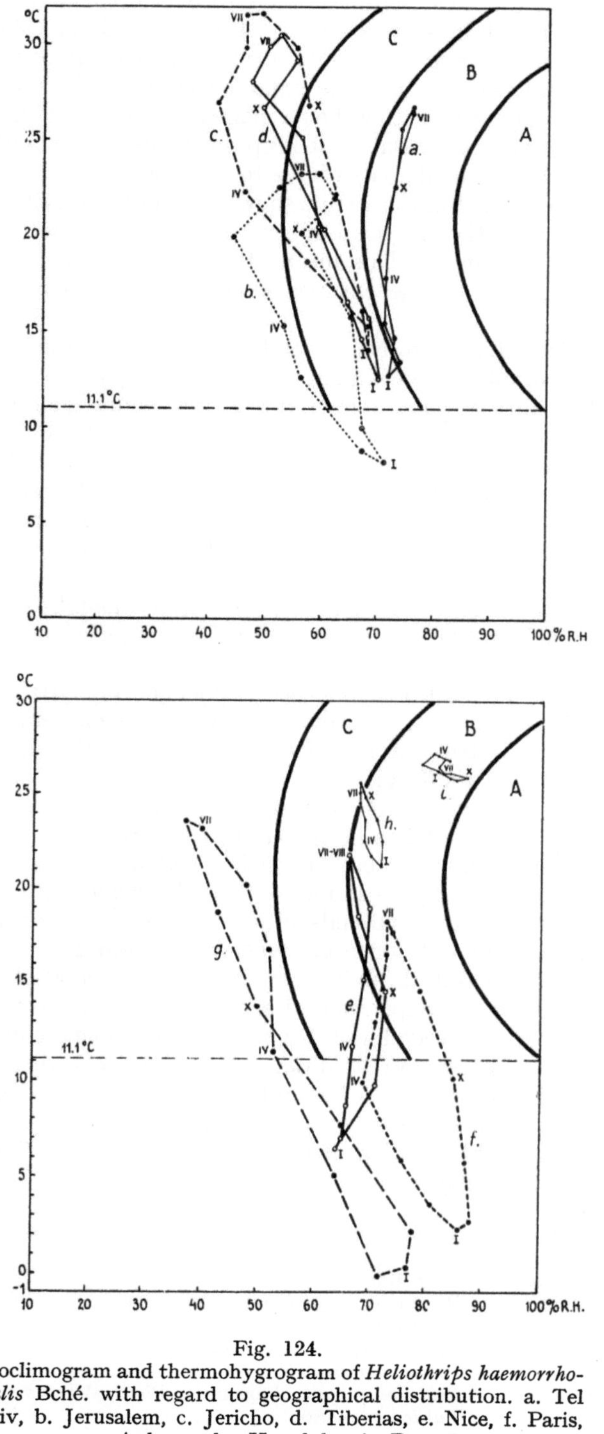

Fig. 124.
Ecoclimogram and thermohygrogram of *Heliothrips haemorrhoidalis* Bché. with regard to geographical distribution. a. Tel Aviv, b. Jerusalem, c. Jericho, d. Tiberias, e. Nice, f. Paris, g. Ankara, h. Honolulu, i. Penang.

100 % of the pupae, the most sensitive stage of development, hatch successfully. In zone B the starving adults survive only 2—4 days and about 1/3 of the pupae die before moulting. In zone C starving adults survive but one day or less and no pupae develop into an adult. Zone A may be regarded as optimal, zone B as tolerable. Conditions of zone C if not preventing the permanent occurrence of *H. haemorrhoidalis* in the place in question entirely, at least keep down the local populations so considerably that no outbreaks or damages have to be expected.

The series of climograms from Palestine show that only Tel Aviv is situated in the tolerable zone, whereas the dry atmosphere of Jerusalem, Jericho and Tiberias prevents any possible damage. This means that Mediterranean types of climate, if not extremely dry, are favorable, whereas desert- and steppe-climates are decidedly unfavorable.

In the northern Mediterranean (Nice), conditions are just at the borderline between tolerable and unfavorable, whereas in the more humid parts of the Euro-Siberian region (Paris) humidity is favorable, but the low winter temperatures prevent permanent establishment outdoors. The Irano-Turanian region (Ankara) is unfavorable both because of the winter cold and the summer drought. In the humid tropics the degree of humidity plays a very decisive role. The more humid Penang approaches optimal conditions, whereas those of Honolulu approach unfavorable conditions. It has to be assumed that under the latter conditions microclimatic differences may play a most important role.

Host-relations: The range of host plants of *H. haemorrhoidalis* in greenhouses is a very wide one, including Azalea, Ficus elastica, Begonia, Cyclamen, orchids, palms, ferns, Euphorbiaceae, Liliaceae.

In nature it has been recorded from azalea, rhododendron, citrus and Cryptomeria in Australia; from Evonymus, Pelargonium and Lagerstroemia in Argentine; from avocado, cherry, citrus, croton, laurel, laurestina, Pellea, Smilax verbana, Viburnum, Vitis, etc. in California.

In the tropics it lives on cacoa, kola, palms, coffee, mango, guava, tea, Garcinia, etc.

In Mediterranean regions it has been found in nature on azalea, rhododendron, Viburnum, Vitis, citrus, myrtus, oak, pistacia, acacia.

In Palestine it was practically restricted to citrus. Bodenheimer (1930) observed it once as quite a common pest in seed-beds of tobacco near Ben Shemen. Rivnay (1934) found it in his excursions only once on avocado in Mikweh Israel, and once in a composite flower near Nazareth. Whereas other suitable food-plants are common in Palestine everywhere and experimental breedings on eucalyptus, date palm and others in the laboratory succeeded well, Rivnay seeks possible ecological explanation. He thinks that under Palestinian conditions, *Heliothrips* can thrive in shade only, and that this optimal requirement is best fulfilled in old, shady, dense citrus groves.

Within the last 15 years this situation hs greatly changed, mainly due to the introduction of many subtropical fruit-trees (Z. Avidov 1950). The additional list of hosts of *H. haemorrhoidalis* in Israel includes also:

Anonaceae	*Anona squamosa, A. cherimolia, Mangifera indica*
Apocynaceae	*Carissa glandifloris*
Caprifoliaceae	*Viburnum sp.*
Diospyridaceae	*Diospyros kaki, D. virginiana*
Iridaceae	*Gladiolus sp.*
Juglandaceac	*Juglans regia, Carya pecan*
Lauraceae	*Persea americana*
Myrtaceae	*Psidium guyava, Eugenia uniflora, Myrtus communis*
Punicaceae	*Punica granatum*
Rosaceae	*Pyrus communis*

Heliothrips is by now a severe pest on avocado (*Persea*) and a minor pest on *Diosporos*, guava, *Carissa*, and amongst ornamental plants on *Myrtus* and *Viburnum*.

Food-habits and damage: On citrus, those leaves are preferred which have gained their full size, but still show the fresh green colour of the young leaf. Neither too young nor too old leaves are normally accepted as food. As is to be expected of a polyphagous species of worldwide distribution, the species is polymeric too. On citrus it develops on leaves as well as on twigs and fruits.

„During spring and early summer the leaves alone serve as food; occasionally, rare individuals are found on the blossom. As the fruit matures towards late June the thrips migrate to the fruit. Too young fruit is not chosen as food. The thrips oviposits where it feeds upon the leaves as well as on the fruit. On the fruit the insect is found, as a rule, in the point of contact between a fruit and a leaf. Quite often there are many other insects in such places. Spider webs and the wax of mealybugs serve the thrips as a place of protection."

The mouth-parts are fit for sucking. Whithin an asymmetric, hollow, conical sheet, which is formed by labrum, second maxilla and labium, move two stinging bristles, the mandible and what is believed to be an epipharyngeal appendix. When the feeding insect jerks, these two bristles pierce the cell-walls of the host-plant. The contents of the injured cell are then sucked in, leaving the cell empty. The chlorophyl, carotin plastids, etc. are all sucked in, upon which depends the colour of the intestinal tract, seen through the dorsum.

Wherever the cell-walls are strong the cell retains its original form, fills itself up with air, getting a silvery shiny appearance. In other cases the empty cells collapse and the punctures remain visible as brown spots. The type of damage occurring depends upon the structure of the plant, and is not conditioned by different types of feeding. The reaction of the plant-tissue may also add a variety of feeding marks. When the damage is recent, inspection of the feeding area with a magnifying lens will invariably show living *H. haemorrhoidalis*. The area is further speckled with the defecations of the insect.

The Greenhouse Thrips attacks both leaves and fruits of citrus in Palestine. Leaves fed on show pale areas, which marks are never intensive enough to cause either leaf-drop or curling. And as full-grown leaves only are attacked, the growth of the injured leaves is not checked as in the California citrus thrips (*Scirtothrips citri;* Horton, 1915). Rivnay observed leaves with thrips markings on trees even 3 years after they had been injured.

The injury on the fruit produced by *H. haemorrhoidalis* in Palestine is easily recognisable as long as it is recent. The pale or silvery coloured irregular feeding area merges gradually into the uninjured peel. Its surface is neither depressed nor elevated. This damage is done to the larger fruit only, when it approaches about the size of an apple. No circular ring-marks around the stalk of the fruit are observed in Greenhouse thrips here. Its areas are irregularly distributed all over the fruit. The original pale and silvery colour of the feeding area changes in time to a yellow brown cork-cover. This cork layer is the reaction of the fruit to the injury and serves as self-protection against the entrance of secondary enemies as fungi, mites, etc. through the damaged surface. In case of *Heliothrips*-injury this cork-layer is thin and uniform.

„When the injury is made in summer, the fruit remains on the tree long enough to allow the formation of such a layer. However, if the injury is caused in spring, when the fruit is ripe, it may be picked from the tree before the cork layer has had a chance to appear over the rind. The size of such thrips markings naturally depends

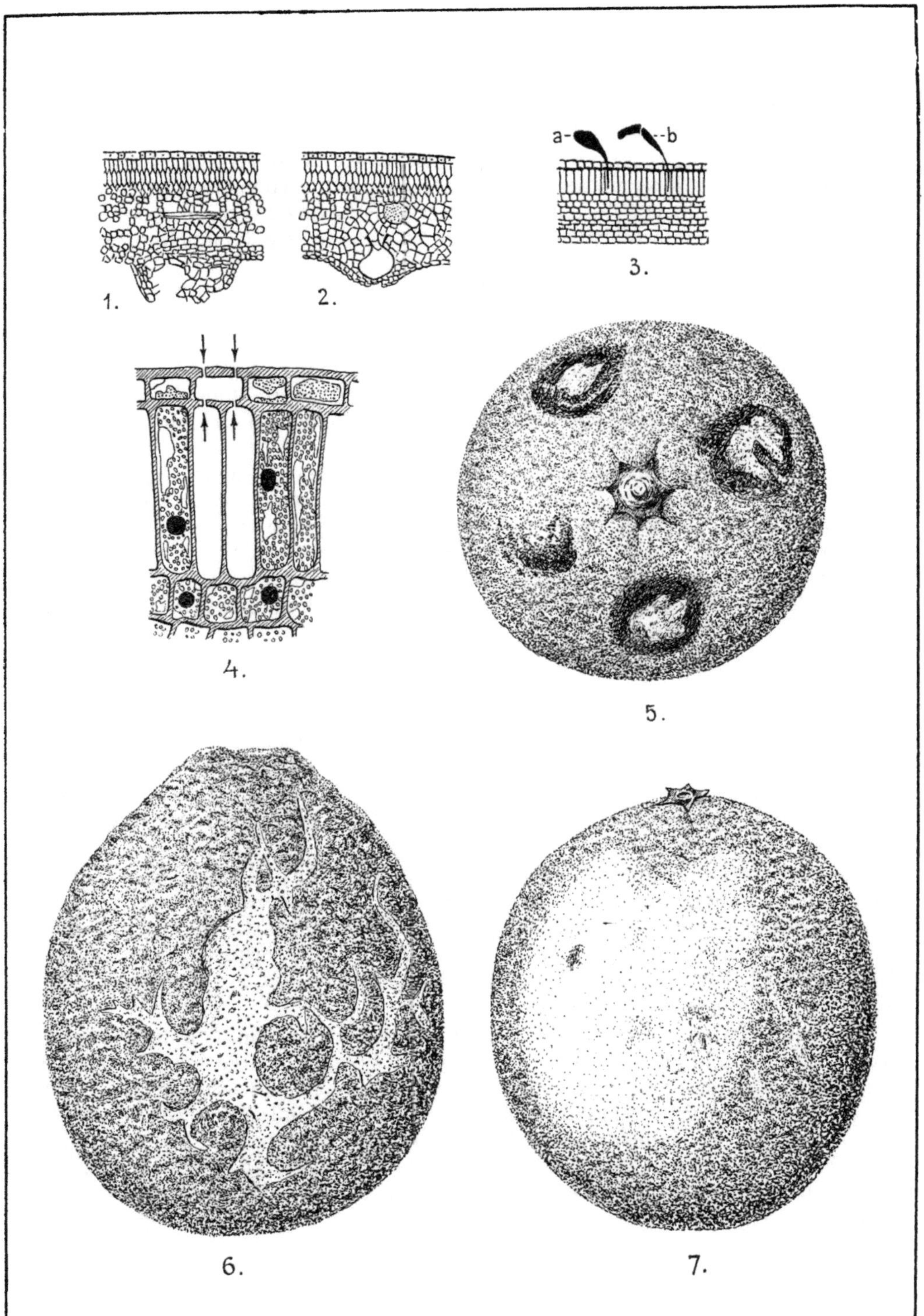

XXXVII. Oviposition- and feeding marks of the Greenhouse Thrips (*Heliothrips haemorrhoidalis* Bché.). **1**) Crosssection through citrus leaf showing cork cells and broken cavity where the eggs were laid; **2**) cross-section through a leaf containing a shriveled egg; **3**) schematic illustration of entrance of stilett (b) and mandible (a) into the cell layers of a *Viburnum*-leaf; **4**) cells of a *Viburnum* leaf sucked empty by *Heliothrips*. The arrows indicate the points where stilett and mandible have been inserted into the leaf; **5**) damage caused by *Heliothrips* on oranges at Cuba; **6**) cork, possibly produced by early damage of *Heliothrips* on orange; **7**) fresh silver (scurf in large extension on orange. (partly adapted frim Buffa, Fawcett & Rivnay).

upon the extent of the infestation and upon the ecologic factors which may check or encourage the injury. Thrips markings are sometimes very small, but they may often cover half of the rind or even the entire fruit. In 1931 a heap of several thousands of oranges in Hedera looked as if every fruit had been wrapped in wax paper."

Reichert and Hellinger (1932) in discussing the orange blemishes describe thrips injury, as follows: ,,The more general form is a whitish opaqueness of small or large scattered surface areas, sometimes involving the greater part of the surface of the fruit. When a piece of the blemished rind is placed in water or in alcohol, the blemish disappears and cannot be distinguished from the normal rind. This rather suggests the removal of cell sap by the insect causing a partial collapse of the cells and larger intracellular air spaces. Thanks to the suction force of the feeding insect coupled with an osmotic force exerted by the cells bordering the insect puncture on the more distant cells, the withdrawal of water will extend over a greater area than actually attacked by the insect. Microscopic examination of the affected rind shows mere air in the superficial cell of the blemish. Severely attacked fruits are very unsightly and become soft and flaccid during storage."

The same authors discuss Silver Scurf, which they define as ,,all markings of fruits which had the form of silvery, greyish or brownish scurf area, mostly very irregular in shape. The surface has a smooth texture from afar, but in actual fact it is composed of small scales. These markings are mostly shallow but may sometimes be very depressed The general cause of all these scurf markings is a liberation of oil which may become very harmful to the cell of the rind. The liberation of oil may be brought about by various agents." Amongst the latter, wind is doubtless the most important. However, puncture of the tissue in or near the oil cells by thrips or other sucking insects may induce it also. ,,The punctures become enlarged with the growth of the fruit and so cause the breakdown of oil cells, thus liberating the oil which kills the surface cells over an area greater than that actually attacked by the insect. The final effect of thrips injury resembles that caused by wind agents. The physical action of wind, together with friction, may stimulate the development of a superficial cork layer which may be formed in small or large irregular spots over the fruit. In such cases the silver scurf does not appear to be the result of inter-cellular reaction of liberated oil causing the death of the tissue around the oil cells. Microscopic examination reveals the formation of cork-like cells in the epidermal layer. The corky development is irregular, producing irregular scales or scurfy patches."

Another type of injury on grapefruits, due to *H. haemoorhoidalis* has been described by Earle and Rogers (1915), from Cuba. On the older fruit there two clustering fruits came in contact with each other or where the leaf is in contact with the fruit, discoloured rings — first gray, then dark brown and finally black — were produced. Such damage has not yet been observed in Palestine.

For a further study of thrips blemishes on orange fruits the reader is referred to Horton (1918), Fawcett (1926) and Hall (1930).

We shall now turn our attention to the intensity of thrips damage on citrus in Palestine. During the author's 15 years stay on Palestine, only one real outbreak occurred in winter 1930/31. Following the statement of Apfelbaum (1931) a similar outbreak occurred in 1911/12. Unfortunately, this determination was not checked by an entomologist, and we may accept it with a certain reserve only.

Indications existing by chance, enable us to give fairly accurate estimates as to the damage done by *H. haemorrhoidalis* in the winter outbreak 1930/31. Reichert and Hellinger (1931) conducted during this year a general survey on orange-fruit blemishes in Palestine, in 3 different groves:

A Orange grove	B Fresh thrips injury	C Silver Scurf	D B % attacked by Penicillium	E C % attacked by Penicillium
Ramath Gan (I.L.G.) .	13.6%	28.0%	1.0%	1.0%
Rishon Le Zion (F.W.)	2.2	30.0	3.8	0.2
Mikweh Israel (O. G.) .	14.4	45.8	—	—

These percentages are the result of random picking of the fruit in different parts of the groves. Column D and E show that penicillium wastage on fruit attacked by thrips or silver scurf was very low. The Silver Scurf was naturally only partially caused by *Heliothrips*, and it is impossible to estimate the latter's part. It may be assumed it was much higher in that special year than in normal years.

The damage caused by *Heliothrips* in normal years is very much below the percentages given above. It may be assumed — taking into consideration the difficulty to differentiate exactly in many cases between silver scurf caused by wind and that caused by thrips — that in normal years it remains below 5% of the total fruit. But not every mark makes the fruit unmarketable, a certain limit being defined by the Plant Inspection Service.

H. haemorrhoidalis plays a more important role in the cultivation of citron. This fruit is grown for ritual purposes only (for use as one of the symbolic fruits in the Tabernacle festival) and every small blemish renders it useless for this purpose. It seems that in normal years the population density of thrips is much higher on cederate than on oranges or grapefruit.

The outbreak in 1931/32 was observed in the Atta-groves near Hedera:

In January 1931 *H. haemorrhoidalis* was first observed on the fruit and in quantity on the leaves. In February in some places up to a third of the fruit showed thrips-injury. In April larvae abounded espec. on the lower side of the leaves which during this, and the two following months, showed heavy thrips-marks. In early July, most of the thrips were killed by heat and no larvae were seen. But late in July, following a few days of cool weather, larvae were seen again. No thrips were seen from August to mid-November, when adults infested the fruit quite generally. Leaves remained practically uninfested, and larvae were not discovered. The heavy infestation of the fruit increased in December, especially where fruits were in contact with each other. The damage was considerable. Many larvae appeared towards the end of the month. This condition remained unchanged until February 1932 when the thrips migrated partially to the leaves. In March it was observed that the larvae tolerated the heavy hamsins better than the adults. A decrease in April and May was followed by a slight increase in June (in older groves only). The summer interval ended in September. The increase from mid-September to mid-October was soon interrupted by a change of weather. During the winter, the thrips damage was moderate in number and the thrips-marks on the fruits were small, originating from the short autumn-increase.

This year of heavy outbreak contrasts certainly with normal conditions. Klein reports no damage (as expressed on culls caused by thrips) for a large orange grove at Rehoboth for 1935/36 and 0.1% for 1936/37. We wish to add, that no further outbreak has been observed in Palestine until 1945.

The distribution on two trees was in July 1932 at Hedera:

height	tree A	tree B	Total	Percentage
0.0—0.5 m	98	27	125	62
0.5—1.0	40	23	63	31
1.0—1.5	5	8	13	6
1.5—2.0	1	—	1	1
2.0—2.5	—	—	—	—

The infestation was heavy in the lowest meter of the tree, inconsiderable above.

It is perhaps early to draw general conclusions from this one count. The same is correct for counts with regard to the compass:

Direction of compass	tree A	tree B	Total	Percentage
North	14	15	29	14
East	6	11	17	8
South	97	12	109	54
West.	27	20	47	23

Control: Owing to the general scarcity of thrips, it proved difficult to perform a regular research program for its control. In small scale experiments we had very satisfactory results by spraying with a nicotine solution, (0.2—0.3%) or a corresponding concentration of nicotine sulphate. However, for extensive scale control pure nicotine is too expensive. Small scale experiments with Californian lime-sulphur did not give very satisfactory results. We shall therefore compile here the last recommandations gained on other species of thrips, injurious to citrus fruit in other countries with a climate similar to that of Palestine. All these measures should be experimented with, here, on the earliest occasion.

Balachowsky and Mesnil (1936) recommend the following formula:

Nicotine (500 gr. per liter)	300 ccm
Soap	1 kg
Water	100 l.

Spraying should be repeated 2—3 times at intervals of 8—10 days.

Hall (1930) uses against the S. African citrus thrips (*Scirtothrips aurantii* Faure) a spray composed of

Lime sulphur (Cap. 36 Baumé) . .	1 l.
Nicotine sulphate (40%)	6 oz.
Capex spreader.	6 oz.
Water make up to.	100 l.

A power sprayer with a constant pressure of 250—300 lb. pressure gives the best results. The lime sulphur in the spray is responsible for killing larvae hatching after the spray, as no 100% killing can be obtained by one spray. In its decomposition it liberates sulphur dioxide, which seems to create an atmosphere unfavorable for

Fig. 125.
Dusting outfit used in California for thrips control in citrus groves. (from McGregor).

the thrips. Two sprayings at a two days interval are recommended for the S. African species of citrus thrips. The first application should be carried out just before the fruit is sufficiently large to be attacked. The annual campaign costs 9.77 d per 10 years old trees. Considering the local importance of the pest, spraying has been included into the regular annual program in S. Rhodesia.

A very high standard of control has been obtained in the dry inland districts of California, where *Scirtothrips citri* Moult. is a permanent menace to citrus groves. Spraying with lime sulphur has been abandoned since about 6 years ago in favour of dusting with sulphur, thanks to the intensive work of Mc Gregor (1931, 1933, 1934, 1936). The results obtained are as follows:

Treatment	Thrips damage to fruit
3 sulphur dustings	3.8%
Lime sulphur (petal time)	9.6
2 sulphur dustings	9.9
Lime sulphur (I/III)	26.1
Lime sulphur (mist spray)	31.5
No treatment	46.6

An even application to all parts of the tree, and powerful dusters have to be used (automobile motors, high volume blowers and fixed two-way fishtail applicator tubes). 4—5 light dustings are preferable to 3 heavy ones, but growers object to the cost of more than 3 sulphurings, (1 pound for first, ¾ for second, ½ a pound per tree for third dusting). During the main period (in California from the end of March to the end of May) a permanent coat of dust should be maintained on the trees. The fineness of the sulphur is of paramount importance. About 90% of the sulphur dust should pass through a 325 mash (per inch) sieve. Crawlers of soft scales are also killed by sulphuring, whereas red scale is not influenced by this treatment.

The control program recently proposed by Boyce & Mabry for California is probably the most feasable for our conditions in case of urgency: ,,The suggested control program for the coming year at Rancho Santa Fe consists of the inclusion of either pyrethrum extract or nicotine sulphate at the rate of 1 to 1200 in the regular late summer oil spray for scale insects and the citrus red mite. This treatment will result in satisfactory control until the following spring when another treatment may be necessary, in which instance pyrethrum extract or nicotine sulphate used alone, or in combination with 0.5% light medium oil may be used . . . Since the insect prefers shaded conditions, it is probably most abundant on the fruit on the inner portions of the tree. Therefore thorough application of materials is very essential for satisfactory results. The problem of satisfactory spray coverage is most difficult when the fruit has reached considerable size, particularly when those fruits growing in clusters are tightly compacted against each other. Because of this fact, the removal of all "off bloom" fruits was urged prior to the present combined treatment for scale insects and greenhouse thrips . . The possibilities of biological control of this insect appear very remote. The tiny Trichogrammatid egg parasite, *Megaphragma mymaripenne* Timb., has been observed attacking the eggs of the thrips in the field; however, they were apparently of little value in reducing the population of thrips."

The duster in fig. 125 is by now rather antiquated and more modern dusting mashinery is at present available throughout Israel. Since 1940 sprays with Tartarum emeticum were used in California, where now a number of new synthetic insecticides are recommanded.

Natural control: The Greenhouse Thrips is apparently very little checked by natural enemies. A predacious thrips, *Franklinothrips myrmicaeformis Zan.* was discovered occasionally during the outbreak in 1930/31 feeding on *Heliothrips*. This species is certainly not a specific predator of the pest and its checking power during the low population interval is practically zero, and not much more so during the outbreak. It may be easily distinguished from other thrips by its ant-like form, its reddish brown colour and the comparatively long legs and antennae.

B. THE ONION THRIPS (*Thrips tabaci* Lindem.)

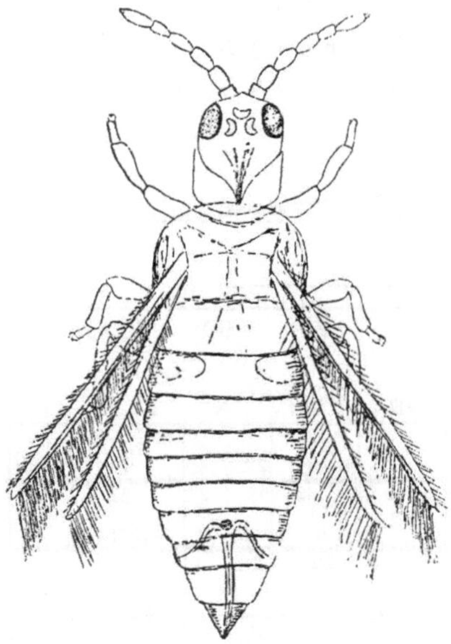

Fig. 126.
The Onion Thrips (*Thrips tabaci* Lind.).

Description: of adult female: Pale yellow to yellow-brown; elongate and flat. Head rectangular, longer than broad. Maxillary palpi 3-jointed; antennae 7-jointed. Wings long, yellowish. Last abdominal segments with long hair. 1.0—1.4 mm long.

Antennal formula: 23 (23), 34 (23), 38 (19), 34 (19), 30 (19), 38 (19), 11 (7) μ.

Male smaller than female. Light yellow. 0.8 mm long.

Egg: White, reniform; 0.26—0.12 mm.

Larvae: First stage white with pink eyes. 0.4 mm long. Second stage bright yellow, legs and antennae greyish. Up to 0.9 mm long.

Pupa: Yellowish to white. Wing pads reaching the 8th abdominal segment. 0.7 mm long.

Distribution: Ubiquistic in Europe, N. America, S. and E. Asia, Australia, etc.

Host-relations: The species is extremely polyphagous. Onions, tobacco, cabbage, cucumbers, apples, pears, carnations serve as food. In Palestine it has been recorded amongst others on: *Citrus, Allium, Cucumis, Solanum, Brassica, Senecio, Lupinus, Pisum.*

Biology: Many data are available on the development of *Thrips tabaci*.

The three hyperbolae based on the breeding data of Sakimura, at Tokio, are:

		A	B	C
development threshold (c.)		10.9	10.9	10.9° C
thermal constant (th.c.)		232	316	410 day-degrees
duration at	30° C	12.2	16.5	21.4 days
	25	16.4	22.4	29.1 ,,
	20	25.5	34.7	45.1 ,,
	15	56.6	77.1	100.0 ,,

The longevity of the different stages may be obtained from the following summary:

Duration of stage	at 25° C	at 20° C
egg	3.1 days	6.0 days
larvae and pupae	11.3	16.3
period of preoviposition	2.0	3.2
period of oviposition	10.7	16.7
period of postoviposition	2.0	2.9
total longevity	29.1 days	45.1 days

The number of annual generations, as far as the coastal plain of Palestine is concerned, is 14½ for shortest, 8 for slowest development and almost 10 as the epidemiological average.

The species may reproduce parthenogenetically, but copulation is the rule. About 80 eggs are laid under favorable environmental conditions, and under those the longevity of the adults is from 2 to 4 weeks.

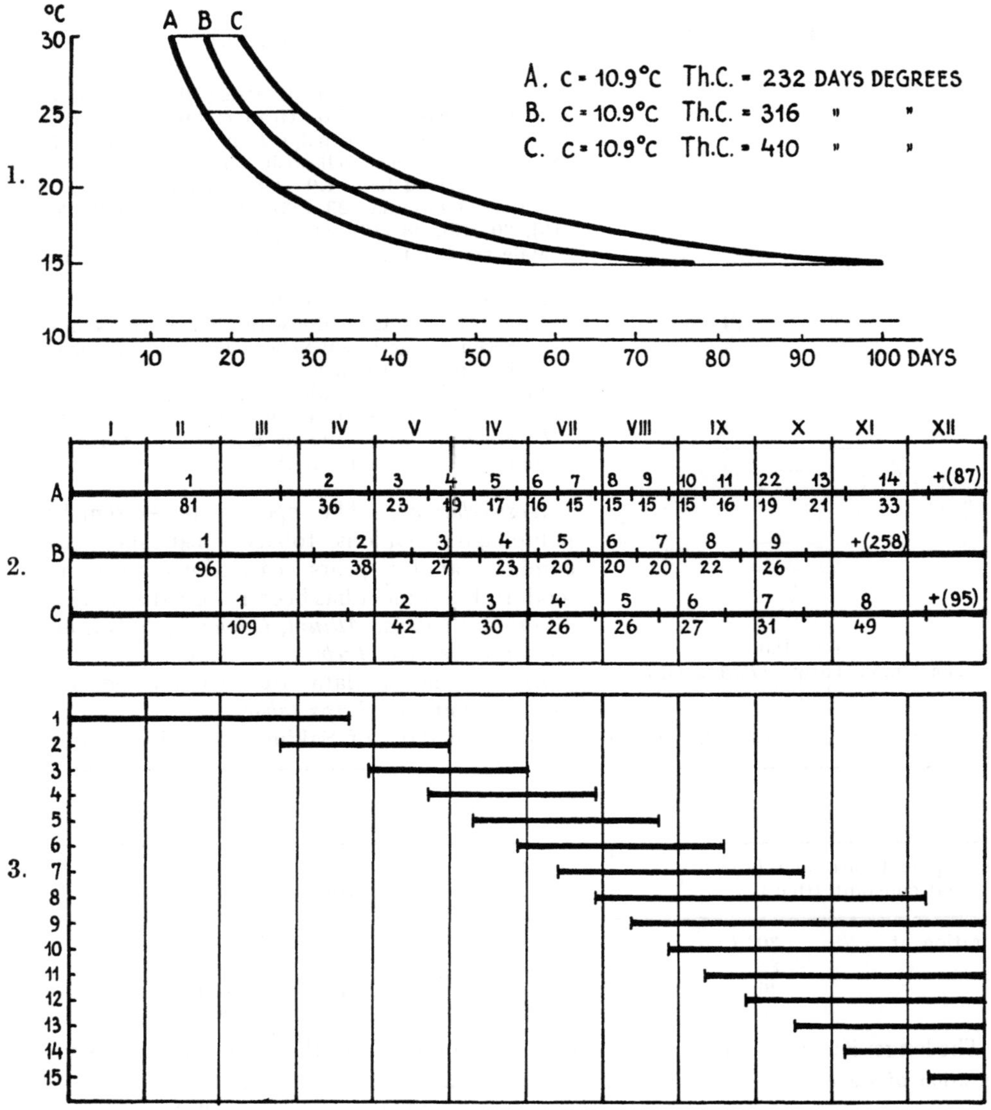

Fig. 127.
Life-history of the *Thrips tabaci* Lind. at Tel Aviv. 1) The three hyperbolae; 2) The life-history as calculated from the hyperbolae; 3) sequence and overlapping of generations.

Damage: We shall restrict ourselves here to a discussion of damage on citrus in Palestine. Rivnay (1933) reports: "In the spring of 1932 it was discovered doing damage to young fruit. Attacking the fruit when it is the size of a nut, at the point of attachment to the stalk, it causes the fruit to become yellow and drop prematurely. Very often, when such a yellow fruit was picked from the tree, *T. tabaci* was found between the fruit and stalk. Damage of this nature was noticed in Benyamina and Hedera in May 1932 and in Benyamina and Mikveh Israel in May 1933. Counts were made of the yellowish fruit which were about to drop from the trees. In Benyamina 7%, in Mikveh Israel 2.5% of the fruit examined were found infested with thrips. Although only a small percentage of these fruit were found with the thrips still feeding on them, the damage is greater because the insects do not linger long on the fruit but wander off in search of other fruit."

The author cannot agree with the above assumptions. During the physiological June-drop in May, which is discussed elsewhere, many fruits drop and grow yellow. It is quite possible, and even probable, that those specimens of *T. tabaci* which infest the young fruits at that season may contribute to the weakening of the fruit and thereby add to the physiological reasons for its dropping down. But there is no reason to regard *T. tabaci* as a primary cause of this phenomenon, and certainly not to regard the percentage of infested fruit as considerably below that which is actually discovered in May. It seems, therefore, that until further details are reported, *Thrips tabaci* should not be regarded as a citrus pest in Palestine.

C. OTHER THRIPS AND PRESUMPTIVE THRIPS DAMAGE.

During 1931/33, a survey was made on the Thysanoptera present on citrus in the laboratory of the author by E. Rivnay (1933) in order to obtain more exact knowledge in this respect. The abundance of species met with was surprisingly high, amounting to 17 species. This list comprises the following species:

A. *Terebrantia.*

1. *Aeolothrips deserticola* Pr. Only one specimen on citrus at Mikveh Israel (V. 1933). Probably a predator and stray visitor.
2. *Limothrips cerealium* Hal. Together with no. 1; one specimen only. Polyphagous on graminaceous plants. Stray visitor.
3. *Taeniothrips discolor* Ka. Polyphagous in the flowers of Cyclamen, Colchicum, Crocus, Senecio, etc. from November to April. Abundant in coastal plain and hills. Stray visitor.
4. *Taeniothrips meridionalis* Pr. Polyphagous on the flowers of wild cruciferae as well as on those of fruit-trees. Common everywhere in April. Stray-visitor.
5. *Odontothrips karnyi rivnayi* Pr. Polyphagous on flowers of *Compositae, Papilionaceae, Eriobotrya, Acacia.* Common everywhere from November to April. Stray-visitor.
6. *Thrips tabaci* Lind. (See above).
7. *Thrips major banaticus* Ka. At Zichron Yakov on flowers of *Eriobotrya, Ruta* and *citrus.* Common from December to April. The economic importance of this species is still unknown.
8. *Heliothrips haemorrhoidalis* Bche. (See above).
9. *Franklinothrips myrmicaeformis* Zan. Predator.
10. *Kakothrips robustus* Uz. Single specimens on citrus in Ramath Gan (IV. 32). Of negligible economic importance.

B. *Tubulifera.*

11. *Haplothrips andresi* Pr. "was found more often, especially in Hedera and Benyamina, during May. Both aldults and larvae frequent the fruit, especially

at the point of its attachment to the stalk. Its allies being predators, it may be that this species also visits such places in search of prey." (Rivnay, **1933**).

12—15. *Haplothrips tritici* Kurdj., *H. gowdeyi* Frankl., *H. cypriotes* Pr. and *H. sp.* These four species of this large genus were found on citrus flowers during April/May at Ramath Gan, Hedera, Rabia and Petah Tikva. Their scarcity on citrus renders them in any case insignificant from the economic point of view.

16. *Hoplandrothrips sp.* One specimen on citrus, both at Ramath Gan (IV. 32) and Hedera (V. 32). Of no economic importance.

17. *Karnyothrips flavipes* Jon. Predator of scale insects.

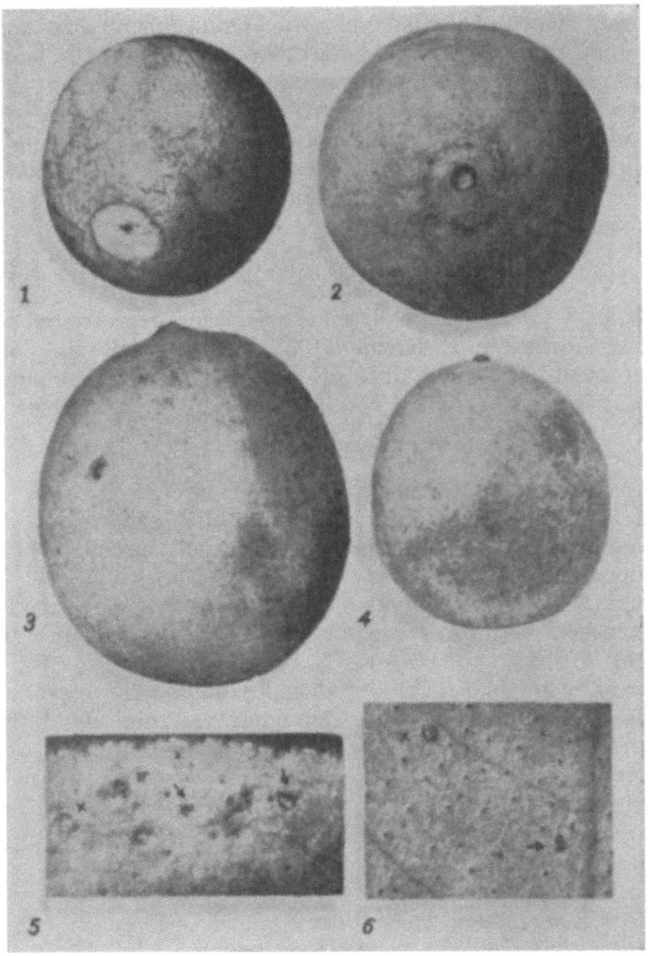

Fig. 128.

Thrips damage. 1 & 2) Types of markings caused possibly by thrips (other than *Heliothrips*). 3 & 4) smooth pale silvery area on fruit characteristic of *Heliothrips*. 5 & 6) microphotograph of upper and lower side of citrus leaf. The arrows indicate oviposition beneath the epidermis, x indicates cork warts often following oviposition. (after Rivnay).

Five out of these 17 species are common polyphagous ones with no specific host affinity to citrus (No. 3—7). Their occasional occurrence is, as a rule, of no economic importance. But this class especially, may be induced to "migrate" to citrus when lacking any other vegetation, conditioned by absence of rains or by agricultural cultivation. We still lack observations as to what degree voluntary mass-migrations take place from the flowers of the lower herbal stratum to those of the higher stratum of the citrus flowers in nature. As a rule, the fluctuations of *Heliothrips* on citrus are certainly due to numeric fluctuations on this plant alone. As far as the author is aware, voluntary host migration of this type has not yet been observed in thrips.

At least two, but probably more species live predaciously (no. 1, 9, 11, 17). No. 2, 10, 15, 16 are of such occasional occurrence as to be of no economic interest. Except No. 8 (and possibly 6) all species are native, and nothing certain can be said about the period of introduction of *Heliothrips*.

In discussing the damage caused by the greenhouse thrips on citrus in this country, Rivnay (1934) points to some other fruit blemishes of citrus: "Among the thrips markings on citrus fruit there may be distinguished two kinds of scars. In one kind the surface is well defined and depressed (Plate XXXVII). Often there is an irregular reticulation around the main scar surface, but it is all well-defined. This kind of injury was made when the fruit was quite immature, and caused by thrips other than *Heliothrips* ... Furthermore, the wing mark around the stalk (Fig. 128), like that caused by the citrus thrips (*Scirtothrips citri*) in California or other thrips in Palestine, is not characteristic of the injury of *Heliothrips*."

The above statement needs urgent verification. Most blemishes which would be regarded by an observer from California or S. Africa as caused by thrips, almost certainly are wind scars. The differentiation between both is not always easy. It is necessary either to discover the thrips which causes these damages or to abandon the theory of a thrips origin. The ring-like blemish around the stalk were suspected to be due to some thrips. Prof. Priesner kindly informed us that he is now satisfied, that these rings are not connected at all with any thrips in our regions, but are caused mechanically.

References.

M. Apfelbaum, Hadar, **4**, 1931, pp. 61—63.

A. Balachowsky et L. Mesnil, Les insects nuisibles aux plantes cultivies. Paris, Vol. 2. 1936. p. 1570—1571.

F. S. Bodenheimer, Schaedlingsfauna Palaestinas. Berlin, 1930. pp. 168—169, 313.

A. M. Boyce and J. Mabry, Calif. Citrograph 1937. p. 19—20, 29—29.

B. Buffa, Redia, **7**, 1911, pp. 71—109.

F. S. Earle, and J. M. Rogers, Ann. Rept. San Pedro Citrus Pathol. Lab. 1915, pp. 5—14.

H. S. Fawcett, Citrus diseases and their control. New York, 1926, pp. 606—509.

W. J. Hall, Brit. S. Africa Company Publ. I, Oxford, 1930, pp. 1—56.

J. R. Horton, U.S. Dept. Agric. Bull. 616. 1918.

H. Z. Klein, Hadar **10**. 1937. pp. 193—195.

A. C. Morgan, Proceed. U.S. Nat. Museum, **46**, 1913, pp. 1—56.

E. A. Mc Gregor, Calif. Citrograph, 1931, p. 358; 1933, p. 128; 1934, p. 232; 1936, p. 436.

I. Reichert and E. Hellinger, Hadar, **5**, 1932, pp. 287—292.

E. Rivnay, Hadar, **6**, 1933, p.p; **7**, 1934, pp.

E. Rivnay, Bull. Ent. Res. **26**, 1935, pp. 267—278.

E. Rivnay, Bull. Soc. Roy. Entom. d'Egypte, 1935, pp. 119—124.

H. M. Russel. U.S. Dept. Agric. Bur. Ent. Bull. 64. 1919.

H. Priesner in Stellwaag, Weinbauinsekten der Kulturländer, Berlin, 1928, pp. 160—162.

H. Priesner, Bull. Soc. Roy. Ent. d'Egypte, 1936, pp. 61—75.

K. Sakimura, Oyo Dobutsugaku Zasshi (Tokyo) **9**, 1937. pp. 1—24.

H. V. Steele, Pamphlet No. 54, Commonwealth Austral. Counc. Sci. and Industr. Res. Melbourne, 1935, pp. 1—59.

Z. Avidov, Pest-phenology of subtropical fruit-trees in Israel (in press).

2. ISOPTERA.

D. THE MEDITERRANEAN SOIL TERMITE
(*Reticulitermes lucifugus* Rossi).

Synonymy: Leucotermes lucifugus Rossi.

Description: The *alate sexuals* are black with mouth and distal part of legs yellow. Wings infuscated, costal and radial nerves black. Antennae 17—18 jointed. Head much larger than pronotum. 6 mm long, wings included 10—12 mm.

The *soldiers* are pale white. Head distinctly longer than broad, with subparallel margins. The slender mandibles are black in their distal half and not dentated at the inner margin. Antennae 17-jointed. 4—5 mm long.

The *workers* differ from the soldiers by a round head with short mandibles.

The body of the *supplementary queen* is similar to that of workers or older nymphs, depending on their origin, which may be from different stages of development of workers or even of soldiers.

Differences between *R. lucifugus* and *Calotermes flavicollis*:

R. lucifugus can easily be distinguished from *C. flavicollis* by the absence of the workers' caste in the latter species and the small size of their colonies, which rarely surpass 400 individuals. Other characters are:

Character	*C. flavicollis*	R. *lucifugus*
Wing of alate	Radius ramified	Radius simple, not ramified
workers	absent	abundant
soldiers	antenna 13-jointed	antenna 17-jointed

Distribution: Mediterranean region; recently introduced into U.S.A.

History of the colony: It is practically impossible to make any accurate estimate of the limits of a nest of *Reticulitermes*. However, a single tree which certainly does not harbour more than one nest may contain at times as much as a liter of termites, i.e. many thousands of individuals (Grassi and Sandias). These indefinite limits of the nest are correlated to the method of propagation of the colonies. Though many thousands of alates leave the nest in the nuptial flight every year, we dispose scarcely of any report, neither from Italy nor from S. France nor from Palestine, as to a sexual colony foundation by an alate pair in nature, nor have true sexuals been discovered in a colony. Instead we find some dozens to a few hundreds of neotenic queens with scattered males (1 ♂: some dozens of ♀). This coexistence of many functional sexuals in one nest is quite remarkable. We find it only in larger colonies. In *R. lucifugus* colony-foundation occurs quite commonly through partition. Part of the workers and soldiers leave the colony, taking along eggs and young larvae to burrow a new nest in the neighbourhood. The new nest remains connected with the mother colony for some time by means of galleries, until at a later period the connection is interrupted and the new nest breeds its own neotenics.

Silvestri thinks that new colonies are often founded by alates which die soon, and leave the further reproduction to the neotenics. What he means is, that high egg production is ascertained in this case by a combined production of many neotenic females, whereas the true queen alone would possess a very low fertility, and not be able to found populus colonies at a rapid rate. Many neotenic females appear, therefore, instead of one queen, with a high egg-laying capacity.

The colony foundation by alates has not yet been observed, whereas that by colony partition is very common. It is very difficult to divide the population of a given territory into nests and to estimate the population of each nest separately.

Under laboratory conditions sexual colony foundation has been observed (Perez, Grassi and Sandias).

The nuptial flight of the alates occurs in winter and spring (in Palestine), but no copulation takes place at this time when huge swarms of alates leave the nest. When they come down, they shed their wings and a male chancing to smell the trace of a female, follows it until it meets the latter, and both remain together. They look for a small cell in the soil, which serves both as nuptial chamber and as a beginning of the new colony. Only very few couples of the alates succeed in finding a nuptial upon leaving the nest and even then many die or may be killed by ants before succeeding to breed the new colony. The first copulation was observed by Perez 5 months after the nuptial flight. Copulation is regularly repeated during the adult life.

Whereas the alates are relatively tolerant towards air-humidity, all other stages are very sensitive not only to dry air, but also to an oversaturated atmosphere. They line the galleries, which may be called the nest of the species, with particles of decayed wood mixed with saliva, which gives them the black, shiny appearance. This lining seems to be primarily a protection against unfavorable low air-humidity. The well-known shyness of termites towards light is probably more due to the strong decline in air-humidity. The building of fragile earthen galleries, wherever *R. lucifugus* makes expeditions, serves also as protection against dry air. These galleries lead from the soil nest to the upper part of the plants and are 20—100 cm long. Their presence is a sure indicator of the presence of these animals.

The colonies of *R. lucifugus* contain:

A. Very young larvae with equal sized heads (1 m long, 11 antennal joints to 2 mm long, 12 antennal joints). They are nourished with saliva and dot not possess the symbiontic Protozoa.

B. Larvae from 2.3—3.8 mm long with 12 antennal joints. Some with larger heads may be regarded as young workers, and may become either adult workers or soldiers.

C. 4 mm long, 14 antennal joints (without wings), soldiers derived from B. Wingless: undifferentiated larvae or larvae of soldiers resp. workers. With wing-rudiments: larvae of sexuals.

D. 4—6 mm long, 15—16 antennal joints. Wingless: young workers, soldiers, larvae of sexuals or neotenics. With wing-rudiments: larvae of sexuals.

E. 5—9 mm long, 17—18 antennal joints. Wingless: adult worker or soldier, larvae of neotenics. With wings: nymphs of sexuals or neotenics.

F. 7 mm long, 17—18 antennal joints. With fully developed wings. Sexuals.

G. Substitute royal forms. 6—11 mm long. 14—16 (17) antennal joints (more intact) pale yellow. Long and thickly spread abdominal hair transversely directed and black maculation only on many parts of the body furnish characteristic points of difference of the substitutes from the sexuals. The latter are of a uniform black and their few abdominal hairs are shorter, and on the sides not pointed transversely, but oblique. These substitute forms may be wingless (derived from early larval-stages), or with winged rudiments (derived from nymphs). However, they may also be derived from unfertilized alates before the nuptial flight, which are slightly paler and smaller than true sexuals, and their wings are torn off irregularly. Also soldier nymphs may be changed into neotenics. (Plate XXXVIII).

The egg stage lasts from 2—3 weeks. Whereas Grassi has observed that hibernation and a prolonged interruption of the reproductive cycle occurs in S. Italy, no prolonged interruption of this kind seems to occur in the coastal plain of Palestine. About 6 months are required for the breeding of workers or soldiers, whereas 18—20 months were needed in S. Italy for breeding alate sexuals. The longevity of the workers and soldiers can be estimated to be between 2—5 years. That of the neotenics is probably of similar duration, whereas the males of the true sexuals, as well as of the neotenics, seem to live for some months only. The supplementary sexuals live in a

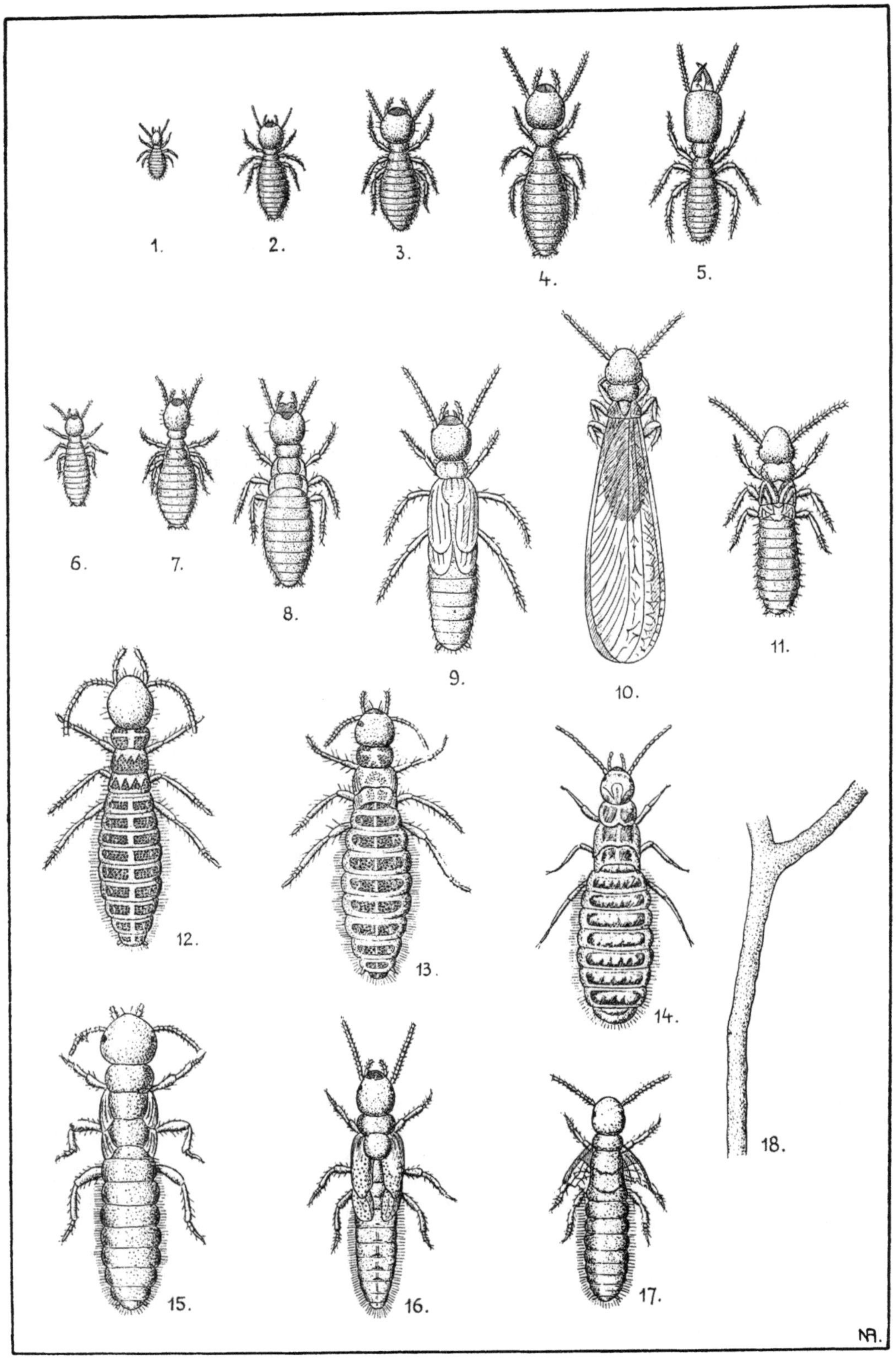

XXXVIII. Development of the Mediterranean Termite (*Reticulotermes eucifugus* Rossi). 1—5) workers and soldiers: 1) larva, (11-antennal joints), 2) large-headed larva (12 joints); 3) dto (14 joints); 4) adult worker (CV); 5) adult soldier (D.V). 6—11) Sexuals: 6) small-headed larvae (12 joints); 7) dto (14 joints); 8) dto (15 joints); 9) nymph; 10) alate sexual; 11) sexual after shedding of wings (A IV). 12—16) supplementary sexuals: 12—14) supplementary sexuals developed from young larval stages, 15—16) supplementary sexuals developed from a nymph. 17) Normal female after shedding of wings. 18) soil gallery covering regular streets above soil, mainly found on stem of plants. (mainly after Grassi and Escherich).

remote part of the nest, but no special royal chamber is built for them as in those species in which the queen grows to an enormous size.

The hind intestine of the workers and soldiers is filled with a series of symbiotic Flagellata, which are supposed to help in the digestion of the cellulose. But it may be that their main importance lies in furnishing the host with vitamins. Also a series of fungal species is connected regularly with *Reticulitermes* and its nests; the role

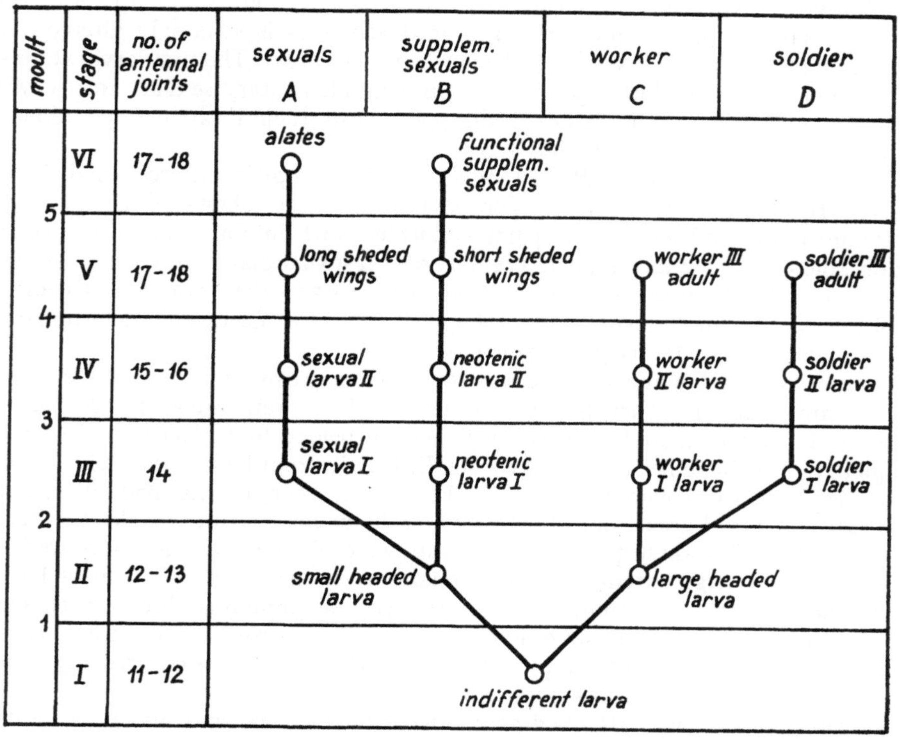

Fig. 129.
Schema of post-embryonic development of *Reticulitermes lucifugus* Rossi (adapted from Grassi and Escherich).

they play in digestion and as food is not yet sufficiently clear. The source of the food is decaying wood. The young larvae, the prospective neotenics, and the sexuals are fed with saliva of old workers.

The social life makes a primitive type of communication indispensable. Besides excited running and turning the attention of brother termites thereby in a certain direction, hearing is specially developed through a tibial tympanic organ. Chirping sounds are produced by rubbing the hind head against the prothorax. The soldiers produce other sounds by rythmical beatings of the head against the soil of the gallery. This sign serves as an alarm signal.

It is fairly generally accepted today, that the young members of a termite colony are capable of growing into the winged form, into a worker, into a soldier, or into a neotenic supplementary sexual. "It depends wholly upon the population of the colony whether they will reach the winged form, or whether they will deviate at an earlier stage and finish the development as a soldier or neotenic." (Kalshoven,

1930). In *Reticulitermes* conditions are not as simple as in *Kalotermes*, where later larval stages appear in natural colonies only. Also Weyer (1931) agrees in a recent review of the problem that apart from the determination of the sex, all freshly laid eggs are equal, and may potentially develop to a worker, soldier, alate or neotenic. External stimuli, mainly regulated by the adult workers and applied in different stages of the postembryonal development, determine the realisation of the castes. Among these external stimuli special nutrition plays probably a very predominant role. Many observations (Grassi and Sandias, Weyer, etc.) tend to show that the intestinal content of larvae changed into neotenics is strikingly altered. The nutritional influence of the saliva on the larvae is very great. The first larval stage — still not containing symbiotic protozoa — and supplementary sexuals are fed with it, and in consequence the protozoa disappear. But their disappearance alone is insufficient to introduce neotenia.

The early development of *R. lucifugus* leads from indifferentiated small larvae to series B ending in adult alates. The deviations are as follows: With the heavy development of head and mouth-parts connected with inhibited growth of sexual glands, soldiers or workers appear. Prematurity of sexual glands in connection with inhibition of development of wings and pigment leads to the formation of neotenic supplementary sexuals (Escherich, 1909). No proof for blastogenic determination of the castes has so far been obtained.

Damage: We dispose of a series of reports on damage caused by termites to young citrus plants. In an earlier publication (1930) these were referred to *Kalotermes flavicollis* F., but re-examination showed them to belong to *R. lucifugus*. The most complete reports have been received by Mr. E. Danin from Kfar Saba: "25.IX.1928. Some weeks ago I discovered the termites by chance, when digging irrigation ditches in a citrus nursery. Patches of soil remained on the main-roots and trunks, producing a sour smell. Below these patches white ants were discovered. They devoured the bark to quite a large extent, and occasionally even the wood. All damaged trees showed heavy gum exudation, as in severe cases of gummosis (brown rot). Only sour lemon was attacked; no sweet lemon plants. 12 trees were attacked in this way (among 6000). Termites reappeared soon, when the roots were covered again and disappeared on the base roots and root-collar."

The same nursery was attacked again the following year and Mr. H. Z. Klein wrote the following report: "7.X.1929. In the older part of the nursery, where lemons (1—1½ m high) were planted last year, many hundreds of termites were produced with every stroke of the how. In the younger part of the nursery, where sour orange (40—50 cm high) seedlings grow, many plants died suddenly. Seedlings which looked green and healthy one day, were dead and dry the next morning. When such plants were removed from the soil, it was seen that from 2—6 cm deep the main root or the main roots were cut by termites. The injured part of the root was black and generally a few individuals of *Reticulitermes* were still connected with it. When the plants which surrounded such dead seedlings and which looked still perfectly healthy were studied, it became obvious that in many of these plants the roots were already partially cut. In one especially strong and healthy plant about 20 termites were observed to cut a root of a diameter of 6 mm by having all fed on the same place in about 30 minutes. All feeding spots show a black discolouration.

Two older trees were very differently damaged, which was recognizable from far. The trunk was surrounded up to 20 cm above the soil surface by a crust of soil which was as hard as stone and 15—20 mm thick. When taken off the trunk, it could be observed that the bark was stripped off in streaks and gum exudation took place, which entered even the soil. The soil was perforated by many irregular tunnels a smooth, dark, shiny surface. Groups of 60—100 termites lived in these galleries. The soil crust around the trunk continues below the soil surface; only it is much thicker there. Feeding marks were observed as deep as 20 cm below the surface, scattered termites as deep as 30 cm. No sexuals were found.

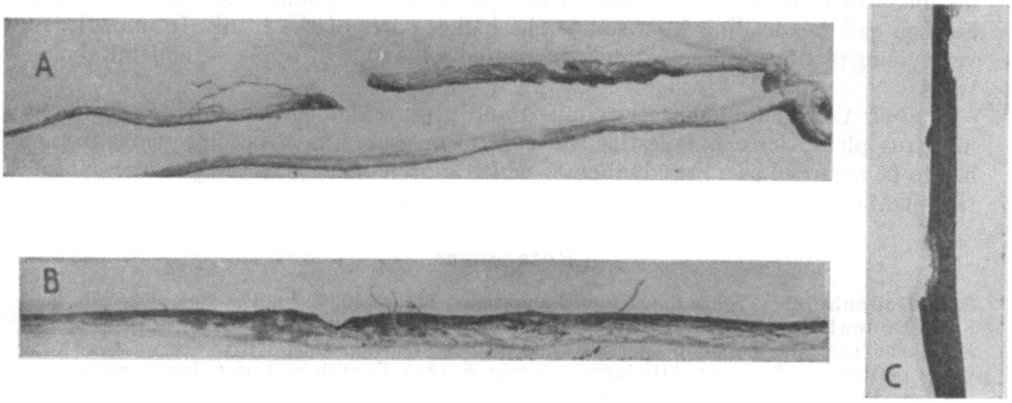

Fig. 130.
Citrus seedlings with damaged bark after attack by *Reticulitermes lucifugus* Rossi (A, B); of *Gryllus bimaculatus* Deg. (C.)

All the four places of attack were 'bur' (Arab fallow) before the nursery was planted. The fallow soil was full of termites, which attacked the very young and tender trees during the first year after the beginning of the plantation."

Probably the same species was concerned in a report by Mr. L. Glassmann from Raananah: "13.VI.1926. This week I opened irrigation ditches and found plenty of termites in the soil. The oculation spots of the trees were covered with a soil layer 2—3 mm, below which termites lived."

No case of attack on older trees has even been heard of. The attack of a certain percentage of the tender plants in the nursery, or in the grove immediately after plantation, is therefore only a quickly passing stage of no long duration. The damage to citrus was never of really alarming proportions.

Setting the young tree too deep is a common cause of termite attack on citrus trees in Florida (*Watson*, 1926). Similar damage was repeatedly reported from young vine-yards, in the weeks and months following plantation. One of the places visited on the occasion of such damage was Ein Harod in October 1926. We were summoned, since the dying off of some dozens of young grape-vines had caused some anxiety. The soil, also in this case fallow land until recently, was heavily infested with *R. lucifugus*. Many plants looked still green, whilst galleries led deeply into the green branches and even into the leaf-stalks. Earthen galleries led from the soil surface 5—20 cm above until the spot where the termites entered the plant. The feeding galleries entered everywhere into the still living tissues. Part of them were filled with gum. In many cases small holes were observed, which the termites had left before leaving deeper into the plant. The black ant *Tetramorium signatum* Men. was an important enemy of the termites in this locality.

Similar damage is quite often reported. But it always occurs in very young plantations and stops everywhere as soon as the plants are definitely rooted to the soil.

R. lucifugus is quite commonly found in the soil, even in the desert districts of Palestine. Their nests are met with in old trees (figs, mulberry, planetree, grapevine), as well as in the old wooden structures of houses in the older settlements. They never touch the external layer of the wood, which looks quite healthy and solid until the thin external layer breaks by some external trauma. Trees are generally invaded from the roots into the higher parts of the trunk, more rarely through earthen galleries leading from soil to the higher parts of the trunk. In nurseries or very young plantations they attacked repeatedly the root of apple — and mirabolan trees.

Control: Considering the very limited and passing damage done by *R. lucifugus* in citrus plantations in Palestine, no control measures have become necessary. It has to be regarded as a „residual pest" which is of no menace to the grove soon ofter the plantation.

References.

F. S. Bodenheimer, Schaedlingsfauna Palaestinas. Berlin. 1930. pp. 174, 261, 269, 278, 356.

F. S. Bodenheimer, Biol. Reviews **12.** 1937. pp. 393—430.

K. Escherich, Die Termiten oder weissen Ameisen. Leipzig, 1909.

B. Grassi and A. Sandias, Atti Accad. Gioenia, **6**, 1893. English in: Quart. Journ. microscop. Sci. **39.** 1897. pp. 246—315; **40.** pp. 1—75.

L. G. E. Kalshoven, Mededeel. Inst. Plantenziekten. Wageningen, 1930. Bull. 76. 154. pp.

J. Perez, R. C. Acad. Sci. **119,** 1894, p. 804, 866.

F. Weyer, Biol. Zentralbl. **51,** 1931, pp. 353—373.

9. LOCUSTS AND CRICKETS (ORTHOPTERA).

A. THE DESERT LOCUST (*Schistocerca gregaria* Forsk., Acrididae).

The Desert Locust is one of the most terrible plagues of Palestine from Biblical times on (Exodus, Joel, Micha), up to our days. Hunger and poverty have followed in its wake for thousands of years. The life history of this locust and the epidemiology of its outbreaks have been obscure until recently. One of the first discoveries of importance was the discovery of phases (Uvarov). When the larvae are bred under gregarious conditions, the morphology, coloring and behaviour of the locusts are different from those bred under solitary conditions. Other studies cleared the question of the permanent home, cycles of migration and other important ecological problems.

Description and theory of phases. — Adult of the Swarming Phase: Large, Pronotum very distinctly constricted in front of the middle, in profile straight, surface almost smooth; hind angle broadly rounded; metazona (i.e. the part behind the hind transverse sulcus), broader than long; colour uniform, without dark pattern. General coloring yellow (when sexually mature) or pink (recently moulted into the adult stage), with small dark spots on elytra. Hind femora unicolored; hind tibiae yellow, spines black on the tips. Prosternum (between the bases of the front legs) with a straight spine with its apex rounded. Male genital plate with a triangular encision at the apex; male cerci compressed laterally with the apex slightly emarginate. Length, male 46—55 mm, female 57 mm. (Uvarov).

Adult of the Solitary Phase: Differ from the swarming phase mainly in its colors and in the structure of the pronotum. The latter is more elongated; median keel more distinct; metazona about as long as broad, hind angle straight, scarcely rounded. General color of freshly moulted adults greenish white; sexually mature adults yellowish-grey, with dark stripes on the pronotum and a pale stripe along its middle. (Uvarov).

Hopper of the first stage: Black. Head black with dirty yellow median line, which continues all over the body. Mouthparts partly yellowish. Sides of the thorax with dirty yellowish bands, which end in the mid-abdomen. The ventral side is dirty grey, with yellow drawings. Antennae black, eyes brown. Hind coxae yellowish. Hind femur black; at the lower margin dirty yellowish. Hind tibia black; at the lower margin yellow; spines black. Hind tarsi black, at the dorsal side yellowish.

Hopper of the second stage: Black. Head pale pink, between the antennae, on its occipital margin, on the distal end of the antennal joints. The median keel of head and thorax mainly black, on the metathorax with pink interruptions and pink dots on the abdomen. The yellowish drawings on the thoracic sides larger and brighter than in the first stage. The sides of the abdomen with many yellow dots. Ventral side as before. Hind-coxa pink. Hind-femur black on both sides, but surrounded by a broad pink band; black on its distal end. Hind tibia pale pink, with broad black band both on the proximal and distal end; spines black. Hind torsi black above, yellowish below.

Hopper of the third stage: Similar to that of second stage, but drawing and colours more distinct. Abdomen with one yellow median and two lateral white lines. The short beginnings of elytra and wings are black.

Hopper of the fourth stage: Very similar to the third stage. The lemon yellow colour dominates greatly over the pale pink, which changes into pink brown. The yellow dots increase on the thorax and abdomen. On the legs, the pink design is replaced by yellow. Tarsi yellowish, only the distal half of the last tarsal joint remaining black. Ventral side yellowish, with black dots. Black, white and yellowish longitudinal lines are present. Elytra and wings yellow with black nervation. The latter reach the posterior margin of the first abdominal segment.

Hoppers of the fifth stage: Besides the size, nearly identical with the preceding stage. The pronotum is more uniform, as the pale bands diappear. The wings reach the posterior margin of the third abdominal segment. The sexual appendices approach their definite shape and both sexes can be easily distinguished.

The above are the discriptions of the gregarious phase. The individuals of the solitary phase are distinctly green with small white drawing and with occasional traces of brown markings. The two phases can be distinguished at first sight.

A few measurements and weights determined for every larval stage (Jerusalem, spring 1929) are compiled in the following table:

Larval stage	Body length	Length of antennae	Length of hind-femur	Length of elytra	Actual weight	Calculated weight	% deviation
I	9.4 mm	4.3 mm	6.0 mm	— mm	0.0232 g	0.0232 g	0.0
II	14.6	5.6	6.9	0.5	0.0481	0.0464	3.5
III	16.2	5.9	8.5	1.8	0.0953	0.0928	2.6
IV	27.9	8.7	13.0	6.2	0.2311	0.3712	
V	41.4	12.0	18.9	16.1	0.5280	0.7124	
Adult ♀	51.4	16.0	25.5	55.4	1.4860	1.4848	0.1
Adult ♀ (mature)		—	—	—	2.8820	2.9696	2.9

Every moult occurs when the female doubles its original weight (besides one latent doubling, which does not lead to a moult). The adult female doubles its weight again before oviposition begins. The mature ovaries are only 17% of the total body weight. In the males such doubling does not occur.

The phases sometimes occur mixed. Green hoppers in a gregarious swarm, however, definitely change into the gregarious phase after the first and second moult. And vice versa: black gregarious hoppers change into or towards the solitary phase, when bred under solitary conditions. At the beginning and at the end of as swarming period, the adults very often show a phasis transiens (aggregans resp. disgregans), which means a transitory stage between both pure phases.

The gregarious phase is connected with changes in the intensity of metabolism. The stimulating effect of large aggregations initiates this change and a moderate increase in numbers is the indispensable factor for a change of a solitary population of *S. gregaria* into a gregarious one. Correlated with the morphological change is one in behaviour. The hoppers aggregate in large swarms, they show a distinct social tendency, in approaching and disappearing in any larger swarm they meet. In the life of the species, both phases have their own importance. In the permanent home, the solitary phase maintains the existence of the species. The gregarious phase coincides with the temporary increase of the species, until the formation of enormous swarms, which finally die.

Distribution, permanent home and migratory routes of S. gregaria. The permanent home of *S. gregaria* has to be looked for in the Sudano-Deccanian Region. The French and Anglo-Egyptian Sudan are now generally recognized as the permanent breeding zones, where the solitary form always persists, where the first nuclei of the migrating swarms are formed. The centres from where the outbreaks originate do not seem to be the permanent home of the solitary. All of these gregarigenous foci seem to be situated on the borderland of the Sudano-Deccanian region, where they meet with a Saharo-Sindian type of early winterrains. Those foci of relatively limited extent seem to exist in the Rio de Oro, in the Red Sea Province of the Sudan, in the Tihama of Yemen and Asir and on the Meccran coast in India. Here the first transit to the gregarious phase preceds emigration. Parts of these territories are the foci, whence Algeria and Morocco, Egypt and Asia Anterior, Kenya and Tanganyika are invaded. The swarms, which invade Baluchistan and Persia seem to have their permanent breeding area in the coastal zone of Baluchistan. All these foci are

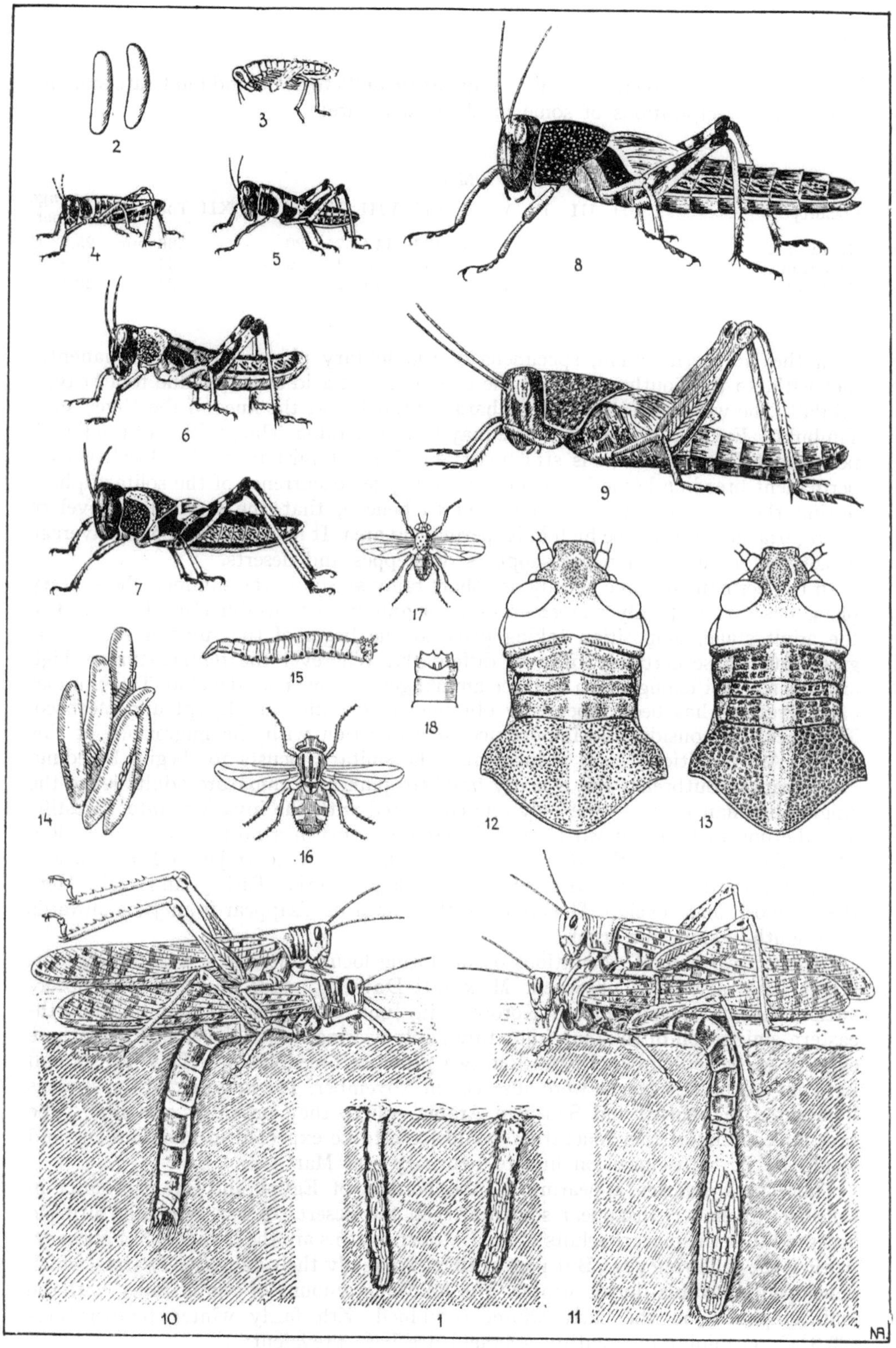

XXXIX. The Desert Locust (*Schistocerca gregaria* Forsk.) 1) Egg-pods in soil. 2) eggs. 3) Hatching vermi ormous larva. 4—7) The first four larval stages of the gregarious phase. 8—9) the last larval stage of the solitary and of the gregarious phase. 10—11) various stages of oviposition a male being attached to the female. 12—13) pronotum of the solitary and of the gregarious phase. 14—16) *Idia lunata* F.: 14) eggs on locust eggs, 15) maggot. 16) fly. 17—18) *Chortophila cilicrura* Rond.: 17) fly, 18) posterior end of maggot (From Kunckel d'Herculais, Vosseler and Bodenheimer).

characterized by an average annual temperature of 25 to 27° C and moderate summer rains. The precipitations of some of these areas are:

Locality	Month I	II	III	IV	V	VI	VII	VIII	IX	X	XI	XII	Total	Average temper.
Obeid	—	—	3	—	5	43	107	**113**	75	20	—	—	366 mm	25.7° C
Khartum	—	—	—	—	3	7	44	**56**	12	9	—	—	131	28.8° C
Timbuktu.	—	—	2	—	7	24	**89**	70	27	10	—	—	224	29.1° C

In the Northern Sudan, specimens of the solitary phase may be permanently met with. In the Southern Sudan, the climate is not arid anymore due to a heavier rainfall. The western focus of the Sahara is probably at the knee of the Niger, near Timbuktu. Further, a possible centre may be in S. Arabia. The position of the fourth permanent focus in India is still unsettled. The establishment of a locality as a permanent breeding home is proven by the regular occurrence of the solitary phase during the intervals. There is no need to believe, that the population level of *S. gregaria* should there be high in its permanent area. It may be low or of the average density of more common grasshoppers of steppes and deserts.

In the permanent home of only one short rainy season in the summer, the solitary phase has certainly only one generation per annum. The winter must be passed in the adult stage, oviposition taking place at the begin of the summer rains. The gregarious phase certainly behaves differently. Wherever the temperature is high and the soil wet enough, two to three annual generations may develop. This number of generations has been repeatedly obtained in breedings in Egypt and Morocco. This fact has considerable importance in connection with the migrations of this locust. Any additional rain induces also the solitary locusts to begin breeding.

In years of outbreaks the recently hatched, pink and immature adults leave the Anglo-Sudanian focus — the only one connected with the invasions into Palestine and the one to be discussed here — in large swarms in a north-eastern direction. They migrate towards the Red Sea shore of the Sudan, to Hejas and Yemen and proceed towards Egypt, Palestine, Transjordania, Syria, Turkey and Irak. Here they oviposit and develop. The adults of these swarms disappear from June onwards in a south-western direction.

At least two annual generations occur during locust outbreaks, but three generations have also been observed. Maxwell-Darling made the following observation: General oviposition in Kordofan in June 1929 was followed by a north-eastern emigration of the pink immature adult swarms, which developed from these eggs, in early September. In the Red Sea province breedings of yellow locusts and hatching of the first larvae were observed in November. The majority of the swarms hibernated in Central and Southern Arabia, where they probably also bred. The first pink adults of this generation would have to be expected in mid-January and would be fit for oviposition in late February and March.

Those pink swarms appearing on the borders of Egypt and Transjordania in October/November, disappear southwards in the desert. It was generally hitherto accepted that the mature adults of the October swarms are the invaders in February/April of the following year. But part of them are really their daughter-swarms, which developed in Arabia during winter. The following amount of winter-rains, with the main emphasis on the early winter combined with fairly winter temperatures (22.9° C, as minimum monthly average) are here prevalent:

	Month													
Locality	I	II	III	IV	V	VI	VII	VIII	IX	X	XI	XII	Total	Average temp.
Port Sudan . . .	11	6	2	1	1	—	10	—	—	8	**42**	17	99 mm	27.8° C
Tokar	**27**	5	1	1	—	4	—	—	—	21	11	24	94	28.1° C

The swarms which arrive in Palestine, Egypt, etc. arrive at a season, when the rainfall is at its peak in the late winter and temperatures are rising (about 15.0° C being the monthly average):

	Month													
Locality	I	II	III	IV	V	VI	VII	VIII	IX	X	XI	XII	Total	Average temp.
El Arish . . .	20	**24**	14	9	2	—	—	—	—	2	18	13	102 mm	18.9° C
Beersheba . .	26	**61**	19	10	—	—	—	—	—	3	12	33	163	20.0° C

On these migrations the recently matured swarms always reach another zoogeographical region, where they find wet soil and a minimum temperature of not below 15° C, generally above 20° C. From the Sudano-Deccanian region (summer-rains) to the northern Saharo-Sindian region on the borders of the Mediterranean region (late winter-rains) and from here back into the Sudan.

These regular migratory cycles seem to have a connection with barometric pressure and wind conditions. When we look at the map of the prevailing winds during the decisive months we find: In June, the prevailing winds drive the swarms directly from Palestine/Egypt into the N. Sudan, west of Obeid. In September, they lead the swarms eastward to Obeid, the Red Sea Province and Hedjas. The winds prevailing in Arabia during late winter are unknown, but it is a well established fact that locusts arrive in Palestine and Transjordania with winds blowing from the south or the south-east.

Life history of Schistocerca gregaria phasis gregaria: The large adult swarms, which invade Palestine from February to April, are sexually mature (yellow coloring) and soon begin to copulate and oviposit. The male approaches the female at the place of rest, strokes it with his feelers and jumps on its back. Supporting itself on the pronotum of the female with the forelegs, it turns the abdomen sidewards and copulates. The copulation act, lasting from 10 minutes to 24 hours, is several times repeated before each oviposition. The female selects a place more or less sheltered from the wind, and begins to dig a hole in the soil. The male is often observed sitting on the back of the ovipositing female. The female presses its abdomen into the soil with great muscular force. The two horny valves at its posterior end then begin to 'drill' their way ahead, until the abdomen is considerably stretched out and has reached the double of its normal length. Ballard observed this process to last for 20 (10—15) minutes. When the hole is ready, the female lifts the posterior tip of ites abdomen slightly from the bottom, and begins to secrete a frothy mucous matter. The eggs are laid into this matter. Secretion and oviposition alternate until all the eggs are deposited. Finally, the abdomen of the female is reduced to its normal size and the hole is filled with the frothy substance. The total process of digging and ovipositing lasted one hour and 20 minutes (1h—2h 47′) (Ballard). The average number of ovipositions was two per female, but as many as six ovipositions have already been observed. The interval is about 7—24 days long. When not disturbed, the latter ovipositions take place in the same area as the former ones. The average number of eggs for each of the first two ovipositions is about 50, decreasing in the further process of egg-laying. On the large oviposition sites contemporaneous coordination by social stimulation shortens duration of pre-copulation as well as of copulation.

The long eggs are golden-brown at the time of oviposition. They are circular in cross-section and round at both ends. Later, the color changes to a dull grey or brown color, with a faint pinkish tinge. The average egg-pod is 40 mm long, 9 mm in diameter. It is straight or slightly curved. The size and weight of the single egg increase during the development of the egg from 7 × 1 to 9 × 2 mm. The change in weight is as follows (after Ballard):

Age in days.	1	2	3	4	5	6	7	8	9	10	12	14	16	18
Weight per egg	10	15	15	10	10	12	12	10	20	20	20	18	18	21 mg

The increase in weight from 5 to 21 mg is induced by the imbibation of water.

Up to 8 egg-pods are laid by one female (3,5 in the average in Egypt) with an total of 334 eggs per female in 1—2 weeks interval. The oviposition period lasts from 1—2 months.

The average incubation, as observed in nature, fluctuates considerably:

Country	Year and Month	Author	Incubation period
East Africa	XI/XII 1903	Vosseler	16—18 days
Egypt	III/IV 1915	Gough	16—30 ,,
Palestine	IV 1915	Buecher	28 ,,
Palestine	V 1928	Bodenheimer	14 ,,
Palestine	III 1929	Bodenheimer	32 ,,

In the laboratory at Rehobot the egg stage of different generations lasted in 1947:

in February/March	43—51 days
March/April	22 ,,
August	16 ,,

The long duration of the development of the egg in the winter is not due to a real diapause, but to the low prevailing temperatures. The eggs develop quicker at higher temperatures.

Oviposition takes place mainly in sandy soil, which is everywhere preferred. Other light soils may be tolerated, occasionally, even heavy soil, but not to a large extent. The female examines the soil with its ovipositor, and when the condition of the soil does not suit it, it retracts the abdomen and attempts to oviposit at another place. A sufficient soil humidity is the main requirement. The upper layer of the soil may be dry, but the eggs are always laid in a thoroughly moist layer. In soils which are wet through and through, oviposition takes place at a relatively high layer, whereas in superficially dry soil it occurs much deeper. The difference can be easily observed on sanddunes in eggpods. In mountains, where deep oviposition is sometimes impossible, oviposition took place between soil and stone, the latter helping to preserve the humidity.

Sandy soil is suitable for the eggs, since they need a well aerated soil. In heavy soil, the large majority of the eggs is destroyed by moulds and by certain parasites, which will be discussed later. The choice of a thoroughly humid soil is of even greater importance for oviposition. Fig. 131 shows, that the *Schistocerca* eggs are very sensitive to low humidity. Their vital optimum falls at 30° C and 100% humidity. During the first day of their development they are so sensitive, that all the eggs removed from the soil die. At the end of the first third of their development a half ellipse touching the combinations 39° C/100% R.H., 30° C/70% R.H. and 24° C/100% R. H. is the extreme border, beyond which no egg develops to hatch. The resistance of the eggs to lower humidity increases considerably with further development. 80—100% of eggs, which are not tampered with and develop under optimal conditions, hatch successfully.

The hatching larva is still envelopped in a membrane, called the amnion, an embryonal cover. During the first 15 minutes after hatching, it gets rid of the amnion. A retractile vesicle in the neck, which had been the locomotion organ of the embryo from the egg to the soil surface, helps to break the amnion in the region of the neck.

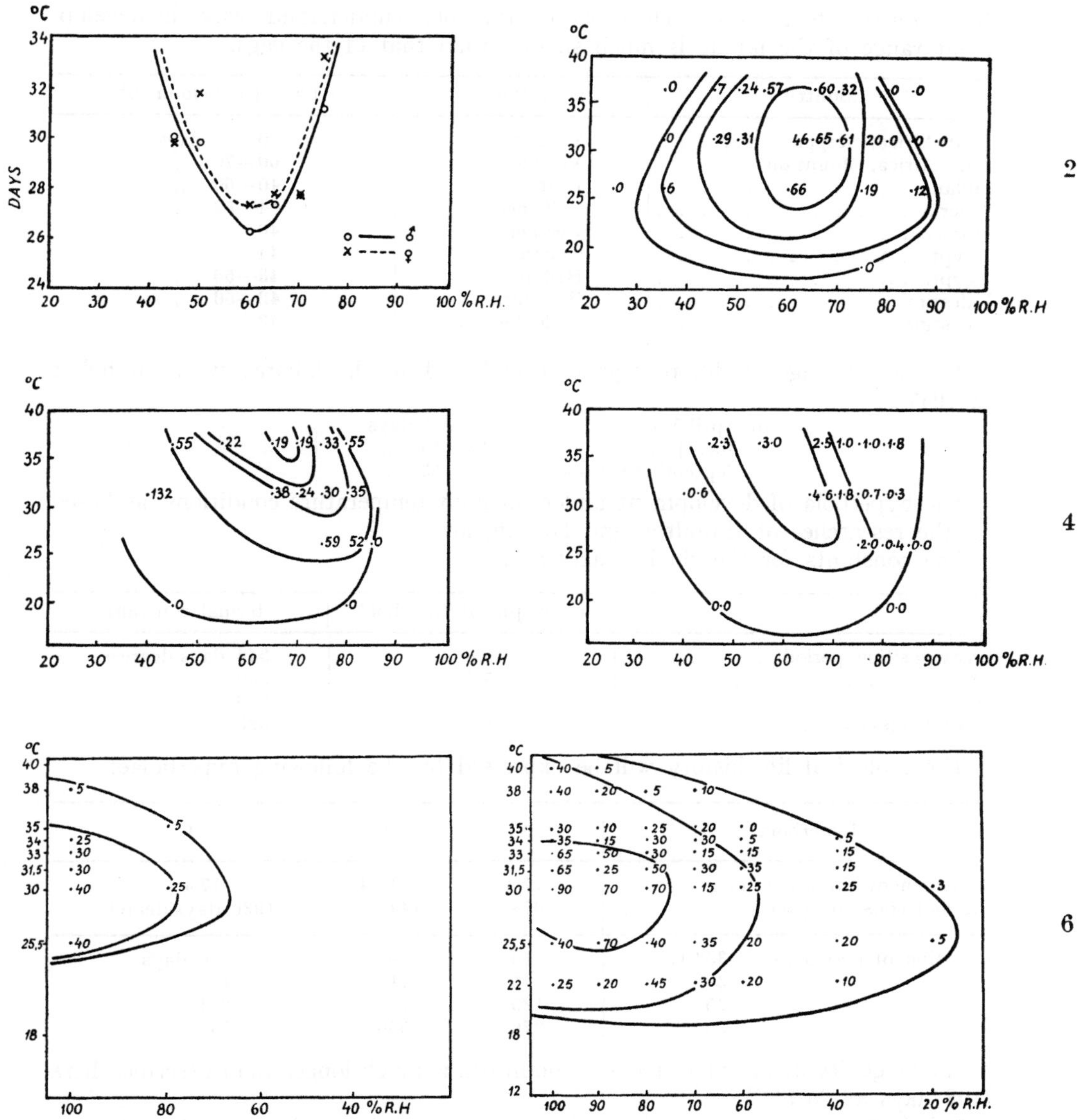

Fig. 131.

Hygrothermograms of different physiological activities of the Desert Locust (*Schistocerca gregaria* Forsk.). 1) Effect of relative humidity at 32.2°C on duration of hopper stage (male and female); 2) effect of temperature and rel. humidity on percentage of survival of hoppers; 3) effect of temperature and relative humidity on lenght of preoviposition period; 4) effect of temperature and relative humidity on number of egg-pods per female; 5) effect of soil temperature and air humidity of soil on survival of eggs in first third of development; 6) as in 5) a the beginning of the last third of embryonal development (after Hamilton & Bodenheimer).

Only now, the transition from embryo to white larva takes place. The larva soon assumes its final color.

The five larval stages have been described above. The body temperature of the larvae, as will be seen later, depends neither on the soil temperature — as that of the eggs — nor on the air temperature, but on the presence of sun radiation, when the latter rises to 40—45° C. Hence we are now able to understand, why the development range of the larvae is much smaller than that of the eggs.

Country	Author	Larval development
East Africa, Steppe	Vosseler	50 days
East Africa, Mountains	Vosseler	60—70 ,,
Sudan	King	40—50 ,,
Algeria	Lallement	44—53 ,,
Morocco	Regnier	40 ,,
Egypt	Gough	44 ,,
Egypt	Ballard	43—56 ,,
Palestine	Buecher	42—56 ,,
Palestine	Bodenheimer	43 ,,

The larval stage of different generation lasted in the laboratory at Rehobot in 1947:

in April/May	46 days
May/June	38—46 ,,
September/October	53 ,,

The hyperbola of development under constant temperature conditions, as based on the researches of Hamilton and Husain, are:

The constants for the single stages are:

	development threshold	termal constant
preoviposition period	15.8° C	344 days-degrees
egg	15.0	240
larva	18.6	386
adult longevity	21.7	523

The ecological life history will be expressed by the following hyperbolae:

hyperbola		A	B	C
development threshold (c)		17.4	17.4	17.4° C
thermal constant (th.c.)		978	1445	1930 days-degrees
Duration of period at	35° C	56	82	110 days
	30	75	115	153
	25	129	191	254
	20	376	556	743

Adult longevity at optimal breeding conditions is much longer than described here (198 days at 26.7° C).

This formula cannot be applied to larvae which develop under natural conditions, because their body temperature rises through sun radiation.

The length of the different larval stages, as observed in the spring of 1929, was as follows:

Larval stage	I	II	III	IV	V	Total
days	7	13	7	6	10	43

Below 18° C the development is only suspended until higher temperatures reoccur.

Development of *Schistocerca* is also influenced by humidity. The optimum humidity for the larvae is at 60 to 70% relative humidity, and development grows longer with deviation to either side (Hamilton).

The mortality of the hoppers as observed in Egypt for 2235 individuals (Ballard) was as follows:

Stage	I	II	III	IV	V	Total
% mortality	42.5	15	10	7.3	2.2	77

This mortality occured in breeding cages. Under optimal natural conditions it is much lower, but under unfavorable conditions it may be much higher, as was observed in Jericho in May 1928. In any case, the recently hatched first larval stage must be regarded as still very sensitive to environmental conditions. Recent breedings of hoppers at different combinations of temperature and humidity showed a vital optimum (50—66% successful development) at 25—38° C and 55—70% relative humidity.

The larvae aggregate as soon as they are hatched. For some days they move irregularly over the hatching area in small swarms. At the end of this period the swarms grow thick and of considerable extent. Regular migrations, directed mainly by the prevailing winds, commence even in the absence of a perfect agreement. The somewhat undirected migrations over plains on calm days with both directions in general agreement are sufficient proof. The gregarious habit is based mainly on a 'social appetite', which has been shown by Allee to be widely distributed among all animal groups. The uniform behaviour of the individuals in the swarms is partly the result of identical environmental conditions (sunning, eating, migration). An equally widely distributed imitation instinct strengthens this tendency. The steady flowing stream of the larval migration is a very impressive picture.

The food ecology of the species is still very imperfectly known. Neither is the exact quantity of food per day and stage known, nor a reliable list of the preferred host plant of this extremely polyphagous species. Cannibalism is occasionally observed, which Uvarov believes to be connected with thirst (need for water).

The longevity of the adults of the gregarious phase has previously been greatly overestimated. In Egypt the longevity of individuals hatched in the respective months was as follows (Ballard):

Month	I	II	III	V	VI	VII	IX	XI	XII	
Males	66	55	39	113	97	54	68	66	89	days
Females	60	43	35	105	81	89	68	76	83	,,

The longevity of the individuals leaving Palestine in June is not known. The latest of those remaining here and dying without oviposition die in November and December. Thus, the average longevity would be about 120 days. This agrees fairly well with the data obtained in Egypt, for our June corresponds climatically to the Egyptian May. The longevity of other generations is shorter. The maximal adult longevity in breeding eages at Rehobot in 1947/48 was 220 to 336 days.

The speed of sexual maturation, easily recognised by the change to a yellow color, depends greatly upon external conditions, which are not yet sufficiently known. In 60 cases, observed by Ballard in Egypt, the range was as follows:

	Average	Minimum	Maximum	
Males	38	13	84	days
Females	56	23	113	,,

The optimum for sexual maturation is at 65% rel. humidity and about 40° C (Hamilton). When sexual maturity in inhibited, as by low humidity, adult longevity is much prolonged.

Warmth and humidity promote sexual maturation, whereas drought has an inhibiting influence. It is still unknown whether humidity affects the maturation process directly or indirectly by its influence on the condition of vegetation.

Under suitable conditions two to three annual generations occur in the gregarious phase. The adults begin to migrate immediately after hatching and settle down to rest only after maturation, to copulate and oviposit. Normally, they remain on this selected area for some weeks, until death.

The density of locusts in these swarms has often been exaggerated. We rarely observed a flight to contain more than one individual per cu.m. of air. But a flight which may sometimes be several kilometers broad and long and some hundred meters deep, leaves a very strong impression and much higher densities are reliably reported.

Body temperature and activity: The Desert Locust belongs, as all grasshoppers do, to the sunloving animals. Its body temperature, when not exposed to sun radiation, equals the temperature of the environment. Under the influence of sun radiation it rises to 35—45° C:

Record of body temperature of fifth hopper stage of *S. gregaria.*

	Temperature of (° C)			
hour	locusts	air (sun)	air (shade)	soil-surface
7:20 h	27.0	23.0	21.2	23.0
7:40	33.6	26.7	22.2	27.3
8:30	35.4	27.3	24.0	30.0
9:00	41.8	31.0	25.0	37.0
9:20	42.8	31.0	25.9	36.5

The body temperature of the locusts was about 5° C higher than the temperature of the soil surface. Many other observations showed that during the day it was generally between 40—45° C.

This fact is of considerable importance, owing to the fact that insects depend greatly upon environmental temperature, as is shown in the following scale of activity:

Beginning of cold torpor	4.6° C
Only slight movements of antennae and legs	7.8
Creeping at intervals	18.4
Beginning of creeping and normal activity	23.5
Beginning of high activity	34.9
Beginning of irritated activity	43.4
Beginning of heat-paralysis	49.6
Instantaneous heat death	50.9

The normal range of activity is at a body-temperature of 23.5—43.4° C. At 18.4° C and higher slow locomotion may commence.

The following time-table of the locusts behaviour (hoppers, April, 1929) will show the locust's activity under field conditions:

Hour and soil-temperature	Behaviour
Before dawn. 5—6:30 h Temperature below 20° C	All hoppers settled on plants, preferably on shrubs. A very small proportion begins to move slowly.
After dawn. 6:30—8:30 h Temperature 20—26° C	Movements grow to be more rapid. When the temperature rises to 20° C, the hoppers assemble on spots, warmed by sun rays, in dense swarms. They turn their maximal front

	(= lateral side) towards the sun in order to get maximum of sun radiation. Moving ceases.
8:30—12h. Temperature between 27—40° C.	Migration begins and continues regularly.
12—14 h. Temperature above 40° C.	When the soil surface temperature rises above 40° C, the migration ceases. The insects settle in such a manner, that their length-axis rund parallel to the direction of sun radiation, so that a minimal surface is turned to the sun. A considerable number climb to the top of herbs and shrubs and thus escape the heat radiation of the soil.
14—16 h. Temperature 40—27° C.	Regular migration recommences. At 15 h many hoppers settle on plants and are busy eating.
16—18 h. Temperature 27—29° C.	The insects assemble on the sunniest spots in dense swarms, as in the morning.
After sunset.	The insects disperse and settle on herbs and shrubs, before going to sleep.
At night.	The insects sleep on shrubs and herbs, only a small proportion having remained on the ground.

This behaviour does not depend on the hour of the day, but on the temperature. In warm hamsin nights they continue to migrate throughout the night, whereas on cool, cloudy days they remain inactive even during the day. The soil surface temperature is a better indicator than the shade temperature of the air, because the former is influenced by sun radiation itself.

The preference temperature of the locust rises with age:

Stage	Ad	L I	L II	L III	L IV	L V	young Ad
Preference temp.	29.4	30.2	28.8	31.6	37.1	36.7	41.3° C

(Ad = adults, L = larvae)

The average speed observed of the early larval migrations is about 60 m per hour or 300—350 m per day.

Natural enemies and diseases: A complete list of the natural enemies in Palestine cannot be given, as only casual attention had been paid to this aspect of the problem.

Two species of flies hatched from the egg-masses in great numbers. *Stomatorrhina lunata* F. (Muscidae), which was previously known from locust eggs in Algeria and Egypt, seems to be restricted to that species, or at least to grasshopper-eggs only. During April, 1929 two generations were bred, each of two weeks duration only. This fly appeared everywhere in large numbers.

Chortophila cilicrura Rond. (Anthomyiadae) is a cosmopolitan ovoviviparous fly, known to develop in eggs of other locusts in Asia Minor and Argentine. Generally the maggot of this extremely polyphagous species develops on decomposing plant material. It attacked the eggs of *Schistocerca* mainly in heavy soil, where it probably always lives, but preferred this change of diet to its normal food. In mid-April it appeared in our breedings in masses.

Occasionally only unidentified mites appeared on the egg masses in numbers and ants, mainly *Myrmecocystus viaticus*, preyed on them.

Species of *Trichodes*, *Zonabris* and *Anthrax*, the adults of which abound in the neighbourhood of the oviposition places in the spring, certainly attack the locust's eggs likewise, but have never been actually bred by us from them.

Many eggs are destroyed by moulds in heavy soil.

The larvae and adults are attacked by a good many species, only few of which are specific enemies of grasshoppers. Among the latter the following may be mentioned:

The larva of a Sarcophagid fly was once discovered in the thoracic musculature of an adult *Schistocerca.* The yellow maggot of a Nemestrinid fly with a very long caudal respiratory tube was found once in the abdomen of an adult (Shulov, 1948).

Among the predacious wasps species of *Sphex, Stizus* and *Ammophila* predominated.

Among the occasional predators, the following may be mentioned: Many ground beetles (Carabidae), such as *Graphipterus serrator, Chlaenius* spp., etc., *Eremiaphila brunneri, Saga ephippigera, Myrmecocystus viaticus,* the hornet, some spiders, lizards (different species) and some birds. Especially common on the oviposition places were storks (*Ciconia alba*), kestrels (*Falco tinnunculus*), rose stars (*Pastor roseus*), and also crows, bee-eaters, wagtails, hoopoes, and others.

The bacterium *Coccobacillus acridiorum* d'Herelle is certainly present in the Palestine swarms. It was used to some extent in the locust campaign of 1915, but has not been used recently anymore. During the dry, warm weather conditions, which generally prevail during our locust season, no artificial plague can be provoked, whereas it would doubtlessly develop at the spot, where local conditions would be favourable for such an outbreak.

Two widely distributed diseases, *Fusarium acridiorum* Girl. and *Empusa grylli,* have not actually been observed in Palestine, but are certainly present.

The united effect of the total biological resistance on the locust potential is relatively unimportant.

The outbreaks of the Desert Locust: Eruption, breaking down and interval. The analysis of the fluctuations in the population of the Desert Locusts which we attempt here, is by no means as thoroughly based as the facts discussed above.

The most sensitive stages are the egg, the young larva and the maturing females.

With regard to the eggs, we have to consider the coupling of

1) the arrival of the migrating swarms in the rainy season,

2) the oviposition of the female in moist soils only, or in other words, maturation, oviposition and development of the eggs as of greatest epidemiological importance. It generally guarantees the optimal development of the eggs.

The adults require temperature conditions, which would secure them a body temperature of about 40° C during 8—12 hours per day, as well as certain conditions of humidity, the exact nature of which is still unknown. Unfavourable conditions delay sexual maturation.

A hot and arid climate is the optimal condition for all other stages. The migrations normally lead the adults into other regions, where conditions are favourable for maturation and development. The number of annual generations therefore rises to 2—3, as soon as the first migration from the permanent home begins. The long imaginal diapause of the solitary phase disappears. At an oviposition rate of 2 egg-masses à 50 eggs per female, this means a potential increase of 5.000 (in two generations), or of 250.000 (in three generations) per pair and year. Actually, during the eruption of an outbreak, a relatively high percentage of the offspring develops to sexual maturity.

It is extremely difficult to understand, how an actual outbreak breaks down, considering the extreme usefulness of those cyclic migrations of the adult swarms, which extend over 2—3 ecologically different regions and which normally lead to suitable breeding places. When we assume a climatic mortality of 30% of the offspring only, a further mortality of 97% is needed simply to reduce it merely to the number of the preceding generation. Human control measures and biotic resistance combined, form only a fraction of this amount. An outbreak, once started, should therefore be expected to go on indefinitely.

The quantity and seasonal distribution of the rains are certainly of importance, as seen before. The more humid south of the Sudan has much larger precipitations, and is not a permanent breeding ground for the solitary phase:

Southern Sudan		N. Sudan (permanent locust home)		Red Sea Province (secondary Locust centre)	
Gallabat	878 mm	Gobeit	114 mm	Port Sudan	99 mm
Malakal	907 mm	Atbara	69	Suakin	195
Wau	1088	Kkartum	121	Tokar	94
Mongalla	987	El Obeid	356		

Small seasonal or quantitative differences in the precipitations must influence the locust breeding very heavily. In the tertiary breeding centres a delay of the late rains may be fatal. In February, 1932, the soil was too dry to permit breeding. In the case of heavy rains in April these swarms may return and start breeding anew. As a matter of fact, no heavy rains occurred, and the swarms died without a chance for normal oviposition. As a consequence, the breeding season in summer 1932 was greatly reduced in the Sudan. Such an interruption is fatal when it occurs over the main area of tertiary breeding.

The breaking down is not necessarily based on a lack or surplus of rains. When the prevailing winds in any one of the migration seasons are decidedly abnormal, and do not lead the locusts into a suitable breeding ground, the cycle is cut and interrupted. The irregular vagrations of the adult swarms in summer 1915 were probably based on such a phenomenon. That outbreak was cut after the first year of eruption. All these swarms perished without having reproduced, and no major swarms returned to the Sudan.

Wind- and rain-conditions are probably the main responsible factors for the breaking down of an outbreak.

Tentatively, we may divide the 13 years cycle of *Schistocerca gregaria* in its Sudanese-Palestinian territory, as follows:

Name	Solitary prodromal period	Gregarious prodromal period	Eruption period	Collapse	Solitary low population level
Length of period in years	2—3	1—2	2—4	1	4—6
length of period in generations . .	4—6	4—6	4—12	1	4—6

Of this cycle about 9 years, equal to 9 generations, are restricted to the permanent breeding centre in the solitary phase. About four years only are passed in the gregarious phase, migrating to secondary and tertiary breeding centres and equaling 8—12 generations.

History of the Locust invasions in Palestine. (1915, 1929/30). Palestine was invaded in 1865, 1878, 1890, 1902, 1915, 1928/30, etc. The parallel reports for Egypt are: 1891, 1904, 1915, 1928/30. However, these reports are incomplete. Aharoni indicates additional swarms of *S. gregaria:*

1892 at Rosh Pinah
1899 on Nahr Rubin
1908 large swarms at Ain Feshcha.

Apart from these minor invasions, a regular cycle of 12—13 years seems to exist for the major locust outbreaks in these regions. Further observations are needed to prove its reliability. The outbreak, which was predicted for 1927/ or 1928, on the assumed basis of such periodicity, was in agreement with the facts. The last outbreak (1945/47) was somewhat delaid.

The outbreak of 1915 is described by Buecher, Sven Hedin, Whiting, Rosenberg, Aaronsohn, and others. The reports of Sven Hedin and Whiting are especially recommended to the reader. At the end of February, 1915, huge swarms of the Desert Locust appeared suddenly in Southern Asia Minor, from the Arabian shores; they copulated immediately and began to oviposit a week later. Oviposition occurred all over Palestine, Syria, and Irak, at suitable localities. The young hoppers commenced to hatch in early April and immediately formed large swarms and began to migrate. The winged adults appeared in June and disappeared in large armies towards the North and North East. Near Adana and in Armenia these swarms were seen in July, in Syria and Irak from September to Decembei. They migrated irregularly here and there, and effected great damage. The orange groves of Hedera were attacked in September, 1915, and greatly injured. at the beginning of the cold season, these swarms died, and disappeared without having oviposited. The severe cold in N. Syria killed the last remainders. Only in the Jordan Valley and in the Negeb several small swarms were still observed in January and February, 1916. Oviposition was observed in these, but none of the larvae hatched.

The cereals were full grown when the locust arrived. Grapes, apricot-, fig-, quince-, pomegranate-trees, etc., fields of tomatoes, cucumbers, melons, sorghum, and others have been destroyed. The adults did not even spare olive-trees, palms and prickly pear. Only Persian Syringe, Charub, sycamore and oleander remained untouched.

The following invasion lasted for at least four years. In February 1928, large swarms appeared in Transjordania, up to the Palestinian border. They were repelled back to the eastern and southern desert by a sudden wave of cold and rains. They reappeared throughout Palestine in April, but not having found suitable conditions for oviposition, anymore, they died. The limited oviposition on irrigated soil near Jericho was easily controlled by flame guns.

In February, 1929, enormous flights of Desert Locusts reappeared in Palestine on route from S. Arabia, and began to oviposit in sandy troughs on the western slope of the hills of Moab, in the same month. The larvae hatched in March, and became adult 5—6 weeks later. Nearly all the swarms of hoppers turned eastwards, and climbed the Transjordanian highplateau by way of the wadis and the flying armies of the adults disappeared in the same direction in late April. In Palestine, oviposition was observed that year only in Galilee, and that only to a small extent.

The largest attack in Palestine occurred in the following year. As early as late October, 1929, some minor swarms invaded the hills, of Samaria temporarily. However, reports from Akaba and Sinai indicated an activity of unpreceeded extent. In January and February, 1930 the whole Negeb proved to be infested by huge ovipositing armies of locusts. Along the coast dunes this area reached the southern border of Rehoboth. The general border of this area was somewhat to the north of Gaza, Beersheba and Tell el Milh. Additional swarms, which invaded the Jordan Valley in early March, fortunately disappeared into Syria. In the main oviposition area near Beersheba, which embraced some 900 square kilometres, the hatching of the larvae started in early March. The direction of the hopper migration was almost invariably towards the north. A very intensive campaign was conducted in this area through April and half May. Minor infestations occurred at Khan Yunis, in the Jordan Valley, near Semakh and in Upper Galilee. The adult locusts disappeared in June in a south-eastern direction.

A considerable locust activity prevailed during 1931 in the neighboring countries, but no oviposition of menacing extent was observed in Palestine. Large flying swarms of *Schistocerca* appeared in February, 1932 on the borders of Palestine and in Transjordania. Owing to a protracted drought these did not find soil suitable for oviposition. They disappeared, and the chain of migrations was broken by this situation. In 1945 to 1947 again invasions occurred. In 1945 they came too late, in May, and only very restricted oviposition took place on irrigated fields. In 1946 the invasion also was late, in early April, and abortive in 1947.

Control measures.

The control of the locust campaign rests wholly in the hands of the government. A special legislation guarantees any possible aid to be given by the population. The plant protection ordinance of 1924 and the locust amendment of 1931 ensure marking and reporting of flying locust swarms, oviposition areas, enforced labour upon the total population, in case control work of any kind is needed. In case of a serious invasion, the Government Entomologist becomes ex officio the Locust Officer. High officials of the Agricultural department are in charge of one special locust area, where a special camp is erected, the necessary control materials stored, and an organization consisting of lower officials of the Government service and the police force is formed. The control work is done by the local population and by paid labour in emergency cases.

The control of *Schistocerca* in Palestine under the mandate was directed mainly against the eggs and larvae. The following measures are used:

1. The arriving swarms may occasionally be destroyed by flame-guns to an appreciable extent. The guns are easily obtained in numbers by furnishing ordinary compressed air back-sprayers with a small additional apparatus in place of the spraying nozzle. The flame reaches over a distance of 5—15 m. It is applied in the evening, when the adults aggregate on a few places, as trees and shrubs, by the thousands.

2. Wherever the oviposition area is carefully marked, and where soil conditions permit, thorough ploughing offers a radical and cheap method for destroying all the eggs. This method is applied easiest in the plain and between cultivated crops. It requires careful work and one well ploughed area is worth more in a campaign than five carelessly ploghed areas. In this type of control, the ploughing must be done speedily, for about 3—4 weeks. On rocky soil, especially in the hills, this method is not possible. Recent research, mainly in Spain, has shown however, that ploughing simply duplicates work, as its effect is much easier obtained by baiting. The possible thinning and dispersal of the hoppers, influences the main baiting campaign unfavourably.

3. The hoppers can be controlled for a period of 6 weeks. Demarkation of the oviposition sites gives additional weeks for the preparations of the baiting campain. The young hoppers may first be attacked by intensive baiting. The bait, which has been first applied in the Sudan, is prepared from a stock solution of

Sodium arsenite	168 lbs (80% As_2O_3)
Water	22 gallons
Molasses	62 gallons

This stock solution is diluted 17 parts of water to one part of solution. Twelve gallons of this diluted mixture are well mixed with 100 pounds of medium-ground wheat bran. The bait is transported from the factory to the place of application in bags, à 5 kilos. It is then scattered amongst the swarms very early in the morning, before the locusts begin their march for the day. The resting hoppers commence to feed on the bait readily, as soon as the bait is scattered in the morning. The bait attracts all stages of hoppers equally (Bodkin, 1930). The killing power of the bait is astonishing. This bait is now entirely replaced by a 2 $^0/_0$ mixture of Agrocide 7 (gammexane) with bran. Hoppers, having fed on the bait, lose their social habits and disperse. The danger of poisoning people or animals is minimal, provided the arsenical bait is scattered under proper supervision. Agrocide is entirely not poisonous.

4. Large swarms of hoppers, especially the third to fifth stages, were controlled by means of the zinc sheet method. Zinc sheets, 33 cm thick and 1 m long, are carefully erected in front of the hopper swarm as a continuous wall of 50—100 m. The erection begins in the afternoon, when the swarm ceases its migration, and continues throughout the night. A number of pits is dug close to the zinc sheets, into which the hoppers fall and where they can be easily destroyed. The number of pits depends on the size of the swarm. The zinc sheets run parallel to the front of the swarm and are somewhat longer at both ends, to form a corner with the main front. The approaching hoppers, not being able to jump over the sheet, turn to the right and to the left, and are finally caught in the pits. The pressure exerted by the following hoppers makes it impossible for the first ones to avoid the pits. The length of one series of sheets may be easily extended to 10 km and even more, in case the size of the swarm requires such an extent. If done properly, even a large swarm may be completely destroyed by one or two consecutive zinc sheet trappings. The method works well in the plain, and is even applied with success in many hilly places, where the rock is not bare. Thus, the swarms may be destroyed for many kilometres before they have a chance to enter cultivated areas. A certain strategic talent is needed to direct a zinc sheet campaign with the maximum of success. This method has been abandoned today altogether in favour of baiting.

5. Minor control measures against hoppers to be applied mainly in broken terrain or against a series of small swarms are: flame-guns, and spraying with mineral oils. The primitive driving methods are more or less abandoned to-day.

6. Important research has been done recently in British Africa. Attempts are made to destroy flying swarms dusting them with arsenical and other dusts from aeroplanes. The first experiments are rather encouraging and if this measure should prove to be practical on a large scale, any future invasion could be successfully prevented by controlling the migrating adult swarms in the Sudan and S. Arabia. Unfortunately the high costs seem still prohibitive to the further development of this method.

7. Recent researches in the Sudan (Johnston, Maxwell-Darling) tend to show definitely that the building up of our locust outbreaks does not occur all over the vast regions of the Sudan, but in restricted territories in the dunes of the Red Sea Coast. A transition from solitary to gregarious phase has been observed there repeatedly, and only there, so far, in our migratory area. The permanent and proper survey of these territories would be a fairly simple matter as well as effectively destructive of all beginning swarms within this area. If it could be proven definitely that the gregarigenous areas are restricted to certain districts in the Sudan and perhaps in the Hedjas, it will be relatively easy to prevent altogether the future building up of large swarms and invasions into Palestine and Egypt. The maintenance of the locust survey and control in the gregarigenous areas would then be a matter of small, but permanent expenses, which all countries concerned in this specific outbreak area should share. The International Locust Conferences and Research under the spiritual guidance of B. P. Uvarov have done very much indeed

to further locust problems. Recent experiences of the writer (1943) in Irak tend to show however, that whilst, initial phase transformation may be restricted to certain "reservations", the huge swarms are probably as a rule be built up in the invasion area, which is too big for regular scouting.

In agreement with the current expert opinion the Govt. of Israel has decided to abandon control of invading adults, except in the rare cases when they should cause considerable danger to cultivated trees, and as well to abandon the ploughing of egg-sites. Baiting with a 2 % bait of Agrocide 7 with bran is at present the main method of control. Experiments on the application of dusting and spraying from aeroplanes and high-power motor mashines are under way, as well a number of additional baits. The molasse component, as entirely useless as locust bait, has been abandonned. The problem, if water is necessary as an ingredient of the bait, is under study.

Locusts and Citrus cultivation.

The invasion of 1915 was of vital importance to the local orange industry, whereas those of 1929/32 and of 1945/47 passed without having left behind serious damage.

In 1915 the invading adult swarms attacked the orange groves of the coast-plain repeatedly and devoured a great deal of leaves. This damage was in itself serious, but only minimal as compared with that done by the hoppers 1—2 months later. Only bare trunks and branches, partly even having stripped off their bark, remained in many invaded orange groves. Most fruits fell off. Only 1600 out of 7665 dunams orange groves were invaded in the Jewish colonies of Judaea and the crop was 200.000 instead of 700.000 orange boxes. However, the real loss was much greater, as the invaded trees remained suffering for many years to come. Even in 1925, the poor conditions of many trees was still attributed to the locust invasion of 1915. The last invasions oviposited mainly in uncultivated districts, and the intensive control measures prevented an attack of the hoppers in the orange district. It should, however, be remembered, that in 1930 the largest oviposition area extended to the very neigbourhood of Rehoboth.

What are the future prospects for our citrus industry, considering the locust menace? It is quite safe to state, that as long as a government of an energetic executive power rules in Palestine, no major damage due to hopper swarms can menace the orange groves or any other intensive crops. The swarms can be easily handled, by the control measures just mentioned, which would prevent the entrance of the swarms into the groves. And even in the case of oviposition in the orchard itself, no major damage should occur, if the ploughing of the soil is combined with intensive baiting.

The only damage which cannot be prevented at present, is that resulting from the feeding of the adult immature locusts. With the improvement of the aeroplane dusting method, even such damage might be avoided in the future.

In connection with the locust menace, the question of locust insurance has been repeatedly discussed among orange growers. According to the opinion of the author, such an insurance is not needed by the farmer anymore. But, if the farmer insists on obtaining that insurance, its costs should be very low, considering

1. the long intervals between locust outbreaks,
2. the present state of control organisation and control prospects.

The present state of government organisation reduces the potential danger of the locust menace for the citrus' grower to a minimum. It is public duty to aid the government in this field to the best of our ability.

References.

A. Aaronsohn, La lutte contre le Criquet palestine. Jerusalem, 1915.
J. Aharoni, Haarbeh. Jerusalem, 1919.
M. J. Atalla and M. S. Hassibi, Bulletin Office Internat. Renseignments sur les Sauterelles de Damas. 1930 et 1931.
E. Ballard, Mistikawi and Zohairy, Min. of Agric., Egypt, Tech, Bull., 110. 1932. 149 pp.
L. Bauer, Zeitschr. D. Pal. Ver. **49.** 1926, p. 168—171.
F. S. Bodenheimer, Hadar **2**, 1929. pp. 136—139. 12 pp.
F. S. Bodenheimer, Biol. Zentralbl. **52**, 1932, p. 598—619.
F. S. Bodenheimer, Direct. Gen. Agric. Iraq. Bull. 29. 1944.
G. E. Bodkin, Bull. Ent. Res. **20,** 1929, p. 123—139.
G. E. Bodkin, Agric. Leaflets Govt. of Palestine Series I, No. 8. 1930, pp. 24.
G. E. Bodkin, The Palestine Locust Campaign 1930. Jerusalem, 1931, 28 pp.
H. Buecher, Die Heuschreckenplage und ihre Bekaempfung, Berlin, 1918.
L. H. Gough, Report on the great invasion of locusts in Egypt in 1915. Cairo, 1916. 72 pp.
S. Hedin, Jerusalem. Leipzig, 1915, pp. 106—112.
K. d'Herculais, Invasions des Acridiens vulgo Sauterelles en Algerie. Alger. 2 Vol. 1893—1905.
J. Rosenberg, Tropenflanzer, **1n**, 1915, pp. 657—669.
B. P. Uvarov, Locusts and Grasshoppers. London. 1929.
B. P. Uvarov, Diverse reports on the locust outbreak and locust situation. London, 1933—38.
J. Vosseler, Berichte Land- und Forstwiss. D. Ostafrika **2,** 1905, pp. 291—374.
J. D. Whiting, National Geograph. Mag. **28,** 1915, pp. 511—550.
M. A. Husain and T. Ahmad, Indian Jenm. Agric. Sci. **6,** 1936. p. 188—262.
A. G. Hamilton, Transact. Raj. Ent. Soc. London n**5**, 1936. p. 1—60.
Various papers submitted to the IV. Internat. Locust Conference at Cairo, 1936 (Bodenheimer, Mistikawy, Hussein, Ramachandra Rao).
A. Shulov, Nature. 162. 1948. p. 255.

B. OTHER ORTHOPTERA IN CITRUS ORCHARDS.

The big Egyptian tree grasshopper (*Anacridium aegyptium* L.), easily identified by the orange keel of its pronotum and its green larvae, is a common insect in the citrus grove. Its occasional feeding on citrus leaves cannot be regarded as damage.

The common bimaculated field-cricket (*Gryllus bimaculatus* Deg.) causes occasionally local damage by feeding on the green bark in young citrus groves (Fig. 130 C).

At the Gan Hasharon grove damage was observed in Juni, 1930. On an area of 350 dunams about 200 recently budded trees were attacked, 25 of them heavily, which required rebudding. The trees were 3—5 mm in diameter. Feeding extended generally over 30—50 mm of the budding, 5—10 mm in the width and 1 mm deep. This means that besides the bark, the wood is also attacked. Also shorter feeding marks were observed, only 20—30 mm long, but deeper; here the wood was destroyed sometimes up to half of the diameter. Thin wildlings have been cut occasionally. The attack was concentrated on three spots, which were situated on elevated parts of the grove on red "chamrah" soil. In the sandy parts of the grove only 2 damaged trees were seen. No weeds were present in the grove. The black crickets have been repeatedly seen eating by the workmen, and many individuals — among them larvae — were found in soil galleries near the attacked trees.

Similar damage, but never to a considerable extent, has been reported from nurseries and young citrus groves from Hedera (IX. 30), Bne Brak (IX. 32), etc.

The European mole-cricket (*Gryllotalpa gryllotalpa* L.) has done a certain, limited damage repeatedly in citrus nurseries. Spreading of a bait, mixed from 100 parts of ground rice, 25 parts of water and 5 parts of zinc-phosphid, offers an easy control measure.

10. MITES (ACARINA).

KEY FOR THE DETERMINATION OF MITES LIVING ON CITRUS IN PALESTINE.

1. Adults with two pairs of legs only. Cephalothorax smooth, abdomen with 28 dorsal half-rings and almost a double number of sternites, also in the form of half-rings. *Phyllocoptes oleivorus* Ashm.
2. Adults with four pairs of legs. 2.
   Adults fairly strong chitinized, brown, on trunk, branches or leaves in great number, more than 1 mm long. *Oribatula plantivaga Berl.*
   All stages with soft exoskeleton. Less than 1 mm long. To be seen with a magnifying lense only. Preferebly on leaves. 3.
3. Fore-legs nearly twice as long as other legs. 0.6—0.8 mm long. Eggs red. *Bryobia praetiosa Koch.*
   Fore-legs not very much longer than other legs. 4.
4. Legs in both sexes distinctly as long as slightly longer than the body. Abdomen dark-brown in female, reddish in male. Length of body 0.45 mm in female, 0,35 mm in male, of legs 0.45 mm in both sexes. Body hair moderately long. Eggs white. *Anychus orientalis Zach.*
   Legs not as long as the body. 5.
5. Abdomen light reddish brown in female, reddish-yellow in male. Body hair moderately long. Length of body 0.45 mm in female, 0.35 mm in male, of legs 0.25 mm in both sexes. Living in self-spun webs. Eggs white. *Epitetranychus althaeae v. Hanst.*
   Abdomen dark reddish brown in female 0.25 mm long. Body hair very long. A non weg-spinning species. Eggs red. (Not yet found in Palestine). (*Paratetranychus pilosus C. & F.*)

A. THE ORIENTAL RED MITE (*Anychus orientalis, Zach., Tetranychidae*).

Synonymy: 1936 *Anychus latus* Sayed, nec Can e Fan.

Whilst Taher Sayed (1942) regarded *Anychus orientalis* Zacher as a mere synonym of *A. latus* C. e F., he restituted both as different in 1946, when he gave good and complete descriptions of both of them. The main differences are:

Character	*A. orientalis*	*A. latus*
Body of female in μ:	436 × 313	400 × 283
Palpus of female:	With short hair on joint I	With no such hair
Body of male in μ:	383 × 206	345 × 204
Dorsal hairs of male:	Spatulate, short	Pointed, longer.

Description: The development of this family is rather complicated. 6-legged larvae, which change into a first restings stage called nymphochrysalis, hatch from the eggs. The nymphochrysalis turns into the first 8-legged nymph, which in turn passes into a deutochrysalis stage. The latter is transformed into the deutonymph, which is followed by a third resting stage, the telochrysalis. This passes into the adult female. The adult male hatches from the deutochrysalis Whereas the female passes from the egg through 3 active and 3 resting stages in its development, the male has only 2 active and two resting stages.

The adult female is broad oval in outline, with legs, as long as the body. Abdomen dark-brown, cephalothorax and legs reddish-yellow. 0.45 mm long, 0.3 mm broad. Legs 6-jointed, with reduced empodium. Pedipalpi 4-jointed. Hairs on body and legs rather short.

The *adult male* is smaller than the female and its body is triangular in shape. Its legs are considerably longer than the body. Abdomen reddish; cephalothorax and legs reddish-yellow. 0.35 mm long, 0.25 mm broad; legs 0.45 mm long.

The *eggs* are round, transparent, light brown. 0.14 mm in diameter.

The other development stages may be briefly described, as follows:

stage	female	male
larva	abdomen bright greenish-brown; cephalothorax light yellowish-green, legs pale yellow. 0.2 × 0.1 mm, 6 legs each o.1 mm long.	abdomen bright reddish, cephalothorax yellow; legs pale yellow. Measurements as those of female larva.
nymphochrysalis	abdomen greenish brown, cephalothorax light red.	abdomen reddish-brown, cephalothorax light red.
I. nymph	abdomen greenish brown, cephalothorax and legs yellow. 8 legs, each 0.15 mm long. 0.3 × 0.2 mm.	abdomen reddish-brown, cephalothorax and legs yellow. 0.3 × 0.2 mm, legs 0.25 mm long.
deutochrysalis	abdomen brown, cephalothorax yellow.	abdomen red; cephalothorax yellowish red.
II. nymph	abdomen greenish-brown, cephalothorax and legs yellow. 0.35×0.25 mm; legs 0.2 mm long.	
telochrysalis	abdomen reddish brown; cephalothorax pale yellow.	

The posterior end of the abdomen is broadly rounded in all female stages, pointed in all male stages.

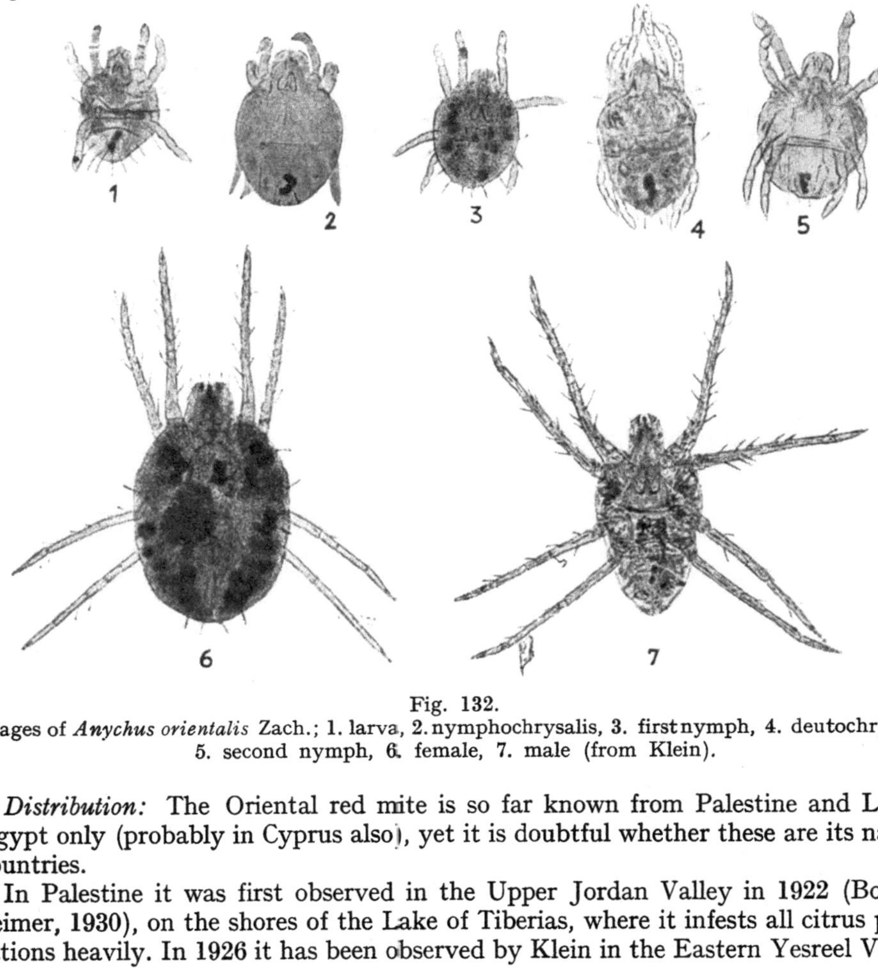

Fig. 132.
Stages of *Anychus orientalis* Zach.; 1. larva, 2. nymphochrysalis, 3. first nymph, 4. deutochrysalis 5. second nymph, 6. female, 7. male (from Klein).

Distribution: The Oriental red mite is so far known from Palestine and Lower Egypt only (probably in Cyprus also), yet it is doubtful whether these are its native countries.

In Palestine it was first observed in the Upper Jordan Valley in 1922 (Bodenheimer, 1930), on the shores of the Lake of Tiberias, where it infests all citrus plantations heavily. In 1926 it has been observed by Klein in the Eastern Yesreel Valley

(Kfar Yehezkiel), where it is still permanently injurious. On the coastal plain it was first reported in 1930 from Hedera; in 1931 from all citrus groves in Samaria around the Sharon-plain until Petah Tikva. In the southern coastal plain (Rehoboth, Rishon lezion) it made its first appearance in 1932, where it has been a permanent pest even since. It is also known from Jericho, being probably recently imported with citrus-stock into the Government's experimental garden.

Life-history: The adults copulate immediately after hatching. Klein (1936) describes the copula: "A male transferred to a leaf on which a female is settled, senses her immediately and moves around her. He climbs over her and circles round her several times at great speed. Finally, he crawls under the posterior end of the female's abdomen, while the latter, leaning forward on her head, lifts her abdomen, spreading her legs widely apart. The male then lifts his abdominal end towards that of the female in order to achieve copulation. Each copulation lasts from 2—3 minutes. A male transferred, following copulation, to another female, copulated again after a short interval. The female requires no second copulation for the rest of her life."

The sex-ratio is fairly constant from April to October, during which time out of 438 adults 70.8 % were females. In late autumn and winter the sex-ratio seems to be different. In November, at 19° C, out of 25 adults only 56.0 % were females. The number of individuals observed is still too small to permit a definite settlement.

During summer, the female lays several (up to 8) eggs per day, mainly along the midrib on the upper side of the leaf. In a heavy infestation eggs are found everywhere, even among the webs of the mite. During winter, oviposition occurs frequently on the lower side of the leaves. The eggs, living and empty shells, may often be seen in long rows along the midribs.

The highest egg rate per female was obtained in summer:

Season	temperature	maximum eggs per female
winter.	16° C	18
autumn, spring. . .	19—20	23
summer	22—27	35

Unfertilized females oviposit normally. The number of eggs is not affected. However, all those parthenogenetically developing eggs develop into males, as usual in red mites.

The very complicated development has been mentioned before. It includes 3 active stages, alternating with 3 resting stages in the female and 2 such stages in the male. The empty exuviae stick to the leaf. The average duration of every stage at the different seasons is compiled in the following table:

Season	Sex	Larva	Nympho-chrysalis	I. nymph	Deuto-chrysalis	II. nymph	Telo-chrysalis	Total
spring . .	♂	5	2	9	4	—	—	20 days
	♀	4	3	4	4	3	2	20
summer .	♂	2	1	2	1	—	—	7
	♀	1	1	1	1	2	1	7
autumn	♂	2	2	4	3	—	—	11
	♀	2	1	2	2	3	1	11
winter . .	♂	4	2	12	5	—	—	23
	♀	3	1	5	2	11	2	24

Despite the lack of two development stages in the male, the latter's total development is not shorter than that of the female. The last active and resting stage of the male is correspondingly prolonged.

The very characteristic webs of this species, as well as of *E. althaeae*, are excreted by two special glands. The two threads excreted by these glands unite into one. The female is more active in spinning than the male. The leaves are often covered with these webs which give them a greyish appearance. (André).

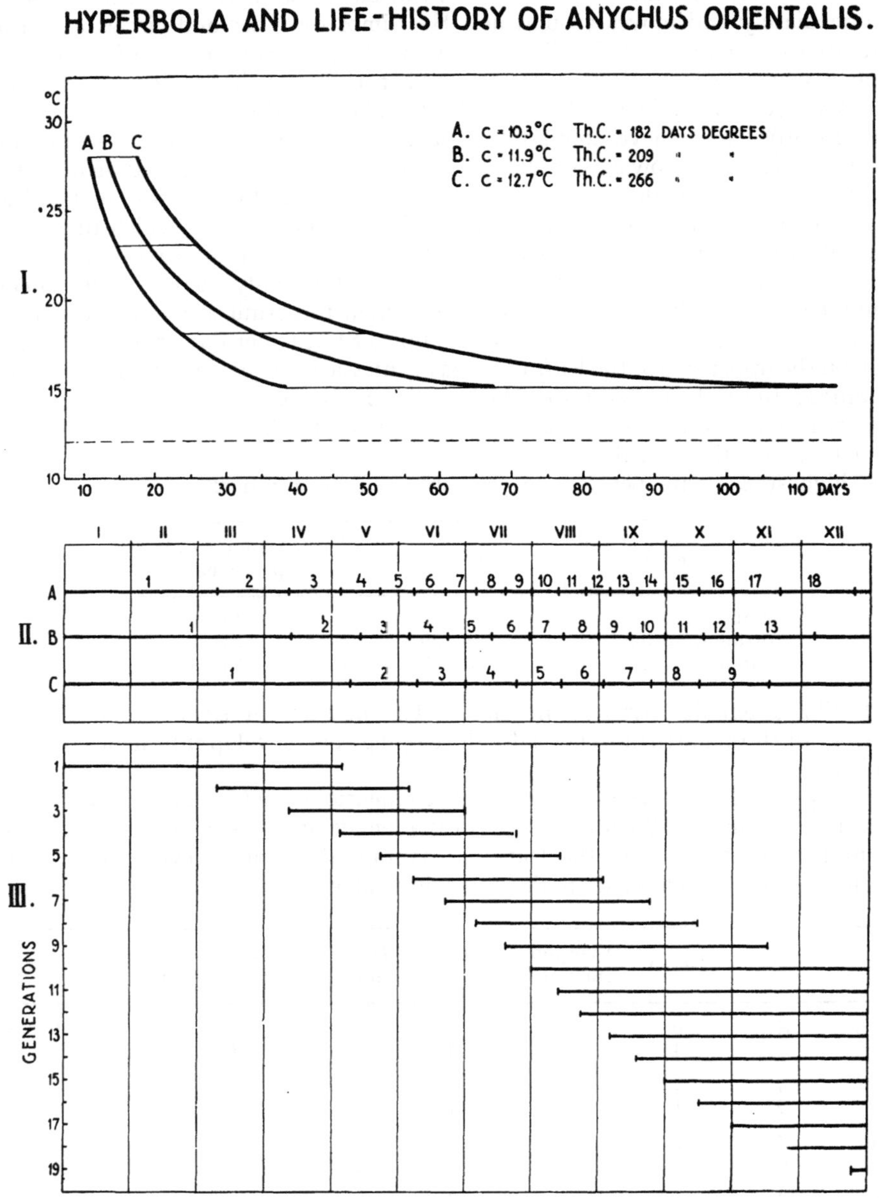

Fig. 133.
Life history of the Oriental Red Mite (*Anychus orientalis Zach.*) at Tel Aviv. I. The three hyperbolae, II. The life-cycle, as calculated from the hyperbolae; III. Sequence and overlapping of generations.

Temperature and development; number of annual generations: The temperature dependency of development is expressed by an hyperbola of the following constants:

development threshold (c) = 11.0° C
thermal constant (th.c.) = 180 days degrees.

The turning point is at about 26° C, above which temperature development slackens (fig. 133).

The preoviposition period lasts 1—2 days in the summer (23° C and above), 2.5—3 days at 20—22° C and 4—8 days at 14—15° C. The longevity of the adults is more or less equal in both sexes. It lasts up to 12 days in summer, 14—18 days in spring and autumn, up to 21 days in winter.

Taking these data into consideration, the hyperbolas A for quickest total development, B for the epidemiological average, C for the latest developing individuals of every generation are computed:

		A	B	C
development threshold		10.3	11.9	12.7° C
thermal constant		182	209	266 days degrees
development at	28° C	10.3	13.0	17.4 days
	23	14.3	18.8	25.8
	18	23.6	34.2	50.4
	15	38.7	67.4	115.7

The difference between C and A yields the adult life span of the individual.

The annual number of generations within the known area of distribution of *Anychus orientalis* is, as follows:

	A	B	C
Tel Aviv	18.1	13.3	9.2
Tiberias	22.6	17.0	13.1
Cairo	21.5	15.9	11.5

A series of breedings, performed by Klein (1936) at Rehoboth agrees fairly well with these calculations:

generation	date (oviposition to adult)	duration of development	temperature	rel. hum.
1	21.VI.—1.VII.	10 days	24.8° C	64%
2	2.VII.—15.VII.	13	25.2	61
3	17.VII.—25.VII.	8	25.8	69
4	26.VII.—7.VIII.	12	26.9	66
5	8.VIII.—18.VIII.	10	26.5	71
6	20.VIII.—1.IX.	12	25.3	66
7	2.IX.—14.IX.	12	25.0	65
8	16.IX.—28.IX.	12	23.7	63
9	1.X.—14.X.	13	22.9	62
10	20.X.—8.XI.	19	21.2	59
11	13.XI.—18.XII.	35	15.8	62
12	26.XII.—4.III.	68	13.0	68
13	8.III.—2.IV.	25	16.5	65
14	4.IV.—22.IV.	18	17.1	55
15	23.IV.—10.V.	17	24.0	53
16	12.V.—24.V.	12	22.4	57
17	26.V.—5.VI.	10	23.4	68
18	7.VI.—18.VI.	11	24.0	63

The scale of activity, following the results obtained by Miss Ben-Ami, is for adult females (n = 24) as follows:

Scale	1	2	3	4	5	6	7	8
Temperature.	6.9	—	11 7	18.0—33.8	36.6	41.8	43.5	47.4° C

The normal activity ranges from 18.0—33.8° C; therefore well suited to the summer temperatures of all parts of Palestine. Below 18° C its normal activity is very much reduced, and many physiological functions are changed at that temperature: development is greatly reduced; the sex-ratio changes; oviposition is very reduced, if continued at all; mortality increases considerably. The limit of cold-torpor (about 7° C) is only for hours reached in the coastal plain. The laboratory experiments agree well with the observations in nature.

Some additional statistical counts on the Oriental Red Mite in different orange groves at Hedera will now be given:

1. Monthly count on 400 leaves during 6 years on trees about 23 years old:

Year	I	II	III	IV	V	VI	VII	VIII	IX	X	XI	XII	Total
1931	78	23	6	10	1	1	—	—	—	2	45	75	244
1932	59	6	8	2	5	3	2	—	2	8	13	3	111
1933	5	—	3	—	3	—	—	—	1	—	—	17	29
1934	11	—	—	—	—	—	2	—	17	10	—	—	40
1935	—	1	5	3	—	1	—	11	14	2	1	3	41
1936	—	—	—	—	4	—	2	—	—	—	—	—	6
Average	24	5	4	2	2	1	1	2	6	4	10	16	

2. Monthly count on 100 leaves during one year on trees 5 years old:

month	I	II	III	IV	V	VI	VII	VIII	IX	X	XI	XII	Total
no. of *A.or.* .	—	3	—	2	—	—	8	9	25	½5	8	—	80

3. Monthly count on a whole branch (about 750 leaves) during one year on a tree 28 years old, in a shady grove.

month	I	II	III	IV	V	VI	VII	VIII	IX	X	XI	XII	Total
no. of *A.or.* .	1	6	9	6	17	41	15	47	140	197	35	19	507

4. 1 year count on 3 other branches:

month	I	II	III	IV	V	VI	VII	VIII	IX	X	XI	XII	Total
no. of *A.or.*	—	6	9	7	7	—	—	15	92	88	80	167	471
dto 2 branches, 1936 .	—	2	—	—	—	—	98	14	12	310	124	—	560

5. 400 leaves in an old (30 years) and young (10 years old) grove:

month	I	II	III	IV	V	VI	VII	VIII	IX	X	XI	XII	Total
1935 old grove	18	79	26	93	(47)	8	20	65	38	49	21	11	475
young grove.	—	11	2	9	1	—	31	35	99	99	30	—	317
1936 old grove	—	—	—	2	—	1	53	84	12	12	—	—	154
young grove (300 leaves)	—	—	—	—	—	—	3	2	67	279	60	—	411

The series 2, 3 and 5 give a good illustration of the seasonal increase in summer and the decrease in late autumn. From September to November *Anychus* was also observed on the fruit.

General epidemiology of the Oriental red mite: The eggs are very sensitive to low humidity, not so the other stages. Not ony mortality, but also duration development are affected in the egg stage by low humidity. All stages show an increased mortality with lowered temperature:

eggs			larvae and nymphs		
temperature	humidity	mortality	temperature	humidity	mortality
24—25° C	63—64%	9.0	24—26° C	40—69%	29.4
16—24	53—64	25.9	19—23	54—60	53.1
13—24	40—67	62.5	13—14	70	63.6

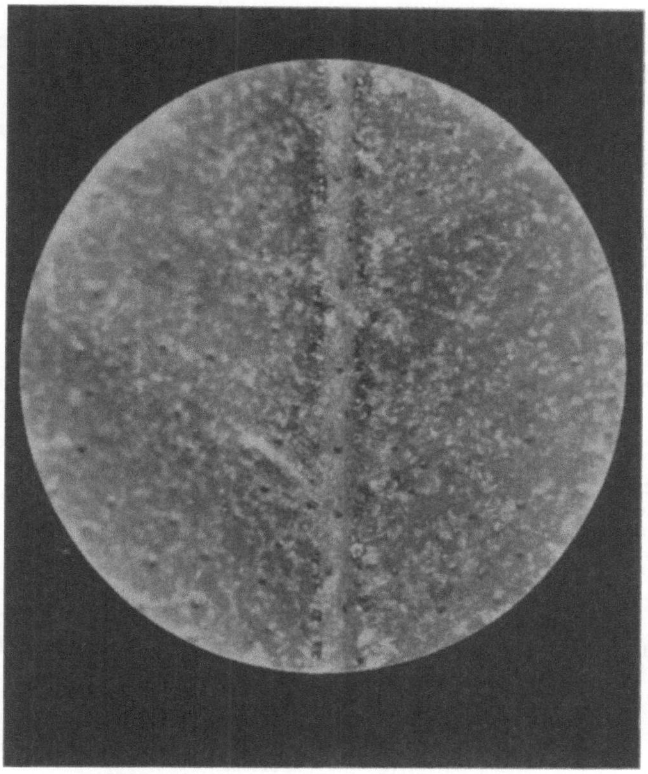

Fig. 134.
Orange leaf with eggs of Oriental red mite (70 ×; from Klein).

The total mortality may be grouped as follows:

mortality	temperature	rel. humidity
20—40%	23—27° C	62—70%
50—75	19—24	40—60
75	13	68—72

High temperature and medium air humidity are optimal conditions for all stages. The most rapid development of each generation may easily be grouped in the thermohygrogram in ellipsoid zones:

duration of development	temperature	rel. humidity
8—10 days	25—27° C	68—72%
10—13	22—27	57—72
17—19	17—24	53—60
25—35	15—17	62—65
68	13	67

The optimal zone for the development of *A. orientalis* is in a zone limited by 21—27° C and 59—70% rel. humidity. Favourable conditions for development are still maintained between 18 and about 30° C and between 35 and about 72% relative humidity. Beyond these limits conditions are decidedly unfavourable for the development of the species.

The optimal zone includes:
1. the shortest development of all stages,
2. the shortest preoviposition period,
3. the highest percentage of females,
4. the maximum number of eggs laid by one female,
5. the lowest mortality of all stages.

Host relations: Only a few hosts of *A. orientalis* have been met in nature in Palestine so far, which is in striking contrast to the extreme polyphagy of *Epitetranychus althaeae*. Following the extended observations of Klein (1936) these few hosts are:

Apocynaceae	*Plumiera alba*
Caricaceae	*Carica papaya*
Euphorbiaceae	*Ricinus communis*
Moraceae	*Morus alba, Ficus macrophylla*
Rutaceae	*Citrus ssp.*

The list of potential hosts is much longer than the actual one. In Egypt it is recorded from citrus, *Albizzia lebbek, Ricinus* and *Melia azederach*. Breeding was attempted on the following plants:

	mortality	shortest development	dto on citrus contem-poraneously
cucumbers .	93.1%	10 days	10 days
eggplants .	82.9	8 10	8 10
beans . . .	75.0	13 16 21	10 13 18

The high initial mortality on a new host is not surprising. It would be of interest to follow up, whether an "adaptation" to the new host takes place during the following generations. The contemporaneous nutritional mortality of *A. orientalis* on citrus was only 16.0%. Duration of development remained unchanged in cucumbers and eggplants, whereas it was prolonged in beans, the latter displaying the lowest mortality among the three experimental hosts.

Among citrus species sour lemon suffers most, so that the degree to which it is attacked may serve as an indicator of the actual intensity of attack.

Epidemiology of Anychus orientalis in Palestine: The annual epidemiology of *A. orientalis* in Palestine is well correlated to the season and length of stay in the optimal zone of the local climogram (Klein, 1936).

In the coastal plain (Rehoboth) the Egyptian red mite appears first in considerable numbers during June. They then increase steadily until reaching the peak in September/October. They decrease in number rapidly in November, and are found throughout the winter in small numbers only, mainly on the lower sides of the leaves. In spring they continue to disappear and the minimum is reached in May.

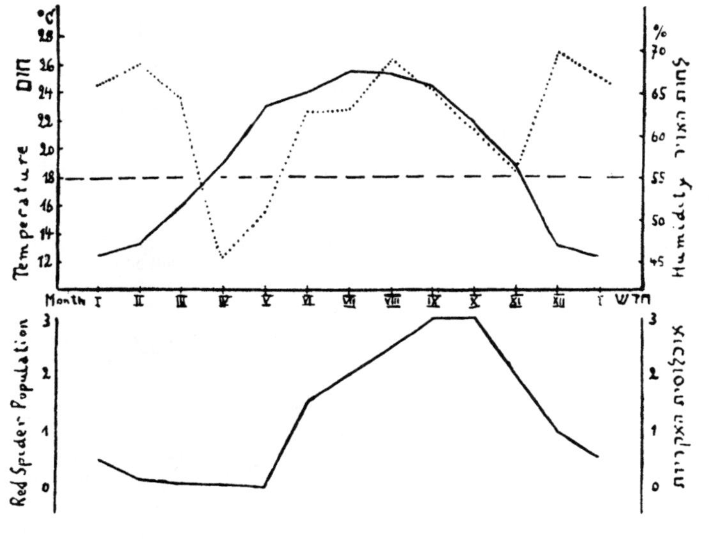

Fig. 135.
Seasonal population trend of *Anychus orientalis Zach.* on citrus leaves at Rehoboth. 0—3 from none to very many red spiders present; —— temperature, ... humidity.

Fig. 135 shows the annual course of increase of the Red Spider at Rehoboth, together with local climatical conditions. It is evident that with a combined rise of temperature and rel. humidity above 18° C and 55% intensive multiplication begins, whereas with a fall of one of these factors below these marks the Red Spider population decreases. Red Spider control begins, as a rule, in August. (Klein, 1936). Low winter temperatures and dryness in spring keep the species down, until summer, when conditions are favorable from June to October.

On the Lake of Tiberias and in the Eastern Yesreel Valley the Egyptian red mite begins to appear in great numbers in April, and maintains a high population level until winter. Sulphur dusting is applied several times during the season. Optimal climatic conditions prevail from April till winter. J. Rosenberg, Superintendent of the Government Horticultural Station at Jericho, informs us: "The Oriental Red

Spider begins to appear in September, and it increases till February/March. During summer it is found in small numbers only. Dusting is applied in March." Severe dryness of the air prevents its increase from April till September, despite the favourable temperatures during this season. Optimal conditions prevail in autumn and towards the end of winter.

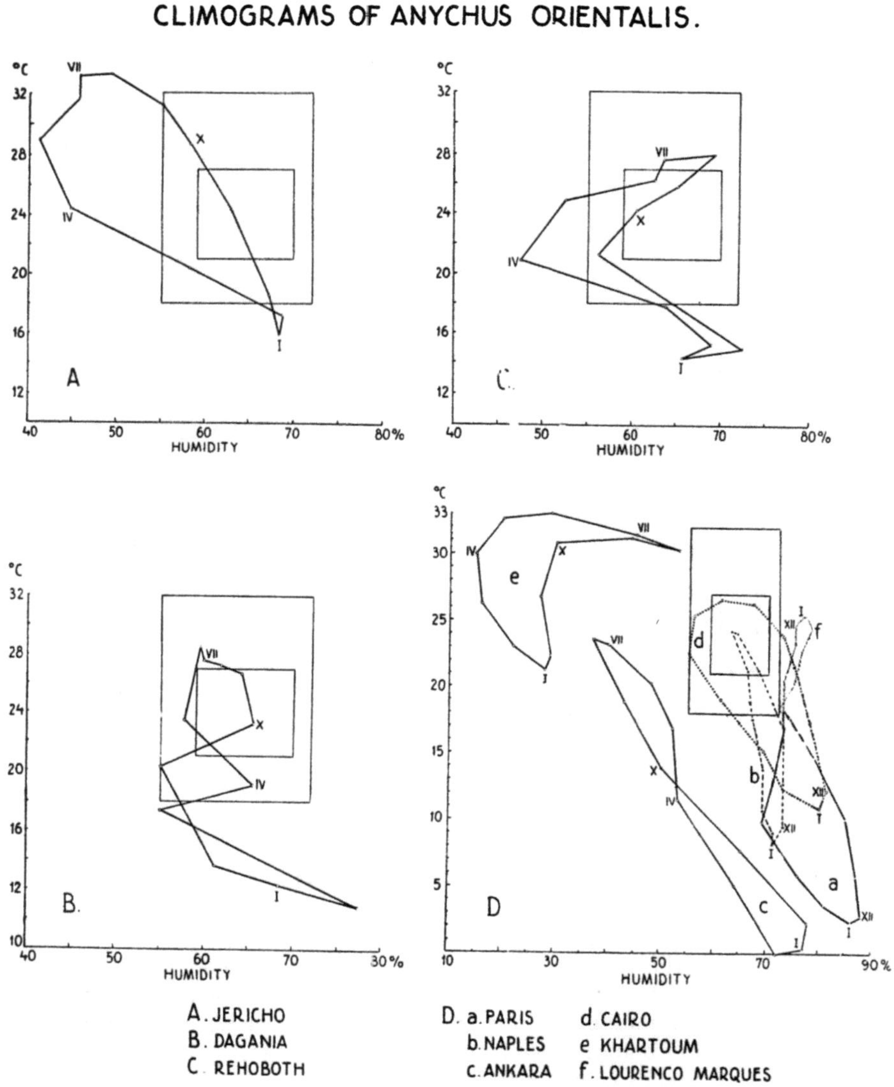

Fig. 136.
Eco-climograms of *Anychus orientalis Zach.* A. Jericho, B. Dagania, C. Rehoboth, D: a. Paris, b. Naples, c. Ankara, d. Cairo, e. Khartoum, f. Lourenco Marques (from Bodenheimer).

The optimal zone, compared with these climograms, explains the epidemiology of *A. orientalis* in the different regions of Palestine very satisfactorily. It explains the seasons of sudden increase, as well as the reasons for the sudden decrease.

Damage: A. orientalis is present on citrus trees all through the year. But its damage is apparent only during summer (beginning with July in the coastal plain, and from April onwards in the Jordan and Yesreel Valleys), when intensive multiplication sets in, and the wind helps its propagation.

A preliminary notice (Klein, 1934) contained a statement to the effect that in the Jordan and Yesreel Valleys the Oriental Red Spider attacks citrus trees of all ages, (beginning with seedlings and ending with old trees), while in the coastal plain it

Fig. 137.
Damage caused by ***Anychus orientalis Zach.*** to young orange trees. Heavy infestation in summer 1930, photographed in late January 1931 at Hedera. A. 2 years old tree, B. 5 years old.

prefers nurseries and young plantations. Now it is possible to affirm, that while in the early years of its propagation the Egyptian red mite did show this preference, lately trees of all ages have been indiscriminately attacked, including some groves thirty years old. Of course, damages in the nurseries and young plantations soon become apparent, for the leaves of the nursling turn chlorotic in a few weeks. But also fruit-bearing trees were severely affected: following the rapid drop of leaves which turned quickly chlorotic, twigs and branches dried up, and fruit, too, dropped.

XXXX. Damage of Oriental red spider on citrus. A. five years old tree showing drop of leaves and drying up of twigs following the heavy outbreak in autumn 1935; B. tree from the same plot, with the twigs, which were dried up, pruned and lying on the ground (to the left of the tree); C. seedling in nursery killed by the Oriental red spider; D. webs of the Oriental red spider on a citrus twig. (from Klein)

But it seems that these are not the only damages. If no control measures are undertaken at the proper time, blossoming, too, seems to be affected, and through it the crop of the following season.

In full grown trees the outer branches are mainly attacked. Drop of top foliage, whereby the top of the tree remained exposed ever since the autumn, was very frequent during the last years. In several cases it was necessary to repeat pruning of dead wood in the autumn following an attack of red mites.

A. orientalis sucks the sap of the leaves on their upper side, the sucked spots turning grey. In a short time the leaf becomes covered with a multitude of such gray spots, when it is reported to have turned chlorotic. The infested leaves weaken and finally drop. This drop of leaves affects the twig, which frequently dries up. In winter, bare nurslings can be observed in nurseries and young plantations, which have been neglected, namely, where no control measures were undertaken against this pest. Of course, the growth of such nurslings is hindered, expecially if the winter happens to be cold, and sometimes such plants shrivel up and die. From a certain grove in Samaria came a report that during October the red mite caused chlorotic colouring in fruit as well (oranges turned gray). It is true that later the fruit developed satisfactorily, and at picking time the colour of the oranges was normal.

Like other members of its order *A. orientalis* spins very fine, grey webs on the leaves and among them: sometimes, webs are spun from to twig; the mites move continuously along these webs, along which they lay their eggs.

One important fact should be put on record here. If irrigation is not in order in the autumn and winter months (in expectation of rains) leaf-shedding is intensified. This phenomenon is a result of the combined adverse action of two factors: A slight attack of red mites combined with partial thirst brings about the same result as a severe attack of red mite or lack of irrigation, each one in itself.

Control: In August—November 1934 and 1935 control experiments were carried out with a view of: 1. establishing the extent to which various materials affect Red Spiders, 2. investigating the difference between the more commonly applied means of control: spraying with emulsions of white oil and dusting with sulphur. These experiments were made on citrus nurslings protected by linen screens. Only nurslings two years old were chosen, and only such of them as were considerably and equally infested. Six nurslings were employed in each experiment, while every month the experiments were repeated with different nurslings. One week after the application of control measures, the nurslings were removed from the "tent" and placed for inspection purposes together with other nurslings, to which no control measures were applied. Observations continued in each case for a period of one month. These experiments, which were conducted by H. Z. Klein (1936), yielded the following results:

The results of control of A. orientalis on citrus nurslings.

Means of control	Origin of the material	The Red Spiders population **)					
		Before control	Following control after				
			2 days	a week	2 w	3 w	4 w
White oil 1.5%	U.S.A.	3	0	0	0	0	0
White oil 1%	U.S.A.	3	0	0	0	Single specimens	1
Rotenon powder	Germany	3	3	3	3	3	3
Rotenon emulsion 1% *)	Germany	3	2	2	2	3	3
Rotenon 0.5% + green soap 0.5%	Switzerland	3	1	2	2	2	3
Pure nicotine 0.15%	Germany	3	3	3	3	3	3
Rotenon 0.5% + green soap 1.5%	Switserland	3	single specimens	single specimens	1	2	2
Sulphate of Nicotine (40%) 0.35%	Great Britain	3	1	1	2	2	3
Essence of Pyrethrine 0.2% *)	Great Britain	3	3	3	3	3	3
Common yellow sulphur, powder	Italy	3	1	0	0	0	0
Fine yellow sulphur, powder	Italy	3	eggs 0	0	0	0	0
Black sulphur, powder	Italy	3	eggs 1	0	0	0	0
Spraying with water		3	3	3	3	3	3
Inspection (no control)		3	3	3	3	3	3

*) According to instructions issued by producers.
**) O = none, 1 = few, 2 = many, 3 = very many red mites.

,,Rotenon, whether as powder or emulsion, liquid Pyrethrine, solution of Nicotine in a concentration of 0.15% showed no effect at all on Red Spiders. A highly concentrated Rotenon emulsion, to which twice as much green soap was added and a 0.35% concentration of Sulphate of Nicotine (40%) reduced the Red Spider population immediately after control, but their effects soon wore off, and a fortnight later the number of Red Spiders increased again. Consequently, control with vegeto-alkaloids gave no satisfactory results. Moreover, it is, as a rule, rather expensive. Water washes (with a sprayer) never achieved any decrease of the population of the mites.

Spraying with emulsions of white-oil, killed almost 100%. A 1.5% concentration gave better results than a 1% concentration. The same complete success followed dusting with sulphur. Dusting with sulphur has been practiced these last six years in groves of the Jordan and Yesreel Valleys. Following the early spreading of the mite in the groves of the coastal plain, spraying with emulsions of white-oil was recommended, based on experience gained at California. Woglum (1932) used to recommend dusting with sulphur only at high temperatures. Parallel experiments carried out during the last two years prove that in this country dusting with sulphur brings about results equal to those achieved by spraying with emulsions of white-oil during all months of the year, in which control of the red mite is necessary. Even in cases of late control (November) dusting with sulphur achieved its purpose. Only an insignificant difference was found between the various kinds of sulphur put to trial (flowers of common sulphur, very fine sulphur, black sulphur), as far as their insecticidal properties were concerned. One tree was especially wetted and dusted with sulphur immediately afterwards, in order to investigate the influence of dusting after a strong dew. No difference whatsoever was apparent in any aspect of the results achieved by dusting on this tree as compared with other trees, which had not been wetted prior to the application of sulphur.

Especial interest was roused by experiments made, both in the laboratory as well as in nature, in order to investigate the effect of control measures on eggs of the mite. Always and everywhere the same results were achieved: spraying with a 1.5% concentration of white-oil emulsions destroyed all eggs, while application of sulphur did not affect them at all. But since the development of eggs in summer and autumn lasts only few days (3—8) while the effect of dusting with sulphur survives for a long enough period of time, all small larvae of the Red Spider (100%) died immediately after hatching.

XXXXI. Damage of Red Spiders on leaves: Above: *Epitetranychus althaeae* on leaves of beans and eggplants; below: *Anychus orientalis* on citrus leaves (from left to right: healthy, medium and heavily attacked leaf). (from Klein)

Control practice: Spraying: each method has its advantages and drawbacks. The drawbacks of spraying are: 1) it requires great practical experience in order to obtain good results, 2) investments in spraying machinery are generally higher in our country than those in dusting-machinery, 3) under local conditions spraying requires more time, and a larger number of workers, than dusting with sulphur. The advantages of spraying are the possibility to work during the major part of the day; then, again, dusting with sulphur is as yet impossible in dense groves, so that spraying is the only method of control available in these cases.

Dusting with sulphur: the great drawback of this method is that it can only be applied when there is no wind blowing, i.e. up to 9 or 10 a.m. Though dusting at night is quite common in this country, it is rather difficult, and does not always give good results. Generally speaking, dusting with sulphur is easier than spraying; then, again, dusting requires fewer workmen, a factor of major importance under conditions obtaining in this country. We have been asked many times already about the advisability of mixing lime with sulphur in order to reduce costs; McGregor has already emphasized some time ago (1931), that an admixture of lime is undesirable, for it does not reduce the burning properties of the mixture, while the deadly effect of sulphur is impaired.

Experiments and observations made during the last few years in groves of various regions prove, that early control, executed immediately on the appearance of the first mites (namely in April in the Jordan and Yesreel Valleys, in August in the Coastal Plain), gives best results, facilitates work, and saves expenses. If control is deferred till the leaves turn chlorotic, it is usually necessary to repeat the treatment at the end of three weeks. The number of sprayings or dustings required during the season varies from district to district, according to the severety of the attack and the local arrangements for control measures. In the Jordan and Yesreel Valleys it is usual to dust the trees with sulphur twice or three times during the season; in some cases up to five dustings were required. In the Coastal Plain there will be no necessity to apply control measures more than twice during the season, provided that control was undertaken in due time, and executed in proper order. In several cases a single dusting was found sufficient in this region.

One winter numerous red mites were observed there late in the season, for the weather was mild; they were especially numerous on lemon trees. The latter were dusted with sulphur in November, following which no Red Spiders were seen on them till the beginning of next summer, whereas on trees which were not dusted Red Spiders were present all though the winter. In this case late dusting, too, was undoubtedly very important.

Damages caused by spraying and dusting with sulphur have already been described in great detail (Klein, 1935). Only one phenomenon is to be especially noticed here: dusting with sulphur executed during the months of July-August in the Jordan Valley has caused in several cases burning of fruit and leaves on all sides of citrus trees. On the burned fruit appeared round, brown spots. This damage is influcted only during the hot summer months, and never at any other season. It is, therefore, most important to introduce early control, so that no repeated action be necessitated during the hot summer months.

Emphasis should be laid here again on damages caused by disorderly irrigation, damages which are usually credited to the attacks of pests, and more especially those of Red Spiders, or to the results of careless application of control methods.

It is as yet impossible to quote exact figures of the expenses involved by spraying or dusting with sulphur respectively. During the last few years prices of materials have changed radically; we have already witnessed the soaring of prices of sulphur during the summer from month to month. As a result of the enormous propagation of the Red Spider in our groves during the last few years, the demand for sulphur has risen so much, that this article was several times unobtainable for a month on end during the most burning season of control.

An example of dusting expenditure is cited below; it is based on figures obtained in control operations at a large grove extending over an area of several hundreds of dunams. A plot, containing trees five years old (after unbudded planting), was dusted with a mechanical duster at a rate of ten kgs. per dunam (42 trees, i.e. 240 grs. per tree); another plot containing trees four years old was dusted with eight kgs. per dunam (i.e. 190 grs. per tree). The work, with a mechanical duster over an area of 150 dunams containing trees five years old, was carried out in nearly nine days, one workman operating the duster and one driving the draft-animal."

B. THE COMMON RED MITE (or SPIDER).

(*Epitetranychus althaeae* v. Hanst.; Tetranychidae).

Description: Adult female: abdomen light red-brown with darker spots along the sides; cephalothorax and legs yellowish, tips of the legs reddish. End of abdomen broadly rounded. 0.45 × 0.30 mm, legs 0.25 mm long.

Adult male: abdomen yellow, cephalothorax and legs yellowish, the tips of the legs reddish. End of abdomen pointed. 0.35 × 0.20 mm, legs 0.25 mm long.

The egg is round, light-brown, 0.1 mm in diameter.

The 6-legged larva is reddish yellow with cephalothorax and legs lighter, almost white. 0.15 × 0.10 mm, legs 0.10 mm long.

The nymphs are 8-legged. Body of the first nymph yellowish with brown spots on the side, tips of the legs reddish. 0.25 × 0.17 mm, legs 0.10 mm long.

Abdomen of the second nymph light red-brown with darker spots on the sides; cephalothorax and legs yellowish, tips of the legs reddish. 0.30 × 0.20 mm, legs 0.20 mm long.

In the three latent stages (Proto-, Deuto- and Telo-nymphs) which alternate with the three active stages of development, the abdomen is yellowish, the cephalothorax lighter.

The coloring of the Common Red Spider changes somewhat with the host. The above given colors agree with our breedings on citrus (Klein, 1936).

Distribution: The species is an extreme ubiquist. It occurs throughout Europe, America and Asia except the extreme north, resp. south, in Australia and at least in N. and S. Africa. On mountains it reaches to a height of up to 2300—2600 m. It develops in the tropics, as well as in a moderate climate.

In Palestine it occurs in numbers in all regions and abounds (Klein, 1936):

in the coastalzone	all year round	
in inland valleys	in spring and autumn	in small number
in the hills	in summer	throughout the year

Life-history: The definite color of the adults passes from greyishyellow to yellow, yellowish pink, pink, light red-brown to the dark redbrown tinge of aged individuals in the female; from grey over yellow to yellowish-brown in the aged specimens in the male. Red ,,winter females" were observed in February on *Solanum nigrum*, which turned to the usual light red brown color in the beginning of March (Klein, 1936).

Copula starts immediately after the adults have hatched. It resembles closely that of *A. orientalis* and lasts 2—3 minutes in summer, up to 10 minutes in winter. The courting (running about the female, etc.) lasts longer than the copula. Klein (1936) frequently observed males which had just moulted waiting near a female telonymph or playing around it. The males quite often copulate with a second female, and even with more. The sex-ratio changed with the seasons:

April/September	65% females	21—26° C	50—69% rel. humidity
March	77	17.5	56
January/February	92	13—14.5	67

The generally high number of females is still higher during winter.

The eggs grow quickly light brown in the summer. In the winter they change colours from transparent violet, over grey-violet and grey to light brown. Shortly before hatching the eggs turn white.

Generally, oviposition takes place in nature on the under side of the leaf, mainly along its ribs. In cases of severe infestations, however, eggs are found also on the upper surface of the leaves and on the webs of *E. althaeae*. This behaviour is the reverse of that of *Anychus orientalis* which prefers the upper surface of the leaves.

The number of eggs laid by one female during 24 hours never exceeded 3 on citrus leaves, 7 on vegetables and *Ricinus*. Mc Gregor (1917) reports 6—9 from cotton.

The total number of eggs per female fluctuates with the seasons (Klein, 1936):

month	eggs per female	average temperature	average rel. humidity
I	6	12.7° C	66%
II	3	13.2	69
III	15	16.0	64
IV	12	18.3	48
V	12	22.8	54
VI	10	24.8	60
VII	11	25.9	64
VIII	9	26.0	69
IX	8	24.4	64
X	8	22.3	60

On citrus leaves the maximal number of eggs per female fluctuates, as follows:

	spring	summer	autumn	winter
nature	**27**	20	15	8
laboratory	39	23	**53**	20

The maximal egg rate was obtained between 16—26° C and 52—69% rel.humidity. Only very few eggs are laid in the winter. The summer climate of the coastal plain surpasses slightly the optimal conditions.

Oviposition was higher on all other tested host-plants than on citrus. On beans it was from twice to three times as high. Beside this range conditioned by different hosts, climatic conditions are very important on one and the same host: Klein quotes 50—100 eggs from cotton in U.S.A. (Mc Gregor), 185 in Uzbekistan (Viontkovskii) and 217 in Tashkent (Tibilova).

The seasonal mortality of the development stages is as follows (Klein, 1936):

temperature	rel. humidity	egg mortality	larval/nymphal mortality	Total mortality
25—26° C	63—69%	10.2%	32.3%	39.2%
24—25	60—63	13.9	35.3	44.3
22—23	53—60	21.7	22.9	39.6
17—19	47—50	33.6	25.5	50.5
13—16	64—69	43.6	25.6	58.0

The egg is fairly sensitive to climatic conditions of the environment. Its mortality in winter is three times as high as in summer. The low air humidity in spring seems to influence the eggs also adversely. Above 22° C, and at medium air humidity 80—90% of the eggs hatch successfully. The larvae and nymphs are considerably less influenced by climate, about 65% reaching maturity above and 75% below 22° C.

These results show that on citrus about half of the individuals perish prematurely in winter and spring, whereas their percentage is somewhat higher in summer and autumn. Citrus, however, is an unfavorable host for this red mite. On more suitable hosts the survival is higher. Total mortality on leaves of beans in the summer was 25.8 %, on leaves of eggplants 29.9 % only.

The experiments on the scale of activity performed by Miss Ben-Ami yielded the following results:

scale	n	1	2	3	4	5	6	7	8
adult females	32	2.7	3.6	8.4	15.9—37.2	39.4	45.3	49.1	51.7
larvae.	8	(5.0)	—	7.3	21.6—37.0	40.3	46.7	49.3	51.9

The range of normal activity extending from 16—37° C is exceptionally wide (21° C.). This feature agrees very well with the ubiquistic distribution, the extreme polyphagy and the general abundance of this extremely polyoecous species. The tolerance towards heat as well as cold is greater than that of *Anychus*.

Temperature and development; number of annual generations: Klein (1936) has calculated the following constants of the hyperbola:

development threshold (c) = 8° C
thermal constant (th.c.) = 170 days-degrees.

Development of both sexes was equally long and lasted

egg		later development stages	
24—26° C	3.1 days	26—27° C	7.4 days
17—23	4.9	23—25	8.9
12—16	9.4	22—23	9.3
		16—17	12.4
		13	17.3

The preoviposition period lasted in

summer (VI/IX)	24—26° C	1 day
spring, autumn	20—23	2
winter	13—19	3—5

The longevity of the adults is equal in both sexes. It was

summer	up to	15 days
spring		19
autumn		17
winter		31.

We shall now calculate the 3 hyperbolae: A for quickest total development; B for the epidemiological average; C for the latest developing individuals of every generation, based on these breedings on citrus under natural conditions:

		A	B	C
development threshold		9.4	7.3	6.6 C
thermal constant.		172	310	460 days-degrees
development duration at	30° C	8.3	13.6	19.6 days
	25	11.0	17.5	25.0
	20	15.9	24.3	34.3
	15	30.7	40.3	54.7

The difference between C and A yields the average adult life span of the individual.

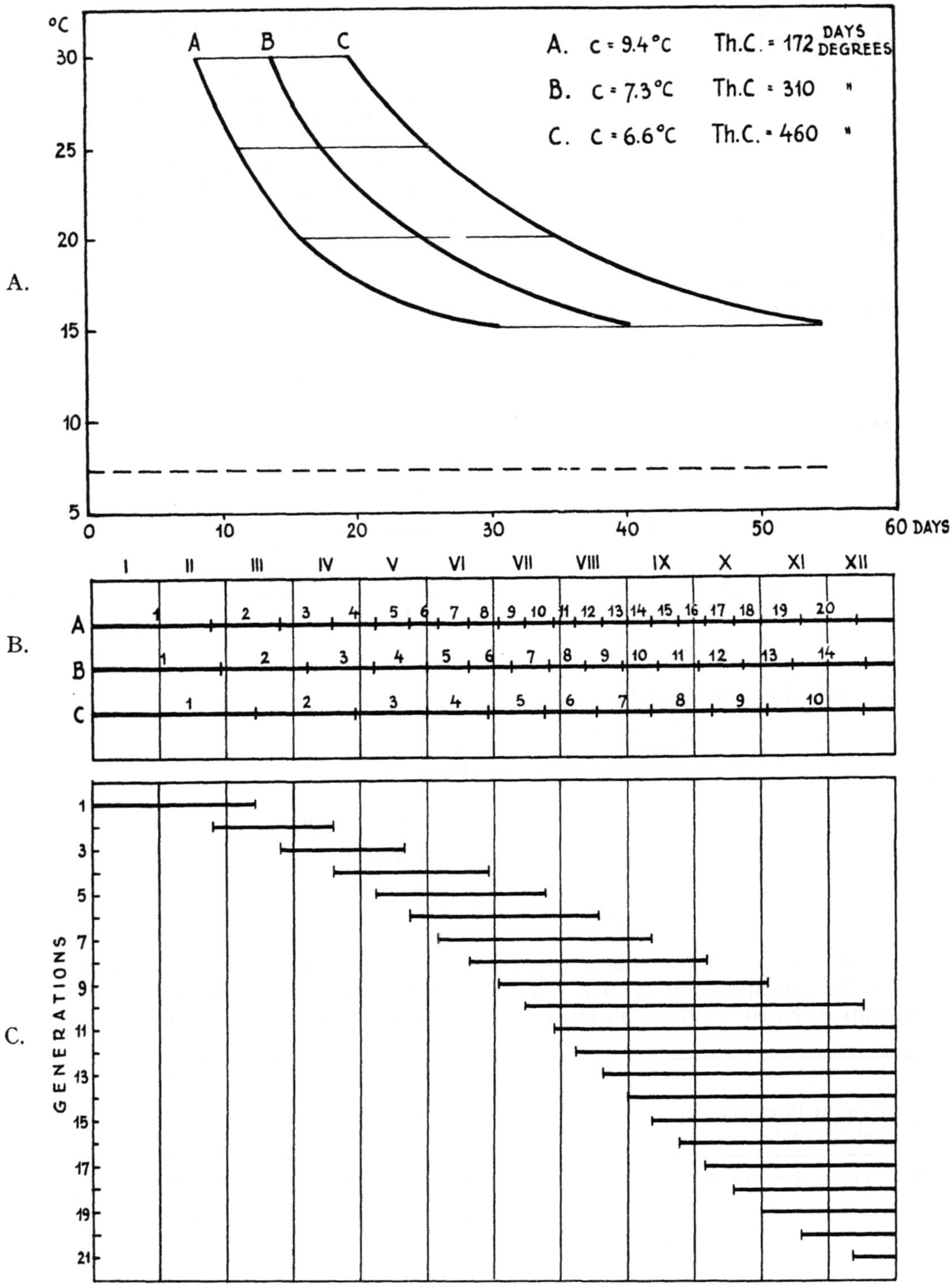

Fig. 138.
Life history of the Common Red Mite (*Epitetranychus althaeae* Hanst.) at Tel Aviv. A. The hyperbolae; B. the life cycle as calculated from the hyperbolae; C. the sequence and overlapping of generations.

A series of breedings agrees rather well with the calculations (Klein, 1936):

Generation	date	development	temperature	rel. humidity
1	16.I.—10.II	26 days	12.5° C	68%
2	13.II.—12.III.	27	14.5	67
3	16.III.—13.IV.	28	16.8	61
4	17.IV—1.V.	14	21.1	53
5	5.V.—17.V.	12	22.2	55
6	19.V.—2.VI.	14	23.0	58
7	4.VI.—16.VI.	12	24.1	58
8	18.VI.—30.VI.	12	24.8	64
9	2.VII.—13.VII.	11	25.3	61
10	14.VII.—24.VII.	10	25.6	67
11	25.VII.—3.VIII.	9	26.8	64
12	4.VIII.—14.VIII.	10	26.6	72
13	14.VIII.—24.VIII.	10	26.4	69
14	25.VIII.—4.IX.	10	25.2	66
15	5.IX.—15.IX.	10	23.0	66
16	18.IX.—28.!X.	10	23.8	61
17	1.X.—11.X.	10	22.7	62
18	13.X.—27.X.	14	22.2	58
19	1.XI.—18.XI.	17	20.0	59
20	21.XI.—23.XII.	22	16.0	62
21	16.XII.—13.I.	28	12.5	64

The breeding of generations no. 19—21 did not succeed under natural conditions. The values have therefore been calculated for these generations. No hibernation occurs in nature, but development of all stages simply slackens.

Laboratory breedings differ somewhat in length from those under natural conditions. In the laboratory, relative humidity, temperature (about 2—3° C) are higher, whereas the temperature range is smaller. Sun radiation is excluded. The most rapid development was:

month	II	III	IV	V	VI	VII	VIII	IX	X	XI	XII
in nature.	27	18	14	12	12	9	10	10	10	(17)	(22)
in laboratory . . .	28	19	18	9	7	9	9	9	10	16	35

Development in the laboratory is generally shorter, except in April when the hamsin winds are more strongly felt in nature and in winter when the lack of sun-radiation is more strongly felt in the laboratory.

The number of annual generations on citrus throughout Palestine would thus be, as follows:

	A	B	C
Tel Aviv	21.0	14.1	10.1
Jerusalem	13.7	9.8	7.6
Tiberias	25.9	16.7	11.8

The shortest development (A) would permit the following number of annual generations at different places:

London	4.9	Palermo	16.4
Berlin.	5.0	Alexandria	22.7
Paris	7.7	Cairo.	24.6
Nice	12.2	Sierra Leone . . .	35.6
Naples	13.6	Honolulu	29.1

However, all these calculations lose greatly in value by the fact that the host influences the length of development considerably. Furthermore, the small differences between the breedings in the laboratory and the contemporaneous breedings under natural conditions cause remarkable differences in the results of both series. These differences are not restricted to length of development alone (I), but refer also to longevity (II), maximal oviposition (III), etc.:

		Host				
		citrus	beans	eggplants	cucumbers	Ricinus
I	nature	9	8	8	10	8 days
	laboratory	7	7	7	7	—
II	nature	15	24	18	14	21 days
	laboratory	22	28	20	—	—
III	nature	20	45	29	25	31 eggs per female
	laboratory	23	71	37	25	—

Host-relations: E. althaeae is extremely polyphagous. Mc Gregor (1917) records 183 host species from N. America, Zacher (1928) about 150 from Germany. The 66 host plants observed so far in Palestine are certainly only a fraction of the number of actual hosts. The following list records the genera only:

Plant family	*Plant genera*
Amarantaceae	*Amarantus*
Araceae	*Zantedeschia*
Bignoniaceae	*Bignonia*
Borraginaceae	*Heliotropium*
Chenopodiaceae	*Beta, Chenopodium*
Compositae	*Sonchus, Cichorium, Scolymus, Silybum, Centaurea Lactuca, Nothobasis.*
Convolvulaceae	*Convolvulus, Ipomaea*
Cruciferae	*Diplotaxis, Eruca, Raphistrum, Raphanus, Brassica, Matthiola.*
Cucurbitaceae	*Cucumis, Citrullus, Cucurbita, Luffa*
Ebenaceae	*Diospyrus*
Euphorbiaceae	*Ricinus, Mercurialis, Acalypha*
Geraniaceae	*Pelargonium*
Labiatae	*Molucella*
Leguminosae	*Trifolium, Medicago, Lathyrus, Phaseolus, Spartium, Ceratonia, Wistaria, Vicia, Pisum, Vigna,*
Loganiaceae	*Buddleya*
Malvaceae	*Hibiscus, Malva*
Moraceae	*Morus, Ficus*
Oleaceae	*Jasminum*
Papaveraceae	*Papaver*
Pittosporaceae	*Pittosporum*
Rosaceae	*Pyrus, Prunus, Rosa, Rubus*
Rutaceae	*Citrus*
Solanaceae	*Solanum, Capsicum, Datura, Petunia*
Umbelliferae	*Ammi, Daucus, Bupleurum*
Urticaceae	*Urtica*
Vitaceae	*Vitis*

Food habits and Damage: "The damage inflicted by the Common Red Spider to citrus is exactly similar to that caused by the Oriental Red Spider. The attack begins near the central vein or other veins on the underside of the leaves. With the help of its mouth parts the Red Spider tears the cells of the leaf open and sucks in the sap of the plant-tissue. As a consequence of this destructive action the leaves turn yellow on their upper sides. They gradually turn a yellowish-grey and drop. In young leaves (especially in wild plants) the edges turn up slightly. In a case of a severe infestation the mites migrate sometimes also to the upper sides of the leaves. The common red mites weave very thin webs, wherever they settle, on which webs they move and lay their eggs. Frequently, such webs cover the whole surface of the leaf and small twigs as well.

Of the 39 cultivated plants, which are subject to the attack of the common red mite, six suffer greatly; these are, in sequence of the importance of injury: beans, egg-plants, ricinus, citrus, pears, *Wistaria*. Six other plants are frequently injured to a greater or lesser extent: tomatoes, cucumbers, squash, *Lathyrus*, *Bignonia* and *Ipomoea*. The other 25 cultivated plants have not so far undergone any injury of economic nature.

Several wild plants are severely attacked by this pest; more especially the black night-shade, Solanum nigrum, which is to be found in large numbers all over the country. Infestation of these wild plants is economically important, since the mite migrates from them to cultivated plants.

The months of heaviest infestation, when most damage is inflicted on the attacked plants, are:

Host	Coastal zone	Inland valleys	Mountains
Citrus	IV—V	—	—
Leguminosae (beans, etc.)	V—X	—	—
Solanaceae (eggplants, etc.)	IV—IX	II—VI, X	—
Cucurbitaceae (cucumbers, etc.)	VI—VIII	IV	—
Ricinus.	I—IX	—	—
Ornamental plants	I—VII, X—XII	III—IV	VIII—IX

The mites continue their activity also during other months of the year, but at such times their attack is not liable to arrest the development of the plants.

It should be added, that every time trees were attacked, (citrus, pears, mulberry, *Diospyrus kaki*, etc.) these were either nurslings or young trees not older than three to four years. The number of host trees in this country is very small and amounts only to 9.5%, out of the total number of hosts.

As was mentioned above, the *E. althaeae* is not to be found on citrus trees except in the coastal zone, and only in spring. During the remainder of the year it is very common on weeds in citrus groves, but does not migrate to citrus leaves. It is quite common in summer and autumn to find only *Anychus* on citrus trees, and nearby the common mite on weeds. Neither of the two would leave its habitat and migrate to the plants occupied by the other.

Why is *E. althaeae* to be found on citrus trees in spring only? In order to attempt a solution of this problem we have to recapitulate at first some of the statements already made above. The most rapid general development of the common red mite sets-in following the rise of temperature above 23° C. Minimal mortality takes place at a temperature of 22—26° C. Maximal oviposition on citrus leaves occurs in March at a temperature of 16° C. Normal temperature in April fluctuates between 17° C and 21° C, in May between 22° C and 23° C.

Till the end of winter every grove is rich in wild plant-life. The common red mite is then to be found in large numbers on various weeds and wild plants. From the beginning of February, and up to the middle of March, ploughing destroys everywhere in Palestine all wild herbs. That part of the population of *E. althaeae* that is not destroyed by ploughing, migrates out of sheer necessity (for lack of their regular plant food) onto the young citrus trees. During this month (March) the fertility of the females is very high. It is true that during March the development of the mite is still slow, but it quickens in April, and still more so in May. Consequently, the population of *E. althaeae* greatly increases on citrus trees during the months of April—May. Just at this season (April/May) the irrigation of groves begins; following each watering wild plants begin to grow anew; so much so, that all through summer and autumn wild plant life is rather plentiful in the groves. As soon as these wild plants begin to grow, the common red mite appears on them. On ricinus trees that were grown as wind-breaks in and round the grove, it developed in quite considerable numbers all the year round. There can be no doubt that during summer months a re-emigration of mites takes place from citrus trees back to the grass-plants. Two more facts help to explain this behaviour of the red mites: the mortality of the Red Spiders on citrus leaves in summer reaches up to 50%, while at the same time on grasses, like beans, it does amount to more than 25% or 30%; the oviposition per female on citrus leaves decreases in summer, it decreases still further in autumn, while on other plants, the oviposition per female is much higher (on beans, e.g. up to 125% and more). It had been already stated above, that the common red mite is never found on trees (deciduous fruit trees, e.g.) during summer months, between July and September. Also systematic observations of Miss Nadel (1936) in light soils of Rehoboth have confirmed these facts, for Miss Nadel, too, found that wild plants are least in number in this vicinity during April and May. This season is an intermediate period, during which plant life is much reduced. Beginning with June, and more especially with July, and up to October wild plant life increases again.

As already mentioned, the common red mite had never been found on citrus in inland valleys (Yesreel and Jordan); apparently, because there are no continuous and large citrus areas in those regions as in the coastal zone. Then again, in inland valleys there are in the nearest proximity of the groves large uncultivated areas growing an abundance of suitable food supply for the mite many months in the year (except during the last months of summer and autumn). Also, the inland valleys grow extensive areas of vegetables. Whenever *E. althaeae* appeared on fruit trees (pears) in the Jordan Valley, they attacked only one or a few single trees.

From the extensive literature should be mentioned only Zacher's (1929) statement that the common red mite settles on vines just on spots exposed to the sun. Of course, citrus trees, like all trees in general, are more shady than grasses. In Florida, Wilson (1931) found, on asparagus, the maximal Red Spider population in dry and hot weather which resembles ours at the time of hot winds (khamsins) in April and May, when Red Spiders multiply here greatly and migrate onto citrus trees.

The same author observed also that rain largely decreases the numbers of Red Spiders, even up to 50%. In Palestine, the decreases caused by humidity and washing off is combined with drop of temperature, reducing thus greatly the damage during winter.

It has already been mentioned above that the Common Red Spider greatly increases in spring and rapidly damages citrus trees to a large extent. Still more important are the injury and economic damages inflicted in summer on vegetables, such as beans, eggplants, etc. Very frequently, large plots of beans can be observed, which have turned white as a result of the sucking of the mites. In such plots any further development of the plants is arrested a short time after the beginning of the attack, and they do not attain fruition. Eggplant seedlings, too, that have been attacked, degenerate and bear but shrivelled fruit. Ornamental plants, and creepers with thin leaves in particular, drop their leaves very rapidly. Such plants, that stand during the summer fully exposed to the rays of the scorching sun, suffer greatly; and even multi-annual plants degenerate and fail to recover fully during the season following infestation, unless timely control measures against the red mite have been instituted." (Klein, 1936).

The author rather wonders as to how far the migrations mentioned by Klein and by other students of *E. althaeae* are real migrations. We believe that transmission from plant to plant by active migration occurs relatively rarely, transport taking place mainly through the wind. Whatmore, the extraordinary potential of increase of the species enables the building up of a high population from a few initial individuals in a surprisingly short time, provided environmental conditions are favorable. It is therefore even improbable that transport by wind is of greater importance than the fluctuations of the population on the same tree. The assumations of Klein on the negative correlation between food-supply (other than citrus) and the development of *E. althaeae* on citrus, must therefore be accepted with some reserve, until more detailed observations are available. Mc Gregor, (1917) on the other hand, observed a female of *E. althaeae* running over a smooth surface at a rate of 125 mm per minute (i.e. 170 meters in 24 hours). The possibility of minor migrations from host to host cannot be denied.

Control: Control measures are the same as against *Anychus:* by dusting with a fine and dry sulphur powder or by spraying with a white oil emulsion of 1.5 %. Control measures should be mainly directed against the under sides of the leaves, where the mites nestle. In the case of citrus and deciduous fruit, trees control should be effected in spring immediately on the appearance of the first Red Spiders, in order to prevent a drop of leaves. Very rapid action is required also in the case of summer vegetables, for the females are very fertile on them at this season of the year; otherwise there is a risk of the loss of fruit. On vegetables and ornamental plants, control measures should usually be repeated twice, and in some cases several times.

It has been observed in some countries that certain varieties are more severely attacked, while other varieties of the same plant are more or less immune against the infestation by the common red mitz. So far no observations of such race immunity have been made in Palestine.

Ecological comparison of both species: An ecological comparison of the two species of Red Spiders, has been given by Klein (1938). Some details of the respective life histories on citrus leaves in the coastal region of Palestine have been compiled, as follows:

	The Oriental mite		The common mite
Number of generations in the year		18	21
Duration of development of one generation	Summer	8—10	9—12 days
	Winter	68	22—28 ,,
% females	Spring to autumn	71%	67—75%
	Winter	56%	92%
Duration of life of adult	Summer	12	15 days
	Spring to autumn	14—18	17—19 ,,
	Winter	21	31 ,,
Maximum number of eggs per female	Summer	35	20
	Autumn	23	12
	Winter	18	8
	Spring	23	27
General mortality rate	Summer	20—40%	42—49%
	Spring to autumn	50—75%	
	Winter above 75%		59—60%
Activity trials:			
Freezing sets in at		6.9° C	2.7° C
Normal activity		18.0—33.8° C	15.9—37.2° C
Death caused by heat at		47.4° C	51.7° C
Zero-temperature of development		10° C	8° C
Thermal constant		180° C	170° C
Number of known plant hosts in Palestine		6	63
Regions of occurence in Palestine	Jericho, Galilean Sea, Valley of Yezreel, Coastal region		All parts of Palestine
Regions of occurence on citrus trees	Jericho, Galilean Sea, Valley of Yezreel, Coastal region		Coastal region
Main season of infestation of citrus:			
Jericho,	September—March		—
Valley of Yezreel, Galilean Sea	April—December		—
Coastal region	August—November		April—May

,,It is clear from the above table that the common species is cosmopolitan, whereas the Oriental species is a subtropic element*. In this connection, the following main differences may be noted: Whereas in the case of the *Anychus*, the development of a generation in winter takes over two months, in the case of the common species the corresponding development is completed within 3—4 weeks. The percentage of winter females of the former species is much smaller; its duration of life is shorter. *Anychus* shows a maximum number of eggs per female in summer, *Epitetranychus* shows a maximum in spring. The seasonal differences in the mortality rate of the common species are not marked, but the seasonal fluctuations of the mortality rate in the case of *Anychus* are tremendous, the value climbing from 20—40% in summer to over 75% in winter. In the activity trials, several differences were marked: — Females of *Anychus* entered a state of freezing already at 7° C; they of the common species did so only below 3° C. Normal activity in females of the common species continued over a greater range (5° C) than was observed in the species. The temperature of death caused by heat was more than 4° lower, the temperature of zero-development 2° C higher in *Anychus* species than the corresponding values for *Epitetranychus*.

The host plants of the Oriental red mite in Palestine, as far as they are known, are either

*) if it is not tropic.

subtropical or tropical elements. Of the hosts of the common species in Palestine, on the other hand only 10% are subtropical whereas fully 90% are temperate zone plants. The common species is spread throughout the country; the oriental species is to be found as yet, only in the regions of citrus cultivation and is therefore altogether absent in the hills. The time of the appearance of the pest on citrus trees in the different regions and seasons varies with the local climate: In a hot dry environment, such as in Jericho, the pest appears in the winter months; in a warm environment such as is to be found in the Galilean Sea district and in the Valley of Yezreel, the season of infestation is from the spring to the beginning of winter; in the coastal region, it extends from the end of summer to the fall. It has been found that two factors hamper the succesful increase of the Egyptian red mite: A temperature of less than 18° C, and a relative humidity of less than 55%.

The common species appears on citrus only in the coastal region and only in the spring months. This peculiarity of behaviour is attributable to the fact that in the various inland regions of the country, numerous plant hosts, both cultivated and wild, are arailable throughout the year,

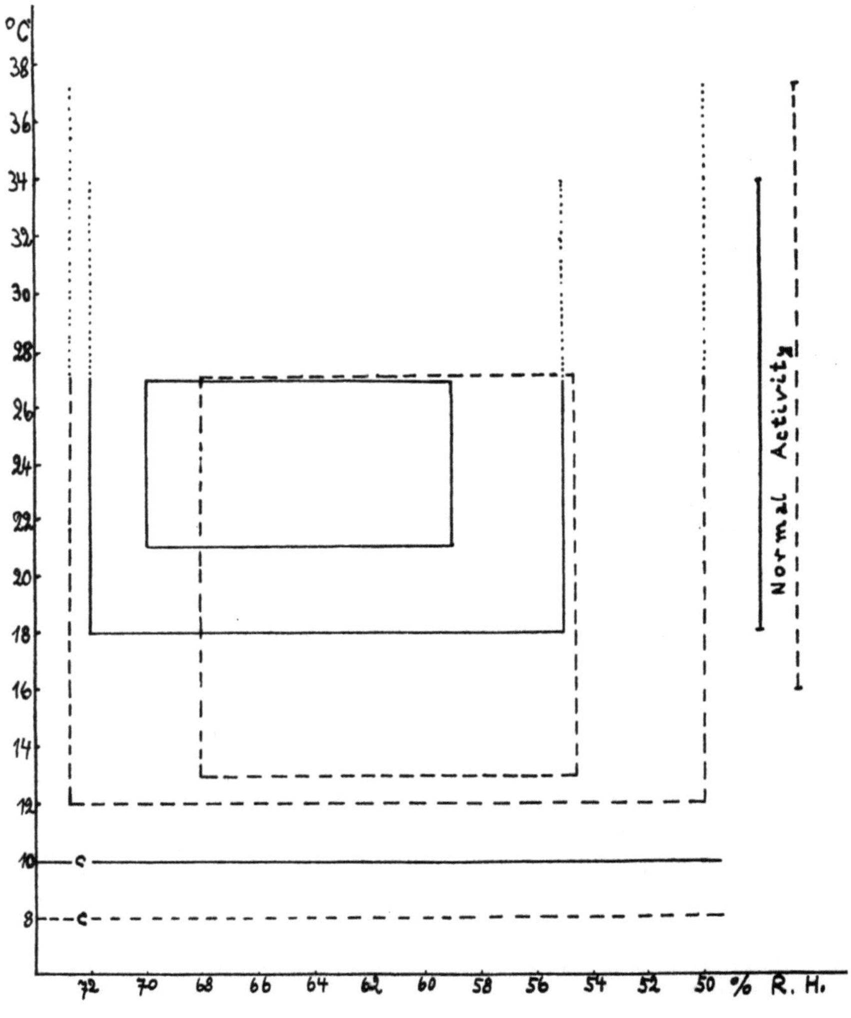

Fig. 139.
Ecoclimograms of red mites: *Anychus orientalis* (full line) and *Epitetranychus althaeae* (broken line). Closed squares: Optimal conditions, open square: tolerable conditions, *c*: threshold of development (from Klein).

whereas in the coastal region where the groves cover large continous areas, only citrus hosts remain in spring when the weeds are destroyed. This circumstance compels the common species to attack the citrus trees, which, however, it leaves immediately the weeds develop after the first irrigation. Some years ago, the pest was common in the spring season. Now, however, that the trees have grown, and neither young groves nor seedbeds are available, the occurrence of the pest is rare, and the species has lost its economic importance for the citrus industry.

The conditions of optimum development (temperature and air humidity) of our two species of red mites are schematically represented in the diagram attached in the form of squares. The range of the conditions still favourable to development and increase indicated by propagation experiments on citrus leaves is also indicated. The limits of normal activity as determined in activity trials is annotated. The marked difference between the two species is here particularly pronounced. The area of optimum conditions for the cosmopolitan common species is three times that of the subtropical *Anychus*. The limits of nearly optimum conditions for increase are a temperature of 12° C and a humidity of 50% in the case of the common red mite, i.e. 6° C, and 5% higher than is the case in the *Anychus*." (Klein 1938).

C. NATURAL ENEMIES OF RED MITES IN PALESTINE.

THE MITE-FEEDING LADYBEETLE

(*Scymnus* (*Stethorus*) *punctillum* Wse., Coccinellidae).

Description: The small *adult beetle* is subcircular. It is black, with yellow-brown antennae mouth and legs. The elytra are covered with thin, light coloured hairs, which are arranged longitudinally. Distance between eyes not much greater than diameter of each eye. 1.2—1.5 mm long, 1.0—1.2 mm broad.

The *larva* is brownish grey, covered with thin, light coloured hair. Each of the eight first abdominal segments has 6 whorls of bristles. Up to 2 mm long and 0.5 mm broad.

The *pupa* is brown, covered with thin, light coloured hair. It is attached to the substrat by its posterior extremities. 1.0—1.25 mm long; 0.6—0.7 mm broad.

Biology: S. punctillum is widely distributed within the Palaearctic region. Klein (1936) met this species quite generally feeding on red mites in the coastal plain. It abounds in March/May and October/December. In summer and winter only scattered individuals are met with. In spring it preys almost exclusively on the common red mite (on weeds, vegetables, or ornamental plants), whereas in autumn it is found on both the common and Egyptian red mites. Development takes about 20 days in April/May, but 26 days in November/December (egg 5, larva 14, pupa 7 days). In the laboratory the adult beetles lived up to 21 days. They oviposit on leaves near the eggs of their prey. Observations on the food-consumption of the adults showed a daily consumption (average of 7 days observation) in spring of 22.8 (15.1—29.4), in autumn 21.5 (7.8—41.6) females of the mites. The larvae apparently prefer the eggs of the mites and therefore devour still more host individuals.

Klein correctly considers the biological control of red mites by *Scymnus* as insignificant despite their high feeding index, especially so, as far as *Anychus* is concerned, which increases in summer when *Scymnus* is absent in nature.

Individual breedings over some generations have not been made, but *S. punctillum* seems to have 2 annual generations in Palestine and the adult beetles estivate resp. hibernate before oviposition. *S. punctillum* is more frequently found together with *Epitetranychus* than with *Anychus*. In March and April and October/December it is met on all plants severely attacked by the common red mite. Its absence in the summer proves that it hides during that season in Palestine. It is of special interest to ascertain where *S. punctillum* passes the summer; the possibility of migrations to the hills should not be overlooked.

Other natural enemies of red mites in Palestine include some species of thrips and spiders, which are still undetermined. Not one of them seems to be of economic importance. The same is probably true of a still undetermined predacious Cecidomyid, which has been observed occasionally.

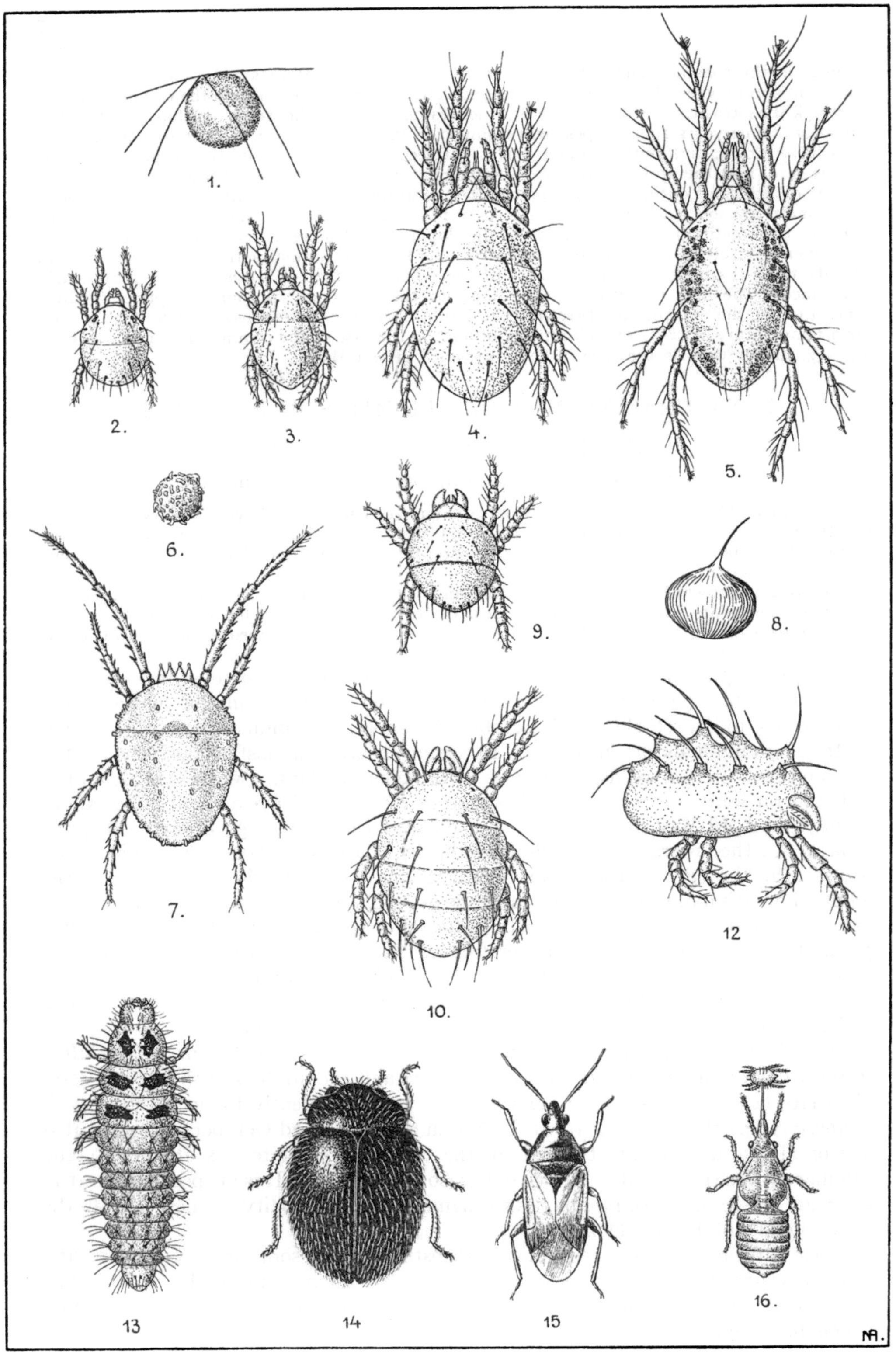

XXXXII. Other Red Spiders injurious to citrus in Palestine, and their predators. 1—5) *Epitetranychus althaeae* v. Hamst.: 1) egg, 2) larva, 3) protonymph, 4) deutonymph, 5) adult female; 6) and 7) *Bryobia praetiosa* Koch: 6) egg, 7) adult female; 8—12) *Paratetranychus pilosus* C. & F.: 8) egg, 9) larva, 10 and 12) adult female from above and in profile; 13—14) *Scymnus punctillum* Wse: 13) larva, 14) beetle; 15) and 16) *Triphleps minuta* L.: 15) adult bug, 16) *Triphleps*-nymph feeding on red mite. (partly from McGregor, Newcomer & Yothers).

The bug *Triphleps minutus* L. (Anthocor.) was met feeding on red mites. This small bug is black, shiny, delicately haired; hemielytra testaceous; in their distal part piceous; legs yellow. 2.2—2.5 mm long. It is a polyphagous predator living on red mites, as well as on aphids, coccids, etc.

References.

M. André, Rev. pathol. veget. et entom. agric. 1933, pp. 8—25.
F. S. Bodenheimer, Die Schaedlingsfauna Palaestinas, 1930. Berlin. p. 165.
H. E. Ewing, Jl. Entom. and Zool; Claremont, Cal. VI, 121—132. 1914. (Rev. Appl. Ent. III, 13. 1915).
H. Z. Klein, Hassadeh, **15,** 1934, pp. **8,** 1935, pp.
H. Z. Klein, Hadar **9,** 1936. pp. 107—111, 126—132, 195—199.
H. Z. Klein, Bull. Ent. Res. **29.** 1938. p. 37—40.
E. A. Mc Gregor, Calif. Citrograph, 1931, p. 358.
E. A. Mc Gregor, U.S.A. Dept. of Agric. Farmer's Bull. 831, 1917 and Revis. 1934.
Y. A. Piontkovskii, Sc. Res. Inst. Cott. Cult. Moscow, 1932 (Rev. Appl. Ent. XX, 599, 1932).
T. Sayed, Bull. Soc. Fouad Ier Ent. **26**. 1942. p. 125—127; idem, ibidem **30,** 1946. p. 143—154.
Tibilova, Bull. Sci. Res. Inst. Cott. Cult. Moscow, No. 1, 1932 (Rev. Appl. Ent. XX, 619. 1932).
J. W. Wilson, Univ. of Florida, Agric. Expt. Sta. Bull. 234, 1931.
R. S. Woglum, Handbook of Citrus Insect Control for 1932, Los Angeles.
E. Zacher, Zool. Anzeiger, 1936 (Anotation not verified! Bd.)
E. Zacher, Mitt. Biol. Reichsanst., Nr. 18, 121—130. 1920.
E. Zacher, Nachrichtenbl. f.d. deutsch. Pfl. Schutzdienst, Nr. 11, 1928.
E. Zacher, Nachrichtenbl. f. d. deutsch. Pfl. Schutzdienst, Nr. 2, 1929.

D. OTHER MITES.

a. THE CLOVER MITE (*Bryobia praetiosa* Koch; Tetranychidae).

Description: Adult female reddish brown with a green tinge. First pair of legs greatly elongated. Body flattened. Abdomen broadly rounded posteriorly. Male slightly smaller and with abdomen pointed posteriorly. Female 0.6—0.8 mm long. 4 lobi protrude from the notum and cover the mouth from above.

The egg is deep red, with a yellow tinge, round.

The 6-legged larva is bright red. Body almost globular. 0.3 mm long.

The two nymphal stages are 8-legged. Color varies.

Distribution: Ubiquist (Europe, America, Africa, Australia).

In Palestine observed once together with *Anychus orientalis* on citrus at Kinnereth. Its life history here is unknown.

Life-history: B. praetiosa is a very polyphagous species, injurious in other regions to many fruit-trees, clover, grasses. In colder regions it passes the winter in the egg stage, in warmer regions the adults clusten behind bark, or below buds. It has some generations per annum.

It induces chlorosis on the infested leaves, which may injure the host seriously. In Palestine no serious damage has yet been observed.

b. THE SOOTY-MOULD FEEDING MITE (*Oribatula plantivaga* Berl.; Oribatidae).

The small, dark red-brown mites of *O. plantivaga* are 0.5 mm long and 0.3 mm broad. Their body is inflated. They are easily to be seen by the naked eye. Throughout the year, especially from late spring to autumn they infest branches, twigs and less commonly leaves of citrus trees in great numbers, often together in big crowds. As far as has been hitherto observed, no damage is connected with these mites, as they seem to feed entirely on sooty mould and other fungal spores developing on honeydew.

E. THE CITRUS RUST MITE *(Phyllocoptrata oleivorus Ashmead)*.

Synonymy: 1879 *Typhodromus oliivorus* Ashmead (misspelling)
1879 *Eriophyes sp.* Pergande
1880 Typhodromus oleivorus Ashmead
1903 Phyllocoptes oleivorus Banks.

Description: The *adult female* is yellow to yellow brown, elongate lancet shaped with the greatest width in the first third. It is about three times as long as broad. The cephalothorax is smooth, the abdomen striated. The number of the dorsal half rings of the abdomen is 28, about the double on the ventral side. With two pair of legs and a pair of lobe-like false feet on the last abdominal segment. 0.10—1.12 broad, 0.04 mm maximal width.

No *males* are known and reproduction is parthenogenetic.

The *eggs* are spherical, smooth, semitransparent to pale yellowish. They are laid singly and usually one or two large eggs are only found in the female.

The *larvae* have two stages which differ from the adult female mainly by smaller size and immature ovaries. They begin immediately after hatching to feed on the cells of the substrate.

Oviposition begins only a few days after the last moult and continues until one or two days before death, mainly into pores and depressions. Yothers and Mason (1930), the monographers of the species, estimate the average egg-production to about 10—15 eggs per female. Usually one or two eggs are laid daily, but up to 5 eggs have been occasionally observed. The rust mites are positively phototactic, aggregating towards dispersed light in the laboratory, scattering during the night.

Origin and distribution: S. E. Asia, the home of *Citrus*, is apparently also the home of this pest. It is now distributed over the southern states of U.S.A., occuring in greenhouses in the northern states. Other records are from the West Indies, Mexico, Venezuela, Hawaii, the Philippines, Australia and Japan.

In Italy it is apparently absent (Silvestri i.l. 1923), also in S. Africa. We have records from Cyprus, Syria and Palestine. Yothers and Mason think, it has hitherto been overlooked in India.

In Florida the citrus rust mite ranks third among the citrus pests (after mussel scale and citrus white fly) and causes heavy damage. Yothers and Mason think that it is present on every citrus tree in Florida, where on an average more than 50 % of the fruit is more or less injured by them.

Host plants: All species of *Citrus* are infested. Most preferred are in their proper sequence: lemon, lime, citron, grapefruit, sweet orange, sour orange, tangerines. Also kumquat (*Fortunella spp.*) and *Chaetospermum glutinosa* are attacked in Florida, where other Rutaceae remained free.

Injury and life cycle in Florida (after Yothers and Mason): The rust mite punctures the epidermal cells of the rind of the fruit. When the mites are abundant, the outer layers of the epidermal cells are largely flattened or destroyed. Slight damage on oranges is called „golden", on young fruits „black russet" or merely „russet". On grapefruits and lemons heavy infestation causes shark skin on young fruits. When thousands of mites swarm over one fruit its growth is stunted. The infested fruits, apart from the blemish and the stunted growth, are not so sweet as the uninfested fruit on the same tree, and its ripening is considerably retarded. Only in spring, after abnormally heavy evaporation, the fruit may be actually sweeter. The devitalising effect of dense rust mite populations on the leaves and on the tender twigs is usually much underestimated.

Climatic factors had the following influence on the rust mite in Florida: In early February 1917 exceptionally heavy night frosts of —3.2 to —9.5° C were observed in the northern counties. Many mites were frozen. Many died in consequence of the

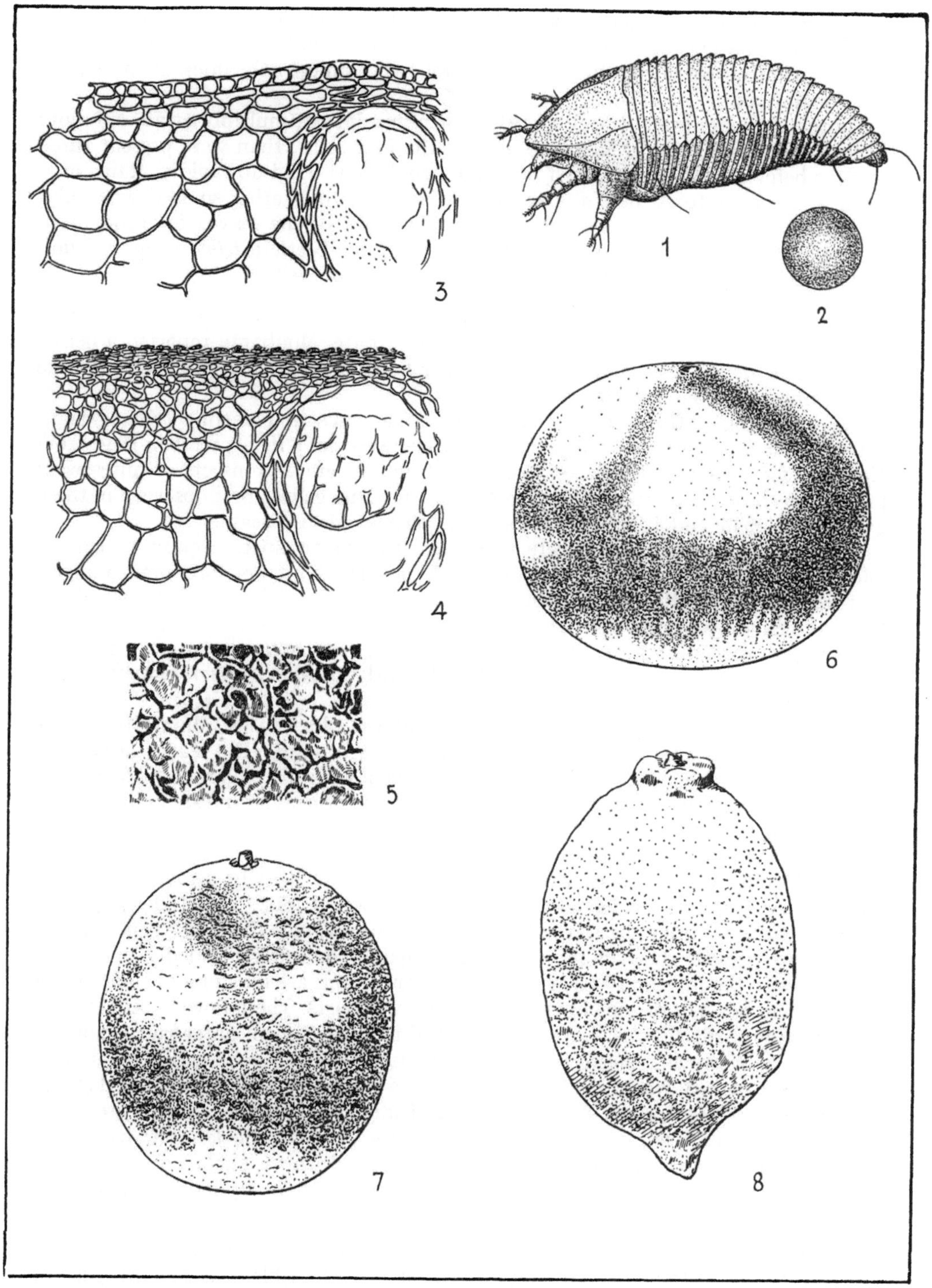

XXXXIII. The Citrus Rust Mite (*Phyllocoptrata oleivorus* Ashm.) 1) female, 2) egg, 3) normal epidermis of orange, 4) epidermis of orange damaged by *Phyllocoptrata*, 6) brown-black discolouration on grapefruit, 7) dto. on orange, 8) silvery discolouration on lemon, and 5) scaly deformation of old lemon blamage caused by *Phyllocoptrata* (1—4 after Yothers and Mason).

shedding of the foliage, as they cannot survive on shedded leaves. The effect on the mite population depended merely on the degree of defoliation which reached 90 to 95 % in some northern counties. Yet in the following summer the infestation reached a normal peak.

During the long drought spell in 1912 rust mites did not multiply at all. No more mites were found in early May, the data of the first rains, than were present three months before. Following the first rains the populations build up rapidly. Rains may wash a few mites from the fruits and the foliage, but to a rather small degree only. In addition, the mites crowled on the lower surface of the leaves for protection.

No parasites have been bred in Florida. A small ladybeetle (*Stethorus nanus* Luc.) and a cecidomyid larva were the only predators observed.

Injury and life cycle in Palestine: Klein obtained about 1935 citrus fruits from Sidon (Lebanon) with the blemishes of the rust mite. Moubarak reported a heavy outbreak all along the coastal citrus area of the Lebanon in 1938. Klein also found the mite some years ago on citrus at Cyprus.

The following is the record of its recent appearance in Palestine (Klein 1946). Citrus fruits with silvery or blackish blemishes were sent in September 1944 from Emek Hefer (Mishmar Hasharon, a.o.) and from Hedera. No insects or mites were present on these fruits. Inquiries revealed that the damage was first observed in these gardens in 1943 and it was regarded as a physiological disease.

The following observations were made in 1945: Silver spots appear on lemons (silver mite), whilst they are dark to black on grapefruit and on some oranges, and rustbrow on most oranges (rust mite). In heavy infested foci a great part of the fruit is covered by the blemish. The silver layer on lemons often cracks and assumes a scaly appearance, which can be easily rubbed off with a knife. In early infested lemons the growth of the fruits stops early. Also early infested fruits of grapefruit and oranges do not reach their normal size.

Leaves and green twigs are also infested, but not older twigs. Mainly the lower side of the leaves are attacked and marked by black spots. The leaves later turn chlorotic, the small ones cease to grow and even wilt. The Florida workers had pointed out their raised transpiration. The lower side of heavily infested older leaves may turn light rust brown. The vitality of strongly infested tender twigs is impaired and black and rusty blemishes are found on them.

In 1945 only citrus groves with wide distances were heavily attacked, such as young groves, groves neglected during the war period, etc., whilst dense groves seriously suffered from the shade loving citrus mussel scale. Yet on open spots in these dense groves, the rust mite abounded again. The mites decidedly prefer the outer rows to the interior of trees with dense crowns and are much more common on the sunny upper half of the crown than in its lower parts. The following sequence in sensitivity towards the rust mite was observed in 1945: lemon, grapefruit, shamuti, Valencia, clementines, tangerine (King, Temple), pomelo. In all cases the young fruits were attacked, whilst the blemish continues to grow with the fruit. The damage during the two last seasons was estimated in badly infested foci of citrus groves to 50—100 %. In all heavily infested citrus fruits the skin was thicker, the fruit-flesh smaller than in normal fruits. Conspicuously blemished fruit is excluded by the fruit inspection service from exportation. They are also unfit for the by-industry of etheric oils.

Seasonal history in 1945: Young lemons from the winter blossom were first seen with silver spots in the middle of May. In late May the first blemishes on other small citrus fruits were observed. The big population rise of the mites began in the middle of June and towards the end of that month many fruits were severely infested. The populations reached their peak in July. They declined markedly in August and almost disappeared from the fruits in September. No mites were ever found on the fruits from autumn to spring. They obviously pass this period in leaves and twigs, in true hibernation for most of the time, as later will be shown. Winter lemons are the earliest fruit attacked in the coming season and they may well serve as indicators for the presence of the mite and the intensity of its early populations. Other red mites (0.25 mm long) belonging to the genus *Tenuipalpus* were observed from September onwards, often in considerable numbers. They caused no apparent damage, and Klein assumes them to feed on the dead bodies of the rust mites.

Three foci in all were discovered in 1945: at Emek Hefer and Hedera, at Petah Tikva, and in the southern citrus area. The first focus was widely scattered and considerable damage was caused. Damage and focus are restricted to an area of heavy soils between Petah Tikva and the Yarkon River, the infestation being fairly strong. In the south very heavy damage was observed in a number of groves in and around Migdal (Mejdel). The mite was almost certainly introduced into this country on nursery stock or on grafting scions.

Seasonal history in 1946*:* Within that year the rust mite spread with surprising rapidity over most of the citrus areas of the coastal plain. A spectacular illustration was the spread from Nizanim to the south of Rishon le-Zion, over 30 km. distance as the crow flies (Klein, 1947). Yet the blemish grew conspicuous in the different districts from early to late June only, i.e. one month later than in the previous season. The main reason for this difference is apparently the spring weather. In 1945 the spring-hamsins were relatively few and mild, in 1946 common and strong:

Year	No. of hamsin days in April	May	June	No. of days with minim. rel. hum. below 20 % April	May
1945	9	10	1	4	2
1946	17	14	5	8	10

June was favorable to rust mite development in both years.

In the areas and plantations in which the pest appeared in 1946 for the first time, infestation was limited to a few, scattered trees, not causing any heavy damage. The upper outer parts of the crown were infested, the lower outer parts only in young orchards with great distances and strong sun-penetration; but not the fruits of the inner parts of the crown, even not in grapefruit. In many places also leaves and green twigs were attacked, eespecially in the late summer months. This lead on light, poor soils with insufficient irrigation, locally to leaf-drop. Generally, the damage remained very low owing to the great scattering of the new infestations, which locally only, on individual trees, lead to a damage of 20 to 40 %.

In the meantime, the citrus rust mite has spread all over our coastal plain.

Physical ecology: Yothers and Mason point out the great difficulties in breeding the mites in the laboratory. From a relatively small number of complete breedings they obtained the following averages:

Season (av. temper.)	duration of stages in days					
	egg	larva I	larva II	preoviposition period	Total development	adult longevity
summer (32.6 °C)	3.1	1.8	1.3	2.7	8.9	6.7 days
winter (25.1 °C)	5.1	4.3	6.4	5.0	20.8	11.3

These temperatures are much higher than the mean monthly averages for these seasons in Florida, and they obviously are microclimatic data of the breedings. From these data we calculated a tentative hyperbola band, as follows:

Hyperbola		A	B	C
Threshold of development (c)		20.0	20.0	20.0° C
Thermal constant (Th.C.)		106	151	197 days-degrees
Duration of development at	35° C	7.1	10.1	13.1 days
	30	10.4	15.1	19.7
	25	21.2	30.2	39.4
	23	35.3	56.3	65.7

The average number of annual generations was calculated for a number of Mediterranean, southern N. American and Indo-pacific localities:

Locality	months above c	average minimum monthly temperature	number of generations with hyperbola		
			A	B	C
Lyon	2	2.4° C	0.4	0.3	0.2 generations
Nice	2	6.5	1.1	0.8	0.6
Genova	4	7.5	3.0	2.2	1.6
Naples	4	7.2	3.1	2.2	1.6
Famagusta	6	11.6	9.2	6.5	5.0
Tel Aviv	6	12.8	7.8	5.8	4.4
Cairo	6	10.9	7.6	5.3	4.1
San Francisco	0	—	—	—	—
Los Angeles	3	12.9	1.2	0.8	0.6
Riverside	4	11.0	3.1	2.3	1.8
Tampa Fla.	7	16.2	11.0	7.7	5.9
Honolulu	12	21.3	12.1	8.6	6.6
Singapure	12	25.5	21.4	15.0	11.5
Manila	12	24.8	22.4	15.7	12.0

It will be impossible, however, to apply these calculations to the epidemiological trend of the mite without further correction.

Yothers and Mason claim that healthy rust mites — inspite of their positive phototaxis — appear to avoid direct sunlight. „It has been observed for many years that rust mites do not attack the outer surfaces of the fruit on the outside of the tree. They are usually found on the side of the fruit which is in semi-shade. Of course they attack the entire surface of fruit on the inside of the tree. It is supposed that the rust mites are not able to endure the direct rays of the hot sun and therefore feed mostly on the sides of the fruit in semi-shade." Mites infected by a fungus disease crawl also in direct sunlight before death in Florida.

The preliminary report on conditions in Palestine shows a very different situation. The shady interior is definitely avoided and the outer rows are preferred. Perhaps, and even probably, direct sun-radiation is avoided at noon hours. But this is easily obtained in the crowns, when the sun is high. This fact is of importance, as mites in the sunny parts of the tree live in a much warmer microclimate than those in the shade. They therefore develop much quicker than can be computed by average shade temperatures. The following table indicates the rapidity of increase, if 15 respectively 5 of the offspring of each female survive:

Generation	Survival 15	Survival 5
P	1	1
Fl	15	5
F2	225	25
F3	3375	125
F4	50625	625
F5	769375	3125.

We now turn to the puzzling seasonal history of the rust mite. Yothers and Mason observed since 1912 that the citrus rust mite remains at a low population level until May. It then rapidly increases to its peak in June, until just after the beginning of the rainy season. In rare instances a single grapefruit was infested at that time with more than 500.000 mites. Shortly after the inset of the rains the mites disappear rapidly as if by magic. By the middle or end of September more than an hour of diligent search is sometimes required to find a single mite. This diminution of numbers is not due to a lack of food, since on an average only about 50 % of the unsprayed fruit is seriously attacked by the rust mites. There is considerable evidence to show that this disappearance of the citrus rust mite is due to a fungous disease. The infested mites aggregate in direct sunlight, contrary to their normal habits, and die. Rust mites always become much more abundant following the use of fungicidal copper sprays than on unsprayed trees or fruits. These sprays apparently eliminate the entomogenous fungi, which normally would have affected the mites. This is in agreement with the experience of Watson on the effect of copper sprays on the entomogenous fungi of the citrus white fly in Florida. These fungi need, as most entomogenous fungi, for their epidemic spread the following conditions: 1) high host density, 2) suitable temperatures, in our case warm weather, 3) a humid atmosphere. The relative scarcity of the rust mite on the excessively humid east coast for the greater part of the year is explained by this humidity which favours the development of the fungus.

In California (Quayle 1938) the rust mite has become established in a restricted southern section of San Diego county for many years. It once appeared in Orange county, but has not been found there since. In 1934 a heavy infestation occurred in North Whittier Heights of Los Angeles county. The damage is apparently restricted to lemons in California, apart from the last mentioned outbreak, on a moderate scale only.

In Honolulu (Bodenheimer 1932) the rust mite exists and causes serious damage in experimental lemon groves. Nothing is known on its seasonal history. The great humidity and rainfall in most parts of the islands throughout the year is favourable to fungous epizootics.

In Palestine the mite reaches its population peak in July and disappears rapidly from the fruits until September. No information is yet available on the influence of the spring hamsins on the mites. But our dry summer is unfavourable to the outbreak of a fungous epizootic. This, however, needs further study. When the fungus is not the cause of the breakdown of the mite population in August, we have to look for other factors. This is to all apparence the age of the fruit. Whilst nothing definite is yet known in this respect, all indications point in the direction that maturing fruits are unfit as food of the mite and that throughout autumn and winter they are restricted to the leaves and twigs. In addition there exists possibly, our observations are too few to deserve great weight, a great population of various, partly predatory mites on the infested fruit.

We come to the following general conclusions: High temperatures, high humidity and dense mite populations together favour heavy fungous epizootics among the rust mites. This is probably the reason, why the pest has been overlooked, so far in the Oriental kingdom. Outbreaks of rust mites are there probably sporadic and local. Nothing is yet known on the influence of prolonged cold on the mite. It is, however, a fact that no permanent rust mite damage has so far been observed where the average of the coldest month is below 10° C. Another point is the high threshold of development (20° C). Only in tropical regions temperature conditions permit development throughout the year. All regions with serious damage have at least six consecutive months above this threshold. Once the seasonal development has set in, i.e. when the spring temperatures surpass 20° C, and after the appearance of the first blemishes, the mite population reaches its seasonal peak rapidly within two or three months. In subtropical regions with summer rains, such as Florida, a fungous disease causes the rapid breakdown of the mite populations. In subtropical regions with summer drought, such as Palestine, the rust mite builds up considerable populations on the fruits in early summer, which break down completely in late summer. The reason for this breakdown is apparently that the fruit grows unfit for the mites at that season, and perhaps the same is true to a minor degree for leaves and twigs. Also predatory mites may help to adduce this breakdown. The blemishes grow and turn more conspicuous at a period, when mites are no more found on the damaged fruits.

Control measures: The gardens have to be scouted for the first blemishes since late April. As soon as the first injury is observed a thorough motor dusting with sulphur has to be performed. It is highly important to perform this dusting as early as possible. The sulphur dust much reach all parts of the tree, especially the upper twigs. Motor dusters are essentially for good penetration of the sulphur into the tree. No dusting should be done on hamsin days or when hamsin days are imminent.

Spraying with Californian lime-sulphur solution (1 %) has also been recommended by the Florida workers. It is apparently during the proper spraying season more expensive in this country. It should be applied, however, when no motor duster is available.

In 1945 hundred dunam of citrus were dusted by motor dusters in early July at Mishmar Hasharon, when severe damage was already done (Klein 1946). A normal tree of about 12 years obtained a dosage of 220 g. of sulphur dust, apparently a generous quantity. Two and five weeks after the dusting the grove was inspected and no living mites were found. On uninfested trees which were marked before the control operation no blemishes were found. The experience of 1946 showed that sulphure-dusting gives a very reliable control. No damage was observed in any grove where dusting was performed, whilst damage was heavy in two untreated orchards.

In 1945, a corner of valencia-oranges in a big plantation had shown infestation. This corner plus a protective girdle of 4 rows of trees in every direction was dusted in 1945 and found to be free from rust mite in 1946. Yet a few infested fruits were found in some single trees of the fifth rows which were not dusted. The costs of dusting were 300 to 500 mils in 1946 per dunam, when most of the work was still done with back-dusters. Costs can probably be much reduced by motor-dusting, and still more, when planned experiments with helicopters will give good results.

Present control: Since sulphure-dusting has been accepted as a routine matter in the infested areas, the economic damage is very small, indeed. Most of it is now done by motor dusters. Groves or corners infested the previous year are dusted in June. The control is performed speedily, as to be terminated by the end of June. In years with few and mild hamsins in the spring, the control will probably have to begin from early May onward. Whereever single trees or small patches of trees are recognised to be infested in late summer (new foci), dusting has to be applied immediately in the patch plus 5 protective rowst of rees in every direction. The sulphure dust needs only to reach the outer part of the crown in its upper half, anyhow not the interior. About 250 g. sulphure per fully grown tree are the dosage to be applied.

References.

W. W. Yothers and A. C. Mason, U.S. Dept. Agric. Techn. Bull. 176. 1930.
H. J. Quayle, Insects of Citrus. Ithaca. 1938. p. 23—26.
F. S. Bodenheimer, Observations on Citrus Insects in many parts of the World. Tel Aviv. 1932. p. 81—82.
H. Z. Klein, Hassadeh, 26, 1946 p. 344—347; ibidem 27. 1947. 2 pp.
E. Moubarak, Rapport sommaire service défense des cultures en 1938. Beyrouth.

11. VARIOUS ANIMALS INJURIOUS TO CITRUS CROPS

A. THE LEMON RAT (*Rattus rattus frugivorus* Rafinesque).

Description: R. frugivorus is usually regarded as a subspecies of the Common rat. It is slender, has long ears. Its tail is longer than the body and has up to 270 rings of scales. It is covered with thick, short, grey to red-brown woolhair and inspersed single, long black hairs. It is easily distinguished from the other rats by its white or yellowish-white belly, which is distinctly demarcated from the upper side. The skull is up to 48 mm long. Feet yellow with grey longitudinal stripes, white below. Tail dark grey. Length of body: 15—20 cm (tail excluded).

Distribution: This species is distributed from India to large parts of Africa, and occurs in Sicily, S. Italy, Syria and Palestine. It has been seen by the author also in Cyprus, Egypt and Cilicia.

The actual taxonomic status of the lemon rat is very obscure, as generally all systematics of the genus *Rattus*. It may finally be shown to be only a biological or an eco-form of *Rattus rattus* L. of no taxonomic value, as well as it may be a separate species. Much more material is needed for a final decision. Unfortunately, we were never able to obtain living material for breeding. The first alternative was much stressed by the greatly increased local damage in orange-groves in various parts of Israel during the rodent outbreak in winter 1949/50, when all specimens feeding on citrus bark, as far as they were obtained for identification proved to be *Rattus rattus alexandrinus*.

Fig. 140.
Feeding marks of *Rattus frugivorus* Rafinesque on a lemon tree at Hedera.

R. r. frugivorus is always of a local appearance. It probably has its origin in swamps, lakes, etc. It is a good swimmer and feeds on fishes and snails in the lake of Antiochia. In some parts of Cyprus and Egypt it is well known as a pest of lemon-trees. Over restricted areas, sometimes, it consumes the bark off entire branches, the latter dying in that or in the following year. As far as could be ascertained, no damage has been done to oranges.

In Palestine, it has been first met in the Atta citrus groves near Hedera. Some old lemon-trees were attacked, which bordered on the old Atta swamp. On 28.X.1930, feeding on the bark of the twigs was observed. Pieces of bark, two mm in size, were seen below the trees. The whole bark was eaten off some branches, 2—3 cm in diameter, along 60—70 cm. This feeding continued throughout November. Two rats were caught in December, upon which the feeding ceased immediately. Besides the bark of the lemon-trees, many pods of the hedge-acacia had been devoured.

Mr. Paker of Dagania (Upper Jordan Valley) informed us in June 1936 about damage caused on grapefruit and, on request, offered the following description:

„The rats nibble within the main branches of the tree (mainly) at a different height from the ground, from about 1.20 m to 2 m height. The eaten area extends to a length of 25—30 cm and a width of about 10 cm. In two isolated cases they also gnawed through the stem, i.e. the oculated stem, below the branching of the main branches. So far, they have nibbled away the bark and in most cases the bark with the cambium up to the xylem, not having touched the xylem itself. I noticed this gnawing, only by accident. At first I could not ascertain as to which animal was doing it. When I was told that it were rats, I distributed under the trees a thallium sulfate bait on paper plates, a few grains on each plate. The activity diminished subequently.

Beneath one tree I also found a few holes which seemed to belong to them. On one tree we found a nest built out of leaves and twigs of small branches and in it a dead rat."

On the publication of Mr. Paker's note the following report was obtained from Mr. Graviner of Gan Haim (Samaria): When I went over a row of lemons in February, I observed on many places that the bark of the main branches was nibbled away some time ago and recently. On about 10 trees the bark and sometimes part of the cambium were missing along a length of 1 m and more, some around the whole branch, some only in longitudinal stripes. Two traps baited with bread and butter were put on the tree, in one of which a rat was found the following morning. Since then no further damage was observed. The attack was restricted to trees of sour lemon.

Similar attacks were reported from Rehoboth, Hedera, a.o. places during 1936/1937. In all these cases the damage was restricted to a few trees only and mainly to sour lemons, caused by a few individuals only. Dead specimens from four places were examined and all proved to belong to this form.

The biology of this rat deserves special study, and living animals are among the urgent desiderata of our laboratory.

The control is best performed by using intensive baiting with grains or paste of thallium-sulfate in the immediate surroundings of the damage. Trapping with simple mouse-traps may be used, if no thallium-sulfate is available.

References.

B. Aharoni, Zeitschr. f. Säugetierkde. **7,** 1932, p. 180—181.
M. Graviner, Hassadeh. **16,** 1936, p. 654. (Hebrew).
G. Neuhauser, Zeitschr. f. Säugetierkde. 11. 1936. p. 174.
M. Paker, Hassadeh **16,** 1936, p. 548—549. (Hebrew).

The Levant vole (*Microtus güntheri* D. et A.), the common vole of the Middle East, a medium sized vole, grey-brown above, greyish-white on the belly, 120 (rump) + 28 (tail) mm long, caused damage for the first time in 1949/50, during a great pullulation on citrus trees. In many parts of the country, in autumn 1949, when no food was available in nature, especially not in the south, yet also near Haifa and elsewhere, voles invaded, usually neglected, orange groves and began to nibble on the bark of orange-trees slightly above the soil. This damage was easily stopped by the dispersal of thalliumsulphate wheat. They invaded from the close neighbourhood from fallow land or from uncultivated soils. This is definitely a very exceptional occurrence.

B. BIRDS.

Birds are only occasionally observed as injurious in citrus groves. The Mediterranean Hooded Crow (*Corvus corone sardonius* Klein.) may be observed during winter picking on the mature fruit. However, as far as our knowledge goes, this damage has never reached an extent to necessitate control measures.

Considerable damage was caused, during the winters 1936/37, to the citrus groves of the Upper Jordan Valley (Gesher, Beth Sera, Dagania B., Menahemia, Afikim) by the Starling (*Sturnus vulgaris* L.). Starlings are well known as winter visitors in the coastal plain and in the Negeb. But late in October 1936 large swarms appeared in the Upper Jordan Valley and chose citrus groves as their roosting places. Evening after evening they settled in the crowns of the trees in enormous quantities. Most of

Fig. 141. Damage caused by starlings at Gesher (winter 1936/37).

the fruits were heavily spoilt by the excrements. The damage done at Gesher alone is estimated at some hundreds of pounds during the season 1936/37, (2000 boxes of grapefruit). Many leaves dropped, many branches remained stripped of all leaves and not a few of these branches died.

The birds were partially driven away by noise and shots. In 1937 they reappeared about the same time in the same district. The damage caused by the starlings is certainly peculiar and locally limited. It is interesting to follow the further development of this new hibernation habit.

References.

E. Smoli, Hatteva vehaaretz **4**, 1937, p. 493—501 (Hebrew).

C. THE CITRUS ROOT NEMATODE (*Tylenchulus semipenetrans* Cobb).

Description: The Citrus Root Nematode is easily distinguished from all other nemas, which may be found on or near citrus roots, by the adult female, embedded with its head in the root tissue. All other instars may be determined by a trained nematologist only.

"The original diagnosis is: Cuticle naked, traversed by 400—500 plain transverse striae. Neck cylindroid, becoming convex-conoid near the continuous head, which is rounded in front. No lips, amphids or eye-spots. Spear and oesophagus typically tylenchoid. Median bulb elipsoidal, with valve; posterior swelling pyriform to elongated, without valve. Male tail conoid to the somewhat blunt terminus. Posterior part of the adult female saccate, with wide blunt tail bent toward the ventral side. Vulva in the midst of a prominent ventral suture." (Cobb).

The species has been discovered, so far, in the subtropical parts of N. and S. America, in many Mediterranean regions and in Australia. It will probably be discovered in all major citrus districts, where it was introduced together with the citrus nursery stock. The species has originally been discovered in California, in 1912, by the agricultural

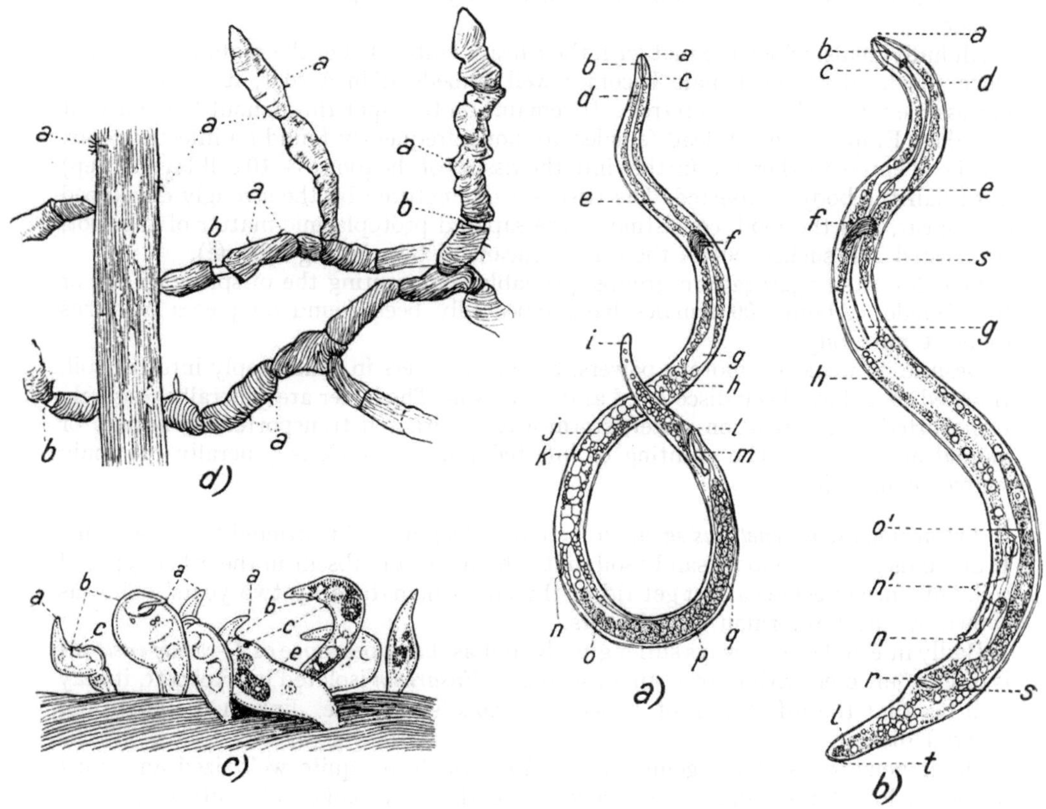

Fig. 142.

The Citrus Root Nematode (*Tylenchulus semipenetrans* Cobb).a) Male (×500); a lip region, b & c spear, c 3-bulbed base of spear, d & e & g oesophagus, f nerve ring, h beginning of intestine, j & k intestinal granula, l anus, m spicula, n excretory pore, o spermatocyte, q spermatozoon, p vas deferens. b) young female, ready for fertilization: a—h as above, o ovary, n excretory pore, r vulva, s wings of the cuticle, l rudimentary anus, t terminus.c) Mature and half-grown females, with their head ends permanently embedded in the feeding-root of a citrus tree: a tail end, b vulva, c excretory pore, e egg in uterus. d) Citrus feeding roots attacked by *T. semipenetrans* (5×). The parasites are shown at a. They are shown black, but in reality are yellowish or brownish. The feeding roots become somewhat enlarged and irregular; the outside portion of the root is somewhat reparated from the axial portion (b). (from Cobb).

inspector of Los Angeles, J. R. Hodges. In Palestine, it has first been discovered in the surroundings of Haifa by the late A. Aaronsohn. Up to date, we have met it in Palestine only occasionally (Petah Tikva, Hedera), which, in all certainty, is due mainly to lack of intensive search.

The eggs of *T. semipenetrans* are large and thin-shelled. Only 1—2 eggs develop in the female uterus at a time. After a few days stay in the uterus, they are deposited into a sticky matter. In time, the female generally deposits the eggs in larger groups, each egg hatching within the first two days after oviposition. The hatching, colorless larvae move slowly and feebly, but are more active than the adults. They are not fit for active migrations over any distance in any of the stages. In the Californian spring, the larval development of the female takes 6—8 weeks.

The adult male decreases in size with age and the characteristic oval spear deteriorates. It is therefore very doubtful, whether the male is ever able to enter tissues of citrus roots. As adults they may not even feed at all, considering their decrease in size and their becoming more and more slender. They develop more rapidly than the females.

All individuals, which are met with their heads embedded in the roots are females. Once the head of the female becomes well embedded in a root, it is practically impossible for the female to retreat. It remains on this spot throughout life, where it also dies. Empty skins of dead females are not infrequently found on infested roots. The head enters further and further into the tissues of the root (8—10 cell-layers deep) until half the body is inserted. The tissues are punctured by the strongly developed oval spear, and the food, consisting of the sap and protoplasmic matter of the root, is ingested and sucked out of the cell by means of the sucking bulb (e).

Females often aggregate in groups, probably representing the offspring of one or two females. About 100 females have repeatedly been found on pieces of citrus roots, 4 mm long.

Despite its weak locomotory powers, the nema enters in time deeply into the soil. Infested roots have been discovered as deep as 4 m. The latter are generally passively transported with irrigation water, storm-water, with soil transported by human or animal agency, etc. The planting of infested nursery stock is generally the main source of infection.

Tylenchulus semipenetrans seems to be monophagous and restricted to citrus roots. It abounds greatly in loose sandy soil, but is by no means absent in the other types of soil. Soil once infested cannot get rid of the citrus nematode by two years' fallow as shown by the Californian experiments.

Badly infested trees always suffer greatly, but as the appearance of *T. semipenetrans* is often connected with that of fungi, such as a *Fusarium* isolated by Fawcett, it may well be, that the infestation of *T. semipenetrans* is only one „link in the chain" of a serial disease.

Other nematodes of the genus *Mononchus*, which are quite well sized and must devour a lot of the smaller *Tylenchulus*, are the only biological controling factor known.

„The above-ground portion of the citrus tree may, or may not, show ill effects from the attacks of the nematode. Trees only slightly infested were found, which to all appearances were in perfect condition. However, among the hundreds of trees examined, every heavily infested tree showed evidence of injury. Badly infested trees have the appearance of being undernourished, the leaves are small and slightly yellow or mottled, the fruits small and often unmarketable, the whole tree being stunted and showing a marked state of deterioration. This condition has been found many times with trees which previously had been healthy and vigorous, and which formerly produced good crops of fruit" (Thomas).

Similar symptoms may be the result of other agencies, as fungi, etc. and slightly attacked trees may show no ill effects at all on the above-ground portions of the tree. When the feeding roots of the tree are examined, a series of symptoms helps to discover the presence of the Citrus Root Nematode. Soil particles commonly cling to infested roots and are not easily removed as they stick to the gelatinous material, in which the eggs are embedded. There, on closer examination, colonies of *T. semipenetrans* females may be revealed. The bark of the rootlet may be easily separated from the inner woody portion. This condition of the bark is not observed in healthy roots. Finally, the number of feeding rootlets is very limited in badly infested trees. Wherever these symptoms, a hand-lens or a microscope generally reveal the swollen bodies of the female *Tylenchulus*, embedded with its head in the bark of the rootlet.

No control has hitherto been necessary in Palestine, where the real extent of damage caused by *T. semipenetrans* is not yet fully determined. An extensive survey of the occurrence and importance of this pest is highly desirable, at the earliest convenience.

So far, no practical method of direct control has been put through anywhere. If the Citrus Root Nematode should prove to be of actual importance in Palestine, the mass breeding of its predator, *Mononchus*, should be attempted. However, even the prospects of this measure are not very promising.

As indirect control some ways of action are recommended:

1) Great care should be taken to keep the nursery stock free from an infestation of *Tylenchulus*. The nursery trees should be grown on uninfested soil, in order that new citrus plantations may be free from the attack of the parasite for a time. Planting infested nursery stock is one of the most important ways in which an orchard may become infested. (Thomas).

2) Fawcett points out that observations in California indicate that, under commercial conditions, fairly healthy trees may be maintained even when the roots are infested with nematodes: Good cultural practices with abundant and proper fertilizing, irrigation and other care will probably maintain profitable trees on good soil, in spite of the presence of nematodes on the roots.

„Their excessive multiplication in certain cases, with bad effects on the tree, followed by periods of apparent diminution and partial or total recovery of the trees under good treatment, is probably due to unknown factors, possibly parasites, influencing or controlling either the number or severity of the nematodes or the resistance of the Citrus host. Many growers have been able to grow successful crops in spite of the presence of the nematode, by building up the apparent resistance of the tree by proper fertilizing and cultural methods. Under California conditions one important factor appears to have consisted in the use of suitable amounts of bulky organic fertilizers" (Fawcett).

References.

F. S. Bodenheimer, Schaedlingsfauna Palaestinas. Berlin, 1930, p. 178—179.
N. A. Cobb, Journ. Washington Acad. Sci. **3**, 1913, p. 288—289.
N. A. Cobb, Journ. Agric. Research **2**, 1914, pp. 217—230.
H. S. Fawcett and H. A. Lee, Citrus diseases and their control. New York, 1926, p. 113—118.
E. E. Thomas, Calif. Agric. Exper. Sta. Circ. **85**, 1913, 14 pp.
E. E. Thomas, Calif. Agric. Exper. Sta. Techn. Paper, No. 2. 1923, 34 pp.

D. SNAILS (*Helix spp.;* Mollusca).

Snails are occasionally met with in citrus groves; but no noticeable damage has yet been observed by us.

Their presence is noted by their very obvious shells and by the broad traces of glue which they leave behind them on the leaves. When they feed, the feeding holes are always very irregular with lacerated borders (never clean cut). Among the few species *Helix joppana* Roth and *H. pisana* Muell., very occasionally *H. seetzeni* are met with mostly. The smaller *H. joppana* is easily distinguished from the others by its rich and dense linear red-brown design. The life-history of *H. pisana* may serve as example for all of them.

The Common White Snail (*Helix pisana* Mueller).

Description: The globulose shell is light buff with lineal brown stripes. Color and stripes are very variable, but the surface of the shell, whether pure white or densely striated, is always shiny. The aperture is pinkish and thickened along the internal margin. During the dry season the aperture is closed by an epiphragma. The snail has 5½—6 whorls and is 13—20 mm in diameter.

Distribution: Mainly Mediterranean. Has been repeatedly introduced into N. America.

Life-history: Copulation takes place after the first rains. The hermaphroditic animals are easily discovered on rainy days in their mutual copulation act. Some weeks later the snail digs a hole in the soil, 3—4 cm deep, which is small at the upper entrance. Here the eggs are deposited through the genital aperture, situated laterally behind one of the antennae. When the oviposition cavity is ready, the eggs are deposited one after another, on an average about 100 per female, and remain in one heap. These eggs imbibe water from the soil eagerly and „grow" until the egg-heap is finally considerably larger then the adult snail. When the oviposition, which may last to late winter, is at an end, the cavity is closed with mucus and soil and left alone. Each female is said to oviposit only once per season (Stephani). This supposition, however, needs to be confirmed, as oviposition actually takes place throughout the season of activity.

After about 2—3 weeks the small snails hatch from the eggs. Their shell is very thin, fragile and contains 1½ whorls. It is strongly carinate and plain on the upperside, being very convex on the lower side. The color is entirely whitish or different numbers of brown bands run along the whitish ground-color. The species apparently develops — at least as a rule — in two years. When born in the early winter, it grows until March. In March it has 3—4 whorls and is somewhat below 10 mm in diameter. Then it ascends on plants, where it may be discovered on the steppe-like landscape of the coast-plain by the millions or on the hedgeplants of the orange groves. It secretes a strong calcarous layer, which closes the entrance of the shell. During the summer-sleep, it may loose up to 50—60 % of its former body weight. In the autumn, it awakes from this summer-diapause with the first rains and begins to feed again. The species is extremely polyphagous. Its feeding marks are very irregular. However, when the mucous traces, which the snails always leave on their way, are discovered, it is plain who the author of these marks is. The snail continues to grow until the shell has 5½—6 whorls and is 15—20 mm in diameter, which occurs as a rule about the end of the second winter. The dry season is passed again on plants in diapause, under the protection of the epiphragma. When the adult snail awakes at the beginning of the third winter, it is ready for copulation and reproduction. It is unknown how many years the average adult continues to live.

Damage: In Sicily, it is reported to be injurious by feeding on the foliage, bark, tender twigs, fruit and blossoms of orange and lemon-trees, as well as on the foliage of almond- and lemon-trees (De Stephani). However, real damage is caused only locally and occasionally. In California, where the species was exterminated by a long campaign, no real damage was done to citrus.

No real damage came to our knowledge in Palestine until the very latest years. Since than repeated and widely spread damage on citrus has been observed all over the coastal plain and in the Emek Yesreel, mainly on light soils. The damage begins

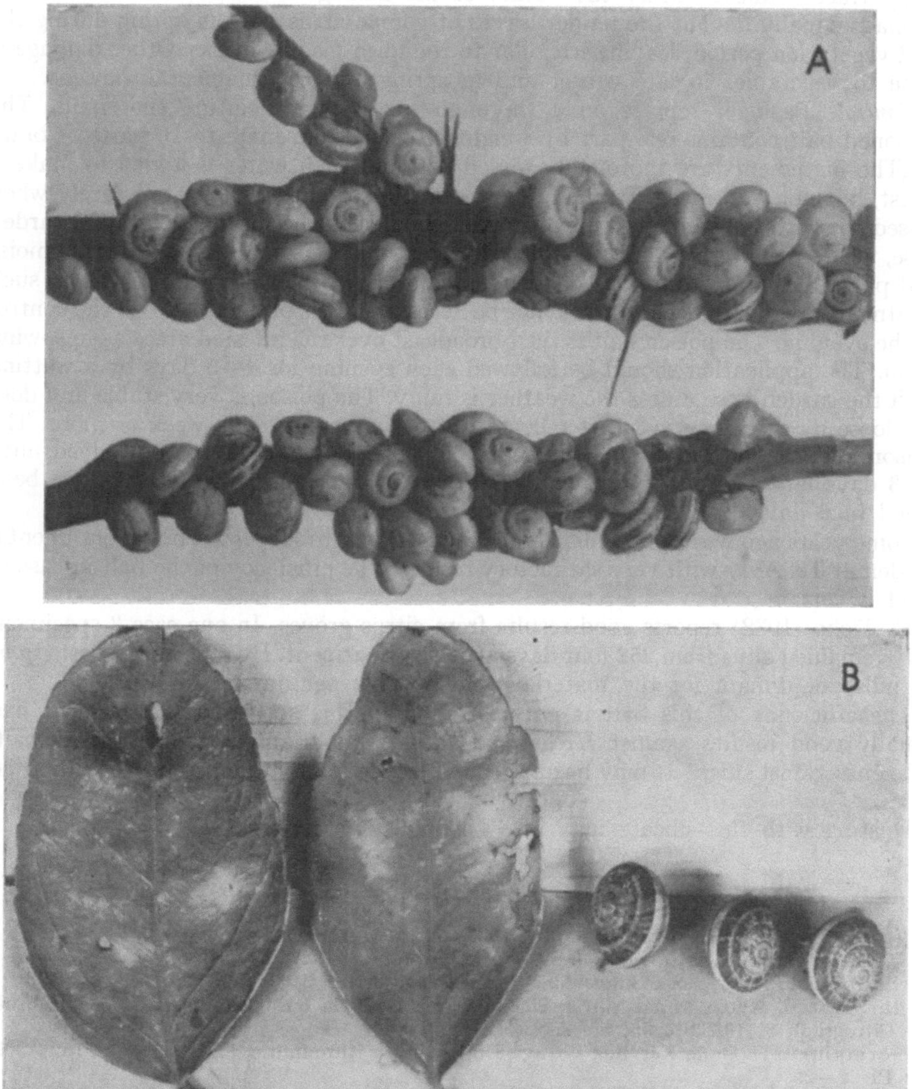

Fig. 143.
Snails in citrus orchards. A. *Helix* (*Euparypha*) *pisana* in summer-sleep on *Acacia farnesiana*-hedge in citrus orchard. B. Feeding marks of snails on orange-leaves. Note the irregular margins of the feeding area.

after the first heavy rains and lasts throughout the picking season. The snails feed almost round holes with irregular margins into the rind of the fruits. The diameter of these feeding marks is up to 1 cm and even more. The rind is usually eaten in its entire depth, but does not attack the pulp. Another type of feeding is more superficial and does not surpass 1 mm in depth. On some fruits such feeding marks may even cover one half and more of the total surface. All species of citrus fruits are equally attacked. All fruits with open and wet wounds, which have not dried, are unfit for packing. In 1938 snails caused 1/3 to 1/4 of all culls on grapefruits at the end of the picking season in the Emek Yesreel. The feeding is mainly nocturnal. But the slimy traces indicate easily the author of the damage.

Snails equally devour the tender leaves of young citrus trees in spring during the first vegetation period, leaving irregular to roundish feeding marks. Other damage is done to vegetables in late winter and in spring and to ornamental gardens.

Control: Basinger reports on a thorough success with baiting the snails. The poisoned bait contains one part by weight of calcium arsenate to 16 parts of bran. ,,The ingredients are thoroughly mixed dry and then water is added to make a moist, but not wet, mash. It must be dry enough so that it will scatter nicely when tossed out with the hand. The infested areas should first be sprinkled with the garden hose, where water is available, as this brings out the snails and keeps the mash moist and palatable. As a control will be made here only in the rainy winter season, such treatment in orchards will as a rule not be necessary. It is preferable to do the control in the evening. The poison bait is then broadcast over the infested areas as in sowing grain. The application should be followed each evening for 4—5 days by a wetting with the garden hose unless the weather is rainy. The poison is very stable and does not loose its efficiency even after being wet each evening for a week or more. The poison acts slowly and the results from an application should not be judged until 2—3 days later. In one experiment 96 % out of 6431 active *H. pisana* have been killed in 6 days." (Basinger).

Some years ago we applied this bait experimentally to *H. pisana* on an ornamental garden at Tel Aviv with very satifactory results. The efficiency of the bait in case of need is certain.

R. Ticho (1938) reports good results from citrus groves. In one case 3 specimens only remained alive from 252 four days after the treatment. He reports the costs to be 65 mils per dunam for the materials and 30 mils per dunam for work.

The efficiency of this bait is not restricted to this species only. Basinger had equally good results against *H. aspersa* and it was adapted from a poison bait, efficient against slugs. It may be applied, wherever snails are injurious to cultivated pests.

Dusting with fluo-silicate may also be applied in nurseries.

References.

A. J. Basinger, Journ. Econ. Ent. **16,** 1923, pp. 456—458, 522—526.

A. J. Basinger, Mo. Bull. State Dept. Agric. California, 1927, pp. 51—77.

F. S. Bodenheimer, Schaedlingsfauna Palaestinas. Berlin, 1930, pp. 53, 164.

T. de Stefani, Nuovi Annali Agric. Siciliana **2,** N.S. 1913, Fasc. 1. translated in California Citrograph **8,** 1922, p. 35; 56—57.

L. Germain, In H. G. de Kerville: Voyage zoologique en Khroumirie, Paris 1908. pp. 182—209. Pl 26—29.

A. Girard, Jorn. Sci. Math. Phys. Nat. Academie R. Sci. Lisboa, 1888, No. 47, pp. 160—166.

R. Ticho, Boustnai X. 1938. p. 30/31 (in Hebrew).

12. CITRUS PESTS IN OTHER COUNTRIES OF THE MIDDLE EAST.

Only Palestine and the Lebanon produce in the Middle East more citrus fruits than are needed for local consumption. In all the other countries vaste possibilities still exist for expansion of citrus cultivation for internal consumption. Only Egypt is a possible competitor on the world market in the near future. Its small, thin-skinned, sweet orange is, however, very different from the Jaffa-orange, which is more estimated by European consumers.

The following short surveys are all based on personal observations, with the exception of that of Transcaucasia. It is highly surprising to learn how the local climates determine the citrus fauna to a much higher degree than usually is assumed for the ubiquistic fauna of a subtropicopolitic host. In none of these countries has the citrusfauna been studied as intensively as in Israel, which is at present in citrus cultivation as well as in citrus pest control, the most progressive area of the Middle East.

A. CITRUS PESTS IN EGYPT.

The area of citrus cultivation has been steadily increasing in Egypt since world war I. There are five times as many citrus trees in that country now, as were in 1920. The following table gives a survey of this increase:

Region	1920 feddans	1929 feddans	1929 trees	1931 trees	1933 trees
Lower Egypt	3578	11219	3.074.400	4.999.073	5.760.322
Middle Egypt	1340	3032	536.100	954.198	1.024.470
Upper Egypt	555	1483	218.700	390.194	401.692
All Egypt	5473	15734	3.829.200	6.343.665	7.186.491

Most of the citrus area is in Lower Egypt. The horticultural conditions of citrus growing are fully discussed by Brown (1924). Experiments with various rootstocks are continued. Our „Jaffa" seems to be a preferred oculation graft. The following table shows the rising trend for export (from 130.000 klo. in 1920 to one million klo. in 1933) and the falling trend of import (from 19.000.000 klo. in 1920 to 4.000.000 klo. in 1933):

Variety	export in kilograms 1920	export in kilograms 1933	import in kilograms 1920	import in kilograms 1933
oranges	—	464.262		2.574.549
sweet lemons	129.247		17.326.307	
mandarines	—	545.029		821
others	—		1.758.757	1.424.688
Total	129.247	1.009.291	19.085.968	4.000.078

Competent surveys on the citrus pests of Egypt have been given by Willcocks (1922) and Hall (1924). Additional information has been obtained by the plant protection service of the Ministry of Agriculture and by three short visits of the writer.

The most important citrus pests of Egypt is the black scale (*Chrysomphalus aonidum* L.). All experts agree that citriculture would be impossible in Egypt, if not a well established system of fumigation would help to check this scale. *C.*

aonidum is much more abundant in Lower than in Upper Egypt, owing to its preference for a humid atmosphere. Heavy leaf dropping and incrustation of the fruit lead to a reduction of the crop in a heavily infested grove to one tenth of its former value, if unchecked. Many orange groves have been uprooted in earlier years owing to this scale.

A number of natural enemies are present:

Coleoptera	Coccinellidae	*Chilocorus bipustulatus* L.
		Exochomus flavipes nigripennis Er.
		Pharoscymnus varius Kirsch.
		Scymnus includens Kirsch.
	Nitidulidae	*Cybocephalus flaviceps* Rtt.
Neuroptera	Chrysopidae	*Chrysops vulgaris aegyptiaca* Nav.
	Coniopterygidae	*Coniopterygidae* spp.
Thysanoptera		*Haplothrips cahirensis* Tryb.
Hymenoptera	Aphelinidae	*Aphytis chrysomphali* Mercet
		Aspidiotiphagus citrinus Craw.
		Aspidiotiphagus lounsburyi B. et P.

The importation of *Comperiella bifasciata* How. is under consideration. Fumigation was organised since 1911 by Dr. L. M. Gough. Originally it was executed by fumigation brigades of the Ministry of Agriculture. Now fumigation is organised on a business basis under governement supervision. The original pot method of fumigation is now largely replaced by cyanic dusts or by liquid HCN. The following table describes the increasing extent of this work:

Season	Gardens fumigated	Trees fumigated	Season	Gardens fumigated	Trees fumigated
1911/12	3	2000	1923/24	4016	738.964
1912/13	31	10000	1924/25	3339	681.216
1913/14	64	31980	1925/26	3400	796.492
1914/15	31	16288	1926/27	4197	966.337
1915/16	46	26320	1927/28	4485	1.084.120
1916/17	79	33750	1928/29	4426	1.153.744
1917/18	167	58877	1929/30	4868	1.286.160
1918/19	86	53817	1930/31	6859	1.706.987
1919/20	270	169738	1931/32	6280	1.748.643
1920/21	1000	388236	1932/33	8754	2.191.669
1921/22	1222	332974	1933/34	10.475	2.675.505
1922/23	3142	686761			

Fumigation is made obligatory by law, and the earlier suspicions of the farmers against this procedure have long since been overcome. In comparison with Palestine the heavy infestation of the twigs should be noted.

Red scale (*Aonidiella aurantii* Mask.) is common everywhere, but more serious in Upper than in Lower Egypt. It infests especially limes and lemons, incrusting the twigs and branches. In consequence of the fumigation campaign against black scale it is generally well under control. Without this check it certainly would be a major pest.

Mycetaspis personatus Comst., a recent introduction into the Alexandria district, is so far an occasional and minor pest on citrus only. The Quince scale (*Hemiberlesia lataniae* Sign.) has been recorded by Hall from orange trees, but is not injurious on this host. The citrus mussel scale (*Mytilococcus beckii* Newm.) is common in the delta, especially along the Mediterranean coast. An attempted eradication campaign was unsuccessful. The black chaff scale (*Parlatoria zizyphi* Luc.) has been once imported on nursery stock, but was eradicated before it could spread.

The Florida wax scale (*Ceroplastes floridensis* Comst.) appears in moderate infestations in the coastal area. Its honeydew excretion causes minor injury. The larger fig wax scale (*Ceroplastes rusci* L.) is only occasionally found on citrus. The common soft scale (*Coccus hesperidum* L.) is spread everywhere and sometimes injurious on young trees.

The citrus mealybug (*Pseudococcus citri* Risso) lives all over the citrus area of Egypt. No outbreaks are known, so far. It is checked by climate and by a number of natural enemies, among which we mention the ladybeetles *Oxynychys marmottani* Fairm., *Scymnus includens* Kirsch. and *S. syriacus* Mars. The ladybeetle *Cryptolaemus montrouzieri*, whose introduction was attempted for the biological control of various mealybugs, failed to become established. The fears of Willcocks, that the newly introduced Hibiscus mealybug (*Phenacoccus hirsutus* Green = *Pseudococcus perniciosus* Newst.) may become a serious citrus pest, did not realise. This polyphagous tropical mealybug is now common all over Egypt, mainly on lebbek, acacias, hibiscus and other trees, but has lost much of its initial impetus, mainly due to a number of indigenous predators and parasites of mealybugs. It shows no preference for citrus, which it infests occasionally.

The citrus fluted scale (*Icerya purchasi* Mask.) is of general occurrence and causes locally sometimes trouble by its abundant honeydew excretion, mainly on lemon trees. It is fairly well under control by the ladybeetle *Novius cardinalis* Muls., which is also an enemy of the Egyptian fluted scale (*Icerya aegyptiaca* Dougl.). This species is only exceptionally met on citrus and not injurious on it.

Aphids are rare on citrus in Egypt. The two most dangerous species, *Toxoptera aurantii* and *Aphis tavaresi* are not yet recorded. Occasional records only refer to the polyphagous cucumber aphis (*Aphis frangulae* Kalt.), the green peach aphis (*Myzodes persicae* Sulz.) and the black *Aphis laburni* Kalt.

The citrus white fly (*Aleurotrachelus citri* P. and H.) is more common in Upper Egypt, where it sometimes causes leaf shedding. No damage is recorded from Middle and Lower Egypt.

The Mediterranean fruitfly (*Ceratitis capitata* Wied.) is well spread all over Egypt. It is, as a rule, no serious danger to the main citrus districts. In small private gardens, where a few citrus trees grow together with peaches, guavas, apricots, etc., infestation on citrus is sometimes high. Major outbreaks may be injurious in some exceptional years, but as a whole the fly is a minor pest on citrus only.

Rose beetles, such as *Tropinota squalida* Scop., visit citrus blossoms not infrequently, but no damage has so far become known. Two Buprestid beetles (*Sphenoptera ardens* Klug and in Upper Egypt *Chrysobothris dorsata* F.) develop in the wood of dying or unhealthy citrus trees. They both are polyphagous and show no special preference for citrus. The larvae feed underneath the bark, separate the bark from the wood and hasten the death of the tree. The only moth recorded so far from citrus in Egypt is the cotton worm (*Prodenia litura* F.). The caterpillar feeds on the seedlings, mainly of sweet oranges. Bishara Effendi showed me also some young oranges into the rind of which cotton worms had eaten holes (½ cm in diameter).

Thrips are very rare on citrus. A few individuals of the polyphagous *Scirtothrips mangiferae* Priesn., common on mango and *Convolvulus*, have been noted in citrus flowers. This lack of thrips is significant, as Prof. Priesner, the eminent specialist of this group, has devoted much attention to the thrips of Egypt.

The mites of Egypt have been revised recently by Taher Sayed (1942). The following species are sometimes met on citrus, without causing serious damage:

the common red mite (*Epitetranychus althaeae* Hanst.), *Tenuipalpus obovatus* Donn. and *Bryobia sp.* The main citrus pest is *Anychus orientalis*, which also infests *Ricinus communis* and sol ornamental trees. „Infested fruits have a russety appearance before they are mature. In some orchards this is the main pests when compared with other citrus pests. Not all citrus species are equally susceptible to this mite. Sweet lemon is more susceptible than orange and mandarine. Bitter lemon is the least susceptible." Sulphur dusting is recommended in winter, oil emulsion (Nan or Volk in 1 %) in summer for its control (Taher Sayed (1940).

In some of the oases (Kharga and Dakle) rats sometimes feed to some extent on the bark of citrus branches, probably *Rattus r. frugivorus* (i.l. Prof. Priesner).

Dropped or damaged citrus fruit is intensively infested by *Drosophila spp.* and by larvae of the beetle *Carpophilus hemipterus* L. The caterpillars of *Ephestia cautella* Walk. live on the dried remains of the decayed fruits.

The quarantine service is well organised and has intercepted a number of potential citrus pests. For some pests an internal quarantine attempts to prevent the spread of some pests into other citrus areas.

References.

T. W. Brown, Min. of Agric. Egypt Techn. Bull. 44. 1924.
W. J. Hall, Min. of Agric., Egypt. Techn. Bull. 45. 1924.
H. Priesner, Min. of Agric., Egypt. Techn. Bull. 117. 1931.
H. Priesner and M. Hosny, Bull. Soc. Fouad. Ier d'Entom. 1940. p. 58—70.
M. T. Sayed, Soc. Fouad Ier d'Entom. 1940. p. 41—45.
M. T. Sayed, Soc. Fouad. Ier d'Entom. 1942. p. 81—132.
F. C. Willcocks, A Survey of the more important economic insects and mites of Egypt. Cairo. 1922. p. 165—170.

B. CITRUS PESTS IN CYPRUS.

Morris (1937) gives a survey of the citrus pests of Cyprus. The writer also visited on two occasions its citrus areas. Citrus has been grown for many centuries on Cyprus. Small, old lemon groves are well scattered all over the coastal plains of the island. Orange cultivation is a recent introduction. It is now growing rapidly, mainly stimulated by the benefit of the imperial preference taxation in the United Kingdom, from which Palestine as a mandate was excluded. Apart from the imperial preference some other factors facilitate the rapid expansion of citrus cultivation on the island: cheap prices of land, cheap labour, cheap water (when present). The loss of markets for other crops, such as potatoes, vegetables, pomegranates were another stimulating factor. The distribution of citrus groves (no. of trees) before the war (in 1934) was:

District	Lemons	Mandarines	Oranges	Sweet limes	Grapefruits
Nicosia	7.548	3.264	65.836	12.442	3.583
Larnaca	7.952	1.391	18.065	523	17
Limassol	25.951	813	3.907	1.201	4.002
Famagusta	6.505	4.944	232.582	1.724	2
Paphos	12.296	3.826	15.461	2.432	—
Kyrenia	27.514	2.919	4.745	1.037	600
Total	87.766	17.157	340.596	19.362	8.204

The export of oranges and lemons (number of fruits; sic!) since 1925 was:

Year	Oranges	Lemons	Year	Oranges	Lemons
1925	17.232.500	1.802.190	1930	17.070.531	1.094.790
1926	14.275.489	1.391.760	1931	21.638.000	2.497.000
1927	20.774.568	1.789.290	1932	17.344.937	2.096.373
1928	16.632.141	4.264.905	1933	20.266.421	3.112.715
1929	13.009.961	2.665.190	1934	28.000.000	not available

In the same year (1934) the total production was:
lemons: 9.386.160, oranges: 40.000.000, mandarines: 2.500.000, sweet limes: 1.500.000, grapefruits: 2.950. Oranges and grapefruit areas have much expanded since, and many young orchards are now in fruit.

The two major pests all over the island are red scale and Mediterranean fruit fly.

Red scale (*Aonidiella aurantii* Mask.) is present in all citrus groves and apparently doing more damage than in Palestine. It mainly infests lemons, causing sometimes heavy defoliation. Host-preference for lemons and less decimating spring winds are probably the main reasons for the heavier infestation. Governement is ready to perform fumigation at the owner's application and cost. The pot method is prevailing (sodium cyanide 1 oz; sulphuric acid 2 oz; water 4 oz). Also some amount of spraying is done.

The citrus mussel scale (*Mytilococcus beckii* Newm.) was in 1934 rather common in the Limassol district. Morris says it was first discovered in that year, but its introduction must date some years back, at least. We found in October 1934 heavy infestation in some orchards on the Jaramaia River. The leaves were heavily encrusted in a 25 years old lemon orchard. No infestation was observed in two neighboured orchards, 3 and 5 years old, the seedlings of which had been taken from the infested orchards. This agrees with our opinion on the epidemiology of *M. beckii* (cf. p. 317). An eradication campaign had been undertaken by the governement in summer 1934, yet in October the same orchards were again heavily infested by living mussel scales. It has, however, to our knowledge, not spread beyond the Limassol district. The absence of *Chrysomphalus aonidum* L. and *C. dictyospermi* Morg. may be mentioned, as well as that of *Parlatoria pergandii* Comst.

Coccus hesperidum L. occurs everywhere, especially on young citrus trees, but does not become a pest. Occasionally only *Saissetia oleae* Bern. is met on citrus. *Ceroplastes rusci* L., a common fig pest, was observed by Morris in one or two localities on orange trees, causing by its sooty mould discoloration on leaves and fruits. We found it at Lefka and at Karavas. Two very recent introductions on seedlings from Palestine into the Limassol district in 1934, *Ceroplastes floridensis* Comst. and *Icerya purchasi* Mask., were apparently eradicated by 1937. *Pseudococcus citri* Risso is common everywhere, without growing abundant. We saw a fairly dense population in an old lemon grove near Lefka. About 1932 Mr. E. E. Green (i.l.) informed us, that he had found *P. comstocki* Mask. on citrus in Cyprus. The species has not been found since. Anyhow, it did not cause any major outbreak, as in its initial phase in Palestine.

Lindinger (1912) mentions *Hemiberlesia lataniae* Sign. on citrus from Cyprus. Hall (1925) collected it from branches and twigs of citrus in one garden near Famagusta and the Egyptian plant quarantine intercepted this scale sometimes from oranges of Cyprus provenience. Morris looked especially for the quince scale on citrus, yet found it to be rare and of little importance. Other interceptions by the

Egyptian plant quarantine service are *Aspidiotus hederae* Vall. on lemons and oranges, *Parlatoria zizyphi* L. on oranges, and *Chrysomphalus aonidum* L. on lemons from Cyprus (Hall 1922) are almost certainly wrong (not the determinations, but the declaration of origin).

The citrus aphis *Toxoptera aurantii* Fonsc. abounds apparently in all citrus districts. We noted the rolled leaves, which it produces, commonly near Famagusta and Lefka. The variegated capper bug (*Stenozygum coloratum* Kl.) causes sometimes minor damages to young shoots of citrus (Morris).

We observed thrips scars on lemon fruits at Lapithos. No thrips has so far been reported from citrus on Cyprus.

The interesting observation of Mr. Morris of caterpillars of the swallowtail (*Papilio machaon sphyrus* Z.) on leaves of orange trees has been mentioned before. The first observations date back to 1933 and 1934 in the Famagusta district. This is beyond Palestine the only Mediterranean region, where this transition occurred.
In addition caterpillars of *Charaxes jasius* L., which usually live on *Arbutus*-shrubs, have been taken on young citrus trees (Morris).

Beetles are of very minor importance only. Morris observed once a good deal of damage to young orange leaves caused by the very common mallow fleabeetle *Podagrica malvae* Ill. On another occasion the common weevil *Lixus algirus* L. fed on orange leaves. *Epicometis hirta* Poda is recorded, probably without reason, as damaging orange blossom, and as causing damage in the larval stage on seedlings in nurseries. Also an unidentified ant was once taken, whilst feeding on orange leaves.

Ceratitis capitata Wied. abounded in every citrus district in Cyprus in October 1934. In some orchards clensel traps contained many hundreds of flies each, over 90 % of them being *Ceratitis*. The intensity of the attack is so high, because other fruit trees, such as apricot-, peach, apple- and other trees are invariably grown in the immediate neighbourhood of or mixed with citrus. In the gardens visited, where two Clensel traps were distributed on every citrus tree, every orange showed signs of multiple oviposition of *Ceratitis*. Under high population pressure of the fly, the bait has obviously little protective value. The spraying of baits was abandonned in consequence of heavy sooty mould development on leaves and fruits. Mandarines and thin-skinned oranges suffer most. Red mites, apparently *Epitetranychus althaeae* Hanst., were fairly common on young plantations in the Famagusta district. They covered the leaves in October 1934 and caused silvery decolorations on the outer side of the fruits. The mites were rather active at that period. Klein also reports the citrus rust mite (*Phyllocoptràta oleivorus* Ashm). as a pest from Cyprus.

Rats (*Rattus frugivorus* Raf.) are locally an important citrus pest. In one 25-years old lemon grove at Kyreni over a dozen trees were derobed of more than half of their bark on stem and branches by these rodents. The damage is well known to the peasants, who told us that the trees die one to two years after the damage is done. Trapping and baiting is often done against the rats, but annually a certain number of citrus trees falls as sacrifice to their activity.

In dropped and damaged citrus fruit the usual association of *Drosophila*-maggots and *Carpophilus*- larvae is found everywhere.

In old lemon groves, where gummosis is fairly common, many of the old trunks are hollow. Larvae of *Rhesus serricollis* Mots. and termites (*Reticulotermes lucifugus* Gr.) contribute to the destruction of these stems. They are anyhow, of secondary importance only.

References.

H. M. Morris, Agric. Dept., Cyprus. Bull. No. 4. 1937.
W. J. Hall, Min. of Agric., Egypt., Techn. Bull. 45. 1924.
K. Lindinger, Die Schildlaeuse Europas. Stuttgart. 1912.

C. CITRUS PESTS IN THE LEBANON AND IN SYRIA.

Citriculture in the Lebanon and in Syria is mainly concentrated in the coastal plains and on the western foothills of the Lebanon Mts. The citrus area (oranges, lemons, mandarines) was in 1942 4.000 dunam for Syria, 50.000 dunam for the Lebanon. In 1940 the citrus area of the Lebanon was 60.000 dunam. War conditions have reduced the area as in Palestine. In 1934 only 26.000 dunam of citrus groves existed in the Lebanon, most of them around Tripoli. Some citrus trees are grown as favoured ornamental trees in the famous house gardens of Damascus and of the other towns of the interior.

In 1934, when we visited the citrus groves of the Lebanon, the methods of citrus cultivation were rather poor. Dead wood, poor growth, heavy scale infestations caught the eye immeditely in the densely planted groves. The leading principle, at that time at least, was to avoid all expenses and to regard any returns as net profit. The Tripoli gardens were mostly at an age of 40 to 50 years at that time. The trees were planted at distances of 2.5 to 3 meters, few were higher than 4 m. The crown was so poor, that fumigation operations could be easily performed in the orchards, in spite of the density of plantation. The hedges consisted mainly of willows, reeds, maples and pomegranates. All these hedge plants are reasonably free from *Chrysomphalus aonidum.* In 1934 the areas of oranges and lemons were almost equal in size, with mandarins not uncommon.

The citrus groves around Tripoli are grouped in three centres. The northern one is on the Bedani River, the southern one on the dunes. Both suffered relatively little from pests in April 1934, with *Chrysomphalus aonidum* and *Mytilococcus beckii* being present in fair numbers. The central group of the Tripoli gardens was so heavily invaded by pests, that it is difficult to describe the situation. *Chrysomphalus aonidum* was most common and first stage larvae abounded everywhere. Almost all females were living and ovipositing at that time. *Mytilococcus beckii* occurs more in patches, but within these foci in high density. *Aonidiella aurantii* was rare. Other citrus scales present were *Parlatoria pergandii, Coccus hesperidum, Ceroplastes floridensis* and *Pseudococcus citri.*

A few, more recently planted gardens were visited in the foothills of Mt. Lebanon at Zghorta (near Tripoli). In two of these gardens moderate infestations of *Chrysomphalus aonidum, Aonidiella aurantii* and *Mytilococcus beckii,* with a few *Icerya purchasi* were found. In two gardens above Zghorta the black chaff scale (*Parlatoria zizyphi* Luc.) dominated. This scale was hitherto entirely unknown from the eastern Mediterranean. Its importance as pest in Africa Minor is not less than that of the black scale. Compared with this scale the infestation with red and mussel scale is unimportant. The two groves were 4 respectively 10 years old. We were told that the nursery stock was obtained from Tripoli, where we could not discover any trace of *P. zizyphi.* It would be of interest if they could be traced to a nursery which obtained part of its stock from Algeria. We have no information how far the black chaff scale has spread in the meantime.

In comparison we report the results of a visit to the orange groves of Antelias near Beirut: *Chrysomphalus aonidum* was abundant, mainly in the shade and almost all females were living. *Mytilococcus beckii* and *Aonidiella aurantii* were fairly abundant and *Parlatoria pergandii*, *Coccus hesperidum*, *Ceroplastes floridensis* and *Pseudococcus adonidum* were also present. On a few rootlets the nema *Tylenchulus semipenetrans* was found, which has, however, no economic importance.

After this report of a visit to the citrus areas we give a short, yet uncomplete list of the citrus pests of both countries.

The most important pest is easily the Egyptian black scale (*Chrysomphalus aonidum* L.), which abounds in all the citrus areas of the coastal plains. It shows heavy incrustations and is common on many ornamental and other plants. Red scale (*Aonidiella aurantii* Mask.) is equally spread, but of little economic importance in the old plantations. *Aspidiotus hederae* Vall. was intercepted by the Egyptian plant quarantine service (Hall 1922) from oranges and lemons of Syrian provenience.

Fig. 144. *Parlatoria zizyphi* Luc. on orange leaves at Zghorta (Lebanon).

Yet in this case also we doubt the validity of the patria declaration. In spite of being quite common, we have not seen the ivy scale on citrus in these states, and nobody else has, so far. *Mytilococcus beckii* Newm., the citrus mussel scale, is not uncommon along the coast, yet does damage in local foci only. The grey chaff scale (*Parlatoria pergandii* Comst.) is common from Saida to Tripoli, but causes no damage. The serious infestation with the black chaff scale (*P. zizyphi* Luc.) near Zghorta has been mentioned before.

Coccus hesperidum L. occurs in all areas, including the Damascus gardens, but is hardly injurious. The Florida wax scale (*Ceroplastes floridensis* Comst.) is a minor pest along the coast. The citrus mealybug (*Pseudococcus citri* Risso) occurs everywhere, but we have no reports of damages. A slight infestation with another mealybug, *Pseudococcus adonidum* L., was observed at Antelias. It can be separated from *P. comstocki* Mask., the only other mealybug with very long caudal wax-threads in the Levant, by microscopical characters. Everywhere is the citrus fluted scale

(*Icerya purchasi* Mask.) which is of no economic consequence. Its enemy *Novius cardinalis* Muls. is introduced since long and well established.

Before the last war the control of citrus scales by fumigation and spraying was encouraged by a liberal export subvention of the governement. At an earlier stage fumigation was subvented by a participation of the governement. Nothing is known yet about the postwar conditions. Yet a serious attempt to eradicate the black chaff scale would be in the interest of the country.

Very little is known about the other citrus pests. The citrus aphis (*Toxoptera aurantii* Fonsc.) occurs. The Mediterranean fruitfly (*Ceratitis capitata* Wied.) is generally present along the coasts and not infrequent in citrus orchards. In fallen and decaying citrus fruit the maggots of *Drosophila* and the larvae of *Carpophilus* beetles are common.

The common red mite (*Epitetranychus althaeae* Hanst.) is widely spread, also on citrus. No damages have been reported. Moubarak (1938) reports a new pest which rapidly spread in October 1938 all along the coast, from Saida to Tripoli. He describes the damage, as follows: ,,One first observes a mealy, dusty layer on the bark of oranges, lemons and mandarines. Later the bark turns brown and almost black. The colour of the lemon bark grows silvery grey and the epidermis cracks slightly. The growth is arrested and the fruits remain small, especially when the disease begins early. Also the leaves and the small twigs are infested and turn blackish." Moubarak has isolated two mites from the infested fruits: *Paratetranychus citri* Mc Gregor and *Phyllocoptrata oleivorus* Ashm. It appears that Mr. Moubarak has made the determinations himself. They should be therefore accepted with reservation. It is, however, obvious that the damage is due to a species of *Phyllocoptes*, the same which has recently appeared in Palestine. Moubarak assumes that dry, warm weather favours the development, whilst high humidity suppresses it. He proposes a limesulfur solution for its control.

References.

W. J. Hall, Min. of Agric., Egypt. Techn. Bull. 22. 1922.

F. S. Bodenheimer, Bull. Soc. entom. France 1926. p. 41—47.

E. Moubarak, Rapport sommaire sur les traveaux du service de la défense des cultures au cours de l'année 1938. Beyrouth.

D. CITRUS PESTS IN TURKEY.

Citrus is commercially grown in four separate districts of the Turkish Republic, all in maritime plains: Cilicia and Antalya on the southern shores, the Egean maritime plain and the eastern corner of the Black Sea shore. The climates varies considerably in these different areas with a very decisive consequence for the insect fauna of the areas. Oranges are more common than all other citrus fruits together. The local orange is a thin-skinned variety corresponding to the ,,Franzawi" of Palestine. It was originally introduced from Algeria. The Jaffa variety has recently been introduced into the southern areas and obtains much higher prices on the local market. The other common citrus fruits are lemons, mandarins and citron (turunj). The entire citrus production of Turkey is yet far from satisfying the home market. The total production in 1940 was:

Varieties	Number of fruits	Fruits per ha.
lemon	34.500.000	109.000
orange	151.196.000	614.000
citron	22.538.000	63.000
mandarine	38.072.000	235.000

This production was distributed over total area by provinces (in percentages of each citrus fruit), as follows:

Fruit	Antalya	Aydin	Coruh	Gireson	Ismir	Isparta	Mugla	Ordu/ Trabzon	Rize	Seyhan
lemon.	61.4	—	—	+	7.9	0.7	0.5	+	1.2	28.1%
orange	20.8	3.0	1.7	0.3	71.	3.0	3.0	3.7	29.8	27.6
citron.	7.4	7.2	—	—	24.5	0.5	1.1	—	—	59.5
mandarine.	48.7	—	11.8	0.4	3.4	2.4	6.2	1.6	21.0	3.8

The great majority of all older trees are grown from seeds and are not budded. In the new plantations of the southern area sweet lemon is the usual root stock. Little cultivation is done in the densely planted older orchards. The trees, nevertheless, look strong. In recent decades many progressive farmers have learned modern methods of citrus cultivation in our settlements in Palestine. The new Citrus Experiment Station at Antalya will certainly do much to improve conditions.

Climate is rather favourable in summer. In spring the ,,foehn'' winds from the Taurus play the same role as our hamsins in Palestine. But in winter all citrus areas are in constant danger of freezing. At Adana frosts occur in January year after year, at Ismir and Rize in four out of five years. Frost can also be expected, as a rule, in these districts in November, February and March. The frosts are never severe enough to destroy the plantations, but they are a definite impediment, and probably will prevent any development of citrus monocultures on the Palestinian scale. From the view of winter frosts, the most favourable citrus districts of Turkey are Antalya and Dörtyöl on the borer of the Hatay. Frosts occur there less often and are also less intense. The citrus areas of Turkey are the most rainy districts. The phenology of citrus is illustrated by the following table:

Locality	Year	Begin of budding	Begin of blossom	Begin of fruit	Mature fruit
Adana	1931	10.IV.	25.IV.	30.IV.	1.XII.
Antalya	1931	25.III.	13.IV.	15.V.	15.X.
Dörtyöl	1931	15.III.	3.IV.	3.V.	1.XII.
Dörtyöl	1932	5.III.	5.IV.	5.V.	1.XII.
Rize.	1932	15.IV.	20.V.	25.VI.	25.XII.

The water for irrigation is usually supplied by rivers or big irrigation channels, and pumped by small motors into the orchard.

Unique in the Middle East are the old orange groves of Dörtyöl. They are usually 50 to 70 years old. The trunks are very often 30 to 50 cm thick in diameter. Those of the oldest trees cannot be embraced by one person. The smooth, clean stem is 150 cm. high, before the well developed crown begins. Most trees are 6 to 8, some up to 10 m high, one even 15 m. This giant yielded annually 10.000 to 15.000 fruits as against 2.000 to 6.000 of the other trees. Cultivation is superficially, manuring poor. Wheat, beans and other vegetables are often grown between the trees. Gummosis and similar diseases are common in the older plantations.

In discussing the citrus pests we begin with a survey of all pests, except the coccids, which will be later discussed by districts, in order to show the influence of the local climate on the distribution of this best known group.

The caterpillars of some polyphagous moths, such as *Prodenia litura* F. and *Laphygma exigua* Hb. feed occasionally in summer on the leaves, especially of young citrus trees. The honeydew moth (*Cryptoblabes gnidiella* Mill.) occasionally feeds in the rind of oranges, where two fruits touch and where the citrus mealybug has previously settled. Up to 20 fruits may drop on some individual trees in September. The damage is usually neglegible. The citrus flower moth (*Prays citri* Mill.) occurs in Cilicia, without doing serious harm.

A nanic specimen of *Apate monachus* F. was once collected at Adana in September 1939. Beetles of the genera *Tropinota, Epicometis* and *Oxythyrea* often abound on citrus blossom, without causing damage. The large Cerambycid beetle *Macrotoma scutellaris* Germ. (ex errore *Ergates faber* L. in 1934) was bred from a very old citrus trunk at Dörtyöl, where also ants (*Crematogaster*) had build their nest. In decaying fruit the larvae of *Carpophilus hemipterus* L. and *C. dimidiatus* F. live together with the maggots of *Drosophila melanogaster* Meig.

The Mediterranean fruitfly (*Ceratitis capitata* Wied.) is widely spread in south-western and central Anatolia. In W. Anatolia it probably occurs, yet no records are available. It does not seem that it is a very old introduction into Turkey. The number of annual generations in various districts is:

Locality	epidemiological average (hyperbola B)	maximum number of generations (hyperbola A)
Adana	3.5	6
Ismir	3.0	6
Istanbul	2.0	4
Rize	2.0	3.5
Ankara	2.0	3.5

The average climate of the eight provinces of Turkey fits as follows into the bioclimograph of *Ceratitis* (cf. p. 123 f.):

Province	Number of months in					Annual sum of effective temperatures
	below threshold of development	optimum	favorable	non-favorable	impossible	
Cilicia	3	—	9	—	—	2500 days-degrees
W. Anatolia	4	—	3	5	—	2100
Marmara	6	—	6	—	—	1500
Thrace	5	—	5	2	—	1500
Black Sea	5	2	5	—	—	1250
C. Anatolia	6	—	2	4	—	1200
E. Anatolia	8	—	—	4	—	500
S. E. Anatolia	5	—	—	3	4	2300

Conditions are thus favorable in Cilicia for *Ceratitis*-development, much less so in W. Anatolia with its drier summer (except perhaps in some coastal districts). The region around the Marmara Sea is more favourable, Thrace less favourable for the fly. In the Black Sea region climate would be very good, indeed, for *Ceratitis*, if not the heavy rains throughout the year would act in the opposite direction. No serious damage needs ever to be expected in Eastern and South-Eastern Anatolia. The presence of *Ceratitis* in Ankara (C. Anatolia) is of no consequence for citrus. It was, however, unknown from that area where in humid summers it is a great pest of peaches in September. The first generation appears in June.

The interesting citrus bug *Dionconotus cruentatus* Brllé., the damage of which was first described by the writer about 25 years ago from Palestine, was observed at Antalya causing the same damage in citrus groves with many weeds. Citrus flowers there from early April to middle of May. The bugs begin to appear in March and disappear in the middle of May. Dr. Yen informed us, that the damage caused by this bug was locally considerable in 1938, but small in 1939. Big cicadas (*Melampsalta* sp.) cause heavy damage in a few orange groves near Antalya. They oviposit in summer in the tender citrus twigs, as well as on many shrubs and herbs. Loose wood fibres

Fig. 145.
Melampsalta sp. on citrus at Adalia: A. *Melampsalta* sp., B. oviposition marks in twigs, C. gum-secretion from oviposition holes in orange twigs (from Bodenheimer).

turn the attention to the oviposition area. A gallery leads downward from the entrance puncture of the ovipositor containing 5 to 10 eggs. Some of these galleries are usually close together. The infested twigs dry and die. The hatching larvae drop to the soil with their forelegs, which are powerful burrowing instruments. There they attach themselves to the roots of trees and herbs from which they suck their food. Large quantities of such larvae have been collected from the roots of grapevine at Mardin, Elazig and Mush. The only recommandation which we could give for the control of these cicadas on citrus was complete clean cultivation, in order to prevent mass development of the larvae on herbs within the grove.

The black citrus aphis (*Toxoptera aurantii* Fonsc.) is apparently common in citrus districts of Turkey. We have not yet seen it from Rize, which generally is the worst known citrus district of the Republic. Many of the tender leaves are curled in late winter and spring. The green *Myzodes persicae* Sulz. has once been collected in Cilicia on a young orange tree.

The greenhouse thrips (*Heliothrips haemorrhoidalis* Bché.) is so far the only thrips collected from citrus flowers in this country. If serious damage occurs, it has been overlooked, so far. The average number of annual generations is 4 at Adana, 3 at Ismir, and 2 at Istanbul and Rize. At Adana and at Ismir the summer is too dry to permit regular damage. The summers are better for the thrips at Istanbul, but almost perfect at Rize, where serious damage should be expected.

The termite *Reticulotermes lucifugus* Rossi builds sometimes galleries into the roots of very young trees and again in the stems of very old trees (Dörtyöl). The mole-cricket (*Gryllotalpa gryllotalpa* L.) does occasional damage in citrus nurseries. The Egyptian tree grasshopper (*Anacridium aegyptium* L.) feeds sometimes a few leaves. The common red mite (*Epitetranychus althaeae* Hanst.) was occasionally observed at Adana. More attention should be paid to it and to other citrus mites.

Among the non-arthropod pests of citrus we first mention the lemon rat (*Rattus r. frugivorus* Raf.), which has girdled many lemon trees in Cilicia. In the same area the citrus root nema (*Tylenchulus semipenetrans* Cobb) has been observed without doing notable damage. The same may be stated for the often abundant snails (*Helicidae*) on citrus trees.

The distribution of the citrus scales in the four main Turkish citrus areas is compiled in the following table (abundance from 0 to 3, + means accidental infestation only or doubtful record):

Species	Cilicia	Antalya	Ismir	Rize
Aonidiella aurantii Mask.	3	3	1	0
Chrysomphalus dictyospermi Morg.	+	+	3	1
Ch. pinnulifer Mask.	0	0	1	1
Mytilococcus beckii Newm.	1	0	0	0
M. gloverii Pack.	0	0	+	2
Parlatoria pergandii Comst.	3	3	0	0
Leucaspis hydrangeae Tak.	0	0	0	2
Coccus hesperidum L.	1	1	1	1
Saissetia oleae Bern.	0	+	0	0
Pulvinaria floccifera Westw.	0	0	0	1
Ceroplastes floridensis Mask.	1	1	0	0
C. rusci L.	0	+	1	0
C. sinensis Del Guerc.	0	0	+	3
Pseudococcus citri Risso	2	2	0	1
Icerya purchasi Mask.	1	1	1	1

The red scale (*Aonidiella aurantii* Mask.) is the most important pest in the two southern districts. On the Egean and the Black Sea shores it is of little consequence. The physical ecology is favourable in the southern areas, much less so in the Egean area, where the summers are too dry, and on the Black Sea shore, where they are too humid. The epidemiological average of annual generations (in brackets the number of months below the threshold of development) is at Adana 3 (3), at Ismir 3 (4), at Rize 2 (6). In W. Anatolia the red scale is replaced by *Chrysomphalus*. Counts performed by the Adana Entomological service on 100 leaves each in 1939 show the following seasonal trend:

	I	II	III	IV	V	VI	VII	VIII	IX	X	XI	XII
Females	1501	1716	1160	861	239	1046	863	969	671	1605	1397	2163
Larvae	1141	1931	1364	1142	767	435	353	1227	968	2530	1042	1418
% living females	10	27		23	0	28	5	19	7	4	23	5

In full agreement with the population trend in Palestine the population ebb lasts from May to middle of July. It reaches rapidly its peak until September/October and remains high during winter. The spring winds exert the same killing influence as our hamsins.

The western red scale *Chrysomphalus dictyospermi* Morg. is the dominating pest in the Egean plains from Nazilli to Bornova. In Rize it is rare, in the southern areas very rare, if it occurs at all. Its predominance in the Ismir area suggests that *C. dictyospermi* prefers, or can tolerate, dry summers better then *Aonidiella*. Counts on 100 orange leaves, performed by the Bornova Entomological service, show the following trend:

	VI	VII	VIII	IX	X	XI
Females	2303	5350	6529	15944	18944	958
Larvae	8364	12047	21063	110079	80379	4502
On 3 fruits	12	117	2078	6939	13960	777

In late October a white oil spray (Volck 1.5 %) was given. This table shows the intensity of infestation which this scale may develop on citrus. The general seasonal trend agrees with that of *Aonidiella* in Cilicia: The population peak is rapidly built up from July/August to September. From Bornova as well as from Rize and Trabzon we obtained *Chrysomphalus pinnulifer* Mask. This species can be separated from *C. dictyospermi* only by careful microscopical examination. It is not abundant. The Egyptian black scale (*C. aonidum* L.) has been introduced into all parts of Turkey abundantly, mainly with citrus fruit from Syria and Egypt. It is almost a miracle that this important pest could not be discovered, so far, in the citrus areas of Turkey, whilst infested imported oranges are sold openly on the markets.

The citrus mussel scale (*Mytilococcus beckii* Newm.) has been observed on orange and lemon trees at Mersin and at Dörtyöl, also on an ornemental lemon tree at Istanbul. It has four annual generations in Cilicia, where it has, apparently not caused damages. The narrow mussel scale (*Mytilococcus gloveri* Pack.) is common in the Rize district, where it mainly infests the young trees. This species flowers only in a warm and humid climate. The same is true for the Oriental *Leucaspis hydrangeae* Tak., a recent introduction into the Rize area, where it is fairly abundant on twigs and leaves of citrus, its only host in this region.

The grey chaff scale (*Parlatoria pergandii* Comst.) is common in Cilicia and Antalya. It has three annual generations and does no economic damage, owing to its lack of conspicuouness on the fruits.

The brown soft scale (*Coccus hesperidum* L.) and the citrus fluted scale (*Icerya purchasi* Mask.) occur at a low population level in all Turkish citrus districts, with rare and minor local outbreaks. The same is true for the citrus mealybug (*Pseudococcus citri* Risso). The olive soft scale (*Saissetia oleae* Bern.) and the fig wax scale (*Ceroplastes rusci* L.) have been taken from citrus at rare occasions only. The Florida wax scale (*Ceroplastes floridensis* Comst.) infests citrus in the two southern areas. It has there three annual generations and has not been, so far, troublesome. The Chinese wax scale (*Ceroplastes sinensis* Del Guerc.) is a recent introduction into the Black Sea area. It has one annual generation only, with oviposition in summer (about 2000 eggs per female). The last larva or the immature female pass the winter. Males mostly hatch in spring. It is fairly common and produces heavy honeydew excretion, followed by sooty mould. Of minor importance, also only in the Rize district, is *Pulvinaria floccifera* Westw.

Natural enemies of the armoured scales are some ladybeetles, mainly *Chilocorus bipustulatus* L., and a few parasitic wasps. They are, however, unable to check the increase of their hosts. Little is known of the parasites and predators of soft scales and mealybugs. The former are preyed upon by the caterpillars of two moths (*Coccidophaga scitula* Z, *Calymma communimaculata* Schiff; Noct.), the later by some ladybeetles (*Scymnus* sp., *Hyperaspis* sp.). The enemy of *Icerya*, the ladybeetle *Novius cardinalis* Muls., has been repeatedly introduced into Turkey from France, S. Italy and from Palestine. It is now permanently maintained at the Adana Experimental Station for distribution. Its permanent establishment in the orchards is often unsuccessful, probably mainly due to the generally low population level of *Icerya*, interrupted by sporadic local outbreaks only.

Entomogenous fungi prevail in summer and autumn in the Rize area with its wet summer. Most armoured scales and soft scales are killed and full with the mycelia. The species infesting the *Diaspididae* are still unidentified, those responsible for epizootics on soft scales are mainly *Cephalosporium lecanii* Zimm. These fungi are doubtless a major controlling factor for most citrus scales in the Black Sea area. Similar entomogenous fungi have been observed in scale insects of the other areas. The dry summer and autumn limits their activity there considerably.

Many experiments with fumigation and spraying were under way when the war broke out. They will probably be taken up again at the first opportunity.

References.

F. S. Bodenheimer, Hadar. 8. 1935. p. 10—14.

F. S. Bodenheimer, Ziraat Dergisi. Mart 1941. p. 1—4.

F. S. Bodenheimer, A survey on the injurious insects of agriculture and their control. Ankara. (In press. Turkish and English).

E. CITRUS PESTS IN TRANSCAUCASIA.

Citriculture in Transcaucasia is mainly restricted to a narrow zone with subtropical climate along the eastern shores of the Black Sea, beginning at the Turkish border and petering out in the north at Sotshi. The climate is similar to that of the neighbouring Rize district of Turkey: a mild climate with rainfall throughout the year. This clima type reminds one of that of C. Japan, and actually many new introductions from that area flower in the subtropical zone od Transcaucasia. Mountains begin to rise at a short distance from the sea, and temperatures grow quickly cooler with rising altitude.

Many plants, among them many citrus varieties, were introduced into this small area from various subtropical regions of the world, such as Japan and Florida. These plant introductions began before 1914, but were much intensified under the new regime. The plant quarantine service obviously was in those initial stages of development, not yet as efficient and as competent as it is today. In consequence, a great number of insects was introduced, which are unique in the Middle East. Today a very strict plant quarantine is maintained and controls all importations as well as the exported plants and fruits. An internal quarantine is, in addition, maintained against a number of important pests, which today are spread over a limited area only. Against initial foci of new or newly spread insect pests often eradication campaigns are undertaken.

A distinctive mark of all control measures in Transcaucasia is, that they are in every village, in every kolchos under the complete supervision and direction of the agricultural authorities. These are thus enabled to launch any control campaign, which they think advisable. This system should finally result in an excellent control system.

Little is known on the citrus pests as such. But fortunately two excellent studies on the coccids of the region have been published (Borkhsenius 1934, Gogiberidze 1938). We compile in the following lines mainly the excellent and detailed information of the later author. His publication is unfortunately published in Russian only and in a little accessible bulletin. We thank Mrs. R. Bodenheimer for its translation.

Our list of citrus coccids begins with a great surprise. Borkhsenius still lists *Aonidiella aurantii* Mask. as the common red scale of the district. Yet at a recent check following the publication of Mc Kenzie's monograph of the genus it became evident that this species is not uncommon in the northern area (Suchum, Gagri), but is not connected with citrus. The common red scale of all citrus areas is the yellow scale (*Aonidiella citrina* Cocq.), a native of Japan and China, and not yet reported from any other area of the Middle East. It infests all citrus species, especially mandarins, from Sotshi south to Maharadzen. It is polyphagous on evergreen plants. Its yellow scale separates it, apart from microscopic characters, easily from the reddish scale of the mature female of *A. aurantii*. Whilst *Chrysomphalus dictyospermi* covers the leaves equally and lives mainly on the upper side of the leaves, *A. citrina* forms groups and lives mainly on the lower side of the leaves. The yellow scale develops in shady places in dense crowns, and on the northern side of the trees. The damage on citrus is considerable. The species is viviparous and the female gives birth to about 100 larvae. It has three annual generations, the last is often facultative. Hibernation occurs in all stages, yet mainly as second larva. Winter cold is badly tolerated. Only 10 to 15 % of the scales survive the winter, as a rule. On mandarines it reaches the population maximum in late autumn. Common predators are the ladybeetles *Chilocorus bipustulatus* and *Ch. renipustulatus*. The following parasites were bred from it in Transcaucasia: *Prospaltella aurantii* How., *Coccophagus lecanii* Fitch, *Aspidiotiphagus citrinus* How. and *Aphelinus mytilaspidis* Le Bar. Biological control is not without importance in some years. White oils are little effective, as the spray rarely covers completely the lower side of the leaves. Fumigation with HCN gives good results.

The western red scale (*Chrysomphalus dictyospermi* Morg.) is a very important citrus pest from the Turkish border to Sotshi. Whilst at Suchum it was abundant in 1934, it has spread and increased at Batum since. At Gagri a few local foci are infested. At Sotshi it is only occasionally harmful. *C. dictyospermi* infests all citrus species, but especially mandarines. The leaves are usually encrusted on their upper side. They turn yellow and drop. Entire hectares are bare in autumn. Twigs and stems are rarely infested. In mandarine-plantations control is needed in order to prevent the dying of the orchards. Encrusted fruits drop prematurely. The young, and subsequently the older twigs die. The infested trees are also very sensitive to cold. Fruits decay after picking in a speed proportional to the infestation. *C. dictyospermi* has three annual generations in the area. It hibernates as female or as larva. The crawlers appear in late May/early June, in the middle of August and from the middle of October to early November. The average number of eggs is 100 per female. This scale also suffers much from the winter cold and only 10 to 15 % survive until the following spring. The population increase begins in midsummer and reaches its peak in the middle of autumn. No parasites or entomogenous fungi have been observed. *Chilocorus bipustulatus* feeds on it, but only in autumn, after the scale has reached its peak of abundance. Fumigation with HCN is the control measure.

A small focus of *Pseudaonidia theae* Mask. (=*P. trilobitiformis* Borkhsenius) was discovered on some citrus trees at Suchum in 1931. By 1932 the scale was eradicated.

The citrus mussel scale (*Mytilococcus beckii* Newm.) occurs everywhere, but on citrus only. In the north it is restricted to greenhouses. Big foci occur at Gonio and

Machinjauri. The scale infests stem, twigs and leaves, mainly of oranges and mandarines. The last generation also infests the fruit. It hibernates as female and as second larva. The female lays 120 to 150 eggs. *M. beckii* has two annual generations in the area. It is regarded as a serious citrus pest and controlled by spraying with white oils or by fumigation with HCN.

The narrow mussel scale (*Mytilococcus gloveri* Pack.) also occurs everywhere but on citrus only, especially at Gagri and Suchum. The hedges of *Poncirus trifoliata* are also heavily infested. It infests mainly stem and twigs, on citrus also the leaves. In slight infestations it is restricted to the lower side of the leaves. Stem and fruits are often encrusted. Whilst *M. gloveri* was abundant in 1932 and 1933, it decreased considerably in the two following years. It has three annual generations. The female lays 80 to 120 eggs. The parasitic wasp *Prospaltella aurantii* How. destroys in August/September sometimes up to 70 and 80 % of the scales. The ladybeetles *Chilocorus bipustulatus* and *Ch. renipustulatus* also feed on it. The control is performed by spraying with white oils or by fumigation with HCN. The later gives complete control.

The black chaff scale (*Parlatoria zizyphi* Luc.) was first recorded in 1916 in the Botanical Garden at Batum. Around 1934 a number of foci were found in Suchum, Batum and Gagri. At Gagri a 20 years old garden of orange and sweet lemon trees was found heavily infested. Only in the coastal area sometimes outbreaks occur. *P. zizyphi* mainly infests the northern side of the trees and the lower half of the crown, settling on the leaves, the tender twigs and the fruits. The economic importance is definitely less than that of the red scale. It is spread with nursery stock. Hibernation takes place as female or as second larva. The female lays 20 to 25 eggs only. It has two annual generations. The eggs of the first generation are laid in late May to early June, those of the second one from middle of August to late September. No parasites have been observed. An eradication campaign has been launched.

The grey chaff scale (*Parlatoria pergandii* Comst.) occurs around Suchum and Batum (Botanical Garden), mainly on citrus, but also on some ornamental plants. It was introduced to Poti with nursery stock from Florida, but rapidly eradicated there. On citrus twigs, bark and branches are attacked. *P. pergandii* is not abundant. It hibernates as female and second larva. The female lays 30 eggs. The scale has two annual generations in Transcaucasia.

The Japanese white scale *Leucaspis japonica* Ckll. was introduced into the Botanical Garden at Batum. There it infests stem, twigs, to a lesser degree the leaves of mandarines, magnolia, roses, etc. The infestation on citrus is heavy. Few foci are known, so far, and an eradication campaign has been ordered.

The brown soft scale (*Coccus hesperidum* L.) is widely spread, yet causes no damage. It has three annual generations. Another recent Far East introduction is the citricola scale (*Coccus pseudomagnoliarum* Kuw.). It can be easily separated from *C. hesperidum* by its symmetrical body which is equally rounded on the anterior and posterior body margins, and by its grey body colour. The citricola scale has one annual generation only. It hibernates in the last larval stage, the females, each of which lays about 2000 eggs, appearing in May and June. By July almost all of them are killed by parasites, among which *Eucomys swederi* Dalm. (also infesting *C. hesperidum*) and the fungus *Cephalosporium lecanii* Zimm. may be mentioned. *C. pseudomagnoliarum* infests all evergreen plants on all parts, preferring the lower side of the leaves. Males are unknown. It is of no economic importance, but widely spread.

The olive soft scale (*Saissetia oleae* Bern.) was found in Batum on citrus, and in a few other places. It is however rare on this host. It has one annual generation.

which oviposits in late May. *Saissetia hemisphaerica* Targ. was once only discovered on *Citrus mitis.*

Pulvinaria floccifera Westw. is a polyphagous species which occasionally only infests citrus. It has one annual generation only. The long and slender ovisacs with up to 2000 eggs appear in May. Hibernation occurs in the first two larval stages. *Pulvinaria aurantii* Ckll., easily separated from the preceeding species by its short ovisac, which is much broader than the body, has two annual generations. Hibernation as in *P. floccifera*. The ovisacs of the first generation with 700 eggs appear in the middle of May, of the second one with 400 eggs in late August. Males develop mainly on the upper, females on the lower side of the leaves. The main hosts are citrus, oleander and *Pittosporum*. It is accompanied by heavy sooty mould development and must be regarded as citrus enemy no. 2 (after *Pseudococcus gahani*). Active spreading occurs slowly only. This new introduction is mainly injurious around Suchum. *Cryptolaemus montrouzieri* Muls. feeds eagerly on the eggs and the fungus *Cephalosporium lecanii* Zimm. causes heavy epizootics in some years. Spraying with white oils has little effect, but fumigation with HCN is satisfactory. Of no economic importance as citrus pest is *Pulvinaria horii* Kuw, which was discovered in 1936 near Batum on *Rosa, Pyrus* and *Citrus*, especially on the former. An eradication campaign has been launched.

The Chinese wax scale (*Ceroplastes sinensis* Del Guerc.) lives all over the littoral, mainly on pomegranate, Feijoa and citrus, with preference on mandarines. It is a minor citrus pest and has one annual generation. It hibernates in the first two larval stages. The female lays almost 2000 eggs. The crawlers appear in July. No parasites have been observed. The ladybeetles prey with minor success on the first larvae only (*Chilocorus bipustulatus* and *Ch. renipustulatus*). *Cephalosporium lecanii* Zimm. mummifies the females and the last larval stage. In 1936 a heavy epizootic followed heavy spring rains. Fumigation with HCN is effective as control measure.

Among the mealybugs *Pseudococcus citri* Risso is not uncommon, but not yet recorded from citrus. *Pseudococcus adonidum* L. with its very long two caudal wax-filaments was recorded by Borkhsenius as an important citrus pest, whilst Gogiberidze mentions it from citrus but not as a pest of economic importance. We believe that the determination should be checked. Possibly *Pseudococcus comstocki* Mask. may be hidden under this name. Another recent introduction is *Pseudococcus maritimus* Ehrh. It is established at Poti, Sotshi, Suchum, etc. It is polyphagous, mainly infesting *Acacia*. It does little damage to citrus, among which it prefers mandarines and lemons.

Pseudococcus gahani Green was introduced in 1930 into Suchum and spread rapidly along the coast. It developed into the most important citrus pest, especially on mandarines. The mealybug lives in dense colonies on the stem, the branches, twigs and fruits, where it settles beneath the sepals and causes fruit drop. It is also accompanied by heavy sooty mould development. The damage was considerable in young plantations in 1930 and continued to be so until 1934. *P. gahani* has three annual generations, with oviposition in spring, in late July and in early October. The eggs of these successive generations are I:600—1000, II:300, III:150 eggs per female. The males swarm before the copula. No parthenogenesis occurs. *P. gahani* continues to develop on the picked citrus fruit. In spring it weakens the young sprouts and leaves, in autumn it leads to the above mentioned fruit drop. Later abundant on the fruits which crack. Mandarines often decay. The species is very resistent against oil sprays as well as against fumigation. It would be of interest to learn, if in the meantime the species has become less abundant in consequence of biological control, as did *P. comstocki* in Palestine.

Good results have been obtained against all these mealybugs by the ladybeetle *Cryptolaemus montrouzieri*. Also predatory Cosmopterygid larvae have been observed. The minor coccinellid predators and the internal parasites have apparently not yet been studied.

The citrus fluted scale (*Icerya purchasi* Mask.) showed considerable increase since 1931 on citrus. Since the introduction in 1933 of *Novius cardinalis* Muls. from Egypt the widely spread scale has lost its economic importance on citrus. It has 2 to 3 annual generations and an average egg production of 1200 eggs per female. Below 12° C the adults show heavy mortality.

References.

N. S. Borkhsenius, Survey of the coccids of the Black Sea shore of the Caucasus. Suchum. 1934.
A. A. Gogiberidze, The Coccidae of the humid subtropics of the Georgian SSR. Suchum. 1938.

F. CITRUS PESTS IN IRAK.

Irak is still the country without agricultural statistics. We have therefore to content ourselves with the following notes. Citrus cultivation is rapidly expanding in Central and Southern Irak. Citrus gardens grow always, at least during their early years, in the shade of date groves. Sun radiation is too intensive to permit young citrus plants to grow without shade. The local orange is very similar to the Egyptian variety. The fruit is small, thin- and smooth-skinned and sweet, with many seeds. Other varieties are under experimentation, but have not yet appeared on the market. The sweet lemon (numi halu) is often served. Guest calls it rightly „a rather insipid table fruit". Limes (numi Basrah, numi hamidh) are much used for lime drinks and the dried fruit is in general use in the Iraki kitchen for flavouring. Lemons (limun), tangerines (naranj) and pomelo (numi hindi) are also cultivated. Grapefruits have recently been introduced, but are not yet on the market. The peel of citron (turunj) is much used for confectionary and for desserts. The total production of citrus fruits is far from being sufficient to supply the demands of the local market. The import of citrus fruits from neighbouring countries is prohibited, in order to prevent the introduction of the Mediterranean fruitfly. This prohibition creates favourable prospects for further expansion of the local citrus industry.

Another favourable factor is the absence of many important citrus pests. This was first pointed out by Ramachandra Rao (1921) in a survey on the insect pests of Irak. The absence of many ubiquistic pests in Irak is almost certainly mainly due to the strict isolation of the country, which is surrounded to the west and east by great deserts. We have to assume, that especially citrus was introduced in earlier days by seeds only, thus avoiding the introduction of pests. This has changed in the last decades. But plant quarantine and the extreme climate of the country have hitherto prevented any important pest introductions, except in the Basra area, where *Aonidiella orientalis* Newst., *Phyllocnistis citrella* St. and an unidentified Aleurodid have become established. The following species have recently and repeatedly been intercepted by the plant quarantine service: *Aonidiella aurantii* Mask., *Chrysomphalus aonidum* L., *Mytilococcus beckii* Newm. and *Parlatoria pergandii* Comst., none of which became established. *Icerya purchasi* Mask. was introduced in 1939 into the Zafraniye farm on citrus nursery stock from Italy, but was speedily exterminated.

The most important citrus pest is *Aonidiella orientalis* Newst. in the Basra area, where it often incrusts the fruits and generally is abundant. The pygidial characters of this species at Basra are: Three well developed pairs of lobes, the first being the largest. Fourth lobe distinctly recognisable as sclerotic triangle. No plates lateral of fourth lobes. Plates between third and fourth lobes with small ensiform processes. Dorso-pygidial ducts in three rows, a single duct in the median interval beginning cephalad of anus. The three prepygidial segments with clusters of submarginal ducts. Genacerores in four groups: anterolaterals: 3(1—7), posterolaterals 3 (1—5). Scales almost circular, slightly convex, pale yellow to yellow-brown, 2.5 to 3.0 mm in diameter. Exuviae subcentric, yellow. Experiments with fumigation with hydrocyanic acid are made.

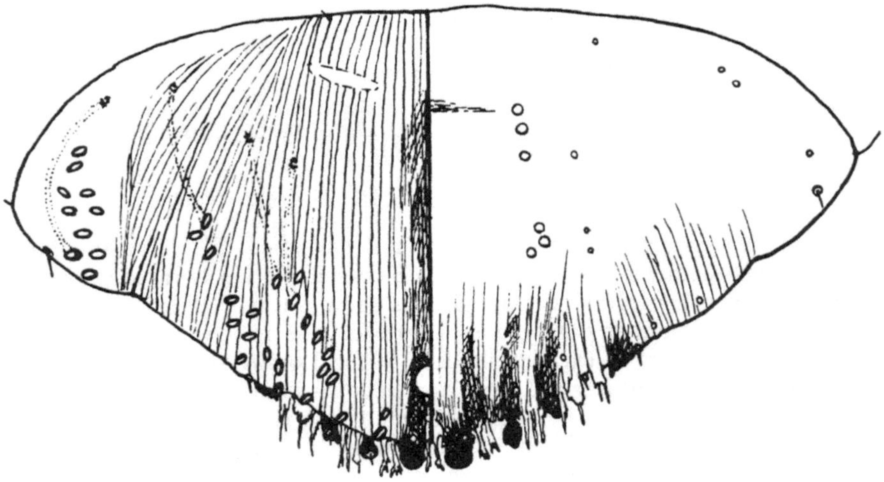

Fig. 146. Pygidium of *Aonidiella orientalis* Newst. at Basra (from Bodenheimer).

The only other coccid on citrus is *Coccus hesperidum* L., which we obtained on citrus from Baquba, Baghdad, Diyale and Diwaniye. It was never abundant and injurious. The citrus mealybug (*Pseudococcus citri* Risso) is widely spread, but was not collected from citrus.

Among aphids *Toxoptera aurantii* Fonsc. is wanting. *Aphis frangulae* Kalt., *Aphis laburni* Kalt. and *Aphis asclepiadis* Pass. occasionally form small colonies on the leaves, without doing any injury. *A. asclepiadis* is apparently even unable to maintain itself or to develop on citrus. The infestation took place from neighbouring oleander shrubs.

The common white fly of C. and S. Irak on Zizyphus sometimes is found in great numbers on citrus leaves in the Basra area. It is very similar to *Aleurotrachelus citri* P. et H. A second unidentified species was also found on citrus leaves in the same area, probably an introduction from the Oriental domain.

Wiltshire (1944) records caterpillars of *Prodenia litura* F. from citrus. The mines of the small citrus miner *Phyllocnistis citrella* St. were first observed by Mr. Dimitri at Basra in 1937. It increased considerably until 1941. The cold winter of 1941/42 decimated it to such a point, that it was first supposed to have been exterminated. In October 1942 we found only old mines. But in May 1943 a few fresh mines were found in one tree. The moth has a number of annual generations, whenever young leaves are present. To a minor degree it also mines in the epidermis of young, green twigs. The infested leaves are curled.

The common red mite (*Epitetranychus althaeae* Hanst.) rarely infests citrus. Yet in November 1942 oranges were received from Diyale which were abundantly infested and spoiled by *Schizotetranychus sp. aff. schizopus* Z. (det. H. Z. Klein). We have not found this mite or its damage again.

Rattus r. frugivorus Raf. does local damage to a few lemon and other trees by girdling the bark.

It is perfectly true until this day, that all citrus groves of Central Irak are practically free from pests. It is an exceptional sight for an experienced citrus entomologist to visit these gardens. The citrus orchards in the Basra area are however so heavily infested by *A. orientalis* Newst. and by a few minor pests, that they have their full load of troubles. So far, the infested fruit is sold at normal prices in the market. As no export of citrus fruit takes place, no economic stimulus exists to enforce control measures.

References.

Ramachandra Rao, A preliminary list of insect pests of Irak. Baghdad. 1921.
E. Guest, Notes on plants and plant products in Iraq. Baghdad. 1933.
E. P. Wiltshire, The Butterflies and Moths of Iraq. Baghdad. 1944.
F. S. Bodenheimer, A first survey of the Coccoidea of Iraq. Baghdad. 1943.
F. S. Bodenheimer, Notes and observations on the agricultural pests of Iraq. Baghdad. (In press)

G. CITRUS PESTS OF IRAN.

The citrus areas of Iran are restricted to the shores of the Caspian Sea and to parts of Khousistan, bordering S. Irak. Citrus are planted as ornamental trees often in the famous Persian gardens in the towns of the interior, whereever the hard winter permits their survival. The climate along the Caspian Sea shore is similar to that of the Rize district, that of Khousistan to the climate of the Basra area.

Unfortunately very little is known on the citrus insects. Prof. J. Afchar of the Karaj Agricultural College and Prof. N. S. Sherbinowsky of the Russian locust delegation to Iran confided a number of samples of citrus scales to the writer. These are about the only citrus pests so far known from Iran.

Aonidiella orientalis Newst., the Oriental red scale, (the same form as described from Baghdad) and the western red scale (*Chrysomphalus dictyospermi* Morg.) are common and in fair infestation on citrus leaves from both districts. They are apparently the most important citrus pests. Afchar records also *Aonidiella aurantii* Mask. from Mazenderan (Caspian Sea). The determination was made by Prof. A. Balachowsky and is therefore doubtless correct. It must however be a species of minor importance.

Restricted to the northern area are also the mussel scales (*Mytilococcus beckii* Newm. and *M. gloveri* Pack.). We have not seen specimens, but the latter species seems to be more common.

The black chaff scale (*Parlatoria zizyphi* Luc.) is common on the Caspian shore. It is doubtless a pest of economic importance. The polyphagous Oriental chaff scale (*Parlatoria morrisoni* Bdhmr.), the scale of which is similar to that of the grey chaff scale, was once collected by Prof. Sherbinowsky from citrus leaves in Khousistan. It is common on various other hosts in C. and S. Irak. The pygidial characters of the species are: Median lobes broad, twice notched. The three lateral lobes narrow and slender, distinctly notched. With macroduct between the median lobes. Three plates between third and fourth lobes, the middle one very broad and occasionally split. Very characteristic is one pair of macroduct openings centrad of the anterior genacerores (dorsally), another pair on each side just covering them from above. Genacerores in four groups; 6—8 anterior, 4—5 posterior genacerores. The economic importance of the species is unknown. The brown soft scale (*Coccus hesperidum* L.) is widely spread, but not common in both citrus areas. *Pulvinaria floccifera* Westw. is not uncommon in the Caspian district., from where also a few samples of *Ceroplastes sinensis* Del Guer. were obtained, which were destroyed by the fungus *Cephalosporium lecanii* Zimm. Restricted to the Caspian shores is *Icerya purchasi* L., which is apparently of little importance on Citrus.

References.

J. Afchar, Insectes nuisibles aux arbres fruitiers en Iran. Teheran. **1316**. (in Persian).
F. S. Bodenheimer, Bull. Soc. Fouad Ier Entom. **28**. **1944**. p. **85—100**.

We obtained once some samples of oranges from Lehaj in the Aden Protectorate of S. Arabia, which were heavily encrusted with *Aonidiella orientalis* Newst. This is, so far, the only record of citrus pests available from Arabia.

13. PRINCIPLES, TECHNIQUE AND ORGANISATION OF CONTROL OF CITRUS PESTS IN PALESTINE.

A. GENERAL REMARKS.

The principles of pest control are dealt with here only briefly, not because of the simplicity of technical and scientific problems of the control, but, on the contrary, because of their intrinsic and extrinsic complexity. The biologically trained entomologist is not an expert in the very complex sphere of insecticides, which are studied to-day by especially trained chemists and technicians, in the interest of the industry mainly. The entomologist can mainly help and further their work by executing routine counts of mortality and by pointing out biological stages or seasons which may be especially suitable to control-actions. The following chapter is therefore written by an entomologist, who reports the position of insecticide research and practice in Palestine from his point of view. A special type of an entomologist with a thorough training in insect physiology and general toxicology begins to appear and will be called on to perform a special task in the future development of insecticides. However, as has just been pointed out, this type of scientist only begins to appear and has scarcely begun to do his work through which our present purely empirical knowledge on insecticides may reach a scientific level. Another reason to remain brief, and to discuss — apart of some historical retrospectives — only principles, is that pest control today is in a revolutionary technical development, where new discoveries and technical improvements chase one the other with cinematographic speed. Almost every recommendation of today is bound to be in less than ten years entirely out ofdate.

Whereas the chemical trend in insecticides is more or less international, considerable differences exist in the various regions with regard to insecticide machinery. With respect to the latter, Europe lags greatly behind the U.S.A. and much remains still to be done in Palestine. Local agricultural practice and requirements differ and, as the following chapter will show, many problems of great economic importance remain to be solved.

The checking of the effect of any treatment is not easy. Our rules for counts of surviving scales, which are the same for checking spraying and fumigation, are:

The first material for checking is taken on the day of the treatment, before the trees have been sprayed or fumigated. The other samples are taken 7 days after the treatment [1]), sufficient material from every plot and material for another control in addition. From 500 to 1000 individuals scales are studied in every count (ab. 80 % adult females are chosen) and the following three categories are recognised:

1) dead, dried individuals (dead for some time)
2) dead, but not yet dried individuals (dead for 8 days at longest)
3) living individuals.

The dried individuals are not included into the final counts for checking purposes. Among 100 scales four may still be living and 10 may have died during the last weeks

[1]) Dr. Ben Amotz kindly informed me, that in Riverside (California) the mortality counts for fumigation are made about 7 days after treatment, those for spraying about a month after treatment.

(compare with check from untreated trees), the percentage of survival is then

$$x = \frac{4 + 100}{100 - 10} = 4.4\ \%$$

In this way only the sometimes heavy fluctuations of recent mortality, espec. after hamsin periods, which are not connected with the control measures, may be eliminated.

Example:

Category	1	2	3	Total (1+2+3)	Total (2+3)	Percentage of survivors (2+3)	Result: real % of survivors
Treated trees	56	369	12	437	381	3.2	3.3
Untreated trees	236	21	616	873	657	96.7	100.0

It is of importance to study also the ovicidal effect of the treatment. About 500 females of the species studied, are preserved in carefully closed test-tubes, having separated these from the control trees from those of each experimental plot. Two to four weeks later the number of crawlers hatched is counted and referred to the average from the check-plots which are put as equal to 100. (An example is given in the chapter on *Mytilococcus beckii* Newm.). Checking can be executed by trained, experienced and reliable persons only. Any statement made in the field (if no special field laboratory has been established) is quite valueless. The necessary equipment is a comfortable table and seat, a binocular microscope, a few small glass-vessels, needles, etc. The analysis should never be executed in a hurry or in the presence of people pressing for the result.

Under the lead of Dr H. Z. Klein we have almost abandoned the mortality count and lay the stress on the inspection of the fruit at harvest time.

Neither laboratory nor specialists have ever been at the disposal of any of the Palestine Research Institutions, to devote special attention to research on insecticides or on biological control. All such research has been done empirically, as by-work, by people fully occupied by intensive work along other lines. We tried, however, to benefit from every experience which farmers or agents of industrial concerns gained in this country. In this way no valuable experience was lost to us. To anyone who approached us we offered any technical help possible, especially by independent checking of experiments performed by agencies such as have just been mentioned.

Fumigation was started and executed by the Department of Agriculture of the Palestine Government about 1921. Some ten thousands of trees were fumigated annually with the pot method in N. Palestine against *Chrysomphalus aonidum* under the plant protection ordinance. Twenty years ago, the Government decided to leave the execution of the fumigation to the private initiative. Up to very recent years prices were very high and prohibitive (6 to 10 LP. per dunam as against about 2½ to 4½ LP. just before the war) and could not compete with the much less costly spraying. The stricter handling of the fruit inspection service and the general opinion that one fumigation equals the value of two sprayings have induced many farmers, after the big drop in fumigation costs, to give it a trial, even when not under obligation to do so by the plant protection ordinance.

Spraying with lime sulphur was attempted in some groves in the Hedera-district in 1929, but was soon replaced all over the country after 1931 by white oil emulsions. During some years the tank mix method has been given extended trials by some farmers, and they are decidedly not encouraging. In 1938 considerable injuries were caused by it. The prices of spraying are still much lower than those of fumigation and there is little doubt that, except in years of heavy outbreaks or in centres of exceptionally high damage, spraying will hold its place side by side with fumigation, as has been the case in California, Spain, Sicily and elsewhere. Recent experiences justify decidedly this view. Thus 94 to 98 % mortality and clean fruit was obtained with 1 to 2 % white oil emulsions against mussel scale and improved oils even gave satisfaction against black scale.

The intensity of control in the future will depend to a very large part on the vigour of the fruit inspection.

Until a few years ago the possibility and advisability of an eradication campaign e.g. against Egyptian black and mussel scale have been considered. At present this question is no longer raised as the areas of these species have become too large, and no way could be found to raise the necessary funds even when those areas were still reasonably restricted.

The plant quarantine service may still be greatly improved, which is discussed in a later chapter.

Better training and qualification of the staff which executes the control measures is urgently needed. A joint short-course of the Agricultural Experimental Station, Rehoboth and the Jewish Farmers Association in summer 1938 for sprayers is a first step in this direction. At present, both fumigation and spraying are performed by independent groups of private workmen, who have no special training and are lent the spraying machinery or the fumigation tents by those firms from which they buy the control materials (oils, HCN).

Biological control is rapidly extending in Palestine. In 1926/28 a small station was at work in Petah Tikva against *Pseudococcus citri*. In 1928/1932 the methods of biological control of Icerya purchasi by *Novius cardinalis* were worked out.

Since 1936 a special parasite-breeding station was opened by the Jewish Farmers Association at Petah Tikva, which has started to put out annually a more or less adequate number of parasites of *Pseudococcus citri*, about 50.000 *Sympherobius* and some thousands of *Scymnus*.

The entomological field-station at Hedera has begun, since early 1938, to breed *Sympherobius* against *P. citri* and has put out in 1939 far over 100.000 individuals. A third breeding station against *P. citri* was later opened at Rehoboth. All these stations were under the writer's supervision. The biological control of the recently introduced *Pseudococcus comstocki* Kuw. is described in its proper place.

Ovisacs of *Icerya* which have been collected by the farmers are kept in all of these field-stations and the *Novius*-beetles, which hatch from them are returned to the farmers for re-introduction in foci of the orchard.

References.

F. S. Bodenheimer, Hadar 6, 1933, p. 285—286.

B. FUMIGATION.

The history of Citrus fumigation is divided by Essig into five periods. It was wholly developed in California.

During the first period (1885—1888) D. W. Coquillet, F. W. Wolfskill and F. W. Morse tried HCN fumigation beneath tents, mainly against *Icerya purchasi*. Many types of tents and tent-pulling-over apparatus were described in these years, but the method remained very crude. The tents in use were not fit to retain the gas sufficiently. The gas was generated in very primitive apparatus beyond the tent and the generated HCN was forced into the tents by means of a small handblower. No accurate dosage was possible and all the work was done during the day.

During the second period (1889—1906) the problem of *Icerya* had lost its importance through the successful introduction of *Novius cardinalis* into California. Red and black scales became now the main objects of the fumigation. The main progress in these years was the developent of the pot generation method under the tent. This method which was long in use in Palestine, allowed a more accurate dosage. Night fumigation replaced the day time work, because during the night the soil is warmer than the air, and this caused a better distribution of the

gas below the tent. Coquillet and Woodworth were the leading personalities of this development.
The third or standardisation period (1907—1912) started with a more thorough investigation by the Bureau of Entomology, in order to develop an exact dosage system, the chemistry of fumigation, the physiological effect of the gas on the citrus-tree and to improve the mechanical equipment. In this important period were laid the foundations for the fumigation on an economic scale. The initiator of these improvements, and of most improvements of fumigation since then was R. S. Woglum, now Entomologist of the Californian Citrus Fruit Exchange. The first requirement was to get an easy method for measuring the space covered by the tent. At the beginning Woglum applied a tape, which was thrown over the tent, thereby measuring the distance over

Fig. 147. Tent fumigation with the pot method (from Bodkin).

and subsequently the distance around. During a short visit in Florida, Woglum became acquainted with Morrill's discovery of the marked tent, which was immediately introduced into California. The use of the marked tent meant progress, because it reduced considerably the time necessary for this measurement. Corresponding to the readings from the marked tent dosage schedules were worked out, which, with very small adaptations, are still in use to-day. A different schedule had to be applied, for large trees and small ones, as the gas leakage from the smaller trees was by far greater. Better tents and higher refined sodium, as well as potassium cyanide as raw materials marked further progresses of this period.

The fourth period (1913—1916) is called by Essig the period of machine fumigation. The old method of generating the gas beyond the tent was improved. The cyanofumer, as the modern type was called, allowed a very exact dosage, and the operation is much quicker than with the pot method. Sulphuric acid spots are to be feared no more, which is very important for the upkeep of the tent. These latter had suffered very much by the sulphuric acid and the short life-duration of the expensive tents added much to the costs of fumigation. The generation of the gas was very quick and its distribution under the tent better. W. Dingle was the first inventor of this improved

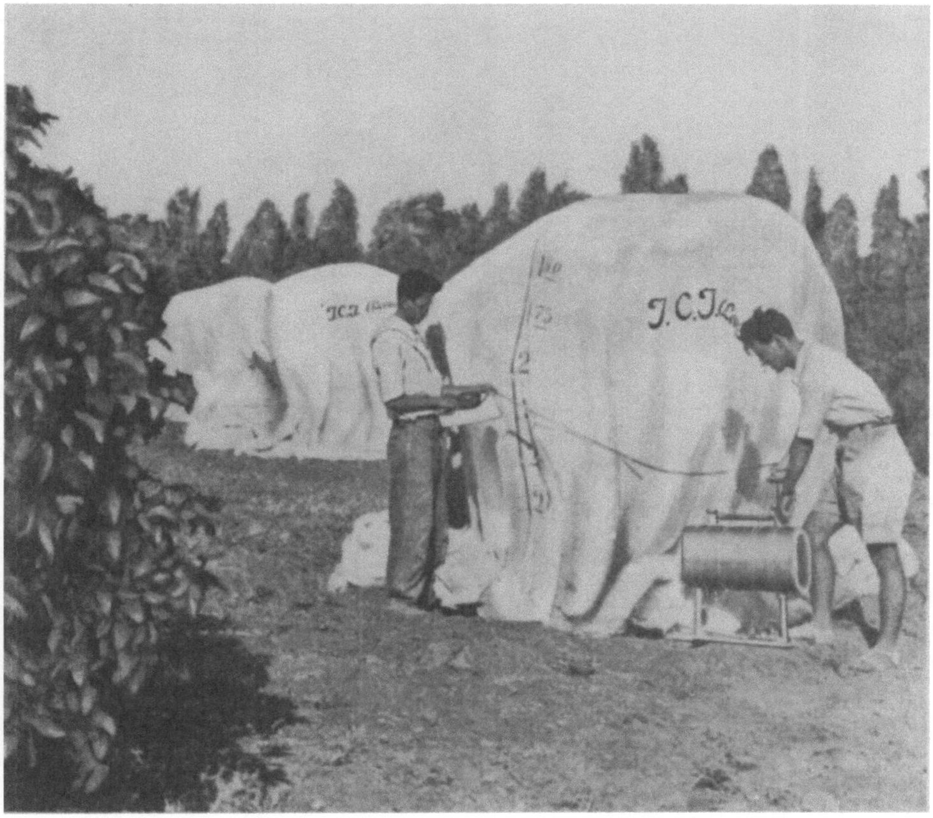

Fig. 148. Tent fumigation with liquid HCN (from Hadar).

machine and Woodworth and Woglum accomplished much in placing the benefits of this method before the farmers.

During the fifth period (1916—1927) the fumigation machine was replaced by the use of liquid hydrocyanic acid gas, which was first discovered by the W. Dingle. The liquid gas is first heated and then a certain quantity of the gas is pressed below the tent. The regulation of the dosage is very easy. The old bulky equipment became replaced by small, easily portable apparatus. Woglum (1923) summed up all the important experiences and prescriptions regarding citrus-fumigation, and those who are interested are referred to this fundamental publication. Quayle and co-workers have recently made some experiments with calcium cyanide dust. Calcium cyanide fumigation under certain conditions is promising as the experiences in Egypt clearly demonstrate. But it is doubtful whether it is able to compete with the liquid gas generation in countries, where the latter is available.

The methods in use in Palestine before the last war were:

1. The old pot-method, in which the hydrocyanic acid is liberated by introducing cyan-natrium into sulphuric acid. This primitive method which develops a warm and wet gas, and shows an equal distribution of the gas beneath the tent, is the less expensive one, but has many drawbacks: fumigation is restricted to night-work in summer and the tents suffer greatly. It is now, in principle, superseded by two more modern methods:

2. Fumigation with liquid hydrocyanic acid is much used, but its application will be restricted until a local factory is erected. The transport from England is not only

Fig. 149. Another view of tent fumigation with liquid HCN.

expensive, but requires a series of special precautions. The advantages of this method are described by Maso as follows:

a) Fumigation can be carried out over a wider range of temperature and humidity than by any other method.

b) The simplicity of this method permits a speed of operation, which is of great importance in Palestine, where the working hours are often limited by high wind and dew. The speed attainable by this method will also allow fumigation to be carried out without interfering with regular irrigation.

c) The risk of damage to the trees is reduced to a minimum, on account of the low temperature at which the gas is liberated, the accuracy of dosage and the absence of moisture.

3. Fumigation by calcium cyanide dust. Its qualifications are summed up by Peters as follows: Summing up we can state that calcium cyanide develops dry and temperatureless hydrocyanic acid gass without the aid of acid or water, but solely dependent upon the moisture of the air.

Moreover there is no question of residues, the dust lying on the plants after losing its HCN content being quite harmless, in particular the high-grade Calcid residues only consist of pure calcium carbonate.

The great storage powers of calcium cyanide must also be referred to.

The climate of Palestine seems to be suitable for the use of calcium cyanide in tree fumigation, even more so, for instance, than in Spain or Sicily. According to the tables of the fumigators, air humidity is always available to a sufficient degree for the liberation of gas from this material, and only very rarely has the air been found too dry for work. Part of the year high winds make fumigation impossible at daytime, but during the greater part of the fumigation season the weather is fairly calm. The condition of the soil in the groves — and this is of extreme importance in daylight fumigation — is favourable both with regard to the moderate moisture and in view of the chemical composition. Jaffa oranges have been found to stand fumigation very well; generally the excellent state of the trees, and the modern planning of the groves, considerably facilitate fumigation.

A fumigation against *Aonidiella aurantii* or *Chrysomphalus aonidum* is required to have at least a 97 % killing effect. In the counts made after properly executed fumigation with HCN the survival was found to be 0.5 to 4.5 %. The reason for a kill lower than 97 % is either fumigation not at the proper season (summer) or the execution of the fumigation by a not adequately trained staff. Inadaquate tents have also been used recently. These sources of undereffect or damage to the citrus-plants are being reduced from year to year. The cost of fumigation, which amounted to 12—20 piasters per tree (6—10 LP. per dunam of 50 trees) till a few years ago, has recently been reduced considerably to 3.5—7 piasters per tree just before the war (depending upon the size, i.e. 1.75—3.50 LP. per dunam). Large scale fumigation has been made possible only by this reduction in price.

Damages caused by fumigation are described by Klein (1935) as follows:

„All those concerned with the growing of citrus know to what extent trees are liable to be damaged by fumigation. If the tree is given an inappropriate dose of toxic material, it drops all its leaves and fruit. Lighter damage causes partial drop, appearances of corky areas on the fruit, signs of partial burning on the leaves etc. The season of the year and the time of the day during which fumigation is carried out, are two very important factors, on which the danger of damage depends to the same extent as an inappropriate dosage. These factors time and dosage, are to be fixed in our country for every district separately. We wish to single out here for illustration one case of great damage: fumigation was started in the Jordan Valley in the evening beginning with the western row of trees. During the night the tents were moved eastwards. Five rows were fumigated in the course of one night. The first three rows showed no damage; in the fourth row slight damages were observed, while the fifth row, which was fumigated just before dawn, showed big damages. The outwardside of the fruit was covered with large whitish spots that mostly spread over half the area of each single fruit. These areas were harder than the green part of the fruit on its inward side (i.e. the side turned towards the tree). Several trees, especially the smaller ones, showed large quantities of burned fruit. The damage in this case was largely confined to grapefruit trees; in the case of blood-orange, shamuti and tangerine trees the damages were much lighter, while the sour lemon trees were not damaged at all. The damaged trees dropped also their leaves.

This damage was caused every night to the rows fumigated just before daylight."

Further details with regard to fumigation are given in the chapters on red scale, Egyptian black scale and mussel scale. The problem of resistant scale is discussed under the heading „red scale".

The Department of Agriculture has been studying for some years the advisability of winter fumigation, espec. in the Upper Jordan Valley. It furthermore studies the advisability of applying two successive fumigations in one year and no control in the following two years.

The practical, satisfactory experiences with tent fumigation in the Upper Jordan Valley in 1937 have recently been fully discussed by Sochovolsky.

References.

F. S. Bodenheimer, Observations on citrus insects in many parts of the world. Tel Aviv 1931, p. 39—47.
E. O. Essig, A history of entomology. New York, 1931, 0. 468—501.
N. Iscander, Min. Agric. Egypt. Bull. 62, 1926.
H. Z. Klein, Hadar, 1935, p. 193—195.
S. Maso, Hadar, 1935, p. 206—207.
G. Peters, Hadar, 1935, p. 208—210.
G. Peters, Blausaeure zur Schaedlingsbekaempfung. Stuttgart, 1933.
H. J. Quayle, Journ. Econ. Entom. **15**, 1922, p. 400; **16**, 1932, p. 327. **20**, 1927, p. 667.
Y. Sochovolsky, Hassadeh **18**, 1938, p. 681—686 (Hebrew).
R. S. Woglum, U.S. Dept. Agr. Bur. Entom. Bull. 79, 1909, 73 pp.
U.S. Dept. Agric. Farmers Bull. 1321. 1929, 58 pp.
Journ. Econ. Ent. **16**, 1923, p. 518; **18**, 1925, p. 593.

C. SPRAYING.

Commercial spraying in citrus orchards has gained sufficient efficiency only since the first war, when oil sprays of improved type became established in California.

Lime sulphur solution and primitive oil sprays were the generally advocated sprays in Palestine until about 1930, when the improved oil sprays of the California-type were first tested in Mediterranean countries, Palestine amongst them. How unsatisfactory the killing effect of the older types of sprays has been, may be gathered from some experiments performed by us in 1928/30 in Palestine. The percentage of survival was:

brand and concentration	Month of experiment					Average survival
	IX	XI	XI	I	IV	
1. Lime-sulphur 1 : 10	—	49.3	—	38.7	58.7	48.6 %
2. Motor oil 2 % — flour	—	39.3	—	19.4	39.1	32.6
3. Motor oil 1 % — flour	—	103.8	—	40.7	—	72.2
4. Petrol-soap emulsion 1 : 7	—	66.7	—	56.4	—	61.5
5. Heptylol 4 %	49.7	52.5	—	—	—	51.1
6. Heptylol 2½ %	62.2	—	—	—	—	62.2
7. Motitush	111.1	68.4	—	73.4	—	84.6
8. Volck's Oil-spray 2½ %	18.6	6.5	4.5	13.8	—	8.9
9. Volck's Oil-spray 1½ %	33.9	29.2	22.0	17.0	4.6	21.3

It is quite obvious that the killing of 15 to 75 % was quite inefficient and that only since the appearance of the improved oil sprays a killing effect of 90—95 % could be obtained. To reach a reliable kill of at least 95 % was the further aim of our experiments.

Before going on to describe the development in Palestine, it will be best to deal briefly with the general properties of these improved oils, which have conquered the Palestine market rapidly.

White oils were first recommended by Volck in about 1911 as safe and effective for spraying citrus orchards. The white oils actually used are characterised by certain chemical properties. The standard distillation at 335.5° C. permits the classification of the oil into the following groups:

% distilled at 350° C	unsulphonated residue	group name
65—100 %	minimum 90	light
50—64	,, 92	light-medium
40—49	,, 92	medium
30—39	,, 92	medium-heavy
0—29	,, 94	heavy

Fig. 150. Spraying with California lime sulphur solution in early days (July 1928 at Hedera).

This classification has been adopted in California since March 1931, and has put an end to the very embarrassing situation, where one manufacturer called a product medium, which another, owing to a different scale, would have called heavy.

It was a very important discovery in 1925 by Gray and De Ong, that the safety of an oil depends on the extent it has been treated with sulphuric acid in the refining process, the treatment by which certain elements regarded as impurities, including the unsaturated hydro-carbons, are removed from the raw distillate. The percentage of unsulphonated residue thus becomes of high importance for the safety of the spray. The higher the unsulphonated residue percentage, the more highly refined the oil. And the higher refined the oil, the less harmful and less dangerous its application to the plants. This is at present the most widely accepted standard of safety. It is proposed by Ralph S. Smith, that for light oils a sulphonation of 90 %, and for medium or heavy oils a sulphonation of 95 % should be regarded as a reasonable satisfactory limit.

The evaporation percentage is, like distillation, a measure of the volatility of an oil. The difference consists in the fact that evaporation is measured at a constant temperature of only 100° C, which is much lower than any of the distillation temperatures.

Analysis of Californian Oil Spray Emulsions by Dr. Marshall, Chemist of the State Department of Agriculture at Sacramento.

		% distilled at 350° C	Oil	Baumé	Viscosity	Unsulfonable residue	Soap	Water	Evaporation 100° C		
									Hours		
									2	6	24
Light	Volck Lt.	69	83.0	32.0	52	95.6	—	15.2	44.2	64.9	86.2
		67	88.0	30.5	56	92.8	—	11.8	41.5	62.7	86.3
Light	Volck Lt Medium. .	56	83.5	31.2	59	97.2	—	15.1	37.3	54.6	75.7
Medium	Med. Oil (Sherw.) . .	55	86.6	30.2	62	91.6	—	13.1	35.0	52.1	74.1
	XXX 60. .	52	88.0	30.4	64	93.2	—	11.4	33.5	50.6	73.9
	Orthol K Med. 70 . .	46	83.7	29.4	70	92.0	—	15.1	29.6	45.7	67.2
	XXX Med..	40	87.6	32.6	63	96.8	—	11.8	27.2	42.1	65.3
Medium Heavy	Med. Oil. . (Sherw.)	38	85.7	29.3	79	92.0	—	14.2	25.0	40.0	58.3
Heavy	Orthol K. . Medium. . .	30	85.0	29.0	87	92.8	—	15.0	17.9	31.8	55.4
	Volck Conc. . . .	24	84.1	29.5	99	100.0	—	15.2	15.9	27.1	55.5
	XXX Heavy. . .	5	87.5	35.1	72	38.0	—	11.7	8.9	11.0	36.7

The viscosity of the oil is measured by the time, taken by a measured quantity of oil to flow through a standard opening. It can be used for classification only with much caution since the volatility correlated to a certain viscosity is different for eastern and western oils. A. C. Browne writes: „It is thought by some that viscosity is not an important factor, and by others that volatilisation is not important. However there are certain data which appear to be definitely connected with one or the other or both. Hindering the natural colouring of ripening fruits, forcing trees into bloom, inducing or injuring foliage or saturating foliage are all phenomena associated with the two above-mentioned physical properties of oils". Those oils, which show the highest viscosity and which remain longest on the tree and consequently on the insect, are generally thought to possess the greatest killing power. Where lighter oils replace heavier oils, the effective dosage has to be increased.

The oil percentage of the preparation as well as its gravity (degrees Baumé) seem to be of little importance for the grower. It will be further observed from the table, that all quoted oils belong to the non-soap type of oil emulsions. Recent analyses have shown, that the size of the oil drops in the spray emulsion is of less importance than was believed at earlier times. The usual California sprays showed a variation from very small and uniform droplets of 2 μ in diameter to emulsions with drops of various sizes, reaching a maximum size of 80 μ (Browne). Smith points out that, aside from the character of the oil they contain, the most important differences among emulsions are the differences in the amount of oil the emulsions deposit on the tree while the spray is being applied. In two different sprays, each containing 2 % oil, the one may deposit 3—4 times as much oil on the tree as the other. In oil deposit a 4 % solution of a light oil may be the equivalent of the same quantity of a 1 % solution of a heavy oil. Even highly refined oils may produce injury, if the oil film, namely the total oil deposit built up, is excessive. And this deposit depends in first line on the kind and amount of emulsifier used in making the emulsion.

The efficiency of the various spray mixtures depends upon the following main factors (Ben Amotz):

1) *The stability* of the emulsion, i.e. its capacity „to stand long periods and various conditions, such as change of temperature and shaking without breaking up; secondly, the degree of its stability when mixed with water of various degrees or hardness; next its compatibility with various insecticides could also be included here. The stability of the emulsion in the above mentioned cases depends upon the physico-chemical structure of the emulsion and the chemical nature of the emulsifier. The emulsifier should be chemically inert in order to remain unaffected by salts in hard waters and other chemicals, with which it might come into contact. The introduction of such emulsifiers as proteins instead of soaps points to some of the work in this direction."

2) *The oil depositing power* or the quantity of oil deposited by the spray emulsion is of great importance. „Both the insecticidal and phytocidal properties of the spray product depend to a very large extent upon this factor. The evaluation of this quality is, therefore, essential in the study of any spray product. A number of methods have been devised for the investigation of this problem. The one now employed a great deal, is to spray a certain measured surface, such as a glass plate or a part of the plant, e.g. a leaf, with the product in question under controlled conditions. The amount of oil deposited upon it, is later determined by weighing or by chemical analysis."

3) The *spreading power* determines the coverage of the plant and to a large extent the oil depositing power as well. It is measured by the 'angle of contact' method. „It consists of placing a drop of the spray liquid in question upon the required surface and measuring the angle of contact, which is formed between the two. The size of the angle is taken as a criterion of the spreading power of the spray liquid. This method appears at the present to give the best results, as compared with others; it is expected that with certain improvements it will develop into a reliable criterion."

Other recent developments of citrus spraying have recently been reviewed by Y. Ben Amotz (1938):

„*Tank mix.* This method was introduced into Palestine several years ago, shortly after it made its appearance in California. The method was mainly developed by Smith (1932) of the University of California. Its main feature is the replacement of the emulsifier (in the oil emulsion) by a spreader, namely a substance which possesses good spreading qualities and in general very slight emulsifying properties.

The oil phase and the water phase are mixed in a special tank under a very strong agitation, in order to emulsify the two together.

The spreader which was recommended by Smith as best for this purpose, was blood albumen: dried blood without the fibrin, mixed together with Fuller's earth in the proportion of 1:3. The product acts mainly on the water phase improving its spreading properties; it serves also to some extent as an auxiliary to the agitation for the emulsification of the oil. The emulsifying power of the blood albumen alone, without the help of tank agitation, is not sufficient; the moment agitation stops, the oil rises immediately to the top of the water.

The spray emulsion formed and applied to the tree is very unstable; the oil phase separates from the water droplets very rapidly (Ben Amotz and Hoskins, 1937). There is a tendency to form a ,,build-up'' of oil-deposit during the process of spraying, especially if drenching spray is applied. Excessive oil deposits, which may thus form, are liable in their turn to injure the sprayed tree. The human and mechanical elements too, play important roles in this method; therefore, very skilled operators and special types of machinery are required if spraying is to be administered by this method.

The method has been used for several years in California; although at the beginning it spread quite rapidly, its application has gradually decreased until it practically went out of use. The reason for this change is, in the opinion of entomological authorities, the damage caused to the trees, and the difficulty of controlling the oil deposit. Similar experiences have been made in Palestine, (Bodenheimer).

Soluble Oil. This product, too, was introduced during the present decade. The main investigations of this product were carried out by Knight and Cleveland (1934). The product is characterised by the fact that the emulsifier is introduced, not into the water phase, as is the case of other sprays, but into the oil phase. The stock product does not contain any water and consists of pure oil, into which is incorporated the specific type of emulsifier. The product will mix freely with water, hence its name: soluble oil.

The main advantages of this method, claimed for it by its inventors, are that in this case the emulsifier actually regulates the spreading properties of the oil, while in the case of the regular oil emulsions it affects only those of the water phase. Consequently it is pointed out that with this product it is possible to obtain a very fine film of oil of high spreading and insecticidal qualities. Furthermore, it is also claimed, that the emulsifier will prevent to a large extent, the penetration of the oil into the tissues of the plant.

Soluble oil has been on the market for a comparatively short time, so that the realization of the above mentioned advantages is still problematic. Scientific authorities in California, where this type of spray was tried, are as yet of divided opinion as to its qualities. It will most likely take still a number of years before its qualities can be definitely ascertained.

Non-Oil Sprays. This is one of the most recent developments in the field of oil sprays. In spite of its name it is actually an oil spray to which an organic insecticide, such as derris, is added in order to increase its efficiency. It is used in lower concentrations than the usual oil emulsions, namely from ½ % to ⅔ %.

This method also has not been in existence long enough to prove its efficiency. Some preliminary results are presented in the Pest Control Circular, California Fruit Growers Exchange (1935, 1936); these show that, as a general rule, they are less effective in the above mentioned concentrations than oil emulsions at the regular concentrations of $1^2/_3$ % to 2 %. It seems that this method requires, as yet, some testing and possibly further improvements before its qualities can be ascertained." (Ben Amotz)

It may be of interest to a specialist of insecticides to read the following report made by Dr. J. Yoffe of the Laboratory for testing materials at Tel Aviv in 1935 of 6 of the most generally used spray-materials on the Palestine market (of each material two samples).

"The samples were brought in test-tubes tightly closed with tin covers or in well closed tins. Each sample weighed from 2 to 3 kilos.

The purpose of the analysis: a) To determine the physical constants of the oils present in the different spraying materials, which will enable a comparison between the spraying materials in different periods, as to whether any changes occurred in their preparation.

b) To learn the nature of several chemical and physical properties of importance in the effect of the spraying materials on the tree and on the pests.

c) To examine several of the practical properties of the spraying materials after they have been mixed with water of different hardness.

To reach the above named purposes, the following analysis were made:

1) the specific weight of the oils, 2) viscosity of the oils, 3) surface-tension of the oils, 4) acidity of the oils, 5) unsulphonable residue, 6) speed of dissolving of the spraying materials in water of different grades of hardness, 7) the value of demulsification in water of different degrees of hardness.

The first four analysis were made of the oils of which the spraying materials are composed. The two last analysis were made on the spraying materials themselves.

Remarks to the analysis. Separation of oils for analysis nos. 1, 2, 3, 4, 5: the spraying materials are an emulsion of oil (from 80 % to 90 % approximately) and water. Besides those, there are substances in the spraying materials known as mulsifiers which serve to keep the spraying material intact and after it has been mixed with water.

In order to separate the oils, we used neutral salts and have kept entirely from using acids, as those may alter the properties of the oils.

This separation met with difficulties, and after some trials, we succeeded in separating all the oils using anhydrous sodium sulphate (Na_2SO_4 siccum). There were cases where the oil was opaque after the separation and filtration, and we then had to treat it which anhydrous calcium chloride ($CaCl_2$) and refilter it. All the oils analysed were clear at the end.

Specific gravity was determined by means of a pycnometer. The results were calculated to 15°.

Viscosity was determined by means of a viscosimeter after Angler in Angler degrees in 25°.

Surface tension. This property of the liquid gives the liquid the capacity to create a tension over its surface. Upon it depends the diffusion of the liquid over the surface of the solid. We measured the surface tension by the stalagmometer method.

From a special pipette (stalagmometer = drop measure), with a disc-like mouth, a definite number of fat drops is let out, weighed and their weight compared with the weight of the same number of distilled water drops let out from the same stalagmometer. The weight of the drop is directly proportional to the surface tension. We calculate the surface tension after the formula $d_1 = \frac{7.61 g_1}{g}$ where g_1 — is the weight of the oil-drop, g-that of the distilled water drop, 7.61 — the surface tension in milligrams of one millimeter of distilled water.

Degree of acidity. The acidity was determined by titration with NaOH N/10 against phenolphthalin. The results are obtained in the "number of acidic units", i.e. mgm of KOH necessary to neutralize the acidity of one gram of oil.

The *unsulphonable residue* gives the per cent of oil which is not attacked by concentrated sulphuric acid. The parts of oil which are attacked by the sulphuric acid are very active and may have a bad effect on the tissue of the tree or the fruit.

5 cc of the oil were mixed with 25 cc of sulphuric acid in apparatus used for determining the oil in cream (creamometer). After this mixture was mixed for 5 minutes and was kept standing for 20 hours, the remaining part of the oil was measured.

Two series of analysis were made: one with concentrated sulphuric acid of 1.84 specific gravity *) the other with fuming sulphuric acid containing 5 % of SO_3; the latter is of course much stronger in its effect.

The experiments made with the spraying materials as they are: The experiments were made with 3 kinds of water: *a*) soft water of 4 Haertegrade, *b*) medium water of 20 Haertegrade, *c*) hard water of 40 Haertegrade.

The speed of dissolution of the material. The practical value of this experiment is to establish which spraying material diffuses quickly in water of different grades of hardness. For this purpose 2½ % (per cent per volume) of the spraying material were added to the water, the liquid was mixed by means of an electrical mixer which makes 100 turns per minute, and is furnished with two cross-like rows of wings, each wing forming an acute angle with the axis of the west (like a propeller of an aeroplane). The mixing was continued and stopped in turn, in order to see each time if the material has diffused in the water; the minute no solid parts were seen, the mixing was stopped and all the liquid was passed through a fine net to make sure that no solid part has remained. We repeated the experiment several times, for each spraying material separately, to determine the exact length of time the material need be mixed to dissolve completely in the water.

We found that while in soft water the spraying materials dissolved after a more or less long time, not exceeding 10 minutes, in medium and hard water small particles appeared in some of the spraying materials which did not dissolve even after prolonged mixing (we tried to mix for two hours). These particles are very scarce (roughly do not exceed 0.1 % of the spraying material) and they were no doubt formed through the effect of lime in the water on the emulgators, or other by-products in the spraying materials. In view of this fact, it was impossible to determine the time of dissolution for the materials mentioned in medium and hard water, and in these cases we put down: "Over an hour". One must, however, keep in mind that this is not a case of undissolution, but a formation of a new product which does not dissolve at all. Thus, for comparison with other spraying materials we must use the results obtained in soft water, in which the above particles were not formed.

The disadvantage of these particles lies in the fact that they may stop the mouth (outlet) of the spraying vessel, and also form centres of attack on the surface of the leaf or the fruit, since they contain the spraying fat in an undiffusible state.

In those spraying materials in which this phenomenon was not observed, the influence of hard water on the speed of dissolution was not recognisable. In several cases the material dissolved faster in hard water than in soft.

The value of demulsification: One of the properties of the emulsions is that they decompose into their different parts after a certain time. In the case of oil emulsions, a layer of fat floats on top (a classical sample is milk). The emulsifiers are substances which serve to preserve the emulsion. The purpose of the demulsification experiments was to see which of the spraying materials are more stable, and especially which of them decompose also in hard water (which contains calcium and magnesium salts). It was clear from the start, that certain emulsifiers are liable to combine with the calcium and magnesium salts to give unsoluble products, or soluble compounds which lack the emulsifying property and the emulsion may hence decompose more rapidly.

In practice, this phenomenon may find its expression in the fact that the emulsion decomposes in the mixing tank already before it is sprayed, and drops of fat and water will fall separately on the tree as well as leaves. The effect of such a solution will not be homogeneous.

The procedure of the experiment. A mixture of spraying material in water of 2½ % (volume) was prepared. This emulsion was mixed by means of the electrical mixer (see "speed of dissolution"), for 10 minutes. In all cases, except for those in which unsoluble particles have remained, the spraying material dissolved, completely during this time. We transferred the liquid immediately following the mixing into a glass 6.5 cm in diameter, and it was then left there for an hour. We then took, by means of a pipette, two samples, one from the bottom of the glass, and right after that, one from 3 mm under the surface of the liquid. The two samples were analysed as to the amount of fat in them, and the difference of the amount of fat calculated in per cent from the amount of fat in the first sample was called the demulsification value.

These experiments were made in soft, medium and hard water. The results obtained in all the analysis and experiments are given in the appended table and figures.

General conclusions. A methodical survey of the properties of oils and spraying materials already shows that there is no relation between the several positive properties in all the spraying

---

*) The method in use in America is to work with sulphuric acid of 1.84 specific gravity.

materials. While one spraying material excels with one property, it lacks another, and on the contrary. We must therefore evaluate the properties separately.

Specific gravity and viscosity: Are directly proportional. Heavy fat is also thicker (see figure no. 1). The advantage of heavy oil lies in the fact that it stays on the tree for a longer time and its effect lasts longer. Its undesirable feature is that it may, for this reason, attack the tissue stronger. One should pay attention here principly to the unsulphonable residue (see below).

Surface tension: This important property is equal in all oils and approches 3.

Acidity: All the oils, except for no. 1, are almost neutral. We must keep in mind that the spraying materials contain ammoniac most of the time, and are therefore slightly alkaline. The spraying material no. 1 was the only acidic one, but a further analysis showed that the acidity is not mineral, which reduces its harm.

Unsulphonable residue: This value approaches 100 % for all oils. Exceptional in this respect are no. 6, with 91.5 % and 87 %, and no. 12 with 90.5 % and 87 %. The low unsulphonable amount of fat is especially liable to harm in the oil no. 12. Judging from its high specific gravity and viscosity, its evaporation is necessarily very small and its effect on the tissue of the tree will therefore be of a very long duration.
The practical experiments were made with soft, medium and hard water. The medium water corresponds to the water used by the orange grower. For example, the hardness of the water of different sources is as follows:

Water from Tel Aviv	from 15 to 32 grades
Water from the Yarkon	18 ,,
Water from Petah Tikva	from 9 to 13 ,,
Water from Bne Brak	from 8 to 10 ,,

Speed of dissolution: Here we shall mention first of all the spraying materials nos. 2 and 7, which leave unsoluble substances even after they have been mixed for a long time with water of medium hardness. No. 8 leaves unsoluble parts only in hard water of 40 grades. Such water is very rare in orange groves. The rest of the spraying materials dissolve in a more or less short time, but after they have been mixed for 7 minutes intensively, they all form a homogeneous solution.

Demulsification value. In figure 2 are given the curves of the demulsification values for all the spraying materials in soft, medium and hard water. According to the values in soft water and according to the construction of the curves, we may divide the spraying materials into the following 3 groups: *a*) those which decompose already in soft water, i.e. their emulsifier is too weak. To these belong the spraying materials: no. 7, 10, 11, 12. *b*) those which do not decompose so quickly in soft water, but do so in medium and hard water. The emulsifier of these is very active in soft water, but is attacked by the calcium and magnesium and does therefore lose its effect. To these belong nos. 1, 3, 8 and 9. *c*) those which contain good emulsifiers which are not attacked even by medium water. To this group belong nos. 2, 5 and 6. We must keep in mind that the experiments on demulsification were made after the materials remained in the water for a whole hour. In practice the time of rest of course in not so long, and the demulsification is smaller. (The numbers correspond to those of the table given below).
The research work on the spraying materials present in this country is not completed as yet."

No.	Specific gravity 15/15°	Viscosity in Angler degrees in 25°	Surface tension α_1	Acidity $\frac{\text{mg KOH}}{\text{gr fat}}$	Unsulfonable residue $H_2SO_4 = 1.84$
1	0.874	4.28	2.99	0.500	100 %
2	0.873	3.76	3.01	0.056	99
3	0.870	3.63	3.00	0.056	98
4	0.853	2.67	3.06	0.028	96.5
5	0.874	3.66	3.03	0.056	97.5
6	0.887	3.44	3.12	0.056	91.5
7	0.865	3.12	3.01	0.140	99
8	0.874	4.00	3.05	0.056	100
9	0.851	2.64	3.04	0.056	98
10	0.871	3.70	3.04	0.056	99
11	0.867	3.07	3.02	0.028	100
12	0.905	4.72	3.10	0.028	90.5

No.	Unsulphonable residue H_2SO_4 + 5 % SO_3	Speed of dissolution in water			Demulsification value in water		
		soft	medium	hard	soft	medium	hard
1	99 %	2′30″	2′40″	1′50″	21.4 %	46.9 %	95.7 %
2	97	20″	over an hour	over an hour	30.0	33.3	66.6
3	98	2′55″	2′30″	3′50″	18.2	83.3	83.0
4	96	4′25″	4′25″	5′5″	7.1	22.2	85.7
5	97	1′45″	1′50″	2′20″	37.9	32.8	60.0
6	87	2′55″	2′15″	2′35″	26.5	31.3	59.9
7	98	6′50″	over an hour	over an hour	66.6	62.5	57.8
8	98.5	1′50″	4′15″	over an hour	46.1	60.0	80.9
9	95	2′50″	2′43″	3′40″	6.66	52.4	95.8
10	98	2′15″	2′20″	2′48″	77.4	84.6	82.2
11	96	3′50″	4′55″	6′30″	68.9	62.5	59.3
12	87	1′40″	1′55″	1′50″	72.0	80.0	98.8

Our experiments for the optimal quantity of spray.

Series of experiments were performed in August and September of 1934 and 1935 in order to evaluate the exact quantity of spraying material needed for an efficient control of red scale on citrus. The material used was a 2½ % spray of Volck concentrate. The same degree of efficiency for other sprays can be taken from the annual recommendations for oil sprays.

We regarded that quantity as optimal at which a minimum mortality of 95 % of red scale was reached, with safety for the tree and at lowest costs.

For comparison different types of spraying machinery were tried: Back sprayers with 4—8 atmospheric pressures (type: four Oaks, Pomonax, etc.), hand sprayers with 8 atmospheric pressures (type: Bean Magic) and motor sprayers with 16 atmospheric pressures (type: Hardy, Bean, etc.). Unfortunately no sprayer of the Bean Magic type was available for the experiments.

The number of survivors are given in the following table (B=Backsprayer, M=Motor sprayer 16 atm.):

Survival of Adult Red Scales one Week after Spraying.

Type of Sprayer	Quantity of spray p. tree	3-year old trees			5-year old trees			7-year old trees			20 budded trees in nurseries, 1½ y. old
		Aug. %	Sept. %	Aver. %	Aug. %	Sept. %	Aver. %	Aug. %	Sept. %	Aver. %	September %
Back Sprayer	5 lit.	14.2	—	14.2	—	—	—	—	—	—	—
	10	11.7	9.3	10.5	7.0	21.7	14.3	—	—	—	22.2
	15	4.6	8.7	6.6	6.3	8.7	7.5	25.0	—	25.0	8.7
	20	4.2	4.1	4.1	5.1	5.4	5.2	9.4	—	9.4	2.1
	25	—	—	—	—	6.2	3.1	—	—	—	—
	30	—	—	—	—	—	—	24.6	—	24.6	—
	40	2.4	—	2.4	5.8	—	5.8	12.9	—	12.9	—
	50	—	—	—	1.5	—	1.5	—	—	—	—
	Unsprayed	98.3	99.3	98.8	98.0	99.3	98.7	99.7	—	99.7	100.0
Motor Sprayer	5 lit.	37.0	—	37.0	—	—	—	—	—	—	—
	10	20.6	17.0	18.9	16.6	12.2	14.4	—	—	—	—
	15	—	10.9	5.5	8.0	8.9	8.4	3.2	8.3	5.7	—
	20	10.3	1.0	5.6	1.7	10.8	6.2	1.7	3.8	2.7	—
	25	—	1.8	1.8	—	2.6	2.6	—	—	—	—
	30	3.0	—	3.0	3.7	—	3.7	—	—	—	—
	40	1.1	—	1.1	2.6	—	2.6	—	—	—	—
	50	—	—	—	1.8	—	1.8	—	9.5	4.7	—
	60	—	—	—	—	—	—	—	—	—	—
	Unsprayed	100	100	100	100	100	100	100	100	100	—

It may be summarized as follows:

1) *3-year old trees* (after grafting).

The 95 % mortality limit is reached with back-sprayer below 20 liters per tree. At 30 liters per tree, leaf fall and discoloration of the fruit occurred. The optimal quantity is 18 liters per tree.

The same quantity is optimal for Hardy.

2) *5-year old trees.*

The 95 % mortality limit is reached at about 21 liters per tree with back-sprayer. At 30 liters per tree, discolorations of leaves occurred.

The optimum for Hardy is at 20 liters or somewhat below. The unsatisfactory result in September (10.8 %) occurred on a tree, which was sprayed in an insufficient manner. It is not advisable to spray more than 20 liters per tree; some occasional discolorations of leaves start in that case. With 8 atmospheres of pressure a very good result was reached at 10 liters per tree.

3) *7-year old trees.*

It was not possible to get satisfactory results with the back-sprayer.

The optimum for Hardy is between 15—20 liters per tree, probably even nearer to 15 liters. 17 liters may be put down as a tentative quantity. At 30 liters and above, discoloration of the leaves was observed.

18 liters per tree is the optimal quantity for all these ages. The gradual reduction of red scale infestation with age is the explanation for this fact.

4) *Nurseries.*

For nurseries back-sprayers are well suited. For inoculated trees, 1½ year old, 2 liters per tree are optimal. The higher kill obtained with that quantity is highly desirable in nursery stock.

In 1935 the percentage of red scale survivors with different types of spraying machinery (B.=Back sprayer, M. 16=Motor sprayer with 16 atmospheres pressure, M 8=dto with 8 atmospheres pressure, B.M.=Bean Magic was rather similar to that obtained in the preceeding year:

		Leaves										Fruit				
liters per	machinery	B.	M.16	B	M.16	B	M.16	M.8	B.M.	M.16	M.8	B.	M.16	B.M.	M.16	M.8
tree	age of tree	2	2	3	3	5	5	5	5	7	7	5	5	5	7	7
4		13.0	7.6	—	—	—	—	—	—	—	—					
7		4.7	3.2	—	—	—	—	—	—	—	—					
10		0.7	0.3	10.2	9.0	10.1	26.5	7.5	14.5	10.7	20.4					
13		2.0	0.0	3.3	( )	12.2	12.5	10.5	(21)	(9.3)	10.6					
16		—	—	3.8	7.4	9.3	5.5	6.0	13.1	2.7	6.7			7	19	28
19		—	—	1.2	5.1	(5.5)	2.4	(4.3)	8.6	1.9	(9.8)	7.0	9.5	11.9		
22		—	—	—	—	2.5	3.3	4.0	8.5	(0.0)	(5.1)	10.9	4.6			

The daily progress of mortality among red scale after spraying with Volck concentrate 2½ % was observed. The maximum effect was observed between the fifth and ninth day. This is in accordance with our rule to perform the count of living red scales 7 days after the control has been made.

Percentage of living Red Scale after spraying (smoothed).

days	3 years (leaves)			5 years (leaves)		5 years (fruit)	
	10 l	16 l	19 l	16 l	19 l	16 l	19 l
0	95.9	92.1	91.2	92.1	88.9	92.7	
1	71.4	67.6	66.0	72.0	63.4	76.3	(68.7)
2	52.7	45.6	38.0	52.0	36.1	62.7	(42.0)
3	28.0	24.1	13.7	27.4	14.1	40.4	(20.0)
4	14.4	17.6	7.0	17.4	7.8	26.1	12.9
5	**5.5**	6.9	2.6	**5.6**	**2.8**	9.5	8.7
6	6.4	3.3	2.5	5.6	6.7	8.7	**6.6**
7	8.3	**2.2**	1.5	6.6	11.3	7.7	6.9
8	6.6	2.3	**0.7**	5.7	11.3	6.9	6.6
9	7.5	2.7	1.0	6.9	6.3	**5.9**	
10	6.2	2.2	0.9	6.0	2.8		

Similar results were obtained in 1935:

days	3 years (leaves)			5 years (leaves)		5 years (fruit)	
	10 l	16 l	19 l	16 l	19 l	16 l	19 l
1	87.6	74.9	73.5	76.8	65.8		
2	46.5	26.5	24.4	34.5	24.4	78.0	
3	34.0	34.0	16.0	40.0	18.0	51.0	23.0
4	3.5	11.9	0.8	3.4	0.0		59.0
5	5.6	7.0	4.2	8.7	5.5	11.2	17.0
6	7.3	1.7	2.7	4.7	3.0	8.0	1.7
7	6.1	1.2	0.6	3.3	7.4	9.2	7.3
8	10.2	3.8	1.2	2.6	5.5	8.9	10.9
9	3.3	2.0	0.4	4.8	3.0	4.8	2.6
10	9.1	2.4	1.2	7.1	2.6	7.0	—

The survival is always higher on the fruit than on the leaves. In experiments during August of 1934 on 5 years old trees the survival was as follows:

Quantity of spray	Back Sprayers		Motor Sprayers	
	Leaves	Fruit	Leaves	Fruits
10 liters	21.7 %	50.8 %	12.2 %	40.0 %
15	8.7	16.7	8.9	17.0
20	5.4	18.0	10.8	23.6
25	6.5	16.1	2.6	19.0
30	—	—	—	23.9
Unsprayed	99.3	100.0	100.0	100.0

Careful counts in the following year revealed the fact that the survivors on fruit were often aggregated in patches, where the liquid had not covered the fruit. This is one main reason for the differential kill, which probably to some degree is natural.

Considering the difference in vitality of scales living on leaves and on fruit (as expressed by fertility), it is quite probable that the scales on the fruit show, in addition, a higher resistance to spraying than those on leaves. No exact data on this physiological difference are available neither from our own experience nor elsewhere in literature.

The average survival of scales on the fruit under orchard conditions is more than double their survival on the leaves. This very unsatisfactory result points further to the advisability of spraying the trees as early as possible, i.e. before the infestation of the fruit. July and early in August seem from this point of view to constitute the optimum spraying season for oranges.

In order to gain an understanding of the important influence of atmospheric pressure on the mortality rate of scales, spraying was carried out with a Hardy motor sprayer comparatively at 16 (normal pressure) and 8 atmospheres. The survival was drawn up for two series, as follows:

Quantity of Spray	5-year old trees		Quantity of Spray	3-year old trees	
	16 atm.	8 atm.		16 atm.	8 atm.
10 litres	16.6 %	0 %	3 litres	10.4 %	4.7 %
15	8.0	0			
20	1.7	0	Unsprayed	100.0	100.0
30	3.7	0			
40	2.6	0			
Unsprayed	100.0	100.0			

The result was in both series in favour of the 8 atmospheres. This is probably due to the fact that the time required for spraying a tree is double at the lower pressure. At 16 atmospheres, 5 litres are sprayed per minute; at 8 atmospheres — only 2½ litres. This allows more time to equally cover all parts of the tree, also a smaller portion of the spray is wasted.

In the following year the results were about equal on leaves at both pressures. The count of fruit showed a decided result in favour of the higher pressure. This question deserves a more detailed and careful study, as all questions connected with machinery and technique of spraying, neglected hitherto to too great an extent.

Our standard material remained for all tests a Volck concentrate solution 2½ %, as this was the first of the standardized improved oils we know of and experimented with. The average kill in 5 years tests was as follows:

Volck concentrate 2½ % Survival

	I	II	III	IV	V	VII	VI	VIII	IX	X	XI	XII	Av.
1931	6.5	15.7	6.4	—	13.6	—	2.3	4.0	0.0	4.6	3.7	5.8	6.3
1932	7.8	12.4	2.9	3.1	—	11.3	0.0	11.6	7.2	5.7	17.7	14.5	8.6
1933	6.2	11.3	1.7	1.9	0.0	9.0	6.5	4.2	6.3	1.3	2.2	2.0	4.4
1934	7.9	5.4	3.8	5.7	13.0	0.6	2.3	17.8	4.7	15.5	2.9		7.2
1935		2.8	—	—	0.0	—	13.2	0.0	0.7	5.3	1.3	3.4	3.3
	(7.1)	9.5	(3.7)	3.6	6.7	7.0	4.9	7.5	3.8	6.5	5.5	(5.9)	6.0

Many other brands appeared on the market, some from foreign countries, some locally produced. As no public control of chemical or insecticidal qualities existed, it seemed necessary to organize some kind of control for the protection of the public. The Citrus Department of the Jewish Farmers Association organized all agents and producers of white oils and submitted the tests to the author, who was in charge of them for four years. The tests aimed mainly at ascertaining the insecticidal qualities and the safety of the spray for the plant. The appropriate concentrations of all brands were published annually by the citrus department and the conclusion was reached that careful spraying following the prescriptions of this department and of the Agr. Exper. Sta. at Rehoboth, if properly executed, insures a 93 to 95 % kill. Most oils in use, including the tank-mix, belong to the medium-heavy or heavy oils. Medium oils, when applied, call for a higher spray-quantity if the same oil deposit it to be obtained on the plant.*)

One oil submitted to our tests, was a light oil highly efficient at a concentration as low as 1½ %. It did not appear on the market owing to its high costs.

It should be mentioned that in addition to the actually counted kill of the spray, additional effects are observed. Ebeling reports that many survivors gave birth to dead crawlers in the subsequent weeks and, what is more important, there is a great reduction in the percentage of crawlers able to settle successfully on the leaves of the sprayed trees for at least one month after spraying.

Cost of spraying: No accurate calculations as to the costs of spraying are yet available. A few calculations may be reported:

1) In one large complex of groves (3—7 years old, average 19 l. per tree, motor sprayer) the costs per dunam of 40 trees were (before the war):

Labour	450 mils
Oil	630 ,,
Amortisation and repairs of machine, for motor	145 ,,
Total	1.225 mils or 30 mils per tree.

2) A young grove in the first year after grafting. One worker sprayed 80 trees (50 trees per dunam) per day with a Bean Magic sprayer. 75 dunam equalled 54 days of work. Work and oil together (no amortisation, etc) costed 4.6 mils per tree or 230 mils per dunam.

3) Similar reports of 200—400 mils per dunam have been reported from many other groves, but in all these cases the quantity of spray per tree was insufficient (5—12 l), no armortisation of machinery was taken into account, etc.

Despite the lowered price of the oils and more economic organisation of labour, it seems highly improbable that any efficient control can be performed below 15 to 20 mils per full grown tree (at pre-war prices).

Damage caused by spraying: During the long experience in California with the many different brands of oil sprays a whole score of injuries came under observation, and their list as compiled by Woglum is quite discouraging to the simple farmer who takes a glance at it. We read as follows:

,,Retarding of blossoming, reduction of blossoming; reduction of crop; retarding fruit colouration; interfering with normal sweatroom colouration; drop of immature fruit; drop of mature fruit; roughened texture mummifying the fruit; increasing crystallisation; producing insipid flavour; decrease in acidity; soluble solids and sugars; increase of dead wood; occurrence of fruit burn and leaf drop; difficulty in cleaning fruit; gumming-up packing house conveyors".

But the actual increase which spraying gains in California over fumigation easily shows that the actual menace cannot be in any relation to the discouraging impression which this list is liable to make on the reader. There is general agreement that

*) Dr. Ben-Amotz makes the following remark (i.l.): "Some recent studies in California showed that in applying medium and heavy oils with the same spray quantity, the same oil deposit is obtained. However, the reason for the fact that in using medium oils it is necessary to use a higher spray concentration or a higher spray quantity, is that the medium oils will volatilise much faster than the heavy oils. The latter will therefore remain longer on the tree and their effect will be more lasting. To compensate for this, when using medium oils in certain cases they are used at higher concentrations than heavy oils."

the large majority of the above mentioned injuries were caused by factors, easily and generally avoided in commercial spraying. The most important reasons of heavy injury are:

1) Application of the spray at a non-advisable season. In California each citrus district has its own spraying season schedule, and various damages are very often the outcome of spraying at a date out of the advised spraying season.

2) The heavier the oil the more important becomes the correct timing of the spray.

3) Spraying with emulsions, which are broken down or separated before spraying, has always serious consequences.

The worst damage to oranges in California resulted from late autumn and early winter spraying with heavy oils. In consequence there was a reduction of blossoming and of the crop (up to more than 50 %). The damages, discoloration, taste, etc. of the fruit also resulted from this untimely spraying.

The more common forms of injury are dead wood in the interior and leaf drop. The latter may take place immediately (1—2 days) after the spraying or may be combined with leaf chlorosis and be protracted over 3 or 4 weeks. Leaf or fruit burning should never occur if the three just mentioned conditions are observed.

As early as 1903 Volck observed that oil causes more damage on the lower sides of the leaves than on their upper side. In connection with this observation the stomata were regarded for a long time as the penetration doors of the oil into the tissues of the plant.

But Knight, Chamberlin, and Samuels (1929) showed, by a very ingenious method, that one hour after spraying the cutinised outer integument covering the epidermal cells absorbs oil rapidly. The oil accumulates in the depressions between the cells. On the lower epidermis the oil pours in through the stomata. At the end of 24 hours the oil enters the leaf from both the upper and under surfaces. The elimination of the oil from the leaves goes through the following steps: After the oil has entered the cell, it penetrates into the vascular system and is translocated down the leaf trace into the stem; from the phloem across the medullary rays. The oil is then deposited in the large storage cells of the pits and in the old wood fibers of the xylem. There it was present 63 days after application. This deposit of oil is supposed to kill the wood and to increase the annual amount of dead wood.

The same authors studied transpiration, photosynthesis and respiration as influenced by petroleum oils. In leaves which were painted on both sides transpiration showed a sharp and abnormal drop. The extent of disturbance varies directly with the amount of oil deposit and with its viscosity. After treatment with light lubricating oils, recovery took place about 1—3 weeks after spraying; with saturated oils of a viscosity of 100—110 secs. according to Sayboldt — from 6—7 weeks after spraying. Other experiments showed that trees which remain 15 days in the dark and give no further starch reactions with iodine, when placed soon after in the light and treated with oils, react as follows:

Treatment	
Both surfaces painted with saturated oil (106 secs Say.)	Starch reaction on 14th day; not yet normal after 40 days.
Dito (50 secs Say.)	Slight trace of strach on 1st. day; Normal in 4 days.
Nearly saturated highly refined kerosene oil	Normal on 1st day
Control	Normal on 1st day

In orchards a marked drop in the starch content was observed during the first 5 weeks followed by a rapid increase until it came to an excess as compared with unsprayed trees (viscosity 106 secs Say; 2 % conc. used).

This excess of starch in the leaves may result from an inability to translocate the starch due to overloading of the conducting cells with oil. Respiration increased considerably, the plants evidently oxidising great quantities of reserve food, whereas plants sprayed with the just mentioned oil returned to normal respiration after somewhat less than one month. In painted leaves this return was not finished after six weeks.

The first and most obvious damage is scorching of leaves and fruit, dropping of fruit and leaves, eventually dying back of twigs. In special experiments in August 1934 with a 2½ % solution of our standard spray, the following damages were experienced:

Age of tree	Quantity of spray	type of damage
3 years	25 liters	many yellow leaves
	30	many yellow leaves and heavy drop of fruit
	40	heavy drop of leaves and fruit
5 years	30	many yellow leaves
	40	many yellow leaves and heavy drop of fruit
	50	many yellow leaves and heavy drop of fruit
7 years	30	some to many yellow leaves
	40	some to many yellow leaves
	50	some to many yellow leaves
	60	some to many yellow leaves and discoloring of fruit.

Dying back of twigs was not observed in any case. Where heavy drop of leaves and fruit occurred, the twigs immediately developed many young shoots and two months later the tree was fully covered with full-grown young leaves. This stimulating influence of oil sprays is observed in normal spraying too, and is used in deciduous fruit-trees to stimulate or advance the blossoming.

Where die-back of twigs is observed, it should first of all be found out whether it is not caused by fungus-attack (*Diplodia*). In all other cases, including the lighter cases of damage, the causes must, as a rule, be ascribed to some mistake in the execution of the spray. Every farmer is warned not to spray thirsty trees. Other common mistakes have been discussed by Klein (1935):

Damage caused by spraying with white oils.

a) Spraying time. Spraying that is carried out not in the appropriate time not only does no good, but is absolutely detrimental. There is a definitely appropriate time for the control of every insect pest. This is to be fixed anew every season by experts. A hamsin occurring after spraying is liable to burn large numbers of fruit and leaves. We have witnessed damage caused by a hamsin that occurred a fortnight after spraying. The dangerous hamsin season is not limited to the period between the end of April and the middle of June, but returns during September/October. After the hamsin, sprayed fruit shows on the side turned away from the tree large brown spots with dark centres. These spots become flattened, and some times sunk. Fig. **151**a shows such fruit. Spraying during the hot noon hours, too, brings about similar damage. In some cases one can observe burnt and even dropped leaves and fruit, a short time after spraying. In such cases investigation has proved that the trees were *thirsty when sprayed.*

Fig. 151.
a. Fruit scorched by a hamsin wind that occured shortly after spraying. b. A cluster of oranges scorched as a result of having been sprayed with the last residue of the spraying solution in the container.

Fig. 152.
Three-years old orange sprayed on 12. VIII. 1934 with 40 l. per tree. 1. Many leaves and some fruit dropped a few days after spraying, 2. Partial picture of young shoots on 15. IX. 1934, 3. The tree has fully recovered on 11. X. 1934.

b) Emulsions. Too concentrated emulsions, or careless preparation of the same, affect the plant adversely: the mixture of the emulsion is not appropriate and the oil becomes easily separated from the water. In order to obtain a good emulsion, the oil is to be poured into the mixing vessel and some water added little by little, while stirring is kept up. The amount of spray applied to every tree is also very important. Of course, every tree requires an appropriate quantity of emulsion, according to its age and development. We wish to emphasize that, generally speaking, our observations have led us to conclude that in common practice too small a quantity of spray is usually applied to every tree. The fixing of an appropriate dose applicable to a given tree taking into consideration its age, under conditions ensuring a maximal kill of insects without damaging the tree, is still in its experimental stage. Dying back of twigs and branches as a result of spraying with white oils is very common in California *So far it has not been observed in this country,* but our experiments have proved to us how detrimental over-doses of spray can be. *Fig.* **151***b shows a tree sprayed with approximately* **40** *litres* of spraying solution. One day after this operation the tree began to drop its leaves and fruit in large numbers. We ought to point out that renewed growth *started some two months later, but the tree was weakened* and lost its entire fruit-crop. Fig. 152 exhibits a five-years old tree, which had been budded in the nursery, and having been sprayed with nearly **30** litres of solution, dropped most of its fruit.

c) Machinery. It is true that those days, when the grower went to his nursery or his young grove and sprayed them with a hand-syringe or a flit-pump, have long passed. And yet we can still witness damage done to groves by the use of inappropriate machinery. It is to be stressed that the success of spraying entirely depends on the operator. In nurseries it is permissable to use portable back-sprayers, but it has been proved that spraying with the last few litres caused burning of seedlings. Such portable back-sprayers usually lack agitators, and the inside pressure is liable to decrease during the operation. Generally speaking, no sprayer that lacks a manometer should be used. Again, sprayers with hand pumps are also in use. It is true that their inside pressure can be raised to 15 atmospheres and more, but their agitators are connected to the handle of the pump, and consequently work to the slow rhythm of the pumping, which is bad for the emulsion. Very frequently the agitator does not reach to the bottom of the container, and the last few litres are sprayed without having been previously stirred. Such machinery sometimes contains up to 200 litres and more. In one case we observed in a large grove that at fixed intervals there were two burned trees every time. Investigation has proved that at every pair of burned trees the mixture had run out of the container. These trees were damaged to such a degree that they died within a few weeks after having been sprayed. The outside half of such fruit was covered with yellow-brownish spots and shrivelled.

There is no doubt that exercising some care and following all prescriptions, no serious damage will be caused by spraying and that, as a rule, even a stimulation of the tree may be expected. The correct quantity for 3—7 years old trees is 18—20 liters of oil solution. It should increase when lower concentrations or lighter oils are used.

This quantity is vastly inferior to that applied in the orchards of California. This difference amounts to 100—200 %. This higher quantity is only partially explained by the use of lighter oils or lower concentrations. It is mainly explained by the use of lighter oils or lower concentrations. It is mainly induced by the necessity to give a full spray to the interior parts of the tree, whereas in Palestine the exterior and central parts are fully covered, but little attention is paid to a full cover spray of the interior. The resulting difference in the total oil-deposit seems to be responsible for the damage observed in California.

Since 1938, experiences have brought out the solution of the different experiences in Palestine and California, namely why die-back has not been observed in Palestine before, as has been stated above by Mr. Klein and much earlier by the writer. While our statements are perfectly correct for young trees up to 7 years and slightly older, the damage is different if older and larger trees are sprayed. Trees of 3 to 7 years age are especially resistent to spray damage. Now, die-back of twigs has been observed, when large trees densily cultivated were selected and sprayed intensively with heavy or even with medium oils. There is a very distinct difference between the damage on young and older trees. Our previous experience referred almost entirely to young trees which formed the bulk of the sprayed area during those $1\frac{1}{2}$ decades. With the aging of the groves planted since 1926 this situation changes rapidly. These conditions are confirmed by the latest experiences of Dr. Ben-Amotz and Dr. Klein.

Dr. Ben-Amotz remarks (i.l.): „My explanation for this is, that younger trees have more space between them and due to more sunlight and better aeration the oil can easily volatilise. The older the trees become, the less space is left between them and due to the decreasing effect of sunlight and less aeration it is more difficult for the oil spray, especially in the inside portion of the tree, to volatilise and as a result these portions of the tree are more susceptible to injury."

The reasons, why heavy-medium oils are avoided in California (excepting the spraying of lemon trees in some coastal districts) *apply by now strictly to all citrus plantations in Palestine older than 7 years.*

The importance of the correct timing of the spraying was equally confirmed. Dr. Ben-Amotz informed the writer (i.l.): „The series of experiments carried out by me last year showed that while drenching sprays in the month of August with 2 % or even higher did not cause damage even under dense cultivation, spraying with a concentration as low as $1^2/_3$ % of oil spray resulted in very definite injury when applied in the second half of the month of October."

As regards the effect of pressure on insect mortality Dr. Ben-Amotz again remarks: „Our experiments here showed that much higher mortality can be obtained with higher pressure. This is especially true in the case of such scales as the mussel scale which prefer the inside and hidden portions of the tree. In the case of larger trees we found out here that it is possible to control the mussel scale only when pressures as high as 23—27 atmospheres are used. I personally attribute to a large extent the good results which we obtained last year in the control of the mussel scale to the high pressure we were using."

Oil sprays may be mixed with Bordeaux-mixture. Insecticidal and fungicidal control can thus be combined wherever this seems advisable. The late Mr. Broadley advocated a combined spray of white oil emulsion and Bordeaux mixture against sooty mould caused by Florida wax scale. Dr. Ben-Amotz obtained satisfactory cleaning of the leaves and the fruit from sooty mould by spraying oils alone.

We shall conclude this chapter by reproducing the instructions published by the Agric. Exper. Sta., Rehoboth in 1938 (H. Z. Klein) for spraying with white oil emulsions:

„*Red scale:* In every case where red scale has appeared in large numbers on leaves, and especially on fruits, during the preceding season, it is recommended to undertake control measures during the summer months. In the coastal plain it is not necessary, as a rule, to fight this pest after the trees have reached their tenth year. In the case of younger trees the grove should be sprayed if numerous fruits show at the end of July or at the beginning of August two scales per fruit, for in such a case a heavy attack is to be anticipated in the autumn. The peel of the shamuti shows minute pits in which the scales settle.

Mussel scale: In every case of heavy damage to fruit during the preceding season, control measures should be undertaken during the summer. During summer mussel scales are to be found more especially on inner branches, on old leaves; towards the end of the summer they take up their abode sometimes under the stem-ends of the fruit. The pest attacks trees of all ages, especially densely planted groves.

Florida wax scale: In every case of considerable sooty-mould on fruit (due to the Florida wax scale) during the preceding season, steps should be taken to control this pest during summer. If during the month of July several leaves harbour four or more larvae per leaf, control is indicated. Considerable secretion of sweet honeydew, and the development of sooty mould as a result, occur during the third stage of the scale. The larvae of this scale are quite conspicuous: white waxy protuberances protrude from the margins of the larva's body. At this stage the body is just as wide as it is long: it is 1.8 mms. long and 1.5 mms. wide (In earlier stages the lenght of the body is nearly twice its width). This pest becomes commercially important from the third year of the tree onwards.

Choosing the most appropriate time for spraying: Larvae are exterminated more easily than adult scales. Therefore, spraying should be executed on the mergence of large numbers of young larvae. The most appropriate time for spraying against red scale is from mid-July to the beginning of September; against mussel scale from mid-August to the beginning of September; against the Florida wax scale all July and the beginning of August. No spraying operations are to be undertaken after the beginning of September for fear of hamsins, which start, as a rule, in mid-September. As is well known a hamsin that occurs within a fortnight after spraying may burn the trees. Spraying should be done early in the morning and in the afternoon hours; during the hot mid-day hours it should be interrupted. Thirsty trees are not to be sprayed; therefore, spraying should be done a short time after watering.

How to spray: Small quantities of emulsion should be prepared for short use. On no account should ready emulsions be left over from one day to the next. Every part of the tree is to be sprayed both above and below, inside and outside. The quantity to be spent on each tree is determined according to age and development. As a rule, a tree in its third year, after budding, requires 8 litres of spraying liquid; in its fifth year 15—20 litres; in its seventh year over 20 litres and so on. Grapefruit trees require larger quantities at all ages. As a general principles concentrations of 2 %—2.5 % are to be used.

Machinery: Motor sprayers should be used in preference to hand sprayers. The pressure required for spraying with emulsions must be at least 15 atmospheres or 225 libres on the manometer attached to U.S.A. machinery. Since large machinery cannot be introduced between the rows of trees, the current practice in this country is to place the sprayer on the road and attach to it rubber hose or iron piping up to 100—150 ms. long. It should be emphasized that the use of such auxiliaries reduces pressure. Measurements have proved that spraying liquids left the orifice of a pipe 100 ms. long under a considerably lesser pressure than that shown by the manometer of the sprayer. The loss of pressure reached up to 25 %. Whenever such auxiliary hose or piping is used, its presence should be taken into consideration, and pressure must be increased, so that the manometer should read over 3oo pounds per inch. Small sprayers with a hand pump are undesirable since their agitator works slowly (in accordance with the action of the pump) and does not agitate the emulsion sufficiently.

Spraying damage: Spraying with white oil emulsions injures thirsty, weak and diseased trees. Wherever machinery with hand pumps is used, the last few litres in the tank should not be sprayed, since their oil content is much larger and is liable to burn the sprayed trees. Too large a quantity of spraying liquids is liable to do the same, causing drop of leaves and fruits. A concentrated emulsion that has been kept a long time under unfavourable conditions (in open containers, exposed to the sun) and lost its homogeneity, should not be used under any circumstances for it is liable to cause injury.

Organisation and supervision: Successful control can be achieved only if it is simultaneously undertaken in all infested groves in a given zone. Since scales are propagated by wind, spraying should commence at the west part of the area to be sprayed and proceed eastwards. The grower should supervise operations to ensure maximum exploitation in best order and avoidance of injury to trees."

References.

Y. Ben-Amotz, Hadar **11**, 1938, p. 211—213, 217.

F. S. Bodenheimer, Observations on citrus insects in many parts of the world. Tel Aviv, 1931, p. 47—59.

E. Broadley, Hadar **1**, 19 p.

A. C. Browne, Mo. Bull. Californ. State Dept. Agric. **1931**, p. 389.

E. R. de Orny, H. Knight & J. G. Chamberlin, Hilgardia **2**, 1927, p. 331—348.

W. Ebeling, Hilgardia **10**, 1936, p. 95—125.

H. Z. Klein, Hadar **8**, 1935, p. 193—196.

11, 1938, p. 219—220.

H. Knight, J. C. Chamberlin & C. D. Samuels, Plant Physiol. **4**, 1929, p. 219.

H. Knight, C. R. Cleveland, Journ. Econ. Ento. **27**, 1934, p. 269—289.

M. Oosthuizen, Plant injury caused by oil sprays. Seminary Paper 1931. Univ. of Minnesota.

R. H. Smith, Calif. Agric. Exper. Sta. Bull. 527, 1932, 48 pp.

R. S. Woglum and coworkers, Handbooks of Citrus Insect Control. Los Angeles (annually).

Further literature is found in the quoted references.

D. OTHER METHODS OF CONTROL.

Other methods of control, such as sulphur-dusting against mites, etc., have been mentioned before. None of these methods have as yet been developed to a great extent. The main reason for the relatively small extent of sulphur dusting in the citrus orchards is the lack of proper equipment before the infestations of the citrus silver mite. Now very modern motor dusters are available, and have been given preference import license.

Damages caused by sulphur dusting as well as by other minor control measures are described by Klein as follows:

„*Sulphur dusting*, which is the regular means for controlling red spider, is sometimes apt to damage trees. Every summer towards the end of July, and during August, one can observe in the Jordan Valley burned leaves and fruit. This burning is to be noticed on all sides of the trees. The outward side of the fruit shows large brown spots. This damage occurs only during the hottest months, but never results during any other season of the year.

Signs of such burning were observed in August at a grapefruit nursery in Samaria following dusting. The tender young leaves showed small and round brown spots, which came out a day or two after dusting. Leaves that were larger showed no spots. This damage greatly troubled the owner of the nursery; however, the damaged leaves continued to develop satisfactorily, and on our visit to the nursery in the autumn the signs of damage had totally disappeared.

It is worth while mentioning that in this case there was an abundant drop of

Fig. 153.
Views of the field stations for breeding *Sympherobius sanctus* against the citrus mealybug at Petah Tikva in 1926 (A, B) and in 1937 (C, D).

leaves in the nurseries and young groves during the autumn, though the attack of red spider was rather mild in that district. Further investigation proved that in all these cases, irrigation was not properly arranged during the end of the summer and in the autumn.

Trees damaged by belts: Every year, during the season of growth, young trees are infested with ants, which climb up to the trunks and nibble off the grafts. Since there

Fig. 154.
Plantation of *Spartium junceum* for mass breeding of *Icerya purchasi* and *Novius cardinalis* at Hedera. a). Shortly after plantation in summer 1929. b). In January 1931. c). In November 1932.

are definitely established methods of controlling these attacks, every grower tries his luck with his own inventions. The usual method is to tie a wide belt round the trunk of the tree, and smear it with one paste or another, such as tangle-foot, cart-grease, etc. In several cases the trunk itself was smeared with these materials; such procedure burned the bark and the cambium. Even, when belts of linen or parchment paper are arranged in four layers, it frequently happens that the paste runs between the belt and the trunk or penetrates to it through the layers of belting.

One spring, hoards of larvae of *Pyrameis cardui* emigrated from wild weeds to young groves. Growers who wished to prevent this attack on their trees smeared the trunks of the trees with machine grease. They choose this grease because they thought that it does not melt quickly and will remain on the trees till the larvae reach the stage of pupae. In these cases, too, the smeared limbs were frequently burned.

Fruit damaged while being cleaned: As a result of a strong attack of the Florida wax scale (*Ceroplastes floridensis*) in some years, sooty mould was very common in the groves. In several groves no attempt was made to control this pest, in others control was unseccessful. Consequently, a large number of fruit was covered with sooty mould at picking time. Such stained fruit is unfit for export (by order). Therefore, attempts were made in several packing houses to wipe the fruit clean with rags before packing. The wiping of grapefruit, the skin of which is smooth, does not present any difficulties, but the wiping of oranges requires great effort. We have seen with our own eyes unexperienced workmen first spit on the orange before wiping it clean, and then try to scrap out the sooty mould with their finger nails from the hollows of the rough skin. Of course such procedure damages the fruit and causes premature decay in large quantities on its long voyage from the producer to the consumer. In order to investigate the extent of the damage caused by cleaning the fruit, one season we instituted a preliminary experiment. Shamouti fruit of equal sizes covered with sooty mould was chosen. Part was wiped in the usual way and packed, and part was packed without wiping. Inspection of the fruit six weeks later showed that the decaying fruit in the unwiped box amounted to 9 %, while the percentage of decay in the box containing wiped fruit rose to 28 %, i.e. as a result of wiping the percentage of decayed fruit was more than trebled. It should be emphasized that the fruit for this experiment was especially chosen, so that under ordinary circumstances damage might be even greater." By now suitable machinery for mechanical cleaning is widely in use.

Biological control has been discussed in the general remarks and in greater detail in the special chapters (cf. espec. mealybugs, *Icerya*, red scale, Egyptian black scale, soft scales, *Ceratitis*). *The principle that the mere introduction or discovery of a beneficial insect is insufficient under our climatic conditions, and that permanent mass-breedings and dispersal in the endangered orchards is required, wherever economic results are to be expected, has been confirmed in every case.*

ADMINISTRATION AND LEGISLATION.

The activities of the fields of administration and legislation are executed by the Department of Agriculture, with regard to citrus, they concentrate mainly on three points:

1. Enforcement of necessary control of citrus pests.
2. Plant quarantine, in order to prevent the introduction of new pests.

3. Inspection of the exported fruit, in order to assure a high standard of grading and to prevent avoidable losses by wastage and decay during the transport.

The first two activities are executed by the Plant Protection Service, the last by the Horticultural Service. The legal frame and background has been created by the Plant Protection Ordinance of 1924 and by the Fruit Export Ordinance of 1928. Both ordinances have been subjected to many changes and amendments. The functions of the Citrus Board have been discussed in the introduction.

Plant Protection Ordi: The Plant Protection Ordinance entitles any plant inspector to inspect any grove, packing house, etc. at any reasonable hour without previous notice. It further entitles the Plant Inspector to force any owner to perform certain control measures within a given period in order to eradicate or control any pest which he has met, and even to destroy plants, infested or not, if such seems necessary. The control may be performed by some person, charged by the Dept. of Agriculture, if such is not done by the owner within the prescribed period at the owner's cost. In addition, the owner is liable to legal prosecution and a fine not exceeding 20 L.P.

7. 1) Any Plant Inspector may by order in writing addressed: —
a) to the owner of any land or plantation, or
b) to any one or more of the owners where the land is held in joint ownership.
direct any such owner to take within a definite time to be specified therein all such measures as may be necessary for the control of any disease or pest including in such measures the total destruction, if necessary, of any plants whether infected with any disease or infested with any pest or not.

3) If any owner shall fail to comply with the directions contained in any order served in accordance with this Ordinance, the Director of Agriculture or any officer authorised by him in writing for that purpose may with or without assistants enter upon the land and take such steps as may be necessary for properly and effectively carrying such order into execution.

8. — 1) The expenses incurred in carrying into effect any order made under this Ordinance shall be paid by the party in default, provided that it shall be lawful for the High Commissioner to dispense with the payment of such expenses or any part thereof.

2) The owners of the land concerned shall be jointly and severally liable to the Government for the said expenses and such expenses may be recovered as a debt from any one or more of such owners. Nothing in this section shall be deemed to prejudice or affect: —
a) the right of recourse of any owner against a co-owner in respect of such expenses, or
b) the liability of any owner to the penalty prescribed by section 11 of this Ordinance of any offence under paragraph (*b*) of section 10 of this Ordinance".

Of great importance in helping the farmer to obtain healthy seedlings and nursery-stock is the following amendment:

1. These rules may be cited as the Plant Protection (Nursery) Rules, 1934.

2. No plants shall on or after a date being three months from the date of these rules, be sold or otherwise distributed except from a nursery licensed under these rules.

3. The owner of any nursery in existence at the date of these rules and the owner of any nursery planted subsequently to such date shall apply in writing for the registration of their nurseries: —
a) in the case of nurseries in existence at the date of these rules, not later than one month from such date, and
b) in the case of nurseries planted subsequently to the date of these rules, not later than one month from the date of such plantation.
either to the Chief Horticultural Officer, Jaffa, or the Fruit Inspector, Haifa, or the Senior Horticultural Officer, Jerusalem.

4. As soon as may be possible after receipt of any application duly made under these rules, the Chief Horticultural Officer or Fruit Inspector, or the Senior Horticultural Officer as the case may be, shall inspect or cause to be inspected the nursery in respect of which application is made, and the Chief Horticultural Officer or Fruit Inspector or the Senior Horticultural Officer, upon being satisfied that any nursery inspected by or on his behalf is free of disease and of the weeds specified in the first schedule to these rules, that the plants are true to name, and that the methods of cultivation employed in such nursery are such as not to impair the vitality of

the plants shall register such nursery upon the payments by the owner of a fee of fifty mils.

5. The owner of any nursery registered under these rules, shall on application to the Chief Horticultural Officer or Fruit Inspector or the Senior Horticultural Officer be granted, upon payment of a fee of fifty mils, a licence to sell or otherwise distribute plants and such licence shall be valid from the date of issue until the thirtieth day of September next following such date:

Provided that the Chief Horticultural Officer or Fruit Inspector or Senior Horticiltural Officer may at any time withdraw a licence granted hereunder if he is of the opinion that the nursery in respect of which it was issued has failed to comply with the conditions upon which it was registered.

Other amendments of interest are:

The fumigation of citrus trees with hydrocyanic acid gas shall be carried out by such persons only who have been authorised in writing to do so by the Director of Agriculture, subject to such conditions as shall be required by the Director of Agriculture.

Schedule.

a)	Fumigation of trees: —	
	For each tree fumigated irrespective of size	35 mils
b)	For each tree felled and uprooted	50 ,,
c)	For each tree pruned or lopped	20 ,,

Notice

The importation of citrus plants and parts of citrus plants from Northern Palestine into Trans-Jordan is prohibited by the Trans-Jordan Government unless the consignment has been fumigated and is accompanied by a certificate of fumigation.

2. All citrus plants and parts of plants for Trans-Jordan dispatched from Haifa by rail can be fumigated before leaving Haifa and a certificate of fumigation will be given by the Plant Inspector.

3. Consignments from other stations and those sent by road will be fumigated at Samakh.

4. The fees charger will be as follows and will be paid by the consignor. In the case of consignments sent by rail the fees must be paid at the time of dispatch.

Each railway truck	300 mils
Each lorry or car	100 mils
Small consignments	50 mils

Plant Quarantine: The more important paragraphs of the Plant Protection Ordinance in force, concerning the citrus industry are:

"Importation of Plants.

1) All plants not included in schedules 1, 11 and 111 of this Order may be imported into Palestine, provided that they are first inspected by a Plant Inspector at the place of entry into Palestine and found free from disease and pests.

2) The importation of all plants included in schedule 1 to this Order shall be prohibited subject to the provisions of paragraph 5 of this Order. All plants included in schedule 11 to this Order may be imported subject to the conditions set out in that schedule".

4) All plants which on inspection are found to be infected with any virus, bacterial, fungus or other disease or harbouring any insect pest whether or not they are accompanied by a free of disease certificate may be destroyed or reconsigned to their place of origin or be subjected to disinfection at the discretion of a Plant Inspector. If re-export is ordered it shall be effected within seven days of the order being given. If exportation is not effected within the time specified the plants may be destroyed. All expenses incurred in the disinfection, destruction of re-exportation shall be paid by the importer.

5. — 1) All plants included in schedules 1 and 3 to this Order which are required for experimental or scientific purposes may be imported into Palestine provided that the written permission of the Director of Agriculture and Forests to import such plants is obtained at least seven days before the date of importation.

Schedule I.

The importation of all plants in this schedule is prohibited, provided, any or all of them may be imported under the provisions of paragraph 5 of the principal order, (as amended by paragraph 4 of this Order), and provided that the prohibition does not apply to preserved, pressed or dried fruits.

All species of citrus other than citrus fruits from Egypt, Syria or Cyprus.

Schedule II.

The importation of plants in this schedule is permitted provided that each consignment is accompanied by a certificate that the plants are free (*a*) from all diseases and pests, and (*b*) in particular, from the diseases or pests indicated opposite the name of each plant respectively in the list below.

Citrus fruit from Egypt, Syria or Cyprus.	*Chrysomphalus ficus* (= *C. aonidum*) Black scale); *Aonidiella aurantii* (Red scale);

Schedule III.

Where any of the plants appearing in the list set out below are imported under the provisions of paragraph 5 of the principal Order (as amended by paragraph 4 of this Order), then in addition to the requirements and conditions set out in that paragraph, every consignment of such plants shall be accompanied by a certificate that the plants are free (*a*) from all diseases and pests, and (*b*) in particular, from the diseases or pests indicated opposite the name of each plant respectively in the list below. The certificate must be signed by an officer of the Phytopathological Service (or any equivalent authority) in the country or origin.

Citrus nursery stock and budwood.	*Pseudomonas citri; Sphaceloma Fawcetti;* all Coccidae.

4. Every plant, article or thing, imported into Palestine in contravention of any Order under this Ordinance or transported within Palestine in contravention of any Regulation made under this Ordinance, or of any of the conditions imposed thereby, may be seized by any Officer of Customs or by any Plant Inspector, and destroyed or otherwise dealt with and no compensation shall be payable to any person claiming any interest in such plants, articles or things.

The legal frame is satisfactory, but in spite of considerable improvements which have taken place in recent years, its execution is still improvable. The professional training of part of the staff in charge is still unsatisfactory, but it is hoped that in the near future it will be above criticism. These remarks do not intend to belittle all that the quarantine service has done for the country hitherto. Among the citrus pests intercepted *Pulvinaria floccosa* Westw. may be mentioned.

A considerable improvement would be achieved if special vacuum fumigation plants could be erected on the main points of import (Haifa, Tel Aviv, Jaffa, Kantara) but financial obstacles seem to have interfered with such installations.

The internal quarantine line which intended to protect Central and S. Palestine from introduction of *Chrysomphalus aonidum* L. from the north, was abolished in 1936 considering 1) the extended focus existing in Jaffa, 2) the assumed small danger of transporting infested fruits into the uninfested area. It seems to the author that the abolishment of this quarantine added much to the speed of dispersal of the Egyptian black scale all over the coastal plain, which was, however, bound to be infested anyhow.

The problems connected with plant quarantaine are masterly exposed in a recent bulletin from California (1933).

Fruit inspection: The third administrative activity of the Government is the inspection of the exported fruit. This is executed by special fruit-inspectors of the Horticultural Department and is not connected with the Plant Protection Service.

The legal frame is the Fruit Export Ordinance originally published in 1927 and quoted here in the edition of August 1937.

The rules about places of inspection requirements with regard to size of boxes, packing, counts, procedure of inspection, boards of reference, etc. may be omitted here. The requirements with regard to quality of fruit are as follows:

5.— (1) All citrus fruit exported shall be of good quality, in sound condition with properly cut seems, fully developed and mature to such an extent as to be not too ripe or too unripe, free from diseases, bruises, mechanical and physiological injuries, open fruit fly punctures, or puffy, or cracked or coarse skins.

(2) Citrus fruit shall be free of scale insects, provided that up to fifteen scales per fruit will be allowed if not more than two per centum of fruit in a case is so affected:

Provided that either the Chief Horticultural Officer or the Chief Fruit Inspector may, at his discretion, pass scalemarked citrus fruit for export, if a process of cleansing has been employed by the exporter which is not likely to cause mechanical injury or the breakdown of such fruit.

(3) Citrus fruit marked with sooty mould or fumagine to the total extent of more than one square centimetre shall not be exported if more than two per centum of the fruit in a case is so affected.

(4) Citrus fruit marked with hail, thrips, or other scare to the total extent of an area equivalent to that covered by a one mil coin shall not be exported if more than two per centum of the fruit in a case is so affected.

The principal provisions of the Citrus Fruit Export Ordinance 1927, are:

(*a*) the prohibition of the export of Citrus fruit without inspection;
(*b*) power to inspectors to refuse to pass fruit for export;
(*c*) provision for Boards of Reference in cases of dispute;
(*d*) power to the High Commissioner to make regulations;
(*e*) penalties for non-compliance.

12. (3) runs: No consignment of citrus fruit shall be refused export unless ten per centum of the boxes forming the consignment have been inspected and found to contain an average of more than two per centum unfit for export: Provided that the Fruit Inspector may reject any boxes of a consignment affected by two per centum of any defect where there are not sufficient grounds under these rules for the rejection of the whole consignment.

During the 1937/38 season the permanent staff comprised: 1 Chief Fruit Inspector, 3 Fruit Inspectors, 11 Asst. Fruit Inspectors, and 4 clerks; and the temporary staff 13 packing Inspectors, 10 Sanitary Inspectors, 81 packers, 58 carpenters and 10 labourers. The inspection service is self-supporting, financed by an export fee of 2 mils per case.

Of the 10,790,110 boxes of the 1936—37 season, 211,800 (1.96 %) were rejected as unfit for export without repacking, and 209,149 boxes (1.94 %) were repacked under the supervision of Fruit-Inspectors, 24,367 boxes being discarded in the process.

The principal reasons for fruit rejections in 1928/29 and in 1936/37 were:

Reason for rejection	Percentage in	
	1928/29	1936/37
Wrongly marked counts	41	—
Waste	30	30
Irregular sizing and loose packing	14	7
Mechanical injuries	—	24
Poor quality	14	—
Other reasons	1	22
Red and mussel scale	—	9
Sooty mould	—	4.5
Mediterranean fruit fly	—	3.5
Total	100	100

The Fruit Inspection Service which at first encountered an open or hidden resistance of a great part of the orange-exporters, has increased its efficiency from year to year, and by patient explanation has won by now the cooperation of the majority of the orange traders. It is by now an important tool in maintaining and improving a high standard in fruit grading as well is in reducing considerable wastage and decay during the transport.

The Ministry of Agriculture of Israel is maintaining and improving all these functions. Some preliminary information is found in the foreword.

References.

F. S. Bodenheimer, Schaedlingsfauna Palaestinas. Berlin, 1930. pp. 22—30.
H. S. Smith and co-workers, The efficacy and economic effects of plant quarantine in California. Agric. Exper. Sta. Berkeley (Calif.). Bull. 553, 1933.
R. O. Williams, Hadar **11**, 1938, pp. 51—53.
Fruit Export Ordinance 1927. Hadar **10**, 1937, p. 203—206.
Plant Protection Ordinance of 1924 and Amendments, Dept. Agric. Jerusalem.

MIX
Papier aus verantwortungsvollen Quellen
Paper from responsible sources
FSC® C105338

If you have any concerns about our products,
you can contact us on
ProductSafety@springernature.com

In case Publisher is established outside the EU,
the EU authorized representative is:
Springer Nature Customer Service Center GmbH
Europaplatz 3, 69115 Heidelberg, Germany

Printed by Libri Plureos GmbH
in Hamburg, Germany